Analysis

ANALYSIS
Volume I

by Einar Hille
YALE UNIVERSITY

ROBERT E. KRIEGER PUBLISHING COMPANY
HUNTINGTON, NEW YORK
1979

Original Edition 1964
Reprint Edition 1979 with corrections

Printed and Published by
ROBERT E. KRIEGER PUBLISHING COMPANY, INC.
645 NEW YORK AVENUE
HUNTINGTON, NEW YORK 11743

Printed in the United States of America

Library of Congress Cataloging in Publication Data

Hille, Einar, 1894-
Analysis.

Reprint of the 1st edition (1964-1966) published by Blaisdell Publishing Co., New York and Waltham, Mass., in series: A Blaisdell book in the pure and applied sciences.
Includes index.
1. Mathematical analysis. I. Title.
[QA300.H53 1979] 515 79-1194
ISBN 0-88275-910-8

Preface

When 'Omer smote 'is bloomin' lyre,
He'd 'eard men sing by land an' sea;
An' what he thought 'e might require,
'E went an' took—the same as me!

RUDYARD KIPLING

Fifty years ago, in preparing for a comprehensive examination, I read C. J. de la Vallée-Poussin's *Cours d'Analyse Infinitésimale*, Vol. I (1909). This treatise left a lasting impression and, when my book was planned, that of the Belgian master served as the model although the final product differs from it in many respects. Let me list some of the aims that dictated the choice of material.

(1) **One variable.** Volume I deals only with functions of one variable, real or complex. Functions of several variables, partial differentiation, and multiple integrals will be treated in Volume II.

(2) **Complex analysis.** Following de la Vallée-Poussin and the French tradition, complex numbers and functions of a complex variable are introduced at an early stage, in Appendix B following Chapter 5. With these tools a proof of the fundamental theorem of algebra can be given, as well as a more satisfactory discussion of rational functions, both in Chapter 6, and, further, a solid foundation is laid for the theory of elementary transcendental functions, Chapter 7.

(3) **Series.** Sequential limits are introduced in Section 1.7 and convergent sequences are used systematically throughout the book. Infinite Series, Chapter 5, form a natural application of this notion and give a background for Taylor's Series in Chapter 6. The latter series figure prominently in Chapter 7.

(4) **The abstract point of view.** Greek mathematics was axiomatic, so is the modern kind. In this book postulates are used to introduce the notions of number, distance, and measure. An effort is made to characterize the function classes used in the Calculus. Stress is laid on the fact that most of these classes are algebras, i.e., invariant under addition, multiplication, and scalar multiplication. Functions continuous at a point or in a fixed interval have this property. So do differentiable functions, while rational functions form a field. Inequalities are stressed and used systematically.

(5) Concrete functions. One may well define a function as a mapping satisfying certain conditions, but unless this abstract definition is illustrated by many examples, the concept might, in the mind of the student, remain "a systematic abuse of a terminology invented *ad hoc*." This is the motivation behind the many concrete examples in Section 2.3 where an attempt is also made to classify the more important types of functions. Continuous functions are studied at length in Chapter 2, differentiable functions in Chapter 3. It is shown in Section C.3 that a continuous function need not be differentiable anywhere. The typical function considered in the differential calculus is piecewise convex. The existence of derivatives of such functions is examined in Section 3.4.

(6) Algebra and geometry. Besides the fundamental theorem, a number of problems are furnished to bring out facts from the elementary theory of equations: roots and coefficients, bounds and geometry of roots. The geometrical concept of convexity figures here and there. Functions of bounded variation are introduced in Section C.1 and used for the discussion of length in the next section. The Stieltjes integral will come in Volume II. Chapter 9 is devoted to Plane Curves, a topic which might well appeal to the student's aesthetic sense. Teachers should like it since nothing provides a better review of the Calculus than systematic curve discussion: change of coordinates, differentiation, maxima and minima, limits, inequalities. In fact, all the tricks of the trade are brought into play.

(7) Historical perspective. The basic problems and some of the methods of the Calculus go back to antiquity. The student gains by knowing about the development of the main ideas involved and about the men who brought these ideas into life. He should also realize that the development is continuing and that old concepts are still steadily revitalized and given new meaning. The many historical notes serve to bring out these facts. For the historical perspective the author is indebted to the fascinating book by Otto Toeplitz *Die Entwicklung der Infinitesimalrechnung*.

(8) Problems. Nobody can learn Calculus without solving problems, many problems, challenging problems as well as routine ones. This volume contains over 2000 problems of varying difficulty. Hard problems are starred. Answers for the odd-numbered problems are given at the end of the book. For the more advanced of these problems, solutions have been worked out. The routine problems are mostly homemade. Among the others are British, French, Russian, and Swedish review and examination problems taken from many sources. I went an' took—the same as 'Omer! The Russian collections of N. M. Günter–R. O. Kusmin and B. P. Demidovič may be accessible in this country. A stay at the University of Uppsala during the fall of 1962 gave me access to a number of Swedish sources.

These are to me the highlights of the book but I realize that some of the topics may raise misgivings in the minds of more experienced teachers. An effort has been made to use fine print for more advanced comments, sidelights, and topics. These parts may be omitted in a first reading without much loss of continuity.

I am indebted to my colleagues at Yale University who encouraged me to brave this task and who let the material be tried in class. The book is intended for honors students and future mathematics majors. It has been tried with some success on selected groups of freshmen. The book starts from scratch but includes much material that traditionally would be given in Advanced Calculus.

A number of referees have read the manuscript at various stages and have made many constructive suggestions. Particular mention should be made of Professors P. M. Cohn, E. E. Moise, and A. E. Taylor who were most helpful. An anonymous critic has willy-nilly contributed much to whatever merits this book may have. I am obliged to Mr. Anders Wik, Uppsala, who read the manuscript and worked many of the problems. I am grateful to my wife who read the whole manuscript carefully, pruned my easy going prose and, in so doing, caught a number of factual errors and inconsistencies. This final revision was made in India, if not under, at least looking at the banyan tree. She has also read the proof with the same vigilance.

I am indebted to U.S. Army Research Office [Durham] and to U.S. Air Force Office of Scientific Research for partial support while this book was in the making.

Finally it is a pleasure to acknowledge the fine cooperation of the staff of Ginn and Company, of the Blaisdell Division, and the consulting editors. The author hopes that the book is relatively free of errors and will appreciate having them brought to his attention.

EINAR HILLE

New Haven, Conn.
January 1964

Contents

Analysis

The Greek Alphabet

Α	α	Alpha	Ν	ν	Nu
Β	β	Beta	Ξ	ξ	Xi
Γ	γ	Gamma	Ο	o	Omicron
Δ	δ	Delta	Π	π ϖ	Pi
Ε	ε	Epsilon	Ρ	ρ	Rho
Ζ	ζ	Zeta	Σ	σ ς	Sigma
Η	η	Eta	Τ	τ	Tau
Θ	θ ϑ	Theta	Υ	υ	Upsilon
Ι	ι	Iota	Φ	ϕ φ	Phi
Κ	κ	Kappa	Χ	χ	Chi
Λ	λ	Lambda	Ψ	ψ	Psi
Μ	μ	Mu	Ω	ω	Omega

1

Introduction

1.1. What is the Calculus† and why? A science is an ordered body of knowledge, a language, and a method. Each science has its own province, though it may be hard to define it exactly; it has its own store of accepted facts, and its own criteria for what constitutes a fact; it employs its own methods in assimilating new facts and in re-evaluating old ones. It refers to the facts and the methods by terms which constitute a separate language. This argot is a mixture of the scientist's everyday language with pseudoclassical words invented *ad hoc*, where many of the common words are used in a sense different from ordinary usage.

All this applies to mathematics and its several sub-divisions. Characteristic of mathematics is *the axiomatic method*, invented by the Greeks for the study of geometry and now made to serve as the basis for work in more and more fields. A brief sketch of this method is in order. A system of objects, also referred to as marks or symbols, and a set of operations which can be performed on these objects constitute the universe of discourse. The objects and the operations are undefined except that they must obey certain rules known as *axioms* or *postulates*. It is preferable that the rules not be redundant; it is necessary that they be consistent. Any statement about such a system obtained by logical conclusions from the axioms is true for that system. The set of all known true statements about the system, organized and logically presented, forms a *mathematical discipline. Mathematics is the set of all such disciplines.* It is obvious that many possible disciplines are of little or no importance. Others are basic for mathematics as well as for its applications. Euclidean geometry is an example of one basic discipline, familiar to the student. Others will be treated in the following.

Mathematicians tried for almost two thousand years to prove that the parallel postulate of Euclid is redundant, that is, a logical consequence of the other postulates. It was finally realized that this is not the case and that, consequently, it is possible to have other geometries with different axioms governing parallels. All such geometries are equally true as logical structures. A totally different question is which of them, if any, holds in the physical world around us. This problem is outside the province of the mathematician. All he can do is to provide astronomers and physicists with mathematical structures needed for the discussion.

† The word *calculus* is the Latin word meaning 'pebble.' The Romans used pebbles in counting and in gambling. See also *calx, calcis* meaning 'stone,' in particular, 'lime stone.' 'Chalk' comes from the same root.

The situation is somewhat similar to the playing of card games. With a deck of 52 cards one can play many different types of games, each with its own rules. None of these games can be said to be true or false, or claim to be the only one to which the cards lend themselves.

The various mathematical disciplines fall into several large subdivisions, from four to eight, according to the classification employed. The reader is already familiar with two subdivisions, *algebra* and *geometry*. He will now learn about a third, known as *analysis*, of which the *Infinitesimal Calculus* is the basic discipline. *Applied mathematics* is often taken as the fourth major subdivision. All of analysis ultimately goes back to the *real number system* for which we can and shall give an axiomatic treatment. There are other basic concepts in analysis, less amenable to such treatment. The two most important are *function* and *limit*. The reader is already familiar with some classes of functions, such as polynomials, logarithms, sines, and cosines. Many others will be studied in the following. The primitive underlying notion is that of correspondence between two classes of objects or, better, the mapping of one class into the other. A graph is a familiar illustration of how such a mapping is done, the objects in this case being numbers.

The notion of a limit is connected with several primitive ideas involving nearness, smallness, relative distance and relative size, changing quantities and their rate of change. Most applications of mathematics to physical problems involve variable quantities whose rate of change is decisive for the problem. Thus, the pressure of water on a dam varies with the depth, the decay of a radioactive substance varies with the concentration, the orbit of a satellite depends upon initial conditions, attraction of the earth, moon, and other bodies, not to mention atmospheric friction. The Calculus and its numerous outgrowths such as the theory of *differential equations* and *celestial mechanics*, have provided the means of handling and solving such problems.

There are also problems of pure mathematics which inspired the early Calculus, such as questions of *tangents*, *lengths*, and *areas*. The Greeks knew how to construct tangents to a conic section, but they were also familiar with curves for which they were not able to construct tangents. This problem grew in importance when the new *analytic geometry*, started by René Descartes (1596–1650), led to a profusion of new and interesting curves. The solution was provided by the Calculus. The problem of area intrigued the ancients because of its agricultural importance. The nonutilitarian approach to the area problem goes back to Archimedes (287?–212 B.C.) who invented highly sophisticated methods of computing areas involving better and better approximations. His methods worked in a number of interesting special cases, but general methods came only with the development of the *integral calculus*. The methods of the Calculus sufficed for two centuries, but it is interesting that the area problem served as prime mover for the new theories of integration which appeared around 1900. The problem of the length of a curve also obtained a solution through the integral calculus, but the question of finding necessary and sufficient conditions for an arc to possess a length was not answered until the end of the nineteenth century.

It remains to say something about the actual discovery of the Calculus and about the men who made it. A certain amount of preliminary work was done during the first half of the seventeenth century (see Section 2.2), but the actual discovery should be credited to Sir Isaac Newton (1642–1727) and to Gottfried Wilhelm Leibniz (1646–1716), an Englishman and a German. Newton's discovery is one of the few good turns that the Great Plague did mankind. The plague forced the closing of Cambridge University in 1665 and young Isaac Newton of Trinity College, who had just got his B.A., returned to his home in Lincolnshire for eighteen months of meditation, out of which grew his *method of fluxions*, his *theory of gravitation* (*Philosophiae Naturalis Principia Mathematica*, 1687), and his *theory of light* (*Opticks*, 1704). The method of fluxions concerns us here. A treatise with this title was written in 1672, but remained unpublished until 1736. The new method was first announced in the *Principia* but in vague general terms without symbolism, formulas, and applications. In a way, all of the *Principia* is an application, but only in posthumous editions is the new technique explained. In view of Newton's great reluctance to publish anything tangible concerning his discovery, it is not surprising that the development on the Continent, in spite of a late start, overtook and went beyond Newton. This led to much resentment, accusations of plagiarism, and animosity between British and Continental mathematicians which soon took the former into splendid but sterile isolation.

Leibniz started his work in 1673, eight years after Newton. In 1675 he initiated the basic notation, dx and $\int$, which is still used. His first publications appeared in 1684 and 1686. These made little stir in Germany but two brothers Bernoulli in Basel, Switzerland, became the dispensers of and added profusely to the new ideas. From 1690 onwards the new Calculus grew rapidly on the Continent. It reached approximately its present state in about one hundred years. A number of new disciplines, initiated by the Calculus, reached independence during the three centuries which have passed since the start of the Calculus. They now form the great subdivision of *Analysis* which is steadily growing.

Finally a few words concerning perspective. Mathematics is a very old science, together with astronomy the oldest of all with an age of over 4000 years. It is also a young and vigorous science. There was more mathematics produced during the 1950's than during the nineteenth century, though the latter probably produced more mathematics than the two preceding centuries. The mathematics presented in the present volume is a very modest part of the whole structure. The results, the formalism, the basic ideas are in the main several hundred years old, but the underlying point of view and the mode of presentation are strongly colored by modern developments. Mathematics is highly stratified. A knowledge of the older, lower strata is necessary for a successful study of the more recent higher ones. Mathematics is neither accounting nor the theory of relativity. Mathematics is much more than the sum total of its applications no matter how important and diversified these may be. It is a way of thinking.

1.2. Sets. The purpose of this section is to make the student familiar with some of the words and signs in the mathematical dictionary.

We often speak of *sets* in mathematics. As a mathematical concept it is less than 90 years old; it became a household word after 1900. Here the noun "set" is used in the everyday sense of assemblage or collection of objects grouped together for one purpose or another. The word "class" is a synonym often used. We are not particular about the nature of the objects—they need not be of the same kind—but we do require exact information about the composition of the set. This may be furnished by explicit membership lists or by a descriptive property which characterizes the members without ambiguity. This excludes statements such as "all men of good will," "all loyal citizens," and so on which involve highly subjective judgments.

The objects which form the set are known as the *elements* or the *members* of the set and the relation between the element a and the set A to which it belongs is written $a \in A$. This is variously read "a is an element of A" or "a belongs to A" or "a is in A." A set B is called a subset of A if each element of B is also an element of A. This is denoted by $B \subset A$ or, equivalently, $A \supset B$. The notation $b \notin A$ means that b is not an element of A. We distinguish between the element a and the subset $\{a\}$ of A which contains a as single element. The set of elements common to two sets A and B is called the *intersection* of A and B and is written $A \cap B$. If A and B have no elements in common we write $A \cap B = \varnothing$. We regard $\varnothing$ as a set having no elements and refer to it as the *empty*, *vacuous*, or *void* set. We avoid the term null set which is sometimes used. The void set is by definition a subset of every set. The *union* of A and B, written $A \cup B$, is the set of elements which are either in A or in B. If B is a subset of A, we write $A \ominus B$ for the subset of A the elements of which do not belong to B. This set is known as the *complement* of B relative to A. If $A \ominus B$ is nonvoid, B is called a *proper subset* of A.

In dealing with sets we suppose we can tell the elements apart, so that if a and b are elements of A we know, if b *is the same element as* a, or, if b is distinct from a. In the first case we say that a is *equal* to b, written $a = b$, and in the second case that a is not equal to b, and we write $a \neq b$. Corresponding to this property of being equal there is a notion of *equality* satisfying the three *laws* or *postulates*:

E_1. $a = a$ (*reflexive property*).
E_2. $a = b$ implies $b = a$ (*symmetric property*).
E_3. $a = b$ and $b = c$ implies $a = c$ (*transitive property*).

More generally, we call a relation between elements of a set an *equivalence*, denoted by $\sim$, if it has the reflexive, symmetric, and transitive properties. If $a \sim b$ we say that a and b are *equivalent* or belong to the same *equivalence class*. Each equivalence class is a subset of the given set A such that the elements belonging to a particular class are equivalent to each other but never equivalent to an element of another class.

It is time to give some illustrations of these concepts. A rather simple one is furnished by the typographical characters on a page of this book, say on page 3. The elements of this set P are the individual letters, punctuation marks, figures, hyphens, etc. which occur on page 3. There are about 2800 members in the set. The characters found on line 10 of the page form a subset of P, and so do the letters which form one of the words on this line or page. The capital letters form a subset of P, the lower case letters form another subset, and the intersection of these two subsets is void. On the other hand, the subset on line 10 has a nonvoid intersection with the subset of lower case types. The set of characters on page 3 which are either defective or belong to the wrong font is also a subset of P; we hope that it is void.

In defining P each character is regarded as an element. Thus, the only possible notion of equality is that $a = a$ and distinct imprints give distinct elements. On the other hand, we have several possible ways of defining an equivalence relation in P. We may say that two elements of P are equivalent if they are imprints of the same character where we have to distinguish between capital, lower case, roman, and italic letters. Thus, all lower case roman a's are made to form an equivalence class, as are the periods, the commas, etc. Another reasonable way of defining equivalence in P would be to group together in the same class all symbols which belong to the same alphabet. This is a cruder subdivision and leads to fewer classes.

These are commonplace illustrations. The sophisticated reader may consider the set of all positive integers, the so-called *natural numbers*, or the set of points on the circumference of a given circle in the plane. These are infinite sets. The set of squares and the set of prime numbers are subsets of the first set. The intersection of these subsets is void. Many other subsets will come to mind. We can separate the natural numbers into two equivalence classes A_0 and A_1 by the following prescription. The integer n belongs to A_0 if it is even, to A_1 if it is odd, that is, according as division by 2 gives the remainder 0 or 1. We get a different system of equivalence classes if we replace the divisor 2 by some other integer m.

In 1889 the Italian logician and mathematician Giuseppe Peano [1858–1932] set himself the task of characterizing the natural numbers by a set of postulates. We reproduce this set for two reasons. First, it is a fairly simple instance of such a system of axioms, one of the early examples which strongly affected the later development. Secondly, the fourth postulate gives the *principle of induction* which is so useful. The basic idea of the system is that of *the immediate successor of an element* of the set of natural numbers denoted by N. The postulates are:

N_1. *Each element x of N has a unique immediate successor x^+ in N.*

N_2. *If $x^+ = y^+$, then $x = y$.*

N_3. $1 \in N$ *and there is no x in N such that $x^+ = 1$.*

N_4. *Any subset of N that contains* 1 *and that contains the successor of each of its elements coincides with N.*

A little cogitation will convince the reader that the set of natural numbers satisfies this system of postulates if $n + 1$ is taken as the immediate successor of n.

We formulate the *principle of induction* as a theorem.

THEOREM 1.2.1. *Suppose that for each natural number n there is an associated statement $E(n)$. Suppose that it is known that $E(1)$ is true and that $E(k^+)$ is true whenever $E(k)$ is true. Then $E(n)$ is true for all n.*

Proof. Let T be the subset of the natural numbers for which $E(n)$ is true. By assumption, T contains 1 and if k is in T so is k^+. By Postulate N_4 we have $T = N$, that is, $E(n)$ is true for all n.

As examples of statements $E(n)$ which may be proved using this theorem we quote:

$$n(n + 1) \text{ is divisible by } 2, \tag{1.2.1}$$

$$n(n + 1)(n + 2) \text{ is divisible by } 6. \tag{1.2.2}$$

We leave the proofs to the reader.

The following is another example of a proposition which can be proved by this theorem.

The sum of the first n natural numbers is

$$S(n) = \tfrac{1}{2}n(n + 1). \tag{1.2.3}$$

This is the statement $E(n)$. It is clearly true for $n = 1$, for then it says that $1 = \frac{1}{2} \cdot 1 \cdot 2 = 1$. Suppose that $E(n)$ is true for $n = k$. Then to verify $E(k^+) = E(k + 1)$ we form

$$S(k + 1) = S(k) + k + 1 = \tfrac{1}{2}k(k + 1) + k + 1 = \tfrac{1}{2}(k + 1)(k + 2)$$

so that $E(k^+)$ is implied by $E(k)$ and the statement must hold for all n. Note that the right member of (1.2.3) is an integer by (1.2.1).

The last new concept for this section is that of a *correspondence* or a *mapping* denoted by T in the following. When we have two sets A and B we can consider a correspondence between them or a mapping of one set into the other. We say that A *is mapped into* B, *if to every element a of A there corresponds a definite element b of B, called the image of a under the mapping.* Here "definite" means "unique and clearly defined." Such a mapping T is fully known if we know all ordered pairs (a, b), $a \in A$, $b \in B$, that is, if we know the image of each object under the mapping. Mappings may be defined in many ways. It is perfectly feasible for all objects a in A to have the same image b in B, so that all of A is mapped on a single element of B. The other extreme is the case where distinct objects give rise to distinct images, so that if (a_1, b_1) and (a_2, b_2) are two pairs of associated object and image, then $a_1 \neq a_2$ always implies that $b_1 \neq b_2$. In this case we say that the mapping is *one-to-one*, often written simply (1, 1). We say that the mapping is "onto" if every element of B is the image of some element of A under the mapping.

A mapping may be one-to-one as well as onto. No finite set B can be the one-to-one and onto map of one of its proper subsets, but that the situation is different for infinite sets is shown by the following example. The set of squares

$$1, 4, 9, 16, \cdots, n^2, \cdots$$

is a proper subset of the set of natural numbers. Nevertheless, the mapping $n^2 \to n$ from the squares to the natural numbers is one-to-one and onto.

The student should now reread the definition of mapping one set into another. If he grasps this idea, he will find that the function concept introduced in Section 2.2 is extremely simple.

EXERCISE 1.2

1. If $A = \{1, 3, 5, 7, 9\}$, and $B = \{2, 4, 6, 8, 10\}$, find $A \cup B$ and $A \cap B$.

2. Let A be the set of natural numbers which divide 30, $A = \{m \mid m$ divides $30\}$, and let $B = \{n \mid n$ divides $72\}$. Enumerate the elements of A, B, $A \cup B$, and $A \cap B$.

3. Let A be the set of even natural numbers, $A = \{m \mid m = 2k\}$, and let $B = \{n \mid n = 4k - 2\}$, where $k = 1, 2, 3, \cdots$. Determine $A \ominus B$.

4. Let $A = \{m \mid m = 6k\}$, $B = \{n \mid n = 4k\}$, $C = \{p \mid p = 3k\}$. Determine $A \cap B \cap C$.

5. Prove (1.2.1) and (1.2.2).

6. Prove by induction that for all natural numbers n
$$1 + 3 + 5 + \cdots + (2n - 1) = n^2.$$

7. Prove that $3^{2n+1} + 2^{n+2}$ is divisible by 7, if n is any nonnegative integer.

8. Prove that $9^n - 8n - 1$ is divisible by 64, if n is any integer $\geqq 2$.

9. Consider the set S with four elements a, b, c, d. It has 16 subsets including the set itself and the empty set. Make out a list of the subsets.

10. How many of these subsets contain the element a? How many of the two elements a and b?

11. Determine the intersection of $\{a, b, c\}$ with each of the subsets of $\{a, b, c, d\}$.

12. Prove that a set of n elements has 2^n distinct subsets including the set itself and the empty set. Try induction!

13. Prove that
$$(A \cup B) \cup C = A \cup (B \cup C) = A \cup B \cup C,$$
$$(A \cap B) \cap C = A \cap (B \cap C) = A \cap B \cap C.$$

14. Prove that
$$(A \cup B) \cap C = (A \cap C) \cup (B \cap C),$$
$$(A \cap B) \cup C = (A \cup C) \cap (B \cup C).$$

In Figure 1.1 five circles are drawn to represent five sets A, B, C, D, E by a so-called *Venn diagram*. The relations between the sets are indicated by the relative position of the circles. Thus, $A \cap B = \varnothing$, $C \subset B$, while $E \cap D \neq \varnothing$ and neither $E \subset D$ nor $D \subset E$ is true.

15. Make Venn's diagram for the sets involved in Problems 13 and 14 and use them to prove the statements.

16. If X is a finite set, let $N(X)$ be the number of elements of X. Prove that
$$N(A \cup B) + N(A \cap B) = N(A) + N(B).$$

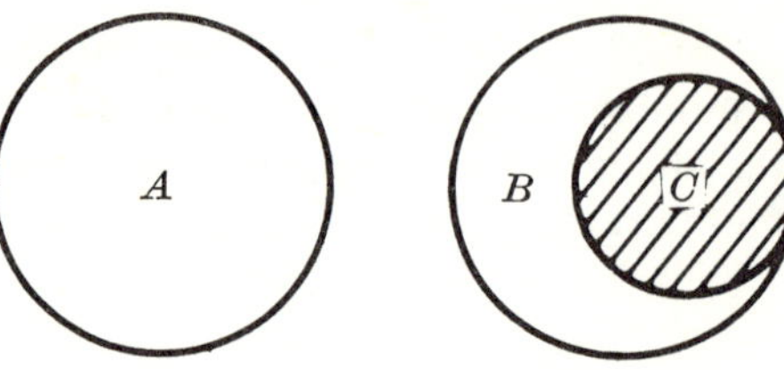

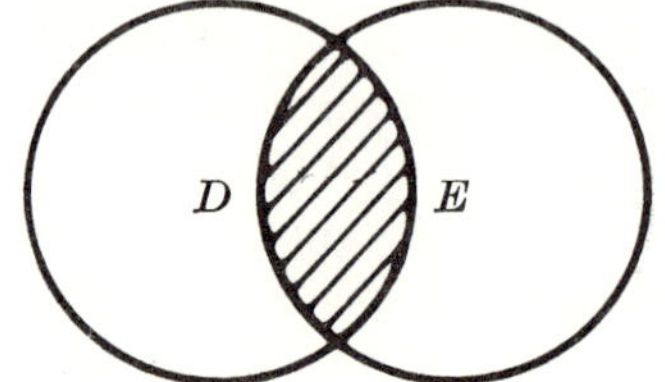

Figure 1.1

17. Let N be the set of natural numbers and let T be the so-called *successor mapping* which takes x into x^+. Prove that T is one-to-one but not onto.

18. Let A_0 and A_1 be the equivalence classes of even and odd natural numbers respectively. If m and n belong to N, prove that $m + n \in A_0$, if m and n belong to the same class, but to A_1, if they belong to different classes. What are the corresponding rules for the product mn?

19. In how many different ways can you map the set $\{a, b, c\}$ into itself? Describe the different mappings. Which are onto, one-to-one, or both?

20. Same question for a set of four elements.

21. In how many different ways can 3 distinct objects be chosen from a set of 5 objects if no attention is paid to order?

22. Restate question 21 in terms of subsets.

23. Prove by induction that the sum of the first n squares is

$$S_2(n) \equiv \tfrac{1}{6}n(n + 1)(2n + 1).$$

1.3. Real numbers, I. The notion of real numbers is basic in analysis. Hence, it is appropriate to start with a discussion of this concept. What the ancient world knew about such things had largely been forgotten during the Middle Ages and it was only after the year 1200 that Western Europe again became conscious of mathematics. Arabic numerals were introduced, Greek learning was revived, and Arabic mathematics became known.

What matters for us is that a concept of numbers arose and developed gradually far beyond that of the Arabic and Greek mentors. Judging from our present-day terminology there was at least passive resistance and much distrust to overcome. Our forefathers were deaf to the tone quality of *quadratic surds* and they pitted *irrational*, *negative*, and *imaginary numbers* against the *rational*, *positive*, and *real varieties*. But inevitably they accepted the new ideas and by the end of the eighteenth century mathematicians could handle real and complex numbers. The critical examination of the underlying concepts started around 1870, is still going on and has become a part in the general revision of mathematics.

It was natural that questioning of the foundations should arise when it did. Mathematics had developed rapidly for over 200 years and the tempo showed every sign of increasing. Much that was accomplished appeared to be of uncertain value since there was too much reliance on geometrical intuition and too little on logic. It was necessary to re-examine the various new theories and replace weak structures. *Number* had also developed in a haphazard manner and it was high time that its properties were codified and analyzed critically. For this purpose two different methods were used, the *genetic* and the *axiomatic*.

The *genetic method* follows more or less the historical development. The point of departure was the notion of *natural number* which was either taken for granted or described by the postulates of Peano or similar device. In the system of natural numbers, N in our previous notation, there are two fundamental operations, *addition* and *multiplication*. N is closed under these operations, that is, sums and products of natural numbers are again natural numbers. On the other hand, the equations

$$m + x = n, \qquad qy = p \tag{1.3.1}$$

for m, n, p, q given elements of N are not always solvable in N. The system N is too small even for the most elementary problems in algebra and must be replaced by a larger one. This is usually done in two steps; the order is immaterial. If we start with the second equation, we introduce the (positive) *rational numbers* as number pairs p/q with agreement as to equality of such symbols and the laws of operation. We shall not specify how this is done. Next we introduce *zero* and the *negative numbers*. This is the so-called *rational field* Q. It is also closed under addition and multiplication, and we can now solve the equations (1.3.1) in Q whenever m, n, p, and q are in Q, the only exception being the second equation when $q = 0$.

The extension from N to Q is a great advance but not sufficient for the needs of algebra. The Greeks knew that the length of the diagonal of a square is not commensurable with the length of the side or, in present-day terms, the square root of two is not a rational number. It is worthwhile to reproduce Euclid's simple proof. If

$$\sqrt{2} = \frac{m}{n}$$

the integers m and n may be assumed to have no factor in common. In particular, if m is even, we may assume that n is odd. Now we have

$$m^2 = 2n^2, \tag{1.3.2}$$

that is, m^2 is an even integer and this requires that m itself is even so that n is odd. But if m is even, there exists an integer k such that $m = 2k$. If this expression is substituted in (1.3.2) and common factors canceled, we see that

$$2k^2 = n^2,$$

so that n must also be even. This is a contradiction of our assumption that $\sqrt{2}$ is rational.

It follows that the equation

$$x^2 = 2 \tag{1.3.3}$$

has no solution in Q. If we want this equation to have a root, a larger system is necessary. In this particular case we could get such a system simply by "adjoining" $\sqrt{2}$ to Q. This would mean replacing Q by the set $Q(\sqrt{2})$ consisting of all numbers of the form

$$r + s\sqrt{2},$$

where r and s are rational numbers. In this enlarged system equation (1.3.3) and many other quadratic equations with integral coefficients have solutions. In modern algebra such adjunctions are commonly used and studied at great length.

For our purposes such measures would not suffice. Even if we enlarged Q to include all (real) roots of algebraic equations with integral coefficients, the resulting field, which is known as the *field of algebraic numbers*, would still not be a sufficiently large domain for the needs of analysis. It would, for instance, not include the number π or logarithms of integers. We are forced to a much more drastic extension which leads to the *field of real numbers*. Various methods of performing the last step were proposed in the 1870's by several mathematicians among whom should be mentioned Georg Cantor [1845–1918], Richard Dedekind [1831–1916], and Karl Weierstrass [1815–1897]. Cantor created set theory, a few concepts of which formed the subject of Section 1.2. Dedekind will be mentioned in Section 1.4 and the name of Weierstrass will occur often in this book. We shall not discuss the genetic method any further. We turn instead to the axiomatic method.

A brief description of this method was given in the Introduction. We proceed to elaborate with special reference to the problem at hand. A system of axioms or postulates is a way of describing or defining mathematical entities. These entities are *a priori* undefined and have merely to satisfy the postulates. The system normally involves certain procedures, also undefined, by means of which two entities or elements can be combined to generate a third element. Finally, we may have relations between two or more elements. Such relations are also undefined terms. The postulates state the desired properties of the undefined terms and any set of objects satisfying the stated conditions is a realization of the system of postulates. It is supposed that the latter is consistent, that is, the conditions are not self-contradictory.

The axioms for the real number system involve the following undefined terms.

(1) *Elements*, later known as *numbers*.

(2) A *class* denoted by R.

(3) A relation of *equality* denoted by $=$.

(4) A rule of *substitution*.

(5) A rule of combination called *addition* and denoted by $+$.

(6) A rule of combination called *multiplication* and denoted by $\cdot$.

(7) A relation of *order* denoted by $<$.

The postulates will be given by groups referring to the different relations and rules of combination. The three axioms of equality have been stated above. The rule of substitution is the old maxim "Equals equal equals" or in a more elaborate form.

S. *If a is an element of R and if $a = b$. then in any relation involving a it may be replaced by b.*

S is a convenient working rule implied by the nature of equality as logical identity.

Before listing the other postulates, we state two definitions as directives for our discussion.

DEFINITION 1.3.1. *The real number system is an ordered field in which nonvoid bounded sets have least upper bounds.*

The term "field" was used above without explanation. It will be defined. The notions of "order" and "least upper bound" will be introduced in Section 1.4.

DEFINITION 1.3.2. *A field F is a set of at least two distinct elements together with the operations of addition and multiplication satisfying the postulates* $A_1 - A_5$, $M_1 - M_5$, *and* D *listed below.*

We start with the *postulates for addition.*

A_1. *For every ordered pair* (a, b) *of elements of* F *there is a unique element* $a + b$ *of* F *called the sum of* a *and* b.

A_2. $a + b = b + a$ (*commutative law*).

A_3. $(a + b) + c = a + (b + c)$ (*associative law*).

A_4. $a + c = b + c$ *implies* $a = b$ (*law of cancellation*).

A_5. *For every pair* (a, b) *of elements in* F *there is an element* x *in* F *such that* $x + a = b$.

Postulate A_4 has an important converse

$$a = b \text{ implies } a + c = b + c \text{ for every } c. \tag{1.3.4}$$

This is simply a consequence of A_1 and S.

Postulate A_5 states that every linear equation

$$x + a = b$$

has at least one root. In particular, this is the case for $b = a$, that is, the equation

$$x + a = a$$

has at least one root. Using E_3, the transitive property of equality, and A_4, the law of cancellation, one shows that there is at most one root of this equation. Hence the root is unique. Moreover, the element which satisfies the equation is independent of the choice of a since

$$x + a = a, \qquad y + b = b$$

implies

$$x + a + b = a + b = y + a + b \quad \text{or} \quad x = y.$$

We denote this element by 0. Thus every field F has a unique *zero element* such that

$$a + 0 = a \tag{1.3.5}$$

for every $a \in F$. In particular, $0 + 0 = 0$ and $z = 0$ is the only solution of the equation

$$z + z = z. \tag{1.3.6}$$

In the same manner we prove that for every $a \in F$ there is a unique element $-a$, called the *negative of* a or *minus* a, such that

$$a + (-a) = (-a) + a = 0. \tag{1.3.7}$$

Combining the last two formulas we see that

$$-(-a) = a. \tag{1.3.8}$$

The *postulates for multiplication* read as follows.

M_1. *For every ordered pair* (a, b) *of elements of* F *there is a unique element* ab *of* F *called the product of* a *and* b.

M_2. $ab = ba$ *(commutative law)*.

M_3. $(ab)c = a(bc)$ *(associative law)*.

M_4. $ac = bc$ *with* $c \neq 0$ *implies* $a = b$ *(law of cancellation)*.

M_5. *For every pair* (a, b) *of elements in* F *with* $a \neq 0$, $b \neq 0$, *there is an element* x *in* F, $x \neq 0$, *such that* $ax = b$.

M_1 together with S gives

$$a = b \text{ implies } ac = bc \text{ for every } c \in F. \tag{1.3.9}$$

We note that M_5 has logical content: since F has at least two elements, there is at least one element $a \neq 0$. We used A_5 to prove the existence of a neutral element under addition. Similarly we can conclude the existence of a neutral element under multiplication from M_5. We call this element *unity* and denote it by 1. It has the property

$$a \cdot 1 = 1 \cdot a = a \tag{1.3.10}$$

for every a in F, except possibly for $a = 0$. We have $1 \neq 0$ and $x = 1$ is a solution of the equation

$$x \cdot x = x. \tag{1.3.11}$$

We shall see below that $x = 0$ is also a solution.

The M postulates also imply that for every $a \in F$ with $a \neq 0$ there is a unique element $a^{-1} \neq 0$, called the *reciprocal* of a, such that

$$a \cdot a^{-1} = a^{-1} \cdot a = 1. \tag{1.3.12}$$

From this we conclude that the equation

$$ax = b, \qquad a \neq 0, \qquad b \neq 0,$$

which has at least one root by M_5, has a unique root given by

$$x = ba^{-1}.$$

The *distributive law* is the final field postulate.

D. *For any three elements* a, b, c *we have*

$$(a + b)c = ac + bc.$$

This postulate has important consequences. Since

$$a \cdot (0 + 0) = a \cdot 0,$$

we see that $z = a \cdot 0$ is a solution of (1.3.6) which has the unique solution $z = 0$. It follows that

$$a \cdot 0 = 0 \text{ for every } a. \tag{1.3.13}$$

In particular, this is true for $a = 0$ and hence the assertion concerning the roots of (1.3.11) holds. We note that

$$ab = 0 \text{ implies either } a = 0 \text{ or } b = 0. \tag{1.3.14}$$

For if $a \neq 0$, then a has a reciprocal a^{-1} and we can multiply both sides of the equation by a^{-1} and obtain $b = 0$. In particular

$$x \cdot x = x \text{ implies either } x = 0 \text{ or } x = 1$$

so (1.3.11) has no other roots than those already noted.

A field may very well be finite. We have assumed the existence of at least two distinct elements and, actually, there are fields with only two elements. We recall the equivalence classes A_0 and A_1 in the system N of natural numbers. Problem 18 of Exercise 1.2 suggests defining addition and multiplication for A_0 and A_1 by the conventions

$$\begin{aligned} &A_0 + A_0 = A_0, \quad A_0 + A_1 = A_1 + A_0 = A_1, \quad A_1 + A_1 = A_0, \\ &A_0A_0 = A_0, \qquad A_0A_1 = A_1A_0 = A_0, \qquad A_1A_1 = A_1. \end{aligned}$$

It is easy to verify that the resulting system is a field of two elements. If we replace the divisor 2 by any prime $p > 2$, the corresponding equivalence classes $A_0, A_1, \cdots, A_{p-1}$ in the system N admit of a definition of addition and multiplication in terms of which these classes form a field of p elements.

EXERCISE 1.3

1. What are the rules for equality of rational numbers, that is, when is $m/n = p/q$?

2. State the rules for addition and multiplication of rational numbers as you know them.

3. Show that the square root of 3 is irrational (= not rational) by a modification of the method used in the text for the square root of 2.

4. Same question for $\sqrt{6}$.

5. The number $\sqrt{2} + \sqrt{3}$ satisfies an algebraic equation with integral coefficients of degree four. Find this equation and determine all its roots.

6. Prove that

$$s = \sqrt{6} - \sqrt{2} - \sqrt{3}$$

is irrational. Show this by expressing $\sqrt{6}$ in terms of s in such a manner that no other square roots occur.

7. Find all quadratic equations with integral coefficients which have roots of the form $r + s\sqrt{2}$, where r and s are rational.

8. Prove that the equation $x + a = a$ has a unique root.

9. Verify that $z = 0$ is the only root of (1.3.6).

10. Prove the existence and uniqueness of unity.

11. Verify the uniqueness of the reciprocal and of the root of $ax = b$.

12. Verify that $(-a)b = -(ab)$.

13. Verify that $(-a)(-b) = ab$.

14. Prove that $(a^{-1})(b^{-1}) = (ab)^{-1}$.

15. Construct a field with three elements. You may use the equivalence classes A_0, A_1, A_2 in the set of natural numbers where $n \in A_j$ if $n = 3k + j$.

16. For real numbers we can define two operations $\oplus$ and $\odot$ by

$$x \oplus y = x + y - 1, \quad x \odot y = x + y - xy.$$

Prove that the reals form a field F under these operations, that 1 plays the role of zero element in F, and 0 that of unity.

1.4. Real numbers, II. According to Definition 1.3.1 R is an ordered field. We now impose a relation of order on our field F. This relation is written $a < b$ or, equivalently, $b > a$ which is read "a is less than b" and "b is greater than a." The ordering is required to satisfy the following four order postulates.

O_1. *For any two elements a and b of F exactly one of the following statements is true:*

$$\text{(i) } a < b; \quad \text{(ii) } a = b; \quad \text{(iii) } a > b \text{ (law of trichotomy)}.$$

O_2. *If $a < b$, then $a + c < b + c$ for all c in F.*

O_3. *If $a < b$ and $0 < c$, then $ac < bc$.*

O_4. *If $a < b$ and $b < c$, then $a < c$ (transitive property).*

DEFINITION 1.4.1. *If $0 < c$, then c is said to be positive.*

O_1 implies that if $a \neq 0$, then either a or $-a$ is positive but not both. From O_2, O_3, and O_4 we conclude that *sum and product of positive elements are positive.*

We note that unity is necessarily positive if the order postulates hold in F. For $1 \neq 0$ and if $0 < -1$ we would have $0 < (-1)(-1) = 1$ by (1.3.8) and this is a contradiction. Hence $0 < 1$. Further, the reciprocal of a positive element is necessarily positive. For if $0 < a$, then $a^{-1} \neq 0$ and the assumption $a^{-1} < 0$ leads to $1 < 0$ if we multiply by a. We just saw that this is impossible so we must have $0 < a^{-1}$.

Adding the positive number 1 to itself, we get a positive number which we denote by 2. Adding 1 again, we get 3 and so on. We refer to these sums of unity as *positive integers*. All these elements are necessarily positive and distinct since $0 < 1$ implies

$$n < n + 1.$$

If we define x^+ as $x + 1$, we see that the set of positive integers satisfies the postulates of Peano for natural numbers, see Section 1.2.

An ordered field must be an infinite set. F contains the positive integers and, of course, also their negatives which we refer to as *negative integers*. More can be proved. We note that the reciprocals of the integers are also in F and if $n^{-1} \in F$ so is

$$n^{-1} + n^{-1} + \cdots + n^{-1} \quad (m \text{ terms}).$$

We denote this element by mn^{-1} and refer to the set $\{\pm mn^{-1}\}$ as the *rational subfield* Q of F. Thus every ordered field contains a rational subfield and may be identical to the latter.

Definition 1.4.2. *If a, b, c are three elements of an ordered field F and if either $a < b < c$ or $c < b < a$, then b is said to lie between a and c.*

Theorem 1.4.1. *If F is an ordered field and if a and b are distinct elements of F, then there is an element of F which lies between a and b.*

Proof. Suppose that $a < b$. Since 2 is positive, so is 2^{-1} and from $a < b$ it follows that $2^{-1}a < 2^{-1}b$. Further

$$1 + 1 = 2 \text{ implies that } 2^{-1} + 2^{-1} = 1.$$

Hence

$$a = 2^{-1}a + 2^{-1}a < 2^{-1}a + 2^{-1}b < 2^{-1}b + 2^{-1}b = b$$

so that $2^{-1}(a + b)$ lies between a and b. The proof shows that there are actually infinitely many elements of F between a and b.

We write $a \leqq b$ if either $a = b$ or $a < b$. The notation $b \geqq a$ is also used.

Note that the real number system cannot be characterized merely as an ordered field. At least one more postulate is necessary. We have to exclude the possibility that $F = Q$, for the field Q is neither adequate for the needs of algebra nor for those of analysis. The field Q is definitely too small for our purposes. On the other hand, there are fields which contain Q as a proper subset whose properties are deficient in other directions. Experience has shown that lengths are comparable and the relations between measure and numbers which are basic in the applications require that numbers also be comparable. This is expressed by the so-called

Axiom of Archimedes. *If $0 < a < b$, then there exists a multiple of a, $a + a + \cdots + a = na$ say, such that $b < na$.*

This property does not hold in every ordered field so our additional assumptions must ensure that the field be Archimedean, that is, that the axiom of Archimedes holds. Actually all we need is *the least upper bound postulate* of Dedekind:

B. *Each bounded nonvoid subset of F has a least upper bound in F.*

This postulate has the following meaning. A subset S of F is said to be *bounded* if there are two fixed elements m_1 and m_2 of F such that $m_1 < x < m_2$ whenever $x \in S$. The postulate then asserts that if this is the case, there is an element b of F, not necessarily in S, such that

(i) $x > b$ is not true for x in S, and

(ii) if $b > d$ then there is an x in S such that $x > d$.

The element b, defined by Postulate B, is uniquely determined for the assumption that (i) and (ii) could hold for two different values of b, say $b = b_1$ and $b = b_2$ where $b_1 < b_2$, leads to a contradiction if we take $b = b_2$ and $d = b_1$. We call b *the least upper bound* of S or *the supremum* of S and write

$$b = \text{l.u.b. } S = \sup S. \tag{1.4.1}$$

If B is applied to the set of negatives of the elements of S, it is seen that S has a uniquely defined *greatest lower bound* β also called the *infimum* of S. We write

$$\beta = \text{g.l.b. } S = \inf S. \tag{1.4.2}$$

THEOREM 1.4.2. *An ordered field R which satisfies the least upper bound postulate is Archimedean.*

Proof. If $0 < a < b$ and we have $na < b$ for all n, then $\{na\}$ is a bounded subset of R and has a least upper bound which we denote by μ. Since $\mu - a < \mu$, there is at least one integer k such that $\mu - a < ka$ or $\mu < (k + 1)a$ which contradicts the definition of μ. It follows that R must be Archimedean.

THEOREM 1.4.3. *In the sense explained below, every element of R is representable by a decimal fraction.*

Proof. It is enough to prove the assertion for the elements between zero and unity. If a is any element of R and R is Archimedean, we can find an integer k such that $k \leqq a < k + 1$. If we can prove the assertion for $a - k$ which lies between 0 and 1, the statement is clearly also true for a. Now, if $0 < a < 1$ and if $10a = a_1$, then $0 < a_1 < 10$. Consider the nonnegative integers less than or equal to a_1. This is a bounded set which contains at least one element, namely 0, and by Postulate B it has a least upper bound, d_1 say. Since the set is finite, the least upper bound is a member of the set, that is, a nonnegative integer. If $a_1 = d_1$, we are through. If not, consider the number

$$a_2 = 10(a_1 - d_1)$$

which also lies between 0 and 10. Let d_2 be the largest integer not exceeding a_2. If $d_2 = a_2$, we are through. If not, consider

$$a_3 = 10(a_2 - d_2)$$

and so on. If this process comes to an end, we have

$$a = d_1 \cdot 10^{-1} + d_2 \cdot 10^{-2} + \cdots + d_n \cdot 10^{-n} \tag{1.4.3}$$

for some choice of n. If this process continues indefinitely, we obtain an infinite sequence of integers

$$d_1, d_2, d_3, \cdots d_n, \cdots, \quad 0 \leqq d_n \leqq 9,$$

such that for each n

$$0 < a - d_1 \cdot 10^{-1} - d_2 \cdot 10^{-2} - d_3 \cdot 10^{-3} - \cdots - d_n \cdot 10^{-n} < 10^{-n}, \tag{1.4.4}$$

$$10^{-k} = (10^k)^{-1} = (10^{-1})^k. \tag{1.4.5}$$

This is the sense of the representation

$$a = 0.d_1\, d_2\, d_3 \cdots d_n \cdots. \tag{1.4.6}$$

We have $0 < 10^{-n-1} < 10^{-n}$, so the degree of approximation in (1.4.4) as measured by the smallness of the error, becomes higher the larger we take n, that is, the more digits we use. Since R is Archimedean, we can make the error as small as we please by taking n sufficiently large. This completes the proof.

Thus an ordered field in which the least upper bound postulate holds, is equivalent to the set of decimal fractions. In this sense R is uniquely defined by the postulates and may be called *the real number system*. The elements of R will be called *real numbers*.

What we have just said does not exclude the possibility of two distinct realizations of the postulates, R and R^* say. It may be shown in such a case, however, that *there is a one-to-one mapping of R^* onto R which takes sums into sums, products into products and preserves order*. Such a mapping is called an *isomorphism*. Examples will be given later.

A few remarks concerning the classification of the elements of R will be in order. We have seen that R contains a subfield Q, the elements of which we shall now call *rational numbers* and denote by the usual symbols. The set $R \ominus Q$ is called the set of *irrational numbers*. It contains numbers like $\sqrt{2}$, $\sqrt{3} + \sqrt{5}$, π, log 2, and so on. Georg Cantor proved in 1874 that $R \ominus Q$ is a much larger set than Q in the sense that no mapping of Q into $R \ominus Q$ can be onto. In this sense "most" real numbers are irrational.

Another important subset of R is the set A of *algebraic numbers. A real number α is said to be algebraic if it satisfies an algebraic equation with integral coefficients*. Thus the number $\sqrt{3} + \sqrt{5}$ is algebraic since it is a root of the equation

$$s^4 - 16s^2 + 4 = 0.$$

We have obviously $Q \subset A$. Actually A is also a field, but we shall not prove this.

The name *transcendental numbers* is given to the elements of the set $R \ominus A$. Their existence was proved by the French mathematician Joseph Liouville [1809–1882] in 1844. Roughly speaking, he found that an unending decimal fraction in which "most" digits are zeros is apt to be a transcendental number. In 1873 another Frenchman Charles Hermite [1822–1901] proved that e, the base of the natural logarithms, is transcendental and in 1882 C. L. F. Lindemann proved the transcendency of π. By that time G. Cantor had already proved that "most" real numbers are transcendental in the sense that no mapping of A into $R \ominus A$ can be onto.

EXERCISE 1.4

1. Which of the following numbers are rational: 1.41, $\sqrt{2}$, $(1 + \sqrt{2})^3$, 3.14, π, π^2, $\sqrt{\pi}$?

2. If a and b are rational, α and β irrational, what can be said concerning $a + b$, $a + \alpha$, $\alpha + \beta$, 2α, $a\alpha$, ab, $\alpha\beta$?

3. The logarithm of 2 to the base 10 is defined by $10^x = 2$. Prove that no rational number can satisfy this equation. Is 0.30103 rational? Is there any discrepancy between these two results?

4. If r and s are rational numbers, show that the reciprocal of $r + s\sqrt{2}$ is a number of the same form.

5. $Q(\sqrt{2})$ is the set of all numbers of the form $r + s\sqrt{2}$, r and s rational. Show that $Q(\sqrt{2})$ is a field.

6. Define an order in $Q(\sqrt{2})$ which satisfies O_1 to O_4.

7. If α is algebraic, prove that $\sqrt{\alpha}$ is also algebraic.

8. If α is algebraic, prove that $\alpha + 1$ is also algebraic.

9. If α is algebraic, prove that α^2 is also algebraic.

10. Show that the reciprocal of an algebraic number is algebraic.

11. An alternate way of defining order in a field is first to define positive elements and then to say that $a < b$ if and only if $b - a$ is positive. Formulate postulates for "positivity" so that the resulting order satisfies O_1 to O_4.

12. Show that the field F of Problem 16, Exercise 1.3 can be ordered by the convention that real numbers less than one are "positive." Show that $x < y$ in F if and only if $y < x$ in R.

13. Show directly that the field of two elements defined at the end of Section 1.3 cannot be ordered.

14. Find supremum and infimum of the set of reciprocals of the natural numbers.

15. Find supremum and infimum of the set $S = \{(\frac{1}{2})^n + (-\frac{1}{3})^n \mid n = 1, 2, 3, \cdots\}$.

16. Same question for $S = \{(\sqrt{2} - 1)^n \mid n = 1, 2, 3, \cdots\}$.

17. Same question for $S = \{2 \sin (n\, 60°) \mid n = 1, 2, 3, \cdots\}$.

18. Consider the set of decimal fractions between 0 and 1 whose only digits are zeros and ones. Find supremum and infimum of the set.

19. Same question if 2's and 3's are the only digits.

20. Verify that R contains the roots of the equation $x^2 = 2$. [*Hint:* Consider the set of rationals whose squares are less than two.]

21. Same question for the equation $x^2 + x - 1 = 0$.

22. Give an example of a bounded set in Q which does not have a least upper bound in Q.

23. Same question for $Q(\sqrt{2})$.

24. Let A and B be nonvoid subsets of Q such that $Q = A \cup B$ and $a \in A$, $b \in B$ implies that $a < b$. Such a partition of Q is known as a *Dedekind cut* $[A \mid B]$. It can serve to define the real number α which is the supremum of A and the infimum of B. Define $\sqrt{3}$ by such a cut. [This was the original form of Postulate B.]

25. Define $\sqrt[3]{2}$ by a Dedekind cut.

1.5. Geometric representation. We shall be concerned with geometric representations of real numbers, number pairs, number triples, and so on. Such representations can be realized in many different ways, the simplest are provided by *analytic geometry* using *Cartesian coordinates*, after Cartesius, the latinized name of the French philosopher and mathematician René Descartes. The reader not familiar with analytic geometry is advised to read Appendix A before going beyond this section.

It is assumed that the reader is familiar with Euclidean geometry, not merely the propositions but also the logical structure so that he has some idea of the nature of the underlying assumptions. It would take us too far afield to list the postulates of Euclidean geometry but we shall recall some of them for the sake of the terminology and as background for the distance postulate stated below.

Euclidean plane geometry deals with an object called a "plane" in which there are other objects called "points" and "lines." The postulates which describe these objects may conveniently be divided into five groups. The first four define the relations "on," "between," "congruent," and "parallel" while the fifth deals with "continuity." We recall that there is one and only one line "on" two given points P and Q and that the points "between" P and Q form the line segment PQ. The line segments form equivalence classes under the relation of "congruence." The latter decides when two line segments are considered to be "equal." The lines in the plane form equivalence classes under the relation of being "parallel." The "continuity postulate" is the

Axiom of Archimedes. *Given three points O, P, Q on a line, with P between O and Q, there exist a natural number n and points $P_1, P_2, \cdots, P_n$ on the line such that the line segments*

$$OP,\ PP_1,\ P_1P_2,\ \cdots,\ P_{n-1}P_n$$

are congruent and Q lies between O and P_n.

See Figure 1.2.

We now introduce a notion of *distance* between points in the plane defined by a postulate.

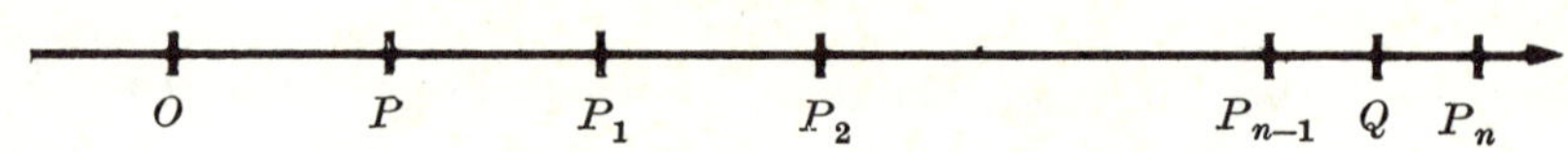

Figure 1.2

Distance Postulate. *To each line segment in the plane can be assigned a number $d(P, Q)$, called the distance from P to Q, satisfying the following conditions:*

(d_1) $d(P, Q) \geqq 0$ *and equals* 0 *if and only if* $P = Q$.

(d_2) $d(P, Q) = d(Q, P)$.

(d_3) *If P_2 lies between P_1 and P_3, then* $d(P_1, P_3) = d(P_1, P_2) + d(P_2, P_3)$.

(d_4) *If PQ and AB are congruent, then* $d(P, Q) = d(A, B)$.

(d_5) $d(P, Q)$ *is uniquely determined once the unit of length has been chosen.*

(d_6) *If α is any positive number, then there is a class of congruent line segments PQ such that $d(P, Q) = \alpha$. In particular, if O is any point in the plane and l is any line on O, then there is a point Q on l such that $d(O, Q) = \alpha$.*

The congruence postulates imply the existence of a class of one-to-one transformations of the plane onto itself which take line segments into line segments and preserve distances. Such a transformation is known as a *rigid motion*. Two line segments belong to the same equivalence class with respect to congruence if and only if there is a rigid motion which carries one segment into the other.

The distance postulate can also be formulated for Euclidean space. The only change in the wording is that "plane" is replaced by "space."

On the basis of these postulates we can introduce Cartesian coordinates. We start with the one-dimensional case and assign to the line a definite direction which is called the *positive orientation*. Normally the line is placed in horizontal position and is described from left to right. See Figure 1.3 where the arrow indicates the positive orientation. This assigns a definite order on the line, the point P precedes the point Q, written $P < Q$, if P lies to the left of Q. This order relation satisfies Postulates O_1 and O_4, the law of trichotomy, and the transitive law, while O_2 and O_3 do not make sense at present. On the line we choose arbitrarily two points O and U, $O < U$, and we set

$$d(O, U) = 1.$$

By condition (d_5) this determines all distances on the line uniquely. We call O the *origin of the coordinate system*. If P is a point to the right of O, $O < P$, and if $d(O, P) = \alpha$, we call α the *coordinate* of P. This number is unique. On the other hand, if $\alpha > 0$ is given, then there is one and only one point P, $O < P$, such that $d(O, P) = \alpha$. That there is one such point follows from (d_6), that there is only one is implied by (d_3). For if $d(O, Q)$ also equals α and if $P < Q$, then

$$\alpha = d(O, Q) = d(O, P) + d(P, Q) > \alpha$$

which is impossible. In the same way we show that Q cannot be to the left of P. Hence the point P with $d(O, P) = \alpha$ is unique.

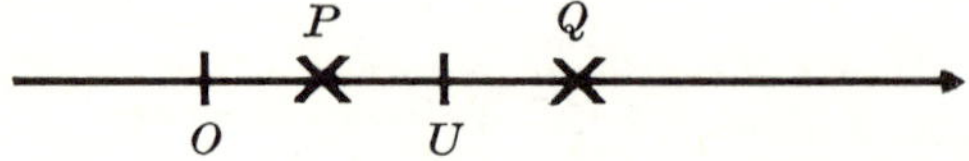

Figure 1.3

So far we have only considered points to the right of the origin. If P^* is a point to the left of O, and if $d(P^*, O) = \alpha$, then we take $-\alpha$ as the coordinate of P^*. In this manner we obtain a correspondence between points on the line and real numbers which defines a one-to-one and onto mapping. The mapping is defined as soon as we have chosen the origin and the unit of length. The correspondence is order preserving: if $P < Q$ then $\alpha < \beta$ where α and β are the coordinates of P and Q respectively. In view of this correspondence we make little or no distinction between the line and the real number system R. We speak of the line R or the *real line* or the *number line* and of the *point* x on this line where x really is the coordinate of the point in question.

Next we define the *absolute value* of x, written $|x|$, by

$$|x| = \max(x, -x). \tag{1.5.1}$$

In other words, $|x|$ is the distance of the point x from 0,

$$|x| = d(0, x). \tag{1.5.2}$$

More generally,

$$d(x_1, x_2) = |x_1 - x_2|. \tag{1.5.3}$$

The absolute value has the following important properties

$$|a + b| \leqq |a| + |b|, \qquad |ab| = |a|\,|b| \tag{1.5.4}$$

which are easily verified.

The set of points on a line between two given points, or, equivalently, the set of real numbers between two given bounds is known as an *interval*. More precisely, we speak of the *open interval* (x_1, x_2) made up of the points x with $x_1 < x < x_2$, also written $\{x \mid x, < x < x_2\}$, and the *closed interval* $[x_1, x_2]$ which is the set $\{x \mid x_1 \leqq x \leqq x_2\}$. Intervals $(x_1, x_2]$ and $[x_1, x_2)$ are also considered. All these intervals have the same length, namely $x_2 - x_1$. Note that the term "interval" presupposes that coordinates have been introduced on the line, if not, we still use the term "line segment."

A useful theorem about *nested intervals* reads:

THEOREM 1.5.1. *If $I_1, I_2, \cdots, I_n, \cdots$ is a set of closed intervals, if $I_n \supset I_{n+1}$ for all n, and if the length of I_n is less than any preassigned positive number for all large n, then there is one and only one point common to all intervals.*

Proof. See Figure 1.4. Let $I_n = [l_n, r_n]$ so that the "nesting" property

$$l_n \leqq l_{n+1} < r_{n+1} \leqq r_n$$

holds for all n. This shows that the set of points which occur as left endpoints of intervals I_n is a bounded set and the same is true of the right endpoints. Let λ be the least upper bound of the first set and ρ the greatest lower bound of the second. Then $\lambda \leqq \rho$ and λ as well as ρ belong to $\cap_n I_n$. But

$$\rho - \lambda < r_n - l_n \quad \text{for all } n.$$

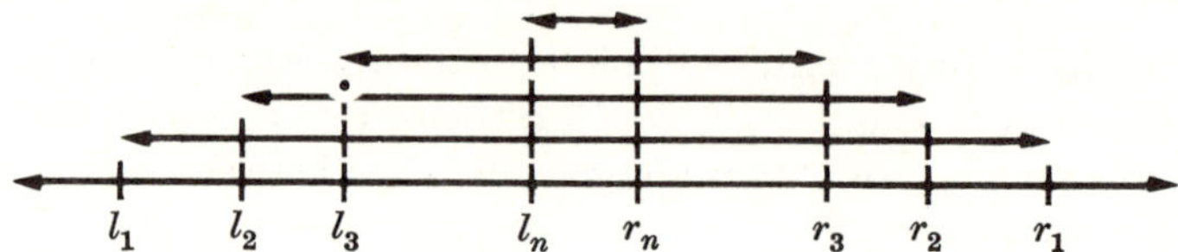

Figure 1.4

Here the right member is the length of I_n and hence arbitrarily small when n is large. It follows that $\lambda = \rho$. This point belongs to all intervals I_n and is the only point with this property since any other point will have a distance from ρ which ultimately exceeds the length of I_n. This completes the proof.

Nested intervals were first used by Georg Cantor to define real numbers when the rational numbers are taken as known. In this case l_n and r_n are rational numbers and the real numbers are defined as equivalence classes of sequences of nested intervals.

A neighborhood of a point x_0 is an open interval containing the point. More precisely, if $\varepsilon > 0$ is given, the *ε-neighborhood* of x_0 is the interval $(x_0 - \varepsilon, x_0 + \varepsilon)$.

We now consider the two-dimensional case, the Euclidean plane. Again we note that once we have chosen a certain line segment to be of unit length, then the lengths of all other line segments are determined. This is utilized as follows. Take an arbitrary point O as origin of the coordinate system and draw through O two perpendicular lines l_1 and l_2, represented by OX and OY in Figure 1.5. It is understood that l_1 is OX produced in both directions and similarly for l_2. The lines are oriented as indicated in the figure. The plane is also oriented in such a manner that a positive 90° rotation of the plane about O will take the positively oriented line l_1 into the positively oriented line l_2. OX is known as the *x-axis*, OY as the *y-axis*.

Next we introduce coordinates on the axes, using the same origin O and the same unit of length. If P is any point in the plane we drop perpendiculars

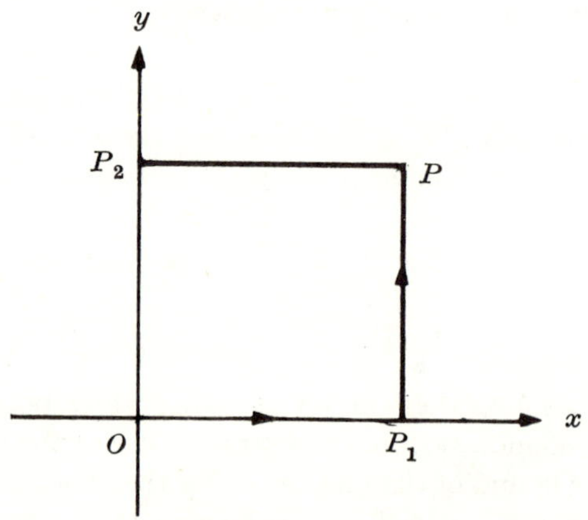

Figure 1.5

from P to the axes, P_1P and P_2P in the figure. If the coordinates of P_1 and P_2 are x_0 and y_0 respectively, then we give P the coordinates (x_0, y_0). The x-coordinate is also known as the *abscissa* of P, the y-coordinate as the *ordinate*. The notation (x, y) should not be confused with that used for open intervals. The context will usually indicate which is meant. We write

$$R_2 = R \times R = \{(x, y)|x \in R, y \in R\}$$

so that R_2 is the set of all ordered pairs of real numbers. It is common practice to refer to R_2 as the *Cartesian product of the system of real numbers with itself.* The correspondence

$$P \leftrightarrow (x, y)$$

defines a mapping of the Euclidean space onto the set R_2. This mapping is also one-to-one. R_2 is frequently used to denote the Euclidean plane as well as the Cartesian product of the real line with itself. The mapping is determined once the origin, the coordinate axes, and the unit of length have been chosen.

If $P_1 = (x_1, y_1)$ and $P_2 = (x_2, y_2)$ are any two points in the plane, then the distance $d(P_1, P_2)$ is given by

$$d(P_1, P_2) = [(x_1 - x_2)^2 + (y_1 - y_2)^2]^{\frac{1}{2}}. \tag{1.5.5}$$

See Section A.1. This quantity satisfies the conditions

D_1. *$d(P_1, P_2) \geqq 0$ and $= 0$ if and only if $P_1 = P_2$.*

D_2. $d(P_1, P_2) = d(P_2, P_1)$.

D_3. $d(P_1, P_3) \leqq d(P_1, P_2) + d(P_2, P_3)$.

Here $D_1 = (d_1)$, $D_2 = (d_2)$. The third property, the *triangle inequality*, states that the sum of the lengths of two sides of a triangle is at least equal to the length of the third. Condition (d_3) states that equality holds if the points are collinear with P_2 between P_1 and P_3. This is evidently the only case where equality holds in D_3.

In the plane rectangles take over the role played by intervals on the line. It is convenient but not necessary to assume that the rectangles have sides parallel to the coordinate axes. Such a rectangle, without its boundary, is a neighborhood for any one of its points. An ε-neighborhood of a point P_0 is usually understood to be the set of points P such that

$$d(P_0, P) < \varepsilon. \tag{1.5.6}$$

This is the interior of a circle with center at P_0 and radius ε. Such a set is often called an *open circular disk*.

A set S in the plane is bounded if the set of distances $\{d(P_1, P_2)\}$, $P_1 \in S$, $P_2 \in S$, is a bounded set of positive numbers. The least upper bound of the latter set is called the *diameter* of S.

The theorem on nested intervals on the line has various two-dimensional analogues. We can consider closed rectangles or closed circular disks, each

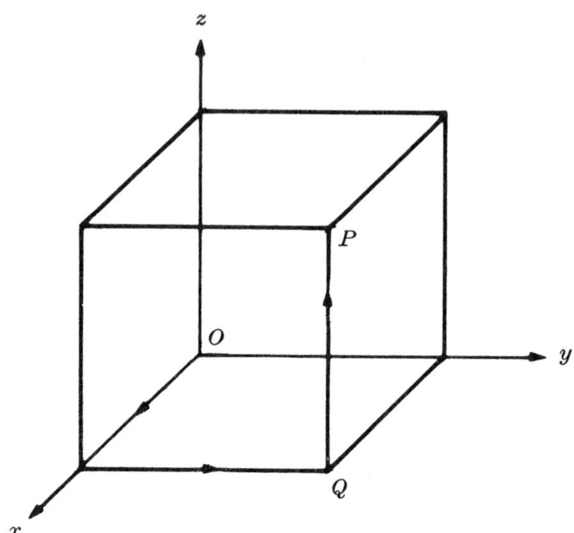

Figure 1.6

containing its successor and with diameters which ultimately become arbitrarily small. Here a closed circular disk is the interior and circumference of a circle. Similarly a closed rectangle is the interior and perimeter of a rectangle. In either case there is one and only one point common to all figures in the sequence.

We can be quite brief concerning the three-dimensional case. Let $R_3 = R \times R \times R$ be the set of ordered triples of real numbers (x_1, x_2, x_3) or, with more common notation, (x, y, z). To represent R_3 geometrically we take a three space, and choose an origin O and three perpendicular axes OX, OY, OZ. The axes, the coordinate planes, and the space are oriented. Let the positive direction of OZ be vertically upward. In Figure 1.6 we have taken OY in the plane of the page with the positive direction to the right. The positive direction of OX is perpendicular to the plane of the page and points toward the reader. The opposite orientation with OX and OY interchanged is also used. The difference is the same as that between a right-handed screw thread and a left-handed one. We introduce coordinates on the axes using O as the origin and the same unit of length on each axis. To get the coordinates of the point P, we project P in the manner indicated in the figure. P has the coordinates (x_0, y_0, z_0) if Q has the coordinates (x_0, y_0) in the xy-plane and if the plane through P, parallel to the xy-plane, intersects the z-axis at the point whose coordinate is z_0.

The distance between two points $P_1 : (x_1, y_1, z_1)$ and P_2: (x_2, y_2, z_2) is given by

$$d(P_1, P_2) = [(x_1 - x_2)^2 + (y_1 - y_2)^2 + (z_1 - z_2)^2]^{\frac{1}{2}}. \tag{1.5.7}$$

See Section 12.1. of Volume II. It satisfies conditions D_1, D_2, and D_3.

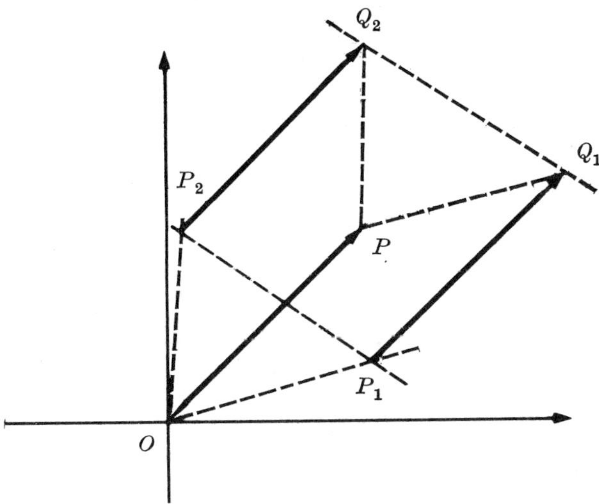

Figure 1.7

Instead of rectangles we now have rectangular *parallelepipeds*. The terms *interval* or *plurisegment* are sometimes used for both concepts. An ε-neighborhood of a point P_0 is usually understood to be the set of points P such that

$$d(P_0, P) < \varepsilon.$$

This is the *interior of a sphere*, often called an *open ball*.

The diameter of a set S is defined as in two dimensions as the least upper bound of the distances $d(P_1, P_2)$ where $P_1 \in S$, $P_2 \in S$. The theorem on nested intervals also extends to three dimensions.

Alternate means of giving geometric representations of real numbers taken singly, in pairs, or in triples are furnished by *vectors. A vector is a directed line segment* in a Euclidean space of one or more dimensions. If P is the initial and Q the terminal point of the segment, we denote the vector by $\overrightarrow{PQ}$. It should be observed that two vectors are defined to be equal if and only if they have the same direction and the same length, that is, if one vector can be carried into the other by parallel displacement. See Figure 1.7. Graphically we represent a vector by an arrow whose length is the magnitude or length of the vector. We normally denote a vector starting at the origin by a single letter, a bold face lower case type, such as $\mathbf{x}$ or y where we reserve italic type for vectors in two space. The length of the vector is denoted by $\|\mathbf{x}\|$ or $\|y\|$.

The representation of numbers, number pairs, and number triples by vectors is illustrated in Figures 1.8, 1.9, and 1.10 which correspond to Figures

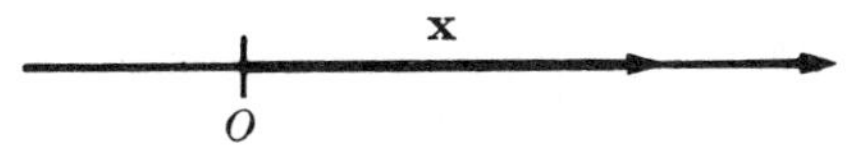

Figure 1.8

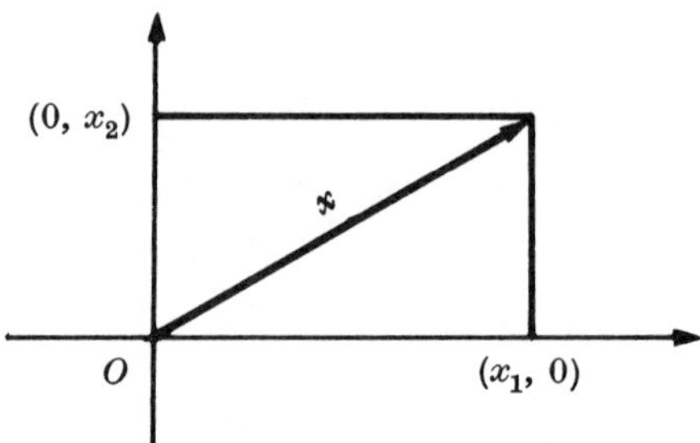

Figure 1.9

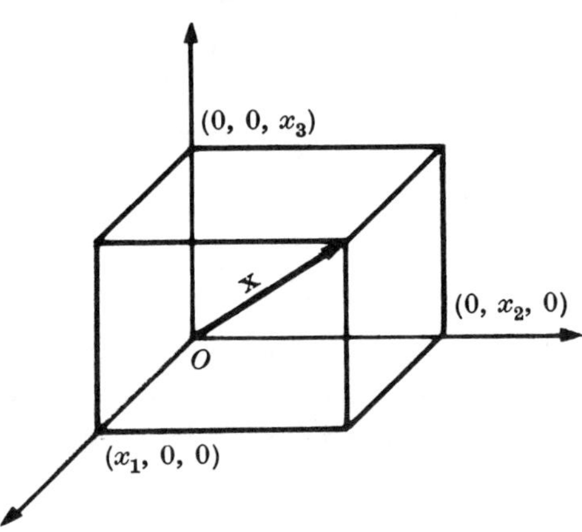

Figure 1.10

1.3, 1.5, and 1.6, respectively. On the line the point P with coordinate x corresponds to the directed line segment $\overrightarrow{OP}$ which is the vector $\mathbf{x} = (x)$. Similarly, in the plane the directed line segment $\overrightarrow{OP}$ is the vector $\mathbf{x} = (x_1, x_2)$ and in three space we have the vector $\mathbf{x} = (x_1, x_2, x_3)$. The numbers x, x_1, x_2, to x_3 are the components of the vector in question. In connection with vectors, numbers are referred to as *scalars*.

EXERCISE 1.5

1. Find the absolute values of 3, -3, $6 - 5\sqrt{2}$, $\pi - 3$, $3\pi - 10$, $\pi - 5\sqrt{2} + 1$.
2. Verify (1.5.4). When does equality hold in the first formula?
3. Show that $|x - a| + |x - b| \geqq |a - b|$ for all x, a, b.
4. Why do congruent segments form an equivalence class?
5. Why do parallel lines form an equivalence class?
6. Sketch the curve $y = |x + 1| - |x - 1|$.
7. In defining coordinates on the line it was asserted that there is a point P, $O < P$, such that $d(O, P) = \alpha$. Condition (d_6) assures that there is one point

on the line such that $d(O, P) = \alpha$, but how can we be sure that there are two such points, one on each side of O? [*Hint:* A rotation is a rigid motion.]

8. Mark in the plane the points

$$(1, 2), (-3, 4), (-1, 1), \text{ and } (3, -3)$$

and find the six distances between these points.

9. Find the midpoints of the diagonals of the quadrilateral whose vertices are given in the preceding problem.

10. If (a, b) is a given point in the plane, characterize geometrically the three points $(a, -b)$, $(-a, b)$, and $(-a, -b)$.

11. If a, b, c are positive numbers, characterize geometrically the eight points $(\pm a, \pm b, \pm c)$ where all possible combinations of signs are taken.

12. What is the diameter of the quadrilateral defined in Problem 8?

13. Find the diameter of a rectangle.

14. Find the diameter of the pyramid with vertices at $(0, 0, 0)$, $(1, 0, 0)$, $(0, 1, 0)$, $(0, 0, 1)$.

15. Find the diameter of a cube.

16. How small must ε be taken in order that the ε-neighborhoods of the points $(0, 0)$ and $(1, 1)$ have no point in common?

17. A set in one, two, or three dimensions is said to be *convex* if whenever P_1 and P_2 are in the set, so is the line segment P_1P_2. Give examples of convex sets.

18. The interior of a circle is convex. Why?

19. A half-plane is the set of points to one side of a given straight line. If three half-planes have a nonvoid intersection, prove that the latter is convex. What different types of sets can occur as intersections?

20. The intersection of two solid spheres is convex. Why?

21. Formulate and prove the theorem on nested circular disks.

22. Show that the vectors $(3, 4)$ and $(-8, 6)$ are perpendicular.

23. When are the two vectors $\mathbf{x}_1 = (a, b)$ and $\mathbf{x}_2 = (c, d)$ perpendicular to each other?

1.6. Inequalities. Manipulations with inequalities play an important role in analysis. We start with some simple observations concerning permissible algebraic operations on inequalities.

We recall the order relation "$<$" introduced in Section 1.4 subject to Postulates O_1 to O_4. One of these, O_2, asserts that

$$a < b \text{ implies } a + c < b + c \text{ for all } c.$$

From this and O_4 it follows that inequalities may be "added" whenever they go the same way. For from

$$x_1 < x_2 \quad \text{and} \quad y_1 < y_2 \tag{1.6.1}$$

we conclude that

$$x_1 + y_1 < x_1 + y_2 < x_2 + y_2,$$

that is,

$$x_1 + y_1 < x_2 + y_2 \tag{1.6.2}$$

which may be regarded as the sum of the inequalities under (1.6.1). This can be extended to any number of inequalities. Note that from

$$u_1 < u_2 \quad \text{and} \quad v_1 > v_2$$

no conclusion can be drawn concerning $u_1 + v_1$.

We recall that O_3 gives the right to multiply an inequality by a positive number: if

$$a < b \quad \text{and} \quad 0 < c \quad \text{then} \quad ac < bc.$$

Multiplication by a negative number, on the other hand, reverses the sense of an inequality: if

$$a < b \quad \text{and} \quad c < 0 \quad \text{then} \quad ac > bc.$$

Inequalities between positive numbers may be "multiplied" by each other for if

$$0 < x_1 < x_2 \quad \text{and} \quad 0 < y_1 < y_2 \tag{1.6.3}$$

then

$$x_1y_1 < x_2y_1 \quad \text{and} \quad x_2y_1 < x_2y_2$$

so that

$$0 < x_1y_1 < x_2y_2 \tag{1.6.4}$$

as asserted. In particular, an inequality between positive numbers may be "squared" or raised to a higher power. Thus

$$0 < x_1 < x_2 \quad \text{implies} \quad x_1^n < x_2^n, \quad n = 2, 3, \cdots. \tag{1.6.5}$$

Conversely, if it is known that x_1 and x_2 are positive numbers, then

$$x_1^n < x_2^n \quad \text{implies} \quad x_1 < x_2. \tag{1.6.6}$$

For we have the algebraic identity

$$b^n - a^n = (b - a)(b^{n-1} + b^{n-2}a + \cdots + ba^{n-2} + a^{n-1}), \tag{1.6.7}$$

whence it follows that $b^n - a^n$ and $b - a$ have the same sign when a and b are positive.

Finally we remind the reader of the elementary fact that the sign of a product of real numbers is $(-1)^k$ where k is the number of negative factors. It is understood that no factor is zero.

We now use these basic principles for the discussion of some elementary inequalities. The general problem is to find what limitations are put on a real number x when it has to satisfy one or more inequalities.

A trivial condition is given by

$$x < 3.$$

This means that $3 - x$ is positive, and the inequality is satisfied by all negative numbers, by zero and by all positive numbers between 0 and 3. A more interesting case is given if we require that simultaneously

$$x < 3 \quad \text{and} \quad x \geqq -1.$$

The second condition can be rewritten $-1 \leqq x$ and we see that the two conditions together give

$$-1 \leqq x < 3$$

that is, all x in the interval $[-1, 3)$ satisfy the conditions imposed on x. We say that the conditions are *compatible*.

On the other hand, the two conditions

$$x < -1, \quad 3 < x$$

are *incompatible*. No x can satisfy more than one of these conditions. Again consider the two sets of conditions

(i) $\quad x < 3, \quad -1 < x < 1, \quad 0 < x,$

(ii) $\quad x > 3, \quad -1 < x < 1, \quad 0 < x.$

Here the first set is compatible and is satisfied if and only if

$$0 < x < 1.$$

The second set is incompatible since the first two conditions are already incompatible. Thus there is no x satisfying (ii).

Let us consider some problems involving absolute values. The condition

$$|x| < 3$$

implies and is implied by

$$-3 < x < 3.$$

An equivalent condition is

$$x^2 < 9.$$

This follows from the definition of the absolute value, see (1.5.1). Similarly

$$|x - 1| < 2$$

is equivalent to

$$-2 < x - 1 < 2$$

or, adding 1 to all three members,

$$-1 < x < 3.$$

A different type of problem is the following:

For what values of a can the inequality

$$|x + 1| + |x - 1| < a$$

be satisfied by some value of x? Find the solutions when they exist.

Here $|x + 1|$ is the distance from x to -1 and $|x - 1|$ is the distance from x to $+1$ so the condition becomes

$$d(x, -1) + d(x, 1) < a. \tag{1.6.8.}$$

This can be satisfied if and only if a exceeds the distance between -1 and $+1$, that is for $a > 2$ and if this condition is satisfied we have either $|x| \leqq 1$ with

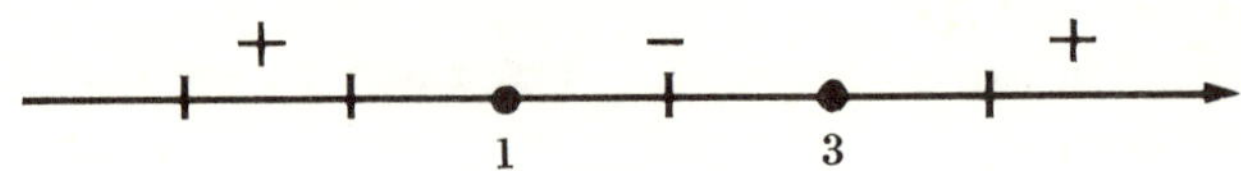

Figure 1.11

$d(x, -1) + d(x, 1) = 2$ or $|x| \geqq 1$ with $d(x, -1) + d(x, 1) = 2|x|$. Thus (1.6.8) holds if $a > 2$ and

$$-\tfrac{1}{2}a < x < \tfrac{1}{2}a.$$

We turn now to a class of inequalities which arise in the applications of the calculus to curve tracing, rectilinear motion, and econometrics. We shall give some instances.

1. For what values of x is

$$3(x^2 - 4x + 3) > 0?$$

The expression factors into

$$3(x - 1)(x - 4).$$

The outside factor 3 is positive and does not affect the result. It was observed above that the sign of such a product is $(-1)^k$ where k is the number of negative factors. Now $x - 1 < 0$ for $x < 1$ and is > 0 if $x > 1$, while $x - 3 < 0$ for $x < 3$ and is > 0 for $x > 3$. The two points $x = 1$ and $x = 3$ are the places where the factors $(x - 1)$ and $(x - 3)$ change signs. These points divide the real line into three intervals

$$(-\infty, 1), \quad (1, 3), \quad \text{and} \quad (3, \infty).$$

Here "∞" is read "infinity" and $x \in (-\infty, 1)$ means simply that x can be any number less than 1. In the first interval both factors are negative and the product positive; in the second, one factor is negative and the product is negative. Finally, in the last interval both factors are positive and so is the product. Hence we arrive at the result that $3(x - 1)(x - 3) > 0$ in the first and the third intervals, that is, if either $x < 1$ or $3 < x$. See Figure 1.11.

2. For what values of x is

$$-2(x + 1)(x - 1)(x - 3) > 0?$$

This is handled in the same way. Note that the factors are written in the order of increasing roots: $-1 < 1 < 3$. Here we have four intervals, separated by the roots,

$$(-\infty, -1), \quad (-1, 1), \quad (1, 3), \quad \text{and} \quad (3, \infty).$$

The outside factor -2 is negative and must be taken into account. There are four negative factors in the first interval, three in the second, two in the third, and one in the fourth. Hence the polynomial is positive in the first and the third intervals, that is, if either $x < -1$ or $1 < x < 3$. See Figure 1.12.

The same technique works in the following problem.

3. For what values of x is

$$|x^2 - x + 1| > 1?$$

This requires either

$$x^2 - x + 1 > 1 \quad \text{or} \quad x^2 - x + 1 < -1.$$

In the first case

$$x(x - 1) > 0.$$

There are three intervals to consider, namely

$$(-\infty, 0), \quad (0, 1), \quad \text{and} \quad (1, \infty).$$

It is found that the inequality holds in the first and in the third intervals. The second case requires

$$x^2 - x + 2 < 0 \quad \text{or} \quad (x - \tfrac{1}{2})^2 + \tfrac{7}{4} < 0$$

which is impossible for a real x. Thus the solution of the problem is that either $x < 0$ or $1 < x$.

A slightly more difficult problem is the following:

4. For what values of x do we have

$$x + 1 > \frac{1}{x} + \frac{2}{x + 1}, \quad x \neq 0, -1. \tag{1.6.9}$$

To solve this inequality for x we must multiply by $x(x + 1)$ to clear fractions. Now inequalities are preserved if the multiplier is positive, but reversed if it is negative. The first case presents itself if either $x < -1$ or $x > 0$. After clearing fractions, we get

$$x(x + 1)^2 > 3x + 1$$

or

$$x^3 + 2x^2 - 2x - 1 > 0.$$

The left member is 0 for $x = 1$ which leads to the factorization

$$(x + \tfrac{3}{2} + \tfrac{1}{2}\sqrt{5})(x + \tfrac{3}{2} - \tfrac{1}{2}\sqrt{5})(x - 1) > 0,$$

where we have to satisfy the condition

$$(x + 1)x > 0.$$

Combining the two conditions we find that either

$$-\tfrac{3}{2} - \tfrac{1}{2}\sqrt{5} < x < -1 \quad \text{or} \quad 1 < x.$$

But we have also to take into account that $(x + 1)x$ may be negative. In this case the cubic polynomial $x^3 + 2x^2 - 2x - 1$ must also be negative and the

Figure 1.12

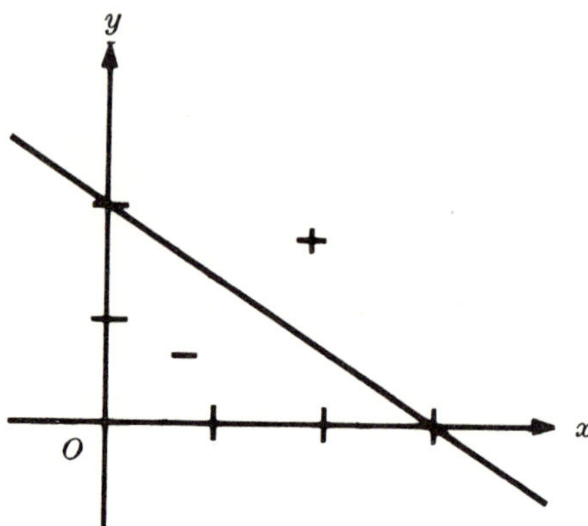

Figure 1.13

two simultaneous inequalities are satisfied for

$$-\tfrac{3}{2} + \tfrac{1}{2}\sqrt{5} < x < 0.$$

Combining the results we see that (1.6.9) holds if x satisfies one of the inequalities

$$-\tfrac{3}{2} - \tfrac{1}{2}\sqrt{5} < x < -1, \quad -\tfrac{3}{2} + \tfrac{1}{2}\sqrt{5} < x < 0, \quad 1 < x.$$

Let us consider some inequalities in the plane. The most common ones usually express that the point (x, y) lies to one side of a straight line or inside a circle or is separated from the origin by some curve or other. A few illustrations will suffice.

5. The straight line

$$Ax + By + C = 0 \tag{1.6.10}$$

separates the plane into two half-planes. See Figure 1.13 where $A = 2$, $B = 3$, $C = -6$. The left member of (1.6.10) is positive in one of these half-planes and negative in the other. To find which is the positive half-plane we need only substitute the coordinates of a point not on the line. In the present case the origin is a good choice and shows that the expression is negative below and positive above the line.

6. Two or more simultaneous inequalities may or may not be compatible. The first case arises for

$$x > 0, \quad y > 0, \quad x + y < 1$$

which is satisfied by the points interior to the triangle with vertices at the points (0, 0), (0, 1), and (1, 0). If in the last inequality we replace 1 by a, we obtain an incompatible set as soon as $a \leqq 0$.

7. The equation

$$(x - \alpha)^2 + (y - \beta)^2 = r^2, \quad 0 < r, \tag{1.6.11}$$

expresses that the distance from the point (x, y) to the point (α, β) equals r.

Hence it is the equation of a circle with center at (α, β) and radius r. It follows that

$$(x - \alpha)^2 + (y - \beta)^2 < r^2$$

represents the interior of the circle and the reversed inequality gives the exterior.

8. The inequality

$$x^2 - y^2 < 1$$

represents that part of the plane which lies between the two branches of the hyperbola

$$x^2 - y^2 = 1.$$

We shall not pursue such questions further. The rest of this section will be devoted to three famous inequalities, all of them known to Euclid.

I. *The triangle inequality. In a triangle the sum of the lengths of two sides equals at least the length of the third.*

This is property D_3 of Section 1.5. It is sufficient to give a proof for the case in which one vertex is at the origin. Suppose that

$$P_1 = (x_1, y_1), \quad P_2 = (x_2, y_2), \quad P_3 = (0, 0)$$

where the numbering is such that

$$d(O, P_2) \leqq d(O, P_1).$$

It is required to prove that

$$d(O, P_1) \leqq d(P_1, P_2) + d(O, P_2)$$

or

$$(x_1^2 + y_1^2)^{\frac{1}{2}} - (x_2^2 + y_2^2)^{\frac{1}{2}} \leqq [(x_1 - x_2)^2 + (y_1 - y_2)^2]^{\frac{1}{2}}.$$

Here the left member is nonnegative by assumption, so the inequality is true if and only if the inequality obtained by squaring both sides is true. We square both sides, expand the squares, cancel terms, and transfer what remains to obtain

$$x_1x_2 + y_1y_2 \leqq (x_1^2 + y_1^2)^{\frac{1}{2}}(x_2^2 + y_2^2)^{\frac{1}{2}}. \tag{1.6.12}$$

This is a special case of what is known as Cauchy's inequality. Now (1.6.12) is certainly true if the left member is negative or zero. If, on the other hand, the left member is positive, the inequality obtained by squaring both sides must be true and vice versa. From

$$(x_1x_2 + y_1y_2)^2 \leqq (x_1^2 + y_1^2)(x_2^2 + y_2^2),$$

by expanding, multiplying out, and collecting terms, we obtain the true inequality

$$0 \leqq (x_1y_2 - x_2y_1)^2. \tag{1.6.13}$$

We can then reverse the steps and conclude that the original inequality must be true.

In order that we shall have the equality

$$d(O, P_1) = d(P_1, P_2) + d(O, P_2),$$

we must clearly have equality at every step and, in particular, in (1.6.13). But if $x_1y_2 = x_2y_1$, then the three points O, P_1, and P_2 are collinear and P_2 lies between O and P_1 since $d(O, P_2) \leqq d(O, P_1)$. In this case equality clearly holds.

II. *The geometric mean is less than or equal to the arithmetic mean.*

For the case of two (positive) numbers a and b this asserts that

$$\sqrt{ab} \leqq \tfrac{1}{2}(a + b), \tag{1.6.14}$$

where the left member is the geometric mean of a and b and the right member the arithmetic mean. We have equality if and only if $a = b$. All this follows from the inequality

$$0 \leqq (\sqrt{b} - \sqrt{a})^2 = a + b - 2\sqrt{a}\,\sqrt{b}.$$

Euclid obtained the result from the construction of the *mean proportional* shown in Figure 1.14. The line segments AB and BC are of length a and b respectively, ADC is the semicircle of diameter AC, while BD is the mean proportional of AB and BC. The similar triangles show that the length of BD is $\sqrt{ab}$. Since the radius of the circle is $\frac{1}{2}(a + b)$ and $BD \leqq OC$, the inequality follows.

The inequality holds also for three or more numbers:

$$(a_1a_2 \cdots a_n)^{1/n} \leqq \frac{1}{n}(a_1 + a_2 + \cdots + a_n). \tag{1.6.15}$$

All quantities are positive, the left member is known as the geometric mean of the n numbers $a_1, a_2, \cdots, a_n$, the right member is the arithmetic mean, and equality holds if and only if all the a's are equal. We omit the proof.

As our last inequality we take one of the oldest problems in maxima and minima.

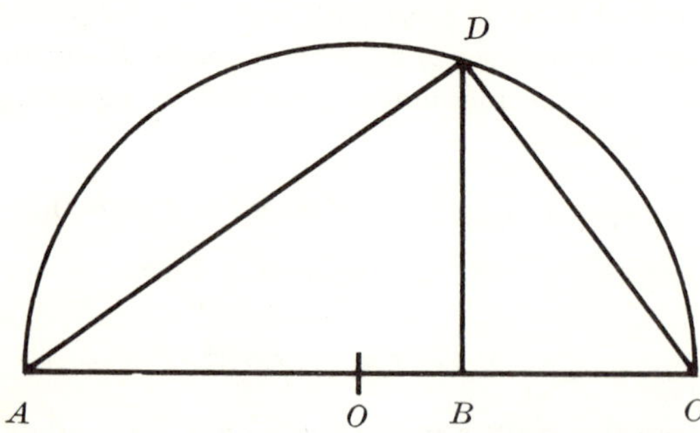

Figure 1.14

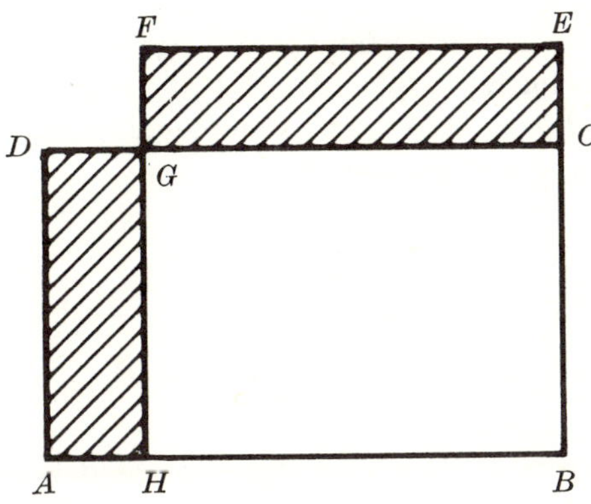

Figure 1.15

III. *Of all rectangles of given perimeter, the square has the largest area.*

The elementary proof of Euclid can be read from Figure 1.15. Here $ABCD$ is a rectangle, $HBCEFG$ a square of the same perimeter. This implies that $DG = GF$. Since $HG < HF = GC$, the area of the left shaded rectangle is less than that of the right one. Adding the area of $HBCG$ to both sides of the inequality, we see that the area of $ABCD$ is less than that of $HBEF$ which is the assertion.

A simple variation of the figure and argument gives:

Of all rectangles of given area, the square has the least perimeter.

These are typical problems in maxima and minima, one of the many topics the Calculus opened for systematic study. The reader will find many such problems in later chapters.

EXERCISE 1.6

1. If A is the set of all persons in a community who have reached the age of twenty-five and if B is the set of those under fifty, what inequalities hold for the age of a person who belongs to (i) $A \cap B$; (ii) $A \cup B$?

2. Two inequalities which determine sets A and B respectively are compatible if $A \cap B \neq \varnothing$. True or false?

3. Given the following sets

$$S_1 = \{x \mid 0 < x \leqq 2\}, \quad S_2 = \{x \mid 1 < x < 4\}, \quad S_3 = \{x \mid -1 < x \leqq 0\},$$

draw a figure and determine the sets $S_1 \cup S_2$, $S_1 \cup S_2 \cup S_3$, $S_1 \cap S_2$, $S_1 \cap S_2 \cap S_3$. Are there any incompatible inequalities?

4. Are the inequalities $|x - 1| < 2$ and $|x + 3| < 3$ compatible, and, if so, what values of x will satisfy both?

5. What is the distance of the point (x, y) from the negative x-axis? Distinguish between $x \leqq 0$ and $x > 0$.

6. For what values of x is $|x + 1|\,|x - 1| < \frac{1}{2}$?

7. Replace $\frac{1}{2}$ by a in the preceding inequality. How large must a be in order that the admissible values of x form a single interval?

8. In what part of the (x, y)-plane is $x^2 + (y - x)^2 < y^2$?
9. Same question for $|x| + |y| < 1$.
10. Prove that $x^2 + xy + y^2 \geqq 0$. Can equality hold if $(x, y) \neq (0, 0)$?
11. Can $2x^2 + 3xy + y^2$ be negative?
12. Verify (1.6.7).
13. If $0 < a < b$, $1 < n$, prove that

$$n(b - a)a^{n-1} < b^n - a^n < n(b - a)b^{n-1}.$$

14. For what values of x is $x^4 - 3x^3 + 2x^2 > 0$?

15. Discuss the inequality $|x + 1| > \left|\frac{1}{x} - 1\right|$.

16. Discuss the inequality $x + 1 > \frac{1}{x} - 1$.

17. If $h > 0$ and m is an integer, $m > 1$, prove that $(1 + h)^m > 1 + mh$. Is it true for $-1 < h < 0$?

18. Show that $a^{1/n} - 1 < (a - 1)/n$ for $1 < a$.

19. Prove that the square has the least perimeter of all rectangles of given area.

20. If $a^2 + b^2 + c^2 = 1$, prove that $ab + bc + ca \geqq -\frac{1}{2}$. If the least upper bound is 1, try to make this plausible.

21. If a, b, and c are positive numbers and $ab \leqq c^2$, show that at least one of the numbers a and b is $\leqq c$. What conclusion can be drawn if $ab \geqq c^2$ instead?

22. If $0 < x < 1$, show that $x(1 - x) \leqq \frac{1}{4}$. Can equality be reached?

23. If the n numbers $x_1, x_2, \cdots, x_n$ lie in the interval $(0, 1)$, show that at least one of the products

$$x_1x_2 \cdots x_n \quad \text{and} \quad (1 - x_1)(1 - x_2) \cdots (1 - x_n)$$

is $\leqq 2^{-n}$.

1.7. Sequential limits. One of the basic new concepts of the Calculus is that of a *limit*. Here it is crucial that the real numbers form an Archimedean system in which the least upper bound property holds. We start with sequential limits, other types will follow.

DEFINITION 1.7.1. *A sequence is a mapping of the set N of natural numbers into a set A.*

This means that to each natural number n corresponds one and only one element of A which we may denote by a_n. We refer to a_n as the nth element of the sequence. The same element of A may be the image of several integers n; in fact, the same value may be repeated infinitely often. We may even have $a_n = a$ for all n. Various symbols are used for sequences such as

$$\{a_1, a_2, a_3, \cdots, a_n, \cdots\},$$

or, in more condensed notation,

$$\{a_n \mid n = 1, 2, 3, \cdots\},$$

or simply

$$\{a_n\}.$$

To begin with, A will be a set of numbers, such as N, Q, or R, but other possibilities will be admitted later.

Let us consider first some special cases where $A = N$. The choice $a_n = n$ gives the identity mapping, and the corresponding sequence is that of the natural numbers in their natural order. Other choices for a_n would be

$$2n,\ 2n + 1,\ n^2,\ n^3,\quad \text{the } n\text{th prime}.$$

A different type of a sequence is the following. Compare the end of Section 1.4. We take A to be the set $0, 1, 2, 3, \ldots, 9$ and let a_n be the nth digit of a given infinite decimal fraction. Here at least one of the elements of A must correspond to infinitely many values of n. If figures are repeated infinitely often, the resulting pattern may be ultimately periodic, but in general the distribution of the digits is completely irregular. Still another type is furnished by the sequence

$$\{\log n \mid n = 1, 2, 3, \cdots\},$$

where the logarithms are to the base 10 and $A = R$, the set of real numbers. These examples could be multiplied without end.

In connection with limits we encounter sequences of a completely different structure.

Definition 1.7.2. *A sequence $\{a_n\}$ of real numbers is said to be a null sequence if, given any $\varepsilon > 0$, there is an integer $N(\varepsilon)$, depending upon the choice of ε, such that*

$$|a_n| \leqq \varepsilon \quad \textit{for all} \quad n \geqq N(\varepsilon). \tag{1.7.1}$$

Geometrically this means that if we choose an ε-neighborhood of the origin, then all but a finite number of the elements a_n of the sequence lie in this neighborhood. None of the sequences described above is a null sequence. On the other hand

$$\{n^{-1}\},\quad \{(-1)^n n^{-2}\},\quad \{[1 - (-1)^n]2^{-n}\},\quad \{(\log n)^{-1}\}$$

are typical null sequences. In the last instance, the logarithms are to the base 10 and $n > 1$.

To see what "all but a finite number of the elements a_n of the sequence" really means in the cases listed, we choose $\varepsilon = 10^{-6}$. We find that $N(\varepsilon)$ is respectively

$$10^6,\ 10^3,\ 20,\quad \text{and} \quad 10^{10^6}.$$

The last number is a 1 followed by a million zeros. This is a finite number even if it dwarfs all astronomical figures.

Definition 1.7.3. *If $\{a_n\}$ is a null sequence, a_n is said to converge to 0 or, equivalently, the limit of a_n is 0.*

This is written

$$a_n \to 0 \quad \text{and} \quad \lim_{n\to\infty} a_n = 0 \tag{1.7.2}$$

respectively.

More generally we have the following

DEFINITION 1.7.4. *Given a sequence* $\{a_n\}$, $a_n \in R$, *we say that* a_n *converges to the number* a *or the limit of* a_n *exists and equals* a, *if* $\{a_n - a\}$ *is a null sequence.*

We write

$$a_n \to a \quad \text{or} \quad \lim_{n\to\infty} a_n = a. \tag{1.7.3}$$

Going back to Definition 1.7.2, we see that (1.7.3) holds if and only if, given any $\varepsilon > 0$, we can find an $N(\varepsilon)$ such that

$$|a_n - a| \leqq \varepsilon \quad \text{for all} \quad n \geqq N(\varepsilon). \tag{1.7.4}$$

How can we tell if a sequence is convergent? It is easy enough to test if it is a null sequence, but the general problem of proving or disproving the existence of an a satisfying (1.7.4) is much harder. We shall take up this problem later; here we discuss a special case in which the solution is easily found.

DEFINITION 1.7.5. *The sequence* $\{a_n\}$ *is bounded if there exists a finite* M *such that* $|a_n| \leqq M$ *for all* n. *The sequence is nondecreasing if* $m < n$ *implies*

$$a_m \leqq a_n. \tag{1.7.5}$$

It is strictly increasing if inequality holds in (1.7.5) *for all* $m < n$. *It is nonincreasing if inequality* (1.7.5) *is reversed and strictly decreasing if* $>$ *always holds. A sequence is monotone or monotonic if it is nondecreasing or nonincreasing.*

The term "increasing" is often used instead of "nondecreasing," and similarly "decreasing" is used where the correct term would be "nonincreasing."

Sequences like

$$\{n\}, \quad \{n^2\}, \quad \{\log n\}$$

are all strictly increasing, but they are obviously not bounded. On the other hand

$$\left\{\frac{n-1}{n}\right\} = \left\{1 - \frac{1}{n}\right\}$$

is a bounded and strictly increasing sequence. In this case obviously 1 is the limit of a_n since

$$\left\{\frac{1}{n}\right\}$$

is a null sequence. The latter sequence is bounded and strictly decreasing. It is of some interest to note that 1 is also the least upper bound of the set $\left\{1 - \dfrac{1}{n}\right\}$.

This will carry over to the general monotonic case, but if the sequence is not strictly increasing we have to consider its range.

DEFINITION 1.7.6. *The range of the sequence* $\{a_n\}$ *is the set of distinct numbers occurring in the sequence.*

Thus in the case of

$$\left\{1 - \frac{1}{n}\right\}$$

the sequence and its range contain the same elements since they are all distinct, but the range of the sequence

$$\{(-1)^n\}$$

is simply the set $\{-1, +1\}$. It is clear that the range of a sequence is bounded if and only if the sequence is bounded. We can now formulate and prove

THEOREM 1.7.1. *A bounded nondecreasing sequence* $\{a_n\}$ *converges to the least upper bound of its range, while a nonincreasing sequence converges to the greatest lower bound of its range.*

Proof. Since the range is a bounded set it has a least upper bound ρ and a greatest lower bound λ. Suppose the sequence is nondecreasing. We have then $a_n \leqq \rho$ for all n. On the other hand, the definition of the least upper bound implies that, given an $\varepsilon > 0$, we can find an integer m such that

$$\rho - \varepsilon \leqq a_m,$$

and this implies that

$$\rho - \varepsilon \leqq a_n \leqq \rho \quad \text{for all} \quad n \geqq m,$$

since the sequence is nondecreasing. It follows that

$$0 \leqq \rho - a_n \leqq \varepsilon, \quad n \geqq m,$$

so that $\{\rho - a_n\}$ is a null sequence and

$$\lim_{n\to\infty} a_n = \rho.$$

This proves the assertion for nondecreasing sequences and the nonincreasing case is proved in the same manner.

Example 1. If $S_1(n)$ is the sum of the first n integers, and

$$a_n = n^{-2}S_1(n), \tag{1.7.6}$$

then the sequence $\{a_n\}$ is strictly decreasing and

$$\lim_{n\to\infty} a_n = \tfrac{1}{2}. \tag{1.7.7}$$

For by formula (1.2.3)

$$a_n = \frac{1}{2} + \frac{1}{2n},$$

whence the assertion follows.

Example 2. The sequence $\{b_n\}$ with

$$b_n = \frac{1}{1\cdot 2} + \frac{1}{2\cdot 3} + \frac{1}{3\cdot 4} + \cdots + \frac{1}{n(n+1)}, \quad n = 1, 2, 3, \cdots, \tag{1.7.8}$$

is strictly increasing and

$$\lim_{n\to\infty} b_n = 1. \tag{1.7.9}$$

Since

$$\frac{1}{k(k+1)} = \frac{1}{k} - \frac{1}{k+1},$$

it follows that

$$b_n = \left(\frac{1}{1} - \frac{1}{2}\right) + \left(\frac{1}{2} - \frac{1}{3}\right) + \left(\frac{1}{3} - \frac{1}{4}\right) + \cdots + \left(\frac{1}{n} - \frac{1}{n+1}\right). \tag{1.7.10}$$

When the parentheses are removed, all terms cancel except the first and the last. The conclusion now follows from

$$b_n = 1 - \frac{1}{n+1}.$$

Example 3. The strictly increasing sequence

$$c_n = 1 + \frac{1}{2^2} + \frac{1}{3^2} + \frac{1}{4^2} + \cdots + \frac{1}{n^2}, \quad n = 1, 2, 3, \cdots \tag{1.7.11}$$

is bounded and hence convergent. Its limit is $\frac{1}{6}\pi^2 = 1.64493$ to five significant decimal places.

Since

$$\frac{1}{k^2} < \frac{1}{(k-1)k}, \quad 1 < k,$$

it follows that

$$c_n < 1 + b_{n-1} < 2$$

for all n. The value of the limit cannot be found with the means available to us.

The next example is of a less obvious nature.

Example 4. The sequence

$$d_n = \frac{1}{n+1} + \frac{1}{n+2} + \cdots + \frac{1}{2n}, \quad n = 1, 2, 3, \cdots, \tag{1.7.12}$$

is bounded and strictly increasing. Its limit is the natural logarithm of 2 or 0.69315.

The difference between two consecutive elements is

$$\begin{aligned} d_n - d_{n-1} = {} & \frac{1}{n+1} + \frac{1}{n+2} + \cdots + \frac{1}{2n-2} + \frac{1}{2n-1} + \frac{1}{2n} \\ & - \frac{1}{n} - \frac{1}{n+1} - \frac{1}{n+2} - \cdots - \frac{1}{2n-2} \\ = {} & \frac{1}{2n-1} + \frac{1}{2n} - \frac{1}{n} = \frac{1}{2n-1} - \frac{1}{2n} = \frac{1}{(2n-1)2n} > 0, \end{aligned}$$

so the sequence is strictly increasing. It is also bounded since d_n is the sum of n terms of which the first is the largest and the last is the smallest. It follows that

$$\frac{1}{2} = n \cdot \frac{1}{2n} < d_n < \frac{n}{n+1} < 1.$$

Since

$$d_n - d_1 = (d_n - d_{n-1}) + (d_{n-1} - d_{n-2}) + \cdots + (d_3 - d_2) + (d_2 - d_1),$$

we obtain

$$d_n = \frac{1}{1 \cdot 2} + \frac{1}{3 \cdot 4} + \frac{1}{5 \cdot 6} + \cdots + \frac{1}{(2n-1)2n}, \tag{1.7.13}$$

which can be used to obtain the actual value of the limit. We shall encounter this sequence in other contexts later.

The last example will be one of the sequences used to define e, the base of the system of natural logarithms. This is the only system used in higher mathematics and is also the oldest system.

Example 5. The sequence

$$e_n = \left(1 + \frac{1}{n}\right)^n, \quad n = 1, 2, 3, \cdots, \tag{1.7.14}$$

is bounded, $e_n < 4$ for all n, and it is strictly increasing. We write

$$\lim_{n\to\infty} e_n = e. \tag{1.7.15}$$

The numerical value of e to twelve decimal places is

$$e = 2.71828\ 18284\ 59 \cdots. \tag{1.7.16}$$

Another thousand digits are available in case of need. We have

$$e_1 = 2, \quad e_2 = 2.25, \quad e_3 = 2.37\cdot, \quad e_4 = 2.44\cdot.$$

The sequence converges fairly rapidly, but we shall later encounter an infinite series for e more suitable for numerical computation. To prove that $e_n < e_{n+1}$ for all n, we use Problem 13 of Exercise 1.6 where we replace n by $n + 1$ obtaining for $0 < a < b$

$$(n+1)(b-a)a^n < b^{n+1} - a^{n+1} < (n+1)(b-a)b^n. \tag{1.7.17}$$

Here we set

$$a = 1 + \frac{1}{n+1}, \quad b = 1 + \frac{1}{n},$$

and use the second inequality in (1.7.17) to obtain

$$\left(1 + \frac{1}{n}\right)^{n+1} - \left(1 + \frac{1}{n+1}\right)^{n+1} < (n+1)\left(\frac{1}{n} - \frac{1}{n+1}\right)\left(1 + \frac{1}{n}\right)^n$$
$$= \frac{1}{n}\left(1 + \frac{1}{n}\right)^n.$$

This simplifies to

$$\left(1+\frac{1}{n}\right)^n < \left(1+\frac{1}{n+1}\right)^{n+1},$$

that is, the desired inequality.

To prove the sequence $\{e_n\}$ bounded, we can also use (1.7.17). Moreover, we can arrange the proof to obtain a strictly decreasing sequence which converges to e from above. For this purpose, choose $k > 1$ as an integer and set

$$a = 1, \quad b = 1 + \frac{1}{kn}$$

in the second half of (1.7.17). After simplification we get

$$\left(1+\frac{1}{kn}\right)^n < \frac{k}{k-1} = \left(1-\frac{1}{k}\right)^{-1}.$$

Taking kth powers we obtain

$$e_{kn} = \left(1+\frac{1}{kn}\right)^{kn} < \left(1-\frac{1}{k}\right)^{-k}. \tag{1.7.18}$$

In particular, for $k = 2$ and $n = m$ we have

$$e_{2m} < 4$$

for all m. Since $e_{2m-1} < e_{2m}$ for all m, $e_{2m} < 4$ implies that $e_n < 4$ for all n. This suffices to prove the existence of $\lim e_n = e$.

We can get more out of a judicious use of (1.7.17). Since $\{e_n\}$ is a strictly increasing sequence, the same is true of any subsequence such as

$$\{e_{mk} \mid m = 1, 2, 3, \cdots\}.$$

Moreover, the two sequences must have the same least upper bound and hence the same limit. Thus

$$e = \lim_{n\to\infty} e_n = \lim_{m\to\infty} e_{mk} \leqq \left(1-\frac{1}{k}\right)^{-k}$$

for each $k \geqq 2$. We now define

$$E_k = \left(1-\frac{1}{k}\right)^{-k}, \quad k = 2, 3, 4, \cdots, \tag{1.7.19}$$

so that

$$e \leqq E_k, \quad \text{for all} \quad k. \tag{1.7.20}$$

Since

$$E_{n+1} = \left(1-\frac{1}{n+1}\right)^{-n-1} = \left(1+\frac{1}{n}\right)^{n+1},$$

we see that

$$E_{n+1} - e_n = \left(1+\frac{1}{n}\right)^{n+1} - \left(1+\frac{1}{n}\right)^n = \frac{1}{n}\left(1+\frac{1}{n}\right)^n < \frac{e}{n}.$$

Hence $\{E_{n+1} - e_n\}$ is a null sequence implying that

$$\lim_{n\to\infty} E_n = e. \tag{1.7.21}$$

Actually the sequence $\{E_n\}$ is strictly decreasing. The proof is left as an exercise for the reader.

Thus e is the common limit of two sequences, one increasing and the other decreasing. Other expressions having e as a limit will be encountered later. John Napier, baron of Merchiston [1550–1617], who published the first table of natural logarithms in 1614, actually did not work with e but with e_n for $n = 10^7$. The difference $e - e_n$ is about 10^{-7}. Logarithms to the base 10 were constructed by Henry Briggs [1556–1631] in 1624 using natural logarithms.

We turn now to *operations with sequences.*

Definition 1.7.7. *If $\{a_n\}$ and $\{b_n\}$ are two sequences and c is a number, then we define the sum of two sequences by*

$$\{a_n\} + \{b_n\} = \{a_n + b_n\}, \tag{1.7.22}$$

the product by

$$\{a_n\} \cdot \{b_n\} = \{a_n b_n\}, \tag{1.7.23}$$

and the scalar product by

$$c \cdot \{a_n\} = \{ca_n\}. \tag{1.7.24}$$

Theorem 1.7.2. *The following relations hold whenever the right members exist:*

$$\lim_{n\to\infty} (a_n + b_n) = \lim_{n\to\infty} a_n + \lim_{n\to\infty} b_n, \tag{1.7.25}$$

$$\lim_{n\to\infty} a_n b_n = (\lim_{n\to\infty} a_n)(\lim_{n\to\infty} b_n), \tag{1.7.26}$$

$$\lim_{n\to\infty} ca_n = c(\lim_{n\to\infty} a_n). \tag{1.7.27}$$

Proof. Suppose that the right members do exist and set

$$\lim_{n\to\infty} a_n = a, \qquad \lim_{n\to\infty} b_n = b.$$

Hence, if we set

$$a_n = a + \alpha_n, \quad b_n = b + \beta_n,$$

then by Definition 1.7.4 $\{\alpha_n\}$ and $\{\beta_n\}$ are null sequences. Since

$$a_n + b_n = a + \alpha_n + b + \beta_n = (a + b) + (\alpha_n + \beta_n),$$

we see that (1.7.25) holds if $\{\alpha_n + \beta_n\}$ is also a null sequence. This is rather self-evident; it says that the sum of two small quantities is also small. A formal proof would go like this. Given an $\varepsilon > 0$, we can find a natural number $N(\varepsilon)$ such that

$$|\alpha_n| \leqq \varepsilon \quad \text{and} \quad |\beta_n| \leqq \varepsilon \quad \text{for all} \quad n \geqq N(\varepsilon).$$

Hence

$$|\alpha_n + \beta_n| \leqq |\alpha_n| + |\beta_n| \leqq 2\varepsilon \quad \text{for all} \quad n \geqq N(\varepsilon).$$

By Definition 1.7.2 this implies that $\{\alpha_n + \beta_n\}$ is a null sequence so that (1.7.25) holds.

In the same way we handle (1.7.27). We have

$$ca_n = c(a + \alpha_n) = ca + c\alpha_n$$

and

$$|c\alpha_n| = |c|\,|\alpha_n| \leqq |c|\,\varepsilon, \quad n \geqq N(\varepsilon).$$

It follows that $\{c\alpha_n\}$ is a null sequence since $\{\alpha_n\}$ is one and this implies (1.7.27).

Finally we have

$$a_nb_n = (a + \alpha_n)(b + \beta_n) = ab + a\beta_n + b\alpha_n + \alpha_n\beta_n.$$

It follows from the previous argument that for $n \geqq N(\varepsilon)$

$$\begin{aligned} |a\beta_n + b\alpha_n + \alpha_n\beta_n| \leqq& \quad |a|\,|\beta_n| + |b|\,|\alpha_n| + |\alpha_n|\,|\beta_n| \\ &\leqq |a|\,\varepsilon + |b|\,\varepsilon + \varepsilon^2 = (|a| + |b| + \varepsilon)\varepsilon, \end{aligned}$$

so that $a_nb_n \to ab$ as asserted.

The idea of the proof is to reduce the problem to the special case of null sequences where the truth of the assertion is intuitively obvious.

Note that the left members of formulas (1.7.25) to (1.7.27) may well exist without the existence of $\lim a_n$ and $\lim b_n$. Thus, if $b_n = -a_n$, the sum $a_n + b_n = 0$ and the resulting trivial null sequence converges to zero regardless of the behavior of a_n. Similarly for the product, for if $a_n \neq 0$ for all n, we may take $b_n = a_n^{-1}$, so that $a_nb_n = 1$. Now the left member of (1.7.26) is 1, while neither a_n nor b_n need have a limit. On the other hand, if $c \neq 0$, the two sides of (1.7.27) coexist, that is, if one limit exists so does the other.

The sequence $\{a_n^{-1}\}$ introduced above, in the case where $a_n \neq 0$ for all n, may be regarded as the *reciprocal sequence*:

$$\{a_n\}^{-1} = \left\{\frac{1}{a_n}\right\}. \tag{1.7.28}$$

We can then also consider the *quotient*

$$\frac{\{b_n\}}{\{a_n\}} = \left\{\frac{b_n}{a_n}\right\} = \left\{b_n \cdot \frac{1}{a_n}\right\}. \tag{1.7.29}$$

THEOREM 1.7.3. *If $a_n \neq 0$ for all n and if*

$$\lim_{n\to\infty} a_n = a \neq 0, \tag{1.7.30}$$

then

$$\lim_{n\to\infty} \frac{1}{a_n} = \frac{1}{a}. \tag{1.7.31}$$

Proof. In our previous notation

$$\frac{1}{a} - \frac{1}{a_n} = \frac{a_n - a}{aa_n} = \frac{\alpha_n}{a(a + \alpha_n)}.$$

We can suppose that ε is given so small that $\varepsilon < \frac{1}{2}|a|$ and that $|\alpha_n| < \varepsilon$ for $n > N$. We have then

$$|a + \alpha_n| \geqq |a| - |\alpha_n| \geqq |a| - \varepsilon > \tfrac{1}{2}|a|.$$

This gives

$$\frac{1}{|a + \alpha_n|} < \frac{2}{|a|},$$

and finally

$$\left|\frac{1}{a} - \frac{1}{a_n}\right| < \frac{2\varepsilon}{|a|^2}.$$

Since ε is at our disposal, this shows that $\left\{\frac{1}{a} - \frac{1}{a_n}\right\}$ is a null sequence and (1.7.31) is proved.

Again we must observe that the left member of (1.7.31) may exist without $\lim_{n\to\infty} a_n$ existing. The choice $a_n = n$ shows this possibility.

Corollary. *If* $\{a_n\}$ *satisfies the assumptions of Theorem* 1.7.3 *and if* $\lim_{n\to\infty} b_n = b$, *then*

$$\lim_{n\to\infty} \frac{b_n}{a_n} = \frac{b}{a}. \tag{1.7.32}$$

This follows by combining the results of the last two theorems.

In the applications we are often concerned with quotient sequences $\left\{\frac{b_n}{a_n}\right\}$ where the corresponding sequences $\{a_n\}$ and $\{b_n\}$ are unbounded. In this case (1.7.32) does not apply. But if the sequences $\{a_n\}$ and $\{b_n\}$ grow at a comparable rate, the limit may well exist. Suppose, for instance, that we can find a positive real number α such that

$$c_n = n^{-\alpha}a_n, \quad d_n = n^{-\alpha}b_n \tag{1.7.33}$$

tend to limits $c \neq 0$ and d respectively. We have then

$$\frac{b_n}{a_n} = \frac{n^{-\alpha}b_n}{n^{-\alpha}a_n} = \frac{d_n}{c_n},$$

whence

$$\lim_{n\to\infty} \frac{b_n}{a_n} = \lim_{n\to\infty} \frac{d_n}{c_n} = \frac{d}{c}. \tag{1.7.34}$$

As an example consider the case

$$a_n = n^2 + n + 1, \quad b_n = 2n^2 - 1.$$

Here we can take $\alpha = 2$ and find that

$$c_n = 1 + \frac{1}{n} + \frac{1}{n^2}, \quad d_n = 2 - \frac{1}{n^2}$$

and

$$\lim_{n\to\infty} \frac{2n^2 - 1}{n^2 + n + 1} = 2.$$

EXERCISE 1.7

1. Show that the digits in the decimal fraction for $\frac{1}{11}$ form the sequence $\{a_n\}$ where

$$a_n = \tfrac{9}{2}[1 + (-1)^n], \quad n = 1, 2, 3, \cdots.$$

2. Let $[u]$ be the largest integer $\leqq u$. Define

$$a_n = \tfrac{5}{2} + (-1)^n + \tfrac{1}{2}(-1)^{[n/2]}, \quad n = 1, 2, 3, \cdots,$$

and form the decimal fraction $0 . a_1 a_2 a_3 \cdots a_n \cdots$. It is periodic in the sense that $a_{n+4} = a_n$ for all n. Find the rational number defined by this decimal expansion.

3. Every ultimately periodic decimal fraction defines a rational number. True or false? Is the converse true?

Verify that $\{a_n\}$ is a null sequence if a_n is given by

4. $\dfrac{1}{\sqrt{n+1}}$. 5. $\dfrac{n}{n^2+1}$. 6. $\dfrac{n^2-1}{n^3+3n}$. 7. $\dfrac{2^n}{4^n+1}$.

[*Hint:* In 5. to 7. use (1.7.34) or similar device.]

8. Verify that

$$a_n = (n+1)^{\frac{1}{2}} - n^{\frac{1}{2}}$$

defines a null sequence. Prove that

$$\lim_{n\to\infty} n^{\frac{1}{2}} a_n = \tfrac{1}{2}.$$

Find an N such that $a_n < 10^{-6}$ for $n > N$.

9. If a_n is defined as in the preceding problem and if

$$A_n = a_1 + a_2 + a_3 + \cdots + a_n, \quad n = 1, 2, 3, \cdots,$$

verify that $\{A_n\}$ is a strictly increasing unbounded sequence.

Use (1.7.34) or similar device to find the limit of q_n if q_n is given by

10. $\dfrac{n}{n+1}$. 11. $\dfrac{n^2-1}{3n^2+1}$. 12. $\dfrac{n}{\sqrt{n^2+1}}$. 13. $\dfrac{n}{n+\sqrt{n^2+1}}$.

14. Find the limit of a_n if

$$a_n = (n^2 + 3n)^{\frac{1}{2}} - n.$$

[*Hint:* Compare with Problem 8.]

15. Prove that the sequence $\{E_n\}$ of (1.7.19) is strictly decreasing. [*Hint:* Use the first half of (1.7.17) with $a = 1 - \dfrac{1}{n}$, $b = 1 - \dfrac{1}{n+1}$, and simplify.]

16. Verify that the sequence $\{n^{1/n}\}$ is strictly decreasing for $n > 3$ and find its limit. [*Hint:* Use the definition of e to show that $n^{n+1} > (n+1)^n$ for $n > 2$.]

17. If a is a fixed positive number, show that the sequence $\{a^{1/n}\}$ is monotonic and find its limit. Note that there are three different cases according as $a < 1$, $a = 1$, or $a > 1$.

18. If $S_2(n)$ is the sum of the squares of the first n natural numbers, show that $n^{-3}S_2(n) \to \frac{1}{3}$. [The geometrical interpretation is that the area bounded by the x-axis, the line $x = 1$ and the parabola $y = x^2$ equals $\frac{1}{3}$. Archimedes.]

19. If $|x| < 1$ and k is a positive integer, prove that

$$\lim_{n\to\infty} n^k x^n = 0.$$

20. If $a_n > 0$ for all n and if $\lim\limits_{n\to\infty} a_n = a$, prove that $\lim\limits_{n\to\infty} \sqrt{a_n} = \sqrt{a}$.

Find the limit of a_n, when it exists, if a_n is:

21. $(2/n)^n$. 22. n^{3-n}. 23. $2^{n^2} \cdot n^{-n}$. 24. $n \cdot 10^{-n}$.

25. Given a convergent sequence $\{a_n\}$ and one of its subsequences $\{b_k\}$. The latter is obtained by omitting some of the elements of the former, possibly infinitely many. Prove that $\lim b_k = \lim a_n$.

26. The sequence $\{a_n\}$ is supposed to be strictly increasing. Define

$$c_n = \frac{1}{n}(a_1 + a_2 + \cdots + a_n),$$

and show that $\{c_n\}$ is also strictly increasing.

27. Define $a_1 = \sqrt{2}$, $a_n = \sqrt{2a_{n-1}}$, $n = 2, 3, 4, \cdots$. Show that $\{a_n\}$ is bounded, strictly increasing and find $\lim a_n$.

Find $\lim a_n$, when it exists, if a_n is

28. $\left(\dfrac{n+1}{n+2}\right)^n$. 29. $\left(1 + \dfrac{1}{n^2}\right)^n$. 30. $\left(1 + \dfrac{1}{n}\right)^{n^2}$.

The binomial coefficient $\binom{n}{k}$, also written $C_{n,k}$, is defined as the coefficient of x^k in the expansion of $(1 + x)^n$. The coefficients for $n = 0, 1, 2, 3, 4, 5$ are shown in the following array, known as *Pascal's triangle:*

```
                1
             1     1
          1     2     1
       1     3     3     1
    1     4     6     4     1
 1     5    10    10     5     1
```

31. The triangle suggests that

(i) $\dbinom{n+1}{k} = \dbinom{n}{k} + \dbinom{n}{k-1}$; (ii) $\dbinom{n}{k} = \dfrac{n(n-1)\cdots(n-k+1)}{1\cdot 2\cdot 3\cdots k}$.

The first property follows from the identity

$$(1 + x)^{n+1} = (1 + x)(1 + x)^n,$$

upon substituting the expansions of the powers, multiplying out, and comparing coefficients. Carry out the details of the proof.

32. It is obvious that $\binom{k}{0} = 1$. Prove that $\binom{k}{1} = k$ by induction on k and verify (ii) by induction on n for $n > k$.

33. The symbol $k!$ is read "k factorial" and is defined as the product of the first k integers. By definition $0! = 1$. Make a table of the first ten factorials. Verify that

$$\binom{n}{k} = \frac{n!}{k!\,(n-k)!}\,.$$

MISCELLANEOUS PROBLEMS FOR CHAPTER 1

1. Show that no point on the line

$$2x + 4y = 11$$

can have coordinates which are both integers.

2. Find all points on the line

$$x + 2y = 6$$

whose coordinates are both integers.

3. Show that each natural number p has a unique *dyadic representation*

$$p = a_n 2^n + a_{n-1}2^{n-1} + \cdots + a_1 2 + a_0,$$

where $a_n = 1$ and a_k is 0 or 1 for $k = 0, 1, 2, \cdots, n-1$. What is the dyadic representation of 1000? Dyadic representations are basic in high-speed computation.

4. Plot the curve

$$y = |x| + |1 - x|.$$

5. Show that $\sqrt{2} + \sqrt{3}$ cannot be rational. You may assume that $\sqrt{6}$ is irrational.

6. Three noncollinear points $P_1 = (x_1, y_1)$, $P_2 = (x_2, y_2)$, $P_3 = (x_3, y_3)$ are given in the plane; $P = (x, y)$ is a variable point. For each position of P there is normally a unique point P_k which is nearest to P. Describe by inequalities and construct geometrically the domain where P_1 is the nearest point. What happens if the points are collinear?

7. Prove that the intersection of two plane convex sets is either a convex set or is void.

8. There are mappings of the natural numbers into the positive rationals which are "onto." Verify that such a mapping $n \to p/q$ is defined by

$$p = n - \tfrac{1}{2}k(k+1), \quad q = k + 2 - p \quad \text{if} \quad \tfrac{1}{2}k(k+1) \leqq n < \tfrac{1}{2}(k+1)(k+2)$$

and $k = 1, 2, 3, \cdots$.

9. If $0 < x < 1$, prove that

$$x(1 - 2x) \leqq \tfrac{1}{8}.$$

Is equality reached? Give a geometric interpretation of the inequality.

10. When is

$$\frac{x+1}{x(1-x)} > 1?$$

11. A sequence is defined by

$$a_0 = 1, \quad a_n = (4 + 3a_{n-1})^{\frac{1}{2}}, \quad n = 2, 3, 4, \cdots.$$

Show that the sequence is monotone and bounded and find its limit.

12. Let $N_1 < N_2 < N_3 < \cdots < N_k < \cdots$ be the natural numbers whose dyadic representations involve only ones. Find

$$\lim_{k\to\infty} \frac{N_{k+1}}{N_k}.$$

13. For $-1 < x, y < 1$ we define

$$x \oplus y = \frac{x+y}{1+xy}.$$

Show that this operation satisfies postulates A_1 to A_5 and that 0 is the zero element.

14. Let F be the ordered field formed by the real numbers under the operations

$$x \oplus y = x + y - 1, \quad x \odot y = x + y - xy,$$

and the order relation $x \prec y$ if $y < x$. Let R be the field of real numbers with the usual operations and order relation. If $z \in R$ and $w \in F$, prove that the mapping $w = 1 - z$ is an order preserving isomorphism. In other words if $z_1 \to w_1$, $z_2 \to w_2$, $z_1 < z_2$, it should be proved that $z_1 + z_2 \to w_1 + w_2$, $z_1 \cdot z_2 \to w_1 \odot w_2$, $w_1 \prec w_2$.

*15. Define a sequence $\{a_n\}$ by

$$a_1 = 1, \quad a_n = 1 + \frac{1}{a_{n-1}}, \quad n = 2, 3, 4, \cdots.$$

Prove that the subsequence $\{a_1, a_3, a_5, \cdots, a_{2k+1}, \cdots\}$ is strictly increasing while $\{a_2, a_4, a_6, \cdots, a_{2k}, \cdots\}$ is strictly decreasing. Show that the two subsequences have the same limit which is the positive root of the equation

$$x^2 - x - 1 = 0.$$

[*Hint:* Consider the difference between two consecutive elements of the original sequence.]

16. For $0 \leqq x \leqq \frac{1}{4}$ define a sequence of polynomials by

$$P_1(x) = x, \quad P_n(x) = x + [P_{n-1}(x)]^2, \quad n = 2, 3, 4, \cdots.$$

Show that for each such x the sequence $\{P_n(x)\}$ is strictly increasing and bounded by $\frac{1}{2}$. Find its limit. Show that $P_n(x)$ is a polynomial of degree 2^{n-1} whose coefficients are positive integers and that $P_n(x) - P_{n-1}(x)$ is divisible by x^n.

17. The harmonic mean of two positive numbers a and b is

$$\frac{2ab}{a+b}.$$

Prove that it cannot exceed the geometric mean and equals it, if and only if $a = b$.

18. Find

$$\lim_{n\to\infty} [(n^3 + n^2 + 2n)^{\frac{1}{3}} - n].$$

19. Show that if $|x - 2| + |x| \geqq 4$, then $|x - 1| \geqq 2$.

20. Which is larger, $\sqrt{5} + \sqrt{51}$ or $7 + \sqrt{7}$?

21. Prove that $(2n - 1)!! < n^n$ for all natural numbers > 1. Here $(2n - 1)!! = 1 \cdot 3 \cdot 5 \cdots (2n - 1)$.

22. The numbers a_n are defined for $n > 2$ by the formula $a_n = a_{n-1} + 3a_{n-2}$ while $a_1 = 1$, $a_2 = 2$. Prove that $a_n > 2^{n-1}$ for $n > 2$.

*23. ABC is a triangle in the Euclidean plane. Its vertices have rational coordinates. Can the triangle be equilateral?

24. If $x_1 < x_2 < x_3$, for what value of x does

$$|x - x_1| + |x - x_2| + |x - x_3|$$

reach its least value?

Appendix A

Plane Analytic Geometry

A.1. Coordinates and vectors. The student should be familiar with Section 1.5 where we introduced coordinates on the line, in the plane, and in three space. Vectors were also defined. We shall be concerned here with the plane case.

We recall that once a frame of reference has been chosen in the plane, that is, two perpendicular lines and a unit of length, then every point P in the plane has a unique set of coordinates. See Figure A.1. The chosen coordinate axes divide the plane into four *quadrants* which are numbered counterclockwise. In the first quadrant both coordinates are positive, in the second x is negative and y positive, in the third both are negative, and in the fourth x is positive and y negative. Thus the points

$$(-1, 2), \quad (1, 3), \quad (-5, -4), \quad (3, -5)$$

lie in the second, first, third, and fourth quadrants respectively.

In the figure we have marked a point $P = (x, y)$ and the vector $\overrightarrow{OP}$ which is known as the *radius vector* of P. Its length r is given by

$$r = +\,(x^2 + y^2)^{\frac{1}{2}}. \tag{A.1.1}$$

We have also marked the *angle of inclination* of $\overrightarrow{OP}$ denoted by θ. This is the angle through which the positive x-axis would have to rotate in order to coincide with the direction of $\overrightarrow{OP}$. It is not uniquely determined; we can add or subtract any multiple of 360° without affecting the considerations. The least

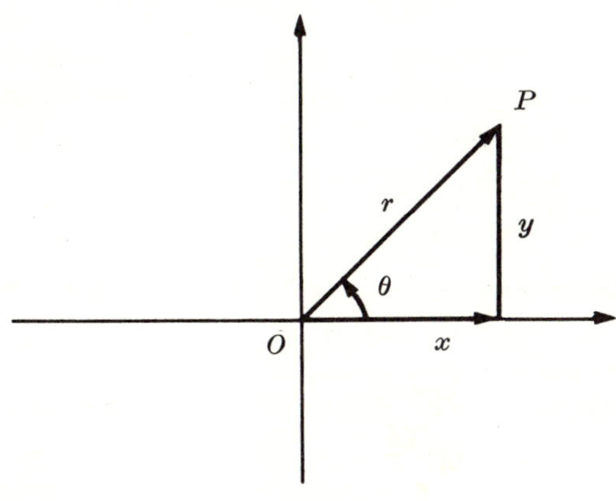

Figure A.1

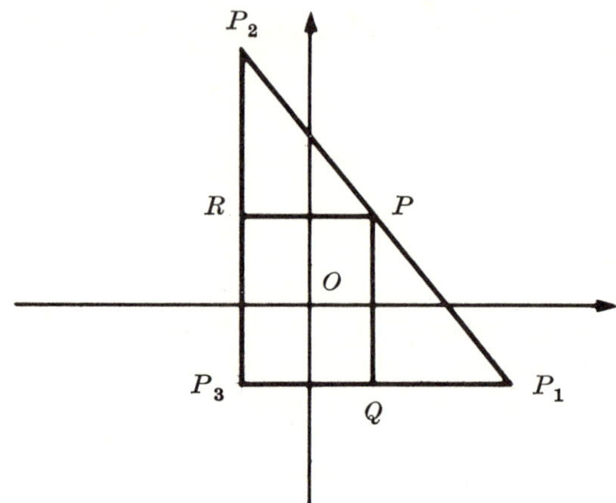

Figure A.2

nonnegative angle is uniquely determined and so are the trigonometric functions of θ to be used. We suppose that the student is familiar with the elementary trigonometric functions, their definition for arbitrary values of the angle, the addition formulas, and so on. From the figure we obtain the relations

$$\sin \theta = \frac{y}{r}, \qquad \cos \theta = \frac{x}{r}, \tag{A.1.2}$$

valid for any position of P. If P is not on the y-axis, we have also

$$\tan \theta = \frac{y}{x}. \tag{A.1.3}$$

Note that the sign of x varies as that of $\cos \theta$ and the sign of y as that of $\sin \theta$.

If P_1 and P_2 are two points in the plane, the midpoint of the line segment joining P_1 and P_2 is

$$P = (\tfrac{1}{2}(x_1 + x_2), \tfrac{1}{2}(y_1 + y_2)). \tag{A.1.4}$$

See Figure A.2 from which the elementary proof can be read. If we draw lines through P parallel to the coordinate axes, the resulting similar triangles show that the point Q is the midpoint of P_1P_3 and R is the midpoint of P_2P_3 where $P_3 = (x_2, y_1)$. If $P = (x_3, y_3)$, the relations

$$x_3 - x_2 = x_1 - x_3, \quad y_3 - y_2 = y_1 - y_3,$$

express the fact that Q is the midpoint of P_1P_3 and R is the midpoint of P_2P_3. Solving for x_3 and y_3 we obtain (A.1.4).

These formulas can be generalized. Let α satisfy $0 < \alpha < 1$ and let P_α be the point which divides the line segment P_1P_2 in the ratio $\alpha:(1 - \alpha)$ so that

$$P_1P_\alpha : P_\alpha P_2 = \alpha:(1 - \alpha).$$

Then

$$P_\alpha = (\alpha x_2 + (1 - \alpha)x_1, \quad \alpha y_2 + (1 - \alpha)y_1). \tag{A.1.5}$$

The proof follows the same lines as the special case $\alpha = \frac{1}{2}$ discussed above. In this formula α is restricted to the interval [0, 1]. It will be shown in the next section that the point P_α lies on the straight line determined by P_1 and P_2 for arbitrary values of α and, in fact, that P_α describes the line when α is given all real values.

Two points P_1 and P_2 are symmetric with respect to a third point P, if P bisects the line segment P_1P_2. The points P_1 and P_2 are symmetric with respect to a straight line l, if l is the perpendicular bisector of the segment P_1P_2.

The following special cases are particularly important for analytic geometry. The point (x, y) is symmetric

to $(x, -y)$ with respect to the x-axis,
to $(-x, y)$ with respect to the y-axis, and
to $(-x, -y)$ with respect to the origin.

For the distance between the points P and Q, the notation $d(P, Q)$ was introduced in Section 1.5. In Figure A.2 we note that

$$d(P_1, P_3) = |x_1 - x_2|, \quad d(P_2, P_3) = |y_1 - y_2|,$$

where $|a|$ is the absolute value of a, defined by (1.5.1). By the Theorem of Pythagoras

$$d(P_1, P_2) = +\,[(x_1 - x_2)^2 + (y_1 - y_2)^2]^{\frac{1}{2}}. \tag{A.1.6}$$

Since the positive square root of a^2 equals $|a|$, the formula is also valid if P_1P_2 is a horizontal or a vertical line segment and not slanting as in the figure.

To the line segment P_1P_2 correspond two directed line segments or vectors, $\overrightarrow{P_1P_2}$ and $\overrightarrow{P_2P_1}$. We set

$$\overrightarrow{P_2P_1} = -\overrightarrow{P_1P_2}. \tag{A.1.7}$$

The set of directed line segments with the same length and the same direction forms an equivalence class in the sense of Section 1.2. This class is represented by any one of its elements. In particular, there is one and only one element of the class which starts at the origin, so that it is of the form $\overrightarrow{OP}$. If $P = (x, y)$, we say that the vector $\boldsymbol{v} = \overrightarrow{OP}$ has the components x and y in the chosen system of coordinates. Figure A.3 shows some vectors belonging to the same equivalence class.

Let the symbol $\|\boldsymbol{v}\|$ denote the length of the vector $\boldsymbol{v}$.

We shall need to determine the angle between two given vectors. By definition this is the angle α in the interval [0°, 180°] formed by the representatives at the origin. We suppose that these are

$$\boldsymbol{v}_1 = (x_1, y_1), \quad \boldsymbol{v}_2 = (x_2, y_2),$$

numbered so that if $\boldsymbol{v}_1$ is rotated through the angle α then its direction will coincide with that of $\boldsymbol{v}_2$. See Figure A.4. Let the angles of inclination of $\boldsymbol{v}_1$ and $\boldsymbol{v}_2$ be θ_1 and θ_2 respectively. We may suppose that these angles have been

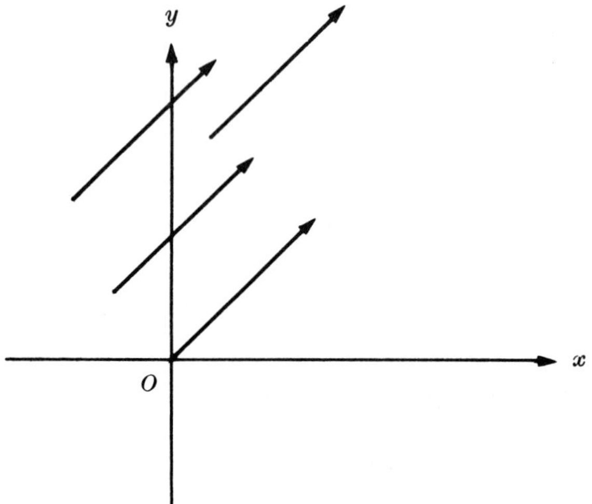

Figure A.3

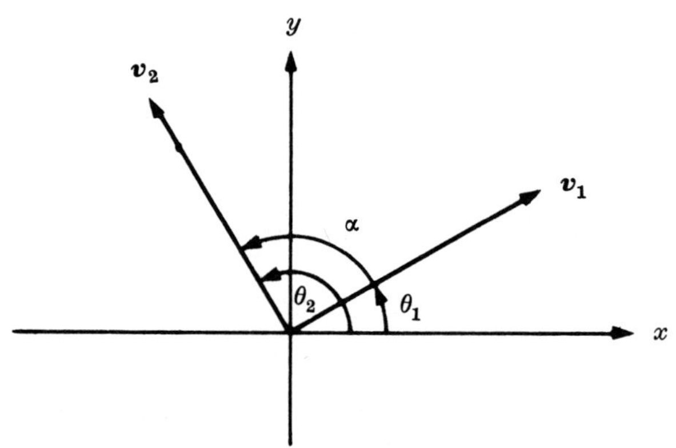

Figure A.4

so chosen that

$$\alpha = \theta_2 - \theta_1.$$

We then have

$$\cos \alpha = \cos (\theta_2 - \theta_1),$$

and by the addition formula for the cosine this becomes

$$\cos \alpha = \cos \theta_1 \cos \theta_2 + \sin \theta_1 \sin \theta_2,$$

and by formula (A.1.2)

$$\cos \alpha = \frac{x_1 x_2 + y_1 y_2}{\|\boldsymbol{v}_1\| \, \|\boldsymbol{v}_2\|}, \tag{A.1.8}$$

or

$$\cos \alpha = \frac{\boldsymbol{v}_1 \cdot \boldsymbol{v}_2}{\|\boldsymbol{v}_1\| \, \|\boldsymbol{v}_2\|}, \tag{A.1.9}$$

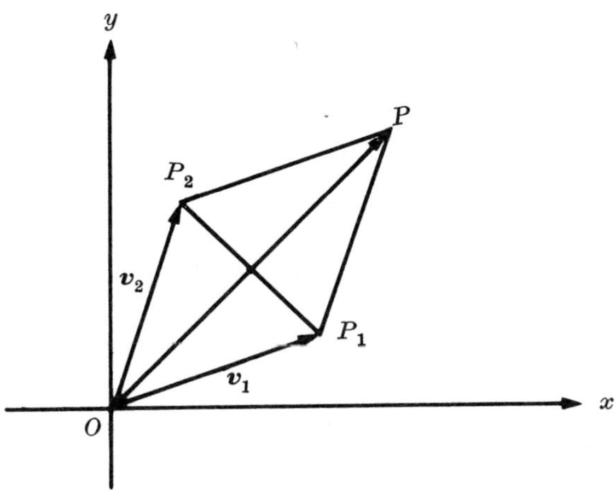

Figure A.5

where

$$v_1 \cdot v_2 = x_1x_2 + y_1y_2 \tag{A.1.10}$$

is known as the *dot product* of v_1 and v_2.

It is clear that this product operation is commutative

$$v_1 \cdot v_2 = v_2 \cdot v_1.$$

We shall define two other operations on vectors. The first is addition: *the sum of two vectors is again a vector*. More precisely, if

$$v_1 = (x_1, y_1), \quad v_2 = (x_2, y_2),$$

then

$$v_1 + v_2 = (x_1 + x_2, y_1 + y_2). \tag{A.1.11}$$

It is clear that addition is *commutative*

$$v_1 + v_2 = v_2 + v_1,$$

since addition of real numbers has this property.

For the geometry of the situation, see Figure A.5 which presupposes that the three points (0, 0), (x_1, y_1), and (x_2, y_2) are not collinear. We construct the parallelogram with three of its vertices at these points. The fourth vertex, P in the figure, gives the vector sum as the vector $\overrightarrow{OP}$. This is one of the diagonals of the parallelogram.

If we have to add two vectors u_1 and u_2 not in standard position, we replace u_1 and u_2 by v_1 and v_2 where v_1 and v_2 start at the origin, v_1 belongs to the equivalence class of u_1, and v_2 belongs to the equivalence class of u_2. Then $v_1 + v_2$ is uniquely defined and we set

$$u_1 + u_2 = v_1 + v_2.$$

Referring to the figure we see that $\overrightarrow{P_1P}$ is equivalent to $\boldsymbol{v}_2$. It follows that

$$\overrightarrow{OP_1} + \overrightarrow{P_1P} = \boldsymbol{v}_1 + \boldsymbol{v}_2 = \overrightarrow{OP},$$

or, in order to construct the sum vector we need only the triangle OP_1P and not the full parallelogram. If we apply this remark to the triangle OP_1P_2 instead, we see that

$$\overrightarrow{OP_1} + \overrightarrow{P_1P_2} = \overrightarrow{OP_2}.$$

Since $\overrightarrow{OP_1} = \boldsymbol{v}_1$ and $\overrightarrow{OP_2} = \boldsymbol{v}_2$, this shows that

$$\overrightarrow{P_1P_2} = \boldsymbol{v}_2 - \boldsymbol{v}_1.$$

It is fairly obvious that vector addition is *associative*, that is,

$$(\boldsymbol{v}_1 + \boldsymbol{v}_2) + \boldsymbol{v}_3 = \boldsymbol{v}_1 + (\boldsymbol{v}_2 + \boldsymbol{v}_3) = \boldsymbol{v}_1 + \boldsymbol{v}_2 + \boldsymbol{v}_3. \qquad \text{(A.1.12)}$$

Since addition is associative for real numbers, we get the components of the sum by adding the components of the summands. A geometric argument is given in Figure A.6 which shows a quadrilateral and its two diagonals. Here

$$\overrightarrow{P_1P_2} = \boldsymbol{v}_1, \quad \overrightarrow{P_2P_3} = \boldsymbol{v}_2, \quad \overrightarrow{P_3P_4} = \boldsymbol{v}_3,$$

$$\overrightarrow{P_1P_3} = \boldsymbol{v}_1 + \boldsymbol{v}_2, \quad \overrightarrow{P_2P_4} = \boldsymbol{v}_2 + \boldsymbol{v}_3,$$

$$\overrightarrow{P_1P_4} = \boldsymbol{v}_1 + \boldsymbol{v}_2 + \boldsymbol{v}_3,$$

and clearly

$$\overrightarrow{P_1P_3} + \overrightarrow{P_3P_4} = \overrightarrow{P_1P_4}, \quad \overrightarrow{P_1P_2} + \overrightarrow{P_2P_4} = \overrightarrow{P_1P_4}$$

so that (A.1.12) holds.

This construction is well known in *statics* where it is used to find the *resultant* of two or more forces acting on the same point. For two forces we get the *parallelogram of forces* the diagonal of which is the desired resultant. Figure A.6 illustrates how to construct the resultant of three forces and extension to a larger number is trivial.

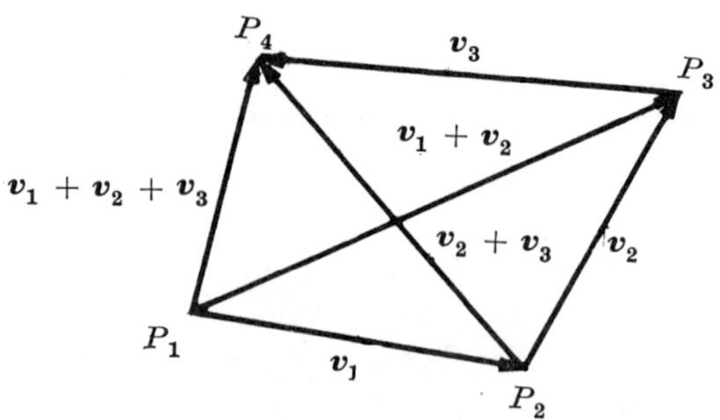

Figure A.6

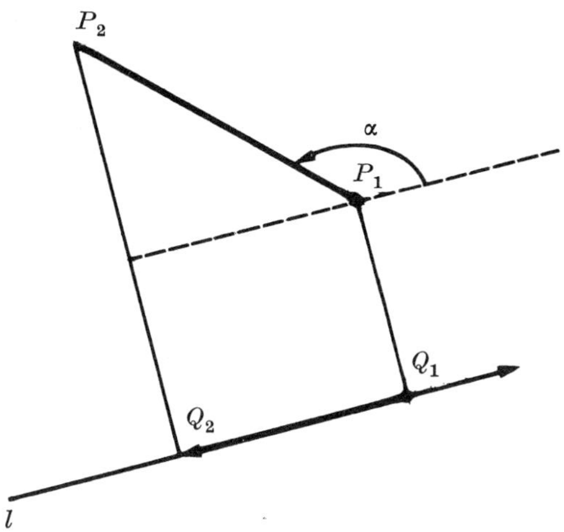

Figure A.7

We note that *vector addition and dot multiplication are distributive operations*, that is,

$$\boldsymbol{u} \cdot (\boldsymbol{v} + \boldsymbol{w}) = \boldsymbol{u} \cdot \boldsymbol{v} + \boldsymbol{u} \cdot \boldsymbol{w} = (\boldsymbol{v} + \boldsymbol{w}) \cdot \boldsymbol{u}. \tag{A.1.13}$$

We also define *scalar multiplication*. If α is a real number and $\boldsymbol{v} = (x, y)$ is a vector, then

$$\alpha \boldsymbol{v} = (\alpha x, \alpha y). \tag{A.1.14}$$

For $\alpha = -1$ we obtain the negative of a vector.

The last notion to be discussed in this section is that of a *projection*. Let $\overrightarrow{P_1P_2}$ be a given vector in arbitrary position and let l be an oriented line. It is desired to define the projection of $\overrightarrow{P_1P_2}$ on l. The geometric construction is shown in Figure A.7. We drop perpendiculars from P_1 and P_2 to l. Let the foot points be Q_1 and Q_2 respectively. Then

$$\text{pr}_l(\overrightarrow{P_1P_2}) = \overrightarrow{Q_1Q_2}.$$

To get an algebraic expression for the projection, let $\boldsymbol{v}$ be the vector starting at the origin, equivalent to $\overrightarrow{P_1P_2}$, and let $\boldsymbol{u}$ be a unit vector starting at the origin with the same direction as l. Let α be the angle between $\boldsymbol{u}$ and $\boldsymbol{v}$ determined in the manner indicated above. This is also the angle between $\overrightarrow{P_1P_2}$ and l. Then

$$\boldsymbol{u} \cdot \boldsymbol{v} = \|\boldsymbol{v}\| \cos \alpha$$

since $\|\boldsymbol{u}\| = 1$ by assumption. The right member is, aside from the sign, equal to the length of $\overrightarrow{Q_1Q_2}$. This gives

$$\text{pr}_l(\overrightarrow{P_1P_2}) = (\boldsymbol{u} \cdot \boldsymbol{v})\boldsymbol{u}. \tag{A.1.15}$$

From this result we get

THEOREM A.1.1. *The projection on a line of the sum of vectors is the sum of the projections of the vectors.*

Proof. This follows from the distributive law. We have

$$\boldsymbol{u} \cdot (\boldsymbol{v}_1 + \boldsymbol{v}_2 + \cdots + \boldsymbol{v}_n) = (\boldsymbol{u} \cdot \boldsymbol{v}_1) + (\boldsymbol{u} \cdot \boldsymbol{v}_2) + \cdots + (\boldsymbol{u} \cdot \boldsymbol{v}_n).$$

If we multiply the vector $\boldsymbol{u}$ by the scalar in the left member, then we get the projection on l of the sum of the vectors $\boldsymbol{v}_1, \boldsymbol{v}_2, \cdots, \boldsymbol{v}_n$. If we use each of the summands in the right member as multiplier of $\boldsymbol{u}$ and add the results, the sum of the projections is obtained. Hence the two must be equal.

In vector analysis it is customary to introduce the two unit vectors

$$\boldsymbol{i} = (1, 0), \quad \boldsymbol{j} = (0, 1). \tag{A.1.16}$$

With this notation the vector

$$\boldsymbol{v} = (x, y)$$

may be written

$$\boldsymbol{v} = x\boldsymbol{i} + y\boldsymbol{j}. \tag{A.1.17}$$

We have

$$\boldsymbol{i} \cdot \boldsymbol{i} = 1, \quad \boldsymbol{i} \cdot \boldsymbol{j} = \boldsymbol{j} \cdot \boldsymbol{i} = 0, \quad \boldsymbol{j} \cdot \boldsymbol{j} = 1. \tag{A.1.18}$$

EXERCISE A.1

1. Plot the points whose coordinates are

$$(1, 3); \quad (-3, 1); \quad (-4, 2); \quad (-3, -2); \quad (4, 5); \quad (5, 4).$$

2. In what quadrants do these points lie?

Find the coordinates of the points symmetric to these points with respect to the:

3. x-axis.
4. y-axis.
5. Origin.
6. Find the distances between the last three points of Problem 1.
7. Find the distances from the origin of the first, third, and fifth points of Problem 1.
8. What relations must hold between the coordinates, if the two points (x_1, y_1) and (x_2, y_2) are symmetric with respect to the straight line which bisects the angle between the positive x- and y-axes?
9. What condition must be satisfied if (x, y) is to lie on this line?
10. If (x_1, y_1) is given, find (x_2, y_2) so that the two points are symmetric with respect to the point $(1, 0)$.
11. Construct the triangle whose vertices are located at the first three points of Problem 1. Find the lengths of the sides and find the coordinates of their midpoints.

12. If $\boldsymbol{v}_1$, $\boldsymbol{v}_2$, $\boldsymbol{v}_3$ are the three vectors from the origin to the vertices of the triangle, find the dot products $\boldsymbol{v}_1 \cdot \boldsymbol{v}_2$, $\boldsymbol{v}_2 \cdot \boldsymbol{v}_3$, and $\boldsymbol{v}_3 \cdot \boldsymbol{v}_1$.

13. Find vectors which start at the origin, are parallel to, and have the same lengths as the three sides of the triangle. Express these vectors in terms of $\boldsymbol{i}$ and $\boldsymbol{j}$.

14. Find the cosines of the angles of the triangle.

15. Find the lengths of the projections of the sides on the line of Problem 8.

16. Draw the three medians of the triangle. On each median find the coordinates of the point which divides the line segment from the midpoint of the side to the opposite vertex in the ratio 1:2. Show that this point is the same for all three medians.

17. When are the vectors $\boldsymbol{v}_1 = (x_1, y_1)$ and $\boldsymbol{v}_2 = (x_2, y_2)$ perpendicular?

18. If the six points in Problem 1 are endpoints of vectors from the origin, are any two vectors perpendicular?

A.2. The straight line. We shall use the tools developed in the preceding section for a study of various geometric configurations starting with the straight line.

If we draw a straight line in a plane where we have a system of Cartesian coordinates, the coordinates of the points on the line are not independent of each other. As we shall see in a moment, if (x, y) lies on the line l, there is a relation of the form

$$Ax + By + C = 0 \tag{A.2.1}$$

between the coordinates where A, B, C are certain constants and the double ratio

$$A:B:C$$

is characteristic for the line in question. To prove this, we consider the various positions which the line can occupy with respect to the coordinate axes.

1. *The line is vertical,* that is, parallel to the y-axis. Suppose that l intersects the x-axis at the point $(a, 0)$. Then $P = (x, y)$ lies on the line if and only if

$$x = a. \tag{A.2.2}$$

The y-coordinate is arbitrary, but the x-coordinate must have the value a and every such point lies on the line. For this reason we call (A.2.2) the *equation of the line.* It is satisfied by every point on the line and by no other points.

2. *The line is horizontal,* that is, parallel to the x-axis. Suppose that l intersects the y-axis at the point $(0, b)$. Then $P = (x, y)$ lies on l if and only if

$$y = b. \tag{A.2.3}$$

Here there is no condition on x. Again, this is the equation of the line for it is satisfied by the coordinates of all points on the line and by no other points.

3. *The line goes through the origin* and does not coincide with either of the axes. Suppose that the angle of inclination of the line is α. We define this angle as the least positive angle through which the positive x-axis should be rotated

in order to coincide with the line l. We have $0 < \alpha < 180°$ and by assumption $\alpha \neq 90°$. Suppose that $P = (x, y)$ is a point on l. If $y > 0$, the angle of inclination of the vector $\overrightarrow{OP}$ as defined in Section A.1 equals α, but if $y < 0$ the angle is $\alpha + 180°$. In either case the tangent of the angle equals $\tan \alpha$. This quantity is denoted by m and called the *slope* of the line l. Formula (A.1.3) now applies and gives

$$\frac{y}{x} = \tan \alpha = m,$$

or

$$y = mx. \tag{A.2.4}$$

This is the equation of the line, for it is satisfied by the coordinates of every point on the line and by no other points.

4. *The line is in general position* with respect to the coordinate axes. Since the line is not vertical, it intersects the y-axis at a point $(0, b)$. Suppose that the angle of inclination of the line is α and set $\tan \alpha = m$. Then for any point (x, y) on the line distinct from $(0, b)$ we have

$$\frac{y - b}{x} = \tan \alpha = m,$$

or

$$y = mx + b. \tag{A.2.5}$$

Here m is again the *slope* and b is called the *y-intercept* of the line.

This completes the survey and shows that in all four cases the equation of the line is of the form (A.2.1). These cases are illustrated in Figure A.8.

We shall prove conversely that each equation of the form (A.2.1) represents a straight line, that is, the set of points (x, y) for which (A.2.1) holds, forms a straight line. We may assume that A and B are not both 0; otherwise C must also be 0 and the equation is satisfied by all points in the plane. Again it is easiest to proceed by cases.

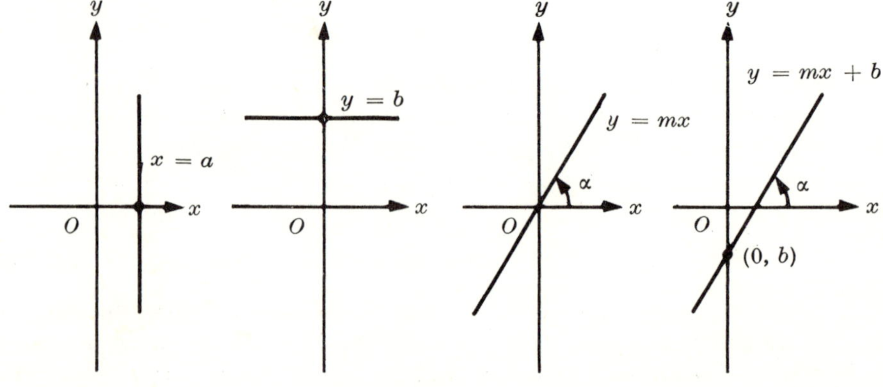

Figure A.8

1. $A = 0$. Since $B \neq 0$,

$$y = -\frac{C}{B}.$$

This is of the form (A.2.3) and represents a horizontal line.

2. $B = 0$. Since $A \neq 0$,

$$x = -\frac{C}{A}.$$

This is of the form (A.2.2) and represents a vertical line.

3. $A \neq 0,\ B \neq 0$. We have then

$$y = -\frac{A}{B}x - \frac{C}{B},$$

This is of form (A.2.5) unless $C = 0$, when we get form (A.2.4).

Thus we see that the general first-order equation in x and y represents a straight line. For this reason (A.2.1) is usually called a *linear equation.*

We note that the straight line defined by (A.2.1) is also defined by the equation

$$cAx + cBy + cC = 0 \tag{A.2.6}$$

where c is any constant different from 0. This shows that only the double ratio

$$A:B:C$$

is determined by the line. Our preceding discussion shows that

$$m = -\frac{A}{B}, \qquad b = -\frac{C}{B}, \tag{A.2.7}$$

so that the individual ratios $A:B$ and $B:C$ have geometrical significance. We have also

$$-\frac{C}{A} = a$$

which is called the *x-intercept* and is the x-coordinate of the point of intersection of the line with the x-axis.

If the equation (A.2.1) is given, we get the x-intercept by setting $y = 0$ and solving for x. We get the y-intercept by setting $x = 0$ and solving for y. Once we have the two points $(a, 0)$ and $(0, b)$, we can construct the line since there is one and only one line on two given points. If both intercepts are different from 0, we can write the equation of the line in the form

$$\frac{x}{a} + \frac{y}{b} = 1. \tag{A.2.8}$$

Since this is a linear equation in x and y, it represents a straight line. Further,

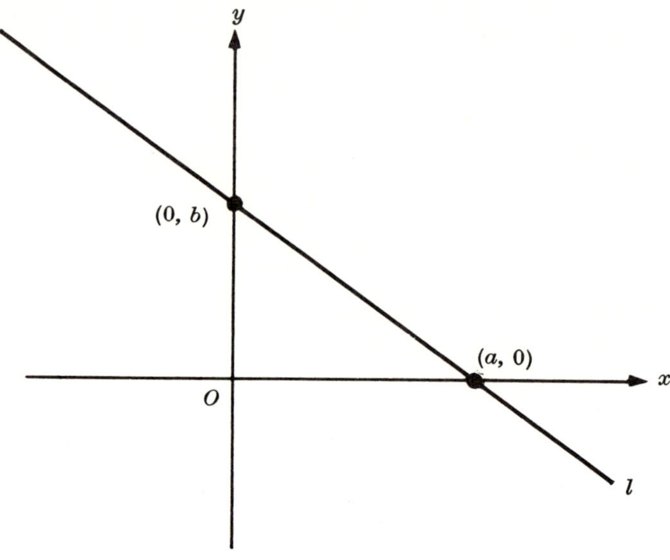

Figure A.9

putting $y = 0$ we see that the x-intercept is a. Similarly we find that the y-intercept is b. Since there is one and only one line with given intercepts, (A.2.8) must be the equation of the original line. (A.2.8) is called the *intercept form* of the equation of a straight line. See Figure A.9.

Formula (A.2.5) is called the *slope-intercept form* of the equation of the straight line. It is obtained by solving the given equation for y. The coefficient of x is the slope and the constant term is the y-intercept.

To illustrate we take the equation

$$3x + 5y = 15.$$

Here

$$a = 5, \quad b = 3,$$

and dividing the equation by 15 we get

$$\frac{x}{5} + \frac{y}{3} = 1,$$

which is the intercept form. Solving the equation for y, we get

$$y = -\tfrac{3}{5}x + 3,$$

so the slope is $-\frac{3}{5}$.

We shall encounter other forms of the equation of a straight line later. Let us consider problems involving two lines

$$\begin{aligned} l_0&: A_0x + B_0y + C_0 = 0, \\ l_1&: A_1x + B_1y + C_1 = 0. \end{aligned} \tag{A.2.9}$$

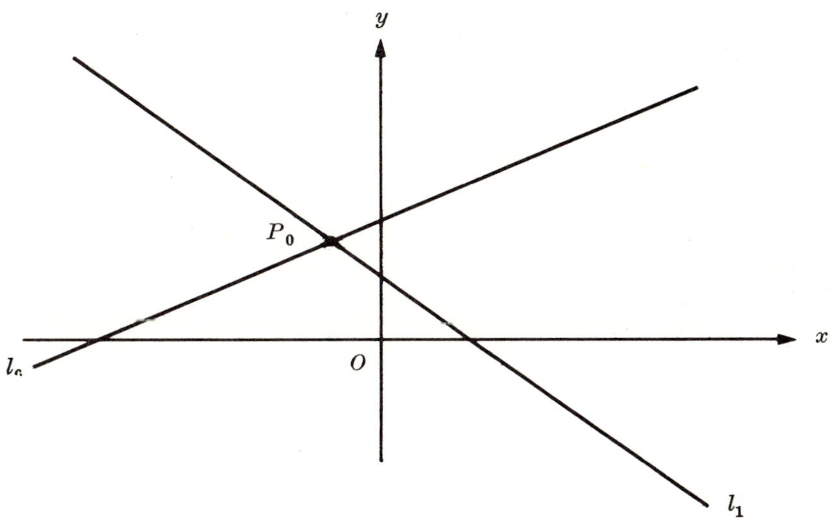

Figure A.10

Two straight lines either intersect or are parallel. They have the latter property, if their slopes are equal. If B_0 and B_1 are different from 0, this condition gives

$$-\frac{A_0}{B_0} = -\frac{A_1}{B_1},$$

or

$$A_0B_1 - A_1B_0 = 0. \tag{A.2.10}$$

This condition is also satisfied if $B_0 = B_1 = 0$ so that the lines are vertical. As an example we see that the three lines

$$x + 2y + 6 = 0, \quad 2x + 4y - 5 = 0, \quad 3x + 6y - 4 = 0$$

are parallel. They all have the slope $-\frac{1}{2}$.

If the lines are not parallel, they intersect and the coordinates of the point of intersection can be found by solving the two equations (A.2.9) as simultaneous equations. See Figure A.10. The point of intersection (x_0, y_0) lies on both lines and this implies that $x = x_0$, $y = y_0$ satisfy both equations. Since obviously $A_0B_1 - A_1B_0 \neq 0$, we can solve the equations and obtain

$$x_0 = \frac{B_0C_1 - B_1C_0}{A_0B_1 - A_1B_0}, \quad y_0 = \frac{A_1C_0 - A_0C_1}{A_0B_1 - A_1B_0}. \tag{A.2.11}$$

Let us use abbreviated notation. We denote the left member of (A.2.1) by L or $L(x, y)$ so that the equation becomes

$$L = 0.$$

Similarly we write the equations of l_0 and of l_1

$$L_0 = 0 \quad \text{and} \quad L_1 = 0,$$

respectively. Let α be a real number, and consider

$$l_\alpha : \alpha L_1 + (1 - \alpha)L_0 = 0. \tag{A.2.12}$$

Substituting the values of L_0 and L_1 we obtain

$$[\alpha A_1 + (1 - \alpha)A_0]x + [\alpha B_1 + (1 - \alpha)B_0]y + \alpha C_1 + (1 - \alpha)C_0 = 0. \tag{A.2.13}$$

This is a linear expression in x and y of type (A.2.1), that is, it represents a straight line no matter how we choose the real number α. If we let α vary we get a *family of straight lines.*

There are two distinct cases. First, suppose that l_0 and l_1 are *parallel.* This means that condition (A.2.10) holds. This implies that *all lines of the family are parallel* for

$$A_0[\alpha B_1 + (1 - \alpha)B_0] - B_0[\alpha A_1 + (1 - \alpha)A_0] = \alpha[A_0B_1 - A_1B_0] = 0.$$

Hence any line of the family is parallel to the first two lines, that is, they are all parallel. By letting α vary we can get every line in the equivalence class of parallel lines determined by l_0. For if $L = 0$ is one of these lines and if it goes through a point (x_1, y_1), we need only substitute $x = x_1$, $y = y_1$ in (A.2.13) to find the value of α. For this value of α the line l_α coincides with the given line $L = 0$.

In the second case l_0 and l_1 intersect. This means that there is a point $P_0 = (x_0, y_0)$ which lies on both lines and this implies that $x = x_0$, $y = y_0$ satisfy both

$$L_0 = 0 \quad \text{and} \quad L_1 = 0.$$

From the structure of L_α it follows that x_0 and y_0 also satisfy the equation

$$L_\alpha = 0.$$

That is, *all the lines of the family pass through* P_0, the point of intersection of l_0 and l_1. The set of all lines through a given point is called a *pencil of lines.* See Figure A.11. Thus the set

$$\{l_\alpha \mid -\infty < \alpha < \infty\}$$

is a pencil of lines on the point (x_0, y_0). For every line in the family goes through this point and, conversely, if we have a line on P_0 we can find a value of α for which l_α coincides with this line. In order to determine α, it suffices to know the coordinates of another point on the line. Substituting these numbers in (A.2.13) we obtain the value of α.

The angle between the intersecting lines is easily found from the equations of the lines, that is, formula (A.2.9). To this end we observe that

$$\boldsymbol{u} = (B_0, -A_0), \quad \boldsymbol{v} = (B_1, -A_1)$$

are vectors at the origin which are parallel to l_0 and l_1 respectively. For the slope of l_0 is $-A_0/B_0$, and this is also the tangent of the angle which $\boldsymbol{u}$ makes with the positive x-axis by (A.1.3). Hence the angle between l_0 and l_1 is the

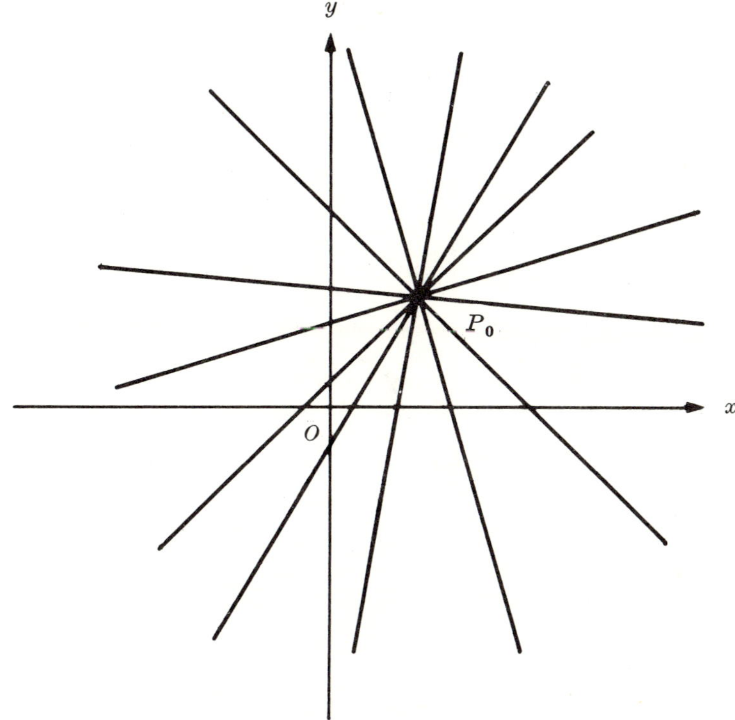

Figure A.11

same as the angle between $\boldsymbol{u}$ and $\boldsymbol{v}$. The latter is obtainable from (A.1.8). We have

$$\boldsymbol{u} \cdot \boldsymbol{v} = B_0B_1 + (-A_0)(-A_1) = A_0A_1 + B_0B_1,$$

$$\|\boldsymbol{u}\| = (A_0^2 + B_0^2)^{\frac{1}{2}}, \quad \|\boldsymbol{v}\| = (A_1^2 + B_1^2)^{\frac{1}{2}},$$

so that

$$\cos \alpha = \frac{A_0A_1 + B_0B_1}{(A_0^2 + B_0^2)^{\frac{1}{2}}(A_1^2 + B_1^2)^{\frac{1}{2}}}. \tag{A.2.14}$$

Another way of finding the angle between the two lines is to use their slopes. See Figure A.12. If the angles of inclination of the lines are θ_0 and θ_1 respectively, we may take

$$\alpha = \theta_1 - \theta_0$$

and use

$$\tan \alpha = \tan (\theta_1 - \theta_0) = \frac{\tan \theta_1 - \tan \theta_0}{1 + \tan \theta_1 \cdot \tan \theta_0}.$$

If the slopes of the lines are m_0 and m_1, both finite by assumption, then

$$\tan \alpha = \frac{m_1 - m_0}{1 + m_1m_0}. \tag{A.2.15}$$

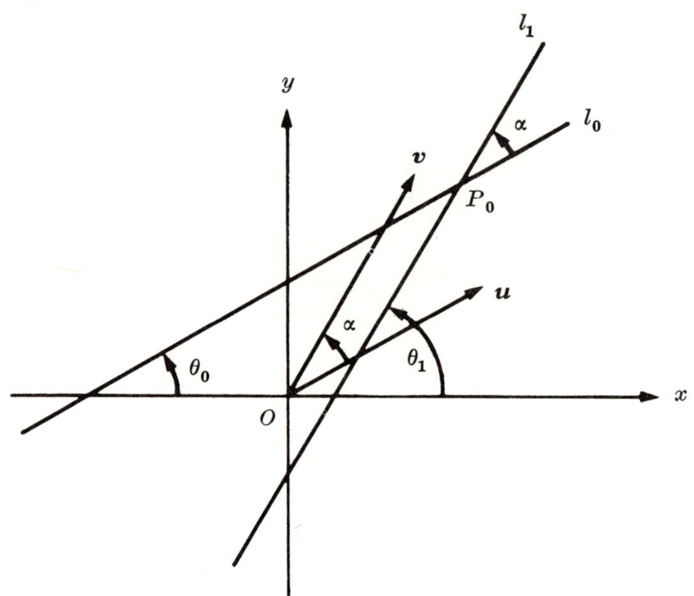

Figure A.12

This formula gives a finite value of tan α unless the denominator is 0. This gives:

Two lines are orthogonal to each other (= perpendicular) if and only if the slope of one is the negative reciprocal of the other.

Formula (A.2.14) remains unchanged if we interchange the subscripts 0 and 1, whereas formula (A.2.15) changes its sign in this case. In other words, one formula may give the acute angle between the two lines while the other gives the supplementary obtuse angle.

As an illustration of the use of the formulas take

$$l_0\colon 3x - 5y + 6 = 0, \quad l_1\colon 2x + 4y - 9 = 0.$$

Here the first formula gives $\cos \alpha = (170)^{-\frac{1}{2}}$ and the second $\tan \alpha = 13$ so that $\alpha = 85.6°$.

We shall now take up the four remaining forms of the equation of the straight line.

1. *The slope-one-point formula.* The equation of the straight line through the point (x_0, y_0) of the slope m is

$$y - y_0 = m(x - x_0). \tag{A.2.16}$$

This is a linear equation and hence it represents a straight line, it is satisfied by (x_0, y_0), and when we solve for y the coefficient of x is m. Hence this is the desired line.

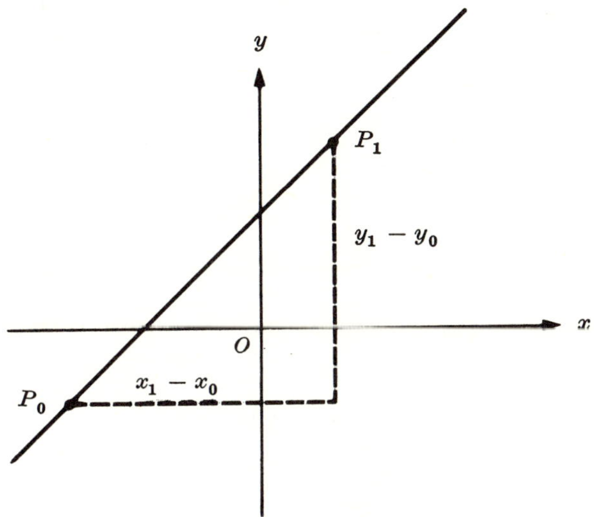

Figure A.13

2. *The two-point formula.* If $P_0 = (x_0, y_0)$, and $P_1 = (x_1, y_1)$ where $x_0 \neq x_1$ and $y_0 \neq y_1$, then the equation of the straight line on P_0 and P_1 is given by

$$\frac{y - y_0}{y_1 - y_0} = \frac{x - x_0}{x_1 - x_0}, \tag{A.2.17}$$

or, equivalently,

$$y - y_0 = \frac{y_1 - y_0}{x_1 - x_0}(x - x_0). \tag{A.2.18}$$

The verification is simple. The equations represent straight lines and are satisfied by the coordinates of the given points. Finally, there is one and only one line on two given points so this must be the line in question. See Figure A.13.

3. *Parametric forms.* Our point of departure is formula (A.1.5) which suggests that the equations

$$\begin{cases} x = \alpha x_1 + (1 - \alpha)x_0, \\ y = \alpha y_1 + (1 - \alpha)y_0, \end{cases} \quad -\infty < \alpha < \infty, \tag{A.2.19}$$

represent the straight line on (x_0, y_0) and (x_1, y_1). We know already that if α goes from 0 to 1 then the point $(x, y) = P_\alpha$ goes from P_0 to P_1. If we rewrite the equations in the form

$$\begin{aligned} x &= x_0 + (x_1 - x_0)\alpha, \\ y &= y_0 + (y_1 - y_0)\alpha, \end{aligned} \tag{A.2.20}$$

we can eliminate α and obtain

$$\frac{y - y_0}{y_1 - y_0} = \frac{x - x_0}{x_1 - x_0},$$

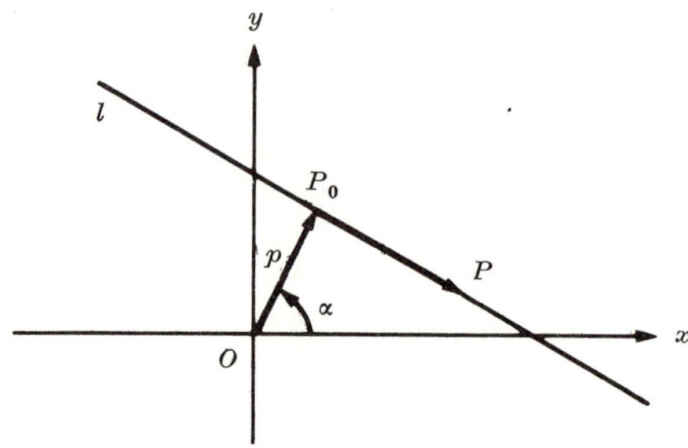

Figure A.14

which we know is the equation of the straight line on P_0 and P_1. We can also define two quantities t and m by the equations

$$t = (x_1 - x_0)\alpha, \quad mt = (y_1 - y_0)\alpha,$$

and rewrite (A.2.20) as

$$\begin{aligned} x &= x_0 + t, \\ & \qquad\qquad -\infty < t < \infty. \\ y &= y_0 + mt, \end{aligned} \tag{A.2.21}$$

These equations give parametric representations of the straight line in terms of the parameters α and t respectively. For each value of α we obtain a point P_α on the line on P_0 and P_1. For $0 \leqq \alpha \leqq 1$ this point belongs to the line segment P_0P_1, for $\alpha < 0$ it lies on the infinite segment adjoining P_0, and for $1 < \alpha$ on the infinite line segment adjoining P_1. In (A.2.21) each value of t gives a point on the straight line through (x_0, y_0) of slope m.

The term "parameter" indicates an "arbitrary constant" or an "auxiliary variable" or an "index." The points P_α defined by (A.2.19) are "indexed" by α and form a "one-parameter family of points." These points in turn form a straight line. This is why we call (A.2.19) the parametric form of the equation of the line. Similarly, the pencil of lines $\{l_\alpha\}$ defined by (A.2.12) is a "one-parameter family of lines," the lines being "indexed" by α. All these lines are on a point (x_0, y_0). Still another example is furnished by formula (A.2.9) which we may regard as defining a "two-parameter family of lines," the two parameters being the intercepts a and b.

4. *The normal form.* We shall obtain a representation of the straight line in terms of two parameters: its distance from the origin, denoted by p, and the angle α which the normal to the line makes with the positive x-axis. See Figure A.14. We take $p > 0$ and $-180° < \alpha \leqq 180°$. If P_0 is the foot point of the perpendicular on the line, then

$$P_0 = (p \cos \alpha, p \sin \alpha)$$

by (A.1.2). If $P = (x, y)$ is any point on the line l, $P \neq P_0$, then the vectors $\overrightarrow{OP_0}$ and $\overrightarrow{P_0P}$ are orthogonal. The latter vector has the components

$$x - p \cos \alpha \quad \text{and} \quad y - p \sin \alpha.$$

By (A.1.8) the condition of orthogonality becomes

$$p \cos \alpha\,(x - p \cos \alpha) + p \sin \alpha\,(y - p \sin \alpha) = 0,$$

which simplifies to

$$x \cos \alpha + y \sin \alpha - p = 0. \tag{A.2.22}$$

This is the normal form of the straight line.

We have supposed $p > 0$. If the line goes through the origin, $p = 0$, and the direction of the normal is defined only up to multiples of 180°. For most purposes it is immaterial which of the possible angles is chosen for α.

If a straight line is given by the equation

$$Ax + By + C = 0,$$

we can always multiply the equation by a constant c so that the resulting equation is the normal form of this particular line. This requires that

$$cA = \cos \alpha, \quad cB = \sin \alpha, \quad cC = -p.$$

If we are only interested in c and not in the actual values of α and p, we square and add the first two equations. This gives

$$c^2(A^2 + B^2) = 1,$$

and

$$c = \pm\,(A^2 + B^2)^{-\frac{1}{2}}.$$

If $C = 0$, $p = 0$ and the choice of the sign is immaterial, but if $C \neq 0$, we have to choose the sign opposite to that of C so that the product $cC = -p$ is negative. This gives

$$\frac{Ax + By + C}{\pm\,(A^2 + B^2)^{\frac{1}{2}}} = 0 \tag{A.2.23}$$

as the normal form of the line with the sign chosen in the manner indicated.

Thus, given the line

$$12x - 5y + 6 = 0,$$

we find

$$(A^2 + B^2)^{\frac{1}{2}} = (144 + 25)^{\frac{1}{2}} = (169)^{\frac{1}{2}} = 13.$$

We must choose the negative square root since $C = 6 > 0$. From this we obtain the following normal form of the line:

$$-\tfrac{12}{13}\,x + \tfrac{5}{13}y - \tfrac{6}{13} = 0.$$

The normal form is used when we want the distance from a line to a given point. See Figure A.15.

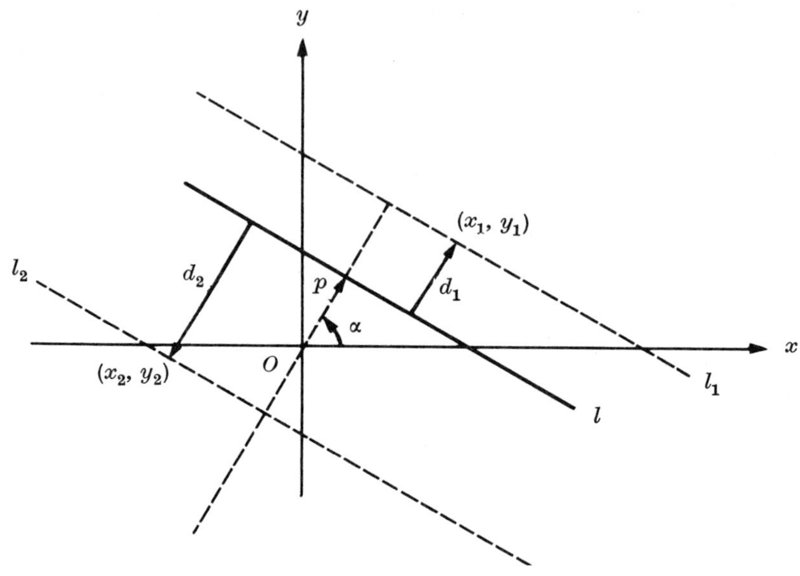

Figure A.15

THEOREM A.2.1. *The (signed) distance from the line* (A.2.22) *to the point* $P_1:(x_1, y_1)$ *is*

$$d = x_1 \cos \alpha + y_1 \sin \alpha - p. \tag{A.2.24}$$

Remark. d is obtained by substituting the coordinates of the point in the equation of the line in normal form. If the line does not go through the origin, d is positive or negative according as the line separates the point from the origin or not.

Proof. If l separates P_1 from the origin, let us draw a line l_1 through P_1 parallel to l. This line is at the distance $p + d_1$ from the origin and it has the same normal as l with respect to the origin. The normal form of the equation of the line l_1 is then

$$x \cos \alpha + y \sin \alpha - p - d_1 = 0.$$

Since P_1 lies on the line, its coordinates satisfy its equation. Substituting $x = x_1$, $y = y_1$, and solving for the unknown quantity d_1, we obtain (A.2.24).

If P_1 lies on the same side as the origin with respect to l, there is nothing to change as long as the origin is to the same side with respect to l and l_1. If, however, the origin lies between the two lines as in the figure for the point $P_2 = (x_2, y_2)$, then the parameter α changes by 180° and the normal form of l_2 becomes

$$-x \cos \alpha - y \sin \alpha + d_2 + p = 0.$$

The left member is the negative of what we obtained in the first case, but the final result is still the same.

If we apply this to our preceding example

$$12x - 5y + 6 = 0,$$

we see that the point (1, 1) is at the distance -1 from the line while $(-1, -1)$ is at the distance $\frac{1}{13}$. Thus the points are separated by the line and the first point lies to the same side as the origin.

We give one more application of the normal form, this time the problem of finding bisectors of the angles formed by two lines l_0 and l_1. We assume that

$$L_0 = 0, \quad L_1 = 0$$

are the normal forms of the equations of these lines. If $P = (x, y)$ is a point in the plane, and if we substitute its coordinates in L_0 and L_1, we obtain the signed distances of P from l_0 and l_1 respectively. Through the point of intersection of l_0 and l_1 there are two bisectors of the angles formed by the lines. See Figure A.16. A point on the bisector is equidistant from the two lines. If P is on one of the bisectors, then the values of L_0 and of L_1 at that point must be equal except for the sign. It follows that the two bisectors are given by

$$L_0 - L_1 = 0, \quad L_0 + L_1 = 0. \tag{A.2.25}$$

These equations represent straight lines through the intersection of l_0 and l_1 by (A.2.12). On the first line $L_0 = L_1$, while on the second $L_0 = -L_1$, which says that points on these lines are equidistant from l_0 and l_1. In other words, they are the desired bisectors.

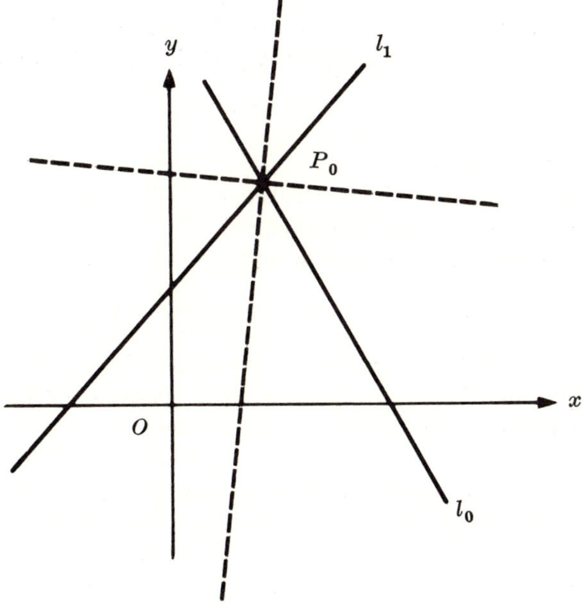

Figure A.16

EXERCISE A.2

1. For the following six lines, determine intercepts and slope and plot the line.

(a) $3x + 4y - 12 = 0$; (b) $x - y + 2 = 0$;
(c) $5x - 6y + 13 = 0$; (d) $x + y - 1 = 0$;
(e) $4x + 3y - 12 = 0$; (f) $2x - y + 4 = 0$.

2. Determine the normal form of the equations in Problem 1. Find also α and p.

3. Find the slope of a straight line on two points if the points are chosen from the following five (ten alternatives!):

$$(-1, 3); \quad (-2, 6); \quad (3, 4); \quad (3, -2); \quad (6, -2).$$

4. Find the equations of the ten lines.

5. Which lines in Problem 1 are orthogonal?

6. Find the cosine of the angle between lines 1(a) and 1(e) and between lines 1(d) and 1(f).

7. Find the tangent of the angle between 1(a) and 1(f), between 1(c) and 1(d).

8. Find the coordinates of the points of intersection of each of the four pairs of lines in the two preceding problems.

9. Find the equation of the line through the point of intersection of the lines

$$3x - 2y + 3 = 0 \quad \text{and} \quad 2x + 9y - 11 = 0$$

which (a) goes through the point (1, 2), (b) has slope 1. [*Hint:* Use (A.2.12).]

10. Find the equation of the line through (1, 2) which is parallel to

$$3x - 2y + 3 = 0.$$

[*Hint:* What is the slope of the line?]

11. Find the equation of the line through (1, 2) which is perpendicular to $3x - 2y + 3 = 0$. [*Hint:* What is the slope?]

12. Find the equation of the bisector of the acute angle formed by the lines

$$4x - y + 3 = 0 \quad \text{and} \quad 2x + 9y - 11 = 0.$$

[*Hint:* Plot the lines and note the sign of the slope of the bisector.]

13. Find the equation of the perpendicular bisector of the line segment joining (—1, 3) with (3, —5). [*Hint:* A point on the bisector is equidistant from the given points.]

14. Prove that the perpendicular bisectors of the sides of a triangle meet in a point. It is permissible to choose the coordinate axes so that the vertices are located at the points $(a, 0)$, $(-a, 0)$, and (b, c) where a, b, c are arbitrary positive numbers.

15. Plot the line $x = 1 + t$, $y = -1 + 3t$.

16. Plot the line $x = \alpha \cdot 1 + (1 - \alpha) \cdot (-2)$, $y = \alpha \cdot 3 + (1 - \alpha) \cdot 1$.

17. Find the equation of a straight line through the point (2, 2.5) which forms a triangle of area 10 with the coordinate axes.

18. Find the equation of a straight line through (2, 2.5) which forms an isosceles triangle with the coordinate axes.

19. After factoring, draw the graph of $4x^2 - 9y^2 = 0$.

20. Prove that the bisectors of the angles of a triangle meet at a point. Choose the coordinates so that the origin is interior to the triangle, let the equations of the sides in abbreviated normal form be $L_1 = 0$, $L_2 = 0$, $L_3 = 0$. Find and discuss the equations of the required bisectors.

A.3. The circle. A circle looks like a very simple thing, but it has given rise to much interesting mathematics and many books have been written on the subject.

Its characteristic property is that all its points have the same distance from a fixed point, the *center* of the circle. The distance is the *radius*. From this definition we can easily obtain the equation of a circle in a Cartesian system of coordinates. Let (x_0, y_0) be the center, r the radius, then the point (x, y) is on the circle if and only if its distance from the center equals r. See Figure A.17. This gives the equation

$$(x - x_0)^2 + (y - y_0)^2 = r^2. \tag{A.3.1}$$

Expanding and rearranging terms we get

$$x^2 + y^2 - 2x_0x - 2y_0y + x_0^2 + y_0^2 - r^2 = 0,$$

that is, an expression of the form

$$x^2 + y^2 + 2Dx + 2Ey + F = 0. \tag{A.3.2}$$

This is an equation of the second degree in x and y but a very special one. The general equation of the second degree involves six arbitrary constants.

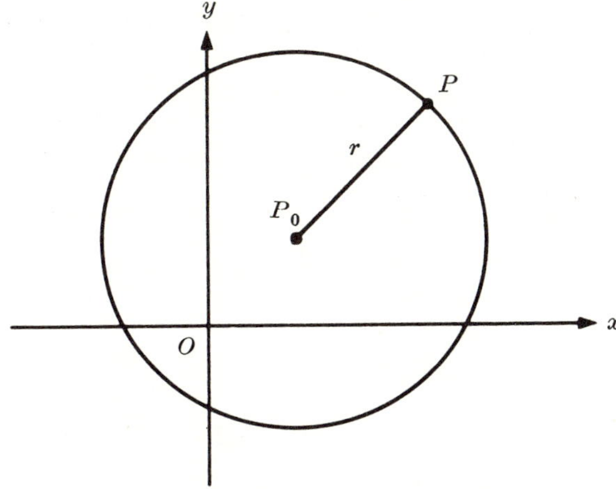

Figure A.17

Thus

$$Ax^2 + 2Bxy + Cy^2 + 2Dx + 2Ey + F = 0, \tag{A.3.3}$$

where the coefficients are arbitrary real numbers and the twos are introduced for the sake of convenience. If we compare the last two equations, we see that (A.3.3) represents a circle if and only if $A = C \neq 0$ and $B = 0$. Without restricting the generality we may take $A = C = 1$ and obtain (A.3.2). However, the remaining coefficients are not completely arbitrary as we shall see. Let us consider equation (A.3.2) and try to find what circle it represents if any. It is natural to complete the squares to obtain

$$(x + D)^2 + (y + E)^2 = D^2 + E^2 - F.$$

Here the left member is the sum of two squares and, hence, not negative. This imposes the condition

$$F \leqq D^2 + E^2 \tag{A.3.4}$$

on the coefficients. If equality holds, the only real values of x and y are

$$x = -D, \quad y = -E$$

and (A.3.2) represents the point $(-D, -E)$. If inequality holds, the equation represents a circle with center at $(-D, -E)$ and radius

$$(D^2 + E^2 - F)^{\frac{1}{2}}.$$

Thus, the following equations represent:

$x^2 + y^2 + 2x - 4y + 1 = 0$, circle with center $(-1, 2)$, radius 2;
$x^2 + y^2 + 2x - 6y + 10 = 0$, point $(-1, 3)$;
$x^2 + y^2 + 1 = 0$, no real graph.

We shall take up various tangent problems for the circle and start with the simplest. Suppose that P_1 is a point on a given circle. Then *the tangent to the circle at P_1 is the straight line through P_1 perpendicular to the diameter ending at P_1*. This is the definition of Euclid. Later we shall recast the definition of the tangent in terms which make sense for more general curves than the circle.

Assuming the circle given by (A.3.1) and taking a point $P_1 = (x_1, y_1)$ on the circle, we want to find the equation of the tangent at P_1. See Figure A.18. If $P = (x, y)$ is a point on the tangent, $P \neq P_1$, then the vector $\overrightarrow{P_0P_1}$ is orthogonal to the vector $\overrightarrow{P_1P}$ by the definition of the tangent. The dot product of these vectors must be zero by the orthogonality condition. Now

$$\overrightarrow{P_0P_1} = (x_1 - x_0, y_1 - y_0), \quad \overrightarrow{P_1P} = (x - x_1, y - y_1),$$

so that

$$(x_1 - x_0)(x - x_1) + (y_1 - y_0)(y - y_1) = 0. \tag{A.3.5}$$

This is the equation of a straight line through (x_1, y_1) and is satisfied by every point on the tangent so it must be the equation of the tangent.

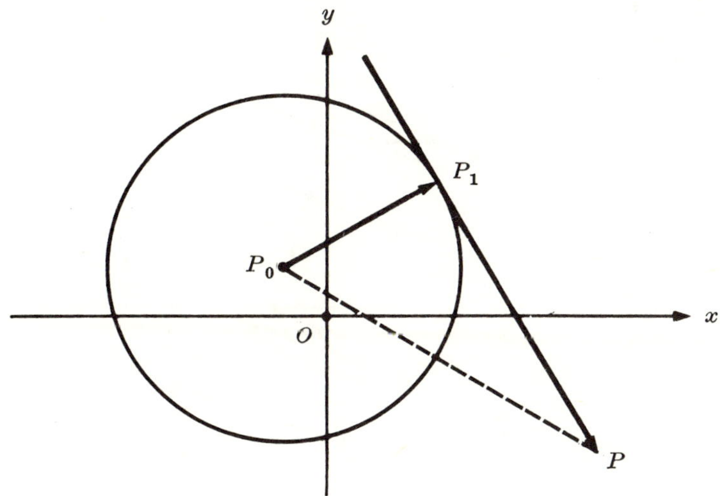

Figure A.18

We could leave the equation in this form, but it is customary to write

$$(x_1 - x_0)(x - x_0) + (y_1 - y_0)(y - y_0) = r^2. \qquad \text{(A.3.6)}$$

This is obtainable by writing

$$x - x_1 = (x - x_0) - (x_1 - x_0),$$

$$y - y_1 = (y - y_0) - (y_1 - y_0)$$

in (A.3.5) and using the fact that (x_1, y_1) satisfies (A.3.1). We omit the details.

From the figure we see that

$$\overline{P_0P}^2 = \overline{P_0P_1}^2 + \overline{P_1P}^2,$$

that is,

$$\overline{P_1P}^2 = (x - x_0)^2 + (y - y_0)^2 - r^2. \qquad \text{(A.3.7)}$$

Let us denote the right member of this relation by $C(x, y)$ or C. Then the equation of the circle (A.3.1) is

$$C(x, y) = 0,$$

and we can express (A.3.7) as follows:

If $P = (x, y)$ is a point outside the circle (A.3.1), *then the square of the length of either tangent from P to the circle is $C(x, y)$.*

It is clear that $C(x, y)$ is negative inside the circle. We shall use this abbreviated notation below in a study of pencils of circles just as we used L or $L(x, y)$ in a study of pencils of lines. First, however, we shall attack the tangent problem from a different angle using less special properties of the circle. This approach can be generalized to other curves.

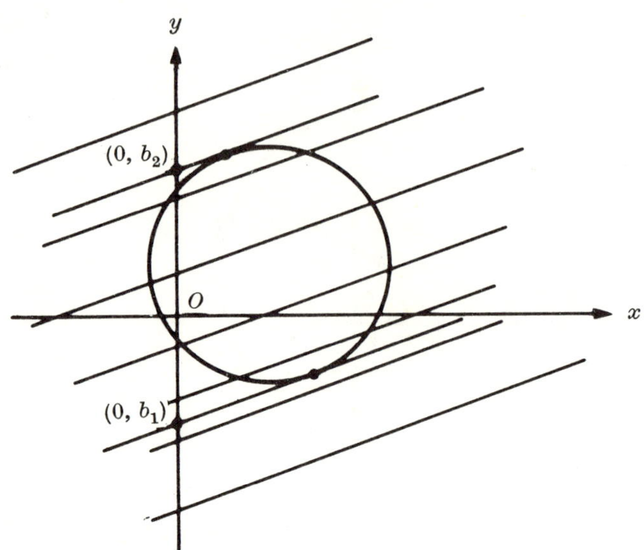

Figure A.19

Let us consider a family of parallel lines

$$y = mx + b, \quad -\infty < b < \infty, \tag{A.3.8}$$

where m is fixed, together with the circle (A.3.1). For large values of $|b|$ the lines do not intersect the circle at all and we see from Figure A.19 that intersection will occur if and only if b lies between the limits b_1 and b_2. The lines

$$y = mx + b_1, \quad y = mx + b_2$$

just touch the circle and are the two tangents to the circle of slope m.

To get the equations of the tangents we must use an algebraic argument. If a straight line of the family intersects the circle, then the simultaneous equations

$$\begin{aligned} (x - x_0)^2 + (y - y_0)^2 &= r^2, \\ y &= mx + b \end{aligned} \tag{A.3.9}$$

have real solutions. To find these, we eliminate one of the variables, y for instance, and obtain a quadratic equation for x whose roots are to be discussed. We can simplify the algebra by a change of variables. We write

$$x - x_0 = s, \quad y - y_0 = t, \quad c = b - y_0 + mx_0,$$

and substitute these values in (A.3.9). This gives a new system of simultaneous equations:

$$\begin{aligned} s^2 + t^2 &= r^2, \\ t &= ms + c. \end{aligned} \tag{A.3.10}$$

The effect of this transformation is to replace the original problem by a similar one for a circle with center at the origin. We now substitute $ms + c$ for t in the first equation and obtain after simplification

$$(1 + m^2)s^2 + 2cms + c^2 - r^2 = 0. \tag{A.3.11}$$

The roots of the quadratic equation

$$\alpha s^2 + 2\beta s + \gamma = 0$$

are

$$\frac{1}{\alpha}[-\beta \pm (\beta^2 - \alpha\gamma)^{\frac{1}{2}}].$$

The roots are real if and only if the *discriminant*

$$\beta^2 - \alpha\gamma \geqq 0.$$

In connection with (A.3.11) we are interested in the case of equal roots which requires that the discriminant be 0. Since

$$\alpha = 1 + m^2, \quad \beta = cm, \quad \gamma = c^2 - r^2,$$

we obtain

$$\beta^2 - \alpha\gamma = c^2m^2 - (1 + m^2)(c^2 - r^2) = 0,$$

or

$$c^2 = r^2(1 + m^2).$$

Hence

$$c = \pm\, r(1 + m^2)^{\frac{1}{2}},$$

and the straight lines

$$t = ms \pm r(1 + m^2)^{\frac{1}{2}} \tag{A.3.12}$$

are tangent to the circle

$$s^2 + t^2 = r^2. \tag{A.3.13}$$

Reverting to the original variables, we see that the lines

$$y - y_0 = m(x - x_0) \pm r(1 + m^2)^{\frac{1}{2}} \tag{A.3.14}$$

are tangents of slope m to the circle (A.3.1).

We can find the coordinates of the points of tangency. If the equation (A.3.11) has equal roots, the root is

$$-\frac{\beta}{\alpha} = -\frac{mc}{1 + m^2} = \mp \frac{mr}{(1 + m^2)^{\frac{1}{2}}}.$$

This is the s-coordinate and since $t = ms + c$, we can easily get the points of contact. For the original circle this means that

$$x_1 = x_0 \mp \frac{mr}{(1 + m^2)^{\frac{1}{2}}}, \quad y_1 = y_0 \pm \frac{r}{(1 + m^2)^{\frac{1}{2}}}, \tag{A.3.15}$$

are the points of contact of the tangents (A.3.14) with the circle (A.3.1)

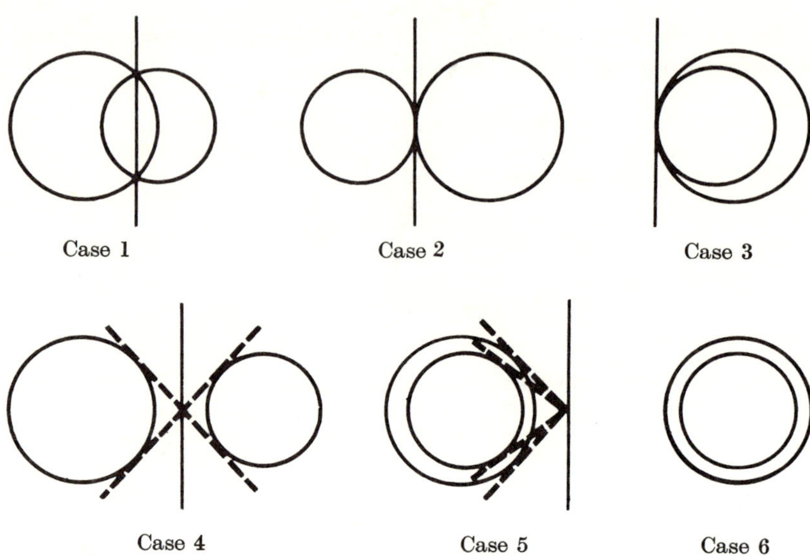

Figure A.20

Formula (A.3.15) expresses the coordinates of the contact points in terms of the slope, but we can use the same formula to get the slope of the tangent from the coordinates of the point of tangency. We write

$$x_1 - x_0 = \mp \frac{mr}{(1 + m^2)^{\frac{1}{2}}}, \quad y_1 - y_0 = \pm \frac{r}{(1 + m^2)^{\frac{1}{2}}}$$

and divide to obtain

$$m = -\frac{x_1 - x_0}{y_1 - y_0}, \tag{A.3.16}$$

which agrees with (A.3.5).

We come now to questions involving more than one circle. Two circles in the plane may or may not intersect. Suppose that the circles are given by their equations:

$$\begin{aligned} C_0(x, y) &= x^2 + y^2 + 2D_0x + 2E_0y + F_0 = 0, \\ C_1(x, y) &= x^2 + y^2 + 2D_1x + 2E_1y + F_1 = 0. \end{aligned} \tag{A.3.17}$$

If the circles intersect, they have two points in common which will satisfy this system of simultaneous equations. Subtracting the two equations we get

$$2(D_1 - D_0)x + 2(E_1 - E_0)y + F_1 - F_0 = 0. \tag{A.3.18}$$

This is a straight line, known as the *radical axis* of the two circles.

In Figure A.20 we have shown the six possible cases: (1) circles intersect; (2) external tangency; (3) internal tangency; (4) circles exterior to each other; (5) one circle interior to the other; and (6) concentric circles. In the last case there is no radical axis, and (A.3.18) becomes $F_1 = F_0$ which is false unless the circles coincide. In all other cases there is a radical axis which is shown in the figure. On the radical axis we have

$$C_0(x, y) = C_1(x, y).$$

If the point $P = (x, y)$ is on the radical axis and exterior to both circles, this equation says that the tangents from P to the two circles have the same length. This can be used to construct the radical axis in Cases 4 and 5.

The family of circles

$$C_\alpha\colon \alpha C_1(x, y) + (1 - \alpha)C_0(x, y) = 0, \quad -\infty < \alpha < \infty, \tag{A.3.19}$$

is called a *pencil of circles*. There are four different types. Cases 2 and 3 give rise to a single type as well as Cases 4 and 5. These types are distinct from each other and from the types corresponding to Cases 1 and 6.

Type 1. C_0 and C_1 intersect, and all circles of the pencil pass through their points of intersection, as does the radical axis which corresponds to the improper value ∞ of the parameter.

Type 2. C_0 and C_1 are tangent to each other and to the radical axis. All circles in the pencil go through the point of tangency and are tangent to the radical axis there.

Type 3. C_0 and C_1 do not intersect and are not concentric. No circles of the pencil intersect and they all have the same radical axis which they do not intersect. All centers are located on the same straight line perpendicular to the radical axis. There are two degenerate point circles, one on each side of the radical axis. Each of these points is an interior point of all the circles on its side of the axis.

Type 4. C_0 and C_1 are concentric and all circles of the pencil have the same center.

We restrict ourselves to this brief description and leave the verification of the statements to the student.

Circles can be represented parametrically in various ways. The best-known representation is in terms of trigonometric functions. It is clear that

$$x = x_0 + r\cos\alpha, \quad y = y_0 + r\sin\alpha \tag{A.3.20}$$

represents a circle with center (x_0, y_0) and radius r for

$$(x - x_0)^2 + (y - y_0)^2 = r^2(\cos^2\alpha + \sin^2\alpha) = r^2.$$

A less well-known representation of the same circle is

$$x = x_0 + r\frac{1 - t^2}{1 + t^2}, \quad y = y_0 + r\frac{2t}{1 + t^2}, \quad -\infty < t < \infty. \tag{A.3.21}$$

This is based on the simple identity

$$(1 - t^2)^2 + (2t)^2 = (1 + t^2)^2.$$

Finally a few words on the possibility of finding circles with preassigned properties. Equation (A.3.2) contains three arbitrary constants, D, E, F. This suggests that it is possible to impose three conditions on the circle and that they determine the constants uniquely. This is often the case, but by no means always. Much depends upon what we mean by "a condition." Some are not compatible and not all have the same compelling character. We shall give some examples.

A. The two coordinates of the center and the length of the radius determine the circle uniquely.

B. There is one and only one circle passing through three given, non-collinear points (six coordinates and an inequality!).

C. There is one and only one circle tangent to a given line at a given point and passing through another point not on the line.

D. There are two and only two circles tangent to two given lines and passing through a given point not on either line with the exception that there is no solution when the lines are parallel and the point is outside the strip bounded by the lines.

E. There are normally four circles tangent to three given lines. If two lines are parallel there are only two circles, and none if all three are parallel.

F. There may be as many as eight circles tangent to three fixed circles.

G. There is no circle of given radius r and tangent to two given parallel lines unless the distance between the lines is $2r$, in which case there are infinitely many such circles.

This list could be continued but we leave it to the reader to invent other cases. We shall show how to obtain the solution in case B, the most important for the applications. If the three points are

$$(x_1, y_1), \quad (x_2, y_2), \quad (x_3, y_3),$$

then the constants D, E, F must satisfy the following system of simultaneous equations:

$$\begin{aligned} x_1^2 + y_1^2 + 2Dx_1 + 2Ey_1 + F &= 0, \\ x_2^2 + y_2^2 + 2Dx_2 + 2Ey_2 + F &= 0, \\ x_3^2 + y_3^2 + 2Dx_3 + 2Ey_3 + F &= 0. \end{aligned} \tag{A.3.22}$$

For these equations express that the three points lie on the circle. The unknowns, D, E, F can be obtained from this system by ordinary methods unless there is a relation between the coordinates of the points which expresses that the points are collinear. We do not enter into further details.

EXERCISE A.3

1. Which of the following equations represent circles? Find center and radius in the relevant cases:

(a) $x^2 + y^2 + x - y = 0$; (b) $x^2 + y^2 - 3x + 4y + 4 = 0$;

(c) $x^2 + y^2 - 2x + 2y + 2 = 0$; (d) $x^2 + y^2 + 5x + 1 = 0$;

(e) $x^2 + y^2 - 11y + 10 = 0$; (f) $x^2 + y^2 - 10x - 8y + 32 = 0$.

2. Find the equation of a circle with center at $(-1, 2)$ and passing through $(1, -2)$.

3. Find the equation of a circle with center at $(a, 0)$ and passing through the origin.

4. What horizontal lines are tangent to the real circles of Problem 1?

5. Same question for vertical lines.

6. Find the tangents of slope 2 to the circle in part (a) of Problem 1. Find also the points of contact.

7. How should a be chosen in order that the line $x + y = a$ be a tangent to the circle $x^2 + y^2 - 2x + 4y = 0$?

8. Find the intersections of the circles

$$x^2 + y^2 - 2y - 1 = 0, \quad x^2 + y^2 - 6x - 2y + 5 = 0.$$

9. Find the intersections of the circles

$$x^2 + y^2 = 4, \quad x^2 + y^2 - 2x + y = 0.$$

10. Find the radical axis of the circles in parts (a) and (b) of Problem 1. Do the circles intersect?

11. Verify that C_α as defined by (A.3.19) actually represents a circle.

12. Verify that if C_0 and C_1 intersect, then all circles C_α pass through the points of intersection.

13. If a varies in Problem 3, show that the corresponding circles form a pencil of type 2 defined above.

14. Construct the pencil of type 3 which has the y-axis as radical axis, the centers being on the x-axis.

15. Find the equation of the circle that goes through the points $(2, 1)$, $(-2, 4)$, and $(-2, -1)$.

16. Find the equation of a circle tangent to the x-axis at $(1, 0)$ and passing through the point $(3, 4)$.

17. Find the equations of the four circles which lie on any three of the points $(0, 0)$, $(1, 2)$, $(4, -2)$, and $(-2, -1)$.

A.4. The conics. In three-dimensional space consider a straight line l passing through the origin and making an angle α, $0° < \alpha < 90°$ with the horizontal plane. If l is revolved about the vertical axis so that its angle of inclination is kept constant, a surface known as a *right circular cone* is generated. The line l is called a *generator* of the cone, the vertical axis is the *axis* of the cone, and the origin its *vertex* or *apex*. If the cone is cut by a plane, the resulting intersection is called a *conic section* or simply a *conic*.

Planes passing through the apex give trivial intersections: the apex itself, one straight line, or two straight lines according as the angle which the plane makes with the horizontal plane is less than, equal to, or greater than α.

Suppose now that the plane does not pass through the origin. Again the character of the intersection will depend upon the angle which the cutting

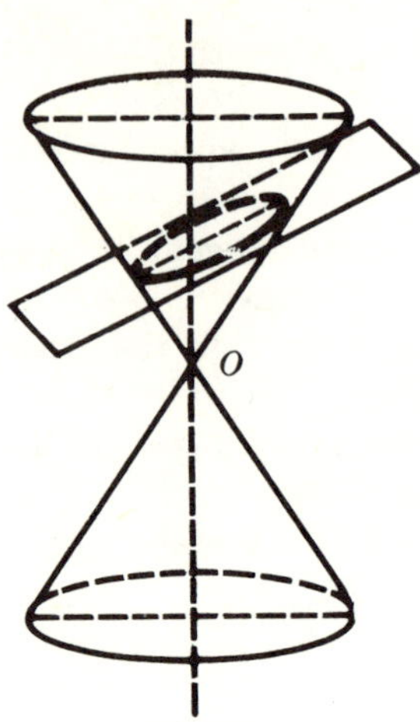

Figure A.21

plane makes with the horizontal. If this angle falls short of α, the intersection is a closed oval known as an *ellipse* (Gr. ἔλλειψις, a defect). If it equals α, an unbounded curve called a *parabola* results (Gr. παραβολή, a comparison). If it exceeds α, the plane intersects both the upper and the lower part of the cone. The resulting unbounded curve consisting of two branches is called a *hyperbola* (Gr. ὑπερβολή, overshooting). Compare the use of "ellipsis," "parable," and "hyperbole" in modern speech. The adjectives "elliptic," "parabolic," and "hyperbolic" serve in many branches of mathematics where the connection with the conics is rather remote. See Figures A.21–A.23 for these three cases. The theory of conic sections goes back to Apollonios of Pergae [about 262–170 B.C.], a Greek who wrote eight books on this subject.

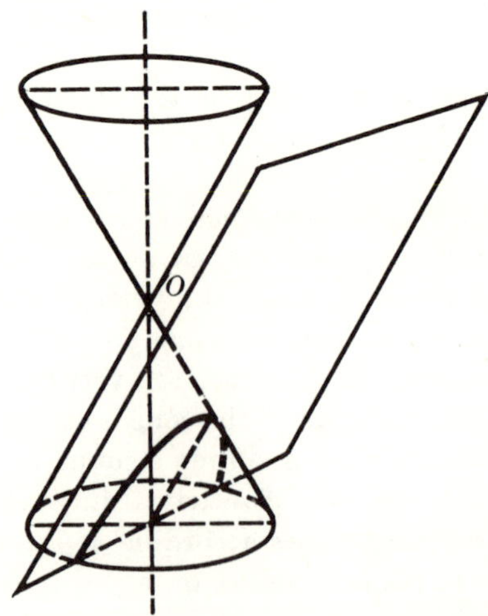

Figure A.22

All conics satisfy second degree equations; that is, if we introduce a system of Cartesian coordinates in the cutting plane, and if (x, y) is a point on the conic, then (x, y) must satisfy an equation of the second degree in x and y which is effectively of the second degree in at least one of the variables.

The generation of the conics sketched above may be used to obtain these equations. It is easier, however, to use a different approach which also leads to the (nondegenerate) conics. We consider the following "locus problem": *Given a straight line in the plane, a point not on the line, and a positive number e, find the locus of a point P which moves in such a way that the ratio of its distance from the fixed point to its distance from the fixed line equals the given number e.*

In this connection the fixed point is known as a *focus*, the fixed line as a *directrix*, and e as the *eccentricity*.

To get the equation of the locus (= curve), we introduce a system of coordinates in the plane. The choice of coordinates is left to our judgment. In such a case it pays to use whatever properties of symmetry the curve has and to choose a line of symmetry as one of the axes. It is clear in the present case that the curve is going to be symmetric with respect to the line through the focus perpendicular to the directrix. See Figure A.24. We take this line as the x-axis. Further we choose as origin the point on the perpendicular between the focus and the directrix which lies on the curve. This is the best choice for $e = 1$ and will serve our needs for the time being also when $e \neq 1$.

After choice of axes, we choose a unit of length. Suppose the equation of the directrix is

$$x + d = 0, \quad d > 0. \tag{A.4.1}$$

Then the focus is $(de, 0)$. This makes the origin a point of the curve. The condition imposed by the problem on the moving point $P = (x, y)$ now becomes

$$[(x - de)^2 + y^2]^{\frac{1}{2}} = e\,|x + d|, \tag{A.4.2}$$

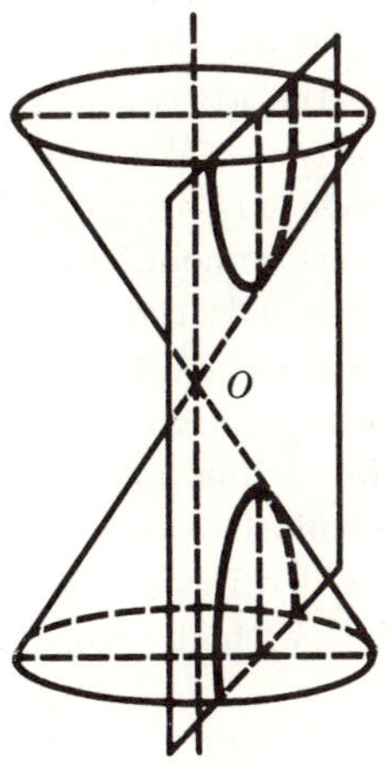

Figure A.23

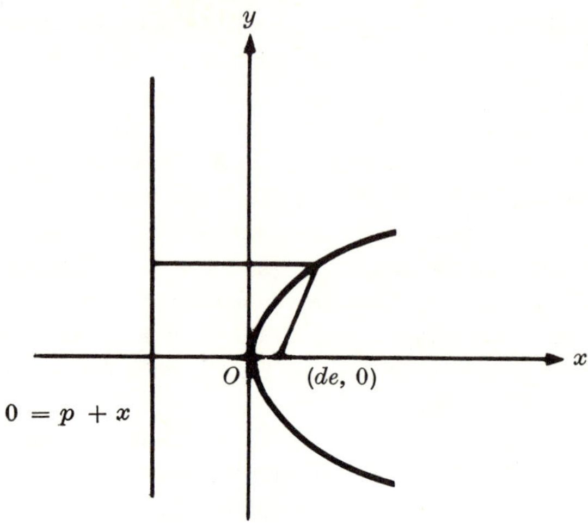

Figure A.24

for the left member is the distance from P to F and $|x + d|$ is the distance of P from the directrix. We square and expand, obtaining

$$x^2 - 2dex + d^2e^2 + y^2 = e^2(x^2 + 2dx + d^2),$$

or

$$(1 - e^2)x^2 + y^2 - 2de(1 + e)x = 0. \tag{A.4.3}$$

This is an equation of the second degree in x and y.

We note that the equation has a different character for different values of e. We distinguish three cases:

(1) $e < 1$, *ellipse*. The quadratic terms have the same sign.

(2) $e = 1$, *parabola*. The quadratic terms form a perfect square y^2.

(3) $e > 1$, *hyperbola*. The quadratic terms have opposite signs.

We shall not justify the use of the same names for these curves as for the conic sections. Rather we take the attitude that an ellipse is defined as the solution of the locus problem for $e < 1$. We prepare the ground for still another shift where an ellipse is regarded as the graph of a second degree equation whose quadratic terms satisfy certain conditions.

Before presenting a more detailed study of the three classes of conics we note that the same curves arise as solutions of other locus problems. The generation of conics as the locus of intersection of related pencils of lines is one of the most important. This is illustrated by a couple of questions in the Miscellaneous Problems at the end of Appendix A.

The rest of this section is divided into three parts, one for each of the three cases. It is advantageous to start with the parabolic case.

I. THE PARABOLA. Here we have $e = 1$ and (A.4.3) gives

$$y^2 - 4dx = 0.$$

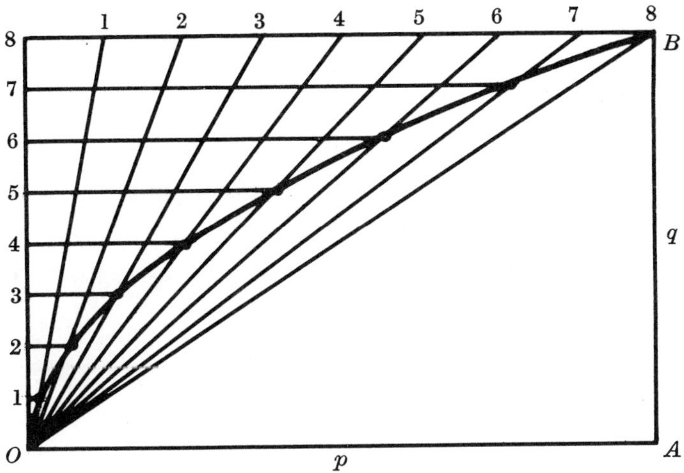

Figure A.25

In this formula we write a instead of d to conform to conventional notation. Thus, the typical form of the equation of a parabola is

$$y^2 = 4ax, \quad a > 0. \tag{A.4.4}$$

We note that the curve lies in the right half-plane except for the origin which in this connection is known as the *vertex* of the parabola. The coordinates of the focus F are $(a, 0)$. The vertical chord of length $4a$ through the focus is called the *latus rectum*. The line through the focus and the vertex is called the *axis* of the parabola and is its line of symmetry.

We can construct any number of points on the parabola using the definition of the latter as the locus of a point which moves so that it is equidistant from the focus and the directrix. Another construction is shown in Figure A.25. This is based on the possibility of generating a parabola as the locus of the point of intersection of corresponding lines of two related pencils, one passing through the vertex, the other being horizontal, namely

$$y = mx, \quad y = \frac{4a}{m}. \tag{A.4.5}$$

Multiplication of the equations eliminates m and shows that lines with the same value of m meet on the parabola (A.4.4).

For the actual construction take a rectangle $OABC$ of width p and height q. Divide the sides CB and OC each into n equal parts. Join the division points on CB with the origin O, draw horizontal lines through the division points on OC, and number the lines from O to C and from C to B. Mark the point where the jth line in the first system meets the jth line in the second system, $j = 1, 2, \cdots, n$. If OA and OC are used as coordinate axes, these points lie on a parabola whose equation is

$$y^2 = \frac{q^2}{p}\, x.$$

It is fairly simple to get the equation of a parabola in any Cartesian coordinate system. If in this system the focus is the point (a, b) and the normal form of the equation of the directrix is

$$x \cos \alpha + y \sin \alpha - p = 0,$$

then the condition that $P = (x, y)$ be equidistant from the focus and the directrix becomes

$$(x - a)^2 + (y - b)^2 = (x \cos \alpha + y \sin \alpha - p)^2,$$

or

$$\begin{aligned} x^2 - 2ax + a^2 + y^2 - 2by + b^2 \\ = x^2 \cos^2 \alpha + y^2 \sin^2 \alpha + 2xy \cos \alpha \sin \alpha + p^2 - 2xp \cos \alpha - 2yp \sin \alpha, \end{aligned}$$

and this becomes

$$\begin{aligned} x^2 \sin^2 \alpha - 2xy \sin \alpha \cos \alpha + y^2 \cos^2 \alpha \\ = 2(a - p \cos \alpha)\, x + 2(b - p \sin \alpha)\, y + p^2 - a^2 - b^2. \end{aligned}$$

Here the left member is a perfect square and we obtain finally

$$\begin{aligned} (x \sin \alpha - y \cos \alpha)^2 \\ = 2(a - p \cos \alpha)x + 2(b - p \sin \alpha)y + p^2 - a^2 - b^2. \end{aligned} \quad \text{(A.4.6)}$$

Thus we see that *in any Cartesian coordinate system the equation of a parabola is of the second degree and the quadratic terms form a perfect square.*

Conversely, a second-order equation in which the quadratic terms form a perfect square represents normally a parabola. The other possibilities are (1) two parallel lines, (2) a double line, and (3) no real graph. The equation

$$(x - y)^2 = b$$

illustrates these three possibilities for $b > 0$, $= 0$, and < 0 respectively.

There are various special cases of interest. If the vertex of the parabola is at the point (x_0, y_0) and the axis is horizontal and points to the right, then the equation is

$$(y - y_0)^2 = 4a(x - x_0), \quad a > 0. \quad \text{(A.4.7)}$$

If instead $a < 0$, the axis points to the left. The equation

$$y - y_0 = 4p(x - x_0)^2 \quad \text{(A.4.8)}$$

represents a parabola with a vertical axis which points up or down according as $p > 0$ or $p < 0$.

Any equation of the form

$$y = a_0x^2 + a_1x + a_2$$

represents a parabola with a vertical axis which points upward if $a_0 > 0$ and downward if $a_0 < 0$.

We come now to tangent problems. The first question is what should be meant by a tangent to a parabola. Here the alternate treatment of tangent problems for the circle points the way. Consider a family of parallel lines of slope m

$$y = mx + b.$$

Such a line intersects the parabola (A.4.4) in real points if the equation

$$(mx + b)^2 - 4ax = 0$$

has real roots. This simplifies to

$$m^2x^2 + 2(bm - 2a)x + b^2 = 0,$$

which has real roots if and only if

$$(bm - 2a)^2 - m^2b^2 \geqq 0,$$

and this condition reduces to

$$b \leqq \frac{a}{m} \quad \text{or} \quad b \geqq \frac{a}{m}$$

according as $m > 0$ or < 0.

The straight line

$$y = mx + \frac{a}{m} \tag{A.4.9}$$

separates the lines which intersect the parabola in two points from those which do not intersect at all. The line in question touches the parabola at the point

$$\left(\frac{a}{m^2}, \frac{2a}{m}\right). \tag{A.4.10}$$

This justifies our calling (A.4.9) the tangent of the parabola of slope m, the point of tangency being (A.4.10). Conversely, if a point (x_0, y_0) is given on the parabola, $y_0 \neq 0$, we conclude from (A.4.10) that there is a tangent at this point and that its slope is

$$m = \frac{1}{2}\frac{y_0}{x_0} = \frac{2a}{y_0}.$$

Substituting this in equation (A.4.9) we get

$$y = \frac{1}{2}\frac{y_0}{x_0}x + \frac{2ax_0}{y_0},$$

or

$$y_0y = \frac{1}{2}\frac{{y_0}^2}{x_0}x + 2ax_0,$$

which reduces to

$$y_0y = 2a(x + x_0). \tag{A.4.11}$$

This is the equation of the tangent to the parabola at the point (x_0, y_0). The equation has important consequences; see Figure A.26.

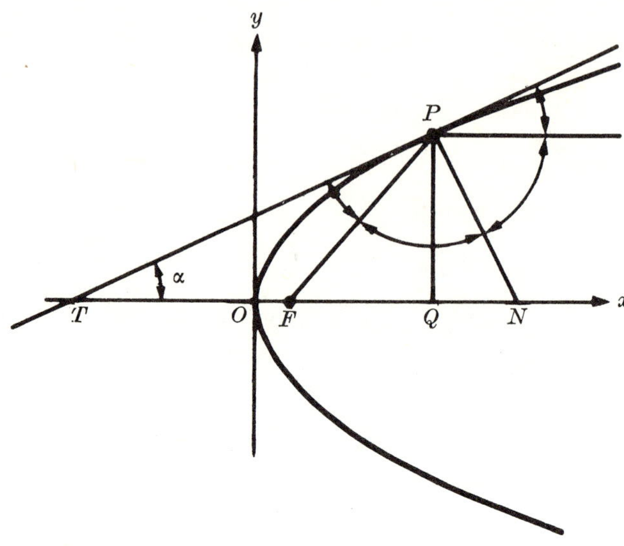

Figure A.26

The equation shows that the tangent intersects the x-axis at the point $(-x_0, 0)$. This means that the tangent to the parabola at the point $P = (x_0, y_0)$ is the straight line on the points P and $T = (-x_0, 0)$.

In the figure we have also marked the focus F, the line PF, and the line PS, the latter being parallel to the x-axis. Further, five angles are shown and the *normal* PN at P, that is, the line through P perpendicular to the tangent. Actually the figure shows the basic principle for the parabolic mirror. The distance of P from the directrix (not shown in the figure) is $x_0 + a$ and since $TO = x_0$, $OF = a$, we see that

$$TF = FP,$$

for P is equidistant from the focus and the directrix. This makes the triangle TFP isosceles so that the angles α and β are equal. Since α and γ are equal, it follows that $\beta = \gamma$ and hence their complements δ and ε are also equal. Now we learn from physics that when light is reflected, *the angle of incidence equals the angle of reflection.* We can take either δ or ε as angle of incidence; the other will then be the angle of reflection. This means that *light from a source at the focus is reflected in a beam parallel to the axis.* It also means that *a beam of light parallel to the axis will be reflected by the parabola through the focus.* The parabolic mirror of a searchlight uses the first principle, that of a reflecting telescope the second.

The parabola plays a role in other practical problems. The cable of a suspension bridge assumes the shape of a parabola if the load and the weight of the cable are uniformly distributed horizontally. Parabolic arches are used in bridge building and in architecture. A projectile from a gun would trace a parabola if acted upon by gravity alone. Certain comets have a parabolic trajectory.

EXERCISE A.4.1

1. A circle is a conic section. Can equation (A.4.3) represent a circle for a suitable choice of e? What happens to the locus problem for this choice?

2. Find the focus and the directrix of the parabolas:

(a) $y^2 = 2x$; (b) $y^2 = -8x$; (c) $y - 3 = 2(x - 1)^2$;

(d) $y = x^2 + 2x$; (e) $y = 4 - x^2$; (f) $y = x^2 + x + 1$.

3. Find the equation of the parabola with focus at $(1, 2)$ and directrix $x + y + 1 = 0$. Where is the vertex?

4. Given the equation

$$(y - x)^2 = x,$$

show that there is a real graph which is located in the right half-plane and is not a pair of parallel lines. Solve for y and sketch the curve.

5. Find the tangent of slope 2 to each of the parabolas in Problem 2.

6. Find the tangent to $y^2 = 2x$ at $(2, -2)$.

7. Find the tangent to $y^2 = -8x$ at $(-2, 4)$.

8. Find the equations of the tangents to $y^2 = 2x$ through the point $(0, -2)$.

9. Consider the parabola $y^2 = 4ax$ and the family of parallel lines $y = mx + b$ where $m > 0$. The parabola cuts off a line segment from each line with $bm > a$. Show that the locus of the midpoints of these line segments is an infinite line segment parallel to the x-axis.

10. A one-parameter family of *confocal* parabolas is defined by $y^2 = 4a(a - x)$ where a is any real number. Show that the origin is the common focus (hence "confocal" family) of all these parabolas. Prove that two parabolas of the family intersect if and only if their parameters have opposite signs. Find where they meet in terms of the parameters and show that they meet at right angles.

Find the equation of the parabola with vertex at the origin which satisfies the additional conditions:

11. Focus at $(-2, 0)$.

12. Directrix $x = \frac{3}{2}$.

13. Focus at $(0, -2)$.

14. Directrix $y = 3$.

15. Focus on the negative x-axis, latus rectum 6.

16. Focus on the positive x-axis, slope at the upper end point of the latus rectum equals 2.

17. Focus on the positive x-axis, $x + y + 1 = 0$ is a tangent.

18. Axis along the x-axis, passing through $(3, 2)$.

19. Axis along the y-axis, passing through $(3, 2)$.

20. Focus on the positive x-axis, latus rectum 8.

Find the equation of the parabola directly from the definition if:

21. Directrix $x = 0$, focus $(b, 0)$.

22. Directrix $y = -1$, focus $(0, p)$.

23. Focus $(-1, -1)$, vertex $(0, 0)$.

24. Directrix $x + y + 1 = 0$, vertex $(0, 0)$.
25. Directrix $x - y = 2$, focus $(5, 4)$.
26. Directrix $x + y = 2$, axis $y = x$, origin exterior to the parabola, latus rectum 6.

II. THE ELLIPSE. We return to the equation (A.4.3) where we now assume that $0 < e < 1$ so that the curve is an ellipse. Our choice of the origin for the locus problem is not the most favorable in the present case because it hides the fact that the curve has two axes of symmetry and not just one as the parabola. The equation shows that the curve intersects the x-axis at

$$x = 0 \quad \text{and at} \quad x = 2\frac{de}{1-e} \equiv 2a.$$

Now

$$x^2 - 2ax = (x - a)^2 - a^2$$

so that the equation may be written

$$(1 - e^2)(x - a)^2 + y^2 = b^2, \quad \text{where } b^2 = (1 - e^2)a^2,$$

or, after division by b^2,

$$\left(\frac{x-a}{a}\right)^2 + \left(\frac{y}{b}\right)^2 = 1.$$

This shows that the ellipse (A.4.3) is symmetric with respect to the x-axis, the line $x = a$, and the point $(a, 0)$.

Hence we can take

$$\left(\frac{x}{a}\right)^2 + \left(\frac{y}{b}\right)^2 = 1$$

as the standard equation of the ellipse referred to its own axes of symmetry. In this form the curve is symmetric with respect to the coordinate axes as well as to the origin. See Figure A.27. We have clearly

$$-a \leqq x \leqq a, \quad -b \leqq y \leqq b.$$

The line segment A_1A_2 is known as the *major axis*, B_1B_2 as the *minor axis*. Their lengths are $2a$ and $2b$ respectively. Owing to the double symmetry, the ellipse has two foci and two directrices which are marked in the figure. Since the original directrix d_1 is d units to the left of the left endpoint of the major axis and the focus F_1 is de units to the right of the same point, we find that the foci are

$$F_1 = (-c, 0), \quad F_2 = (c, 0), \quad c = ae,$$

and the directrices are

$$d_1\colon x + \frac{a}{e} = 0, \quad d_2\colon x - \frac{a}{e} = 0.$$

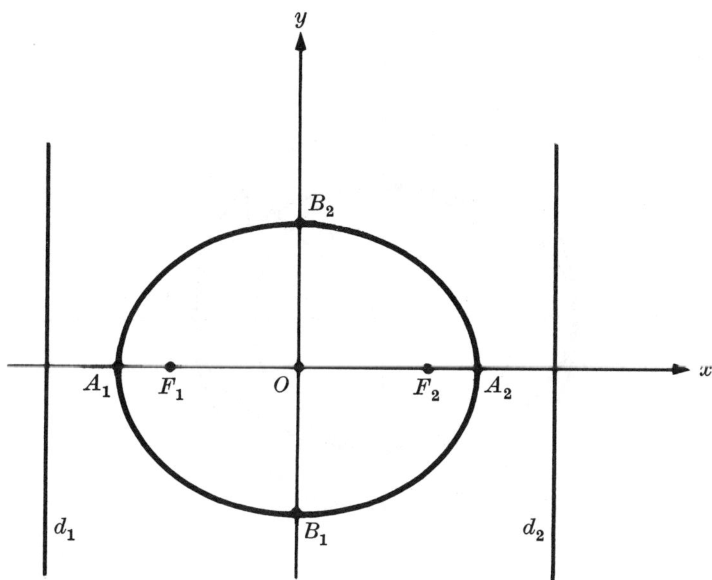

Figure A.27

Here we have used that

$$a + d = \frac{de}{1 - e} + d = \frac{d}{1 - e} = \frac{de}{1 - e}\frac{1}{e} = \frac{a}{e},$$

$$a - de = \frac{de}{1 - e} - de = \frac{de^2}{1 - e} = ae.$$

The distance between the foci equals $2c$. We note that

$$a^2 = b^2 + c^2.$$

There are several simple ways of constructing an ellipse or an arbitrary number of points on an ellipse. One of these is based on the parametric representation of the ellipse

$$x = a \cos \theta, \quad y = b \sin \theta, \tag{A.4.13}$$

which is an immediate consequence of the identity

$$\cos^2 \theta + \sin^2 \theta = 1.$$

If we draw in the plane two circles with center at the origin and radius b and a respectively, then the ellipse is inside the outer circle, except for the endpoints of the major axis, and outside the inner circle, except for the endpoints of the minor axis. See Figure A.28.

If OQR makes the angle θ with the positive x-axis, then we have $Q = (b \cos \theta, b \sin \theta)$, $R = (a \cos \theta, a \sin \theta)$. Hence if we draw a vertical line through R and a horizontal line through Q, then the point where they intersect:

$$P = (a \cos \theta, b \sin \theta)$$

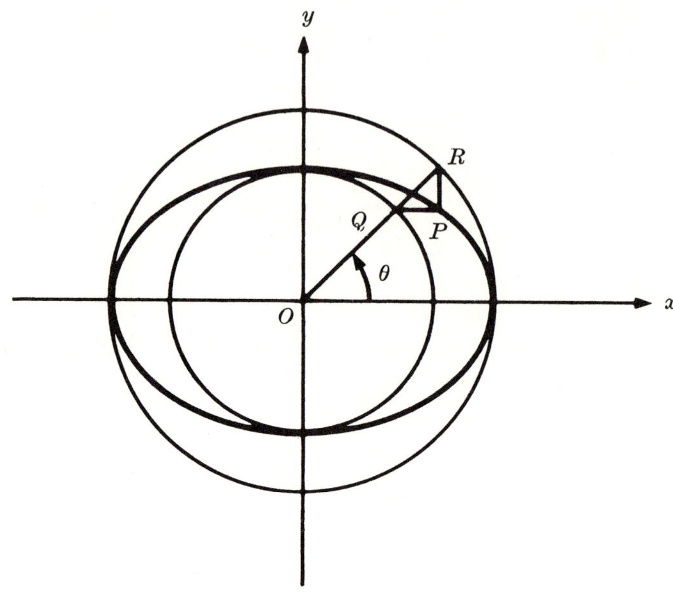

Figure A.28

is on the ellipse. This enables us to find any number of points on the ellipse.

There is also a string construction which is based on the following locus problem:

The locus of a point which moves in such a way that the sum of its distances from two fixed points is a constant, is an ellipse with foci at the fixed points.

We take a system of coordinates in which the fixed points are $(-c, 0)$ and $(c, 0)$. Let the constant be denoted by $2a$. We must have

$$c < a$$

in order for the problem to have a solution. See Figure A.29. If we express the condition

$$F_1P + PF_2 = 2a$$

in terms of coordinates, we get

$$[(x - c)^2 + y^2]^{\frac{1}{2}} + [(x + c)^2 + y^2]^{\frac{1}{2}} = 2a.$$

Here we transfer the first term to the right and square both sides, obtaining

$$(x + c)^2 + y^2 = 4a^2 + (x - c)^2 + y^2 - 4a[(x - c)^2 + y^2]^{\frac{1}{2}}.$$

This reduces to

$$4a[(x - c)^2 + y^2]^{\frac{1}{2}} = 4(a^2 - cx).$$

Canceling the factor 4 and squaring we obtain

$$a^2[x^2 - 2cx + c^2 + y^2] = a^4 - 2a^2cx + c^2x^2,$$

which reduces to

$$(a^2 - c^2)x^2 + a^2y^2 = a^2(a^2 - c^2).$$

Setting $a^2 - c^2 = b^2$ we obtain

$$b^2x^2 + a^2y^2 = a^2b^2,$$

and this is equivalent to (A.4.12).

To construct an ellipse, one can loop a string of length $2(a + c)$ over tacks placed at the foci. A pencil placed in the loop and moved so as to keep the string taut will then describe the ellipse.

The distances of the point $P = (x, y)$ from the directrices are

$$\frac{a}{e} - x \quad \text{and} \quad \frac{a}{e} + x.$$

It follows that the focal radii are

$$PF_1 = a - ex, \quad PF_2 = a + ex,$$

which adds up to $2a$.

The ellipse also has interesting reflecting properties with practical applications, if whispering galleries can be called practical. To describe these we take up the tangent problem for the ellipse. We proceed as in the case of the parabola. Consider a family of parallel lines

$$y = mx + q,$$

where m is fixed. A line of this family will have a real intersection with the ellipse (A.4.12) if

$$-(a^2m^2 + b^2)^{\frac{1}{2}} \leqq q \leqq (a^2m^2 + b^2)^{\frac{1}{2}},$$

as we see by eliminating y between the two equations and writing down that the discriminant of the resulting second-order equation be nonnegative. In

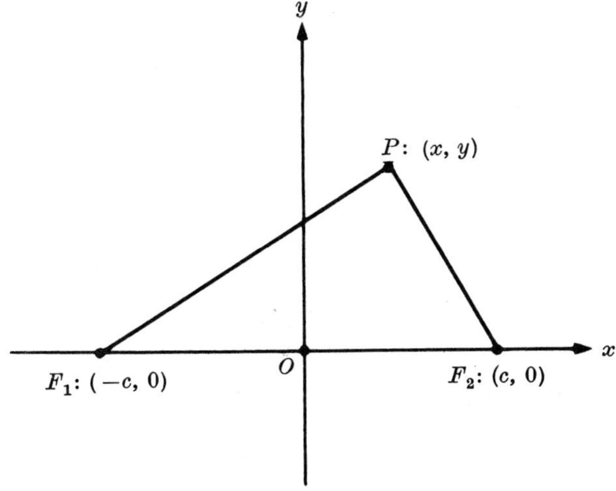

Figure A.29

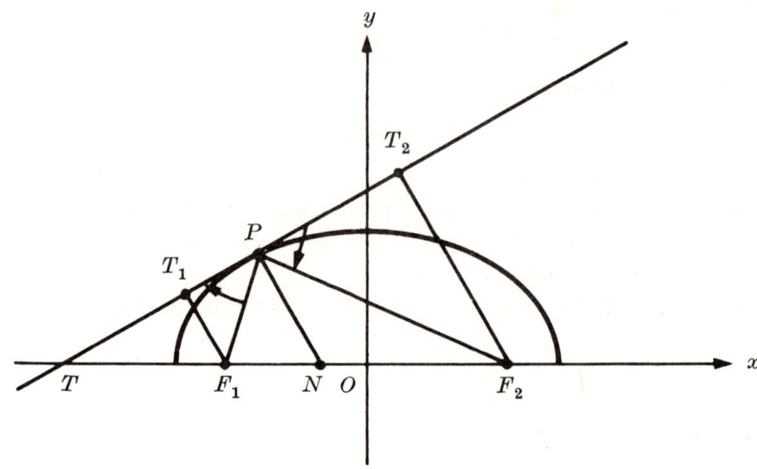

Figure A.30

particular, we have equal roots if the sign of equality holds, that is, the two lines

$$y = mx \pm (a^2m^2 + b^2)^{\frac{1}{2}} \tag{A.4.14}$$

are tangents to the ellipse with slope m. Their points of tangency are

$$\left(-\frac{a^2m}{(a^2m^2 + b^2)^{\frac{1}{2}}}, \frac{b^2}{(a^2m^2 + b^2)^{\frac{1}{2}}}\right) \quad \text{and} \quad \left(\frac{a^2m}{(a^2m^2 + b^2)^{\frac{1}{2}}}, -\frac{b^2}{(a^2m^2 + b^2)^{\frac{1}{2}}}\right). \tag{A.4.15}$$

Conversely, if (x_0, y_0), $y_0 \neq 0$ is a given point on the ellipse, then we can set

$$x_0 = -\frac{a^2m}{(a^2m^2 + b^2)^{\frac{1}{2}}}, \quad y_0 = \frac{b^2}{(a^2m^2 + b^2)^{\frac{1}{2}}}.$$

We determine m by dividing the first equation by the second and simplifying. We get

$$m = -\frac{b^2x_0}{a^2y_0} \tag{A.4.16}$$

as the slope of the tangent to the ellipse at the point (x_0, y_0). This leads to the equation

$$y - y_0 = -\frac{b^2x_0}{a^2y_0}(x - x_0)$$

for the tangent which simplifies to

$$b^2x_0x + a^2y_0y - a^2b^2 = 0. \tag{A.4.17}$$

We can now prove:

The focal radii at P make equal angles with the tangent.

For the proof see Figure A.30. If T_1 and T_2 are the foot points of the perpendiculars from the foci to the tangent, it suffices to show that the triangles PT_1F_1 and PT_2F_2 are similar. This requires that

$$T_1F_1 : PF_1 = T_2F_2 : PF_2.$$

Now T_1F_1 and T_2F_2 are proportional to the values of the left member of (A.4.17) at the points F_1 and F_2 respectively. These values are

$$-b^2x_0c - a^2b^2 = -b^2(cx_0 + a^2) = -b^2a(a + ex_0)$$

and

$$b^2x_0c - a^2b^2 = b^2(cx_0 - a^2) = -b^2a(a - ex_0).$$

The last factor is PF_1 in the first case and PF_2 in the second. This proves that the triangles are similar and hence that the angles in question are equal. This also makes the angles with the normal equal, and we have the following result:

An elliptical mirror will reflect light or sound rays, originating at either focus so that the rays converge at the other focus.

The orbit of a planet with respect to the sun is approximately an ellipse, the sun being at one focus.

EXERCISE A.4.2

1. Find axes, foci, directrices, and eccentricity of each of the following ellipses:

(a) $x^2 + 4y^2 = 4$; (b) $4x^2 + 9y^2 = 16$; (c) $x^2 + 5y^2 = 5$;

(d) $3x^2 + 2y^2 = 12$; (e) $4x^2 + y^2 = 4$; (f) $x^2 + 3y^2 = 4$.

2. The following ellipses are not in standard position, but their axes are parallel to the coordinate axes. In each case find the center, the axes, the foci, and the eccentricity, and give the equation of the ellipse when referred to its own axes:

(a) $\frac{3}{4}x^2 + y^2 - 3x = 0$; (b) $4x^2 - 8x + y^2 - 2y = 0$;

(c) $4x^2 + y^2 - 6y + 5 = 0$.

3. Find the equation of the ellipse of eccentricity $\frac{1}{2}$ with a focus at $(1, 1)$ and corresponding directrix $x + y = 4$. Find also the length of the major axis.

4. The equation $x^2 + xy + y^2 = 3$ represents an ellipse with center at the origin and tilted axes. Show that if (x, y) is on the curve, then $-2 \leqq x \leqq 2$, $-2 \leqq y \leqq 2$. Determine the four points where one of the signs of equality holds.

5. Find the equations of the tangents of slope 2 to the ellipses in Problem 1.

6. Find the tangent of slope 3 to the ellipse in part (b) of Problem 2.

7. Find the tangent at $(\sqrt{3}, \frac{1}{2})$ of the ellipse in Problem 1(a).

8. Find the tangent at $(\sqrt{2}, \sqrt{3})$ of the ellipse in Problem 1(d).

9. Given the ellipse $b^2x^2 + a^2y^2 = a^2b^2$ and the family of straight lines $y = mx + q$, m fixed. The ellipse cuts off a line segment of each of the lines which intersects it. Prove that the midpoints of these segments form a chord through the origin of slope $-b^2/(a^2m)$.

10. Prove that the ellipses $px^2 + (p + 1)y^2 = p(p + 1)$, $p > 0$, form a confocal family with foci at $(-1, 0)$ and $(1, 0)$.

11. Prove that the following is a parametric representation of an ellipse

$$x = a\,\frac{1 - t^2}{1 + t^2}, \quad y = b\,\frac{2t}{1 + t^2}.$$

Find the equation of the ellipse satisfying the given conditions:

12. Vertices at $(\pm 5, 0)$, foci at $(\pm 4, 0)$.
13. Vertices at $(0, \pm 4)$, foci at $(0, \pm 2)$.
14. Vertices at $(\pm 5, 0)$, eccentricity $\frac{1}{2}$.
15. Vertices at $(0, \pm 4)$, eccentricity $\frac{2}{3}$.
16. Foci at $(\pm 4, 0)$, directrices $x = \pm 8$.
17. Axes along the coordinate axes, passing through $(1, 2)$ and $(\sqrt{3}, -1)$.
18. Vertices at $(\pm 5, 0)$ passing through $(1, 2)$.
19. Foci at $(0, \pm 2)$, passing through $(1, 2)$.
20. One directrix $x + y = 2$, corresponding focus $(0, 0)$.
21. Axes along $y = x$ and $y = -x$, passing through $(0, 6)$ and $(4, -4)$.

III. THE HYPERBOLA. We return once more to the equation (A.4.3) where we now take $e > 1$. We set

$$a^2 = \frac{de}{e - 1}, \quad b^2 = (e^2 - 1)a^2,$$

so that the equation becomes

$$(e^2 - 1)(x + a)^2 - b^2 - y^2 = 0,$$

or

$$\left(\frac{x + a}{a}\right)^2 - \left(\frac{y}{b}\right)^2 = 1.$$

This shows that the hyperbola also has a center and two perpendicular axes of symmetry. Referred to its own axes, the standard form of the equation of the hyperbola becomes

$$\left(\frac{x}{a}\right)^2 - \left(\frac{y}{b}\right)^2 = 1. \tag{A.4.18}$$

Here $2a$ is known as the *transverse axis*, $2b$ as the *conjugate axis*. The end-points of the transverse axis are on the hyperbola and are called the *vertices* of the latter; the conjugate axis is not a chord of the hyperbola. The foci are at the points $(-c, 0)$ and $(c, 0)$ where now

$$c^2 = a^2 + b^2$$

so that $c > a$. The directrices are

$$x + \frac{a}{e} = 0 \quad \text{and} \quad x - \frac{a}{e} = 0. \tag{A.4.19}$$

If $P:(x, y)$ is a point on the hyperbola and $x > 0$, the left members of the equations (A.4.19) give the distance from P to the directrices, and if we multiply by e we obtain the focal radii

$$PF_1 = ex + a, \quad PF_2 = ex - a, \tag{A.4.20}$$

which shows that

$$PF_1 - PF_2 = 2a. \tag{A.4.21}$$

Hence:

The locus of a point P which moves in such a way that the difference of its distances from two fixed points is a constant, is a hyperbola.

In the case of the ellipse, the major axis is by definition longer than the minor axis. There is no such relationship between the transverse and the conjugate axes.

The hyperbola is an unbounded curve consisting of two separate branches whose minimum distance is $2a$. The behavior of these branches for large values of the coordinates is fundamental for the theory of the hyperbola. First we note that if $P:(x, y)$ lies on the hyperbola, then

$$\left(\frac{x}{a}\right)^2 = 1 + \left(\frac{y}{b}\right)^2 \geqq 1,$$

that is,

$$|x| \geqq a.$$

There is no upper bound for $|x|$, and y can take on all real values. Moreover, it is clear that $|x|$ is large when $|y|$ is large for we have

$$\left(\frac{x}{a}\right)^2 > \left(\frac{y}{b}\right)^2,$$

that is,

$$|x| > \frac{a}{b}|y| \quad \text{or} \quad |y| < \frac{b}{a}|x|.$$

This is for P on the hyperbola. We have equality instead of inequality if P lies on either of the straight lines

$$bx - ay = 0, \quad bx + ay = 0 \tag{A.4.22}$$

which are known as the *asymptotes* of the hyperbola. The first step in sketching a hyperbola should always be to draw the asymptotes. See Figure A.31. It is convenient to start by drawing the rectangle of sides $2a$ and $2b$ whose diagonals when extended give the asymptotes.

The inequalities proved above show that in the first quadrant the hyperbola lies to the right of the line $x = a$ and below the asymptote $bx - ay = 0$. We shall prove that as P proceeds along the hyperbola it gets closer and closer

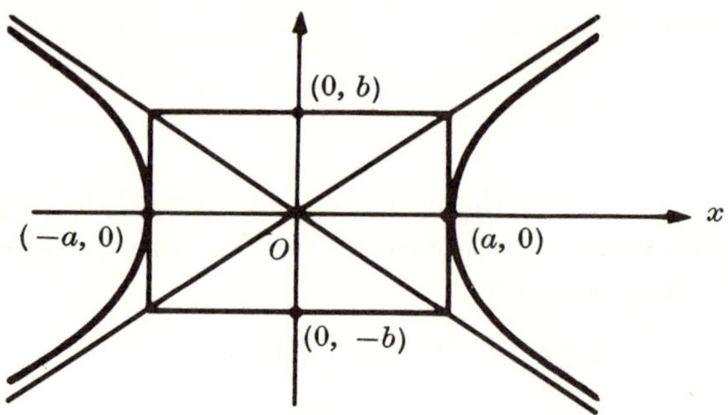

Figure A.31

to the asymptote and the distance of P from the asymptote tends to zero. We can factor the equation of the hyperbola as follows:

$$(bx - ay)(bx + ay) = a^2b^2.$$

Here we divide by $a^2 + b^2$ and obtain

$$\frac{bx - ay}{(a^2 + b^2)^{\frac{1}{2}}} \cdot \frac{bx + ay}{(a^2 + b^2)^{\frac{1}{2}}} = \frac{a^2b^2}{a^2 + b^2}. \tag{A.4.23}$$

On the left we have the product of the distances from $P{:}\,(x, y)$ to the asymptotes. This product then is a constant and this implies that if one factor is large, the other must be small. Now as P recedes to the distant part of the first quadrant, its distance from the other asymptote becomes arbitrarily large, and this forces P to approach the asymptote in the first quadrant. The further out P is, the closer it gets to the asymptote.

It is easy to get bounds for the distance of P from the nearest asymptote. We have

$$\frac{b}{a}x - y = \frac{1}{\dfrac{b}{a}x + y}$$

from the equation of the hyperbola. Here the first term on the left is the ordinate on the asymptote, the second the ordinate on the hyperbola. The difference is the vertical (not perpendicular!) distance from the hyperbola to the asymptote. Since $\frac{b}{a}x > y$, we have

$$2y < \frac{b}{a}x + y < 2\frac{b}{a}x,$$

so that

$$\frac{a}{2bx} < \frac{b}{a}x - y < \frac{1}{2y}.$$

This shows that the vertical distance tends to zero and also gives its bounds.

Owing to the symmetry of the curve, what has been said about the asymptote in the first quadrant applies with obvious changes to all quadrants.

If $b = a$, the equation becomes

$$x^2 - y^2 = a^2, \tag{A.4.24}$$

which is known as an *equilateral hyperbola*. Here the asymptotes are at right angles to each other. The equation

$$xy = p^2 \tag{A.4.25}$$

also represents an equilateral hyperbola. Here the coordinate axes are the asymptotes.

Tangent problems for the hyperbola can be handled by the same methods as used for the ellipse. We note minor differences and final results. The slope of a tangent to the hyperbola satisfies the inequality

$$|m| > \frac{b}{a}. \tag{A.4.26}$$

If this condition is satisfied, the equation of the tangent is

$$y = mx \pm (m^2a^2 - b^2)^{\frac{1}{2}}, \tag{A.4.27}$$

and the corresponding points of tangency are

$$\left(\mp \frac{a^2m}{(a^2m^2 - b^2)^{\frac{1}{2}}}, \mp \frac{b^2}{(a^2m^2 - b^2)^{\frac{1}{2}}}\right). \tag{A.4.28}$$

On the other hand, if the point of tangency (x_0, y_0), $y_0 \neq 0$, is given, the slope is

$$m = \frac{b^2x_0}{a^2y_0}, \tag{A.4.29}$$

and the equation of the tangent is

$$b^2x_0x - a^2y_0y = a^2b^2. \tag{A.4.30}$$

The tangent again occupies a special position with respect to the focal radii. Using the same method as for the ellipse we get:

The tangent to the hyperbola at the point P bisects the angle between the corresponding focal radii.

An application of the hyperbola is to *range finding*. This is based on formula (A.4.21). If two listening posts, at a known distance apart, time the firing of a gun, the time difference multiplied by the velocity of sound gives the value of $2a$ and hence determines a hyperbola on which the gun must be located. A third observation post gives a second hyperbola and the gun placement is at one of the intersections of the hyperbolas.

EXERCISE A.4.3

1. Find the axes, foci, directrices, and asymptotes of the following hyperbolas:

(a) $x^2 - 4y^2 = 4$; (b) $4x^2 - y^2 = 4$; (c) $3x^2 - 2y^2 = 12$;
(d) $x^2 - 5y^2 = 5$; (e) $y^2 - 4x^2 = 4$; (f) $5y^2 - x^2 = 5$.

2. The following hyperbolas do not have their centers at the origin, but their axes are parallel to the coordinate axes. Find the axes, foci, directrices, and asymptotes. Find also the equations of the hyperbolas referred to their own axes.

(a) $-3x^2 + y^2 - 6x = 0$; (b) $x^2 - 6x - y^2 + 2y = 0$;
(c) $4x^2 - 8x - y^2 + 6y - 1 = 0$; (d) $4x^2 - y^2 + 2x + 2y + 10 = 0$.

3. Show that the curve $xy = 1$ is symmetric with respect to the lines $y = x$ and $y = -x$ and also to the origin. Assuming that the equation represents a hyperbola, find vertices, foci, and lengths of the axes.

4. Find the equation of the hyperbola of eccentricity 2 whose foci are at the points (3, 3) and (−1, −1).

5. Do all the hyperbolas in Problem 1 have tangents of slope 1? Give the equations of the tangents and the points of contact for those that do.

6. Prove formulas (A.4.26) to (A.4.30).

7. Prove that the tangent to the hyperbola bisects the angle between the focal radii.

8. The slope of a tangent to $xy = p^2$ is necessarily negative. Find the equation of the tangent of slope m (< 0) and its point of contact. Find also the equation of the tangent at the point (x_0, y_0).

9. Given the hyperbola $b^2x^2 - a^2y^2 = a^2b^2$ and the family of straight lines $y = mx + q$ with m fixed, $m > b/a$. The hyperbola cuts off segments from the lines that intersect it. Show that the midpoints of these line segments lie on a straight line through the origin and find its slope.

10. Show that the hyperbolas

$$(\sin \alpha)^2x^2 - (\cos \alpha)^2y^2 = (\sin \alpha \cos \alpha)^2$$

form a confocal family with foci at the points (−1, 0) and (1, 0). [Actually this family is orthogonal to the confocal family of ellipses in Problem 10 of Exercise A.4.2.]

Find the equation of the hyperbola satisfying the following conditions:

11. Vertices at $(\pm 4, 0)$, foci at $(\pm 5, 0)$.
12. Vertices at $(0, \pm 3)$, foci at $(0, \pm 4)$.
13. Directrices $x = \pm 2$, foci at $(\pm 4, 0)$.
14. Vertices at $(\pm 6, 0)$, passing through (7, 8).
15. Axes along the coordinate axes, conjugate axis 4, passing through (3, 4).
16. Axes along the coordinate axes, passing through (3, 4) and (4, 5).
17. Axes along coordinate axes, passing through (1, 2) where the slope is 3.

18. Asymptotes $y = \pm 2x$, passing through (1, 2).

19. Directrix $x + y = 2$, corresponding focus (0, 0), and passing through (−3, −4).

20. Axes along the lines $y = \pm x$, passing through (1, 2) and (3, 4).

21. Axes along the lines $y = \pm x$, vertices at (1, 1) and (−1, −1), eccentricity 2.

22. Asymptotes the coordinate axes, passing through (2, 3).

23. Prove that a parametric representation of the hyperbola is furnished by

$$x = a \sec \theta, \quad y = b \tan \theta, \quad \sec \theta = \frac{1}{\cos \theta}.$$

What θ-interval corresponds to the right branch, the left branch?

24. Prove that

$$x = a\,\frac{1 + t^2}{1 - t^2}, \quad y = b\,\frac{2t}{1 - t^2},$$

also gives a parametric representation of the hyperbola. What t-intervals correspond to different parts of the curve?

A.5. Transformation of the axes. At the beginning of the preceding section the question of the geometric meaning of the general second-order equation was raised. This question will now be discussed.

Let us set

$$Q(x, y) \equiv Ax^2 + 2Bxy + Cy^2 + 2Dx + 2Ey + F, \tag{A.5.1}$$

and ask what the equation

$$Q(x, y) = 0 \tag{A.5.2}$$

represents when x and y are Cartesian coordinates of a point in the plane.

It is quite feasible that the equation cannot be satisfied by any real values of x and y so there is no real representation. This is the case, for instance, for

$$x^2 + y^2 + 1 = 0$$

which clearly cannot hold for real values of x and y.

We have seen above in the study of the special conics that a great deal depends upon the adroit choice of the coordinate system and that much simplification can usually be gained by using the properties of symmetry of the curve considered. This brings up the question of how to pass from one coordinate system to another. We assume that both have the same unit of length. Geometrically we can think of this passage as affected by a *rigid motion* composed of a *translation* and/or a *rotation*.

Let us denote the two systems of coordinates by (x, y) and (s, t) respectively. Let the origins be O and O^* where the coordinates of O^* in the (x, y)-system are (h, k). This means that the coordinates of O in the (s, t)-system are $(-h, -k)$. If the axes of the two systems are parallel and similarly oriented, we have

$$x = s + h, \quad y = t + k. \tag{A.5.3}$$

This means that we pass from one system to the other by a pure translation.

Here it is important to consider the effect of a translation on the quadratic polynomial $Q(x, y)$. We have

$$\begin{aligned} Q(s+h, t+k) &= A(s+h)^2 + 2B(s+h)(t+k) + C(t+k)^2 \\ &\quad + 2D(s+h) + 2E(t+k) + F \\ &= As^2 + 2Bst + Ct^2 \\ &\quad + 2(Ah + Bk + D)s + 2(Bh + Ck + E)t \\ &\quad + Q(h, k). \end{aligned} \tag{A.5.4}$$

Hence a translation does not affect the second-order terms at all. By a suitable translation we can modify the first-order terms. We can always get rid of one of them and, usually, both. If both are to be eliminated, we must be able to find numbers h and k such that

$$\begin{aligned} Ah + Bk + D &= 0, \\ Bh + Ck + E &= 0. \end{aligned} \tag{A.5.5}$$

This is possible if and only if

$$\Delta \equiv B^2 - AC \neq 0. \tag{A.5.6}$$

If this condition holds, and if h and k satisfy (A.5.5), then equation (A.5.2) becomes

$$As^2 + 2Bst + Ct^2 = M, \quad M = -Q(h, k). \tag{A.5.7}$$

This equation is unchanged if we replace (s, t) by $(-s, -t)$, that is, the graph is symmetric with respect to O^*. The term "graph" may be a misnomer for the equation need not have a real representation. If it has, then the symmetry property holds. In any case we say that O^* is the *center* of the (real or imaginary) curve.

If $B = 0$, a translation will always suffice to transform the given equation (A.5.2) into one of the standard forms of Section A.4. If $\Delta \neq 0$, we choose h and k as the solution of the system (A.5.5) and obtain

$$As^2 + Ct^2 = M. \tag{A.5.8}$$

If A, C, and M all have the same sign, the equation represents an ellipse where the major axis is horizontal if $|A| < |C|$ and vertical if the inequality is reversed. It is a circle when $A = C$. If M is 0, the ellipse reduces to a single point, O^*, and if the sign of M is opposite to that of A and C, then there is no real graph.

If $AC < 0$ and $M \neq 0$, equation (A.5.8) represents a hyperbola and reduces to two intersecting lines if $M = 0$.

On the other hand, if $\Delta = 0$ and $B = 0$, then either A or C is 0. For the second case the equation is

$$Ax^2 + 2Dx + 2Ey + F = 0.$$

If $E \neq 0$, a translation can be found to take the equation into

$$As^2 + 2Et = 0.$$

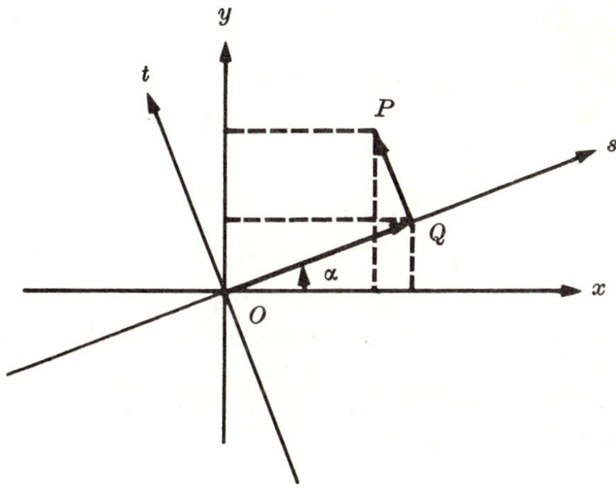

Figure A.32

This is a parabola with a vertical axis pointing down if $AE > 0$, up if $AE < 0$. If $E = 0$, we can reduce the equation to the form

$$As^2 + G = 0,$$

which is either a pair of parallel lines or the t-axis or no real graph. The case $A = 0$, $B = 0$ is handled in the same way.

Now suppose that $B \neq 0$. In this case it is necessary to use a rotation to get rid of the term in xy. Figure A.32 shows a rotation of axes where the two systems have the same origin. Suppose that a rotation through an angle α will take the positive x-axis into the positive s-axis. If we project the sum of the vectors $\overrightarrow{OQ}$ and $\overrightarrow{QP}$ on the positive x- and y-axes, we obtain

$$\begin{aligned} x &= s \cos \alpha - t \sin \alpha, \\ y &= s \sin \alpha + t \cos \alpha. \end{aligned} \tag{A.5.9}$$

Such a transformation takes a quadratic polynomial in x and y into a quadratic polynomial in s and t, but in this case the coefficients of the second degree terms are not invariant (= unchanged). There are some numerical *invariants*, however, to be noted below. It is enough to consider the second degree terms. Thus

$$\begin{aligned} Ax^2 + 2Bxy + Cy^2 = {} & A(s \cos \alpha - t \sin \alpha)^2 \\ & + 2B(s \cos \alpha - t \sin \alpha)(s \sin \alpha + t \cos \alpha) \\ & + C(s \sin \alpha + t \cos \alpha)^2. \end{aligned}$$

Expanding this and collecting like terms, we obtain

$$\begin{aligned} & (A \cos^2 \alpha + 2B \sin \alpha \cos \alpha + C \sin^2 \alpha)s^2 \\ & \quad + 2[(C - A) \sin \alpha \cos \alpha + B(\cos^2 \alpha - \sin^2 \alpha)]st \\ & \quad + (A \sin^2 \alpha - 2B \sin \alpha \cos \alpha + C \cos^2 \alpha)t^2 \\ & \quad \equiv A^*s^2 + 2B^*st + C^*t^2. \end{aligned} \tag{A.5.10}$$

Here B^* simplifies to

$$B^* = B \cos 2\alpha - \tfrac{1}{2}(A - C) \sin 2\alpha. \tag{A.5.11}$$

We note that

$$A + C = A^* + C^*, \tag{A.5.12}$$

so this sum is invariant under a rotation. Another invariant is furnished by Δ:

$$\Delta = B^2 - AC = (B^*)^2 - A^*C^*. \tag{A.5.13}$$

The proof is elementary but rather time-consuming and so it is left to the reader. Repeated use is made of the trigonometric function of twice an angle.

It is now clear that we can remove the term in st from the transformed equation by a suitable choice of the angle α. The condition $B^* = 0$ gives the trigonometric equation

$$B \cos 2\alpha - \tfrac{1}{2}(A - C) \sin 2\alpha = 0.$$

If $A \neq C$, we choose an angle 2α between $-90°$ and $+90°$ such that

$$\tan 2\alpha = \frac{2B}{A - C}. \tag{A.5.14}$$

If $A = C$, we take $\alpha = 45°$.

Once α has been chosen, we can express A^* and C^* algebraically in terms of A, B, and C. For if $\tan 2\alpha = a$, then

$$\sin 2\alpha = a(1 + a^2)^{-\frac{1}{2}}, \quad \cos 2\alpha = (1 + a^2)^{-\frac{1}{2}},$$

where the positive square roots are taken. Then

$$\begin{aligned} A^* &= A \cos^2 \alpha + B \sin 2\alpha + C \sin^2 \alpha \\ &= \tfrac{1}{2}(A + C) + \tfrac{1}{2}(A - C) \cos 2\alpha + B \sin 2\alpha, \end{aligned}$$

where we substitute the values of $\sin 2\alpha$ and $\cos 2\alpha$ in terms of a and express a in terms of A, B, and C. The result is an expression for A^* in A, B, C which involves nothing more complicated than square roots. The upshot of this is that the trigonometric functions play an intermediary role in the reduction problem which is purely algebraic in character. Unless the problem calls explicitly for the angle α, there is no need for trigonometric tables; in fact, they should be avoided.

Let us summarize the results of the discussion so far.

Case 1. $\Delta \neq 0$. We first use a translation to transform the equation to the form (A.5.7). Then we remove the st term by a rotation to obtain an equation of the form

$$A^*u^2 + C^*v^2 = M, \tag{A.5.15}$$

where M is the same numerical constant as in (A.5.7), that is, $M = -Q(h, k)$ where (h, k) is the center. Further

$$A^* + C^* = A + C, \quad \Delta = B^2 - AC = -A^*C^*. \tag{A.5.16}$$

Now equation (A.5.15) has the same form as (A.5.8) and the various cases are discussed in the same way. We are led to the following subcases:

1.1. $\Delta < 0$. *Elliptic curves.*

(a) $A + C > 0$, $M = -Q(h, k) > 0$. Ellipse.

(b) $A + C > 0$, $M = 0$. Point (h, k).

(c) $A + C > 0$. $M < 0$. No real graph.

If $A + C < 0$, change the sign.

1.2. $\Delta > 0$. *Hyperbolic curves.*

(a) $M \neq 0$. Hyperbola.

(b) $M = 0$. Intersecting pair of lines.

If $\Delta = 0$, we start with a rotation unless B is already 0. Suppose that the original equation is

$$(ax + by)^2 + 2Dx + 2Ey + F = 0, \tag{A.5.17}$$

where

$$a^2 = A, \quad ab = B, \quad b^2 = C.$$

By assumption $ab \neq 0$ and we can take $b > 0$. Here we can read off a suitable transformation by inspection. We take

$$s = \frac{bx - ay}{(a^2 + b^2)^{\frac{1}{2}}}, \quad t = \frac{ax + by}{(a^2 + b^2)^{\frac{1}{2}}}. \tag{A.5.18}$$

It is clear that $s = 0$ and $t = 0$ are orthogonal lines in the (x, y)-plane. This gives a transformed equation

$$t^2 + 2Ks + 2Lt + F = 0. \tag{A.5.19}$$

There are two subcases according as $K \neq 0$ or $K = 0$. We can read off the results from our preceding discussion and can sum up this case as follows:

Case 2. $\Delta = 0$. *Parabolic curves.*

(a) $K \neq 0$. Parabola.
(b) $K = 0$, $F \leqq L^2$. Two parallel lines.
(c) $K = 0$, $F > L^2$. No real graph.

EXERCISE A.5

Classify the following conics as to type and reality of graph:

1. $x^2 - 4xy + 2y^2 + 1 = 0$.
2. $x^2 - 2xy + y^2 = 4$.
3. $2x^2 + 2xy + y^2 = 0$.
4. $x^2 - 4xy + 4y^2 + x + y = 0$.
5. $xy + y^2 - x - y = 0$.
6. $x^2 - xy + y^2 = 1$.
7. $x^2 - 2xy + y^2 - 2x + 2y + 3 = 0$.
8. $x^2 - 2xy + 2y^2 + 2x + 2y - 6 = 0$.

9.–12. Transform to their own axes the equations of the conics in Problems 1, 4, 6, and 8.

Transform to their own axes the equations of the following (real) conics:

13. $x^2 + xy + y^2 = 3.$

14. $xy = 4.$

15. $2x^2 + 2xy + y^2 = 5.$

16. $xy + x + y = 0.$

17. $x^2 - 2xy + y^2 = 2x.$

18. Without transforming the axes, find the asymptotes of the hyperbola in Problem 16.

Without transforming the axes, find the asymptotes of the hyperbolas:

19. $x^2 - 4xy - 5y^2 + 9 = 0.$

20. $xy + y^2 - x - y = 1.$

[*Hint:* Try to factor and argue as in Section A.4.III.]

21. Prove formula (A.5.13).

22. Let $C = 0$, $|B| + |E| \neq 0$ in equation (A.5.2). Then, for all x with at most one exception, we can solve the equation for y in terms of x. Show that the conic cannot be of elliptic type and determine when it is of hyperbolic type and when of parabolic. Illustrate by numerical examples.

A.6. Polar coordinates. So far we have only considered Cartesian coordinates. Now the purpose of coordinates is to fix position with respect to some frame of reference. Latitude and longitude are used on the globe. Cartesian coordinates serve in the plane, but this choice is obviously not the only way of achieving the desired result. One important alternative is that of *polar coordinates.*

Here we choose an origin O in the plane, usually referred to as the *pole* of the system, and a ray from the origin called the *polar axis.* This is the frame of reference. We give the position of a point P with respect to the pole and the axis by measuring the length of the radius vector OP and the angle which the latter makes with the polar axis. It is preferable that coordinates are pure numbers or, at least, are measured in the same units. One number and so and so many degrees is not a good combination. This and other considerations make it desirable to introduce another method of measuring angles, namely in terms of *radian measure.*

Here we measure an angle by the length of the arc which the angle subtends on the unit circle (circle of radius one, center at the origin) when the angle is placed at the center of the circle. See Figure A.33. We postpone the discussion of how to define and measure lengths of circular arcs until Chapter 7. In the meantime we shall use the naive, intuitive notion of length.

The unit of angle subtends an arc of length one on the unit circle. It is called *one radian* and is approximately $57°17'$. We have the correspondence

$$2\pi \text{ radians} = 360°, \tag{A.6.1}$$

so that

$$30° = \frac{\pi}{6}, \quad 45° = \frac{\pi}{4}, \quad 60° = \frac{\pi}{3}, \quad 90° = \frac{\pi}{2} \text{ etc.}$$

Hereinafter all angles are measured in radians unless otherwise stated.

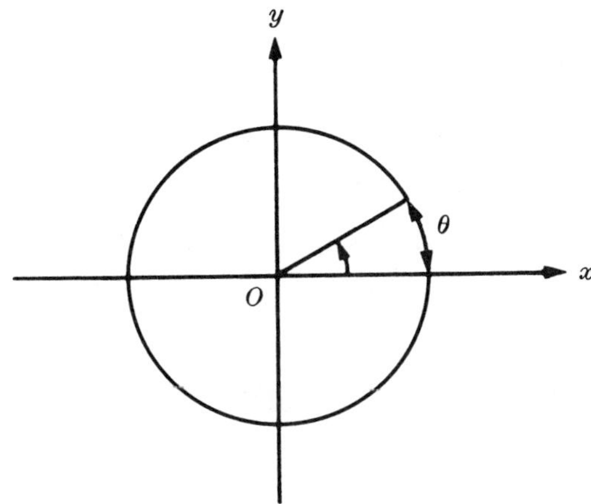

Figure A.33

The polar coordinates of a point P are $[r, \theta]$, where r is the *length of the radius vector* OP and θ is the *vectorial angle*, measured in radians, which the radius vector makes with the polar axis. See Figure A.34. Here r is by definition nonnegative (see below, however); r is uniquely determined while the vectorial angle is determined only up to additive multiples of 2π. We often have to use simultaneously Cartesian and polar coordinates. We assume that the origin = the pole is the same in both systems and that the positive x-axis is the polar axis. The relations between the two systems are then (see Section A.1)

$$x = r \cos \theta, \quad y = r \sin \theta,$$
$$r = (x^2 + y^2)^{\frac{1}{2}}, \quad \tan \theta = \frac{y}{x}. \tag{A.6.2}$$

Later we shall study some curves whose equations are given in polar coordinates. At this juncture we restrict ourselves to straight lines and conics. Here are some simple examples.

(1) *Vertical line*: $x = a$ becomes

$$r \cos \theta = a \quad \text{or} \quad r = a \sec \theta$$

in polar coordinates.

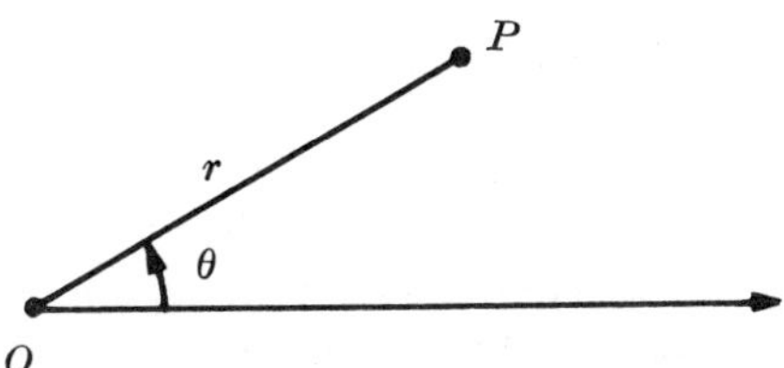

Figure A.34

(2) *Horizontal line*: $y = b$ becomes

$$r \sin \theta = b \quad \text{or} \quad r = b \operatorname{cosec} \theta.$$

(3) *Straight line*: $Ax + By + C = 0$ becomes

$$r(A \cos \theta + B \sin \theta) + C = 0.$$

This also holds if $C = 0$ so that the line goes through the origin.

(4) *Circle with center at the origin*: $x^2 + y^2 = a^2$ reduces to

$$r = a.$$

(5) *General circle*: $(x - x_0)^2 + (y - y_0)^2 = a^2$ becomes

$$r^2 - 2r(x_0 \cos \theta + y_0 \sin \theta) = a^2 - x_0{}^2 - y_0{}^2.$$

If the circle goes through the origin, this reduces to

$$r = 2(x_0 \cos \theta + y_0 \sin \theta). \tag{A.6.3}$$

(6) *Equilateral hyperbola*: $x^2 - y^2 = a^2$ becomes

$$r^2 \cos 2\theta = a^2. \tag{A.6.4}$$

There are some interesting equations in polar coordinates which represent general conics. Thus the equation

$$r = m(b - a \cos \theta)^{-1}, \tag{A.6.5}$$

where all the constants are positive, represents

an *ellipse* if $a < b$,
a *parabola* if $a = b$,
a *hyperbola* if $a > b$.

To prove this we rewrite the equation in the form

$$br - ar \cos \theta = m,$$

or

$$br = m + ax.$$

Squaring gives

$$b^2(x^2 + y^2) = m^2 + 2max + a^2x^2,$$

or

$$(b^2 - a^2)x^2 + b^2y^2 - 2max = m^2.$$

This equation differs from (A.4.3) by a translation and by a different notation. It represents a *conic having a focus at the origin*. The nature of the conic is obviously as stated above.

If $b > a$, every value of θ is admissible and gives a unique finite value of r which lies between $m(b + a)^{-1}$ and $m(b - a)^{-1}$. For $b = a$ the values $\theta = 0$ or multiples of 2π are exceptional and give an infinite value for r. All other values are admissible and give finite values. We conclude that, for $b > a$, equation (A.6.5) represents the ellipse completely, that is, there is a one-to-one correspondence between the points of the ellipse and the pairs $[r, \theta]$ which satisfy (A.6.5). The same holds for the parabola when $b = a$.

We have a different situation when $b < a$. Here the values of θ for which

$$\frac{b}{a} \leq \cos\theta \leq 1$$

are inadmissible, the left bound gives $r = \infty$ and all other values make r negative. As a consequence equation (A.6.5) represents only the right branch of the hyperbola and the left is omitted. Now in using polar coordinates the rule that r be nonnegative is often disregarded, and if $r < 0$ for some value of θ, a point with coordinates $[|r|, \theta + \pi]$ is added to the graph, that is, negative values of r are laid off on the continuation of the radius vector beyond the origin. If this is done in the present case, we obtain the missing left branch of the hyperbola.

EXERCISE A.6

Plot the curves:

1. $r = \sin\theta$.
2. $r(\sin\theta + \cos\theta) = 1$.
3. $r = \cos^2\theta$.
4. $r = (2 - \cos\theta)^{-1}$.
5. $r^2 \sin 2\theta = 1$.
6. $r^2 = \sin\theta$.

7.–9. Transform the equations in Problems 3, 5, and 6 into Cartesian coordinates.

10. Transform the equation $(x^2 + y^2)^2 = x^2 - y^2$ into polar coordinates and use the result to plot the curve. [This is a *lemniscate*.]

11. Show that the following is a central conic and determine its type:

$$r^2[\cos^2\theta - 2\cos\theta\sin\theta + 4\sin^2\theta] = 1.$$

MISCELLANEOUS PROBLEMS FOR APPENDIX A

1. Find the intersections of the conics:

(a) $$x^2 + y^2 = 4, \quad (x-1)^2 + (y+1)^2 = 4;$$

(b) $$x^2 + 2y^2 = 12, \quad x^2 - 2y^2 = 10.$$

2. Discuss the number of intersections of two general conics. If two equations $Q_1(x, y) = 0$ and $Q_2(x, y) = 0$ are given, show that they admit a linear combination which involves y either to the first power or not at all. In the first case, solve for y in terms of x and substitute in one of the original equations. The result is an equation in x of degree four which has four or two or no real roots. Interpret geometrically! Carry through the details and discuss also the more elementary second case where the linear combination does not involve y.

3. The circles $x^2 + y^2 = 4$, $(x - 4)^2 + y^2 = 1$ have four common tangents. Find their equations. [*Hint:* Use formula (A.3.12) and choose m so that the line also touches the second circle.]

4. Find the common tangents to the circle $x^2 + y^2 = 1$ and the parabola $y^2 = 4x$.

5. Find the equation of a conic with center at the origin passing through the points $(-1, -1)$, $(-3, 0)$, and $(0, 1)$.

6. Find the equation of a parabola with vertex at $(1, 1)$ that is symmetric with respect to the line $y = x$ and passes through the point $(0, -2)$.

7. Find the equation of the conic that passes through the five points $(1, 1)$, $(1, -1)$, $(-1, 1)$, $(-1, -1)$, and $(2, 0)$.

8. Let four points P_1, P_2, P_3, P_4 be given in the plane no three of which are collinear. Let $L_{jk}(x, y) = 0$ be the equation of the straight line through P_j and P_k where $j, k = 1, 2, 3, 4$, and $j \neq k$. Let α be any real number. Show that the family of curves

$$Q_\alpha: \alpha L_{12}(x, y)L_{34}(x, y) + (1 - \alpha)L_{23}(x, y)L_{41}(x, y) = 0$$

is a pencil of conics passing through the four given points.

9. If a fifth point P_5 does not lie on any of the lines $L_{jk} = 0$, show that there is a unique conic in the pencil that passes through P_5. Hence: *There is a conic that goes through five given points no three of which are collinear.*

10. Show that the conic is uniquely determined by the five points. [*Hint*: Two distinct conics passing through five distinct points would have five or more intersections. Is this possible? Show that it is not possible for the two conics to degenerate and have a line in common. What condition would this violate?]

11. Use this result to prove that the pencil of conics in Problem 8 contains all conics through the points P_1, P_2, P_3, P_4.

12. What is the locus of a point which moves in such a way that the sum of its distances from two given lines is a constant?

13. What is the locus if the product of the distances is a constant?

14. What is the area of a triangle in terms of the coordinates of the vertices?

15. Given two pencils of straight lines

$$y = m(x + a) \quad \text{and} \quad y = k(x - a),\ a > 0,$$

find the locus of the point of intersection of a line of the first pencil with a line of the second pencil, if the slopes of the two lines satisfy a fixed relation

$$Am + Bk + C = 0,$$

where A, B, C are given constants. Note in particular the case $B = -A$.

16. Solve the same question if $km = A$ instead.

17. Discuss the curve $2r \sin^2 \dfrac{\theta}{2} = m$.

18. Given a straight line l and a point O at the distance a from l. Through O draw a line l_1 intersecting l at the point Q. Mark on l_1 the points P_1 and P_2 such that $P_1Q = QP_2 = b$ where $b < a$. Find the loci of the points P_1 and P_2 when l_1 varies. Use O as the pole and the perpendicular to l from O as the polar axis and find the equations in polar coordinates. Transform to Cartesian coordinates.

19. Find the equations of the common tangents of the curves $y = x^2$ and $2y = 4x - x^2 - 2$.

20. Find the equation of a straight line such that the point $(3, \frac{5}{2})$ divides the line segment between the axes in the ratio 2:1.

21. A parabola has its focus at the point (0, 1) and passes through (7, 0) and (—1, 0). Find the equations of the parabola and of its directrix.

22. Through the point (5, 7) on the hyperbola

$$xy - 4x - 3y + 6 = 0$$

an arbitrary line is drawn which intersects the asymptotes in two points. Through these points draw straight lines parallel to the asymptotes. These lines intersect in a point P. Find the locus of P when the line first mentioned turns about the point (5, 7).

23. Show that the equation in the preceding problem actually represents a hyperbola by solving the equation for y and discussing the result.

24. Show that the equation

$$(2x - y + 3)^2 = x + 2y - 1$$

represents a parabola which lies above the line $x + 2y - 1 = 0$ except for the point (—1, 3) where the line is tangent to the parabola.

25. Show that $x^2 + xy + y^2 = 3$ represents an ellipse by solving for y in terms of x. For what values of x is the result real?

2

Functions

2.1. Sets and limits. As a preparation for the discussion of functions in this chapter, we must develop some concepts relating to sets and limits not mentioned in Chapter 1.

DEFINITION 2.1.1. *A point set is a set of points in a Euclidean space. The nonempty point set S is called bounded if the set of distances $d(O, P)$, $P \in S$, is bounded, where O is a fixed origin.*

If the set is on the line, we can quite as well speak of a set of real numbers. In the plane we have sets of ordered number pairs, in three space ordered number triples, and so on.

In a bounded set of points the coordinates form bounded sets of real numbers. For these coordinate sets least upper bounds and greatest lower bounds exist.

On the line there are already great variations in structure between sets, and this becomes still more pronounced in higher dimensions. To illustrate we start with some simple examples. We take

1. S_1 is the set of reciprocals of the natural numbers

$$S_1 = \left\{1, \tfrac{1}{2}, \tfrac{1}{3}, \tfrac{1}{4}, \cdots, \frac{1}{n}, \cdots\right\}.$$

2. S_2 is the set of proper fractions

$$S_2 = \left\{\frac{p}{q} \,\middle|\, \begin{array}{l} p \text{ and } q \text{ relatively prime} \\ \text{positive integers, } p < q \end{array}\right\}.$$

The first set is made up of *isolated points*; to every point $\dfrac{1}{n}$ there is an ε-neighborhood of the point which contains no other point of the set. This holds as soon as

$$\varepsilon < \frac{1}{n} - \frac{1}{n+1} = \frac{1}{n(n+1)}.$$

The larger the n, the smaller we have to take the neighborhood. The reader may ask about 0. Every ε-neighborhood of 0 contains infinitely many points of the set. This is certainly true, but 0 is not the reciprocal of an integer and is not a member of the set. We shall call 0 a *cluster point* of the set.

The set S_2 is totally different. There are no isolated points, and every point of the set has the property that all its ε-neighborhoods contain infinitely many points of the set. Thus all points of the set are cluster points. Moreover, every real number in $[0, 1]$, rational or irrational, is a cluster point.

It is time to make a formal definition.

DEFINITION 2.1.2. *Let S be an infinite point set. A point P_0, not necessarily in S, is called a cluster point of S if every ε-neighborhood of P_0 contains infinitely many points of S.*

The examples considered may suggest that every infinite point set has at least one cluster point. This is not true, however, as is shown by the set of natural numbers. No interval of length <1 can contain more than one natural number, and so there can be no cluster points. On the other hand, the proposition becomes true if we add the condition of boundedness. This is the sense of the *Bolzano-Weierstrass theorem.* Weierstrass has already been mentioned. Bernard Bolzano [1781–1848] was professor of the philosophy of religion at Prague; this particular result of his comes from a tract on the paradoxes of infinity, published posthumously in 1850 and unnoticed for about twenty years.

THEOREM 2.1.1. *An infinite bounded point set has at least one cluster point.*

Proof. We shall base the argument on the mathematical version of the classical maxim: "Divide et impera!" We suppose that the set S is linear; the device works equally well in higher dimensions. Since the set is bounded, there exists a closed interval I_1 which contains S. What is important here is that I_1 contains infinitely many points which belong to S. If we bisect the interval, at least one of the parts will have the same property. If both have it, we select the left subinterval. In any case we can find an interval I_2, $I_2 \subset I_1$, whose length is $\frac{1}{2}l$ if l is the length of I_1, and I_2 contains infinitely many points of S. See Figure 2.1. We take I_2 to be closed. We can now subdivide I_2 in the same manner and obtain an interval I_3 such that $I_3 \subset I_2$, the length of I_3 is $\frac{1}{4}l$, and I_3 contains infinitely many points of S. This procedure can be continued indefinitely and leads to a sequence of nested closed intervals $\{I_n\}$ each containing infinitely many points of S. The length of I_n is $2^{1-n}l$. By Theorem 1.5.1 there is one and only one point P_0 common to all the intervals I_n.

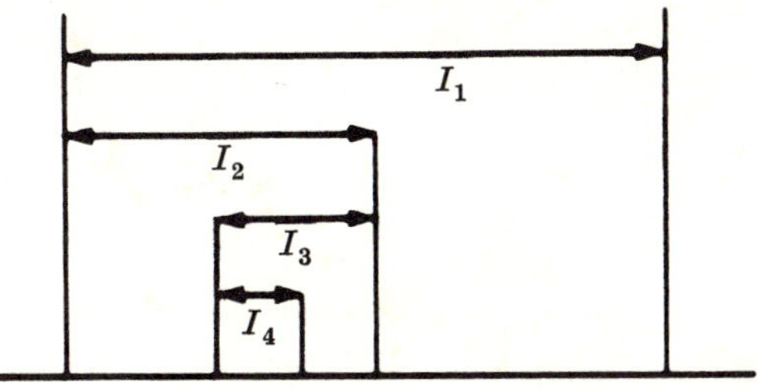

Figure 2.1

This point P_0 is a cluster point of S. For if $\varepsilon > 0$ is given, we can choose n so large that $2^{1-n}l < \varepsilon$. This says that I_n is contained in an ε-neighborhood of P_0, that is, there are infinitely many points of S in the ε-neighborhood. Since this holds, no matter how small the ε, we conclude that P_0 is a cluster point of S and the theorem is proved.

We have seen that a set may, but need not, contain any of its cluster points.

Definition 2.1.3. *The set of cluster points of the set S is called the derived set of S and is denoted by S'* (read "S prime"). *S is called closed if*

$$S' \subset S. \tag{2.1.1}$$

The union of S and S' is called the closure of S and is denoted by $\bar{S}$

$$\bar{S} = S \cup S'. \tag{2.1.2}$$

A finite set is said to be closed since $S' = \varnothing$ and (2.1.1) *holds.*

Thus for a closed set $\bar{S} = S$. Neither of the special sets considered is closed. For the set S_1 we have $S_1' = \{0\}$; for S_2 we have $S_2' = [0, 1]$. Incidentally, a closed interval is a closed point set in the sense here defined. Note that $\bar{S}$ is always closed regardless of the nature of S.

If S' is also an infinite point set, it has a derived set S'', called the *second derived set* of S, and so on. It may be shown that S' is closed, that is, $S'' \subset S'$. For a linear set the extreme cluster points are particularly important.

Definition 2.1.4. *Let S be a linear bounded infinite set. Let α be the greatest lower bound of S' and β the least upper bound. Write*

$$\alpha = \liminf S, \quad \beta = \limsup S. \tag{2.1.3}$$

The abbreviations stand for "limes inferior" and "limes superior" respectively, but it is common to say *inferior limit* and *superior limit.*

It is clear that S has a single cluster point if and only if $\alpha = \beta$. This is the case for the set S_1. For S_2 we have $\alpha = 0$, $\beta = 1$.

In the applications of point set theory to real analysis one must remember that a sequence is not a point set so that the preceding definitions do not apply as they stand. On the other hand, the range of the sequence, introduced in Definition 1.7.5, is a point set to which the definitions do apply. Here we run into a difficulty, since infinitely many elements of the sequence may correspond to the same element of the range. Hence we need other conventions.

Definition 2.1.5. *If $\{a_n\}$ is a bounded sequence and if there exists a number a such that for every $\varepsilon > 0$ the inequality*

$$|a_n - a| < \varepsilon \tag{2.1.4}$$

holds for infinitely many values of n, then a is called a cluster point of the sequence or a limiting value of a_n.

The reader should note the great difference between this definition and Definition 1.7.4 where the limit of a_n is defined. It is clear that $\lim a_n$ is a limiting value of a_n, in fact the only one, when it exists. The converse is not true, for in Definition 2.1.5 we require only that (2.1.4) shall hold infinitely often. It is not required that it hold for all large values of n. Thus the sequence $\{(-1)^n\}$ has the limiting values -1 and $+1$ but not a unique limit.

As in the point set case, the least and the largest limiting values are important.

DEFINITION 2.1.6. *We set*

$$\alpha = \liminf a_n = \inf \{\lambda \mid \lambda \textit{ a limiting value of } a_n\}, \tag{2.1.5}$$

$$\beta = \limsup a_n = \sup \{\lambda \mid \lambda \textit{ a limiting value of } a_n\}. \tag{2.1.6}$$

From this definition it follows that the inequality

$$\alpha - \varepsilon < a_n < \beta + \varepsilon \tag{2.1.7}$$

holds for all large values of n, and each of the inequalities

$$a_n < \alpha + \varepsilon, \quad \beta - \varepsilon < a_n \tag{2.1.8}$$

holds for infinitely many values of n.

In the case of the sequence $\{(-1)^n\}$ we see that -1 is the inferior limit and $+1$ is the superior one.

We have the following analogue of Theorem 2.1.1.

THEOREM 2.1.2 *Every bounded sequence has at least one limiting value.*

Proof. If the range of $\{a_n\}$ is infinite, then Theorem 2.1.1 applies. If a is a cluster point of the range of $\{a_n\}$, then a is also a limiting value of a_n in the sense of Definition 2.1.5. On the other hand, if the range is finite, it contains at least one point a which corresponds to infinitely many values of n. But if $a_n = a$ infinitely often, then a satisfies (2.1.4) and is a limiting value of a_n. This completes the proof.

THEOREM 2.1.3. *We have*

$$\lim_{n\to\infty} a_n = a$$

if and only if

$$\liminf a_n = \limsup a_n = a.$$

Thus there is only one limiting value which is then the limit of a_n.

We can give a different criterion for the existence of a limit. This criterion is known as *Cauchy's Convergence Principle* named after Augustin-Louis (baron de) Cauchy [1789–1857], the father of modern analysis.

THEOREM 2.1.4. *A necessary and sufficient condition that the sequence* $\{a_n\}$ *have a limit is that, given any* $\varepsilon > 0$, *we can find an* $N(\varepsilon)$ *such that*

$$|a_m - a_n| \leqq \varepsilon \quad \textit{for all} \quad m \textit{ and } n \geqq N(\varepsilon). \tag{2.1.9}$$

Proof. (1) *The condition is necessary.* For if $\lim a_n = a$, then by Definition 1.7.4 there exists an $N(\varepsilon)$ such that for m, $n \geqq N(\varepsilon)$ we have

$$|a - a_m| \leqq \tfrac{1}{2}\varepsilon, \quad |a - a_n| \leqq \tfrac{1}{2}\varepsilon,$$

so that

$$|a_m - a_n| = |(a - a_m) - (a - a_n)| \leqq |a - a_m| + |a - a_n| \leqq \tfrac{1}{2}\varepsilon + \tfrac{1}{2}\varepsilon = \varepsilon.$$

Hence, if $\lim a_n$ exists, then (2.1.9) holds.

(2) *The condition is sufficient.* For suppose that it holds. Then, in particular, for $m = N$ we have $|a_N - a_n| \leqq \varepsilon$ when $n > N$. Thus there is only a finite number of elements of the sequence outside of $[a_N - \varepsilon, a_N + \varepsilon]$, an interval of length 2ε. From this we conclude that any and all limiting values of the sequence must lie in this interval. This implies that

$$\limsup a_n - \liminf a_n \leqq 2\varepsilon.$$

Here ε is at our disposal. Since the inequality holds for every $\varepsilon > 0$, we must have

$$\limsup a_m = \liminf a_n,$$

and by Theorem 2.1.3 their common value is the limit of a_n. This completes the proof.

This theorem is basic for the theory of infinite series to which Chapter 5 is devoted. The student's attention is called to the phrase "a necessary and sufficient condition." He should carefully distinguish between a "necessary condition," a "sufficient condition," and a "necessary and sufficient condition" for the validity of some assertion of the form: "A has the property P." A condition C is necessary for A to have the property P, if all objects A which have P also satisfy C, that is, P implies C; it is still possible that some objects which satisfy C do not have the property P. On the other hand, C is sufficient if C implies P; in this case it is possible that P does not imply C. Finally, C is necessary and sufficient if C implies P and C is implied by P. If P holds "only if" C holds, then C is a necessary condition. If P holds "if" C holds, then C is sufficient, and, finally, if P holds "if and only if" C holds, then C is necessary and sufficient.

EXERCISE 2.1

1. A set S_1 is made up of all (distinct) numbers of the form

$$\frac{1}{m} + \frac{1}{n} \quad \text{where} \quad m, n = 1, 2, 3, \cdots.$$

Find its supremum, lim sup, lim inf, and infimum.

2. Find the first and second derived sets of S_1. Is S_1 closed?

3. Show that the sequence $\{a_n\}$ with

$$a_n = (n+1)^{\frac{1}{2}} - n^{\frac{1}{2}}, \quad n = 1, 2, 3, \cdots$$

is strictly decreasing. What is its limit?

4. With a_n defined as in the preceding question, consider the set $S_3 = \{a_m + a_n \mid m, n = 1, 2, 3, \cdots, m \leqq n.\}$. Find its supremum, lim sup, lim inf, and infimum. Find also its first and second derived sets.

5. If a sequence $\{a_n\}$ has several limiting values, one of which is a, then there is a subsequence which converges to a. True or false?

6. Give an example of an unbounded sequence with $\lim (a_{n+1} - a_n) = 0$.

7. Why is a convergent sequence bounded?

8. Prove that S' is closed if it is an infinite set.

9. Prove that $\bar{S}$ is closed.

10. Given

$$U_n = 1 - \tfrac{1}{2} + \tfrac{1}{3} - \cdots + (-1)^{n-1}\frac{1}{n},$$

use Theorem 2.1.4 to prove that the sequence $\{U_n\}$ converges.

11. A set S consists of all the points

$$a_{m,n} = (n^2 + m)^{\frac{1}{2}} - n, \qquad n = 1, 2, 3, \cdots, \quad m = 1, 2, 3, \cdots, 2n.$$

Prove that (i) $S \subset [0, 1]$; (ii) $S' = [0, 1]$. If α is a fixed number, $0 < \alpha < 1$, how should $m(n)$ be chosen in terms of n in order that

$$\lim_{n\to\infty} a_{m(n),n} = \alpha?$$

12. Let S be the set of all decimal fractions involving only zeros and ones. Verify that no point of S lies in any of the intervals

$$(\tfrac{1}{9}10^{-k}, 10^{-k}), \quad k = 0, 1, 2, 3, \cdots$$

Are there other gaps? Prove that $S' = S$. Verify also that every point of S is a cluster point of points which are not in S. To prove the last proposition it suffices to show that every interval in $(0, 1)$ contains a subinterval which contains no point of S.

13. Find $\lim a_{n+1}/a_n$ if $a_n = n!\, n^{-n}$.

2.2. The function concept. This central and basic notion in analysis can be traced back to antiquity. The sine is a function and the first tables of sines were constructed by Hipparch from Nicaea around 150 B.C. Logarithms are also functions. We have learned that the first table of logarithms, constructed by Napier, appeared in 1614. Various things happened in the intervening millennium and a half, but very little that has any bearing on our present quest.

It is in the middle of the seventeenth century that we first find anything that could be called a general function concept. Actually we would then find two such concepts, one geometrical, the other analytical. These developed very

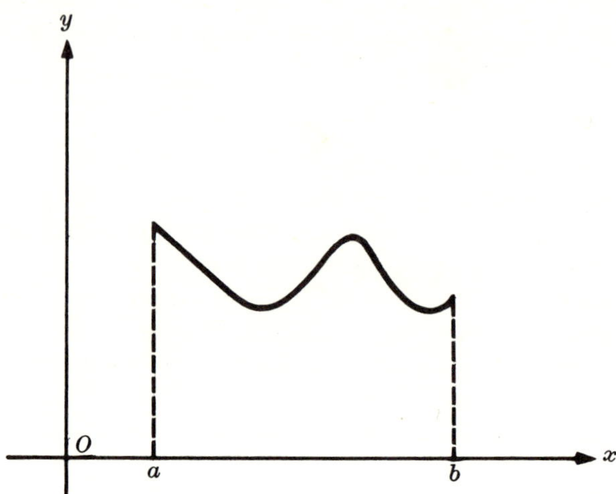

Figure 2.2

gradually. Sometimes one was in vogue, sometimes the other, and their unification took place after 1900.

When we draw a graph in the (x, y)-plane such that each vertical line cuts it in at most one point, then the graph defines a function. For each abscissa in some interval there is a unique corresponding ordinate. See Figure 2.2. This is a simple example of the geometrical function concept: the ordinate is a graphically given function of the abscissa. The reader is familiar with this type of graphical representation. It may represent barometric pressure, temperature readings, inventories, stock prices, vital statistics, or anything with variations considered worth recording.

This concept is familiar, but it may not be associated with the word "function" in the mind of the student. Graphic representation of scientific observations, demographic or economic data, and so on, is essentially a creation of the

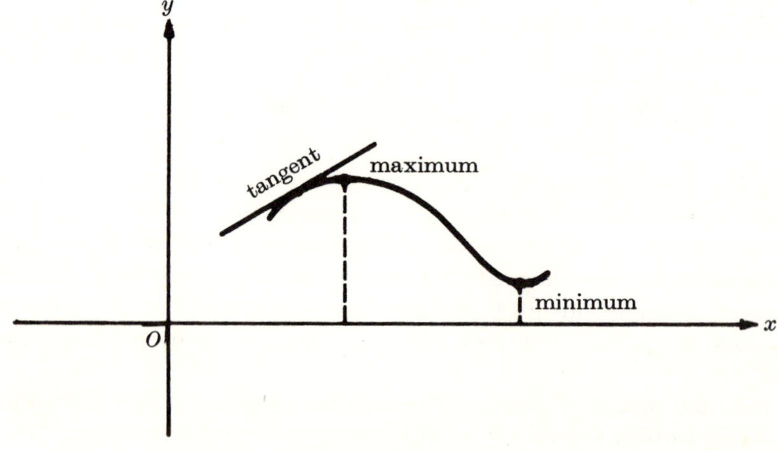

Figure 2.3

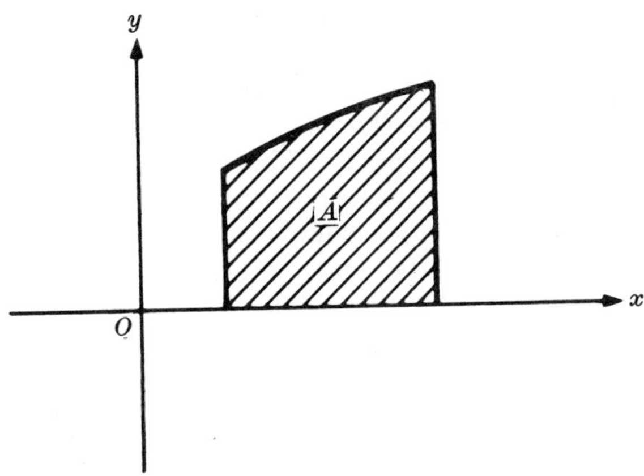

Figure 2.4

nineteenth century though there are a few instances of a much earlier date. The geometric notion of a function is of a venerable age. In the main, it presupposes knowledge of analytic geometry which started with Descartes in 1637. Once this tool was mastered, which took a fairly short time, mathematicians found themselves studying, not without success, a wide class of geometrical problems involving maxima and minima, tangents, and areas. See Figures 2.3 and 2.4 for orientation.

In our terminology this means that they could discuss, and in certain cases also evaluate, the derivative or the integral of an ordinate function with respect to the abscissa. Here should be mentioned in particular the Italian mathematician Bonaventura Cavalieri (1591–1647), who as early as 1630 could integrate powers of x ultimately up to and including x^9. He realized that the process was a linear one, so that if the result is known for two functions it is also known for all their linear combinations. Other great names from this precalculus period are Pierre de Fermat (1601–65), Blaise Pascal (1623–62), Christiaan Huygens (1629–95), James Gregory (1638–75), John Wallis (1616–1703), and Isaac Barrow (1630–77). Barrow had a clear idea of what later became known as the fundamental theorem of the integral calculus. He was Newton's teacher at Cambridge; in 1669 he resigned his chair in favor of his illustrious pupil and henceforth devoted himself to theology. It has been remarked that the unseemly fight between the founders of the Calculus ought not to have been between Newton and Leibniz, who in the main thought alike on basic matters, but between Barrow and Newton who did not.

The analytical function concept swept in with Newton and Leibniz and reigned supreme for almost a hundred years. The masters of the Calculus studied elementary functions such as powers, polynomials, rational functions, roots of rational functions, power series, exponential functions, logarithms, trigonometric functions and their inverses, in brief, functions defined by explicit analytic expressions. In the long run, the restriction to such functions is not adequate, not even for the needs of the Calculus. While it is true that the derivative of an elementary function is also elementary, this need not be the case for the integral. Moreover, these functions give a distorted view of the domain of the operations of the Calculus. Every continuous function has an integral, but need not have a derivative, whereas the elementary functions suggest that the opposite is true. This was of course realized only much later.

Actually it was the demands of the applications which created a "comeback" for the geometric function concept. The mathematical treatment of the motion of a vibrating string set off an animated debate between leading mathematicians two hundred years ago as to what could be accepted as a function. In part the debate was over the question: can analytic expressions be found for functions defined geometrically?

The student who objects to functions defined geometrically by broken lines or verbally by a sentence (examples will be found in the next section) would have found sympathy and understanding from eminent mathematicians two hundred years ago, when such newfangled and scatterbrained ideas were first aired by men who ought to have known better. As usual the opposition ultimately died a natural death and the new ideas won the day. Gradually the two concepts moved closer to each other and mathematicians became more skilled in devising explicit formulas to describe empirical functions.

The term "function" is found in the writings of Leibniz as early as 1692, but was then used for various geometric quantities associated with a curve such as abscissa, ordinate, normal, tangent, radius of curvature, and so on. The term was later given a more analytical meaning. Thus, in his "*Introductio in Analysin Infinitorum*" of 1748 Leonard Euler (1707–83), the greatest mathematician of that century, gave a definition of "function" which is essentially that of an elementary function as described above. But the very same year Euler admitted "arbitrary" geometrically defined functions as initial values for position and velocity of a vibrating string. When the smoke of the ensuing battle finally cleared, another great mathematician, Peter Gustav Lejeune Dirichlet (1805–59), in 1837 formulated the following

DEFINITION 2.2.1. *y is a single-valued function of the variable x, in the interval $[a, b]$, when a definite value of y corresponds to each value of x such that $a \leqq x \leqq b$, no matter in what form this correspondence is specified.*

Dirichlet's definition stood unchallenged until the beginning of this century; in fact it is still frequently used and serves a useful purpose. Mathematicians do not change definitions at frequent intervals, but a more modern formulation seems desirable. We shall define and use concepts which are closer to the twentieth century models.

DEFINITION 2.2.2. *If X and Y are two sets, then the set of all ordered pairs (x, y) where $x \in X$, $y \in Y$, is called the Cartesian product of X and Y and is denoted by $X \times Y$. Here $(x_1, y_1) = (x_2, y_2)$ if and only if $x_1 = x_2$, $y_1 = y_2$. We refer to x as the first coordinate of the pair (x, y) and y as the second coordinate.*

Note that here X and Y are arbitrary sets. In Figure 2.5 we have depicted the case in which x and y are real numbers. X is a set of three closed intervals

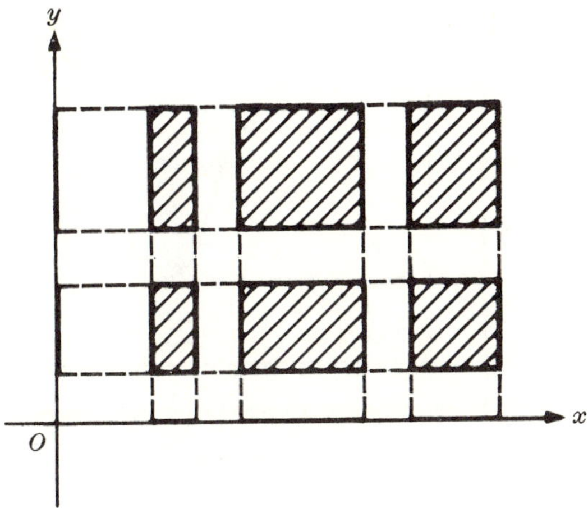

Figure 2.5

on the x-axis. Y a set of two closed intervals on the y-axis, and $X \times Y$ a set of six closed rectangles.

DEFINITION 2.2.3. *A function f on X to Y is a nonvoid subset of pairs (x, y) in $X \times Y$, such that no two distinct pairs have the same first coordinate.*

It is clear that the product set in Figure 2.5 does not define a function on reals to reals. On the other hand, Figure 2.2 defines such a function.

DEFINITION 2.2.4. *The sets*

$$D[f] = \{x \mid x \in X, (x, y) \in f\}, \tag{2.2.1}$$

$$R[f] = \{y \mid y \in Y, (x, y) \in f\}, \tag{2.2.2}$$

are called the domain of f and the range of f, respectively.

In Figure 2.6 the domain of f is the interval $[a, b]$, and the range is the interval $[c, d]$. These sets are evidently the projections of f on the x-axis and the y-axis, respectively. Quite generally we shall refer to $D[f]$ as the *projection* of f on X, and $R[f]$ as the projection of f on Y. The reader should note the difference in the nature of the mappings involved here. We have a mapping of X into Y, more precisely of $D[f]$ onto $R[f]$ which in general is not one-to-one. The projection of f on X is a one-to-one mapping of f onto $D[f]$, while the projection of f on Y is onto $R[f]$ but normally not one-to-one.

The term "range" has already been used in connection with sequences. See Definition 1.7.6. *We can now define an infinite sequence $\{a_n\}$ of real numbers as a function on natural numbers to reals.* Here f is the set of all pairs (n, a_n). We have $D[f] = N$, and $R[f]$ is the set of all real numbers α such that $a_n = \alpha$ for some value of n. This set is identical with the range of the sequence as previously defined.

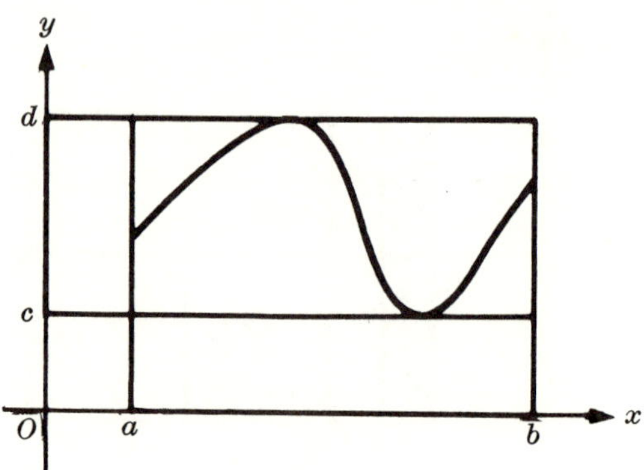

Figure 2.6

Definition 2.2.5. *If f is a function on X to Y and if $(x, y) \in f$ then y is called the value of f at x and is denoted by $f(x)$* [read "f of x"].

Purists make a great point of distinguishing between the function f and its value $f(x)$ at x. People of more slipshod habits still speak of the function $f(x)$ and in so doing often make life easier for themselves and their readers. Even purists retrogress occasionally and we shall make no excuse for suiting ourselves. In this connection it should be mentioned that the geometric representation in the (x, y)-plane of the set

$$\{(x, f(x)) \mid x \in D[f]\}$$

is frequently called the *graph* of the function f or $f(x)$.

We shall give a more systematic discussion of examples in the next section, but we insert a few here as a drill in notation and terminology.

Example 2.2.1. We start with the "square function" defined by

$$\{(x, x^2) \mid x \in R\}.$$

The graph of this function is the parabola whose equation is

$$y = x^2.$$

Its range is the set of all real nonnegative numbers which we denote by R^+. If $f(x) = x^2$, then

$$f(-3) = 9, \quad f(-1) = 1, \quad f(0) = 0, \quad f(2) = 4, \quad f(\pi) = \pi^2,$$

and so on. This is the simplest instance of a function with the property

$$f(x) = f(-x) \quad \text{for all } x. \tag{2.2.3}$$

Such a function is said to be *even*.

Example 2.2.2. We take

$$f(x) = x(1 - x^2)^{\frac{1}{2}},$$

and ask for what values this gives a function on reals to reals. This restricts the domain to the closed interval $[-1, +1]$, and to complete the definition we agree that, as usual, $a^{\frac{1}{2}}$ means the positive square root of a. The range turns out to be the interval $[-\frac{1}{2}, \frac{1}{2}]$. Every value in the range is assumed by the function for two distinct values of x with three exceptions. The value 0 is taken on three times, namely at $x = -1, 0, +1$. The extremal values $-\frac{1}{2}$ and $\frac{1}{2}$ are assumed once only, the former for $x = -2^{-\frac{1}{2}}$ and the latter for $x = 2^{-\frac{1}{2}}$. The function is *odd* in the sense that

$$f(-x) = -f(x) \quad \text{for} \quad -1 \leqq x \leqq 1. \tag{2.2.4}$$

Example 2.2.3. We define

$$f = \{(x, y) \mid x \in [0, 1], \quad y = 1 - |2x - 1|\}.$$

This is an example of a function which it is much simpler to define by an explicit formula, in this case

$$f(x) = 1 - |2x - 1|, \quad 0 \leqq x \leqq 1. \tag{2.2.5}$$

The graph of the function is drawn in Figure 2.7, which shows that the range is the interval $[0, 1]$, every value of which is taken on twice by the function except the value one which corresponds only to $x = \frac{1}{2}$. This function also has a property of symmetry, analogous to (2.2.3) and (2.2.4). In this case

$$f(1 - x) = f(x).$$

It should be noted that formula (2.2.5) defines a function for all values of x and not merely in the interval $[0, 1]$. In such a situation the function defined on a proper subset of the maximal domain of definition is regarded as the *restriction* to the subset of the function defined on the maximal domain.

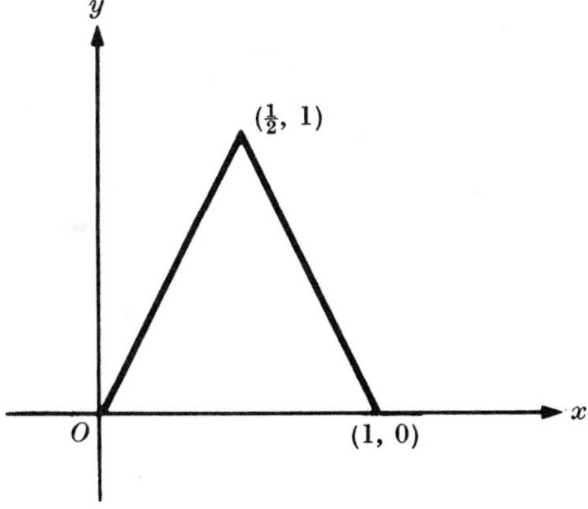

Figure 2.7

We now turn to the problem of combining functions in one way or another. At this stage we are only interested in algebraic operations. Suppose that f and g are two functions on X to Y which have the same domain $D[f] = D[g] = D$. Here D can be any set, but Y must be such that we can add and multiply elements as well as multiply by a constant without leaving the set. This requirement will be met if we take Y to be the real field R. Thus f and g are real-valued functions on D. We define their *sum*, *product*, and *scalar multiples* by the following convention.

DEFINITION 2.2.6. *Let f and g be functions on X to R and let α be a real number. Let $D[f] = D[g] = D$. Then $f + g$, fg, and αf are also functions on X to R defined for all x in D by*

(i) $(f + g)(x) = f(x) + g(x)$,

(ii) $(fg)(x) = f(x) \cdot g(x)$,

(iii) $(\alpha f)(x) = \alpha f(x)$.

We say that the real-valued functions of domain D form an *algebra over the real field*, denoted by $F[D]$. We shall not give a detailed definition of what we mean by an algebra. It is enough to remark that it is a set closed under addition, multiplication, and scalar multiplication by real numbers. The operation of addition is supposed to satisfy Postulates A_1 to A_5 of Section 1.3, multiplication satisfies M_1 and M_3, and all three operations are distributive while scalars commute with the elements. The algebra is called *commutative* if multiplication is commutative. An algebra always has a zero element but need not have a unit element. We can then state

THEOREM 2.2.1. *The real-valued functions f defined in a domain D form a commutative algebra $F[D]$ over the real field which has a unit element.*

This is merely a paraphrase of the definitions. The unit element is the function

$$e = \{(x, 1) \mid x \in D\};$$

the zero element is

$$0 = \{(x, 0) \mid x \in D\}.$$

DEFINITION 2.2.7. *If f and g belong to $F[D]$ and if $g(x) \neq 0$ for all x in D, then*

$$\frac{f}{g}(x) \equiv \frac{f(x)}{g(x)}$$

is also an element of $F[D]$.

Normally the domain of definition of the quotient is not D but a proper subset of D, namely

$$D \ominus \{x \mid g(x) = 0\}.$$

There is one more, commonly used operation by means of which we can sometimes compound two given functions to obtain a third one. This operation requires that the functions to be compounded be rather carefully adjusted to each other.

DEFINITION 2.2.8. *Let f and g be real-valued functions such that $R[g] \subset D[f]$. Then $f(g)$ is well defined on $D[g]$ by*

$$[f(g)](x) \equiv f[g(x)]. \tag{2.2.6}$$

This is a type of "product," and sometimes the symbol $f \circ g$ [read "f circle g"] is used. The operation is associative when it is definable, but normally it is not commutative. Moreover, $f \circ g$ may not exist while $g \circ f$ does. The expression $f[g(x)]$ is known as a *function of a function* or a *composite function*.

Other operations will be encountered later.

EXERCISE 2.2

1. Functions on reals to reals are defined by:

$$\text{(a)}\ \frac{1+x}{1-x};\quad \text{(b)}\ x^3 - x;\quad \text{(c)}\ (1-x)^{\frac{1}{2}};\quad \text{(d)}\ x^{\frac{1}{3}};\quad \text{(e)}\ \frac{2+(1-x)^{\frac{1}{2}}}{2-(1+x)^{\frac{1}{2}}}.$$

In each case determine the domain of the function and the values $f(0)$, $f(-\frac{1}{2})$, $f(-1)$.

2. If the functions defined in Problem 1 are restricted to the interval $[-1, 1)$, what are the corresponding ranges? Sketch the graphs.

A function on X to reals is said to be *bounded* if its range is a bounded set. It is said to be *positive* if the range is a set of positive numbers.

3. Examine the functions defined in Problem 1 for boundedness in the interval $[-1, 1)$.

4. Which of them are positive in $[-1, 1)$?

5. Verify that the function defined in Example 2.2.3 has the property that $f(x)$ is twice the distance of x from the nearest endpoint of the interval $[0, 1]$. Find a function g such that $g(x)$ equals this distance.

6. If C is a constant and $f(x) = Cx$, verify that

$$f(x) + f(1-x) = f(1).$$

7. If $f(x) = x^2$, show that $f(x+1) - 2f(x) + f(x-1) = 2$ for all x.

8. If $f(x) = 4x^2 - 4x + 1$, find $f(1-x)$, $f(\frac{1}{2}x)$, and $f(x+1)$.

9. If $f(x) = x^3$, find $f(x+1) - f(x)$.

10. (i) Find a function f such that $f(x)$ is the area of an equilateral triangle whose sides have the length x. (ii) Find a function g such that $g(x)$ is the length of the side of an equilateral triangle whose area is x. What are the domains and ranges of these functions?

11. Find the range $R[f]$ if f is given by
(i) $x^2 - 4x + 7$, $2 \leqq x \leqq 3$; (ii) $x^2 + 4x + 7$, $0 < x < 3$.

12. In the preceding problem, do inf $R[f]$ and sup $R[f]$ belong to $R[f]$?

13. The length of the perimeter of a rectangle is a units, $a > 0$. Find a function f such that $f(x)$ is the area of this rectangle when its width is x. Find $D[f]$ and $R[f]$.

14. If $f(x) = x^3 - x + 2$, $g(x) = \dfrac{1}{x+1}$, find the functions $f + g$, $f - g$, fg, $\dfrac{f}{g}$, $f \circ g$, and $g \circ f$. Determine their domains (on R to R).

15. Let $f(x) = x^2 + 1$, $g(x) = x^3$. Verify that $f \circ g \neq g \circ f$.

16. If $f(x) = x^2 - 1$, form the two functions $f \cdot f$ and $f \circ f$.

17. Let $f(x) = x^{\frac{1}{2}}$, $g(x) = 1 - x^2$, and determine the domains of $f \circ g$ and of $g \circ f$.

18. Show that the real-valued functions, defined and *bounded* on the set D, form an algebra $B[D]$. How is this algebra related to $F[D]$?

19. Do the positive functions on D form an algebra?

20. Do the polynomials with real coefficients form an algebra over R?

In the following problems $P = (x, y)$ is a point in the Euclidean plane R_2 (see Section 1.5) and f is a function on R_2 to R.

21. If $f(x, y) = x^2 + 4y^2 - 5$, find the range of f. Where is $f(x, y) = 0$? Find $f(0, \frac{3}{2})$, $f(1, 2)$, $f(2, 1)$.

22. Determine f so that $f(x, y)$ is the distance of P from the origin. What is the range of f? Where is $f(x, y) = 2$?

23. Determine f so that $f(x, y)$ is the distance from P to the nearer of the two points $(-1, 0)$, $(1, 0)$. Plot the curve where $f(x, y) = 2$.

24. If f and g are two functions on R_2 to R, can $f \circ g$ or $g \circ f$ be defined?

25. The graph of a function on R_2 to R is an entity in the Cartesian product $R_2 \times R = R_3$, that is, in three space. What kind of entity? Represent the case $f(x, y) = x^2 + y^2$ by a figure.

2.3. Various functions. This section is used to introduce various classes of functions, to illustrate definitions, to provide examples for later use, and to extend the student's vocabulary. The student should review the mapping concept (Sections 1.2 and 2.2).

These classes fall into larger groups. We start with elementary functions given by explicit formulas. Since we do not want to use trigonometric functions and logarithms at this stage, we are restricted to algebraic functions. We start with powers, polynomials, rational functions, and end with simple irrational algebraic functions. The second group deals with piecewise linear functions such as absolute values and step functions. Such functions are possibly nonelementary but they are definitely simple. The third group is a miscellaneous collection including set functions and sequences.

I. Elementary Functions

1. *Powers.* Here we start with the constant function

$$f(x) = a. \tag{2.3.1}$$

The domain is R and the range is the set $\{a\}$ consisting of the single element a. The graph is the straight line $y = a$ parallel to the x-axis.

The simplest function which actually varies is

$$f(x) = x. \tag{2.3.2}$$

Here domain and range are both R and the graph is the straight line

$$y = x,$$

which bisects the first and the third quadrants of the (x, y)-plane. The mapping

$$x \to x$$

maps R onto R and leaves every point invariant. It is known as the *identity mapping* and is usually denoted by I. In recent years this notation has also been used for the corresponding function, then referred to as the *identity function*.

The square $f(x) = x^2$ was discussed in the preceding section. It maps R onto R^+ and the mapping is not one-to-one. For the cube, $f(x) = x^3$, the domain and range are both R, and the mapping is onto and one-to-one. For odd powers of x the range is R and the mapping is onto and one-to-one, while for even powers the range is R^+ and the mapping is onto but not one-to-one. Figure 2.8 represents the graphs of

$$y = x^n$$

for $n = 1, 2, 3, 4$ and $0 \leqq x \leqq 1$. For $n = 1$, the graph is the diagonal of the

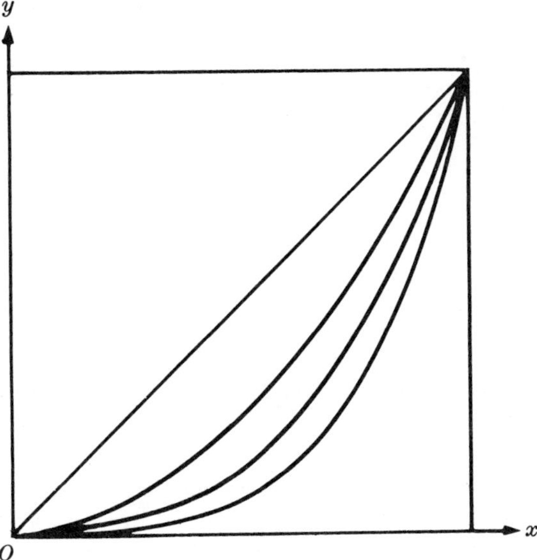

Figure 2.8

unit square; for $n > 1$ the graph lies below the diagonal except at the endpoints. At the left endpoint the curve is tangent to the x-axis for $n > 1$, and the larger the n, the flatter the curve at the origin. For large n the curve is quite close to the sides of the square. Thus, for $n = 100$ the distance of a point on the graph to the nearest side of the square nowhere exceeds 0.034.

2. *Polynomials.* A polynomial is an expression of the form

$$P(x) = a_0 + a_1x + a_2x^2 + \cdots + a_nx^n, \tag{2.3.3}$$

where the a's will be assumed to be real numbers. If $a_n \neq 0$, the integer n is called the *degree* of the polynomial. Here the domain of P is R. The range is also R if n is odd, but an interval $(-\infty, a]$ or $[a, \infty]$ if n is even where the first alternative goes with $a_n < 0$ and the second with $a_n > 0$.

The set of polynomials with real coefficients is a commutative algebra with unit element over the reals. The unit element is the function that is one for all x. Here we take addition, ordinary multiplication $(f \cdot g)(x) = f(x)\, g(x)$, and multiplication by real numbers as the three operations. It is interesting to see what happens when we replace multiplication by the circle operation, $(f \circ g)(x) = f[g(x)]$, and keep the other operations unchanged. The polynomials form a nondistributive algebra over the reals with respect to addition, circle product, and scalar multiplication. The new algebra is no longer commutative, since ordinarily $f \circ g \neq g \circ f$ and the distributive law does not hold. It has a unit element which, however, is not $f(x) = 1$ but $f(x) = x$.

3. *Rational functions.* The quotient of two polynomials may be a polynomial, but is usually a rational function. The general rational function is of the form

$$R(x) = \frac{a_0 + a_1x + a_2x^2 + \cdots + a_mx^m}{b_0 + b_1x + b_2x^2 + \cdots + b_nx^n}, \tag{2.3.4}$$

where the a's and b's are real numbers. The domain of such a function is R with omission of the points, if any, where the denominator vanishes so the quotient is not defined. The range may or may not be R. The four functions

$$x + \frac{1}{x}, \quad x^2 + \frac{1}{x^2}, \quad \frac{x}{1 + x^2}, \quad \frac{1}{1 + x^2}$$

are all rational. Their ranges are

$$(-\infty, -2] \cup [2, \infty), \quad [2, \infty), \quad [-\tfrac{1}{2}, \tfrac{1}{2}], \quad (0, 1],$$

respectively. Their graphs are sketched in Figure 2.9. The first is a hyperbola.

We can apply the operations of arithmetic, addition, subtraction, multiplication, and division, to rational functions and the results will always be rational functions. We express this by saying that *the rational functions form a field over the real numbers.* The student is urged to verify that the field postulates of Section 1.3 hold. The circle product can also be formed, and it is easy to see that *a rational function of a rational function is also a rational function.*

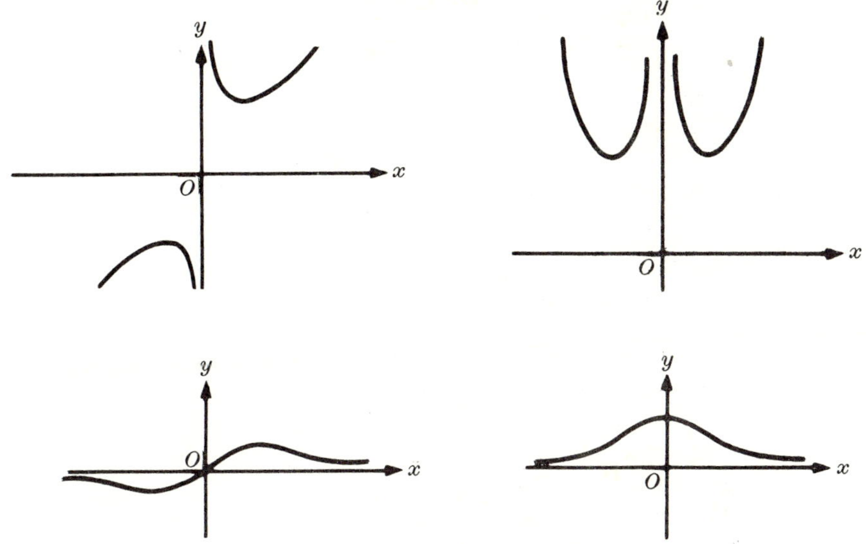

Figure 2.9

4. *Algebraic Functions.* The simplest algebraic functions are roots of rational functions. As examples we give

$$x^{\frac{1}{2}},\ (1-x)^{-\frac{1}{2}},\quad (1-x^2)^{\frac{1}{4}},\quad \left(\frac{1+x}{1-x}\right)^{\frac{1}{3}}. \tag{2.3.5}$$

We want these roots to define functions on reals to reals. This requirement determines the domains to be

$$R^+,\quad (-\infty, 1),\quad [-1, 1],\quad R\ominus\{1\},$$

respectively. We choose the positive determination of the root in the first three cases and the real root in the fourth. The ranges are then

$$R^+,\quad (0, \infty),\quad [0, 1],\quad R\ominus\{1\}.$$

If we denote the four functions by f_1, f_2, f_3, and f_4, then the following identities hold:

$$[f_1(x)]^2 = x,\quad (1-x)[f_2(x)]^2 = 1,$$

$$[f_3(x)]^4 = 1-x^2,\quad (1+x)[f_4(x)]^3 = 1-x,$$

for each x in the domain of the function in question. This means that for fixed x in the domain, $f_j(x)$ satisfies an algebraic equation of degree 2, 2, 4, and 3, respectively, namely

$$y^2 = x,\quad (1-x)z^2 = 1,\quad u^4 = 1-x^2,\quad (1+x)v^3 = 1-x.$$

This explains why these functions are said to be *algebraic.*

Quite generally, we say that a function f, defined in some domain $D \subset R$, is an algebraic function of x if we can find an integer n and $n + 1$ polynomials

$$Q_0(x), Q_1(x), \cdots, Q_n(x),$$

not all zero, such that for each x in D we have identically

$$Q_0(x)[f(x)]^n + Q_1(x)[f(x)]^{n-1} + \cdots + Q_{n-1}(x)f(x) + Q_n(x) = 0. \quad (2.3.6)$$

On the other hand, given an equation of the form

$$Q_0(x)y^n + Q_1(x)y^{n-1} + \cdots + Q_{n-1}(x)y + Q_n(x) = 0, \quad (2.3.7)$$

where the Q's are polynomials, it is always possible to find a function f whose values $f(x)$ satisfy the equation for each x in some domain. This is not necessarily true, however, if we restrict ourselves to real-valued functions. For a more detailed study of algebraic functions it becomes necessary to let the domain and the range be sets of complex numbers.

Algebraic functions form a field over complex numbers (over the reals if we restrict ourselves to real-valued functions). This is fairly difficult to prove and definitely beyond the scope of this text. Similarly, *algebraic functions of algebraic functions are algebraic.* This normally requires the consideration of complex-valued functions and is harder to prove than the field theorem.

II. Piecewise Linear Functions

5. *Absolute Values.* In Section 1.5 we defined

$$|x| = \max(x, -x). \quad (2.3.8)$$

We can also write

$$|x| = (x^2)^{\frac{1}{2}}, \quad (2.3.9)$$

that is, the positive square root of x^2, for this is x when x is positive but $-x$ if x is negative. The domain of the function is R, its range is R^+, and its graph is shown in Figure 2.10.

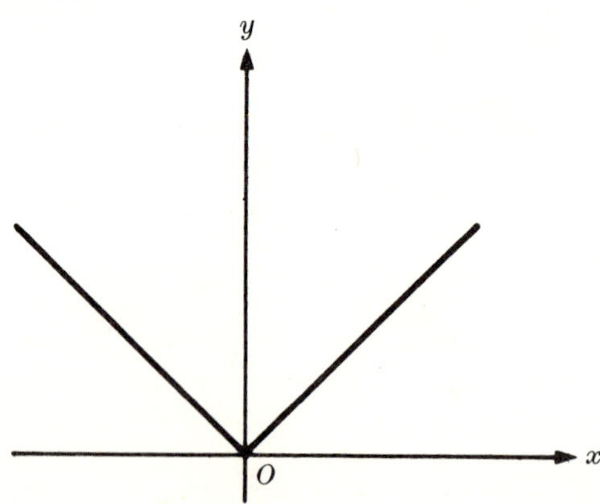

Figure 2.10

At this juncture we shall use absolute values to discuss *piecewise linear functions*. A function f is *linear* in an interval (a, b) if

$$f(x) = Ax + B, \quad a < x < b, \tag{2.3.10}$$

where A and B are two constants. This means that the part of its graph which projects onto this interval is a line segment. A function is piecewise linear if its domain is a finite or infinite interval which is the union of nonoverlapping subintervals in each of which the function is linear. At the common endpoint of two contiguous subintervals there are two possibilities according as the line segments join or not. Here we consider the first case; the second will appear later.

We start with the function L defined by

$$L(x) = c_1 |x - a_1| + c_2 |x - a_2| + \cdots + c_n |x - a_n|, \tag{2.3.11}$$

where the c's and the a's are real numbers and

$$a_1 < a_2 < \cdots < a_n.$$

The domain of L is R which is the union of the nonoverlapping subintervals

$$(-\infty, a_1], \quad (a_1, a_2], \cdots, (a_{n-1}, a_n], \quad (a_n, \infty).$$

$L(x)$ is linear in each of these subintervals. As an illustration we show the graph of

$$y = \frac{b}{2a}\{|x + a| - 2|x| + |x - a|\} \tag{2.3.12}$$

in Figure 2.11.

Conversely, if a piecewise linear function is given graphically with domain R or some subinterval (a, b), then we can represent $f(x)$ by an expression of the form

$$f(x) = L(x) + Ax + B, \tag{2.3.13}$$

where $L(x)$ is of type (2.3.11). Here it is understood that the line segments join

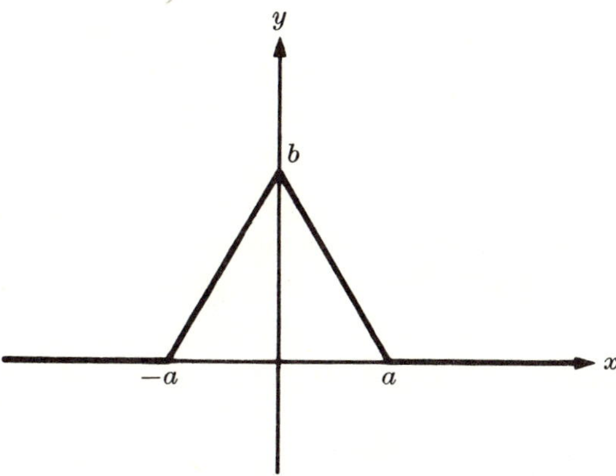

Figure 2.11

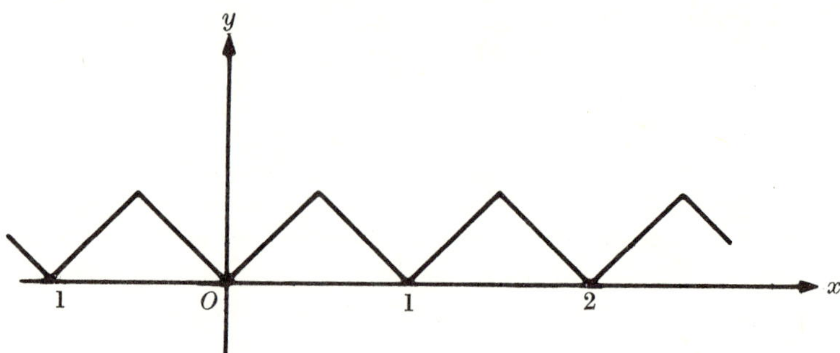

Figure 2.12

without gaps and that the vertices are finite in number and located at the points

$$(a_1, b_1), (a_2, b_2), \cdots, (a_n, b_n),$$

where the a's are the a's of $L(x)$. If (a, b) is finite, then $(a, b) = (a_1, a_n)$. We can take $A = B = 0$ and then the coefficients c_k of $L(x)$ are uniquely determined by the n conditions

$$L(a_k) = b_k, \qquad k = 1, 2, \cdots, n.$$

If the interval is infinite, we also must use the equation of the infinite line segment or segments. We still find enough conditions to determine the coefficients.

Thus far we have been concerned with piecewise linear functions whose linear pieces are finite in number. The following is an example where there are infinitely many pieces. Nonetheless it is a very simple function with many applications. We define:

$$(x) \textit{ is the distance from } x \textit{ to the nearest integer.} \tag{2.3.14}$$

The domain of (x) is R, its range is the interval $[0, \frac{1}{2}]$, and its graph is shown in Figure 2.12. In the interval $[0, 1]$ we have

$$(x) = \begin{cases} x, & 0 \leqq x \leqq \frac{1}{2}, \\ 1 - x, & \frac{1}{2} < x \leqq 1. \end{cases}$$

For all x we have

$$(x + 1) = (x).$$

We express this by saying that (x) *is a periodic function of* x *whose period is* 1. In a way it is the simplest of all periodic functions.

The function $(2x)$ has the period $\frac{1}{2}$; more generally (nx) has the period $\dfrac{1}{n}$. Thus, the larger the n the faster the function (nx) oscillates. Its graph is again a sawtooth curve, but with a finer saw (more points per unit of length) than that of (x).

6. *Step Functions.* We now consider a piecewise linear function all of whose line segments are horizontal. Such a function is called a step function and is illustrated in Figure 2.13. Here we represent $[x]$, "bracket x" or "the

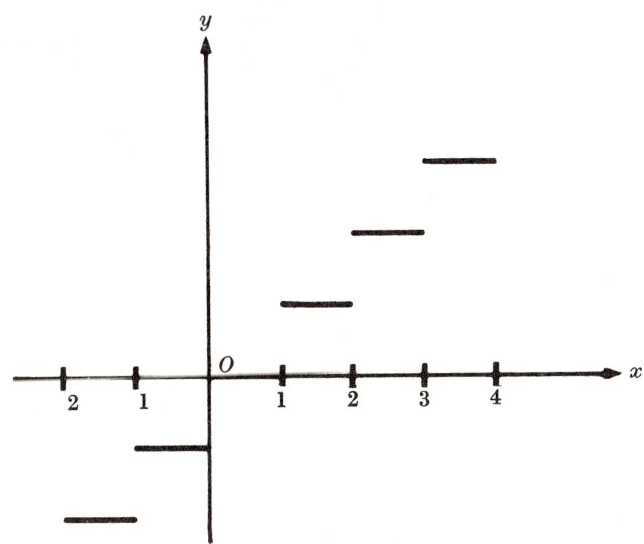

Figure 2.13

integral part of x" defined by

$$[x] \text{ is the largest integer } \leqq x. \tag{2.3.15}$$

The domain is R, and the range Z, the class of integers. This function has a "jump" or "saltus" at each of the integers. Let us take an integer n and inspect the two adjacent intervals $(n - 1, n)$ and $(n, n + 1)$. In the left interval $[x] = n - 1$; in the right $[x] = n$. If x increases and passes through $x = n$, the value of $[x]$ suddenly increases by one unit. We say that $[x]$ is *discontinuous* at the integers.

Similarly we define

$$[x) \text{ is the largest integer } < x. \tag{2.3.16}$$

This function differs from $[x]$ only at the integers where

$$[n] = n \quad \text{while} \quad [n) = n - 1.$$

The function $[x)$ has its uses. Thus

$$p = 5[x + 1), \quad 0 < x \leqq 1120,$$

is first class postage in cents on a letter which weighs x ounces.

A charge for services rendered which is based on a fixed price per unit or fraction thereof is represented graphically by a step function. Utility charges and income taxes are familiar illustrations.

The simplest of all step functions is the so-called *characteristic function of an interval*. The graph of such a function is shown in Figure 2.14. Let the interval be denoted by I and define

$$\chi(x, I) = \begin{cases} 1, & x \in I, \\ 0, & x \notin I. \end{cases} \tag{2.3.17}$$

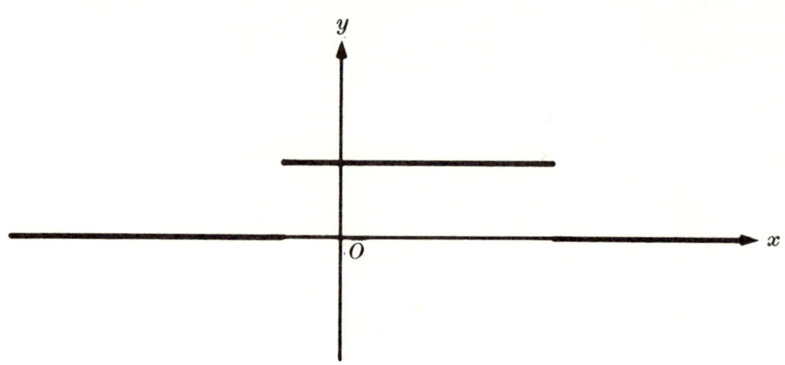

Figure 2.14

We recall that $\notin$ means "not a member of." The interval I may be open at either end or at both. The range in this case is the set $\{0, 1\}$ of the two elements 0 and 1.

More generally, if S is any set of points on the real line, we call $\chi(x, S)$ the characteristic function of S if its value for x in S equals 1 and it takes on the value 0 everywhere else. As early as 1837, Dirichlet considered the function with value 1 on the rationals and 0 on the irrationals, that is, the characteristic function of the set of rationals $\chi(x, Q)$. This function has a graph, the set $(x, \chi(x, Q))$ in the plane, but there can be no question of a pictorial representation in any practical sense.

We return to step functions. Every step function can be written as a linear combination of characteristic functions of intervals. For suppose that $f(x)$ takes the value a_n throughout the interval I_n and that the domain of the function is the union of the nonoverlapping intervals I_n which may be open. Then we have

$$f(x) = \sum_n a_n \chi(x, I_n). \tag{2.3.18}$$

The function f defined by

$$f(x) = x - [x] \tag{2.3.19}$$

is known as "the fractional part of x." It is not a step function but it is piecewise linear and discontinuous at the integers. The domain is R, the range is the half-open interval $[0, 1)$, and its graph is the "rip saw" shown in Figure 2.15. Note that the dotted vertical lines do not belong to the graph. A table of

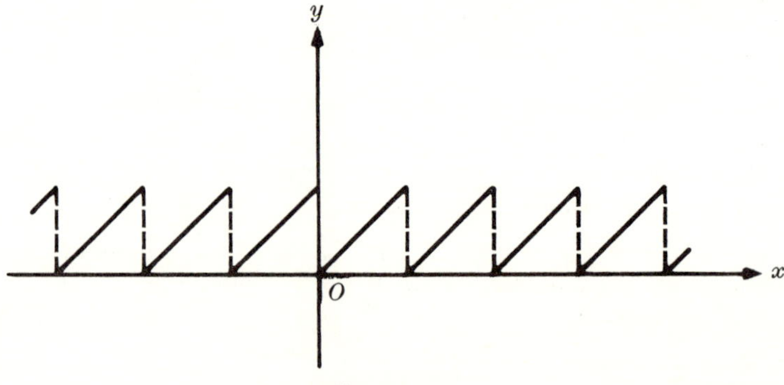

Figure 2.15

ordinary logarithms gives the fractional part of the logarithm, in this connection called the "mantissa," while the integral part is known as the "characteristic."

There is a large class of step functions on R^+ to Z known as "enumerative functions." They answer the question: How many positive real numbers $\leq x$ have a given property P? We can ask, for instance, how many integers lie in the interval $(0, x)$, and how many squares, cubes, or primes? In the first three cases the answer is elementary and may be expressed in terms of the "bracket function" of (2.3.15). The fourth case is different. Here we are dealing with a very irregularly increasing function which has been studied by the world's leading mathematicians during more than a century with only moderate success.

III. Miscellaneous

7. *Sequences.* We can consider a sequence $\{a_n\}$ as a function on N (more generally on Z) to R. We refer the reader to Sections 1.7 and 2.1 for specific instances. In Section 1.7 the reader will also find a discussion of algebraic operations on sequences from which it may be concluded that *sequences form an algebra over the reals*. The algebra is commutative and

$$\{1, 1, \cdots, 1, \cdots\}$$

acts as unit element.

An important generalization of the binomial coefficients is furnished by the sequence $\{C_{\alpha,k}\}$ where

$$C_{\alpha,k} = \binom{\alpha}{k} = \frac{\alpha(\alpha-1)(\alpha-2)\cdots(\alpha-k+1)}{k!} \tag{2.3.20}$$

and α is a fixed real number. If $\alpha = m$, a positive integer, this reduces to an ordinary binomial coefficient as defined in Exercise 1.7.

8. *The Functions of Leibniz.* To illustrate, let us consider an ellipse of major axis $2a$ and minor axis $2b$. In Figure 2.16 we have placed the ellipse in the usual position with respect to the coordinate axes. Let $P = (x, y)$ be a point on the ellipse but not an endpoint of either of the axes. We may take $0 < x < a$, $0 < y < b$. Mark the foot point Q of the perpendicular from P to the x-axis, draw the tangent and the normal to the ellipse at P, and produce them until they intersect the x-axis at T and N respectively. Finally draw a perpendicular from the origin to the tangent which it meets at M. Then the lengths of the line segments OM, OQ, OT, QP, PN, and PT are functions of P and hence of x. We can take $(0, a)$ as the domain of each of these functions. The ranges vary from one function to the next; they are trivial for OQ and QP, but the other cases require careful consideration. Such functions will be considered in later chapters in connection with applications of Calculus to geometry.

9. *Set Functions.* We have already considered the characteristic function of a set. Here we should merely add that finite linear combinations of characteristic functions of sets in the same space are often called *simple functions*.

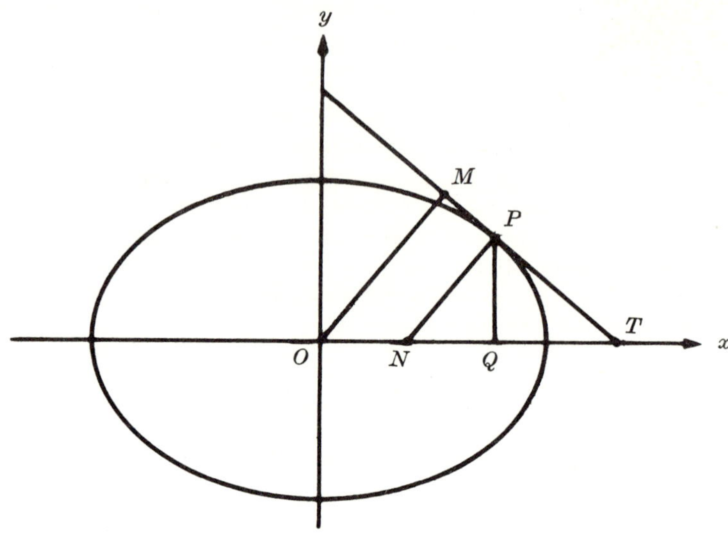

Figure 2.16

It is the range which is simple; it consists of a finite number of real numbers. The domain is the whole space.

The term "set function" is used for real-valued functions whose domain of definition is a set of point sets on the line or in the plane or in three space. We have encountered examples of such functions. The diameter $d(S)$ of a set S in the plane, defined in Section 1.5, is a set function. Its domain consists of all sets in the plane, and its range is R^+.

The length of an interval is a positive "interval function." It is possible to extend the notion of length to other sets on the line, but we shall not consider such questions. On the other hand, we shall be concerned with the notion of "length of a curve." This is a set function definable for certain classes of curves to be specified later.

Areas of triangles, squares, and circular disks are further examples of set functions. One of the main tasks in the applications of the Calculus is to extend the notion of "area" from these elementary figures to more general two-dimensional sets.

This ends our survey of special functions. It contains too many unrelated facts to be assimilated in a first reading. The wide scope of the function concept is unmistakable, and particular details can be studied whenever the need arises.

EXERCISE 2.3

1. Show that the equation

$$\frac{x-1}{x+1} = a$$

has a uniquely determined root for every finite $a \neq 1$. Show that for $a = 1$ the difference of the two sides of the equation becomes arbitrarily small for $|x|$ sufficiently large. How small is the difference if $x = 999{,}999$?

2. Discuss the equation

$$\frac{1 - x^2}{1 + x^2} = a.$$

For each of the following functions on reals to reals determine the domain and the range and sketch a graph.

3. $\dfrac{x-1}{x+1}$. 4. $\dfrac{2x-1}{x+1}$. 5. $\dfrac{1-x^2}{1+x^2}$.

6. $\frac{1}{2}x + x^2$. 7. $\frac{1}{3}x - x^3$. 8. $\frac{1}{3}x + x^3$. 9. $x^2 - x^4$.

10. $x^{\frac{1}{3}}$. 11. $2(1 - x^2)^{-\frac{1}{2}}$. 12. $(1 + x^2)^{\frac{1}{2}}$. 13. $x^{\frac{1}{2}} + (1 - x)^{\frac{1}{2}}$.

14. Show that for $a > 3$ the equation

$$x^2 + \frac{2}{x} = a$$

has a root in each of the intervals

$$(-\infty, -a^{\frac{1}{2}}), \quad \left(\frac{2}{a}, 1\right), \quad (1, a^{\frac{1}{2}}).$$

What happens for $a = 3$? for $a < 3$? Sketch the curve

$$y = x^2 + \frac{2}{x}.$$

15. If f and g are rational functions, what can be said concerning the set $D[f] \cap D[g]$? Prove that $f + g$, fg, αf, and βg are defined in this set. Is there a nonvoid set on R where all real-valued rational functions are defined?

16. Prove that the set of rational functions forms a field in the sense of Section 1.5 and that this field $\mathscr{R}$ contains R as a subfield.

17. The functions defined in Problems 11, 12, 13 are algebraic. Construct algebraic equations of degree 2, 2, and 4, respectively, which are satisfied by their values for fixed x in the domain in question.

18. Denote the three functions of Problem 17 by f, g, and h. There are nine possible circle products involving two factors, but several of them are not functions on reals to reals. Test $f \circ g$, $g \circ f$, $f \circ h$, and $h \circ g$ from this point of view.

19. Draw the curve $y = x + |x|$.

20. A piecewise linear function L is identically zero outside the interval $(-3, 3)$ and its graph has vertices at the points $(-3, 0)$, $(-2, 3)$, $(2, 3)$, and $(3, 0)$. Express $L(x)$ in terms of absolute values. See formula (2.3.11).

21. The weight limit for third class mail is 8 ounces. If the rate is 2 cents for the first 2 ounces or fraction thereof, plus 1 cent for each additional ounce or fraction thereof, plot the graph of this step function.

22. What is the range of $x - \frac{1}{100}[100\, x]$?

23. Plot the curve $y = (x) - \frac{1}{2}(2x)$. See formula (2.3.14).

24. Express the number of cubes $\leq x$ in terms of the bracket function.

25. Plot $x^{\frac{1}{2}}$, say for $0 \leqq x \leqq 9$. Plot also $x^{\frac{1}{2}} - [x^{\frac{1}{2}}]$ and compare the two graphs.

26. Prove that $0 \leqq [x + y] - [x] - [y] \leqq 1$. Are the bounds reached?

Do the ordered number pairs (x, y) which satisfy one of the following equations form a function?

27. $x^2 + y^3 = 0$.
28. $|x| - |y| = 0$.

29. Given $S_n(x) = 1 + x + x^2 + \cdots + x^n$, prove that for $x \neq 1$ we have

$$S_n(x) = \frac{1 - x^{n+1}}{1 - x}.$$

For what values of x does $S_n(x)$ tend to a (finite) limit when $n \to \infty$? Find the limit. $S_n(x)$ is the nth partial sum of the so-called geometric series, the simplest of all power series.

30. Compute $C_{-2,k}$ and $C_{-3,k}$.

31. Express OT in Figure 2.16 as a function of the abscissa and find its range.

2.4. The limit of a function. The basic problem of this section is to assign a meaning to the formula

$$\lim_{x \to c} f(x) = C$$

(read "the limit of f of x as x approaches c is C"), to analyze its significance and some of its applications, and to generalize the relation. This is a fundamental problem in Calculus and is apt to puzzle the student.

To fix the ideas, suppose that $D[f]$ contains the intervals (a, c) and (c, b) so that f is defined to both sides of $x = c$ but not necessarily at $x = c$.

DEFINITION 2.4.1. *If for every $\varepsilon > 0$ there exists a $\delta = \delta(\varepsilon)$ such that*

$$|f(x) - C| < \varepsilon \quad \textit{for all } x \textit{ with} \quad 0 < |x - c| < \delta, \tag{2.4.1}$$

then

$$\lim_{x \to c} f(x) = C. \tag{2.4.2}$$

Alternate notation is

$$f(x) \to C \quad \text{when} \quad x \to c. \tag{2.4.3}$$

Let us illustrate the various possibilities by some examples.

(1) $f(x) \to C$, when $x \to c$, $c \in D[f]$, and $C = f(c)$.

This is the ideal case. A simple example is furnished by $f(x) = x^2$, c arbitrary. Here we have

$$\lim_{x \to c} f(x) = c^2 = f(c)$$

for any choice of c. For in this case

$$|f(x) - f(c)| = |x^2 - c^2| = |x + c|\,|x - c| \leqq (|x| + |c|)\,|x - c|.$$

Since we are interested in values of x close to c, let us suppose that at the outset we restrict x to $|x - c| < 1$. This gives $|x| < |c| + 1$ and $|x| + |c| < 2\,|c| + 1$

so that

$$|f(x) - f(c)| < (2\,|c| + 1)\,|x - c|.$$

If we choose $|x - c|$ so small that the right member of this inequality is $< \varepsilon$, then the left member is also $< \varepsilon$. Hence

$$|f(x) - f(c)| < \varepsilon \quad \text{if} \quad |x - c| < \frac{\varepsilon}{2\,|c| + 1}.$$

The last quantity can be taken as $\delta(\varepsilon)$. We note, however, that δ depends upon c as well as upon ε; the larger we take $|c|$ the smaller we must choose δ.

With a view to later applications we note that if $|x|$ is bounded, say $|x| \leqq M$, then we have

$$|f(x) - f(c)| < \varepsilon \quad \text{for} \quad |x - c| < \frac{\varepsilon}{2M + 1},\ |c| \leqq M,$$

regardless of the particular value of c. We say that $f(x)$ tends to its limit $f(c)$, when x approaches c, *uniformly* with respect to c in any fixed finite interval.

We now turn to the other possibilities.

(2) $f(x) \to C$ when $x \to c$ and f is not defined at $x = c$.

In this case we may extend the domain of f by adjoining $x = c$ to $D[f]$ and defining $f(c) = C$. For the extended function case (1) now holds. This situation would arise, for instance, if we defined $f(x)$ to have the value x^2 for all $x \neq 1$ and omitted to make any provision for $x = 1$.

(3) $f(x) \to C$ when $x \to c$, c belongs to $D[f]$ but $f(c) \neq C$.

As an example we take the characteristic function (in the sense defined in Section 2.3: 6, step functions) of the set $\{c\}$ which consists of the single point c. Here we have $f(c) = 1$ while $f(x) = 0$ for all $x \neq c$. In this case

$$\lim_{x \to c} f(x) = 0 \neq 1 = f(c).$$

(4) $f(x)$ does not tend to any limit as x approaches c.

In this case it is immaterial if $f(c)$ is defined. As an example we may take the characteristic function of the rationals, $\chi(x, Q)$, which is 1 when x is rational and 0 when x is irrational. Since every neighborhood of $x = c$ contains both rational and irrational numbers, $f(x)$ cannot tend to any limit when x approaches c.

THEOREM 2.3.1. *A necessary and sufficient condition that* (2.4.2) *shall hold is that for each sequence* $\{x_n\}$ *such that*

$$\lim_{n \to \infty} x_n = c, \tag{2.4.4}$$

the sequence $\{f(x_n)\}$ *converges to* C, *that is,*

$$\lim_{n \to \infty} f(x_n) = C. \tag{2.4.5}$$

Proof. (1) The condition is necessary. For if (2.4.2) holds, then condition (2.4.1) holds and (2.4.4) implies the existence of an $N = N(\delta)$ such that

$$|x_n - c| < \delta \quad \text{for} \quad n > N(\delta).$$

Combining this with (2.4.1) we see that

$$|f(x_n) - C| < \varepsilon \quad \text{for} \quad n > N(\delta),$$

and this implies (2.4.5).

(2) The condition is sufficient. Suppose that (2.4.4) implies (2.4.5) for every choice of the sequence $\{x_n\}$. Suppose that this does not imply the validity of (2.4.2). We shall show that this assumption leads to a contradiction. Let us choose an $\varepsilon > 0$ and a corresponding $\eta > 0$ such that the set $\{x \mid 0 < |x - c| < \eta\}$ belongs to $D[f]$. This set will then contain a point x_1 such that $|f(x_1) - C| > \varepsilon$. We can then find a point x_2 such that $x_2 \neq x_1$, $0 < |x_2 - c| < 2^{-1}\eta$, and $|f(x_2) - C| > \varepsilon$. In this way we can proceed and find a sequence $\{x_n\}$ of distinct points such that for each n we have $0 < |x_n - c| < 2^{-n}\eta$ and $|f(x_n) - C| > \varepsilon$. For if the procedure would stop after we have found m points, say, then we must have $|f(x) - C| \leq \varepsilon$ for $0 < |x - c| < 2^{-m-1}\eta$ and this would imply (2.4.1) and hence (2.4.2), against the hypothesis. We have now an infinite sequence $\{x_n\}$ which clearly converges to c and we have also for all n

$$|f(x_n) - C| > \varepsilon,$$

and this in its turn contradicts (2.4.5). Hence the assumption that (2.4.2) does not hold proves to be untenable, the condition is indeed also sufficient and the proof of the theorem is finished.

It is often necessary to consider limits when x is allowed to approach c only from the left or only from the right.

DEFINITION 2.4.2. *We say that*

$$\lim_{x \to c-0} f(x) = C, \tag{2.4.6}$$

if, given any $\varepsilon > 0$, there exists a $\delta = \delta(\varepsilon)$ such that $|f(x) - C| < \varepsilon$ for every x such that $c - \delta < x < c$.

This is the *left-hand limit*. We leave it to the student to formulate a corresponding definition for the *right-hand limit* which is written

$$\lim_{x \to c+0} f(x). \tag{2.4.7}$$

To illustrate these concepts we recall that the function $[x]$ has right- and left-hand limits everywhere and these limits are distinct if (and only if) x is an integer for

$$\lim_{x \to n-0} [x] = n - 1, \qquad \lim_{x \to n+0} [x] = n = [n].$$

Since it is possible for a function to fail to have a limit as x approaches some value c, it is natural to consider superior and inferior limits, found to be quite useful in the case of sequences.

DEFINITION 2.4.3. *Let f on reals to reals be defined for $0 < |x - c| < \eta$. We say that*

$$\liminf_{x \to c} f(x) = A, \quad \limsup_{x \to c} f(x) = B, \tag{2.4.8}$$

where A and B are finite, $A \leqq B$, if, given any $\varepsilon > 0$, there exists a $\delta = \delta(\varepsilon)$ such that when $0 < |x - c| < \delta$ we have

$$A - \varepsilon < f(x) < B + \varepsilon \text{ for all such } x,$$
$$f(x) < A + \varepsilon \text{ for infinitely many such } x, \text{ and}$$
$$f(x) > B - \varepsilon \text{ for infinitely many such } x.$$

As an illustration we consider the characteristic function of the rationals. Here for any c we have

$$\liminf_{x \to c} \chi(x, Q) = 0, \qquad \limsup_{x \to c} \chi(x, Q) = 1.$$

Similarly for the distance to the nearest integer

$$\liminf_{x \to 0} \left(\frac{1}{x}\right) = 0, \quad \limsup_{x \to 0} \left(\frac{1}{x}\right) = \tfrac{1}{2}.$$

We can also define one-sided superior and inferior limits. All we have to do is to restrict x to one side of c in the preceding definition.

So far c has been supposed to be finite, but in studying the Calculus it is not desirable to restrict oneself to finite quantities only. In fact, terms like "infinite," "infinitely many," "infinitely large," "infinitely small," and "infinity" are likely to appear in almost any problem. The student must learn to accept such expressions, to use them correctly, and to avoid being diverted by speculations concerning the possible philosophical implications of the terms.

We have used "∞" (read "infinity") as a matter of convenient notation without thus far attaching any intrinsic meaning to the symbol. Thus, the interval (b, ∞) is simply the set of all real numbers greater than b, while $(-\infty, a)$ stands for the set of real numbers less than a, and $(-\infty, \infty)$ is the set of all real numbers.

The idea that there are "infinitely many" integers and hence also infinitely many rational and real numbers should not lead to any conceptual difficulties. Similarly, the idea of an infinite sequence should by now be firmly established in the student's mind. In connection with such sequences we defined (see Definition 1.7.4) the meaning of

$$\lim_{n \to \infty} a_n = a, \tag{2.4.9}$$

and we have also defined $\limsup_{n \to \infty} a_n$ and $\liminf_{n \to \infty} a_n$ (see Definition 2.1.6). Now an infinite sequence is actually a function on natural numbers to reals which is defined for large integers. In the preceding sections we have encountered many functions on reals to real which are defined for all large values of x. It is then natural to try to attach a meaning to symbols like

$$\lim_{x \to \infty} f(x), \quad \liminf_{x \to \infty} f(x), \quad \text{and} \quad \limsup_{x \to \infty} f(x).$$

We arrive at the following

DEFINITION 2.4.4. *Let f be defined and real-valued for x in (b, ∞). If there exists a number C such that for any $\varepsilon > 0$ there exists an $\omega = \omega(\varepsilon)$ with the property that*

$$|f(x) - C| < \varepsilon \tag{2.4.10}$$

for all x with $x > \omega(\varepsilon)$, then, by definition,

$$\lim_{x\to\infty} f(x) = C. \tag{2.4.11}$$

We can get illustrations from the first five problems of Exercise 2.3. Problems 1 and 3 lead to

$$\lim_{x\to\infty} \frac{x-1}{x+1} = 1.$$

Similarly Problems 2 and 5 show that

$$\lim_{x\to\infty} \frac{1-x^2}{1+x^2} = -1.$$

In Problem 4 the corresponding limit is found to be 2.

There is of course a similar definition for

$$\lim_{x\to-\infty} f(x),$$

when $D[f]$ contains an interval $(-\infty, a)$. Further we note the following analogue of Theorem 2.4.1 which we state without proof.

THEOREM 2.4.2. *A necessary and sufficient condition for the validity of* (2.4.11) *is that for every strictly increasing unbounded sequence $\{x_n\}$ the sequence $\{f(x_n)\}$ shall converge to C,*

$$\lim_{n\to\infty} f(x_n) = C. \tag{2.4.12}$$

Using Definitions 2.4.3 and 2.4.4 we can easily define relations like

$$\lim_{x\to\infty} \inf f(x) = A, \quad \lim_{x\to\infty} \sup f(x) = B. \tag{2.4.13}$$

We leave this task to the reader. As an illustration we observe that

$$\liminf_{x\to\infty} \{x - [x]\} = 0, \quad \limsup_{x\to\infty} \{x - [x]\} = 1.$$

There is one more step that should be taken and this is to admit infinite limits. In connection with Definition 2.4.1 we gave a classification of possible behavior of $f(x)$ when x approaches c. The discussion is completely satisfactory and exhaustive as long as we deal with bounded functions. But it is impossible to exclude unbounded functions without serious loss of generality, and for the unbounded case it is desirable to split (4) into several subcases. Consider the three functions f, g, and h defined for $x \neq 0$ by

$$f(x) = \frac{1}{x^2}, \quad g(x) = \frac{1}{x}, \quad h(x) = \frac{1}{x}\left\{\tfrac{1}{4} - \left(\frac{1}{x}\right)\right\}.$$

None of these functions has a limit in the sense of Definition 2.4.1 when x approaches zero, but their behavior for small values of $|x|$ is strikingly different. The first function assumes only large positive values when x is small. For instance, we have $f(x) > 10^{12}$ when $|x| < 10^{-6}$. Here it is quite natural to say that $f(x)$ tends to $+\infty$ as x approaches 0. The second function is odd and assumes large positive values for small positive x and large negative values for small negative x. In this case we still have no limit, but we can accept $+\infty$ as right-hand limit and $-\infty$ as left-hand limit. The third function shows a much more complicated behavior. The first factor is $g(x)$, the behavior of which we have just described. The second factor is bounded. Since $\left(\frac{1}{x}\right)$ oscillates between 0 and $\frac{1}{2}$; the second factor oscillates between $-\frac{1}{4}$ and $\frac{1}{4}$. The combined effect of the two factors as x approaches zero is an oscillation where both frequency and amplitude are unbounded. More precisely, if n is any integer, then

$$h\left(\frac{1}{n}\right) = \tfrac{1}{4}n, \quad h\left(\frac{1}{n + \frac{1}{2}}\right) = -\tfrac{1}{4}n.$$

Here there can be no question of a limit or of right-hand or left-hand limits. We can accept $+\infty$ as superior limit and $-\infty$ as inferior limit of $h(x)$ as x approaches 0 and here there is no difference between one-sided and two-sided approach. We shall complete this discussion by giving a formal definition for $+\infty$ as the limit of $f(x)$.

Definition 2.4.5. *We say that*

$$\lim_{x \to c} f(x) = +\infty, \tag{2.4.14}$$

if, given any positive ω, there exists a $\delta = \delta(\omega)$ such that $\omega < f(x)$ for $0 < |x - c| < \delta(\omega)$.

With this and previous definitions as a pattern it is easy to lay down definitions for $-\infty$, for one-sided limits, for superior and inferior limits, and, finally, for infinite limits at $+\infty$ or at $-\infty$. We shall not press these matters any further.

Assuming now that we have a clear idea of different types of limits, we still have to discuss how to work with limits. We have seen that functions with a common domain of definition D form an algebra $F[D]$. Further we saw that the bounded functions form a subalgebra $B[D]$. This encourages us to expect that "limitable" functions also form an algebra. This is indeed the case provided we restrict ourselves to functions whose limits are finite.

Let us say that f belongs to the set $L(c)$ if (i) for some $\eta > 0$ the point set $\{x \mid 0 < |x - c| < \eta\}$ belongs to $D[f]$ and (ii) $\lim_{x \to c} f(x)$ exists and is finite. We have then

THEOREM 2.4.3. *The set* $L(c)$ *is an algebra. Moreover, if* f *and* g *belong to* $L(c)$, *then*

$$\lim_{x\to c} (f+g)(x) = \lim_{x\to c} f(x) + \lim_{x\to c} g(x), \tag{2.4.15}$$

$$\lim_{x\to c} (\alpha f)(x) = \alpha \lim_{x\to c} f(x), \tag{2.4.16}$$

$$\lim_{x\to c} (fg)(x) = [\lim_{x\to c} f(x)][\lim_{x\to c} g(x)]. \tag{2.4.17}$$

Proof. This result is completely analogous to Theorem 1.7.2 and the same technique applies. We set

$$\lim_{x\to c} f(x) = C, \quad \lim_{x\to c} g(x) = D,$$

and define

$$F(x) = f(x) - C, \quad G(x) = g(x) - D,$$

so that

$$\lim_{x\to c} F(x) = \lim_{x\to c} G(x) = 0.$$

This means that to every $\varepsilon > 0$ there is a corresponding $\delta = \delta(\varepsilon)$ such that

$$|F(x)| < \varepsilon, \quad |G(x)| < \varepsilon \quad \text{when} \quad 0 < |x-c| < \delta(\varepsilon). \tag{2.4.18}$$

We now turn to (2.4.15). We have

$$\begin{aligned}(f+g)(x) - (C+D) &= f(x) + g(x) - C - D \\ &= [f(x) - C] + [g(x) - D] = F(x) + G(x).\end{aligned}$$

Here we can apply (2.4.18), which implies that the absolute value of the last member does not reach 2ε whenever $0 < |x-c| < \delta(\varepsilon)$. Since ε is arbitrary, it follows that

$$\lim_{x\to c} [F(x) + G(x)] = 0,$$

and this is equivalent to (2.4.15).

The next relation is fairly obvious since

$$(\alpha f)(x) - \alpha C = \alpha[f(x) - C] = \alpha F(x),$$

and for $0 < |x-c| < \delta(\varepsilon)$ the absolute value of the last member of the equality does not reach $|\alpha|\varepsilon$. Here α is a fixed finite number and ε is at our disposal so we conclude that $\alpha F(x) \to 0$ when $x \to c$.

Finally we have

$$\begin{aligned}(fg)(x) - CD &= f(x)g(x) - CD = [F(x) + C][G(x) + D] - CD \\ &= DF(x) + CG(x) + F(x)G(x),\end{aligned}$$

and for $0 < |x-c| < \delta(\varepsilon)$ the absolute value of the last member does not reach

$$[|C| + |D| + \varepsilon]\varepsilon.$$

Hence the limit of the last member in the equality is 0 and (2.4.17) holds. This completes the proof.

THEOREM 2.4.4. *If f and g belong to L(c) and if*

$$\lim_{x\to c} g(x) = D \neq 0,$$

then

$$\lim_{x\to c} \frac{f}{g}(x) = \frac{C}{D}. \tag{2.4.19}$$

Proof. With the notation used in the proof of the preceding theorem we have

$$\frac{f}{g}(x) - \frac{C}{D} = \frac{C + F(x)}{D + G(x)} - \frac{C}{D} = \frac{DF(x) - CG(x)}{D[D + G(x)]}.$$

We can assume ε so small at the outset that $\varepsilon < \frac{1}{2}|D|$ so that

$$|D + G(x)| > |D| - \varepsilon > \tfrac{1}{2}|D| \quad \text{for} \quad 0 < |x - c| < \delta(\varepsilon).$$

This ensures the existence of the quotient f/g and gives the estimate

$$\left|\frac{f}{g}(x) - \frac{C}{D}\right| < \frac{|C| + |D|}{|D|^2} 2\varepsilon,$$

which is as small as we please. Hence (2.4.19) must hold and the theorem is proved.

These theorems are used whenever we work with limits. Let us list some of the many consequences.

First we note that (2.4.15) can be extended to any finite number of summands. Thus, if $f_1, f_2, \cdots, f_n$ belong to $L(c)$, so does $f_1 + f_2 + \cdots + f_n$, and *the limit of the sum is the sum of the limits*. Similarly, $f_1 f_2 \cdots f_n \in L(c)$, and *the limit of the product is the product of the limits*. This is proved by mathematical induction on n.

It should be noted that $f + g$ may belong to $L(c)$ without f and g having this property. Thus, if $g = -f$, then $f + g = 0$, and the limit of the sum exists regardless of the behavior of f. A similar remark holds for the product.

The following observation is an immediate corollary of Theorems 2.4.3 and 2.4.4 together with the remark on the limit of a product, in this case of equal factors, when the limits of the factors exist.

THEOREM 2.4.5. *For each c and each nonnegative integer n*

$$\lim_{x\to c} x^n = c^n. \tag{2.4.20}$$

fI $c \neq 0$, this holds also for negative integers.

Going back to Theorems 2.4.3 and 2.4.4 again we obtain the

COROLLARY. *For each polynomial $P(x)$ we have*

$$\lim_{x\to c} P(x) = P(c). \tag{2.4.21}$$

Further, if $Q(x)$ is also a polynomial and $Q(c) \neq 0$, then

$$\lim_{x\to c} \frac{P}{Q}(x) = \frac{P(c)}{Q(c)}. \tag{2.4.22}$$

The following is a useful remark in connection with quotients. If $g(c) = 0$ and if $f(c) \neq 0$, then $\frac{f}{g}(x)$ cannot have a finite limit as $x \to c$, for it is clear that $\frac{f}{g}(x)$ must take on arbitrarily large values in any neighborhood of $x = c$. The best we can get here is an infinite limit either for a one-sided approach or for a two-sided. It is of course understood that f and g belong to $L(c)$. The problem of finding the limit of a quotient $\frac{g}{f}(x)$ when x approaches a value c such that $g(c) = 0$ is very important in the Calculus. From what has just been said, we see that a finite limit cannot exist unless $f(c)$ is also 0. In this case Theorem 2.4.4 does not apply and formula (2.4.19) becomes completely meaningless. If $f(x) = P(x)$, $g(x) = Q(x)$ are polynomials, and $P(c) = Q(c) = 0$, then by long division

$$P(x) = (x - c)P_1(x), \quad Q(x) = (x - c)Q_1(x),$$

where P_1 and Q_1 are also polynomials in x of degree one unit less than the degrees of P and Q, respectively. For $x \neq c$

$$\frac{P(x)}{Q(x)} = \frac{P_1(x)}{Q_1(x)},$$

and the problem of finding the limit of the left member is the same as that of finding the limit of the right one. If $Q_1(c) \neq 0$, we can apply Theorem 2.4.4 and get the desired result. Otherwise we have to continue the factorization. Since P and Q are polynomials, this process comes to an end after a finite number of steps.

In conclusion it should be observed that Theorems 2.4.3 and 2.4.4 obviously hold for a one-sided approach of x to c, and there are also analogous theorems for the approach of x to $+\infty$ or to $-\infty$. We can leave the formulation of such theorems to the reader. Finally we note that

$$\lim_{x\to c} f(x) = +\infty, \quad \lim_{x\to c} g(x) = +\infty \tag{2.4.23}$$

imply that

$$\lim_{x\to c} (f + g)(x) = +\infty, \quad \lim_{x\to c} (fg)(x) = +\infty. \tag{2.4.24}$$

Here we can replace "plus" by "minus" in the hypothesis and in the first conclusion. If only one plus is replaced by a minus in the hypothesis, then the sign should be changed in the second conclusion and no conclusion can be drawn in the first case.

EXERCISE 2.4

Which of the following functions tend to a limit when $x \to 2$? Find the limits.

1. $x^3 + 3x^2 - 9x + 1$. 2. $8x^3 - 57x$. 3. $\dfrac{x^2 + x + 1}{x^2 + 2x}$.

4. $\dfrac{x^2 - 4}{x + 2}$. 5. $\dfrac{x^2 - 4}{x - 2}$. 6. $\dfrac{x^2 - 3x + 2}{x^2 + x - 6}$.

7. $\dfrac{1 - x^2}{1 + x^2}$. 8. $\dfrac{x^3 - 3x^2 + 4}{x^2 - 3x + 2}$. 9. $x - \dfrac{4}{x}$.

10. $[x]$. 11. $(x^2 - 4)\,[x]$.

12. Which of the preceding functions tend to finite limits when $x \to +\infty$? Find the limits.

It is known that

$$\lim_{x\to 0} f(x) = C, \quad \lim_{x\to 0} g(x) = D, \quad \lim_{x\to 0} h(x) = E$$

exist and are finite. Find the limits of the following functions when $x \to 0$.

13. $(f + g + h)(x)$. 14. $(fg + gh + hf)(x)$. 15. $f^2(x) = [f(x)]^2$.

16. $(f + gh)(x)$. 17. $(f + g)^2(x)$. 18. $(f - 2g + 3h)(x)$.

19. If $f \in L(c)$, prove that so does $|f|$.

20. If $f(x) < 0$ in $D[f]$ and $c \in D[f]$, find

$$\lim_{x\to c} [f(x) + |f(x)|].$$

Verify that the following three functions tend to the limits 0, 0, and 1, respectively, when $x \to 0$. Show first in each case that formula (2.4.17) does not apply.

21. $[x]\,[x + 1]$. 22. $x\left(\dfrac{1}{x}\right)$. 23. $x\left[\dfrac{1}{x}\right]$.

24. Formulate the analogue of Definition 2.4.2 for right-hand limits.
25. Formulate the analogue of Definition 2.4.3 for $x \to +\infty$.
26. Prove Theorem 2.4.2.
27. Prove the analogue of Theorem 2.4.3 for $x \to +\infty$.
28. Prove (2.4.23).

Find the inferior and the superior limits as $x \to +\infty$ of the following functions:

29. $1 - 4(x)$. 30. $x^{\frac{1}{2}} - [x^{\frac{1}{2}}]$. 31. $(-1)^{[x]}$.

32. $x + \left[\dfrac{1}{x}\right]$. 33. $2(x^2)$. 34. $\dfrac{x + [x]}{x}$.

2.5. Continuous functions. We are going to develop the concept of continuity for functions on reals to reals and start with the case where $D[f]$ contains an open interval (a, b). We want to examine the local behavior of f, that is, its properties in a small neighborhood of an arbitrary point c in (a, b). It is reasonable to ask if nearness of x to c implies nearness of $f(x)$ to $f(c)$. A brief review of the functions considered in the two preceding sections shows this need not be the case and makes it desirable to single out the cases where the property does hold.

DEFINITION 2.5.1. (i) *A function f on (a, b) to reals is said to be continuous at the point $x = c$, $a < c < b$, if, given any $\varepsilon > 0$, there exists a $\delta = \delta(\varepsilon, c)$ such that $|x - c| \leqq \delta$ implies $|f(x) - f(c)| \leqq \varepsilon$. If this condition is not satisfied, f is said to be discontinuous at $x = c$.*

(ii) *f is said to be continuous in (a, b), if it is continuous at each point of (a, b).*

Comparing this with Definition 2.4.1 and the various possibilities discussed afterward, we see that continuity corresponds to case (1). We can state this conclusion as

THEOREM 2.5.1. *A function f is continuous at $x = c$, if and only if* (i) $c \in D[f]$ *and* (ii)

$$\lim_{x \to c} f(x) = f(c). \tag{2.5.1}$$

Geometrically the situation is described in Figure 2.17. The part of the graph $(x, f(x))$ which corresponds to the interval $c - \delta \leqq x \leqq c + \delta$ is confined to a small rectangle in which

$$f(c) - \varepsilon \leqq y \leqq f(c) + \varepsilon.$$

The height of this rectangle can be made as small as we please by making the width sufficiently small.

The reader should realize that normally the "graph" of a continuous function cannot be plotted. We can plot points, a finite number of points, and join them by a smooth curve, but it is likely that this curve has no bearing on the problem. A smooth curve has direction and its direction changes continuously. Neither property pertains to the graph of a continuous function. Thus when we draw a diagram to illustrate a property of a continuous function, we have to be aware of the limitations of this device and not trust it too far. What is significant in Figure 2.17 is merely that the graph is confined to the rectangle

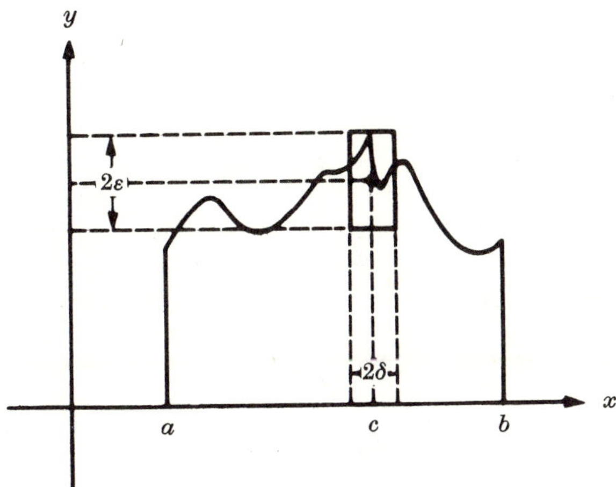

Figure 2.17

when x is in $[c - \delta, c + \delta]$ and that the graph passes through the center of the rectangle. The rest is sheer invention.

We can also define *right-hand continuity* as well as *left-hand continuity.*

DEFINITION 2.5.2. *f is continuous to the right at $x = c$ if* (i) $c \in D[f]$, *and* (ii) *if, given any $\varepsilon > 0$, there exists a $\delta = \delta(\varepsilon, c)$ such that $c < x \leqq c + \delta$ implies that $|f(x) - f(c)| \leqq \varepsilon$. Similarly, f is continuous to the left at $x = c$ if* (i) $c \in D[f]$, *and* (ii) *if, given any $\varepsilon > 0$, there exists a δ such that $c - \delta \leqq x < c$ implies that*

$$|f(x) - f(c)| \leqq \varepsilon.$$

It is clear that *if f is continuous both to the left and to the right at $x = c$, then f is continuous at $x = c$.*

The simplest type of discontinuity is where the function has left- and right-hand limits at the point, but these either do not agree with each other or differ from $f(c)$. We define

$$f(c - 0) = \lim_{x \to c-0} f(x), \quad f(c + 0) = \lim_{x \to c+0} f(x), \tag{2.5.2}$$

when these limits exist. We obviously have continuity if

$$f(c - 0) = f(c + 0) = f(c).$$

If there is inequality in either place, we say that f has a *discontinuity of the first kind* at $x = c$. Any discontinuity, not of the first kind, is said to be of the *second kind.*

The step functions studied in Section 2.3 are simple examples of functions whose discontinuities are only of the first kind. Thus $[x]$ has such discontinuities at the integers and is continuous everywhere else. Dirichlet's function $\chi(x, Q)$ (see Section 2.3:6, step functions) enjoys the distinction of being discontinuous everywhere and all its discontinuities are of the second kind. The function $(1/x)$ has a discontinuity of the second kind at $x = 0$ and is continuous everywhere else. See below.

We can easily extend our definitions to functions defined in a closed interval $[a, b]$. All we need is to add a convention for the endpoints. We have to be satisfied with right-hand continuity at $x = a$ and left-hand continuity at $x = b$. If these conditions hold and if f is continuous at all interior points, then f is continuous in $[a, b]$.

It happens sometimes that a function is defined and continuous in an open interval (a, b), but is not defined at one of the endpoints, say at $x = a$. Suppose that $f(a + 0)$ exists as a finite number (the reason for this restriction will be clear in a moment). We can then define f "by continuity" at $x = a$ by the convention

$$f(a) \equiv f(a + 0).$$

The reader should note this use of the sign "$\equiv$" to denote "equality by definition." The function as extended is continuous in $[a, b)$.

It is common practice to say that functions like

$$\frac{1}{x}, \quad \left[\frac{1}{x}\right], \quad \text{and} \quad \left(\frac{1}{x}\right)$$

are discontinuous at $x = 0$. Strictly speaking this is incorrect, for these functions are not defined at $x = 0$. What is true, is the following assertion: No matter how we extend the definition of such a function so as to include a value for $x = 0$, the resulting function will be discontinuous at $x = 0$. The first and the second functions have $+\infty$ as right-hand limit and $-\infty$ as left-hand limit at $x = 0$. There is no way of bridging this gap. Since $f(0 + 0)$ and $f(0 - 0)$ exist, even though they are infinite, we may say that these functions have a discontinuity of the first kind at $x = 0$. The third function is bounded, but we have

$$\liminf_{x \to 0} f(x) = 0, \qquad \limsup_{x \to 0} f(x) = \tfrac{1}{2},$$

so in whatever way we define $f(0)$, we cannot escape a discontinuity of the second kind.

Note that it is not sufficient for continuity that $f(c - 0)$ and $f(c + 0)$ exist as finite limits and be equal. Their common value must also agree with $f(c)$.

We turn now to the behavior of continuous functions under algebraic composition. Here the lengthy discussion in Section 2.4 will pay its first dividend.

THEOREM 2.5.2. *If f and g are functions on (a, b) to R, if $a < c < b$, and f and g are continuous at $x = c$, so are $f + g$, fg, and αf. If, further, $g(c) \neq 0$, then there exists an open subinterval in which f/g is defined, and this function is continuous at $x = c$.*

Proof. We know that $f(c)$ and $g(c)$ exist as finite numbers. Then $f + g$, αf, and fg are also defined at $x = c$ and have the values

$$f(c) + g(c), \quad \alpha f(c), \quad \text{and} \quad f(c)g(c),$$

respectively. Further we know that

$$\lim_{x \to c} f(x) = f(c), \quad \lim_{x \to c} g(x) = g(c).$$

By Theorem 2.4.3 this implies that

$$(f+g)(x) \to f(c) + g(c), \quad (\alpha f)(x) \to \alpha f(c), \quad (fg)(x) \to f(c)g(c),$$

whence it follows that the functions $f + g$, αf, and fg are continuous.

In the discussion of the quotient we appeal to Theorem 2.4.4. From its proof follows the existence of the quotient in some open interval about $x = c$ and the limit relation

$$\frac{f}{g}(x) \to \frac{f(c)}{g(c)}.$$

This completes the proof.

Corollary 1. *The functions on reals to reals whose domains contain $x = c$ and which are continuous at $x = c$, form an algebra.*

Corollary 2. *The functions which are continuous in $[a, b]$ and have values in R, form an algebra denoted by $C[a, b]$.*

Corollary 3. *The polynomials in x with domain R and range in R are continuous for all x. The same is true for the rational functions of x on R to R provided that we exclude for each function those values of x for which the denominator is zero.*

In Corollary 3 we use Theorem 2.4.5 and its Corollary.

There is still another rule of composition, based on the circle operator, which may be applicable to continuous functions and which always leads to a continuous result when it applies.

Theorem 2.5.3. *Suppose that $g(x)$ is a function on (a, b) to R whose range is a subset of (A, B), and suppose that $g(x)$ is continuous at $x = c$, $a < c < b$. Suppose also that $f(X)$ has domain (A, B) and range in R and that it is continuous at $X = C = g(c)$. Then the composite function $(f \circ g)(x) = f[g(x)]$ is defined in (a, b) and it is continuous at $x = c$.*

Proof. The assumptions have obviously been so chosen that $f[g(x)]$ be defined for x in (a, b). To prove continuity at $x = c$ we note that if $\varepsilon > 0$ is given, we can find first an η and then a δ such that

$$|f(X) - f(C)| \leqq \varepsilon \quad \text{if} \quad |X - C| \leqq \eta,$$

and

$$|g(x) - g(c)| \leqq \eta \quad \text{if} \quad |x - c| \leqq \delta.$$

It follows that

$$|f[g(x)] - f[g(c)]| \leqq \varepsilon \quad \text{if} \quad |x - c| \leqq \delta,$$

and, hence, that $f[g(x)]$ is continuous at $x = c$. This completes the proof.

We may also consider compositions involving three or more such functions. The analogue of Theorem 2.5.3 is valid for such compositions.

As an example let us take

$$f(X) = X^{\frac{1}{2}}, \quad 0 \leqq X, \quad g(x) = 1 - x^2, \quad -1 \leqq x \leqq 1.$$

When x is restricted in this manner, the range of g is the interval $[0, 1]$ which is a subset of the domain of f. We note that g is a polynomial in x and hence continuous for all x. The positive square root of X is that nonnegative number whose square is X. The existence of such a number and its uniqueness follow from the postulates of Section 1.4. Now the square root is clearly continuous to the right at $X = 0$, and continuous for any positive value C since

$$\sqrt{X} - \sqrt{C} = \frac{X - C}{\sqrt{X} + \sqrt{C}}$$

is small when $|X - C|$ is small. Note that the denominator exceeds $\sqrt{C}$ regardless of the value of $X \geqq 0$. Thus Theorem 2.5.3 applies and shows that

$$(1 - x^2)^{\frac{1}{2}}$$

is continuous in $[-1, 1]$.

Similar considerations apply to any fractional power of a rational function.

Finally we introduce the concepts of *oscillation of a function in an interval and at a point*. As the name indicates, these quantities measure, how the function varies. It is a fairly crude measure, but it suffices to distinguish between continuous and discontinuous functions.

DEFINITION 2.5.3. *Let f be defined in (a, b) and let I be a subinterval of (a, b). The oscillation of f in I is by definition*

$$\omega(f, I) = \sup_{x \in I} f(x) - \inf_{x \in I} f(x). \tag{2.5.3}$$

It should be observed that if f is unbounded in I then $\omega(f, I) = +\infty$. We can define the oscillation of f in a set S by simply replacing I by S in (2.5.3).

THEOREM 2.5.4. *Let $c \in (a, b)$, let $\delta < \min(c - a, b - c)$ and let I_δ be the interval*

$$\max(a, c - \delta) \leqq x \leqq \min(b, c + \delta).$$

Let $\omega(f, I_\delta)$ be defined as above. Then

$$\omega(f, c) = \lim_{\delta \to 0+0} \omega(f, I_\delta) \tag{2.5.4}$$

exists as a finite number or $+\infty$ and is called the oscillation of f at $x = c$.

Proof. There are two possibilities. Either $\omega(f, I_\delta) = +\infty$ for all admissible values of δ or there is a δ_0 such that $\omega(f, I_\delta)$ is finite for $0 < \delta < \delta_0$. In the first case we have used the limit symbol improperly, and (2.5.4) is to be understood as meaning that $\omega(f, c) = +\infty$. In the second case we have to show the existence of a proper finite nonnegative limit.

To this end we note that if $0 < \delta < \eta < \delta_0$, then $I_\delta \subset I_\eta$ so that $\omega(f, I_\delta) \leqq \omega(f, I_\eta)$, that is, shortening the interval does not increase the oscillation. This implies the existence of the desired limit by Theorem 2.7.1. If we do not want to anticipate this result, we can argue as follows. Let $\{\delta_n\}$ be a strictly decreasing null sequence and form the corresponding sequence $\{\omega(f, I_{\delta_n})\}$. It is nonincreasing as we have just seen, and by Theorem 1.7.1 it has a unique limit which we denote by L. If we take any other strictly decreasing null sequence $\{\eta_n\}$, then the corresponding sequence $\{\omega(f, I_{\eta_n})\}$ is also non-negative and nonincreasing, so it has a limit L_1. Using the two sequences $\{\delta_n\}$ and $\{\eta_n\}$ we form a new sequence $\{\zeta_n\}$ by intercalating the elements of the second sequence between those of the first, so that the third sequence is also strictly decreasing and contains the first and the second sequences as proper subsequences. Then $\{\omega(f, I_{\zeta_n})\}$ is also nonincreasing and has a limit L_2. But any subsequence of a convergent sequence has the same limit as the latter and $\{\omega(f, I_{\delta_n})\}$ is a subsequence of $\{\omega(f, I_{\zeta_n})\}$. It follows that $L = L_2$ and by the same token $L_1 = L_2$, so that all three limits are equal. This shows that

$$\lim_{n \to \infty} \omega(f, I_{\delta_n}) = L \tag{2.5.5}$$

for any choice of the null sequence $\{\delta_n\}$. We can regard $\omega(f, I_\delta)$ as a function of δ defined for $0 < \delta < \delta_0$, and by Theorem 2.4.1 the existence of (2.5.5) for any choice of $\{\delta_n\}$ implies the existence of (2.5.4) and that the desired limit is L. It is clear that Theorem 2.4.1 holds also for one-sided approach which is the case that we need. This completes the proof.

THEOREM 2.5.5. *f is continuous at $x = c$, if and only if $\omega(f, c) = 0$.*

Proof. The condition is necessary. For if f is continuous at $x = c$, then $|f(x) - f(c)| \leqq \varepsilon$ for $|x - c| \leqq \delta(\varepsilon)$ and this gives

$$\begin{aligned} |f(x_1) - f(x_2)| &= |[f(x_1) - f(c)] - [f(x_2) - f(c)]| \\ &\leqq |f(x_1) - f(c)| + |f(x_2) - f(c)| \leqq 2\varepsilon \end{aligned}$$

for x_1 and x_2 in $I_\delta = [c - \delta, c + \delta]$ and hence $\omega(f, I_\delta) \leqq 2\varepsilon$. Since ε is arbitrary, this requires $\omega(f, c) = 0$.

The condition is sufficient. For if it is satisfied, then $\omega(f, I_\delta) \leqq \varepsilon$ where I_δ is any interval of length $2\delta < 2\delta(\varepsilon)$ centered at c. In particular, $|f(x) - f(c)| \leqq \varepsilon$ for any x in I_δ. This shows that $f(x)$ is continuous at $x = c$. Thus the condition is also sufficient and the theorem is proved.

EXERCISE 2.5

The following functions are defined in (0, 1). Examine their properties of continuity. In what cases is it possible to define $f(0)$ and $f(1)$ by continuity and what are the corresponding values?

1. $\dfrac{x + 3}{x^2 + 1}$. 2. $(1 - x)^{\frac{1}{2}}$. 3. $(1 - x)^{-\frac{1}{2}}$. 4. $[x]$. 5. (x).

For the following functions determine the domain where the function is defined and is real. Discuss continuity.

6. $\dfrac{3x + 2}{x^3 - 8}$. 7. $\dfrac{1 + x^{\frac{1}{2}}}{1 - x^{\frac{1}{2}}}$. 8. $\dfrac{1 + (1 - x^2)^{\frac{1}{2}}}{1 - (1 - x^2)^{\frac{1}{2}}}$. 9. $\dfrac{x^{\frac{1}{2}}}{2 - x^{\frac{1}{2}}}$.

10. Prove that $x^{\frac{1}{4}}$ is continuous for $x \geqq 0$. [*Hint:* Use (1.6.7) with $n = 4$, $b^4 = x$, $a^4 = c$.]

11. If $x^{1/n}$ is continuous for $x \geqq 0$, how do you prove that $x^{m/n}$ has the same property? Here m and n are positive integers relatively prime to each other.

Discuss the continuity of:

12. $\left(\dfrac{1 + x^2}{1 - x^2}\right)^{\frac{1}{4}}$. 13. $\left(\dfrac{1 - x}{1 + x}\right)^{\frac{1}{5}}$.

Sketch the graphs and discuss the discontinuities of:

14. $[x] - [x - \frac{1}{2}]$. 15. $[2x] - 2[x]$.

16. A function $f(x)$ is defined to be 0, if x is irrational, and $\dfrac{1}{n}$, if $x = \dfrac{m}{n}$, where m and n are relatively prime. Show that $f(x)$ is continuous at each of the points of the first set but admits the points of the second set as discontinuities of the first kind.

17. A function f is defined by

$$f(x) = \lim_{n \to \infty} \frac{1}{1 + n^2x^2}.$$

Show that $f(x)$ is the characteristic function of the set $\{0\}$.

18. If $f(x)$ is $\left(\dfrac{1}{x}\right)$ for $x \neq 0$ and $f(0) = 0$, find $\omega(f, 0)$.

19. Let $f \in C[a, b]$, $a < c < b$, and let $f(c) > 0$. Prove the existence of an open interval (α, β), containing c, in which $f(x)$ is positive. If (α, β) is the largest such interval prove that if $a < \alpha$, then $f(\alpha) = 0$ and if $\beta < b$, then $f(\beta) = 0$.

20. Find the oscillation of x^2 in $(c - \delta, c + \delta)$. You may take $0 \leqq c$ but allow for the possibility that $c - \delta$ may be negative.

21. Same question for the function x^3.

22. If f and g are defined in (a, b) and if $I \subset (a, b)$ prove that

$$\omega(f + g, I) \leqq \omega(f, I) + \omega(g, I).$$

23. Prove also that

$$\omega(fg, I) \leqq \|f\|\,\omega(g, I) + \|g\|\,\omega(f, I).$$

Here $\|F\|$ denotes the supremum of $|F(x)|$ in (a, b). This quantity is known as the "sup norm" of F. [*Hint:* Use the identity

$$AB - CD = A(B - D) + D(A - C).]$$

24. Use Problem 23 together with Theorem 2.5.5 to give an alternate proof of the continuity of fg when f and g are continuous.

25. Determine the point set where the function $1 - 4\left(\dfrac{1}{x}\right)$ is positive.

26. The following is an elaboration of Problem 19. If f is continuous in (a, b) and takes on positive values somewhere in (a, b), then the set of points

x for which $f(x) > 0$ is the union of nonoverlapping open intervals not necessarily finite in number (as shown by the preceding problem).

27. Formulate a definition of continuity for functions on R_2 to R. You may suppose that $f(x, y)$ is defined in an open rectangle with sides parallel to the coordinate axes.

2.6. Properties of continuous functions. In this section we are concerned with properties of continuous functions on reals to reals. The properties depend in a decisive manner on the domain of definition. We shall concentrate on the case in which $D[f]$ is a finite closed interval. Our point of departure is Problem 19 of the preceding exercise which we restate in an equivalent form and prove as Lemma 2.6.1.

LEMMA 2.6.1. *If $f \in C[a, b]$ and $f(a) > 0 > f(b)$, then there is an α in $[a, b]$ such that $f(\alpha) = 0$.*

Proof. Since f is positive at $x = a$, we can find an interval $[a, \xi)$ in which f is positive. There is such an interval since $f(x) > f(a) - \varepsilon$ for $a \leqq x < a + \delta$ and since f is continuous in $[a, b]$. Consider the intervals $[a, \xi)$ in which f is positive. The set of points $\{\xi\}$ for which this holds is bounded and has a greatest lower bound $\alpha < b$. Now we cannot have $f(\alpha) > 0$, for then there would be an interval $[a, \xi)$ with $\alpha < \xi$ in which f is positive and this contradicts the definition of α. Similarly we cannot have $f(\alpha) < 0$, for then f would be negative in an interval extending to the left of α and this is again a contradiction. It follows that $f(\alpha) = 0$, and the lemma is proved.

We obtain the same conclusion if the hypothesis is changed to

$$f(a) < 0 < f(b).$$

The proof is the same.

From this simple proposition we get a surprising number of important consequences. We start with Theorem 2.6.1.

THEOREM 2.6.1. *If $f \in C[a, b]$ and if two numbers A and B belong to $R[f]$, where $A < B$, then every point of the interval $[A, B]$ belongs to $R[f]$.*

Proof. By assumption there are points x_1 and x_2 in $[a, b]$ such that

$$f(x_1) = A, \quad f(x_2) = B.$$

To clarify the ideas we may assume $a \leqq x_1 < x_2 \leqq b$. We now take a number C with $A < C < B$ and form the function

$$g(x) = f(x) - C.$$

The restriction of this function to the interval $[x_1, x_2]$ belongs to $C[x_1, x_2]$ and we can use the lemma with a, b replaced by x_1, x_2, respectively. The function $g(x)$ is positive at $x = x_2$ and negative at $x = x_1$. By the lemma there exists an x_3 such that $g(x_3) = 0$, that is, where

$$f(x_3) = C.$$

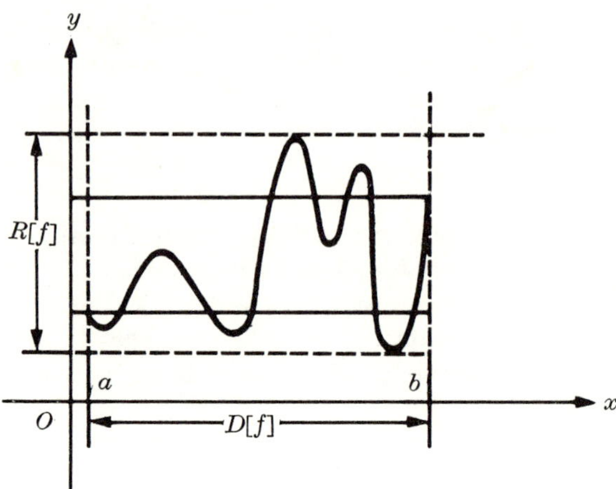

Figure 2.18

Moreover, the proof of the lemma shows that x_3 lies between x_1 and x_2. This completes the proof of the theorem.

COROLLARY 1. *If $f(a) \neq f(b)$ then $f(x)$ assumes every value between $f(a)$ and $f(b)$.*

COROLLARY 2. *The range $R[f]$ is an interval.*

We refer to Figure 2.18 for an illustration of this theorem and its corollaries.

We do not yet know if the range is bounded, and if so, if the interval is open or closed or half-open. Both questions are fairly easy to settle.

THEOREM 2.6.2. *The range $R[f]$ is a bounded closed interval, if $f \in C[a, b]$.*

Proof. We prove first that $R[f]$ is a bounded set. The argument is indirect. Suppose that $R[f]$ has no finite least upper bound. This means that f takes on arbitrarily large positive values and this implies that f takes on all large values since $R[f]$ is an interval. In particular, f takes on all large positive integers, say $n \geqq N$. We can then find a sequence $\{x_n \mid n \geqq N\}$ of points in $[a, b]$ such that $f(x_n) = n$. This is a bounded infinite point set and has at least one cluster point x_0 by the Bolzano-Weierstrass theorem. We can then find a subsequence $\{x_{n_k}\}$ such that

$$\lim_{k \to \infty} x_{n_k} = x_0.$$

The point x_0 belongs to the closed interval $[a, b]$ and $f(x)$ is continuous at $x = x_0$. We choose $\varepsilon = \frac{1}{3}$ and can then find a corresponding δ such that for $|x - x_0| \leqq \delta$ we have $|f(x) - f(x_0)| \leqq \frac{1}{3}$. We can also choose an integer p such that $|x_0 - x_{n_k}| \leqq \delta$ for $k \geqq p$. For such values of k

$$|f(x_{n_k}) - f(x_0)| \leqq \tfrac{1}{3},$$

or

$$|n_k - f(x_0)| \leqq \tfrac{1}{3}.$$

Such an inequality, however, can hold for at most one integer n_k and not for infinitely many. This contradiction shows that $R[f]$ must be bounded above and the same type of argument shows that $R[f]$ is bounded below. Thus $R[f]$ is a bounded interval.

It remains to prove that this interval is closed. Let its left endpoint be m, its right M. We have to prove the existence of points ξ_1 and ξ_2 such that $f(\xi_1) = m$, $f(\xi_2) = M$. The same method applies in both cases so it is enough to prove, say the second one.

Since M in any case is the supremum of $R[f]$, there exists a sequence $\{x_n\}$ of points in $[a, b]$ such that

$$f(x_{n-1}) < f(x_n) \quad \text{and} \quad \lim_{n\to\infty} f(x_n) = M.$$

This is a bounded infinite sequence of points x_n so there is at least one cluster point ξ_2. There is at least one convergent subsequence, but we may assume that we have already eliminated the points that we cannot use, so that

$$\lim_{n\to\infty} x_n = \xi_2.$$

Since $f(x)$ is continuous at $x = \xi_2$, we have

$$|f(\xi_2) - f(x_n)| \leqq \varepsilon \quad \text{for} \quad n \geqq N(\varepsilon).$$

But $x_n \to \xi_2$ and $f(x_n) \to M$, so we must have

$$f(\xi_2) = M.$$

In the same way we prove that there is a point ξ_1 in $[a, b]$ where $f(\xi_1) = m$. This completes the proof.

What we have just proved is so important that we restate part of it as a

COROLLARY. *A function which is continuous in a finite closed interval takes on both its supremum and its infimum in the interval.*

It is important to realize that the properties which we have proved may well be lost if we modify the assumptions made on $D[f]$.

(1) *Suppose that* $D[f]$ *is not connected*. As an example we may take $D[f]$ to be the union of two disjoint closed intervals $[a, b]$ and $[c, d]$. We give f the constant value A in the first interval and the value B in the second. The range of this continuous function consists of two values, A and B, and Theorem 2.6.1 does not hold.

(2) *Suppose that* $D[f]$ *is open*. To illustrate what may happen we take the interval $(0, 1)$ and the function $\dfrac{1}{x}$ which is continuous there. Here Theorem 2.6.1 does not hold, this function is not bounded, and it assumes neither its supremum nor its infimum in $(0, 1)$.

The last important property to study is that of *uniform continuity*. We start with

DEFINITION 2.6.1. *$f(x)$ is uniformly continuous in $[a, b]$ if given any $\varepsilon > 0$, there exists a $\delta = \delta(\varepsilon)$ such that for any points x_1 and x_2 in $[a, b]$ the inequality*

$$|x_1 - x_2| \leqq \delta \quad \textit{implies} \quad |f(x_1) - f(x_2)| \leqq \varepsilon. \tag{2.6.1}$$

We note that *if f is uniformly continuous in $[a, b]$, then f is continuous in $[a, b]$*, that is, $f \in C[a, b]$. We shall prove that the converse is also true. Again this basic property depends upon the domain of definition. A function which is continuous in an open interval is not necessarily uniformly continuous. Counter-examples will be given below.

Uniform continuity is easily expressed in terms of oscillation. The definition implies that $\omega(f, I)$ does not exceed ε if the length $l(I)$ of I does not exceed $\delta(\varepsilon)$ regardless of where I is placed in the interval $[a, b]$. Or, we can say that the inequality $\omega(f, I) \leqq \varepsilon$ holds uniformly with respect to all intervals I, provided $l(I) \leqq \delta(\varepsilon)$.

The term "uniform" will occur in other contexts in later chapters. It indicates that the property under consideration holds for all values of whatever is variable in the situation on hand. In the present case it is the placement of the interval I which is variable, and we want the oscillation in I to be small wherever I is placed.

Figure 2.19 gives a geometric characterization of the situation. A small rectangle of height 2ε and width 2δ is centered at a point on the curve. We have uniform continuity, if we can choose the width of the rectangle, whenever its height is given, so that while the rectangle slides along the curve remaining parallel to itself, the curve projects from the vertical sides but does not penetrate the top or the bottom of the rectangle. This rectangle is a sort of gauge, measuring the uniformity of the "product," in this case the graph $(x, f(x))$. It is clear that a particular gauge is apt to get stuck if the curve is very steep, or oscillates rapidly, or is unbounded.

THEOREM 2.6.3. *If $f \in C[a, b]$, then f is uniformly continuous in the interval $[a, b]$.*

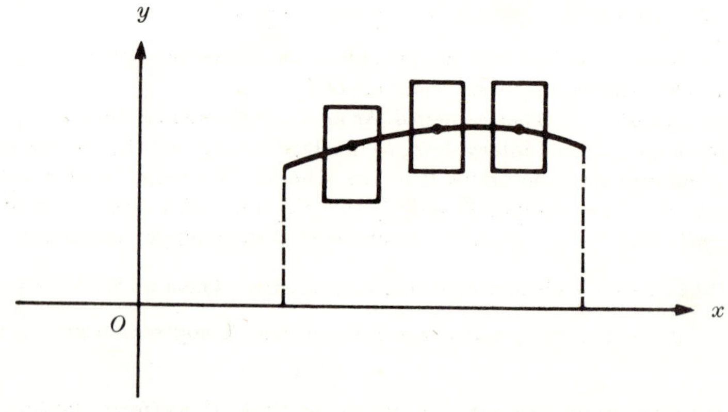

Figure 2.19

This result is an immediate consequence of the following

THEOREM 2.6.4. *If $f \in C[a, b]$, $b - a < \infty$, and if $\varepsilon > 0$ is given, then there exists a subdivision of $[a, b]$ into a finite number of equal parts in none of which the oscillation of f exceeds ε.*

Proof. We give an indirect proof, that is, we assume that there is a function $f \in C[a, b]$ and an $\varepsilon > 0$ for which no finite partition of the desired type exists, and we show that this assumption leads to a contradiction. If f is such a function, then subdividing $I = [a, b]$ into two equal parts will not improve matters. The partition property may possibly hold in one of the subintervals but not in both for then it would also hold in I. Suppose that I_1 is one of the subintervals in which the partition property does not hold. We then subdivide I_1 into two equal parts and pick one of the subintervals I_2 in which the partition property fails. Proceeding in this manner we get an infinite sequence of nested intervals $\{I_n\}$ in each of which the partition property fails to hold. In particular, this implies that $\omega(f, I_n) > \varepsilon$ for every n. Now the length of I_n is $2^{-n}(b - a)$ which tends to 0. By Theorem 1.5.1 there is one and only one point x_0 common to all the closed intervals I_n and $x_0 \in [a, b]$. By Theorem 2.5.4 this implies that $\omega(f, x_0) \geqq \varepsilon$ and this in turn means that f is discontinuous at $x = x_0$. This conclusion is also valid in case x_0 should be one of the points a or b. But $f \in C[a, b]$ and cannot have any discontinuities. Thus our assumption that Theorem 2.6.4 does not hold has led to a contradiction. It follows that this theorem must be true and this also implies the validity of Theorem 2.6.3. This completes the proof.

Examples of continuous functions, defined in an open interval, which are not uniformly continuous, are easy to find. We merely take a function defined and continuous in an interval (a, b) such that there is no way of extending the definition to the closed interval $[a, b]$ and obtaining a continuous function by suitable choice of $f(a)$ and $f(b)$. Two functions with such properties used earlier are

$$\frac{1}{x} \quad \text{and} \quad \left(\frac{1}{x}\right), \qquad 0 < x < 1. \tag{2.6.2}$$

In the first case, suppose that $0 < x_1 < x_2 < 1$. Then

$$\frac{1}{x_1} - \frac{1}{x_2} = \frac{x_2 - x_1}{x_1 x_2} > \frac{x_2 - x_1}{(x_2)^2}.$$

This shows that we cannot make $\omega(f, I)$ small merely by making $l(I)$ small. The location of I is essential. If a is a fixed positive number and if I is located in $[a, 1]$, then $\omega(f, I) < \varepsilon$ provided $l(I) < \varepsilon a^2$, that is, we have uniform continuity in $[a, 1]$. But this does not extend to $(0, 1)$. The reason for the failure in this case is that f is unbounded in $(0, 1)$.

The second function in (2.6.2) is of a different nature. We recall that (t) is the distance of t from the nearest integer. The function is bounded in $(0, 1)$, but if I is an interval of the form $(0, a)$, then $\omega(f, I) = \frac{1}{2}$ no matter how small a is. This is true since f oscillates infinitely often between 0 and $\frac{1}{2}$ in such an interval. This prevents uniform continuity in $(0, 1)$. Naturally we have uniform continuity in $[a, 1]$, for this is a closed interval and f is continuous there.

In connection with uniform continuity the notion of the *modulus of continuity* of f plays an important role. We denote it by $\mu(\delta, f)$ and refer to

DEFINITION 2.6.2. *If $f \in C[a, b]$, we set*

$$\mu(\delta, f) = \sup \omega(f, I) \tag{2.6.3}$$

for all intervals I of length $\leqq \delta$ and located in $[a, b]$.

It may be shown that this is a continuous nondecreasing function of δ which tends to the limit 0 as δ decreases to 0. The least value of the quantity $\delta(\varepsilon)$ which figures in Definition 2.6.1 is simply the root of the equation

$$\mu(\delta, f) = \varepsilon. \tag{2.6.4}$$

Further properties of the modulus of continuity can be found among the Miscellaneous Problems for Chapter 2.

EXERCISE 2.6

1. If f is continuous in an open interval and f is not a constant, show that $R[f]$ is an interval.
2. Show by examples that this interval need not be finite or open at either end if it is finite.
3. If $f \in C[a, b]$, if $f(a) < 1$, and $f(b) > 1$, then there is a point c, $a < c < b$, where $f(c) = 0$. True or false?
4. Show that a cubic equation

$$x^3 + a_1x^2 + a_2x + a_3 = 0$$

with real coefficients has at least one real root. Generalize to equations of higher order. Does the parity of the degree matter?
5. The equation $x^3 - 3x + 1 = 0$ has three real roots. Find the integral part $[x]$ of each root.
6. If $y = x^3 + x - 1$, show that the pair (y, x) defines a function with domain and range R. Show that to each y in R there is one and only one x. Prove that if $y_1 > y_2$, then $y_1 - y_2 > x_1 - x_2 > 0$. Does this prove that x is a continuous function of y?
7. Prove that the modulus of continuity of x^2 in $[0, 1]$ is $1 - (1 - \delta)^2$, $0 \leqq \delta \leqq 1$. Determine $\mu(\delta, f)$ for $f(x) = x^n$.
8. Prove that the modulus of continuity of $x^{\frac{1}{2}}$ in $[0, 1]$ is $\delta^{\frac{1}{2}}$. Determine $\mu(\delta, f)$ for $x^{1/n}$.
9. Prove that $|f(x)|$ is continuous wherever $f(x)$ is.

2.7. Monotone functions. We have found the notion of monotone sequences very useful, and shall now consider *monotone functions.* The motivation for our study is that most functions encountered in the Calculus are at least piecewise monotone, that is, we can divide the (finite) interval of definition of f into a finite number of subintervals in each of which the function is monotone. Whenever we draw a graph of a function as an illustration, the graph usually shows a piecewise monotone function.

DEFINITION 2.7.1. *Let f be a function on reals to reals defined on an interval I which may be open at either end. If for each choice of x_1 and x_2 in I, the inequality $x_1 < x_2$ implies that $f(x_1) \leqq f(x_2)$, then f is said to be nondecreasing in I. It is nonincreasing if $f(x_1) \geqq f(x_2)$ instead. A function having either of these properties in I is said to be monotone or monotonic in I.*

If $x_1 < x_2$ always implies that $f(x_1) < f(x_2)$, we speak of a "strictly increasing" function. Similarly, if $f(x_1) > f(x_2)$ holds instead, f is "strictly decreasing." Figure 2.20 depicts a nondecreasing function. Here there is an interval of constancy and also two subintervals in which f is strictly increasing. Further we have exhibited a point of discontinuity.

The function x^2 is strictly decreasing for $-\infty < x \leqq 0$, and strictly increasing in $[0, \infty)$. It is a piecewise monotone function. The function (x), so often used, is piecewise monotone in any finite interval, while $[x]$ is nondecreasing.

Some of these examples show that a monotone function need not be continuous everywhere, but, as we shall see, the points of discontinuity are of a rather special nature.

THEOREM 2.7.1. *A monotone function has right- and left-hand limits everywhere in I. There are one-sided limits also at the endpoints of I but these may be infinite unless I is closed.*

Proof. To clarify the ideas, suppose that f is nondecreasing and let $c \in I = (a, b)$ with $a < c < b$. Consider the range of f, restricted to the open interval (a, c). This range has a least upper bound α. Similarly the range of f, restricted

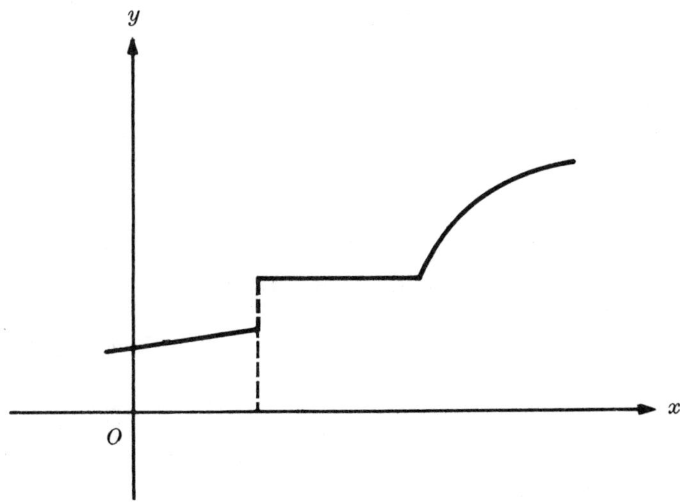

Figure 2.20

to the open interval (c, b), has a greatest lower bound β and

$$\alpha \leqq f(c) \leqq \beta.$$

Now if $\{x_n\}$ is a strictly increasing sequence converging to c, the sequence $\{f(x_n)\}$ is nondecreasing and has a limit $\leqq \alpha$. On the other hand, from the definition of the least upper bound it follows that there must be such sequences which converge to α and the monotone character of the function then implies that every such sequence converges to α. This implies the existence of $f(c-0)$, and in the same manner we prove that $f(c+0)$ exists and we have

$$-\infty < \alpha = f(c-0) \leqq f(c) \leqq f(c+0) = \beta < +\infty. \tag{2.7.1}$$

The same argument shows the existence of $f(a+0)$ and $f(b-0)$, but if $f(a)$ and $f(b)$ are not defined, these limits may be infinite.

The oscillation of monotone functions has remarkable additive properties. Suppose that $I = [a, b]$ and f is nondecreasing. Then $f(a)$ and $f(b)$ are finite numbers and

$$\omega(f, I) = f(b) - f(a). \tag{2.7.2}$$

This holds for any closed subinterval $I_0 = [a_0, b_0]$, that is,

$$\omega(f, I_0) = f(b_0) - f(a_0).$$

Suppose now that I is the union of nonoverlapping intervals

$$I_k = [c_{k-1}, c_k], \quad c_0 = a, \quad c_n = b, \quad k = 1, 2, \cdots, n.$$

Since

$$f(b) - f(a) = f(b) - f(c_{n-1}) + f(c_{n-1}) - f(c_{n-2}) + \cdots + f(c_1) - f(a),$$

we have, reversing the order,

$$\omega(f, I) = \omega(f, I_1) + \omega(f, I_2) + \cdots + \omega(f, I_n). \tag{2.7.3}$$

Here all terms are nonnegative. It follows that if some intervals I_k^* are simply nonoverlapping, but do not necessarily exhaust I, then

$$\omega(f, I_1^*) + \cdots + \omega(f, I_m^*) \leqq \omega(f, I). \tag{2.7.4}$$

Suppose that a_k is an interior point of I_k^*. We have then

$$f(a_k + 0) - f(a_k - 0) = \omega(f, a_k) \leqq \omega(f, I_k^*). \tag{2.7.5}$$

Combining this with the preceding inequality we get

$$\omega(f, a_1) + \cdots + \omega(f, a_n) \leqq \omega(f, I). \tag{2.7.6}$$

This holds for any choice of the points a_k. It states that *the sum of the oscillations of f at points in I does not exceed the oscillation of f in I*. We recall that f is discontinuous at $x = a$, if and only if $\omega(f, a) > 0$. We note that (2.7.6) holds also if f is nonincreasing.

The discontinuities of f are all of the first kind by Theorem 2.7.1. Formula (2.7.6) imposes another important restriction on the discontinuities. We can have infinitely many points of discontinuity in $[a, b]$, but (2.7.6) implies that we can enumerate them, say in the order of decreasing value of the oscillation. We shall not carry out this in detail. We merely note that if $\varepsilon > 0$ is given and if $\omega(f, I) = \omega$, then the number of points where the oscillation exceeds ε cannot be greater than ω/ε. This provides a basis for the enumeration.

As an example of a nondecreasing function with infinitely many discontinuities in a finite interval, we may take the step function

$$f(x) = 1 - 2^{-n+1}, \quad 1 - 2^{-n+1} \leqq x < 1 - 2^{-n}, \quad n = 1, 2, 3, \cdots, f(1) = 1.$$

We shall now examine what happens to the monotonic properties under algebraic operations.

THEOREM 2.7.2. *If f and g are nondecreasing functions in $[a, b]$, then so is $f + g$ and $\alpha f + \beta g$, provided α and β are $\geqq 0$. If, in addition, f and g are positive, then their product is also nondecreasing. If f and g are positive, if f is nondecreasing and g is nonincreasing then f/g is nondecreasing. These statements remain true when "increasing" and "decreasing" are interchanged throughout.*

Proof. If $x_1 < x_2$ implies that

$$f(x_1) \leqq f(x_2) \quad \text{and} \quad g(x_1) \leqq g(x_2),$$

then (see Section 1.6) we may add the inequalities, obtaining

$$f(x_1) + g(x_1) \leqq f(x_2) + g(x_2).$$

We may multiply by the positive numbers α and β before adding to obtain the more general inequality. If f and g are positive, then the inequalities may also be multiplied which gives

$$f(x_1)g(x_1) \leqq f(x_2)g(x_2).$$

This proves the first two statements.

If f and g are positive in I, then f/g is defined there. If $x_1 < x_2$ implies that

$$f(x_1) \leqq f(x_2) \quad \text{and} \quad g(x_1) \geqq g(x_2),$$

then the second inequality gives

$$\frac{1}{g(x_1)} \leqq \frac{1}{g(x_2)}.$$

We can now multiply by the first inequality and obtain

$$\frac{f}{g}(x_1) \leqq \frac{f}{g}(x_2)$$

as asserted. The reversal of the inequalities is obvious. This completes the proof.

The functions of the Calculus usually have a much stronger property than piecewise monotony: they are also *piecewise convex or concave. In an interval where the function f is convex* (downward) *the graph always lies below or on the chord*. See Figure 2.21. Analytically this property is expressed by

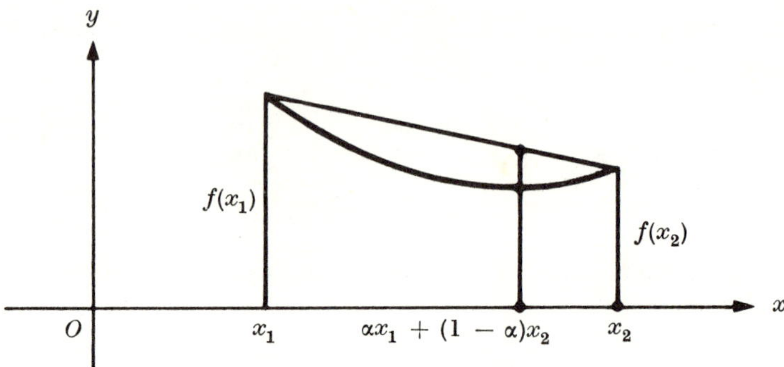

Figure 2.21

DEFINITION 2.7.2. *A function f on reals to reals is convex in $[a, b]$ if for any choice of x_1 and x_2, $a \leqq x_1 < x_2 \leqq b$, and any α with $0 < \alpha < 1$, we have*

$$f(\alpha x_1 + (1 - \alpha)x_2) \leqq \alpha f(x_1) + (1 - \alpha)f(x_2). \tag{2.7.7}$$

f is concave in $[a, b]$ if $-f$ is convex.

It may be proved that if f is bounded above in some subinterval of $[a, b]$ and convex in $[a, b]$, then f is continuous in $[a, b]$.

The function x^n is convex in $[-b, b]$, b arbitrary, if n is an even integer; it is convex in $[0, b]$ and concave in $[-b, 0]$ if n is odd.

EXERCISE 2.7

In what intervals are the following functions strictly increasing or decreasing:

1. $x^2 + 2x$. 2. $x^3 + 3x$. 3. $x + \dfrac{1}{x}$. 4. $x - \dfrac{1}{x}$. 5. $(1 - x^2)^{\frac{1}{2}}$?

6. Given that $f(x)$, defined in $(-a, a)$, is even if $f(-x) = f(x)$ and odd if $f(-x) = -f(x)$, which of the functions 1 to 5 are even and which are odd?

7. Prove that if f is odd and monotonic in $(0, a)$, then it is also monotonic in $(-a, 0)$ and it varies in the same sense in both intervals. Is the function necessarily monotonic in $(-a, a)$? Prove that this is true if f is continuous at $x = 0$. What is the value of $f(0)$ in the latter case?

8. Suppose that f is even and monotonic in $(0, a)$. What can be said of f in $(-a, 0)$?

9. Prove that the function

$$\frac{1}{x} + \frac{1}{x - 1}$$

is monotone in $(0, 1)$ and takes on every real value once and only once there.

10. Do Theorems 2.6.1 and 2.6.2 hold for strictly increasing functions? If not, give counter-examples.

11. Prove directly that x^2 satisfies the inequality (2.7.7). [*Hint:* Use (1.6.14).]

12. Examine the convexity properties of the function in Problem 1.

13. Discuss the monotonic properties of fg if (i) f is positive and nondecreasing while g is negative and nonincreasing; (ii) f and g are negative and nondecreasing.

14. Discuss the monotonic properties of the circle product $f \circ g = f(g)$ assuming f and g to be monotonic, possibly in opposite sense, for the values under consideration.

2.8. Inverse functions. We have defined a function f on X to Y as a set of ordered pairs

$$\{(x, y) \mid x_0 \in X_0, \quad y_0 \in Y_0\}, \tag{2.8.1}$$

such that no two pairs have the same first element x. Here $X_0 \equiv D[f]$ is the domain of f in X and $Y_0 = R[f]$ is the range of f in Y. To every point $y_0 \in Y_0$ there is at least one corresponding point $x_0 \in X_0$ such that $y_0 = f(x_0)$ and, normally, more than one such point. Here we are interested in the case where the mapping is one-to-one.

DEFINITION 2.8.1. *If the mapping defined by f of X_0 onto Y_0 is one-to-one, then the inverse function of f, denoted by $\check{f}$ (read "f inverse"), is defined by*

$$\{(y, x) \mid y \in Y_0, x \in X_0\}. \tag{2.8.2}$$

Remark. American and British authors normally use the notation f^{-1} for the inverse of f, but continental European authors have usually avoided this practice and not without reason. In the case of numerically valued functions, the case of main interest to us, it is necessary to avoid confusion between the reciprocal of a function and the inverse of a function. Moreover, for such functions we have the general problem of defining powers of a function and this means that we must make a choice between different function algebras over the reals (or over the complex numbers as the case may be). We have encountered two such algebras, one based on ordinary multiplication for the product operation and the other based on the circle product. The former algebra is commutative, the product can always be formed, and $f(x) \equiv 1$ is the unit element. The second algebra is nondistributive and noncommutative, the product operation severely limits the admissible elements, and $f(x) \equiv x$ is the unit element.

In the former algebra it is natural to define f^n by

$$f^n(x) = [f(x)]^n \tag{2.8.3}$$

for all n, and this gives

$$f^{-1}(x) = [f(x)]^{-1}, \tag{2.8.4}$$

the reciprocal of f. The domain of f^{-1} is then the subset of $D[f]$ where $f(x) \neq 0$. In the circle product algebra, on the other hand, we have

$$f^k(x) = (f \circ f \circ \cdots \circ f)(x), \quad k \geqq 1, \tag{2.8.5}$$

where there are k factors in the symbol on the right. We can then try to define f^{-k} by

$$f^{-k}[f^k(x)] = x. \tag{2.8.6}$$

We prefer to define the powers f^n by (2.8.3) and to use f^{-1} for the reciprocal of f. This requires a different notation for the inverse and we have settled for $\check{f}$ which does not carry other connotations.

THEOREM 2.8.1. *The domain of $\check{f}$ is the range of f and vice versa. More precisely,*

$$\check{f}[f(x)] = x \quad \text{for all } x \text{ in } D[f], \tag{2.8.7}$$

$$f[\check{f}(y)] = y \quad \text{for all } y \text{ in } R[f]. \tag{2.8.8}$$

This is simply a paraphrase of the definition.

So far nothing has been said about the nature of X and Y. As indicated above, however, we are mainly interested in $X = Y = R$. We restrict ourselves to the case: $D[f]$ is an interval I, but we allow $R[f]$ to be a more general set than an interval. If the mapping is one-to-one, then $x_1 \neq x_2$ implies $f(x_1) \neq f(x_2)$. Suppose that $I = [a, b]$ and $a \leqq x_1 < x_2 \leqq b$. Then we have either

$$f(x_1) < f(x_2) \quad \text{or} \quad f(x_1) > f(x_2).$$

Actually there are several possibilities. The inequality $x_1 < x_2$ may imply

$$f(x_1) < f(x_2) \quad \text{for all } x_1 \text{ and } x_2, \tag{2.8.9}$$

or

$$f(x_1) > f(x_2) \quad \text{for all } x_1 \text{ and } x_2, \tag{2.8.10}$$

or, finally, the first inequality for some choices of x_1, x_2 and the second for others. The third possibility cannot be excluded. In Figure 2.22 is shown a

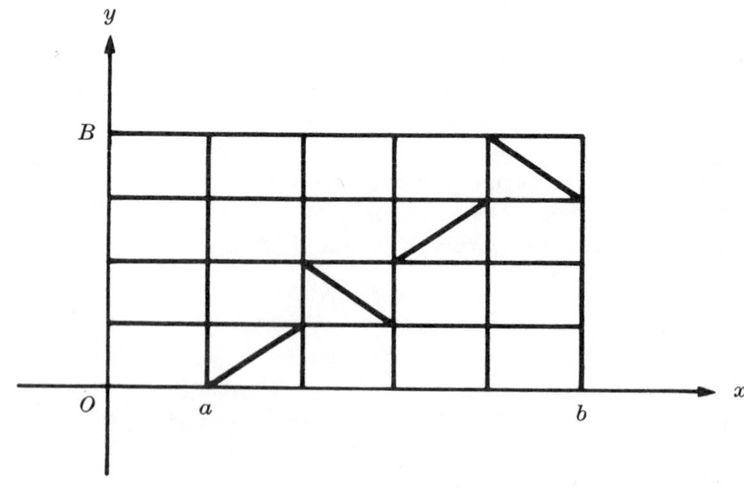

Figure 2.22

function whose domain and range are both finite intervals, the mapping is one-to-one, but f is not monotone in $[a, b]$ though it has this property in each of the four subintervals. It is clear that this example can be modified so as to obtain as many distinct subintervals of monotony as desired, even infinitely many. This function is discontinuous and, in fact, such a phenomenon cannot occur for continuous functions.

THEOREM 2.8.2. *If f is continuous in $[a, b]$, and if $a \leqq x_1 < x_2 \leqq b$ implies $f(x_1) \neq f(x_2)$, then f is strictly monotone in $[a, b]$, that is, either* (2.8.9) *or* (2.8.10) *holds.*

Proof. This is essentially a consequence of Theorem 2.6.1. Since $f(a) \neq f(b)$, we have either $f(a) < f(b)$ or the reversed inequality. We shall show that

$$f(a) < f(b) \tag{2.8.11}$$

implies that (2.8.9) holds. Suppose, contrariwise, that we could find $x_1 < x_2$ such that $f(x_1) > f(x_2)$. We have then necessarily $a < x_1 < x_2 < b$. Various inequalities can hold between the four quantities $f(a), f(x_1), f(x_2), f(b)$ in addition to those already listed. First we note that if

$$f(x_2) < f(a),$$

then $f(x)$ would have to assume the value $f(a)$ somewhere between x_2 and b. This is impossible since $f(x) \neq f(a)$ for $a < x \leqq b$. In the same way we exclude

$$f(x_1) < f(a), \quad f(b) < f(x_2), \quad \text{and} \quad f(b) < f(x_1).$$

This leaves us with

$$f(a) < f(x_2) < f(x_1) < f(b).$$

But this is also impossible, for it would require that $f(x)$ take the value $f(x_1)$ somewhere between x_2 and b. This contradiction shows that (2.8.11) implies (2.8.9). The reversed inequality $f(a) > f(b)$ is handled in the same way. This completes the proof.

It remains to discuss the continuity properties of the inverse function. In this direction we have

THEOREM 2.8.3. *If f is continuous and strictly increasing (decreasing) in $[a, b]$, and if $f(a) = A$, $f(b) = B$, then $\check{f}$ is defined, continuous, and strictly increasing in $[A, B]$ (decreasing in $[B, A]$).*

Proof. The existence and monotonic properties of $\check{f}$ follow from the preceding discussion. We have only to prove that $\check{f}$ is continuous. To clarify the ideas, suppose that f is strictly increasing in (a, b). This implies that $\check{f}$ is strictly increasing in $[A, B]$. From this it follows that $\check{f}$ has right- and left-hand limits at all interior points of $[A, B]$. Thus if $A < C < B$, then

$$\check{f}(C - 0) \leqq \check{f}(C) \leqq \check{f}(C + 0).$$

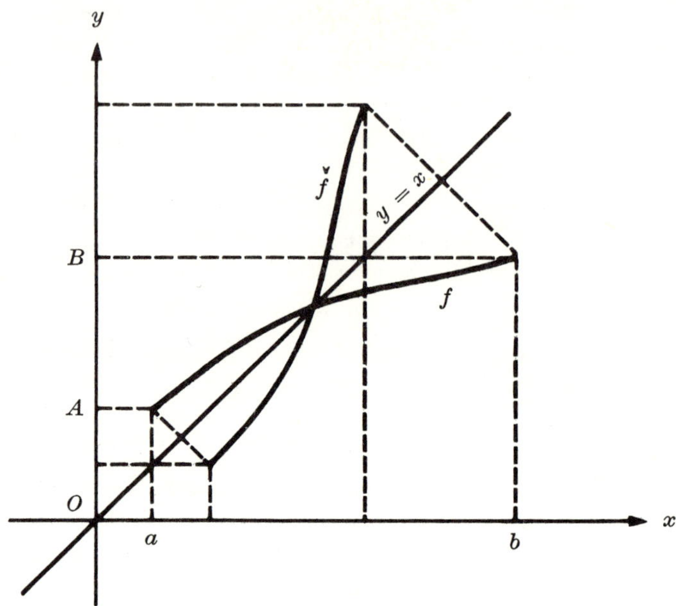

Figure 2.23

It is required to prove that equality holds in both places. Suppose, contrariwise, that

$$\check{f}(C+0) - \check{f}(C) = \delta > 0.$$

By the definition of $\check{f}$, we can find a value c, $a < c < b$, such that

$$\check{f}(C) = c;$$

moreover, this value c is unique. Hence, for any $\varepsilon > 0$,

$$\check{f}(C+\varepsilon) \geqq \check{f}(C+0) = \check{f}(C) + \delta = c + \delta.$$

From this it follows that

$$f[\check{f}(C+\varepsilon)] \geqq f(c+\delta),$$

or

$$C + \varepsilon \geqq f(c+\delta)$$

for every $\varepsilon > 0$. This would give

$$f(c) = C \geqq f(c+\delta),$$

and this is absurd since f is strictly increasing. Hence $\delta = 0$ and $\check{f}(C+0) = \check{f}(C)$. This argument also applies at $y = A$. In the same manner we prove continuity on the left at all points of $(A, B]$. This completes the proof.

There is a simple way of constructing the graph of the inverse function from the graph of the given function. See Figure 2.23. This consists in *reflecting the graph of* $y = f(x)$ *in the line* $y = x$. This depends upon the simple and well-known fact that the points $P = (x, y)$ and $Q = (y, x)$ are symmetric with respect to this line. This is what we mean by saying that Q is obtained from P by reflection in the line. Thus, if $(c, f(c))$ is a point on the graph of f, and if $f(c) = C$, then $(C, \check{f}(C))$ is a point on the graph of $\check{f}$ and, clearly, all points of the graph of $\check{f}$ are obtainable by this method.

EXERCISE 2.8

Five functions are defined on $[0, 1]$ by:

1. x^2. 2. $(1 - 2x)^3$. 3. $(2 - x)^4$. 4. $\dfrac{x-1}{x+1}$. 5. $\dfrac{1}{1+x} + x$.

Find the inverse functions and their domains of definition.

6. What is the maximum domain of definition of each of the five inverse functions in the preceding problems when the restriction of x to the interval $[0, 1]$ is dropped?

7. The function $-y^{\frac{1}{2}}$, $0 \leqq y$, is also an inverse of x^2. What is its range? Determine if the functions in Problems 2–5 have inverses other than those found above and, if so, their ranges.

8. If

$$f(x) = \left\{1 + \frac{x}{n}\right\}^n,$$

find $\check{f}(y)$.

9. In Figure 2.22 let $a = 3$, $b = 15$, $B = 8$. The domain and the range are each divided into four equal parts and the graph is made up of straight line segments. Give equations representing $f(x)$ in the various subintervals. Be careful with the definition of $f(x)$ at the points of discontinuity in order that the inverse exist. Express the inverse analytically.

10. If $y = (1 - x^2)^{\frac{1}{2}}$, $0 \leqq x \leqq 1$, find the inverse function and determine its domain of definition.

11. We define $f(x)$ to be x if x is rational, otherwise $-x$. Does this function have an inverse?

12. Is the product of two strictly increasing functions also strictly increasing?

13. Given that f is continuous with domain $[a, b]$ and range $[A, B]$. It is known that f has an inverse, does it follow that f^2 has the same property? Here $f^2(x) = [f(x)]^2$.

14. Use the theory of inverse functions to prove that there exists a unique function $y = f(x)$ with domain and range R, which satisfies the equation $y^3 + 3y + x = 0$.

15. Same question for the equation $y^5 + 5y + x = 0$.

MISCELLANEOUS PROBLEMS FOR CHAPTER 2

1. For $0 < x$ a sequence of functions $\{f_n(x)\}$ is defined by

$$f_1(x) = x^{\frac{1}{2}}, \quad f_{n+1}(x) = [x + f_n(x)]^{\frac{1}{2}}, \quad n = 1, 2, 3, \cdots.$$

Prove that each f_n is a strictly increasing function of x. Prove also that for each fixed x the sequence $\{f_n(x)\}$ is strictly increasing as a function of n and is bounded by $f_1(x) + 1$. Find its limit $f(x)$ and show that f is strictly increasing as a function of x.

2. Find the inverse of the function f of the preceding problem.

3. Given

$$S_n(x) = 1 + 2x + 3x^2 + \cdots + (n + 1)x^n,$$

verify that

$$(1 - x)^2 S_n(x) = 1 - (n + 2)x^{n+1} + (n + 1)x^{n+2}.$$

Show that $S_n(x)$ tends to a limit as n becomes infinite, if and only if $|x| < 1$. Find the limit. $S_n(x)$ is the nth partial sum of a certain power series which is the square of the geometric series.

4. If f is defined for all x in R by

$$f(x) = \max\{x, 1 - x, 2x - 3\},$$

plot $y = f(x)$ and prove that f is continuous and convex.

5. Express this f in the form

$$f(x) = Ax + B + a\,|x - \tfrac{1}{2}| + b\,|x - 3|.$$

6. If a function f is continuous and convex in $[a, b]$, where does it reach its maximum?

7. If f is continuous and convex in $[a, b]$, and if f reaches its minimum value A for $x = x_1$ and $x = x_2$, $x_1 < x_2$, prove that $f(x) = A$ for $x_1 \leqq x \leqq x_2$.

8. If f and g are convex in $[a, b]$ so is their sum. Why? Show by examples that their product may be convex or concave or piecewise one or the other.

9. If $f \in C[a, b]$, prove that f^+ defined by

$$f^+(x) = \max\{f(x), 0\}$$

also does.

10. If f and $g \in C[a, b]$, prove that h defined by

$$h(x) = \max\{f(x), g(x)\}$$

also does.

11. In what intervals is the function $3x - x^3$ (a) convex; (b) concave?

12. Given an interval I: $[a, 2a]$, $a > 0$; how small must a be in order that the oscillation of $1/x$ in I will exceed a given positive number ω?

13. Given

$$f(x) = \frac{1}{x} + \frac{1}{x - 1}, \quad x \neq 0, 1,$$

where is f strictly increasing, strictly decreasing? Is f decreasing in the set $(-1, 0) \cup (0, 1)$?

14. Find the inverse function $x = \check{f}(y)$ if x is restricted to $(0, 1)$ and f is defined as in Problem 13.

15. What other inverses exist and what are their ranges?

16. Given

$$g(x) = \frac{1}{x} + \frac{1}{x-1} + \frac{1}{x-2}, \quad x \neq 0, 1, 2,$$

show that g is negative when $x < 0$ and that g takes on every negative value once and only once for $x < 0$. How many times does g take on a given negative value in its domain and in what intervals are the roots located? What can be said about positive values of g?

The following four problems deal with the modulus of continuity.

17. Prove that $\mu(\delta_1 + \delta_2, f) \leqq \mu(\delta_1, f) + \mu(\delta_2, f)$.

18. Prove that this inequality together with $\lim_{\delta \to 0} \mu(\delta, f) = 0$ implies that $\mu(\delta, f)$ is a continuous function of δ.

19. Prove that f is a constant if $\mu(\delta, f) \equiv 0$.

20. Prove that $\mu(\delta, f)$ is not decreasing.

21. Let $a > 0$ be given, and define a sequence $\{a_n\}$ by setting $a_1 = a$, $a_n = a + \sqrt{a_{n-1}}$ for $n > 1$. Find the limit of a_n.

22. How should A be chosen in order that

$$(x^2 + x + 1)^{\frac{1}{2}} - Ax$$

will tend to a finite limit when $x \to \infty$, and what is then the limit?

23. A function f is defined by

$$f(x) = \begin{cases} x^2 + a, & 0 \leqq x, \\ 2(x^2 + 1)^{\frac{1}{2}} + b, & x < 0. \end{cases}$$

Determine the constants a and b so that $f(x)$ is continuous at $x = 0$ and $f(1) = 1$.

24. A function f is defined by

$$x^{-2}[(1 + x^2)^{\frac{1}{2}} - 1], \quad x \neq 0, \quad f(0) = \tfrac{1}{2}.$$

Show that f is continuous for all x. [*Hint:* Multiply and divide by $(1 + x^2)^{\frac{1}{2}} + 1$.]

25. Show that $x(1 - x)\,[x]$ is continuous in $(-1, 2)$ where $[x]$ is the largest integer $\leqq x$.

26. If f is continuous in (a, b) and $f(x) \neq 0$, show that f keeps a constant sign in (a, b).

Do the following functions on reals to reals have inverses? If so, define $\check{f}$ and give its domain and range when f is given by:

27. $x + (x^2 - 4)^{\frac{1}{2}}$.

28. $[1 + (4 + x)^{\frac{1}{2}}][1 - (4 + x)^{\frac{1}{2}}]^{-1}$.

29. $x^{\frac{1}{4}} + x^{\frac{1}{2}}$.

30. $x[x + 1], \quad 0 \leqq x$.

3

The Differential Operator

3.1. The derivative. The terms "change," "growth," and "speed" are common words in modern society. They are not merely words. We all know that it is often possible to obtain quantitative measurements of the corresponding phenomena. This presupposes that the changing quantity can be measured and is known as a function of time or whatever variable is involved. There is rapid growth if during a small time interval there is a substantial increase of the measured quantity. The ratio between the two increments, that of the measured quantity to that of the time, gives an average rate of growth during the time interval. If the latter is $[t_0, t_0 + h]$, then it may happen that the ratio of the increments tends to a limit as h decreases to zero. If this is the case, we call the limit the *rate of change* or the *rate of growth* at the time t_0. We shall make this observation more precise and start with a very simple special case.

Suppose that the phenomenon to be studied leads to the function x^2. This is the case, for instance, if we consider the area of a square as a function of the length of the side. Then the problem would be to find the rate of change of the area in terms of the side. See Figure 3.1.

For $x = a$ the area is a^2. Let h be a small quantity which to begin with is taken to be positive, and let $x = a + h$. We say that we give the side the *increment* h. The corresponding area is

$$(a + h)^2 = a^2 + 2ah + h^2,$$

and the increment of the area is the difference

$$(a + h)^2 - a^2 = 2ah + h^2,$$

which is the area of the shaded border in Figure 3.1. The ratio of the increment

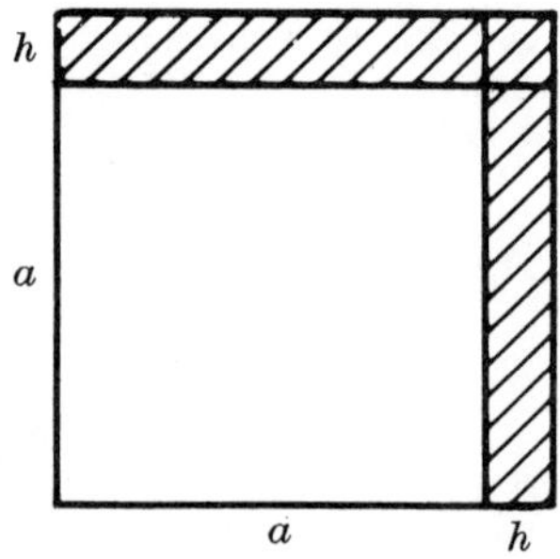

Figure 3.1

of area to the increment of side is

$$\frac{2ah + h^2}{h} = 2a + h. \tag{3.1.1}$$

This expression tends to the limit $2a$ as $h \to 0$. We say that the rate of change of the area of a square with respect to the length of its side is twice the length of the side.

This result implies that the increment of area is approximately $2ah$ when the side is a and its increment is h. More precisely stated, the increment is $2ah$ with an error which is small in comparison with h when h itself is small. This is easily verified for we know the exact value of the increment in question, namely $2ah + h^2$, so that the error is h^2 which is indeed small in comparison with h for small values of h. It is clear that in this discussion we are not restricted to positive values of h. The limit $2a$ obtained above exists regardless of how h tends to zero.

It is desirable to use this opportunity to discuss the question of making a comparison of small quantities, which is an ever-recurring problem in the calculus, and to introduce appropriate notation. We said above that h^2 is small in comparison with h when h itself is small. This follows from

$$h^2 = h \cdot h \quad \text{so that} \quad \frac{h^2}{h} = h \to 0 \text{ with } h.$$

Similarly

$$h^3 = h^2 \cdot h = h \cdot h^2,$$

which shows that h^3 is small in comparison with h^2 as well as with h. A convenient notation for these relationships is

$$h^2 = o(h), \quad h^3 = o(h^2), \quad h^3 = o(h) \tag{3.1.2}$$

Definition 3.1.1. *Suppose that f and g are functions on reals to reals defined for $|x| < \delta$ and suppose that $g(x) \neq 0$ for $0 < |x| < \delta$. Then*

$$f(x) = o[g(x)] \quad \textit{for} \quad x \to 0 \tag{3.1.3}$$

shall mean that

$$\lim_{x \to 0} \frac{f(x)}{g(x)} = 0. \tag{3.1.4}$$

Similarly,

$$f(x) = O[g(x)] \quad \textit{for} \quad x \to 0 \tag{3.1.5}$$

shall mean that there exists a constant K such that

$$|f(x)| \leqq K\,|g(x)| \quad \textit{for} \quad |x| < \delta. \tag{3.1.6}$$

In the last relation $g(x)$ may be allowed the value 0.

The symbols are read "small oh of g of x" and "capital oh of g of x" respectively. We see that

$$|h|^\beta = o(|h|^\alpha) \quad \text{as} \quad h \to 0 \text{ if and only if } \beta > \alpha. \tag{3.1.7}$$

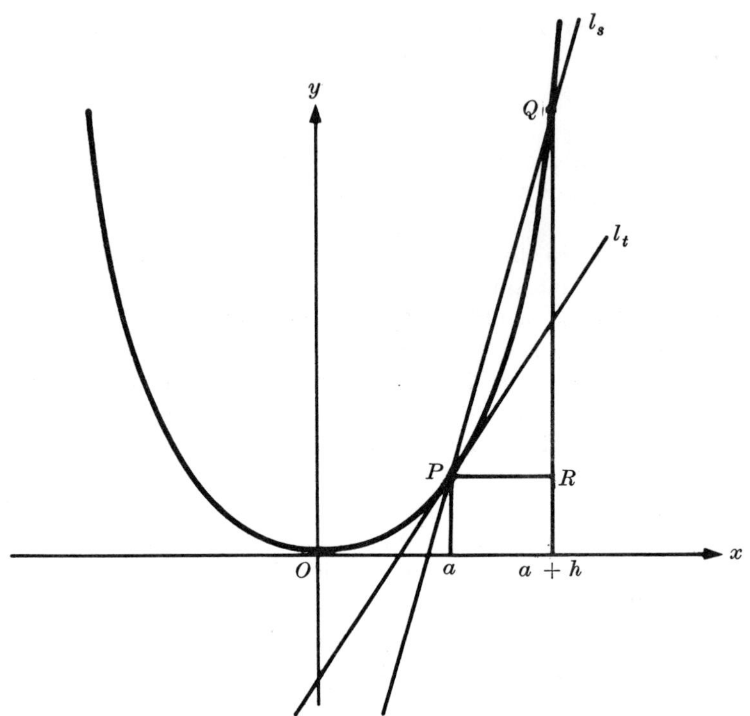

Figure 3.2

In particular, note that $f(h) = o(1)$ means that $f(h) \to 0$ with h, while $f(h) = O(1)$ means merely that $f(h)$ is bounded for small $|h|$. The "oh"-notation is also used to make comparisons between functions when the variable tends to other limits than 0. Thus

$$3x^2 - 10x + 4711 = O(x^2) \quad \text{as} \quad x \to \infty,$$

and the same expression is also $o(x^3)$. The notation is very useful when we are interested in showing only that a quantity is small but do not need to know how small it is as long as it is small in comparison with some standard gauge function. In working with this notation one observes that oh-relations can be "added" and "multiplied." Thus, for instance

$$o(1) + o(1) \Rightarrow o(1), \quad O(1) + O(1) \Rightarrow O(1) \tag{3.1.8}$$

simply state that the sum of two small (bounded) quantities is small (bounded). See also Theorem 2.4.3.

We now return to the rate of change of the area of the square. Another aspect of this problem is brought out in Figure 3.2. Here we have plotted the curve whose equation in rectangular coordinates is

$$y = x^2,$$

which the student will recognize as a parabola. Let

$$P = (a, a^2), \quad Q = (a + h, (a + h)^2).$$

Then

$$h = PR, \quad (a + h)^2 - a^2 = RQ,$$

and the ratio of the increments is

$$\frac{RQ}{PR}. \tag{3.1.9}$$

This is simply the slope of the chord PQ. In this connection the straight line l_s passing through P and Q is known as a *secant*. Letting $h \to 0$ means keeping P fixed and letting Q slide along the parabola towards P. Since (3.1.9), which is the same as (3.1.1), tends to the limit $2a$ when $Q \to P$ or, equivalently, $h \to 0$, we see that the secant tends to a limiting position, namely to the straight line l_t through P of slope $2a$. The equation of this line is

$$y - a^2 = 2a(x - a) \quad \text{or} \quad y = 2ax - a^2.$$

If we eliminate y between the two simultaneous equations

$$\begin{cases} y = x^2, \\ y = \ 2ax - a^2, \end{cases}$$

we obtain

$$x^2 = 2ax - a^2 \quad \text{or} \quad (x - a)^2 = 0,$$

that is, $x = a$ is a double root. Thus the line and the parabola have only one point in common, namely (a, a^2). We say that *the line is the tangent of the parabola at the point* (a, a^2). This is of course in agreement with our previous result in Section A.4.

What we have done here for the special function x^2 can be generalized very considerably. Suppose that

$$y = f(x) \tag{3.1.10}$$

is a real-valued continuous function defined in an interval (a, b). Let $a < x_0 < b$ and consider the ratio of the increments

$$\frac{RQ}{PR}, \tag{3.1.11}$$

also known as the *difference quotient* (of f at $x = x_0$). We are interested in what happens to this quotient when $h \to 0$. In the case of $f(x) = x^2$ we could assert that the quotient tends to a limit, namely $2x_0$, when $h \to 0$. In the general case we have no such assurance, and instead of a theorem we have to state a definition. As an orientation, see Figure 3.3.

Definition 3.1.2. *If*

$$\lim_{h\to 0} \frac{f(x_0 + h) - f(x_0)}{h} = f'(x_0) \tag{3.1.12}$$

exists as a finite number, then $f'(x_0)$ is called the derivative of $f(x)$ at $x = x_0$ and

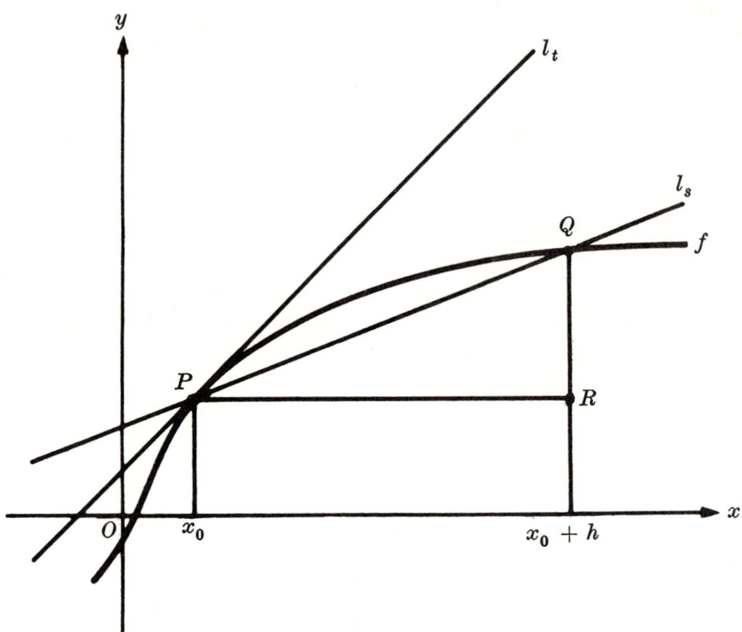

Figure 3.3

$f(x)$ is said to be differentiable at $x = x_0$. If $f(x)$ is differentiable for all x in (a, b), then $f(x)$ is said to be differentiable in (a, b) and its derivative $f'(x)$ is a function defined for all x in (a, b).

The derivative is sometimes referred to as the *derived function*. The operation by means of which $f'(x)$ is obtained from $f(x)$ is called *differentiation*. The corresponding verb is to *differentiate*.

It was supposed above that $f(x)$ is continuous. It is clear from (3.1.12) that the existence of $f'(x_0)$ presupposes that

$$\lim_{h\to 0} [f(x_0 + h) - f(x_0)] = 0, \tag{3.1.13}$$

that is, that $f(x)$ is continuous at $x = x_0$. But it must be understood that this condition is merely necessary and far from sufficient for the existence of a derivative.

Let us first note some functions which are differentiable. It is clear from (3.1.1) that x^2 is differentiable for all x and has the derivative $2x$. Since

$$\frac{(x + h)^3 - x^3}{h} = 3x^2 + 3xh + h^2 = 3x^2 + O(h),$$

we see that x^3 is also differentiable and its derivative is $3x^2$. In the next section we shall show that all powers of x are differentiable and that this implies the differentiability of every rational function of x. Other classes of differentiable functions will be encountered later.

It is easy to give examples of continuous functions which do not possess a derivative at certain points. The simplest example is furnished by the function $|x|$ at the origin. Here

$$\frac{f(h) - f(0)}{h} = \frac{|h|}{h} = \begin{cases} +1 & \text{if} \quad h > 0, \\ -1 & \text{if} \quad h < 0. \end{cases}$$

Thus the difference quotient does not tend to a unique limit as $h \to 0$. This function fails to have a derivative at one point only, but it is not difficult to construct continuous functions which lack a derivative at infinitely many points. To construct a continuous function which is nowhere differentiable is much more difficult. The existence of such a function was proved by Weierstrass in 1861 and many other examples have been found later. We shall consider such a function in Section C.3. We may feel inclined to regard nondifferentiability as a pathological phenomenon. Actually this is the rule for continuous functions rather than the exception but even a rigorous formulation of this assertion is already beyond the scope of this text.

Let us return to the case in which $f(x)$ is differentiable at $x = x_0$ so that the derivative $f'(x_0)$ exists. We can then give the same geometric interpretation of the derivative as in the case of $f(x) = x^2$. See Figure 3.3. The difference quotient

$$\frac{RQ}{PR}$$

is again the slope of the secant through the points P and Q and when $Q \to P$ along the curve, the slope of the secant tends to the limit $f'(x_0)$. We call *the straight line*

$$y - f(x_0) = f'(x_0)(x - x_0) \tag{3.1.14}$$

the tangent of the curve $y = f(x)$ *at the point* $(x_0, f(x_0))$ *and the slope of the curve at the point is*, by definition, *the slope of its tangent, that is* $f'(x_0)$. In this case we cannot assert that the tangent has a single intersection with the curve; actually there may be infinitely many intersections between the curve and its tangent at $x = x_0$ in any neighborhood of x_0. For an example, see Problem 51 in Exercise 3.1.

The value of $f'(x_0)$ shows what the function $f(x)$ does in a small interval $(x_0 - \delta, x_0 + \delta)$, whether it is increasing or decreasing or practically stationary. We shall discuss such problems in detail later; here it is only a question of general orientation.

DEFINITION 3.1.3. *The rate of change (or rate of growth) of a differentiable function* $f(x)$ *at* $x = x_0$ *is by definition the value of its derivative.*

The notion of "speed" or, more precisely, of "velocity" also leads to a derivative, in this case the time derivative of the distance traveled. We shall give general definitions later; it is enough that we treat a special case here. A body which falls from rest *in vacuo* under the influence of gravity travels the distance

$$s = \tfrac{1}{2}gt^2 \tag{3.1.15}$$

during the first t seconds. If cgs units are used, then g is approximately 981 cm/sec² which is equivalent to about 32 ft/sec². The time derivative of this function is

$$v = gt, \tag{3.1.16}$$

and this is the velocity of the body attained after t seconds.

EXERCISE 3.1

1. Verify (3.1.16).
2. If a is a constant, show that the derivative of ax equals a.
3. Find the derivative of $(x - 1)^3$ by two different methods. (i) Work directly with the function and simplify the difference quotient with the aid of formula (1.6.7). (ii) Expand $(x - 1)^3$ by the binomial theorem before replacing x by $x + h$. Do the results agree?
4. Apply the same techniques for finding the derivative of $(x + 2)^4$.

Find the derivatives of the following functions:

5. 1. 6. $1 - 2x$. 7. $3x^2 + x$. 8. x^5. 9. gt.

10. $x^2 - 2x + 1$. 11. $x^3 + 3x^2$. 12. $x^4 - 5x^3 + 10x^2$.

13. $\dfrac{1}{x}$. 14. $x + \dfrac{1}{x}$. 15. $x^2 - \dfrac{1}{x}$. 16. $\dfrac{1}{x^2}$.

17. $1 - \dfrac{1}{x^2}$. 18. $x^2 - \dfrac{1}{x^2}$. 19. $x^3 - 3x + \dfrac{1}{x}$. 20. $(x + 1)^5$.

21. Express the area of a square in terms of the length of the diagonal and find the rate of change of the area with respect to the diagonal.
22. What is the rate of change of the area of a circular disk with respect to the radius? Interpret the result.
23. What is the rate of change of the volume of a sphere with respect to the radius? Find also the rate of change of the area of the surface of the sphere with respect to the radius. Interpret! [*Hint:* $V = \frac{4}{3}\pi r^3$, $A = 4\pi r^2$.]
24. What is the rate of change of the area of an equilateral triangle with respect to the length of the side?
25. What is the rate of change of the volume of a cube with respect to the length of the side? What is the rate of change of the area of the surface?
26. A particle P is moving along a straight line under the influence of a force. At the time t its distance from a fixed origin O is given by $s = t^2 + 6t + 8$. Find the velocity of P at the time t (velocity = rate of change of distance with respect to time).
27. More generally, let $s = t^2 + at + b$. Where is P at the time $t = 0$ and what is its velocity?

28. The rate of change of the velocity with respect to the time is known as the *acceleration*. Determine the acceleration of P in the two preceding problems.

29. Find the velocity and the acceleration if the distance of P from O is given by $s = t^3$.

30–39. Find the equations of the tangents to the curves $y = f(x)$ at $x = 2$, if f runs through the functions of the above Problems 7, 11, 12, 13, 14, 15, 17, 18, 19, and 20.

40. Prove that for a differentiable function f

$$\lim_{h\to 0} \frac{f(x_0 + h) - f(x_0 - h)}{2h} = f'(x_0).$$

41. More generally, if α and β are two fixed positive numbers and $f(x)$ is differentiable, prove that

$$\lim_{h\to 0} \frac{f(x_0 + \beta h) - f(x_0 - \alpha h)}{(\beta + \alpha)h} = f'(x_0).$$

42. The limits listed in the two preceding problems may exist even if $f'(x_0)$ does not exist. To prove this, take $f(x) = |x|$, $x = 0$, and show that the corresponding limit is

$$\pm \frac{\beta - \alpha}{\beta + \alpha} \text{ depending on whether } h\downarrow 0 \text{ or } h\uparrow 0.$$

43. Show how to construct geometrically the tangent to the parabola $y = x^2$ at the point (a, a^2).

44. Does the function $x^{\frac{1}{3}}$ have a derivative at $x = 0$? Why not? Does the curve $y = x^{\frac{1}{3}}$ have a tangent at the origin, that is, does the secant through OQ tend to a limit when $Q \to O$?

45. Same questions for the function $x^{\frac{2}{3}}$.

46. Let (x) as usual denote the distance of x from the nearest integer. Let n be a positive integer. Find the derivative of the function (nx) wherever it exists. Show that the derivative is a step function and find its points of discontinuity.

47. Show that the sum of the squares of the first n natural numbers is $\frac{1}{3}n^3 + O(n^2)$.

48. Show that the relation (3.1.3) is transitive, that is, $f = o(g)$ and $g = o(h)$ implies $f = o(h)$. Does it have any other of the properties of an equivalence relation? Which of the order postulates hold?

49. If $x \to 0$, which of the statements "$f(x) = O(x^2)$" and "$f(x) = o(x)$" implies the other? What is the answer if $x \to \infty$ instead?

50. A function $f(x)$ is the sum of three functions and when $x \to 0$ these are

$$O(|x|^{\frac{3}{2}}), \quad O(x^2), \quad \text{and} \quad o(|x|^{\frac{1}{2}})$$

respectively. What is the best estimate for $f(x)$?

51. Show that the x-axis is tangent to $y = x^2\left\{\left(\frac{1}{x}\right) - x\right\}$ at $x = 0$ and has infinitely many intersections with the curve.

3.2. Rules of differentiation. We have already encountered a number of differentiable functions but it is now time for a more systematic treatment. It is obviously impossible to verify for each individual function whether or not it is differentiable. What we need are general results which will enable us to draw conclusions for large classes of functions. We can do this by a study of the structure of the class of functions which are differentiable either at a point or in an interval. We found this procedure quite helpful in the case of continuous functions and we shall obtain similar results here. It will be shown that the class of differentiable functions is an algebra, that is, if it contains two functions f and g, and if C is an arbitrary number, then it also contains $f + g$, Cf, and fg. Moreover, quotients f/g belong to the class at all points where they are defined. Finally we shall consider composite functions. The results are quite similar to those obtained in Section 2.5.

Actually we could proceed as in Section 2.5 translating Definition 3.1.2 in the δ,ε-language. We would then say that $f(x)$ is differentiable at $x = x_0$ if there exists a number $f'(x_0)$ and, for every $\varepsilon > 0$ a corresponding $\delta = \delta(\varepsilon, x_0)$, such that

$$\left|\frac{1}{h}[f(x_0 + h) - f(x_0)] - f'(x_0)\right| < \varepsilon \quad \text{for} \quad |h| < \delta. \tag{3.2.1}$$

A less elaborate procedure is to write

$$f(x_0 + h) - f(x_0) = hf'(x_0) + o(h). \tag{3.2.2}$$

We recall that this means

$$\frac{1}{h}[f(x_0 + h) - f(x_0) - hf'(x_0)] \to 0 \quad \text{when} \quad h \to 0, \tag{3.2.3}$$

and this is clearly equivalent to (3.2.1) and to Definition 3.1.2. We have already indicated that oh-symbols can be combined by algebraic operations. We shall need some results in addition to formula (3.1.8) and the following lemma gives what is needed.

LEMMA 3.2.1. *The sum of two variable quantities is $o(h)$ as $h \to 0$, if each of the summands is $o(h)$. The product of two variable quantities is $o(h)$, if one factor is bounded, that is, $O(1)$ and the other is $o(h)$. Such a product is $O(h^2)$ if each factor is $O(h)$ as $h \to 0$ and this implies that the product is $o(h)$. Symbolically we have the implications*

$$o(h) + o(h) \Rightarrow o(h), \tag{3.2.4}$$

$$O(1) \cdot o(h) \Rightarrow o(h), \tag{3.2.5}$$

$$O(h) \cdot O(h) \Rightarrow O(h^2) \Rightarrow o(h). \tag{3.2.6}$$

Proof. In the first case we have two functions g_1 and g_2, defined for small values of h, and we form their sum g_3,

$$g_3(h) = g_1(h) + g_2(h).$$

Knowing that

$$\frac{g_1(h)}{h} \to 0, \quad \frac{g_2(h)}{h} \to 0 \quad \text{as} \quad h \to 0, \tag{3.2.7}$$

we can obviously conclude that

$$\frac{g_3(h)}{h} = \frac{g_1(h)}{h} + \frac{g_2(h)}{h} \to 0,$$

which is the assertion in (3.2.4). A formal proof may be based on Theorem 3.4.3, formula (2.4.15).

In the second case there is a product, one factor of which is bounded and the other is $o(h)$, say

$$B(h)g_1(h) \equiv g_4(h),$$

where $g_1(h)$ satisfies (3.2.7) and $|B(h)| < K$. Here

$$\frac{g_4(h)}{h} = |B(h)| \left| \frac{g_1(h)}{h} \right| < K \left| \frac{g_1(h)}{h} \right|,$$

and the limit of the last member is 0 by Theorem 2.4.3, formula (2.4.16).

Finally, in (3.2.6) we have

$$g_5(h) \cdot g_6(h) \equiv g_7(h).$$

We assert the boundedness of

$$\frac{g_7(h)}{h^2} = \frac{g_5(h)}{h} \cdot \frac{g_6(h)}{h},$$

which is obvious since each factor on the right is bounded by assumption. This completes the proof.

We can now prove the main rules of differentiation.

THEOREM 3.2.1. *If f and g are differentiable at $x = x_0$, and if C_1 and C_2 are numbers, then $C_1f + C_2g$ is also differentiable at $x = x_0$ and*

$$(C_1f + C_2g)'(x_0) = C_1f'(x_0) + C_2g'(x_0). \tag{3.2.8}$$

Proof. We are given that

$$\begin{aligned} f(x_0 + h) - f(x_0) &= f'(x_0)h + o(h), \\ g(x_0 + h) - g(x_0) &= g'(x_0)h + o(h). \end{aligned} \tag{3.2.9}$$

Multiplying by C_1 and C_2 respectively and adding, we obtain

$$C_1[f(x_0 + h) - f(x)] + C_2[g(x_0 + h) - g(x_0)]$$
$$= C_1f'(x_0)h + C_2g'(x_0)h + C_1 \cdot o(h) + C_2 \cdot o(h).$$

Here the left member is simply

$$C_1f(x_0 + h) + C_2g(x_0 + h) - C_1f(x_0) - C_2g(x_0)$$
$$= (C_1f + C_2g)(x_0 + h) - (C_1f + C_2g)(x_0),$$

and by Lemma 3.2.1

$$C_1 \cdot o(h) + C_2 \cdot o(h) \Rightarrow o(h) + o(h) \Rightarrow o(h).$$

Combining we get

$$(C_1 f + C_2 g)(x_0 + h) - (C_1 f + C_2 g)(x_0) = [C_1 f'(x_0) + C_2 g'(x_0)]h + o(h),$$

and this implies (3.2.8) and proves the theorem.

Thus we see that the elements of the set of functions differentiable at $x = x_0$ can be combined by addition and scalar multiplication. The next theorem asserts that we can also combine elements by multiplication, that is, the set is actually an algebra.

THEOREM 3.2.2. *If f and g are differentiable at $x = x_0$, then so is their product and*

$$(fg)'(x_0) = f'(x_0)g(x_0) + f(x_0)g'(x_0). \tag{3.2.10}$$

Proof. Our point of departure is the identity

$$AB - ab = (A - a)b + a(B - b) + (A - a)(B - b).$$

This expression shows clearly that AB is close to ab if A is close to a and B to b. Here we set

$$A = f(x_0 + h), \quad a = f(x_0), \quad B = g(x_0 + h), \quad b = g(x_0)$$

and use (3.2.9). We then have

$$A - a = f(x_0 + h) - f(x_0) = hf'(x_0) + o(h) \Rightarrow O(h),$$
$$B - b = g(x_0 + h) - g(x_0) = hg'(x_0) + o(h) \Rightarrow O(h),$$

so that

$$(A - a)(B - b) = O(h) \cdot O(h) \Rightarrow O(h^2) \Rightarrow o(h)$$

by Lemma 3.2.1, formula (3.2.6). Substitution of these expressions in the identity then gives

$$f(x_0 + h)g(x_0 + h) - f(x_0)g(x_0)$$
$$= [hf'(x_0) + o(h)]g(x_0) + f(x_0)[hg'(x_0) + o(h)] + O(h^2).$$

Here

$$g(x_0) \cdot o(h) + f(x_0) \cdot o(h) + O(h^2) \Rightarrow o(h) + o(h) + o(h) \Rightarrow o(h).$$

Thus finally

$$f(x_0 + h)g(x_0 + h) - f(x_0)g(x_0) = [f'(x_0)g(x_0) + f(x_0)g'(x_0)]h + o(h),$$

which is equivalent to (3.2.10) and proves the theorem.

COROLLARY. *The set of functions which are differentiable at $x = x_0$ forms an algebra. The same is true for the functions which are differentiable in an interval (a, b).*

Quotients are handled in essentially the same way as products.

THEOREM 3.2.3. *If f and g are differentiable at $x = x_0$ and if $g(x_0) \neq 0$, then their quotient is differentiable and*

$$\left(\frac{f}{g}\right)'(x_0) = \frac{f'(x_0)g(x_0) - f(x_0)g'(x_0)}{[g(x_0)]^2}. \tag{3.2.11}$$

Proof. We start with the identity

$$\frac{A}{B} - \frac{a}{b} = \frac{(A-a)b - (B-b)a}{b^2}\left\{1 - \frac{B-b}{B}\right\},$$

and make the same substitutions as in the preceding theorem. On the left we have then the increment of f/g. The expression on the right is the sum of two terms, a leading term to be denoted by $L(h)$ and a remainder $R(h)$. Here

$$L(h) \equiv \{[f(x_0+h) - f(x_0)]g(x_0) - [g(x_0+h) - g(x_0)]f(x_0)\}[g(x_0)]^{-2}.$$

We use (3.2.9) and obtain

$$L(h) = \{[hf'(x_0) + o(h)]g(x_0) - [hg'(x_0) + o(h)]f(x_0)\}[g(x_0)]^{-2}$$
$$\Rightarrow Q(x_0, f, g)h + O(1)\cdot o(h) \Rightarrow Q(x_0, f, g)h + o(h) \Rightarrow O(h),$$

where $Q(x_0, f, g)$ denotes the right member of (3.2.11). To complete the proof, we have to show that $R(h) = o(h)$. Now

$$R(h) \equiv -L(h)\frac{g(x_0+h) - g(x_0)}{g(x_0+h)}.$$

We note that $g(x)$ is necessarily continuous at $x = x_0$ and, by assumption, $g(x_0) \neq 0$. It follows that $|g(x_0+h)|$ is bounded away from 0 for small values of $|h|$ so that its reciprocal is bounded. Hence

$$L(h) = O(h), \quad g(x_0+h) - g(x_0) = O(h), \quad [g(x_0+h)]^{-1} = O(1),$$

and

$$R(h) = O(h)\cdot O(h)\cdot O(1) \Rightarrow O(h^2) \Rightarrow o(h).$$

Combining we see that

$$\frac{f(x_0+h)}{g(x_0+h)} - \frac{f(x_0)}{g(x_0)} = Q(x_0, f, g)h + o(h),$$

and the theorem is proved.

The discussion of composite functions is postponed until the next section. Before going further into the general theory, let us insert some remarks concerning notation. So far we have exclusively used a prime to denote differentiation, but there are many other notations in use with which the student should be familiar.

Newton, who spoke of fluxions rather than of derivatives, denoted them by a dot placed over the letter. The dot notation is still used, but only in very special cases, namely to denote time derivatives in kinematics. Thus, if $x(t)$ denotes the position of a particle traveling along a straight line, then its velocity at the time t is the derivative of $x(t)$ with respect to t and is written $\dot{x}(t)$.

The following are more common notations. The operation of differentiation with respect to x is often denoted by one of the symbols

$$\frac{d}{dx},\ D,\ D_x. \tag{3.2.12}$$

Corresponding notations for the derivative of

$$y = f(x)$$

are

$$\frac{dy}{dx},\ \frac{df}{dx},\ \frac{df(x)}{dx};\qquad Dy,\ Df,\ Df(x);\qquad D_xy,\ D_xf,\ D_xf(x). \tag{3.2.13}$$

The double-d notation goes back to Leibniz. It is still very popular even if it has probably caused more misunderstandings and trouble than any other single item in the Calculus. The other notations go back to French writers: the prime notation to Joseph Louis Lagrange (1736–1813), and the capital D notations to Louis Arbogeste (1759–1803) and to Cauchy.

In this connection it should be mentioned that the increment of x, for which we use the single letter h, is usually denoted by Δx. The increment of $y = f(x)$ is then denoted by Δy or by Δf and the relation between difference quotient and derivative is written

$$\lim_{\Delta x \to 0} \frac{\Delta y}{\Delta x} = \frac{dy}{dx}. \tag{3.2.14}$$

Let us now return to the discussion of the class of differentiable functions. Theorems 3.2.1 to 3.2.3 tell us something about the structure of this class. It is an algebra and quotients are included, provided they are well-defined. These results enable us to assert that rational functions are differentiable at all points where they are defined.

The argument goes as follows. A constant is differentiable and its derivative is identically zero,

$$D_xC = 0. \tag{3.2.15}$$

This follows from first principles and the formal proof is left to the reader. The function x is differentiable and

$$D_xx = 1. \tag{3.2.16}$$

Since x^2 is the product of the differentiable function x with itself, x^2 is also differentiable by Theorem 3.2.2. Using an induction argument (see below in the proof of Theorem 3.2.4) we show that x^n is differentiable for all x and all positive integers n. Using Theorem 3.2.1 we see that any linear combination of integral powers of x is differentiable for all x. In other words, any polynomial

$$P(x) \equiv a_0x^n + a_1x^{n-1} + \cdots + a_{n-1}x + a_n \tag{3.2.17}$$

is differentiable. Now a rational function is the quotient of two polynomials

$$R(x) = \frac{P(x)}{Q(x)} . \tag{3.2.18}$$

Theorem 3.2.3 then asserts that $R(x)$ is differentiable for all values x_0 of x such that $Q(x_0) \neq 0$. This proves the assertion.

It was stated above that x^n, n positive integer, is differentiable. We shall need the explicit expression for the derivative.

THEOREM 3.2.4. *The derivative of x^n, n positive integer, exists for all x and*

$$D_x x^n = n x^{n-1}. \tag{3.2.19}$$

Proof. We shall give an induction argument. Other methods of proof are suggested in Exercise 3.2. Formula (3.2.19) is certainly true for $n = 1$, provided we understand x^0 to be 1 for all x including $x = 0$. Suppose then that the formula holds for $n = k$. Let us prove that it is valid for $n = k + 1$. We have

$$x^{k+1} = x \cdot x^k,$$

whence by Theorem 3.2.2

$$D_x(x^{k+1}) = xD_x(x^k) + x^k D_x(x) = x \cdot kx^{k-1} + x^k \cdot 1 = (k + 1)x^k.$$

It follows that (3.2.19) holds for positive integers n.

Formula (3.2.15) shows that (3.2.19) is also true for $n = 0$. Next we prove that it is valid for negative integers provided $x \neq 0$. For if $n = -k$, $k > 0$, then

$$D_x(x^{-k}) = D_x \frac{1}{x^k} = \frac{x^k \cdot D_x 1 - 1 \cdot kx^{k-1}}{x^{2k}},$$

$$= -kx^{k-1-2k} = -kx^{-k-1},$$

which was to be proved.

For $x > 0$ and k a positive integer, we define

$$\sqrt[k]{x} = x^{1/k}$$

as that positive number whose kth power is x. We shall show that this function is also differentiable and that its derivative is given by formula (3.2.19) with n replaced by $\frac{1}{k}$. For this purpose we go back to first principles and consider the difference quotient

$$\frac{1}{h}[(x + h)^{1/k} - x^{1/k}], \quad x > 0, \quad x + h > 0.$$

To simplify we write

$$x^{1/k} = u, \quad (x + h)^{1/k} = v,$$

and use formula (1.6.7) where we set

$$a = u, \quad b = v, \quad n = k,$$

so that

$$b^n - a^n = v^k - u^k = x + h - x = h.$$

This gives the identity

$$h = (v - u)(v^{k-1} + v^{k-2}u + v^{k-3}u^2 + \cdots + vu^{k-2} + u^{k-1})$$

and the following expression for the difference quotient under consideration

$$\frac{1}{h}(v - u) = [v^{k-1} + v^{k-2}u + v^{k-3}u^2 + \cdots + vu^{k-2} + u^{k-1}]^{-1}.$$

There are k terms inside the bracket. When h tends to 0, v tends to u and for any j with $1 \leqq j \leqq k$,

$$\lim v^{k-j}u^{j-1} = u^{k-1}.$$

It follows that the difference quotient has a limit and this equals

$$[ku^{k-1}]^{-1} = \frac{1}{k}u^{-k+1} = \frac{1}{k}(x^{1/k})^{-k+1} = \frac{1}{k}x^{1/k-1},$$

so that

$$D_x(x^{1/k}) = \frac{1}{k}x^{1/k-1}, \quad x > 0, \tag{3.2.20}$$

as asserted.

The argument given above shows that any integral power of $x^{1/k}$ is differentiable for $x > 0$ and, consequently, that rational functions of $x^{1/k}$ have this property. We shall obtain further results along these lines in the next section.

EXERCISE 3.2

Find the derivatives of the following functions:

1. $2 - 3x$. 2. $3x^2 + 4x - 3$. 3. $2x^5 - 10x^4 + 5x^3$.

4. $(x - 1)^6$. 5. $(x + 1)^3 - (x - 1)^3$. 6. $x^{2n} - nx^{-n}$.

7. $\dfrac{1}{x}$. 8. $2x - \dfrac{5}{x}$. 9. $x^2 - \dfrac{1}{x}$. 10. $x^3 + 3x - \dfrac{2}{x}$.

11. $x^2 + \dfrac{3}{x}$. 12. $x^3 - \dfrac{2}{x^2}$. 13. $\dfrac{x^3 + 1}{x}$. 14. $\dfrac{ax + b}{cx + d}$.

15. $\dfrac{ax^2 + b}{cx^2 + d}$. 16. $\dfrac{x}{x^2 + a^2}$. 17. $\dfrac{x^3 + 3}{x^3 - 3}$. 18. $\dfrac{1}{x} + \dfrac{1}{x^2}$.

19. $x^n + \dfrac{1}{x^n}$. 20. $\dfrac{x^2}{(x^2 + a^2)^2}$. 21. $x^{-1} + x^{-2} + x^{-3} + x^{-4}$.

Find the derivatives of the following functions for $x > 0$;

22. $\sqrt{x} + \dfrac{1}{\sqrt{x}}$. 23. $x^{\frac{1}{3}} + x^{\frac{2}{3}}$. 24. $\dfrac{\sqrt{x} - 1}{\sqrt{x} + 1}$. 25. $\dfrac{\sqrt{x}}{(1 + \sqrt{x})^2}$.

26. $\dfrac{a\sqrt{x} + b}{c\sqrt{x} + d}$.
27. $\dfrac{(1 + \sqrt{x})^3}{(\sqrt{x})^3}$.
28. $\dfrac{4x^{\frac{1}{4}} + 1}{3x^{\frac{1}{3}} + 1}$.

29. $\dfrac{5x^{\frac{1}{5}} + 3x^{\frac{1}{3}}}{2x^{\frac{1}{2}} + 1}$.
30. $\left(\sqrt{x} + \dfrac{1}{\sqrt{x}}\right)^3$.
31. $1 + x^{-\frac{1}{2}} + x^{-\frac{1}{3}} + x^{-\frac{1}{4}}$.

Find the derivatives of the following functions from first principles. In Problem 32, $x > -1$; in Problem 34, $x > 1$.

32. $(x + 1)^{\frac{1}{2}}$.
33. $(x^2 + 1)^{-\frac{1}{2}}$.
34. $\left(\dfrac{x - 1}{x + 1}\right)^{\frac{1}{2}}$.
35. $(x^3 + 1)^{-\frac{1}{3}}$.

Find the derivative of x^n, n positive integer:

36. From the difference quotient, using the binomial theorem to expand $(x + h)^n$.

37. From formula (1.6.7) with $a = x, b = x + h$, proceeding as in the proof of (3.2.20).

38. If k is an odd integer, $x^{1/k}$ is also defined for $x < 0$ as the negative number whose kth power is x. Is formula (3.2.19) valid in this case?

39. If F is the product of three functions f_1, f_2, f_3, prove that

$$F' = f_1'f_2f_3 + f_1f_2'f_3 + f_1f_2f_3'.$$

Generalize to n factors by induction.

40. If

$$P(x) = (x - x_1)(x - x_2) \cdots (x - x_n),$$

prove that

$$\frac{P'(x)}{P(x)} = \frac{1}{x - x_1} + \frac{1}{x - x_2} + \cdots + \frac{1}{x - x_n}.$$

41. If $P(x) = (x - a)^kQ(x)$, where k is a natural number and $Q(x)$ is a polynomial with $Q(a) \neq 0$, then the polynomial $P(x)$ is said to be divisible by $(x - a)^k$ and the equation $P(x) = 0$ has $x = a$ as a k-fold root. Show that $P'(x)$ is divisible by $(x - a)^{k-1}$.

42. Choose the constant c so that the equation

$$x^3 - x^2 - x + c = 0$$

has a double root.

43. Same question for the equation

$$x^3 - 2x^2 + x - c = 0.$$

44. What is the shape of the graph of $y = P(x)$ for x near to a double root of the equation $P(x) = 0$? Same question for a triple root. Illustrate by taking $x = 0$, $P(x) = x^2$ and x^3.

45. Find the points on the curve whose equation is

$$y = \tfrac{1}{3}x^3 + 4x^2 + x,$$

where the slope equals 10, and give the equations of the tangents to the curve at these points.

46. Same question for $y = x(x - 1)^2$ and the slope 1.

47. A function f is defined to be x^2 for $x \geqq 0$ and 0 for all $x < 0$. Show that $f'(x)$ exists for all x including $x = 0$.

3.3. Composite functions. In Section 2.5 composite functions defined by the circle product are considered with regard to continuity. Here we are concerned with differentiability instead. The basic result is

THEOREM 3.3.1. *Given two functions f and g with the following properties:* (i) *$D[g]$ contains a neighborhood of $x = x_0$ and $g(x_0) = u_0$;* (ii) *$D[f]$ contains a neighborhood of $u = u_0$;* (iii) *$R[g] \subset D[f]$; and* (iv) *$g'(x_0)$ and $f'(u_0)$ exist. Then $F = f \circ g$ is well defined for $x \in D[g]$, F is differentiable at $x = x_0$ and*

$$F'(x_0) = f'(u_0)g'(x_0). \tag{3.3.1}$$

Proof. Let $|h|$ be so small that $x_0 + h \in D[g]$ and set

$$p(h) = g(x_0 + h) - g(x_0).$$

Further, define

$$\varepsilon(h) = \begin{cases} \dfrac{f(u_0 + p(h)) - f(u_0)}{p(h)} - f'(u_0) & \text{if} \quad p(h) \neq 0, \\ 0, & \text{if} \quad p(h) = 0. \end{cases}$$

Then

$$\frac{1}{h}[F(x_0 + h) - F(x_0)] = [\varepsilon(h) + f'(u_0)]\frac{p(h)}{h}$$

for all values of h under consideration. Note that if $p(h)$ happens to be zero for some particular value of h, we have $\varepsilon(h) = 0$ and both sides of the equation are zero. From this relation we get

$$\frac{1}{h}[F(x_0 + h) - F(x_0)] - f'(u_0)g'(x_0) = f'(u_0)\left[\frac{p(h)}{h} - g'(x_0)\right] + \varepsilon(h)\frac{p(h)}{h}.$$

The left member is the difference between the difference quotient and the proposed derivative of F at $x = x_0$. We have to prove that this difference tends to 0 with h. To this end we note that

$$\frac{p(h)}{h} \to g'(x_0) \quad \text{as} \quad h \to 0$$

by the definition of the derivative of g. Since $p(h) \to 0$ with h, we have $\varepsilon(h) \to 0$, now by the definition of the derivative of f. Formula (3.3.1) follows from this argument and the theorem is proved.

Formula (3.3.1) is often called the "chain rule."

If f and g are differentiable in their domains of definition and if (iii) holds, then F is differentiable for all x in $D[g]$, and (3.3.1) is valid for all such x_0.

The formula may also be written

$$D_x(f \circ g)(x) = (f' \circ g)(x) \cdot g'(x). \tag{3.3.2}$$

In the notation of Leibniz we write

$$y = f(u), \quad u = g(x), \quad \frac{dy}{dx} = \frac{dy}{du} \cdot \frac{du}{dx}. \tag{3.3.3}$$

This is very compact and easy to remember. One should also remember to substitute $u = g(x)$ in the result. This variant of the formula appears self-evident and scarcely worth such a complicated proof as that given above. One simply multiplies numerator and denominator of the left member of the last equation by "du" and regroups the factors so that the right member results. What is wrong with that? Briefly, the argument presupposes that the symbols "dx," "du," and "dy" stand for definite numbers. Actually they have not been defined at all; $\dfrac{du}{dx}$ is merely another way of writing $g'(x)$. True, it is written as a fraction, but neither the "numerator" nor the "denominator" has been defined so far. The proposed "proof" has no real basis; it is just a loose speculation encouraged by the suggestive notation of Leibniz. A slightly cruder way of "proving" (3.3.3) is to cancel all the d's in the last formula. The remaining letters form an algebraic identity which is obviously true. Hence (3.3.3) is also true.

In a more serious vein, let us pass to more complicated composite functions. Suppose that we have three functions $F(v)$, $f(u)$, and $g(x)$. We suppose that g is defined and differentiable for g in an interval (a, b) and that the mapping

$$u = g(x)$$

takes (a, b) into an interval (c, d) which is a subset of the domain of f. The latter function is supposed to be differentiable for u in (c, d) and to map this interval into an interval (A, B) where $F(v)$ is defined and differentiable. Then for $x \in (a, b)$

$$F\{f[g(x)]\} = (F \circ f \circ g)(x) \tag{3.3.4}$$

is well-defined and differentiable. This can be proved by an argument similar to that used in the proof of Theorem 3.3.1 or by an induction proof. Its derivative equals

$$F'\{f[g(x)]\} \cdot f'[g(x)] \cdot g'(x) = (F' \circ f \circ g)(x) \cdot (f' \circ g)(x) \cdot g'(x), \tag{3.3.5}$$

or, in Leibniz's notation,

$$\frac{dy}{dx} = \frac{dy}{dv} \cdot \frac{dv}{du} \cdot \frac{du}{dx}. \tag{3.3.6}$$

This formula and its extension to more "factors" are also referred to as the "chain rule."

Let us apply these methods to find the derivative of

$$y = \left(\frac{1 - x^2}{1 + x^2}\right)^{\frac{1}{3}}.$$

Here

$$y = F(v) = v^{\frac{1}{3}}, \quad v = f(u) = \frac{1-u}{1+u}, \quad u = g(x) = x^2.$$

Now g maps R onto R^+, and f is certainly defined, continuous and differentiable for $u \geqq 0$. The only point where $g(u)$ becomes infinite is $u = -1$ and this does not correspond to a real value of x. The function f maps R^+ onto the interval $(-1, 1]$, and $F(v)$ is well defined for such values of v. It is also differentiable except at $v = 0$ which corresponds to $u = 1$ and $x = \pm 1$. Thus we find that y is differentiable for all $x \neq \pm 1$. We have

$$F'(v) = \tfrac{1}{3}v^{-\frac{2}{3}}, \quad f'(u) = \frac{-2}{(1+u)^2}, \quad g'(x) = 2x.$$

Combining and substituting we obtain

$$\frac{dy}{dx} = -\tfrac{1}{3}\left(\frac{1-x^2}{1+x^2}\right)^{-\frac{2}{3}} \frac{2}{(1+x^2)^2} \cdot 2x = -\tfrac{4}{3}x(1-x^2)^{-\frac{2}{3}}(1+x^2)^{-\frac{4}{3}}$$

In using the chain rule, the student is advised to write out all intermediary steps carefully. Once used to the process, he can of course write the result directly. Care should be taken to make the substitutions in formulas (3.3.3) and (3.3.6), $u = g(x)$ and so on. It is of paramount importance that the right substitutions be made. It is not true, for instance, that

$$(f \circ g)'(x) = (f' \circ g')(x).$$

The reader is also warned that the derivative of a product is not the product of the derivatives.

Finally a few words about the use of the formulas. It is clear that a fraction can be written as the product of the numerator times the reciprocal of the denominator, that is,

$$\frac{f}{g} = f \cdot g^{-1}.$$

Note that g^{-1} denotes the reciprocal of g and not the inverse. It follows that we can get the derivative of a quotient using the product formula and the formula

$$D_x[g(x)]^\alpha = \alpha[g(x)]^{\alpha-1}g'(x), \tag{3.3.7}$$

which is a special case of (3.3.1). Taking $\alpha = -1$ we get

$$D_x\{f(x)[g(x)]^{-1}\} = f'(x)[g(x)]^{-1} - f(x)[g(x)]^{-2}g'(x). \tag{3.3.8}$$

This is of course the same as (3.2.11) with x instead of x_0. Sometimes this procedure has its advantages and it is worth keeping in mind. In particular, the case in which $f(x)$ is a constant should never be handled by the fraction formula. It calls for (3.3.7) with $\alpha = -1$.

We note that (3.3.7) has been proved for the case in which α is rational. It is also true if α is irrational, but the definition of $[g(x)]^\alpha$ for α irrational has to wait until we have developed the exponential function and the logarithm.

EXERCISE 3.3

1–6. Do Problems 22–27 of Exercise 3.2 using Theorem 3.3.1. Take $g(x) = x^{\frac{1}{2}}$ except in Problem 23 where $g(x) = x^{\frac{1}{3}}$ instead.

7–10. Solve Problems 32–35 of Exercise 3.2 by the same method.

11. Find the derivative of

$$[x + (x^2 + 1)^{\frac{1}{2}}]^{-\frac{1}{2}}.$$

12. Find the derivative of

$$\frac{x^3 + 1}{x} = x^2 + x^{-1},$$

and compare the amount of work involved in using (3.2.11) on the left and the power formula on the right.

Find the derivatives of the following functions.

13. $(ax + b)^n$.
14. $(x^2 - 1)^5$.
15. $\dfrac{1}{(1 + x^2)\sqrt{1 + x^2}}$.
16. $(1 + x^3)^{\frac{1}{3}}$.
17. $x\sqrt{x^2 + 1}$.
18. $\sqrt{x + \sqrt{x}}$.
19. $\sqrt[3]{(2x + 1)^2}$.
20. $(x + x^{\frac{1}{3}})^{\frac{2}{3}}$.
21. $\sqrt{a^2 - x^2}$.
22. $\sqrt{x\sqrt{x\sqrt{x}}}$.
23. $\dfrac{2x - 1}{x^7}$.
24. $(x^2 + x + 1)^{\frac{3}{2}}$.
25. $(x^2 - 1)^{\frac{1}{4}}(6x^2 - 2x + 3)^{-\frac{2}{3}}$.
26. $\left(\dfrac{x^2 - 1}{x^2 + 1}\right)^{\frac{5}{2}}$.

Find the derivatives of the following functions with respect to the variable under consideration.

27. $(s^{\frac{1}{3}} + 3)^{\frac{3}{5}}$.
28. $\dfrac{t^2 + t + 1}{t + 1}$.
29. $\dfrac{u^{19}}{1 - u}$.
30. $\sqrt{v(1 - v)}$.
31. $\dfrac{(w - 1)^3}{(w + 1)^2}$.
32. $\left(\dfrac{z}{z^2 + 1}\right)^{\frac{1}{3}}$.

Find the coordinates of the points on the following curves where the tangent is horizontal. (*Hint:* What is the slope at such a point?)

33. $y = x^2 - 4x + 5$.
34. $y = x^2 + \dfrac{1}{x}$.
35. $y = x^2 + \dfrac{1}{x^2}$.
36. $y = \sqrt{x(1 - x)}$.
37. $y = \dfrac{(x - 1)^3}{(x + 1)^2}$.
38. $y = \dfrac{x}{1 + x^2}$.

3.4. Differentiability of convex functions.

It was remarked in Section 2.7 that the typical function considered in the Calculus is a piecewise convex-concave function. If the reader will look at the figures in this book, he will find that all the continuous functions represented are of this type. The domain of the function shown splits up into subintervals in each of which either

f or $-f$ is convex in the sense of Definition 2.7.2. The obvious importance of such functions for mathematics, pure and applied, makes it desirable to examine their differentiability properties in some detail. Simple examples show that a convex function need not have a derivative at all points. We can think of $|x|$, for instance. It turns out, however, that convex functions possess the slightly weaker property of having right- and left-hand derivatives at all interior points of the intervals where they are defined. We proceed to define one-sided derivatives.

DEFINITION 3.4.1. *Let f be defined in an interval $[a, a + \delta]$ and let f be continuous to the right at $x = a$. The right-hand derivative of f at $x = a$ is*

$$\lim_{h\to 0} \frac{f(a + h) - f(a)}{h} \equiv f_+'(a), \tag{3.4.1}$$

if the limit exists as a finite number and h is restricted to be positive. Similarly, if f is defined in $(b - \delta, b]$ and continuous to the left at $x = b$, then the left-hand derivative is defined as

$$\lim_{h\to 0} \frac{f(b) - f(b - h)}{h} \equiv f_-'(b), \tag{3.4.2}$$

if the limit exists as a finite number and h is restricted to be positive.

Such one-sided derivatives may exist at points where the ordinary derivative does not. Thus $|x|$ has right- and left-hand derivatives everywhere. We have $f_-'(0) = -1, f_+'(0) = +1$. The following theorem is obvious:

THEOREM 3.4.1. *A necessary and sufficient condition that f be differentiable at $x = x_0$ is that $f_-'(x_0)$ and $f_+'(x_0)$ exist and are equal.*

We recall that f is convex in $[a, b]$ if, for any three points P_1, P_2, P_3 on the graph of f with P_2 between P_1 and P_3, the point P_2 lies on or below the chord P_1P_3. This suggests that some elementary facts concerning the geometry of the triangle may have a bearing on our problem. In Figure 3.4 a triangle $P_1P_2P_3$ is shown. The

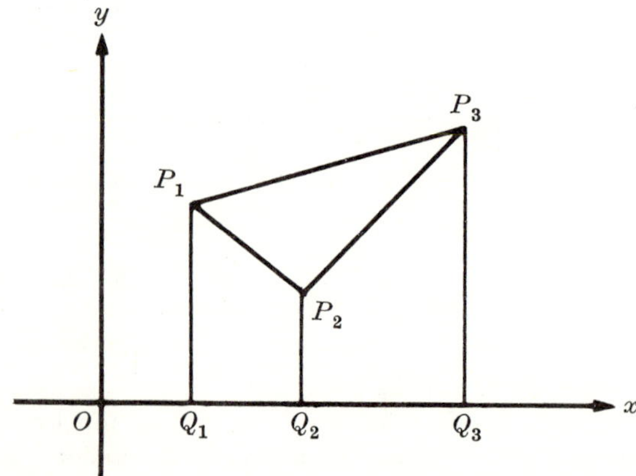

Figure 3.4

coordinates of P_j are (x_j, y_j), $j = 1, 2, 3$, where

$$x_1 < x_2 < x_3.$$

Moreover, P_2 lies below P_1P_3. Without restricting the generality we may suppose that $y_j > 0$, $j = 1, 2, 3$. The foot point on the x-axis of the perpendicular from P_j is denoted by Q_j.

The figure shows three trapezoids in addition to the triangle. The fact that P_2 lies below the side P_1P_3 is equivalent to the statement that the area of the largest trapezoid $Q_1Q_3P_3P_1$ equals the sum of the areas of the two smaller trapezoids plus the area of the triangle. From this we can compute the area of the triangle in terms of the coordinates of the vertices. The area of $Q_1Q_3P_3P_1$ equals

$$\tfrac{1}{2}(y_3 + y_1)(x_3 - x_1).$$

Computing the areas of the other trapezoids and subtracting we find the area of the triangle which we denote by A. We have

$$2A = y_3(x_2 - x_1) + y_1(x_3 - x_2) + y_2(x_1 - x_3). \tag{3.4.3}$$

This expression is positive unless the three points P_1, P_2, P_3 are collinear and the triangle degenerates. The inequalities we are interested in are a consequence of $A \geqq 0$. As an example we give

$$(x_3 - x_1)y_2 < (x_3 - x_2)y_1 + (x_2 - x_1)y_3, \tag{3.4.4}$$

which states that P_2 lies below P_1P_3. If P_1, P_2, P_3 are three points on the graph of a convex function, we have $y_j = f(x_j)$ and (3.4.4) becomes

$$(x_3 - x_1)f(x_2) < (x_3 - x_2)f(x_1) + (x_2 - x_1)f(x_3), \tag{3.4.5}$$

which is equivalent to (2.7.7). Note that if the three points are collinear, "$<$" is to be replaced by "$=$" in (3.4.5).

Another consequence of $A > 0$ is inequalities for the slopes of the sides of the triangle. Let us denote the slope of the side P_jP_k by $\{j, k\}$. Then

$$\{j, k\} = \{k, j\},$$

and we have

Lemma 3.4.1. *Under the assumptions made*

$$\{1, 2\} < \{1, 3\} < \{2, 3\}. \tag{3.4.6}$$

Proof. A proof can be read off from Figure 3.5. The student who prefers an analytical proof can use the condition $A > 0$ to obtain one.

Again, if P_1, P_2, P_3 are three noncollinear points on the graph of a convex function, the inequalities (3.4.6) become inequalities between difference quotients. We need one more such inequality before we can proceed to the proof of the existence of one-sided derivatives for convex functions. We can extend what we proved for a triangle to a convex quadrilateral. See Figure 3.6.

Lemma 3.4.2. *Suppose that $P_1P_2P_3P_4$ is a convex quadrilateral such that*

$$x_1 < x_2 < x_3 < x_4;$$

P_2 and P_3 lie below the side P_1P_4. Draw the diagonals P_1P_3 and P_2P_4. Then P_2 lies

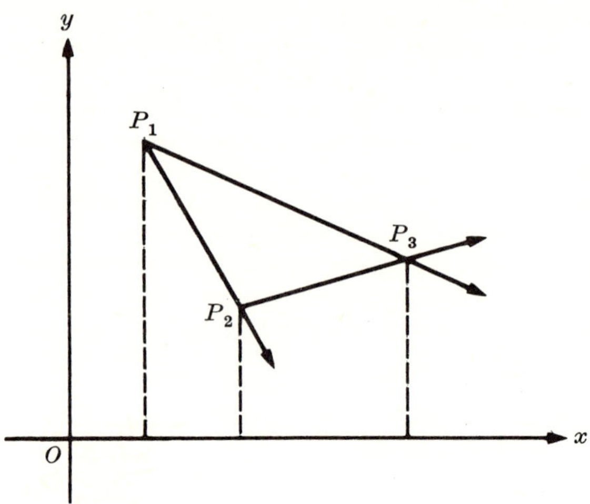

Figure 3.5

below P_1P_3 *and* P_3 *below* P_2P_4. *Further, if* $\{j, k\}$ *denotes the slope of* P_jP_k, *the following inequalities hold:*

$$
\begin{gathered}
\{1, 2\} < \{1, 3\} < \{2, 3\} < \{2, 4\} < \{3, 4\}, \\
\{1, 3\} < \{1, 4\} < \{2, 4\}.
\end{gathered}
\tag{3.4.7}
$$

Proof. The notion of a convex set was defined in Exercise 1.5, Problem 7. We recall that a set is convex if, whenever it contains two points P and Q it also contains the line segment joining them. That the quadrilateral is convex then means

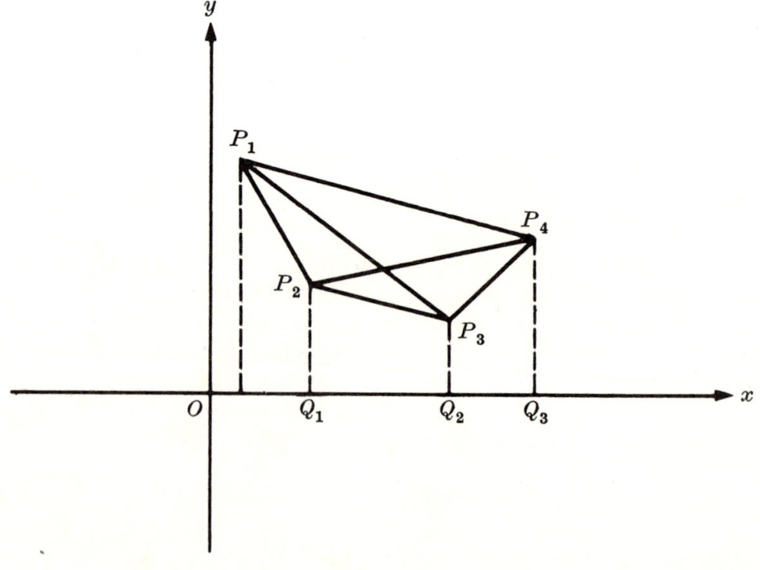

Figure 3.6

that the diagonals belong to the interior of the quadrilateral, except for the endpoints and this requires that P_1 and P_3 be on opposite sides of P_2P_4 and P_2 and P_4 be on opposite sides of P_1P_3. We now have four triangles which satisfy the conditions of Lemma 3.4.1 and which are bounded by sides P_jP_k. To get the listed inequalities we apply Lemma 3.4.1 to these triangles and combine the results.

We have assumed tacitly that the quadrilateral is nondegenerate. If three or possibly all four points are collinear, the corresponding inequalities become equalities.

We now have all the material we need for the discussion of differentiability of convex functions.

THEOREM 3.4.2. *If $f \in C[a, b]$ and is convex in $[a, b]$, then $f_-'(x)$ and $f_+'(x)$ exist for $a < x < b$ and are nondecreasing functions. When $x \to b$, $f_-'(x)$ either goes to a finite limit $f_-'(b)$ or to $+\infty$. When $x \to a$, $f_+'(x)$ either goes to a finite limit $f_+'(a)$ or to $-\infty$. For each x in (a, b) we have*

$$f_-'(x) \leqq f_+'(x). \tag{3.4.8}$$

Proof. We introduce the one-sided difference quotients

$$\begin{aligned} \Delta_-(x, h, f) &= \frac{1}{h}[f(x) - f(x - h)], \\ \Delta_+(x, h, f) &= \frac{1}{h}[f(x + h) - f(x)]. \end{aligned} \tag{3.4.9}$$

These are functions of two variables x and h. We take $h > 0$ and restrict x and h so that

$$a \leqq x - h < x < x + h \leqq b.$$

We want to prove that the quotients are nondecreasing in their dependence upon x and h and that they are bounded when x and h are suitably restricted.

All this can be shown using Lemma 3.4.2. The method of proof will consist in choosing points P_1, P_2, P_3, P_4 on the graph of f and in noting what the corresponding inequalities imply for the difference quotients. Usually we need the inequality $\{1, 2\} \leqq \{3, 4\}$, but we shall also have occasion to appeal to $\{1, 2\} \leqq \{1, 3\}$. We start by proving boundedness of the difference quotients.

We choose a small positive δ and take four points P_1, P_2, P_3, P_4 whose x-coordinates are such that

$$a \leqq x_1 < x_2 < x_3 = b - \delta < x_4 = b.$$

The inequality $\{1, 2\} \leqq \{3, 4\}$ gives

$$\frac{f(x_2) - f(x_1)}{x_2 - x_1} \leqq \frac{1}{\delta}[f(b) - f(b - \delta)].$$

The right member is $\Delta_-(b, \delta, f)$. Once δ is chosen, this is a fixed finite positive number which we denote by $M_1(\delta)$. The left member is a difference quotient. If we take $x_1 = x - h$, $x_2 = x$, the difference quotient is $\Delta_-(x, h, f)$. Taking instead $x_1 = x$, $x_2 = x + h$ we get $\Delta_+(x, h, f)$. This shows that

$$\begin{aligned} \Delta_-(x, h, f) &\leqq M_1(\delta), \quad a + h \leqq x \leqq b - \delta, \\ \Delta_+(x, h, f) &\leqq M_1(\delta), \quad a \leqq x \leqq b - \delta - h. \end{aligned} \tag{3.4.10}$$

In the same manner we show that the difference quotients are bounded below: there exists an $M_2(\delta)$ such that

$$\begin{aligned} M_2(\delta) &\leqq \Delta_-(x, h, f), \quad a + \delta + h \leqq x \leqq b, \\ M_2(\delta) &\leqq \Delta_+(x, h, f), \quad a + \delta \leqq x \leqq b - h. \end{aligned} \tag{3.4.11}$$

Next we choose three points P_1, P_2, P_3 whose abscissas satisfy

$$a \leqq x_1 = x_0 < x_2 = x_0 + h_1 < x_3 = x_0 + h_2 \leqq b,$$

and use the inequality $\{1, 2\} \leqq \{1, 3\}$. We obtain

$$\frac{1}{h_1}[f(x_0 + h_1) - f(x_0)] \leqq \frac{1}{h_2}[f(x_0 + h_2) - f(x_0)],$$

that is,

$$\Delta_+(x_0, h_1, f) \leqq \Delta_+(x_0, h_2, f), \quad 0 < h_1 < h_2.$$

Hence, for any fixed x_0 the quotient $\Delta_+(x_0, h, f)$ is a nondecreasing function of its second argument h. We can now let h decrease to 0. If $x_0 \geqq a + \delta$, we have

$$\Delta_+(x_0, h, f) \geqq M_2(\delta),$$

so that the left member tends to a finite limit. By Definition 3.4.1 this limit is $f_+'(x_0)$ so that

$$\lim_{h \to 0} \Delta_+(x_0, h, f) = f_+'(x_0) \tag{3.4.12}$$

exists.

The same argument applies to Δ_-. In this case we let the abscissas of the three points be

$$x_0 - h_2, \quad x_0 - h_1, \quad x_0,$$

and use the inequality $\{1, 2\} \leqq \{1, 3\}$.

Now to prove that the one-sided derivatives are monotone in x, we prove first that the difference quotients have this property. We consider the quadrilateral whose vertices have the x-coordinates

$$x_1 = x_0, \quad x_2 = x_0 + h_1, \quad x_3 = x_0 + \eta, \quad x_4 = x_0 + \eta + h_2$$

where $\eta > h_1 > 0$. The inequality $\{1, 2\} \leqq \{3, 4\}$ gives

$$\Delta_+(x_0, h_1, f) \leqq \Delta_+(x_0 + \eta, h_2, f).$$

Here we keep η fixed and let $h_1 \to 0$. The left member tends to the limit $f_+'(x_0)$. We then let $h_2 \to 0$. The right member tends to the limit $f_+'(x_0 + \eta)$. It follows that

$$f_+'(x_0) \leqq f_+'(x_0 + \eta), \tag{3.4.13}$$

so that $f_+'(x)$ is nondecreasing. In the same manner we handle $f_-'(x)$. The behavior of $f_+'(x)$ as $x \to a$ and of $f_-'(x)$ as $x \to b$ follows from the monotonic properties of the derivatives.

Finally, use of the inequality $\{1, 2\} \leqq \{1, 3\}$ in the triangle whose vertices have the x-coordinates

$$x_1 = x_0 - h_1, \quad x_2 = x_0, \quad x_3 = x_0 + h_2$$

gives

$$\Delta_-(x_0, h_1, f) \leqq \Delta_+(x_0, h_2, f).$$

Letting $h_1 \to 0$, $h_2 \to 0$, we get (3.4.8). This completes the proof.

We have not touched on the question of the existence of the ordinary derivative $f'(x)$ which requires equality in (3.4.8).

THEOREM 3.4.2. *Let f be continuous and convex in $[a, b]$. Then the set of points where $f_-'(x)$ and $f_+'(x)$ are discontinuous can be enumerated. They are the vertices of the graph of the convex function. At all other points in (a, b) the derivative $f'(x)$ exists and we have*

$$f_-'(x) = f_+'(x) = f'(x). \tag{3.4.14}$$

Proof. Since the one-sided derivatives are nondecreasing, they have right- and left-hand limits everywhere in (a, b). It was indicated in Section 2.7 that the discontinuities of such a function can be enumerated $\{x_1, x_2, x_3, \cdots\}$, say in decreasing order of the saltus. Suppose now that

$$a < x_0 - \delta < x_0 < x_0 + \delta < b.$$

Using suitable quadrilaterals we can add the following to our list of inequalities:

$$f_+'(x_0 - \delta) \leqq f_-'(x_0) \leqq f_+'(x_0) \leqq f_-'(x_0 + \delta). \tag{3.4.15}$$

Suppose now that $f_+'(x)$ is continuous to the left at $x = x_0$. Then

$$\lim_{\delta \to 0} f_+'(x_0 - \delta) = f_+'(x_0)$$

by the continuity. On the other hand, the first inequality under (3.4.15) requires that the limit does not exceed $f_-'(x_0)$ which, however, cannot exceed $f_+'(x_0)$. This requires that (3.4.14) holds at all points where $f_+'(x)$ is left-continuous. Similarly, if $f_-'(x)$ is continuous to the right at $x = x_0$, then

$$\lim_{\delta \to 0} f_-'(x_0 + \delta)$$

is equal to $f_-'(x_0)$; on the other hand it should be $\geqq f_+'(x_0)$ and this requires that (3.4.14) holds at all points where $f_-'(x)$ is right-continuous. This completes the proof.

EXERCISE 3.4

1. A function f is defined by $f(x) = (x - 1)^2$ for $x \geqq 0$ and $f(-x) = f(x)$ for $x < 0$. Show that f is convex in each of the intervals $(-\infty, 0)$ and $(0, +\infty)$ but not in $(-\infty, +\infty)$. Plot the curve. Find the right- and left-hand derivatives at $x = 0$.

2. If f is defined as $\left(\dfrac{1}{x}\right)$ for $x \neq 0$, find for what values of x the right- and the left-hand derivatives differ and find their values at these points.

3. Give an analytical proof of (3.4.6).

3.5. Differentiation of inverse functions. In this brief section we shall examine the differentiability of inverse functions.

For the geometry of the situation we refer to Figure 3.7. Here we have represented the graphs of the function f and of its inverse $\check{f}$. At a point (x_0, y_0) on f we assume the existence of f' and have drawn the tangent l_t to the curve at this point. Let λ_t be the straight line symmetric to l_t with respect to the line $y = x$. It is clear that λ_t is the tangent to $\check{f}$ at (y_0, x_0). If $f'(x_0) \neq 0$, the

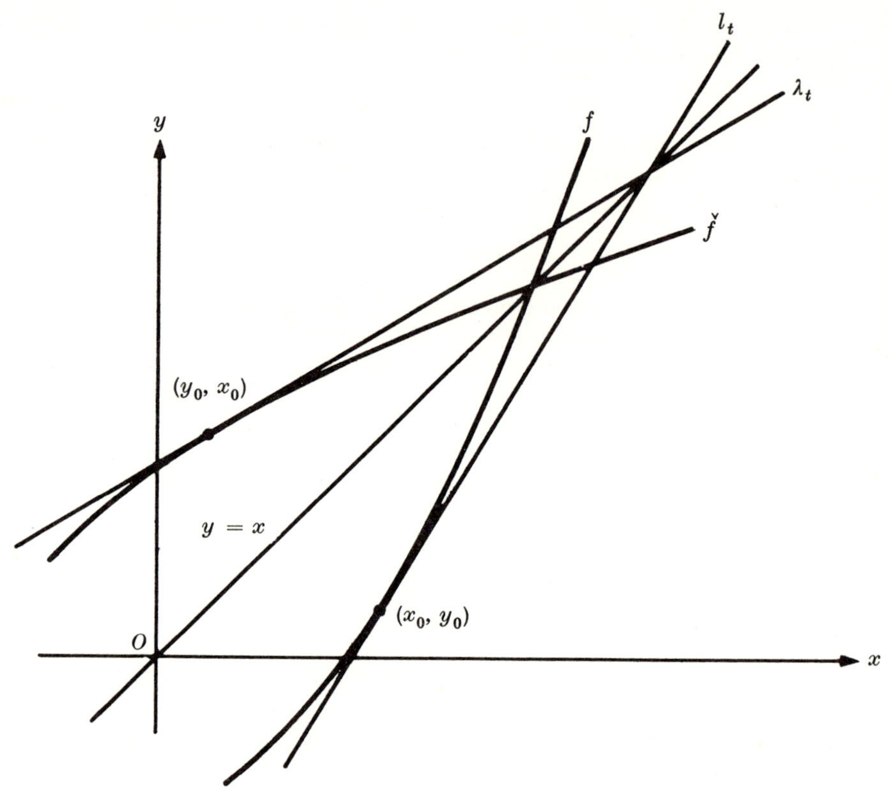

Figure 3.7

line λ_t is not vertical. This would then imply that $\check{f}$ has a derivative at $y = y_0$. Moreover, the product of the slopes of l_t and of λ_t is 1 so we should have

$$f'(x_0)\check{f}'(y_0) = 1. \tag{3.5.1}$$

If on the other hand $f'(x_0) = 0$, then λ_t is vertical and $\check{f}$ cannot have a finite derivative at $y = y_0$.

These geometric considerations suggest what we should try to prove and what assumptions are desirable to make for the proof.

THEOREM 3.5.1. *Suppose that f is continuous and strictly monotone in $[a, b]$ and differentiable at some point x_0, $a < x_0 < b$, where $f'(x_0) \neq 0$. Then $\check{f}$ is differentiable at $y = y_0$ and* (3.5.1) *holds.*

Proof. The existence of the inverse function follows from Theorem 2.8.3. To fix the ideas we may suppose that f is strictly increasing. Then $f'(x) \geqq 0$ wherever it exists. In particular, $f'(x_0) > 0$. Let $x_0 < x_1 < b$, set $f(x_1) = y_1$, and consider the difference quotient

$$\frac{f(x_1) - f(x_0)}{x_1 - x_0} = \frac{y_1 - y_0}{\check{f}(y_1) - \check{f}(y_0)}. \tag{3.5.2}$$

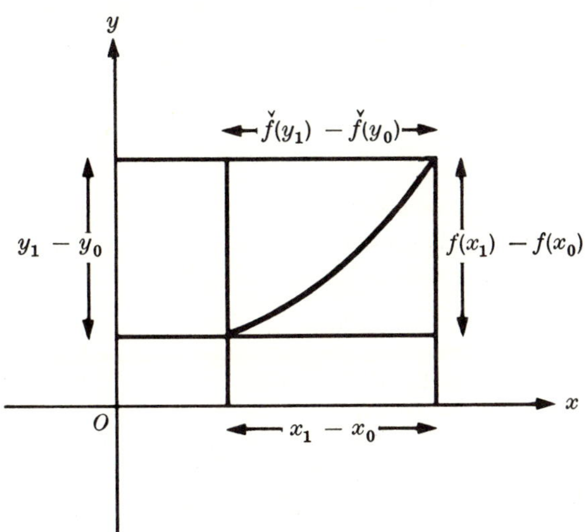

Figure 3.8

See Figure 3.8. The left member tends to the limit $f'(x_0)$ as $x_1 \to x_0$. Since $f(x)$ is strictly increasing and continuous, $y_1 = f(x_1)$ decreases to $y_0 = f(x_0)$ when x_1 decreases to x_0. This means that

$$\lim_{y_1 \to y_0} \frac{y_1 - y_0}{\check{f}(y_1) - \check{f}(y_0)}$$

exists and equals $f'(x_0)$, that is, neither 0 nor infinity. Under such circumstances the reciprocal also has a limit which is finite and not 0. This implies that

$$\lim_{y_1 \to y_0} \frac{\check{f}(y_1) - \check{f}(y_0)}{y_1 - y_0} \equiv \check{f}'(y_0) \tag{3.5.3}$$

exists and that (3.5.1) holds. Strictly speaking we have only shown the existence of a right-hand limit, but the left exists by the same type of argument. This completes the proof.

In the notation of Leibniz we can write

$$\frac{dx}{dy} = \frac{1}{\dfrac{dy}{dx}}, \tag{3.5.4}$$

where on the left the derivative is taken at $y = y_0$ and on the right at $x = x_0$. Again we have highly suggestive formulas which are easy to remember.

As an application let us consider the case

$$y = x^n, \quad x = y^{1/n},$$

where x and y are positive and n is a natural number. Here we have

$$D_x(x^n) = nx^{n-1} = ny^{1-1/n},$$

so that

$$D_y(y^{1/n}) = \frac{1}{n}\, y^{1/n-1}$$

in agreement with previous results.

A less obvious situation is presented by the following problem. In Exercise 2.8, Problem 14 we observed that the cubic equation

$$y^3 + 3y + x = 0 \tag{3.5.5}$$

established a one-to-one mapping of R onto R. Here x is a polynomial in y and obviously differentiable. We have

$$D_y x = -3(1 + y^2).$$

By Theorem 3.5.1 y must be a differentiable function of x and

$$D_x y = -3^{-1}(1 + y^2)^{-1}. \tag{3.5.6}$$

This formula looks rather different from the expressions we normally obtain for a derivative of y with respect to x inasmuch as $D_x y$ is expressed in terms of y instead of in x. In this particular case it would still be possible to express y in closed form in terms of x and hence also to obtain such an expression for the derivative. If, however, y^3 is replaced by y^5 in (3.5.5), such a possibility no longer exists, and we can only express $D_x y$ in terms of y.

The relation (3.5.6) may be regarded as a functional equation, more precisely, a differential equation, which is satisfied by the function $y = f(x)$ defined by (3.5.5).

EXERCISE 3.5

1. When is the tangent to $y = f(x)$ at (x_0, y_0) perpendicular to the tangent of $x = \check{f}(y)$ at (y_0, x_0)?

2. If $y = \left(\dfrac{1-x}{1+x}\right)^{\frac{1}{2}}$, $-1 < x < 1$, find the inverse function. Determine $D_y x$ and use it to find $D_x y$.

3. If (x, y) describes the hyperbola $xy = 1$, prove that y satisfies the differential equation $xD_x y + y = 0$.

4. Find the area of the triangle bounded by the tangent to $xy = 1$ at (x_0, y_0) and the coordinate axes.

5. For $-1 \leqq x \leqq 1$ two continuous functions f_1 and f_2 satisfy the equation $x^2 + y^2 = 1$. Show that they also satisfy the differential equation

$$yD_x y + x = 0.$$

6. If $f(x) = x + \dfrac{1}{x}$, $x > 1$, find the inverse function and its derivative.

3.6. Implicit differentiation. Our point of departure is a particular problem similar to those considered in the preceding Exercise 3.5.

Figure 3.9 shows the graph of the ellipse

$$x^2 + xy + y^2 = 3. \tag{3.6.1}$$

It also shows two horizontal and two vertical tangents which form a square in which the ellipse is inscribed. How do we find the tangent at a point (x_0, y_0) of the ellipse? We have solved such problems in Appendix A using elementary algebra. This approach works well with conics but leads to unsurmountable difficulties for curves of higher order than the second. It is natural to turn to the Calculus for a simple method which will work also for curves of higher order.

We have seen that the Calculus transforms a tangent problem into the question of finding the derivative of the ordinate with respect to the abscissa. In the present case this is a continuous function $f(x)$ defined for $-2 \leqq x \leqq 2$ such that

$$[f(x)]^2 + xf(x) + x^2 = 3. \tag{3.6.2}$$

Actually we have two such functions, an upper and a lower. For the following it is immaterial which of these we consider. To make clear the ideas, let $f(x)$ be the upper ordinate function such that

$$f(x) > -\tfrac{1}{2}x.$$

Once we know the existence of a continuous ordinate function $f(x)$, it is not difficult to prove that it is differentiable and to express the derivative in terms of x and $f(x)$.

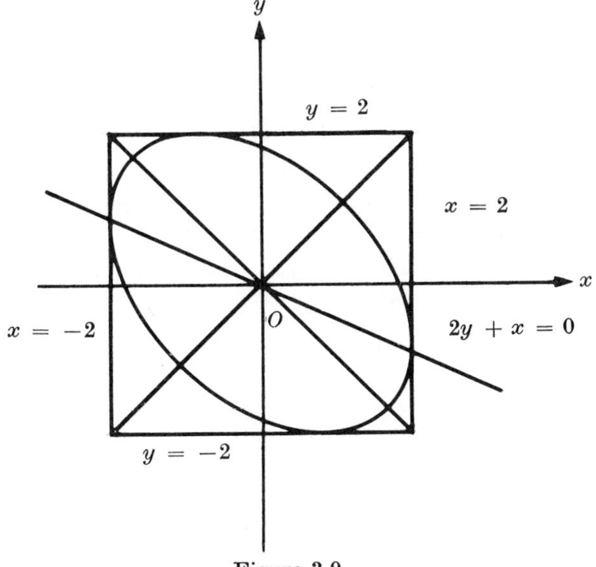

Figure 3.9

Actually we do not need to assume continuity; the equation takes care of this property automatically. First we note that the equation (3.6.2) can have a real solution only when x is restricted to a fairly short interval and that $f(x)$ is necessarily bounded there. In fact, from (3.6.1) we get

$$x^2 + y^2 = 3 - xy \leqq 3 + |xy| \leqq 3 + \tfrac{1}{2}(x^2 + y^2)$$

by (1.6.14). This gives

$$x^2 + y^2 \leqq 6, \tag{3.6.3}$$

and we have equality for $y = -x = \pm\sqrt{3}$. Hence, no point of the ellipse can have a distance from the origin which exceeds $\sqrt{6}$, and we also have $|x| \leqq \sqrt{6}$ and $|y| \leqq \sqrt{6}$ for real solutions. In the last two inequalities we can actually replace $\sqrt{6}$ by 2 as is shown by a more refined analysis.

Next, boundedness of the ordinate function implies continuity in view of (3.6.2). To prove this we take two points (x_1, y_1) and (x_2, y_2) on the ellipse where we assume

$$y_1 > -\tfrac{1}{2}x_1, \quad y_2 > -\tfrac{1}{2}x_2,$$

so that $y = f(x)$ is the upper ordinate function. We have

$$y_2^2 + x_2y_2 + x_2^2 = 3,$$

$$y_1^2 + x_1y_1 + x_1^2 = 3,$$

and, by subtraction

$$y_2^2 - y_1^2 + x_2y_2 - x_1y_1 + x_2^2 - x_1^2 = 0,$$

or

$$(y_2 - y_1)(y_2 + y_1) + x_2(y_2 - y_1) + (x_2 - x_1)y_1 + (x_2 - x_1)(x_2 + x_1) = 0.$$

Collecting terms, we get

$$(y_2 - y_1)(y_1 + y_2 + x_2) + (x_2 - x_1)(x_1 + x_2 + y_1) = 0. \tag{3.6.4}$$

We want to prove that $y_2 \to y_1$ when $x_2 \to x_1$. We note that

$$y_1 + y_2 + x_2 > y_1 + \tfrac{1}{2}x_2 \to y_1 + \tfrac{1}{2}x_1 > 0$$

when $x_2 \to x_1$. Since all the numbers x_1, x_2, y_1, y_2 are bounded, the second product in (3.6.4) tends to 0 when $x_2 \to x_1$. This shows that the first product must tend to zero, and we have just seen that the second factor of this product has a positive lower bound when $x_2 \to x_1$. Hence the first factor must tend to zero and we have proved that

$$\lim_{x_2 \to x_1} f(x_2) = f(x_1)$$

for x_1 and x_2 in the domain of definition of $f(x)$. This is the asserted continuity.

We can now go one step further and prove that continuity implies differentiability. We go back to (3.6.4) and divide by $x_2 - x_1$. This gives

$$\frac{y_2 - y_1}{x_2 - x_1}(y_1 + y_2 + x_2) + x_1 + x_2 + y_1 = 0.$$

We saw above that the quantity inside the parentheses has a positive limit when $x_2 \to x_1$, namely $2y_1 + x_1 = 2f(x_1) + x_1$. For x_2 close to x_1 we can then divide the equation by $y_1 + y_2 + x_2$ and obtain

$$\frac{y_2 - y_1}{x_2 - x_1} = -\frac{x_1 + x_2 + y_1}{y_1 + y_2 + x_2},$$

or, equivalently

$$\frac{f(x_2) - f(x_1)}{x_2 - x_1} = -\frac{x_1 + x_2 + f(x_1)}{f(x_1) + f(x_2) + x_2}.$$

When $x_2 \to x_1$

$$x_1 + x_2 + f(x_1) \to 2x_1 + f(x_1),$$

$$f(x_1) + f(x_2) + x_2 \to 2f(x_1) + x_1 > 0,$$

whence it follows that the difference quotient has a limit and

$$f'(x) = -\frac{2x + f(x)}{2f(x) + x}, \quad -2 < x < 2. \tag{3.6.5}$$

Once we know that $f(x)$ is differentiable, we can find the derivative in a much simpler fashion. We go back to (3.6.2) and apply the rules of the Calculus. Since $f(x)$ is differentiable, so are $xf(x)$ and $[f(x)]^2$. Further, 0 is a differentiable function of x and its derivative is also 0. This gives the identity

$$2f(x)f'(x) + f(x) + xf'(x) + 2x = 0. \tag{3.6.6}$$

If we solve for $f'(x)$, we obtain (3.6.5).

This argument presupposes that $f(x)$ is the upper ordinate function, but the lower one requires only trivial modifications. Formula (3.6.5) holds in both cases.

We can now easily solve the tangent problem which served as the background of this lengthy discussion. Suppose that (x_0, y_0) is a point on the ellipse and that $x_0 + 2y_0 \neq 0$. This means that (x_0, y_0) is not one of the points $(2, -1)$ and $(-2, 1)$, where the tangent is vertical. Then by (3.6.5) the slope of the tangent equals

$$-\frac{2x_0 + y_0}{x_0 + 2y_0} \neq \infty,$$

and the equation of the tangent is

$$y - y_0 = -\frac{2x_0 + y_0}{x_0 + 2y_0}(x - x_0).$$

Clearing fractions and remembering that (x_0, y_0) satisfies (3.6.1), we can reduce this to

$$(2x_0 + y_0)x + (x_0 + 2y_0)y = 6. \tag{3.6.7}$$

We note that the tangent is horizontal if $y_0 + 2x_0 = 0$, that is, at the points $(-1, 2)$ and $(1, -2)$. The four vertical and horizontal tangents are marked in Figure 3.9. It is clear that they bound the ellipse and justify the assertion made earlier that $|x| \leqq 2$ and $|y| \leqq 2$ on the ellipse.

We have discussed this problem in great detail not so much on account of its own importance but because of the many important questions of general interest that are involved in the discussion.

This is not the first time that the student has encountered a functional equation, but previously we have gone from the function to the equation. In this section the approach is different. Here the equation is given and we discuss the function or functions which satisfy it. This can be interpreted in various ways:

(1) We may ask for an explicit solution in closed form involving only known elementary functions. Or

(2) we may prove the existence of a solution, usually by a limiting process. Or

(3) we may appeal to the equation for information concerning the properties of the solutions without the use of any explicit representation.

In the case of equation (3.6.2) there is no difficulty in finding explicit solutions. The equation is quadratic and the two ordinate functions are simply

$$f_1(x) = \tfrac{1}{2}\{-x + \sqrt{3}(4 - x^2)^{\frac{1}{2}}\}, \quad f_2(x) = \tfrac{1}{2}\{-x - \sqrt{3}(4 - x^2)^{\frac{1}{2}}\}.$$

Continuity and differentiability follow directly from these representations and we can also use them for numerical computation if so desired.

Algebraic resolutions are available only for equations of degree less than five in y and their usefulness is limited in the case of cubics and almost nil for quartics. It is consequently necessary to use nonalgebraic methods both to show the existence of solutions and to study their properties.

Actual existence proofs will be given later in Chapter 17, Vol. II. At this stage it is important that the student realizes that the lack of an explicit representation of a function does not put an end to all further discussion. In particular, he should understand that a wealth of information can be hidden in a functional equation and may be had for the asking. This possibility was illustrated above where we gave a detailed discussion of the continuity and differentiability properties of the solutions of (3.6.2) without the use of any representation.

The time has come to make a definite convention. If a curve C is given by an equation of the form

$$F(x, y) = 0, \tag{3.6.8}$$

which is not solved for y, then the student may assume until further notice that in some interval or other there is at least one function f such that

$$F(x, f(x)) = 0 \tag{3.6.9}$$

holds for all values of x in the interval in question. Moreover, if the problem refers to derivatives, then it is to be understood that F is a differentiable function of each of its two arguments, that f is a differentiable function of x, and that the derivative of f can be found by straightforward differentiation of (3.6.9). This does not exclude the possibility that there are points where $f(x)$ has no derivative or an infinite one. This simply means that C may have a vertical tangent somewhere.

The process by which $f'(x)$ is computed is called *implicit differentiation*, and $f(x)$ is known as an *implicit function* since f is not given explicitly.

To illustrate further, let us consider the so-called *leaf of Descartes* whose equation is

$$x^3 + y^3 = 3axy, \tag{3.6.10}$$

where a is a fixed positive number. The graph is shown in Figure 3.10. The straight line

$$x + y + a = 0,$$

also shown in the figure, is an asymptote. The student is familiar with this term from the theory of the hyperbola. The figure shows that there is always at least one ordinate function, sometimes even three. We want to find the derivative of one of these functions and use implicit differentiation to obtain the relation

$$3x^2 + 3y^2 D_x y = 3a(xD_x y + y),$$

so that

$$D_x y = \frac{ay - x^2}{y^2 - ax}.$$

In general this equation gives a finite definite value for $D_x y$, but the expression becomes meaningless at $(0, 0)$ and gives an infinite value at a point where the denominator is 0 but the numerator is different from 0. Elimination of x between the equations

$$x^3 + y^3 = 3axy, \quad y^2 = ax$$

gives the points $(2^{\frac{2}{3}}a, 2^{\frac{1}{3}}a)$ and $(0, 0)$. At the former the tangent is vertical; at the latter there are actually one vertical and one horizontal tangent corresponding to the two branches that pass through the origin.

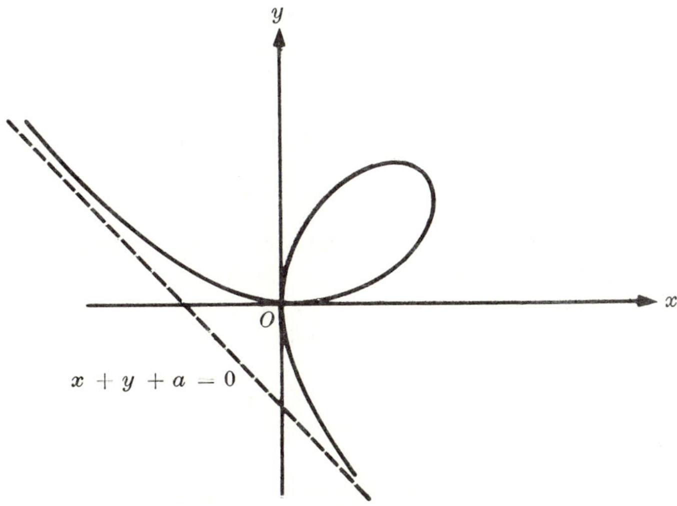

Figure 3.10

EXERCISE 3.6

1. Show that

$$b^2x_0x + a^2y_0y = a^2b^2$$

is the equation of the tangent at (x_0, y_0) of the ellipse

$$b^2x^2 + a^2y^2 = a^2b^2.$$

2. Find the equation of the tangent at (x_0, y_0) of the hyperbola

$$b^2x^2 - a^2y^2 = a^2b^2.$$

3. Find the equation of the tangent at (x_0, y_0) of the parabola

$$y^2 = 4ax,$$

using the method of implicit differentiation.

4. The straight line $y = tx$, t arbitrary real number, intersects the leaf of Descartes in three points, counting the origin as two. Show that the third has the coordinates

$$x = \frac{3at}{1 + t^3}, \quad y = \frac{3at^2}{1 + t^3}.$$

This is a *parametric representation* of the curve in terms of the *parameter* t. The term "parameter" signifies an arbitrary constant or an auxiliary variable. The equation of the ellipse involves the two parameters a and b which are the lengths of the semiaxis. In the present case t is the slope of the radius vector from $(0, 0)$ to (x, y). In dynamical problems time is an often-used parameter for representation of path curves.

5. Use the parametric representation to discuss the behavior of the leaf of Descartes in the neighborhood of the origin. Show that there are two branches going through $(0, 0)$, one corresponding to small values of t and the other to large values.

6. Show that x and y tend to $+\infty$ $(-\infty)$ when $t \to -1 + 0$ $(t \to -1 - 0)$ and that $x + y + a$ decreases to 0 in either case. What is the geometric interpretation?

7. Show that the straight line $x + y = s$ intersects the leaf in two points if and only if $-a < s \leqq 3a$. The two points coincide for $s = 0$ and for $s = 3a$.

8. The curve

$$(x^2 + y^2)^2 = a^2(x^2 - y^2)$$

is known as a *lemniscate* (Latin "lemniscatus" adorned with ribbons, Greek $\lambda\eta\mu\nu\iota\sigma\kappa o\varsigma$ from $\lambda\eta\nu o\varsigma$ wool). It looks like the symbol for infinity. Find the slope at a point (x_0, y_0) distinct from $(0, 0)$, $(a, 0)$, and $(-a, 0)$.

9. Verify that a parametric representation of the right loop of the lemniscate is given by

$$x = a\frac{\sqrt{1 - t^2}}{1 + t^2}, \quad y = a\frac{t\sqrt{1 - t^2}}{1 + t^2}, \quad -1 \leqq t \leqq 1.$$

The parameter is again the slope of the radius vector. Discuss values of t close to ± 1. How is the left loop represented?

Find $D_x y$ by implicit differentiation if:

10. $(x + y)^2 + x = 0$.
11. $x^2 + xy + y^2 - x - y = 0$.
12. $x^3 + y^3 - y = 0$.
13. $x^3 + x^2y + xy^2 + y^3 + y = 0$.
14. $(x^2 + y^2)^2 - 2xy = 0$.
15. $x^5 + y^5 - xy = 0$.
16. At what point or points does the parabola in Problem 10 have a vertical tangent?
17. Same question for the ellipse in Problem 11.
18. Where does the cubic in Problem 12 have a horizontal tangent?
19. Same question for the cubic in Problem 13.
20. The curve in Problem 15 has two branches through the origin, one with a vertical, the other with a horizontal tangent. Is there any other point on the curve where the tangent is horizontal? If so, find it.

3.7. Derivatives of higher order. Let f be defined in (a, b) and differentiable there. Then f' is a function of x defined in (a, b) and may possibly have a derivative there, at isolated points or everywhere.

DEFINITION 3.7.1. *A function $f \in C(a, b)$ is said to be twice differentiable at the point x_0 in (a, b), if f is differentiable in some neighborhood of $x = x_0$ and if f' has a derivative at x_0. The derivative of f' is called the second derivative of f and its value at x_0 is denoted by $f''(x_0)$. If this situation holds at all points x_0 in (a, b), then f is said to be twice differentiable and to have a second derivative in (a, b). We then write f'' for the function defined by $f''(x)$.*

As an example we can take

$$f(x) = x^3 - 5x^2 + 3x - 7$$

for which

$$f'(x) = 3x^2 - 10x + 3, \quad f''(x) = 6x - 10.$$

Here we can obviously continue. There is a third derivative

$$f'''(x) = 6,$$

and all higher derivatives are zero.

The prime notation is not used beyond the third order. One writes

$$f^{(4)} \quad \text{or} \quad f^{(\mathrm{iv})}, \tag{3.7.1}$$

and similarly for fifth and higher orders. These notations are also used for the letter y so that the first four derivatives of $y = f(x)$ are written

$$y', y'', y''', y^{(4)}. \tag{3.7.2}$$

In addition we have the notation of Leibniz and that of Arbogeste-Cauchy. In Leibniz's notation we write

$$\frac{dy}{dx}, \frac{d^2y}{dx^2}, \frac{d^3y}{dx^3}, \frac{d^4y}{dx^4} \tag{3.7.3}$$

and similarly with y replaced by f or by $f(x)$. The student should note carefully the positions of the numerals in these symbols, they follow the "d" in the numerator but the "x" in the denominator. In the D notation we write

$$D^2y,\ D^2f,\ D^2f(x),\ D^3y,\ D^3f,\ D^3f(x),\ \cdots. \tag{3.7.4}$$

We define the nth derivative of y by recursion as the first derivative of the $(n-1)$st derivative:

$$f^{(n)} = D^nf = D[D^{n-1}f]. \tag{3.7.5}$$

As a consequence we have

$$D^{m+n}f = D^m[D^nf] = D^n[D^mf]. \tag{3.7.6}$$

Polynomials have derivatives of all orders, but these are zero as soon as the order exceeds the degree of the polynomial. The basic formulas are

$$D^kx^n = \begin{cases} n(n-1)(n-2)\cdots(n-k+1)x^{n-k}, & k \leqq n, \\ 0, & k > n. \end{cases} \tag{3.7.7}$$

Rational functions have derivatives of all orders which are again rational functions and do not reduce to zero for large n unless the original function is a polynomial. Here the basic formula is

$$D^n(x-a)^{-1} = (-1)^n n!\,(x-a)^{-n-1}, \tag{3.7.8}$$

which implies

$$D^n(x-a)^{-k} = (-1)^n k(k+1)\cdots(k+n-1)(x-a)^{-n-k}. \tag{3.7.9}$$

Actually these formulas enable us to differentiate any rational function as often as we please, for we shall see later that each rational function is a linear combination of negative powers $(x-a)^{-k}$ plus possibly a polynomial. Here the a's run through the roots of the denominator so that some or all of the a's may be complex numbers.

As an example of a function which fails to have higher order derivatives everywhere we can take any odd power of $|x|$. Thus if

$$f(x) = |x|^3,$$

we have

$$f'(x) = \begin{cases} 3x^2,\ x \geqq 0, \\ -3x^2,\ x < 0, \end{cases} \qquad f''(x) = \begin{cases} 6x,\ x \geqq 0, \\ -6x,\ x < 0, \end{cases} \qquad f'''(x) = \begin{cases} 6,\ x > 0, \\ -6,\ x < 0. \end{cases}$$

The higher derivatives are 0 for $x \neq 0$ and $f^{(n)}(0)$ fails to exist for $n \geqq 3$.

A more complicated example is

$$f(x) = \left(\frac{1}{x}\right), \quad 0 < x < 1.$$

Here

$$f(x) = \frac{1}{x} - n, \quad \frac{2}{2n+1} \leqq x \leqq \frac{1}{n},$$

$$f(x) = n + 1 - \frac{1}{x}, \quad \frac{1}{n+1} < x < \frac{2}{2n+1}, \quad n = 1, 2, 3, \cdots.$$

It follows that $f(x)$ has derivatives of all orders except at the endpoints of the subintervals where no derivatives exist. At all other points we can use formula (3.7.8).

If f and g have derivatives of order n, so do their sum, product, and quotient. We have

$$D^n(f + g) = D^nf + D^ng, \tag{3.7.10}$$

$$D^n(fg) = D^nf \cdot g + \binom{n}{1} D^{n-1}f \cdot Dg + \binom{n}{2} D^{n-2}f \cdot D^2g + \cdots + f \cdot D^ng, \tag{3.7.11}$$

where, as usual,

$$\binom{n}{k} = \frac{n!}{k!\,(n-k)!}\,.$$

The first of these formulas is an immediate consequence of Theorem 3.2.1 and the definition of the nth-order derivative. The second formula can be obtained from Theorem 3.2.2 by an induction argument, but the following device is more interesting. It is fairly obvious from Theorem 3.2.2 that we must have a formula of the type

$$D^n(fg) = D^nf \cdot g + C_{n,1}D^{n-1}f \cdot Dg + C_{n,2}D^{n-2}f \cdot D^2g + \cdots + f \cdot D^ng,$$

where the coefficients $C_{n,k}$ are natural numbers independent of the choice of f and g. The values of the coefficients can be obtained by substituting suitable special functions. We take, for instance,

$$f(x) = x^{n-k}, \quad g(x) = x^k,$$

so that the left-hand side becomes

$$D^n(fg) = D^n(x^{n-k}x^k) = D^nx^n = n!.$$

On the right-hand side the derivatives of f occur in decreasing orders while those of g are in increasing orders. This gives

$$D^{n-j}x^{n-k} = \begin{cases} 0 & \text{if } j = 0, 1, \cdots, k-1, \\ (n-k)! & \text{if } j = k, \end{cases}$$

and

$$D^jx^k = \begin{cases} k!, & j = k, \\ 0, & j > k. \end{cases}$$

This means that all terms on the right drop out with one exception only, so that we get the equation

$$n! = C_{n,k}(n-k)!\,k!.$$

Hence $C_{n,k}$ has the value asserted in (3.7.11). This is known as the *formula of Leibniz.*

There is no simple formula for the nth derivative of the quotient f/g when it is known that f and g have nth-order derivatives. It is not difficult to show, however, that the nth derivative of the quotient f/g is also a quotient where the denominator is the $(n + 1)$st power of g and the numerator is the sum of a certain number of products each involving $n + 1$ factors and being of *weight* n. Here the weight is obtained by adding the orders of the derivatives involved. In particular, no derivative of order greater than n can occur.

Next we examine composite functions for higher derivatives. *We assume that f and g have derivatives of order $\leqq n$ and we want to prove that $F(x) = f[g(x)]$ also has derivatives of order $\leqq n$.* This is proved by induction. The general formula is complicated though it is not too difficult to analyze its structure. We restrict ourselves to the first four derivatives. If

$$F = f[g],$$

then

$$\begin{aligned} F' &= f'[g]g', \\ F'' &= f''[g](g')^2 + f'[g]g'', \\ F''' &= f'''[g](g')^3 + 3f''[g]g'g'' + f'[g]g''', \\ F^{(4)} &= f^{(4)}[g](g')^4 + 6f'''[g](g')^2g'' + f''[g]\{3(g'')^2 + 4g'g'''\} + f'[g]g^{(4)}. \end{aligned} \tag{3.7.12}$$

In the special case $g(x) = x^2$, the general formula is sufficiently simple to be quoted. We have

$$\begin{aligned} F^{(n)} = (2x)^n f^{(n)}(x^2) &+ A_{n,1}(2x)^{n-2} f^{(n-1)}(x^2) \\ &+ A_{n,2}(2x)^{n-4} f^{(n-2)}(x^2) + \cdots + A_{n,k}(2x)^{n-2k} f^{(n-k)}(x^2) \end{aligned} \tag{3.7.13}$$

where $2k$ is the largest integer $\leqq n$ and

$$A_{n,m} = \frac{1}{m!} n(n-1)(n-2)(n-3)\cdots(n-2m+2)(n-2m+1).$$

It is fairly obvious that the formula must have this structure. The numerical coefficients can be found by substituting special functions such as

$$f(x^2) = \frac{2}{x^2 - 1} = \frac{1}{x-1} - \frac{1}{x+1}.$$

We leave the details to the reader.

Finally let us consider the inverse function. *Suppose that $y = f(x)$ is strictly monotone and twice differentiable in (a, b) and suppose that $f'(x) \neq 0$ for $a < x < b$. Then $x = \check{f}(y)$ exists and is twice differentiable with respect to y.* For we have

$$\check{f}'(y) = \frac{1}{f'(x)}$$

which is a differentiable function of x. On the other hand,

$$\check{f}'(y) = \check{f}'[f(x)]$$

is a composite function. It is differentiable with respect to x and so is $f(x)$. It follows that $\check{f}'(y)$ must be a differentiable function of y. Hence

$$\check{f}''(y)f'(x) = -\frac{f''(x)}{[f'(x)]^2},$$

or

$$\check{f}''(y) = -\frac{f''(x)}{[f'(x)]^3}, \tag{3.7.14}$$

which is the desired result. The method extends to higher derivatives.

EXERCISE 3.7

For the following functions find the derivative of the indicated order:

1. x^7, 7. 2. x^5, 6. 3. $(4x - 5)(2x^2 + 9)(x + 7)^3$, 6. 4. $\sqrt[5]{x^3}$, 3.

5. $\dfrac{x^3}{x-1}$, 3. 6. $x^2 + \dfrac{1}{x^2}$, 2. 7. $x^{\frac{5}{3}}$, 4.

8. Find the nth derivative of

$$\frac{ax + b}{cx + d},$$

where a, b, c, d are real constants and $ad - bc \neq 0$.

9. Find the third derivative of $\dfrac{ax^2 + b}{cx^2 + d}$.

Find the nth derivative of:

10. $\dfrac{1}{x(x+1)}$. 11. $\dfrac{1}{x^2 - 1}$.

12. Verify that

$$\frac{x^4 + 1}{x^3 - x} = x - \frac{1}{x} + \frac{1}{x-1} + \frac{1}{x+1}$$

and find the nth derivative, $n > 1$.

13. Find, and simplify as far as possible, the second derivative of $(1 - x^2)^{\frac{1}{2}}$.

14. Verify that

$$fg'' - f''g = D[fg' - f'g].$$

This is a special case of the so-called *identity of Lagrange*.

15. If $f'' > 0$ in (a, b), present some reasons for believing that f is convex.

16. Prove that $X_n(x) \equiv D^n(x^2 - 1)^n$ is a polynomial of degree n which is an even or an odd function of x according as n is even or odd. Use Leibniz's formula to show that $X_n(x)$ is positive for $x \geqq 1$.

17. Find $D^3f(x^3)$. 18. Find $D^4f(x^4)$. 19. Find $D^2f(\sqrt{x})$.

20. Find $D^4f(x - x^2)$. 21. Find $D^3f(1 + x + x^2)$.

22. If $f(x)$ is strictly monotone and three times differentiable, prove that $\check{f}(y)$ is three times differentiable and that

$$\check{f}'''(y) = \frac{3[f''(x)]^3 - f'(x)f'''(x)}{[f'(x)]^5}.$$

23. A point $P = (x, y)$ describes a curve C in the plane. At the time t the position of P is given by the equations

$$x = t^2, \quad y = t^3.$$

Sketch C. Express y in terms of x. Note that the curve C corresponds to two functions of x according to the sign of y.

24. By going back to first principles, show that the slope of this curve C at the time t is given by

$$\frac{D_t y}{D_t x} = \tfrac{3}{2}t.$$

Express the slope as a function of x (two cases!).

25. Find the second derivative of y with respect to x, in terms of t and in terms of x.

26. Express the third derivative of y with respect to x in terms of t.

3.8. Differentials. The term Calculus is an abbreviation of the classical title Differential and Integral Calculus. This indicates that some constructs called *differentials* played a sufficiently important role in the development to give their name to the subject. This is unfortunately true. Differentials are one of the diffuse concepts which Leibniz originated and they have caused endless confusion and misunderstanding. In the case of functions of one variable there is little that differentials can do that derivatives cannot do and usually better. It is only for functions of several variables that the concept is really important. On the other hand, differentials have played an important role, the notation is used everywhere, and differentials are popular in the applications for purposes of approximation. Hence it is necessary to make brief mention of the facts without obscurities.

Suppose that f is defined and differentiable in an interval (a, b). We write

$$\Delta f(x, h) \equiv f(x + h) - f(x) \tag{3.8.1}$$

for $x, x + h \in (a, b)$. This is the *first difference of f at the point x with span or increment h.* We recall formula (3.2.2) which says

$$\Delta f(x, h) = f'(x)h + o(h). \tag{3.8.2}$$

Definition 3.8.1. *The differential of f at the point x with increment h is*

$$\delta f(x, h) \equiv f'(x)h. \tag{3.8.3}$$

These three formulas give

Theorem 3.8.1. *We have*

$$\lim_{h \to 0} \frac{1}{h} \{\Delta f(x, h) - \delta f(x, h)\} = 0. \tag{3.8.4}$$

For the geometry of the situation, see Figure 3.11. Here we have marked the arc PQ of the curve f between the abscissas x and $x + h$ and the tangent

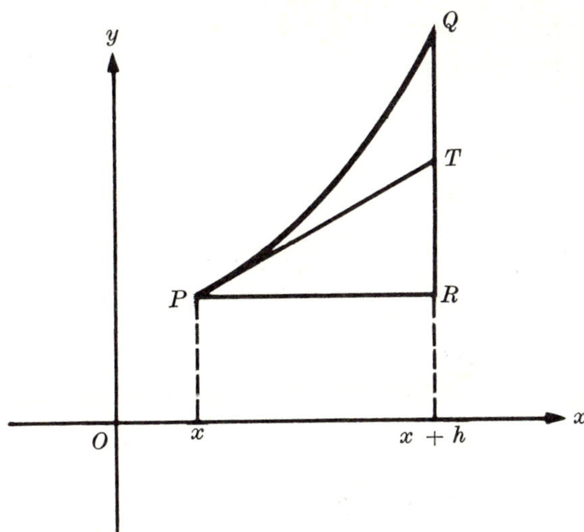

Figure 3.11

PT at P. Further

$$\Delta f(x, h) = RQ, \delta f(x, h) = RT,$$

so that

$$\Delta f(x, h) - \delta f(x, h) = TQ.$$

The theorem then asserts that TQ is small in comparison with PR in the sense that the ratio tends to 0 with the length of PR:

$$\frac{TQ}{PR} \to 0 \quad \text{when} \quad PR \to 0.$$

Using the sign "$\sim$" in the sense "is approximately equal to" we can write

$$\Delta f(x, h) \sim \delta f(x, h). \tag{3.8.5}$$

This is the vague way that differentials are used in the applications. The student may very well ask: What does it all mean? It should mean Theorem 3.8.1, but that is normally not the way in which the formula is used. When we replace differences by differentials, we hope to get a reasonably good approximation, that is, that the error committed be small, absolutely or relatively. The formula and the derivation tell us nothing about the size of the error, only that it is $o(h)$. This can often be improved to $O(h^2)$, but as long as we do not know anything about the magnitude of the bound which lies hidden in the "O," we are not much better off. This is obviously an important problem and it will be attacked later. See Section 6.2 for the case of functions having higher order derivatives. There it will be shown that a good approximation of the error is

$$\tfrac{1}{2}f''(x)h^2. \tag{3.8.6}$$

In the meantime we give a numerical example to illustrate the situation. We take

$$f(x) = x^3 - 2x^2 + 5x - 10, \quad x = 10.$$

The following table lists the results for four different values of the span h. The last column gives the relative error in percent, and the preceding column the absolute error.

h	$\Delta f(10, h)$	$\delta f(10, h)$	$\Delta - \delta$	$100 \dfrac{\Delta - \delta}{\Delta}$
1	294	265	29	9.87
0.5	139.625	132.5	7.125	5.1
0.1	26.781	26.5	0.281	1.04
0.01	2.653	2.65	0.003	0.11

The improvement in the absolute and relative errors for decreasing values of h is striking. Whether or not the use of the approximate value for Δ is justified will depend upon the degree of accuracy that the problem requires.

The notation $\delta f(x, h)$ that we have used for the differential is customary in advanced mathematics, but it is not the traditional notation which goes back to Leibniz. The conventional notation is

$$dy \quad \text{or} \quad df, \tag{3.8.7}$$

so that

$$dy = df = f'(x)h. \tag{3.8.8}$$

Now if $f(x) = x$, we have $f'(x) = 1$ and the last formula becomes

$$dx = h. \tag{3.8.9}$$

This is normally used to rewrite (3.8.8) in the form

$$dy = f'(x)\, dx. \tag{3.8.10}$$

We have now obtained a definition of the symbols dx and dy introduced in an obscure manner by Leibniz. Since dx and dy are real numbers and $dx \neq 0$, we can divide by dx and obtain

$$\frac{dy}{dx} = f'(x)$$

in agreement with previous notation. When the left member was first introduced in formula (3.2.13) as notation for the derivative, neither dy nor dx had any independent meaning. They are now defined: dx is the increment of x denoted by h or Δx, and dy is defined by (3.8.10).

During the first hundred years of the development of the Calculus, dx and dy were regarded as infinitely small quantities, often called *infinitesimals*. In this realm of ideas the tangent line to the curve $y = f(x)$ passed through the two points (x, y) and $(x + dx, y + dy)$ which were considered to be infinitely close to each other. Nevertheless, the formula for the slope of a straight line through two given points was used and gave

$$\frac{dy}{dx}$$

as the slope of the tangent and of the curve. The great analysts of the nineteenth century such as Cauchy, Bernhard Riemann (1826–66), and Weierstrass brought order in this chaos, but far too much obscurity survives to this day in the teaching of elementary calculus "made easy" by avoiding limits.

We have also differences and differentials of higher order. Thus, the second difference of f at x with span h is

$$\Delta^2 f(x, h) = f(x + 2h) - 2f(x + h) + f(x). \tag{3.8.11}$$

We note that

$$\Delta^2 f(x, h) = \Delta f(x + h, h) - \Delta f(x, h). \tag{3.8.12}$$

If f is twice differentiable, we define the second differential by

$$\delta^2 f(x, h) = f''(x)h^2. \tag{3.8.13}$$

It may be shown that

$$\Delta^2 f(x, h) = \delta^2 f(x, h) + o(h^2), \tag{3.8.14}$$

whence it follows that

$$f''(x) = \lim_{h\to 0} h^{-2}\Delta^2 f(x, h). \tag{3.8.15}$$

The classical notation for $\delta^2 f(x, h)$ is

$$d^2y \quad \text{or} \quad d^2f.$$

With $h = dx$, we can then write

$$d^2f = f''(x)\,(dx)^2.$$

Here it is customary to omit the parentheses and to write simply

$$d^2f = f''(x)\,dx^2. \tag{3.8.16}$$

Division by dx^2 gives

$$f''(x) = \frac{d^2f}{dx^2},$$

in agreement with previous notation. This explains the difference in the placement of the superscripts in the numerator and in the denominator which was observed earlier.

Corresponding definitions hold for differences and differentials of higher order. Thus we have

$$\Delta^n f(x, h) = \sum_{k=0}^{n} (-1)^{k+n}\binom{n}{k} f(x + kh), \tag{3.8.17}$$

$$\delta^n f(x, h) = f^{(n)}(x)h^n. \tag{3.8.18}$$

It will be shown later that

$$\Delta^n f(x, h) = \delta^n f(x, h) + o(h^n). \tag{3.8.19}$$

We shall now take up some questions involving applications of differentials. Three different uses of varying importance will be discussed: (1) *approximate computation;* (2) *approximate errors;* and (3) *differential notation.*

Example 1. As an example of approximate computation let us find the value of $\sqrt[4]{17}$ without the use of tables and let us also try to get some idea of the accuracy of the approximations. We can do this with the use of differentials. Formula (3.8.5) can be used to get a first approximation, and formula (3.8.6) will give some idea of the error. We take

$$f(x) = x^{\frac{1}{4}}, \quad x = 16, \quad h = 1.$$

First we get

$$f'(x) = \tfrac{1}{4}x^{-\frac{3}{4}}, \quad f''(x) = -\tfrac{3}{16}x^{-\frac{7}{4}}.$$

The values of f, f', and f'' for $x = 16$ are

$$2, \ \tfrac{1}{32}, \quad \text{and} \quad -\tfrac{3}{2048},$$

respectively. We have

$$\Delta f(16, 1) = \sqrt[4]{17} - 2, \quad \delta f(16, 1) = \tfrac{1}{32}.$$

Hence

$$\begin{aligned} f(17) &= f(16) + \Delta f(16, 1) \\ &\sim f(16) + \delta f(16, 1) = 2\tfrac{1}{32}. \end{aligned}$$

This gives as a first approximation

$$\sqrt[4]{17} \sim 2\tfrac{1}{32} = 2.031250.$$

This value is too large and formula (3.8.6) says that the error is approximately

$$-\frac{3}{4096} = -0.000732.$$

Adding this correction to the first approximation we get a second and better approximation, namely

$$\sqrt[4]{17} \sim 2.030518.$$

This time the approximation is too small, but it may be shown that the error does not reach

$$\frac{1}{3!}f'''(16) = \frac{7}{2^{18}} < 0.000027.$$

This would give a third approximation

$$\sqrt[4]{17} \sim 2.030545$$

and here the error is less than one digit in the last decimal place. It is clear that in the discussion of the problem we have anticipated methods and results which will be proved later. Only the first approximation is a bona fide application of differentials as such. Any estimate of errors must go beyond formula (3.8.5).

Example 2. *Estimating errors* is a topic of interest to the applications of Calculus to physical measurements. Let us take a simple problem. We want to find the volume of steel balls, assumed to be perfect spheres, by measuring the diameter. The machine used for this purpose does not give the true value d for the diameter, but an approximate value $d + h$. The relative error in this measurement is

$$\frac{h}{d}.$$

The volume of the sphere is

$$V(d) = \tfrac{1}{6}\pi d^3,$$

so that

$$\delta V(d, h) = \tfrac{1}{2}\pi d^2 h.$$

The relative error in the volume is

$$\frac{\Delta V(d, h)}{V(d)} \sim \frac{\delta V(d, h)}{V(d)} = 3\frac{h}{d}.$$

Thus the relative error in the volume measurement is (approximately) 3 times the relative error in the original measurement cf the diameter. In the same manner one shows that the relative error in the computed area of the surface of the sphere is (approximately) 2 times the relative error in the diameter.

Finally we shall have something to say about the use of *differential notation* to express differentiation formulas. This is useful in various connections, in integration for instance, and in the elementary theory of differential equations. In the following, u and v are thought of as differentiable functions of some auxiliary variable, say x. This means that the differentials du and dv are well defined:

$$du = u'(x)\,dx, \quad dv = v'(x)\,dx.$$

We can then express the basic formulas of differentiation given in Section 3.2 in differential form as follows:

$$\begin{cases} dC = 0, \quad d(Cu) = C\,du, \\ d(u + v) = du + dv, \\ d(uv) = u\,dv + v\,du, \\ d\left(\dfrac{u}{v}\right) = \dfrac{v\,du - u\,dv}{v^2}, \\ d(u^n) = n\,u^{n-1}\,du. \end{cases} \tag{3.8.20}$$

EXERCISE 3.8

In the following six problems, find $\Delta f(a, h)$, $\delta f(a, h)$, and $\Delta f(a, h) - \delta(a, h)$, as well as the approximate error given by (3.8.6).

1. $f(x) = x^2 + x + 1$, $a = 1$, $h = 0.01$.
2. $f(x) = x^2 + x - 2$, $a = 1$, $h = -0.02$.
3. $f(x) = x^3 - 3x^2 + x - 3$, $a = 3$, $h = 0.01$.
4. $f(x) = x^{-1}$, $a = 3$, $h = -0.1$.
5. $f(x) = x^{\frac{1}{2}}$, $a = 4$, $h = 0.05$.
6. $f(x) = x^{-\frac{1}{2}}$, $a = 4$, $h = -0.05$.

In the following six problems, make approximate computation using the differential and estimate the error as in Example 1.

7. $\sqrt{101}$. 8. $\sqrt[3]{28}$. 9. $(0.96)^{-\frac{1}{2}}$.

10. $\sqrt[4]{82}$. 11. $33^{\frac{1}{5}}$. 12. $37^{\frac{1}{2}}$.

Prove the following approximation formulas which all are of the form $f(1 + h) \sim f(1) + f'(1)h$.

13. $(1 + h)^2 \sim 1 + 2h$.
14. $(1 + h)^{-1} \sim 1 - h$.
15. $(1 + h)^{\frac{1}{2}} \sim 1 + \frac{1}{2}h$.
16. $(1 + h)^{-\frac{1}{2}} \sim 1 - \frac{1}{2}h$.
17. $(1 + h)^{-2} \sim 1 - 2h$.
18. $(1 + h)^{\frac{3}{2}} \sim 1 + \frac{3}{2}h$.
19. $(1 + h)^{\frac{1}{3}} \sim 1 + \frac{1}{3}h$.
20. $(1 + h)^{-\frac{1}{3}} \sim 1 - \frac{1}{3}h$.

In each of Problems 13–20 the difference D between the left and the right members is $O(h^2)$. It is to be proved by elementary means that D satisfies the following inequalities for $0 < h < 1$.

21. In Problem 13 $D = h^2$.
22. In Problem 14 $0 < D < h^2$.
23. In Problem 15 $0 > D > -\frac{1}{8}h^2$.
24. In Problem 16 $0 < D < \frac{1}{3}h^2$.
25. In Problem 17 $0 < D < 5h^2$.
26. In Problem 18 $0 < D < \frac{7}{8}h^2$.
27. In Problem 19 $0 > D > -\frac{4}{27}h^2$.
28. In Problem 20 $0 < D < \frac{2}{9}h^2$.

[*Hint:* Simplify D by clearing fractions and/or using formula (1.6.7). Note that $1 < 1 + h < 2$ and $h^2 < h$.]

29. How should the preceding estimates be modified if $-\frac{1}{2} < h < 0$. Omit Problem 28.

30. Use Problem 23 to show that

$$10.049875 < \sqrt{101} < 10.05.$$

31. If a is a fixed positive number, show that

$$(x + a)^{\frac{1}{3}} - x^{\frac{1}{3}} \to 0$$

as $x \to \infty$. Find an approximate value for the difference.

In analogy with Example 2 in the text, determine the approximate relative error

$$\frac{\delta f(a, h)}{f(a)},$$

if f, a, and h are given by

32. $f = x^3$, $a = 10$, $h = 0.01$.
33. $f = x^{-1}$, $a = 3$, $h = -0.1$.
34. $f = x^{\frac{3}{2}}$, $a = 4$, $h = 0.05$.
35. $f = x^2 + x + 1$, $a = 3$, $h = 0.02$.
36. $f = x^{\frac{1}{2}}$, $a = 4$, $h = -0.02$.

37. State a problem involving physical measurements which leads to Problem 32.

38. Write formula (3.8.17) in full for $n = 5$.

39. If f is a polynomial of degree two, show that f'' is a constant. If this constant is $2a$, show that

$$\Delta^2 f(x, h) = 2ah^2.$$

40. If $f(x)$ is a polynomial of degree two, show that

$$f(x+h) = f(x) + \frac{h}{1!} f(x) + \frac{h^2}{2!} f''(x).$$

41. What is the corresponding formula for a third degree polynomial?

MISCELLANEOUS PROBLEMS FOR CHAPTER 3

1. Find the equation of the tangent to the parabola $3y = x^2 - 3x + 3$, parallel to the line $y = x$.

2. For what value of a does the curve $4y = ax - x^2$ intersect the x-axis at an angle of 45°?

3. Find the equation of the tangent of slope 3 to the parabola $y = x^2 - 4x$.

4. At what angle do the two parabolas

$$4y^2 = 1 - 4x, \quad 4y^2 = 1 + 4x$$

intersect? [*Hint:* The angle is the angle between their tangents at the point of intersection.]

5. A straight line of slope 1 is normal (= perpendicular) to the ellipse $4x^2 + y^2 = 4$. Find its equation (two solutions!).

6. Find the equation of the normal of slope 2 to the parabola $y = x^2 - 4x$.

7. Find the derivative of $x(1 - x^2)^{-\frac{1}{2}}$.

8. Find the second derivative of $\left(x + \frac{1}{x}\right)^6$.

9. Same question for $(x^3 + 1)^2$.

10. Find the third derivative of $(3x - 1)/x^5$.

11. Same question for $(x^2 + 1)^3$.

12. Show that if f is defined as $x\left(\frac{1}{x}\right)$ for $x \neq 0$ and $f(0) = 0$, then f does not have a derivative for $x = 0$.

13. Show that if $g(x) = x^2\left(\frac{1}{x}\right)$ for $x \neq 0$ and $g(0) = 0$, then $g'(0) = 0$.

Find a function $f(x)$ whose derivative has the value:

14. $2x$. 15. $x^3 + 1$. 16. $(x+1)^{-2}$. 17. $x - \frac{1}{x^2}$.

18. $x^5 + \frac{1}{x^5}$. 19. $\sqrt{x}$. 20. $\frac{1}{\sqrt{x}}$. 21. $x^{\frac{1}{3}}$.

Such a function is called the *antiderivative* or the *indefinite integral* of the given function. Is it uniquely determined?

22. Show that the derivative of a rational function is also a rational function. Is it true that each rational function is the derivative of a rational function?

Which of the following functions is the derivative of a rational function? If it has this property, find the function of which it is the derivative.

23. $\frac{1}{x^2}$. 24. $\frac{1}{(x-a)^3}$. 25. $\frac{1}{x+1}$. 26. $\frac{1}{x(x-1)}$.

27. $\dfrac{x+1}{x}$.

28. $\dfrac{1}{x^2} + \dfrac{1}{x-1}$.

29. $\dfrac{3}{x^4} + \dfrac{1}{(x-1)^2} - x$.

30. $\dfrac{x^2+1}{x^2}$.

31. Show that the derivative of

$$f(x) = \frac{2}{x} + \frac{1}{x+1} + \frac{4}{x+3}$$

is negative wherever it exists. Use this to show that the equation $f(x) = a$ has a root in each of the intervals $(-3, -1)$ and $(-1, 0)$. If $a > 0$, there is a root in $(0, \infty)$; if $a < 0$, there is one in $(-\infty, -3)$.

32. Let $f(x)$ be defined and differentiable for $-1 < x < 1$ and let its range be the interval $(-1, a)$ where $0 < a < 1$. Let $f(0) = 0$, $f'(0) = \alpha > 0$. Form successively

$$p(x) = \frac{a - f(x)}{1 - af(x)}, \; q(x) = [p(x)]^{\frac{1}{2}}, \; g(x) = \frac{a^{\frac{1}{2}} - q(x)}{1 - a^{\frac{1}{2}}q(x)}.$$

Determine the range of each function. What is $g(0)$? Express $g(x)$ in terms of $f(x)$ and find $g'(x)$ by the chain rule. Compute $g'(0)$ and show that $g'(0) > f'(0)$.

33. If n and k are positive numbers, show that

$$(n^2 + k)^{\frac{1}{2}} < n + \frac{k}{2n}.$$

Find an upper bound for the difference.

34. If $f(x)$ is a rational function, show the existence of numbers a and b such that $f(x)$ is monotone for $x < a$ and for $b < x$.

35. If $f(x)$ is convex in (a, b) and is twice differentiable there, show that $f''(x) \geqq 0$.

36. $P(x)$ is a polynomial of degree three in x and $y^2 = P(x)$. Express

$$y^{-2}D[y^3D^2y]$$

in terms of P and its derivatives and show that its value is a constant.

37. If a function $f(x)$ exists which is differentiable and satisfies the equation $f'(x) = f(x)$, show that $f(x)$ has derivatives of all orders and that all derivatives are equal.

38. Show that the polynomial $X_n(x)$ of Exercise 3.7, Problem 16, satisfies the differential equation

$$D[(1 - x^2)DY] + n(n+1)Y = 0.$$

This is known as Legendre's equation after A. M. Legendre (1752–1833). $X_n(x)$ is a constant multiple of the nth polynomial of Legendre.

39. Show that if a function $f(x)$ satisfies the preceding differential equation for $-1 < x < 1$, then $f(x)$ must have derivatives of all orders. Note that $f(x)$ is not necessarily a polynomial so the question is not trivial.

40. If $f(x)$ satisfies Legendre's equation, what differential equation is satisfied by $f^{(k)}(x)$?

41. Show that the zeros of the derivative of the polynomial

$$x(x-1)(x-2)(x-3)$$

are real and find in what intervals they are located.

42. For the following problem, see the end of Section 3.6 and Problem 4 of Exercise 3.6. Show that three points P_1, P_2, P_3 on the leaf of Descartes lie on a straight line if and only if their parameters satisfy $t_1t_2t_3 = -1$. The points are assumed to be distinct.

43. Compute the second derivative of y in terms of x and y if (x, y) is a point on the leaf of Descartes.

44. $f(x)$ is strictly monotone and three times differentiable. Show that

$$f'(x)f'''(x) - 3[f''(x)]^2 = 0$$

implies that

$$\check{f}'''(y) = 0.$$

From this it follows that $\check{f}$ is a polynomial in y of degree $\leqq 2$.

4

Elements of Integration

4.1. Measure of area. There are two different aspects of integration, both highly important for analysis as well as for the applications. On the one hand, *integration may be considered as the inverse operation of differentiation.* In this context we ask what function has a given function f as its derivative. The required function is called the *antiderivative* or the *indefinite integral* or the *primitive* of f. Thus, x^2 is a primitive of $2x$, but not the only one since we can clearly add any constant. More generally, if n is an integer, $n \neq -1$, then

$$\frac{x^{n+1}}{n+1} + C \text{ is a primitive of } x^n.$$

On the other hand, *with the given function f is associated a family of interval functions, the definite integral of f over the interval $[\alpha, \beta]$ which is any subinterval of the domain of definition of f.* The connection between the two concepts is that *the interval function for the interval $[\alpha, x]$ is a differentiable function of x and that its derivative equals $f(x)$* under rather general assumptions on f.

To the beginner the second aspect sounds more forbidding than the first. Actually it is quite easy to visualize and it ties in with the age-old problem: *What is the area of a region with known boundary?*

The Babylonians and the Egyptians could answer this question for triangles and simple regions that can be triangulated, that is, subdivided into a finite number of triangles. The Greeks as usual went much further than their forerunners and sometime mentors. No longer do practical considerations dominate but logical philosophical ones. Their feeling for abstract concepts and for mathematical rigor was unsurpassed until our century. Among the many names that could be mentioned, we select two: Eudoxos from Knidos (408–355 B.C.) and Archimedes of Syracuse, the foremost mathematician produced by Greek civilization. To Eudoxos apparently we owe the *method of exhaustion* (see below) which can be described as the idea of approximating a geometrical figure whose length, area, or volume is to be found, by inscribed or circumscribed figures whose measures can be computed. This requires a limiting process and a definition of the measure of the limit as the limit of the measures. Using such considerations Eudoxos found the volumes of the pyramid and of the circular cone. Archimedes used this method to find the circumference of a circle, the areas of the ellipse and the parabolic segment, the center of gravity of a triangle, the volume and surface area of a sphere, etc. On his own evidence we know that the proofs did not spring full-fledged from his head and that he often resorted to mechanical considerations. Once he knew what to prove, he would produce strictly rigorous and highly ingenious arguments to demonstrate his propositions. He varied his approach from one problem to the next, however, and failed to see the underlying principle.

Later generations had little to add to this problem and it was over 1880 years before any new significant development took place. Among the forerunners of the inventors of the Calculus during the middle of the seventeenth century, Cavalieri and Barrow were outstanding in developing the geometrical function concept. Cavalieri used the exhaustion principle with great success in the study of area problems for curves of the form

$$y = P(x),$$

where $P(x)$ is a polynomial of degree $\leqq 9$. Beyond this stage the algebra became too complicated. What he did, was to find the definite integral of x^n, $n \leqq 9$, by the method of exhaustion. Barrow had the general concept of the definite integral.

The invention of Calculus with its emphasis on explicit analytic expressions brought the indefinite integral into the foreground and gave a new form to Barrow's fundamental theorem. Gone were the days of painstaking rigor. Cavalieri had carefully followed the high logical standards of Archimedes. Not so the new masters: their methods seemed to handle all problems with the greatest of ease and it took almost a century before problems were encountered which required a real understanding of infinitary processes and ultimately led to a reexamination of all basic concepts.

"Area" is one of these. In the terminology of Section 2.3 it is a set function defined for certain plane sets. We avoid inessential difficulties by not restricting ourselves severely to the sets which are of immediate interest to the Calculus. Under these circumstances it is preferable to use a more neutral term than "area" and we shall speak of (two-dimensional) *measure* instead. Seven postulates list the properties that this measure should have. The last one is the exhaustion principle; all the others follow directly from our intuitive ideas of area.

Two sets S_1 and S_2 will be said to be *directly congruent* if there is a rigid motion which takes S_1 into S_2, and *reversedly congruent* if there is a rigid motion followed by a reflection in a straight line which takes S_1 into S_2.

We can now state the measure postulates.

B_1. *There exists a class B of measurable plane sets. To each such set S corresponds a nonnegative number $\mu(S)$, called the measure of S.*

B_2. *Each rectangle belongs to B. If its sides have the length a and b, then its measure is ab.*

B_3. *The union of two measurable sets is measurable and, if they are disjoint,*

$$\mu(S_1 \cup S_2) = \mu(S_1) + \mu(S_2). \tag{4.1.1}$$

B_4. *The intersection of two measurable sets is measurable.*

B_5. *If $S_1 \subset S_2$ and both belong to B, then*

$$\mu(S_1) \leqq \mu(S_2). \tag{4.1.2}$$

B$_6$. *If $S_1 \subset B$ and if S_2 is directly or reversedly congruent to S_1, then $S_2 \in B$ and* $\mu(S_2) = \mu(S_1)$.

B$_7$. *Let S be a bounded set, let $\{I_n\}$ and $\{C_n\}$ be two infinite sequences of measurable sets such that*

$$C_1 \supset C_2 \supset \cdots \supset C_n \supset \cdots \supset S \supset \cdots \supset I_n \supset \cdots \supset I_2 \supset I_1,$$

and

$$\lim_{n\to\infty} [\mu(C_n) - \mu(I_n)] = 0. \tag{4.1.3}$$

Then S is measurable and

$$\mu(S) = \lim_{n\to\infty} \mu(C_n) = \lim_{n\to\infty} \mu(I_n). \tag{4.1.4}$$

"Rectangle" in this case means the two-dimensional figure with or without the boundary. The terms "triangle," "parallelogram," and "polygon" are used in the same manner below.

As a consequence of the postulates it is easy to show that parallelograms and triangles are measurable and that their measures are given by the classical area formulas. If a polygon of n sides is convex, the diagonals from one vertex to the other vertices divide the polygon into $n - 3$ triangles. Hence, such a polygon is measurable. In particular, all regular polygons have this property. In topology it is proved that every simple closed polygon can be triangulated so every polygon is actually measurable. We shall not need this fact. Most of the polygons used below will be the union of vertical rectangles and obviously measurable.

We shall show two applications of the exhaustion principle. One is to the area of a circle, say of radius 1. Here we can take a system of inscribed and circumscribed regular polygons with

$$3, 6, 12, 24, 48, 96, \cdots, 3 \cdot 2^n, \cdots$$

sides. Each such polygon is measurable and intuition tells us that condition (4.1.3) is satisfied. Actually the proof is not elementary. It is equivalent to showing that

$$\lim_{k\to\infty} k \tan \frac{\pi}{k} \sin^2 \frac{\pi}{k} = 0.$$

For the values of k under consideration $k \tan \dfrac{\pi}{k} \leqq 3\sqrt{3}$ (this requires a proof!); since the sine factor tends rapidly to 0, the conclusion follows. The exhaustion principle then gives the area of the unit circle as

$$\lim_{k\to\infty} k \tan \frac{\pi}{k},$$

which has the value π as we shall see in Section 7.3.

Archimedes deduced the area of the segment of a parabola cut off by a straight line in two different ways. The first, which appears to have been his very first mathematical discovery, uses deeper properties of the parabola than the second. We shall sketch enough of the argument to bring out the ingenious idea. Figure 4.1 shows an arc of the parabola

$$y^2 = 4ax$$

and a straight line which intersects the parabola at the points A: (x_1, y_1) and B: (x_2, y_2). The problem is to find the area of the segment S between the parabola and the line. There is a tangent to the parabola parallel to the line AB which touches at the point C whose y-coordinate is

$$\tfrac{1}{2}(y_1 + y_2).$$

See Problem 9 of Exercise A.4.1. Archimedes set himself the task of computing the area of the triangle $T_0{:}ABC$. He had no coordinates at his disposal but the Greeks worked with proportions and his definition of the parabola gave him a relation which we would state in the form

$$x_1 : x_2 = (y_1)^2 : (y_2)^2.$$

This is equivalent to our equation. Imitating Archimedes' procedure we find that

$$\mu(T_0) = \frac{(y_2 - y_1)^3}{32a} \equiv \frac{\delta^3}{32a}\,. \tag{4.1.5}$$

We omit the calculations. Now the triangle T_0 is a first approximation of S. If we remove T_0 from S, we have left two smaller parabolic segments. One of these contains the triangle $T_{11}{:}ACD$, the other the triangle $T_{12}{:}CBE$. Here the tangent at D is parallel to AC and the tangent at E is parallel to CB. These triangles have the same area which we can compute from formula (4.1.5) replacing δ by $\tfrac{1}{2}\delta$. We now remove the triangles T_{11} and T_{12} and are left with

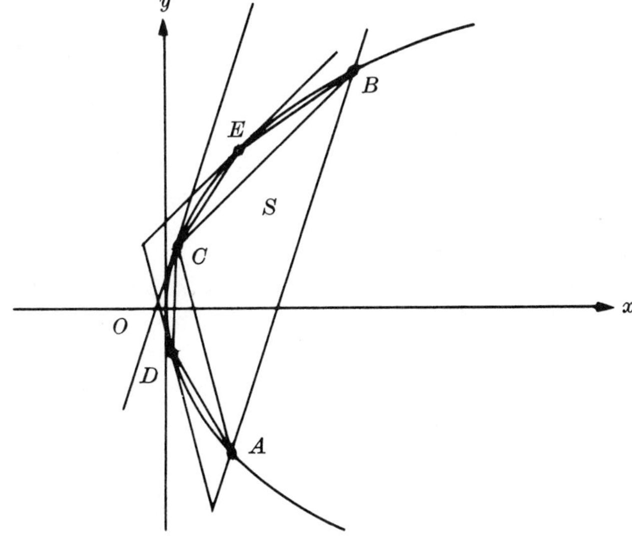

Figure 4.1

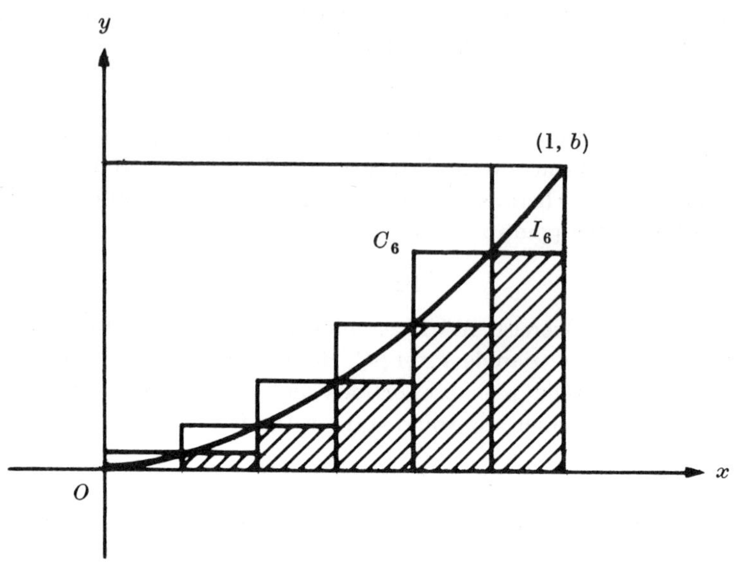

Figure 4.2

four still smaller parabolic segments which can be handled in the same way. The process can be continued indefinitely. At the nth stage we have left 2^{n+1} tiny parabolic segments and the areas of the triangles removed add up to

$$\frac{\delta^3}{32a}\left\{1 + 2 \cdot \tfrac{1}{8} + 2^2 \cdot \frac{1}{8^2} + \cdots + 2^n \cdot \frac{1}{8^n}\right\}$$
$$= \frac{\delta^3}{32a}\left\{1 + \frac{1}{4} + \frac{1}{4^2} + \cdots + \frac{1}{4^n}\right\} = \frac{\delta^3}{32a}\frac{1 - 4^{-n-1}}{1 - 4^{-1}},$$

and as n becomes infinite this converges to

$$\frac{\delta^3}{24a}. \tag{4.1.6}$$

This then is the area of the parabolic segment.

Here we take $y_1 = -y, \quad y_2 = y > 0$. Then

$$\delta^3 = 8y^3 = 8y^2 \cdot y = 32axy$$

so that

$$\mu(S) = \tfrac{4}{3}xy. \tag{4.1.7}$$

This shows that the area of the segment cut off by a vertical line is two-thirds of the area of the smallest rectangle in which the segment can be inscribed. This was Archimedes' discovery.

His second proof was simpler and basically more important. It represents the first use of approximation of areas by rectangles which is standard practice in Calculus. This time we tilt the parabola so that the axis points upward. We have to prove that the area under the parabola is one-third of the area of the circumscribed rectangle. See Figure 4.2. We take the parabola

$$y = bx^2 \quad \text{for} \quad 0 \leq x \leq 1,$$

divide the interval $[0, 1]$ into n equal parts ($n = 6$ in the figure), and note that

$$\left(\frac{k-1}{n}\right)^2 b \leqq y \leqq \left(\frac{k}{n}\right)^2 b \quad \text{for} \quad \frac{k-1}{n} \leqq x \leqq \frac{k}{n}, \quad k = 1, 2, \cdots, n.$$

We now construct the polygons C_n and I_n shown in the figure. Both are unions of rectangles. The upper horizontal boundary of C_n in the kth interval is the line segment

$$\frac{k-1}{n} \leqq x \leqq \frac{k}{n}, \quad y = \left(\frac{k}{n}\right)^2 b,$$

while that of I_n is the line segment

$$\frac{k-1}{n} \leqq x \leqq \frac{k}{n}, \quad y = \left(\frac{k-1}{n}\right)^2 b.$$

We note that the width of each of the rectangles is $1/n$. Adding the areas of the rectangles we obtain

$$\mu(C_n) = b\left\{\left(\frac{1}{n}\right)^2 + \left(\frac{2}{n}\right)^2 + \cdots + \left(\frac{n-1}{n}\right)^2 + \left(\frac{n}{n}\right)^2\right\}\frac{1}{n}$$

$$\mu(I_n) = b\left\{\left(\frac{1}{n}\right)^2 + \left(\frac{2}{n}\right)^2 + \cdots + \left(\frac{n-1}{n}\right)^2\right\}\frac{1}{n}.$$

The student should note that the rectangles in the two sums are pairwise congruent except for the largest one. It follows that

$$\mu(C_n) - \mu(I_n) = \frac{b}{n} \to 0 \text{ as } n \to \infty.$$

Thus condition (4.1.3) is satisfied and

$$\mu(S) = \lim_{n\to\infty} \mu(C_n).$$

Now, by Problem 23 of Exercise 1.2,

$$1^2 + 2^2 + 3^2 + \cdots + (n-1)^2 + n^2 = \tfrac{1}{6}n(n+1)(2n+1)$$

so that

$$\mu(C_n) = \frac{b}{6}\frac{(n+1)(2n+1)}{n^2} \to \tfrac{1}{3}b = \tfrac{1}{3}\mu(C_1)$$

as asserted.

This beautiful result was admired, but its possibilities remained hidden until Cavalieri attacked and solved the corresponding problem for the cubic parabola $y = x^3$ around 1630.

EXERCISE 4.1

1. Find the area of the triangle bounded by the lines

$$y = 0, \quad y = \frac{h}{b}x, \quad x = b,$$

using the second method of Archimedes.

2. Find the area bounded by

$$y = 0, \quad y = x^3, \quad x = 1.$$

Verify by induction or otherwise that the sum of the first n cubes is

$$S_3(n) = \tfrac{1}{4}n^2(n+1)^2$$

and use this in finding the area.

3. Consider the equilateral hyperbola $y = \dfrac{1}{x}$ and let $A(a, b)$ be the area between the curve, the x-axis, and the vertical lines $x = a$, $x = b$, $0 < a < b$. Prove by comparing rectangles that $A(1, 2) = A(2, 4)$. Generalize.

4. What is the area enclosed by the parabola $y = x^2$, the x-axis, and the lines $x = a$, $x = b$, $0 < a < b$?

5. What is the corresponding area for the cubic parabola $y = x^3$?

6. Knowing that

$$S_4(n) = 1^4 + 2^4 + 3^4 + \cdots + n^4 = \tfrac{1}{5}n^5 + O(n^4) \quad \text{as} \quad n \to \infty,$$

show that the area enclosed by $y = x^4$, $y = 0$, and $x = b > 0$ is $\frac{1}{5}b^5$.

7. Find the area enclosed by $y = x^2 + x$, $y = 0$, $x = b > 0$.

4.2. The definite integral. Without perhaps being aware of it, the student has already computed some definite integrals. In the preceding section including the Exercise, we have been concerned with

$$\int_0^b t^n\, dt \quad \text{for} \quad n = 1, 2, 3, \text{and } 4.$$

It is time to make a more systematic approach.

We start by defining the definite integral for step functions. Suppose that f is defined in some interval $[a, b]$ and that $[a, b]$ breaks up into a finite number of subintervals in each of which f has a constant value. Suppose that the subintervals are

$$(\alpha_1, \alpha_2), (\alpha_2, \alpha_3), \cdots, (\alpha_n, \alpha_{n+1}), \tag{4.2.1}$$

where

$$\alpha_1 = a, \quad \alpha_{n+1} = b,$$

and that

$$f(t) = u_j, \ \alpha_j < t < \alpha_{j+1}, \quad j = 1, 2, \cdots, n. \tag{4.2.2}$$

The value of $f(t)$ at $t = \alpha_j$ is immaterial for our purposes.

Definition 4.2.1. *We set*

$$\int_a^b f(t)\, dt \equiv u_1(\alpha_2 - \alpha_1) + u_2(\alpha_3 - \alpha_2) + \cdots + u_n(\alpha_{n+1} - \alpha_n)$$

$$= \sum_{j=1}^{n} u_j(\alpha_{j+1} - \alpha_j). \tag{4.2.3}$$

$f(t)$ is called the integrand, and (a, b) the interval of integration.

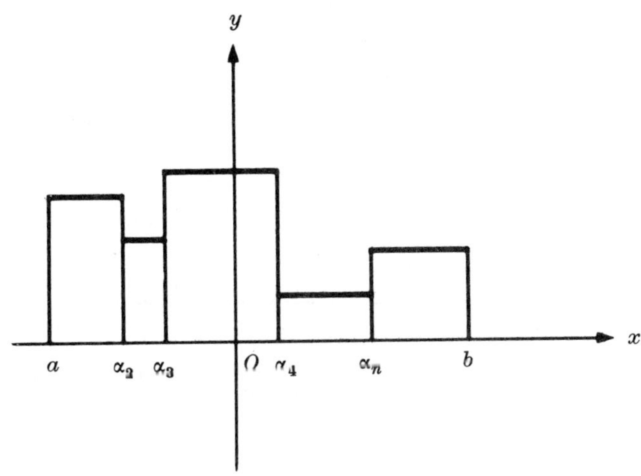

Figure 4.3

The geometrical interpretation is immediate when each $u_j \geqq 0$. See Figure 4.3. Then the sum in the second member is the sum of the areas of the rectangles corresponding to the n intervals of constancy, that is, the integral is the measure of the set of points between the graph, the x-axis, and the lines $x = a$, $x = b$.

This integral has various properties which here are almost trivial but which carry over to much wider classes of functions. We state the results in the form of a theorem.

THEOREM 4.2.1. I. *The integral is an additive function of intervals: if $a < c < b$, then*

$$\int_a^c f(t)\,dt + \int_c^b f(t)\,dt = \int_a^b f(t)\,dt. \tag{4.2.4}$$

II. *The integral is an additive function of the integrand: if $f(t)$ and $g(t)$ are step functions, so is $f(t) + g(t)$ and*

$$\int_a^b [f(t) + g(t)]\,dt = \int_a^b f(t)\,dt + \int_a^b g(t)\,dt. \tag{4.2.5}$$

III. *The integral is a homogeneous function of the integrand: if C is any constant, then*

$$\int_a^b [Cf(t)]\,dt = C\int_a^b f(t)\,dt. \tag{4.2.6}$$

IV. *The integral is a bounded operation on functions to real numbers: if*

$$\|f\| = \sup_{a \leqq t \leqq b} |f(t)|, \tag{4.2.7}$$

then

$$\left| \int_a^b f(t)\,dt \right| \leqq \int_a^b |f(t)|\,dt \leqq \|f\|\,(b - a). \tag{4.2.8}$$

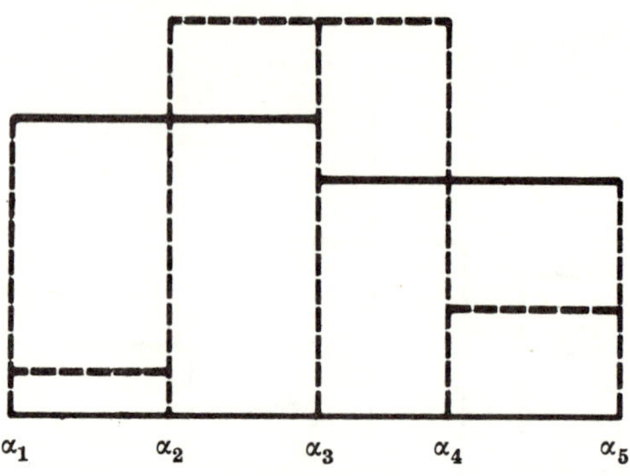

Figure 4.4

Proof. Re property I. This is completely trivial if c is one of the points α_j. If, on the other hand, $\alpha_j < c < \alpha_{j+1}$, then the integral $\int_a^c$ gives a term $u_j(c - \alpha_j)$ which adds to the term $u_j(\alpha_{j+1} - c)$ from the integral $\int_c^b$ to give the term $u_j(\alpha_{j+1} - \alpha_j)$ of $\int_a^b$.

Re property II. That $f + g$ is also a step function is obvious. Thus both sides of (4.2.5) are well defined, and it remains to prove the equality. Without restricting the generality, we may assume that f and g have the same intervals of constancy. For in the definition of a step function it is not required that the function assume distinct values in adjacent intervals of constancy. Hence it is possible to divide $[a, b]$ into a finite number of subintervals (α_j, α_{j+1}) in each of which both f and g assume constant values. See Figure 4.4 for a typical case. Here the solid line gives the graph of f while the broken one represents g.

Suppose that

$$f(t) = u_j, \quad g(t) = v_j, \quad \alpha_j < t < \alpha_{j+1}, \quad j = 1, 2, \cdots, n.$$

We also have then

$$f(t) + g(t) = u_j + v_j, \quad \alpha_j < t < \alpha_{j+1}.$$

Hence

$$\int_a^b f(t)\,dt = \sum_{j=1}^n u_j(\alpha_{j+1} - \alpha_j), \quad \int_a^b g(t)\,dt = \sum_{j=1}^n v_j(\alpha_{j+1} - \alpha_j),$$

$$\int_a^b [f(t) + g(t)]\,dt = \sum_{j=1}^n (u_j + v_j)(\alpha_{j+1} - \alpha_j),$$

and property II follows.

Re property III. The left member of (4.2.6) is

$$\sum_{j=1}^n Cu_j(\alpha_{j+1} - \alpha_j) = C \sum_{j=1}^n u_j(\alpha_{j+1} - \alpha_j)$$

which equals the right member. Hence (4.2.6) holds.

Properties II and III are often combined into a single statement:

II + III. *The integral is a linear function of the integrand.*

Re property IV. We have

$$\left|\int_a^b f(t)\,dt\right| \leqq \sum |u_j|\,(\alpha_{j+1} - \alpha_j) = \int_a^b |f(t)|\,dt$$
$$\leqq \|f\| \sum (\alpha_{j+1} - \alpha_j) = \|f\|\,(b - a)$$

and this completes the proof.

We now proceed to define the class of functions that are *integrable in the sense of Riemann over the interval* $[a, b]$. We denote this class by $R(a, b)$. We proceed by several stages.

Definition 4.2.2. *Let f be a bounded real-valued function defined in $[a, b]$. Let $\underline{f}$ and $\bar{f}$ be two step functions defined in $[a, b]$ such that*

$$\underline{f}(t) \leqq f(t) \leqq \bar{f}(t) \tag{4.2.9}$$

for all t. Then $\underline{f}$ is called a lower step function of f and $\bar{f}$ is an upper step function. Let $L[f]$ be the set of all lower step functions of f and $U[f]$ the class of all upper step functions.

We have obviously

$$\int_a^b \underline{f}(t)\,dt \leqq \int_a^b \bar{f}(t)\,dt \tag{4.2.10}$$

for every $\underline{f} \in L[f]$ and every $\bar{f} \in U[f]$. This inequality shows that the two sets of real numbers

$$L = \left\{\int_a^b \underline{f}(t)\,dt, \underline{f} \in L[f]\right\},$$
$$U = \left\{\int_a^b \bar{f}(t)\,dt, \bar{f} \in U[f]\right\} \tag{4.2.11}$$

are bounded above and bounded below, respectively.

Definition 4.2.3. *We set*

$$\sup L = \underline{\int_a^b} f(t)\,dt, \quad \inf U = \overline{\int_a^b} f(t)\,dt \tag{4.2.12}$$

and call them the lower and the upper integrals of f, respectively.

Definition 4.2.4. *The function f belongs to $R(a, b)$ and is said to be integrable over (a, b) in the sense of Riemann if*

$$\underline{\int_a^b} f(t)\,dt = \overline{\int_a^b} f(t)\,dt. \tag{4.2.13}$$

The common value is called the integral of f over the interval (a, b) and is denoted by

$$\int_a^b f(t)\,dt. \tag{4.2.14}$$

We omit "in the sense of Riemann" and say simply "integrable." Riemann's definition of 1854 has been superseded in advanced analysis by more powerful definitions.

The following condition of integrability is obvious.

THEOREM 4.2.2. *$f \in R(a, b)$ if and only if for each $\varepsilon > 0$ there is an $\underline{f}_\varepsilon \in L[f]$ and an $\bar{f}_\varepsilon \in U[f]$ such that*

$$\int_a^b \bar{f}_\varepsilon(t)\, dt - \int_a^b \underline{f}_\varepsilon(t)\, dt < \varepsilon. \tag{4.2.15}$$

This implies and is implied by (4.2.13).

We have considered monotone sequences of numbers in earlier sections beginning with Definition 1.7.5. We say that a sequence of functions $\{f_n\}$ is monotone in the interval (a, b) if, for each x, $a < x < b$, the numerical sequence $\{f_n(x)\}$ is monotone in the same sense for all x. Thus the sequence $\{x^n\}$ is monotone and strictly decreasing in $(0, 1)$. It is also monotone in $(1, 2)$ but now it is strictly increasing. It is not monotone in $(0, 2)$

With this terminology we can reformulate Theorem 4.2.2 as follows.

THEOREM 4.2.3. *If $f \in R(a, b)$, then there exists a nondecreasing sequence of lower functions of f and a nonincreasing sequence of upper functions whose integrals converge to that of f, from below in the first case and from above in the second. Conversely, if two such sequences of integrals converge to a common limit, then $f \in R(a, b)$ and the limit is the integral of f.*

The class $R(a, b)$ contains all step functions with a finite number of "steps" so it is not void. We proceed to a closer examination of the structure and membership of the class. We prove first that $R(a, b)$ is a linear system and in the process we verify that the integral possesses the properties II and III of Theorem 4.2.1.

THEOREM 4.2.4. *If f and g are in $R(a, b)$ and if C is any real constant, then $f + g$ and Cf are in $R(a, b)$ and formulas (4.2.4) and (4.2.5) hold.*

Proof. We can use Theorems 4.2.1 and 4.2.3. Let $\{\underline{f}_n\}$ and $\{\underline{g}_n\}$ be ascending sequences of lower functions of f and g respectively and let $\{\bar{f}_n\}$ and $\{\bar{g}_n\}$ be descending sequences of upper functions such that

$$\int_a^b \underline{f}_n(t)\, dt \to \int_a^b f(t)\, dt, \quad \int_a^b \underline{g}_n(t)\, dt \to \int_a^b g(t)\, dt,$$

$$\int_a^b \bar{f}_n(t)\, dt \to \int_a^b f(t)\, dt, \quad \int_a^b \bar{g}_n(t)\, dt \to \int_a^b g(t)\, dt.$$

Then $\{\underline{f}_n + \underline{g}_n\}$ is an ascending sequence of lower functions for $f + g$, while

$\{\bar{f}_n + \bar{g}_n\}$ is a descending sequence of upper functions. Further

$$\int_a^b [\underline{f}_n(t) + \underline{g}_n(t)]\,dt = \int_a^b \underline{f}_n(t)\,dt + \int_a^b \underline{g}_n(t)\,dt$$
$$\to \int_a^b f(t)\,dt + \int_a^b g(t)\,dt,$$

and

$$\int_a^b [\bar{f}_n(t) + \bar{g}_n(t)]\,dt = \int_a^b \bar{f}_n(t)\,dt + \int_a^b \bar{g}_n(t)\,dt$$

has the same limit. But this shows that $f + g \in R(a, b)$, and since the common limit is by definition

$$\int_a^b [f(t) + g(t)]\,dt,$$

we see that the integral of the sum equals the sum of the integrals. Thus (4.2.5) holds and formula (4.2.6) is proved by the same type of argument.

We can use the same technique to verify properties I and IV.

THEOREM 4.2.5. *Formulas* (4.2.4), (4.2.7), *and* (4.2.8) *are valid for all functions of* $R(a, b)$.

Proof. With the same notation as in the preceding theorem we have

$$\int_a^b \underline{f}_n(t)\,dt = \int_a^c \underline{f}_n(t)\,dt + \int_c^b \underline{f}_n(t)\,dt,$$
$$\int_a^b \bar{f}_n(t)\,dt = \int_a^c \bar{f}_n(t)\,dt + \int_c^b \bar{f}_n(t)\,dt.$$

Here the left members converge to

$$\int_a^b f(t)\,dt.$$

The four terms in the right members are bounded monotone functions of n. They consequently tend to finite limits, A and B in the first line, C and D in the second. Here

$$A \leqq C, \quad B \leqq D, \quad A + B = C + D = \int_a^b f(t)\,dt.$$

The assumption that $A < C$ would clearly imply $B > D$ and this is a contradiction. Hence, $A = C$, $B = D$. This implies that f is integrable over each of the intervals (a, c) and (c, b) and that the integrals are related by (4.2.4).

Let us define

$$f^+(t) = \max[f(t), 0], \quad f^-(t) = \max[-f(t), 0], \tag{4.2.16}$$

so that

$$f = f^+ - f^-, \quad |f| = f^+ + f^-. \tag{4.2.17}$$

We use similar notation for other functions. Now if $\{\underline{f}_n(t)\}$ is an ascending sequence of lower step functions of f, and if $\bar{f}_n(t)$ is a descending sequence of upper step functions, then $\{\underline{f}_n^+(t)\}$ and $\{\bar{f}_n^+(t)\}$ are similarly related to $f^+(t)$, and from the inequality

$$\int_a^b \bar{f}_n^+(t)\,dt - \int_a^b \underline{f}_n^+(t)\,dt \leqq \int_a^b \bar{f}_n(t)\,dt - \int_a^b \underline{f}_n(t)\,dt,$$

we conclude that the left member tends to 0 as n becomes infinite whenever the right member does. This implies that f^+ belongs to $R(a, b)$ and the function f^- is handled in the same manner. Hence $|f| \in R(a, b)$ as the sum of two integrable functions.

By (4.2.17)

$$\begin{aligned}\left|\int_a^b f(t)\,dt\right| &= \left|\int_a^b [f^+(t) - f^-(t)]\,dt\right| \\ &= \left|\int_a^b f^+(t)\,dt - \int_a^b f^-(t)\,dt\right| \\ &\leqq \int_a^b f^+(t)\,dt + \int_a^b f^-(t)\,dt = \int_a^b |f(t)|\,dt.\end{aligned}$$

This is the first half of the inequality (4.2.8). To obtain the second half we note that the function which is constantly equal to $\|f\|$ in $[a, b]$ is an upper step function with respect to $|f|$. Its integral then cannot be exceeded by the integral of $|f|$ and this is the second half of the inequality. This completes the proof.

Thus far we have not had any examples of integrable functions that are not step functions. It is time to fill this gap.

THEOREM 4.2.6. *Every function which is continuous in a closed finite interval is integrable over this interval, that is,*

$$C[a, b] \subset R(a, b). \tag{4.2.18}$$

Proof. Let $f \in C[a, b]$. We recall that a function which is continuous in a closed interval is uniformly continuous (see Definition 2.6.1 and Theorem 2.6.3). This means that the oscillation of f in an interval (α, β), $a \leqq \alpha < \beta \leqq b$, is small as soon as $\beta - \alpha$ is small regardless of the location of (α, β). This fact can be used for the construction of suitable lower and upper functions. Let $\varepsilon > 0$ be given and choose $\delta = \delta(\varepsilon)$ so small that

$$|f(s_1) - f(s_2)| \leqq \varepsilon$$

for any choice of s_1 and s_2 with $|s_1 - s_2| \leqq \delta$. Then take n so large that $b - a < n\delta$ and divide the interval $[a, b]$ into n equal parts with the aid of the points

$$t_k = a + \frac{k}{n}(b - a), \quad k = 0, 1, 2, \cdots, n.$$

Set

$$m_k = \min f(t), \quad M_k = \max f(t), \quad t_k \leqq t \leqq t_{k+1}. \tag{4.2.19}$$

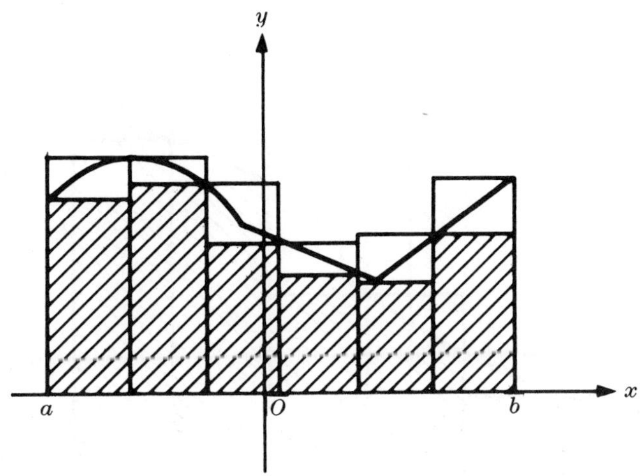

Figure 4.5

Then by the choice of n

$$0 \leqq M_k - m_k \leqq \varepsilon$$

for each k. Next we define step functions (see Figure 4.5)

$$\begin{aligned} \underline{f}_\varepsilon(t) &= m_k, \\ \bar{f}_\varepsilon(t) &= M_k, \end{aligned} \quad t_k \leqq t \leqq t_{k+1}, \quad k = 0, 1, 2, \cdots, n-1. \tag{4.2.20}$$

Then

$$\underline{f}_\varepsilon(t) \leqq f(t) \leqq \bar{f}_\varepsilon(t)$$

for all t and

$$\begin{aligned} \int_a^b \bar{f}_\varepsilon(t)\,dt - \int_a^b \underline{f}_\varepsilon(t)\,dt &= \int_a^b [\bar{f}_\varepsilon(t) - \underline{f}_\varepsilon(t)]\,dt \\ &= \sum_{k=0}^{n-1} (M_k - m_k)(t_{k+1} - t_k) \leqq \varepsilon \sum (t_{k+1} - t_k) = \varepsilon(b-a). \end{aligned}$$

Since ε is arbitrary, Theorem 4.2.3 applies and shows that $f \in R(a, b)$. This completes the proof.

Another important subclass of $R(a, b)$ is given by

THEOREM 4.2.7. *A bounded monotone function is integrable over any finite interval.*

Proof. Let f be monotone in $[a, b]$. To fix the ideas we assume that f is nondecreasing. We use the same notation and construction as in the proof of the preceding theorem. See Figure 4.6. In this case

$$m_k = f(t_k), \quad M_k = f(t_{k+1}), \tag{4.2.21}$$

and

$$\sum (M_k - m_k)(t_{k+1} - t_k) = \frac{1}{n}(b-a) \sum [f(t_{k+1}) - f(t_k)],$$

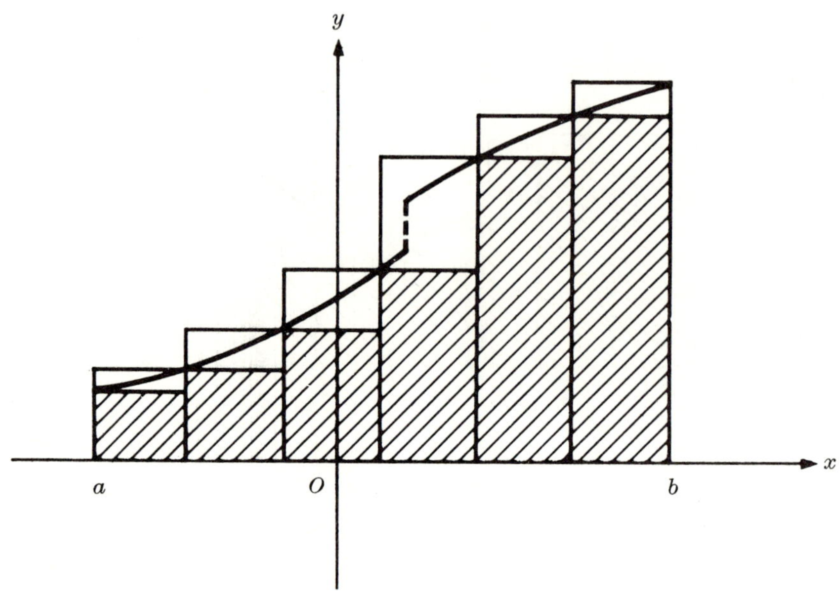

Figure 4.6

since all intervals have the same length $(b - a)/n$. The sum "telescopes" into $f(b) - f(a)$ since all terms cancel except $f(t_{n+1})$ and $-f(t_0)$. Hence

$$\sum (M_k - m_k)(t_{k+1} - t_k) = \frac{1}{n}(b - a)[f(b) - f(a)].$$

This tends to 0 as n becomes infinite so that $f(t) \in R(a, b)$. If $f(t)$ is nonincreasing instead, we have

$$m_k = f(t_{k+1}), \quad M_k = f(t_k), \tag{4.2.22}$$

and in the final expression for the difference of the integrals we get $f(a) - f(b)$. The conclusion is the same.

It is clear that any linear combination of bounded monotone functions is also integrable. In Section C.1 we shall encounter such combinations under the name of *functions of bounded variation.*

The functions of $R(a, b)$ are limits of monotone sequences of step functions. This fact imposes certain restrictions on the discontinuities of f if $f \in R(a, b)$. It is clear that f can have any number of discontinuities and also infinitely many. What is of importance to us is that f cannot be discontinuous everywhere. It may be shown that the points of discontinuity can be enclosed in subintervals, possibly infinitely many and overlapping, but such that the sum of their lengths is less than any preassigned positive number. We shall not need this, but the following simple consequence will be used. We state it without proof.

THEOREM 4.2.8. *If $f \in R(a, b)$, then every subinterval $(\alpha, \beta) < (a, b)$ contains a point of continuity of f.*

We note the

COROLLARY. *If $f \in R(a, b)$, and if $f > 0$ in (a, b), then there exists a subinterval (α, β) and a positive number c such that $f(x) > c$ for all x in (α, β).*

For if x_0 is a point of continuity of f, then there exists a $\delta > 0$ such that $f(x) > \frac{1}{2}f(x_0)$ for $x_0 - \delta < x < x_0 + \delta$. We take $\alpha = x_0 - \delta$, $\beta = x_0 + \delta$, $c = \frac{1}{2}f(x_0)$.

For the applications we can usually restrict ourselves to *piecewise continuous functions*. By this we mean that $[a, b]$ can be broken up into a finite number of subintervals in which f is continuous (continuous to the right at left endpoints, continuous to the left at right endpoints). Every such function is in $R(a, b)$.

Finally we have to connect the main concept of this section with that of the preceding one.

THEOREM 4.2.9. *Let $f \in R(a, b)$, and let f be nonnegative in $[a, b]$. Let S be the set*

$$S = \{(x, y) \mid a \leqq x \leqq b, 0 \leqq y \leqq f(x)\}, \tag{4.2.23}$$

the so-called ordinate set of f. Then

$$\mu(S) = \int_a^b f(t)\, dt. \tag{4.2.24}$$

Proof. We can always find an ascending sequence of lower step functions $\{\underline{f}_n(t)\}$ and a descending sequence of upper step functions $\{\bar{f}_n(t)\}$. We may assume that $\underline{f}_n(t) \geqq 0$ for all t. The ordinate set of $\underline{f}_n(t)$ is a set I_n, and the ordinate set of $\bar{f}_n(t)$ a set C_n in the notation of Postulate B7 (the exhaustion principle). Both I_n and C_n are unions of nonoverlapping rectangles (see Figure 4.3), and by Definition 4.2.1 we have

$$\int_a^b \underline{f}_n(t)\, dt = \mu(I_n), \quad \int_a^b \bar{f}_n(t)\, dt = \mu(C_n).$$

Here the approximation functions can be chosen in such a manner that

$$\mu(C_n) - \mu(I_n) \to 0,$$

since $f \in R(a, b)$. By B7 the common limit of $\mu(C_n)$ and of $\mu(I_n)$ is $\mu(S)$. This implies that (4.2.24) holds. The student should note how basic the exhaustion principle is for this argument.

In conclusion a few words concerning notation and terminology. The sign $\int$ goes back to Leibniz (1675). It is a deformed capital S and bears witness to the name *calculus summatorius* originally used by Leibniz. The term "integral" was introduced by Jacob (= James) Bernouilli (1654–1705) in 1690 and the name *calculus integralis* was agreed upon in 1696. Note that the terms "somme" and "sommable" are still used by French mathematicians.

The numbers a and b are known as the *limits of integration*, a as the *lower limit*, b the *upper*. Normally $a < b$, and in any case we have the convention

$$\int_a^b f(t)\,dt = -\int_b^a f(t)\,dt \tag{4.2.25}$$

which goes back to Definition 4.2.1.

The letter "t" used in our formulas is called the *variable of integration*. It is a "dummy" in the sense that it can be replaced by any other letter which is not already used in the formulas. Thus we cannot replace t by a, b, d, or f. In this section we have avoided using x as variable of integration, because in the next section x will be used as an upper limit in an important context. The dt is nowadays often omitted in writing integrals. We do not follow suit. It will be seen that dt does serve a useful purpose.

EXERCISE 4.2

1. Show that (i) $\int_a^b dt = b - a$, (ii) $\int_a^b t\,dt = \frac{1}{2}(b^2 - a^2)$.

2. Show that $\int_0^b t^2\,dt = \frac{1}{3}b^3$. What is the value of the integral if the lower limit is a instead of 0?

For what value of α is

3. $\int_0^1 (t - \alpha t^2)\,dt = 0$.

4. $\int_{-1}^1 (t + \alpha t^2)\,dt = 0$.

If (t) denotes the distance of t from the nearest integer find the value of:

5. $\int_0^1 (t)\,dt$.

6. $\int_{-1}^1 t(t)\,dt$.

If $[u]$ is the integral part of u, compute

7. $\int_0^1 t[3t]\,dt$.

8. $\int_0^2 t[2t]\,dt$.

9. Show that

$$\int_{-a}^a f(t)\,dt = 2\int_0^a f(t)\,dt$$

if $f(t)$ is an even function of t, but 0 if f is odd. [*Hint:* Consider partitions of $[-a, a]$ into $2n$ equal parts, use upper and lower functions of the same parity as f, and compare contributions from subintervals symmetric with respect to the origin.]

10. This is an example of a function which is not Riemann integrable. Take $f(t) = 0$ or 1 according as t is rational or irrational. Show that the lower integral is 0 while the upper integral equals the length of the interval of integration. This function is discontinuous everywhere.

4.3. The fundamental theorem of the integral calculus. We continue the examination of the properties of the definite integral. This will include various estimates for the integral, the first mean value theorem, a study of the integral as a function of its upper limit, the fundamental theorem of the integral calculus, and the existence of primitives. We start with the following result which is an immediate consequence of Theorem 4.2.8, Corollary and of Theorem 4.2.9.

THEOREM 4.3.1. *A nonnegative integrable function which is actually positive in some subinterval of the interval of integration has a positive integral.*

This theorem has a number of important consequences.

THEOREM 4.3.2. *If $f_1(t)$ and $f_2(t)$ belong to $R(a, b)$, and if $f_1(t) \leqq f_2(t)$ in (a, b) where the strict inequality holds in some subinterval, then*

$$\int_a^b f_1(t)\,dt < \int_a^b f_2(t)\,dt. \tag{4.3.1}$$

For $f_2(t) - f_1(t)$ satisfies the condition of Theorem 4.3.1. Hence: *An inequality between integrable functions can be integrated.*

The following very useful theorem is nothing but a particular case of Theorem 4.3.2.

THEOREM 4.3.3. *If $f(t)$ belongs to $R(a, b)$, and if the constants m and M are so chosen that*

$$m \leqq f(t) \leqq M \tag{4.3.2}$$

for all t, then

$$m(b-a) \leqq \int_a^b f(t)\,dt \leqq M(b-a). \tag{4.3.3}$$

Proof. We use Theorem 4.3.2 twice, first with $f_1 = m$, $f_2 = f$ and secondly with $f_1 = f$ and $f_2 = M$. We have inequality between the first and the second member of (4.3.3) if $m < f(t)$ in some subinterval. There is inequality in the second place if $f(t) < M$ in some subinterval. If $f(t)$ is actually continuous, we can make a sharper statement:

THEOREM 4.3.4. *If $f \in C[a, b]$, then there exists at least one number c such that $a \leqq c \leqq b$ and*

$$\int_a^b f(t)\,dt = f(c)(b-a). \tag{4.3.4}$$

Proof. For by the preceding theorem the value of the integral is $\mu(b-a)$, where

$$m = \min f(t) \leqq \mu \leqq M = \max f(t).$$

Since f is continuous by assumption, Theorem 2.6.2 shows that the range of f

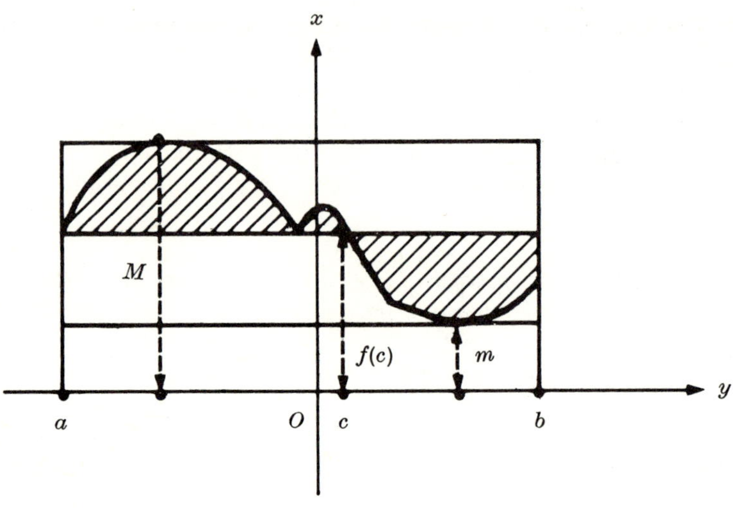

Figure 4.7

is the closed interval $[m, M]$ and every value in the range is assumed at least once. Hence there is at least one number c in $[a, b]$ where $f(c) = \mu$. See Figure 4.7 for this and the preceding theorem. This completes the proof.

This is known as the *first mean value theorem of the integral calculus*. For the generalization of this theorem we need the following result which is of independent interest.

THEOREM 4.3.5. *The product of two (bounded) integrable functions is integrable. Hence $R(a, b)$ is an algebra.*

Proof. To prove that $f g \in R(a, b)$ when f and g do, it is sufficient to prove that we can find lower and upper functions of $f g$ whose integrals differ arbitrarily little. These functions need not be step functions; it is enough that they are integrable. The following construction will lead to the desired result. Since f is integrable, we can find a partition of $[a, b]$ into subintervals (t_k, t_{k+1}) such that for a given $\varepsilon > 0$

$$\sum (M_k - m_k)(t_{k+1} - t_k) < \varepsilon(b - a), \tag{4.3.5}$$

where

$$m_k = \inf f(t), \quad M_k = \sup f(t) \quad \text{in } [t_k, t_{k+1}].$$

Now $g \in R(a, b)$, so that each of the integrals

$$\int_{t_k}^{t_{k+1}} g(t)\, dt$$

exists. It is clear that we can add any constant to g without affecting the integrability of the product: if $f g$ is integrable so is $f(g + C)$ and vice versa. It is then no restriction to assume that g is positive in the interval $[a, b]$. We then define a lower function $\underline{h}$ and an upper function $\bar{h}$ for $f\,g$ by the formulas

$$\begin{aligned} \underline{h}(t) &= m_k g(t), \\ \bar{h}(t) &= M_k g(t), \end{aligned} \qquad t_k \leqq t < t_{k+1}, \quad k = 0, 1, \cdots, n-1.$$

This gives, using (4.3.5),

$$\begin{aligned} \int_a^b \bar{h}(t)\,dt - \int_a^b \underline{h}(t)\,dt &= \sum (M_k - m_k) \int_{t_k}^{t_{k+1}} g(t)\,dt \\ &< B \sum (M_k - m_k)(t_{k+1} - t_k) \\ &< \varepsilon B(b-a), \end{aligned}$$

if B is an upper bound for g in $[a, b]$. Since ε is at our disposal, we see that the product of f and g is integrable as asserted.

We come now to a generalization of the mean value theorem.

THEOREM 4.3.6. *If $f(t)$ and $g(t)$ belong to $R(a, b)$, and if*

$$m \leqq f(t) \leqq M, \quad 0 \leqq g(t), \tag{4.3.6}$$

then

$$m \int_a^b g(t)\,dt \leqq \int_a^b f(t)\,g(t)\,dt \leqq M \int_a^b g(t)\,dt. \tag{4.3.7}$$

If $f(t)$ is continuous, then there is at least one value c in $[a, b]$ such that

$$\int_a^b f(t)\,g(t)\,dt = f(c) \int_a^b g(t)\,dt. \tag{4.3.8}$$

Proof. For all values of t we have the inequality

$$mg(t) \leqq f(t)\,g(t) \leqq Mg(t).$$

By Theorem 4.3.2 we can integrate an inequality over an interval where it is valid. This gives (4.3.7). To obtain (4.3.8) we argue as in the proof of Theorem 4.3.4. The details are left to the reader.

Theorem 4.3.6 gives an estimate of the integral of a product. A much more important estimate goes under the name of the *Bouniakovsky-Schwarz inequality* [Victor Jacovlevich Bouniakovsky (1804–1889) of St. Petersburg and Hermann Amandus Schwarz (1843–1921) of Berlin]. It corresponds to Cauchy's inequality for sums of products; a special case is formula (1.6.12).

THEOREM 4.3.7. *If $f(t)$ and $g(t)$ are bounded integrable functions, then*

$$\left[\int_a^b f(t)\,g(t)\,dt\right]^2 \leqq \int_a^b f^2(t)\,dt \cdot \int_a^b g^2(t)\,dt. \tag{4.3.9}$$

If f and g are continuous, then equality holds if and only if f and g are proportional.

Proof. Let α be a real number. From the obvious inequality

$$0 \leqq [\alpha f(t) + g(t)]^2 = \alpha^2 f^2(t) + 2\alpha f(t)g(t) + g^2(t) \tag{4.3.10}$$

we get by integration (permitted since f^2, $f\,g$, and $g^2 \in R(a, b)$)

$$0 \leqq \alpha^2 \int_a^b f^2(t)\,dt + 2\alpha \int_a^b f(t)\,g(t)\,dt + \int_a^b g^2(t)\,dt. \tag{4.3.11}$$

This is a quadratic polynomial in α with constant coefficients and it is never negative. This requires that the discriminant be negative or zero, or precisely the inequality (4.3.9). Suppose now that equality does hold in (4.3.9). This means that there is a real value of α for which equality holds in (4.3.11). If f and g are continuous, this requires that equality holds in (4.3.10) for *all* values of t, and this says that $g(t)$ is a constant multiple of $f(t)$. This completes the proof.

We come now to the dependence of the integral upon the limits of integration. We concentrate on the upper limit. If $f \in R(a, b)$, we define

$$F(x) = \int_a^x f(t)\,dt\,, \quad a \leqq x \leqq b. \tag{4.3.12}$$

THEOREM 4.3.8. *$F(x)$ is continuous in $[a, b]$ and $F(a) = 0$. At every point where $f(x)$ is continuous, $F(x)$ is differentiable and*

$$F'(x) = f(x). \tag{4.3.13}$$

Moreover, the right-hand derivative of $F(x)$ equals $f(x + 0)$ wherever this limit exists. Similarly the left-hand derivative equals $f(x - 0)$ wherever it exists.

Proof. Suppose that $a \leqq x < x + h \leqq b$. Then

$$\begin{aligned} F(x + h) - F(x) &= \int_a^{x+h} f(t)\,dt - \int_a^x f(t)\,dt \\ &= \int_a^x f(t)\,dt + \int_x^{x+h} f(t)\,dt - \int_a^x f(t)\,dt \\ &= \int_x^{x+h} f(t)\,dt = h\mu(x, h), \end{aligned}$$

where $\mu(x, h)$ is a number which lies between the infinum and the supremum of $f(t)$ in the interval $[x, x + h]$. This is also valid if h is negative. Set

$$B = \sup |f(t)|, \quad a \leqq t \leqq b.$$

Then $|\mu(x, h)| \leqq B$ and

$$|F(x + h) - F(x)| \leqq B\,|h|. \tag{4.3.14}$$

This is a so-called *Lipschitz condition* [Rudolf Lipschitz (1832–1903) of Bonn] and clearly implies continuity.

Suppose $\delta > 0$ and that

$$\lim_{\delta\to 0} f(x+\delta) \equiv f(x+0)$$

exists. This means that for small positive values of h the quantity $\mu(x, h)$ is close to $f(x+0)$ and we have

$$\lim_{h\to 0} \frac{1}{h}[F(x+h) - F(x)] = f(x+0) \tag{4.3.15}$$

as value of the right-hand derivative. The left-hand derivative is handled the same way. In particular, if $t = x$ is a point of continuity of $f(t)$, we have $f(x+0) = f(x-0)$, and $F(x)$ has $f(x)$ as its derivative. This completes the proof.

A special case of this theorem was discovered by Isaac Barrow in 1667 who proved what in our notation would be written

$$\frac{d}{dx}\int_a^x f(t)\,dt = f(x), \tag{4.3.16}$$

assuming $f(t)$ to be continuous and monotone in $[a, b]$. To clarify the ideas, suppose that f is nondecreasing. Figure 4.8 then shows that

$$f(x) \leqq \frac{1}{h}[F(x+h) - F(x)] \leqq f(x+h).$$

When $h \to 0$ we obtain (4.3.16) since $f(x+h)$ tends to $f(x)$ when h tends to 0.

Barrow's Fundamental Theorem, that is (4.3.16), was the summit of the development of the geometrical function concept during the seventeenth

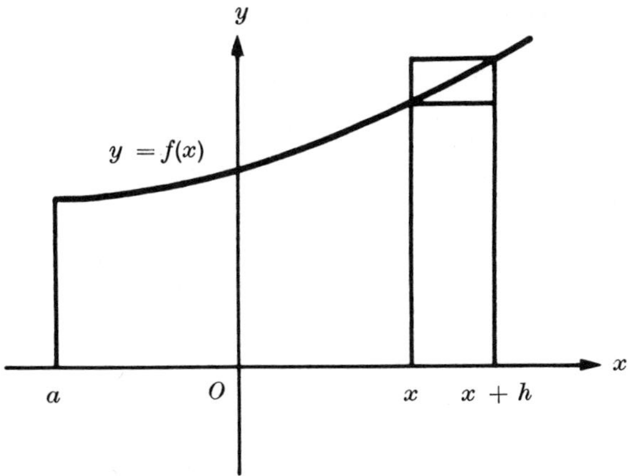

Figure 4.8

century. The result was known to Newton and Leibniz but their point of view was different. They worked with explicit analytic expressions rather than with geometrical concepts and consequently their formulation of the fundamental theorem was somewhat different. This is our next topic.

DEFINITION 4.3.1. *Let $f(x)$ be a given function, continuous in the interval $[a, b]$. A function $F(x)$ such that*

$$F'(x) = f(x), \quad a < x < b, \tag{4.3.17}$$

is called an antiderivative or a primitive or an indefinite integral of $f(x)$. If the last name is used one writes

$$F(x) = \int f(x)\,dx. \tag{4.3.18}$$

Theorem 4.3.8 shows the existence of at least one primitive of a continuous function, and it is clear that we can add any constant to the function of (4.3.12) and still obtain a primitive of $f(x)$. To show that all primitives are of this form we need

THEOREM 4.3.9. *If $G(x)$ is differentiable for all x in (a, b), and if $G'(x) = 0$ for all x, then $G(x)$ is a constant.*

Proof. This theorem is usually proved using the mean value theorem of the differential calculus, Theorem 6.1.2. To avoid anticipating a later result, we give the following direct argument.

Without restricting the generality we may suppose that $G(t)$ is continuous in the closed interval $[a, b]$. Now take a point c with $a < c < b$ and suppose that $G(c) = C$. Since $G'(c) = 0$, there exists for every given $\varepsilon > 0$ a $\delta_1 = \delta_1(\varepsilon)$ such that

$$|G(c + h) - G(c)| \leqq \varepsilon\,|h| \quad \text{for} \quad |h| \leqq \delta_1.$$

Let $[x_1, x_2]$ be the largest closed subinterval of $[a, b]$ in which

$$|G(x) - G(c)| \leqq \varepsilon\,|x - c|. \tag{4.3.19}$$

Such an interval exists, is not void, and may possibly coincide with $[a, b]$. If not, suppose that $c < x_2 < b$. Since $G'(x_2) = 0$ we can find a $\delta_2 = \delta_2(\varepsilon)$ such that

$$|G(x_2 + h) - G(x_2)| \leqq \varepsilon h \quad \text{for} \quad 0 \leqq h \leqq \delta_2.$$

Now let x_3 be the largest value of x such that

$$|G(x) - G(x_2)| \leqq \varepsilon(x - x_2) \quad \text{for} \quad x_2 \leqq x \leqq x_3.$$

For such values of x

$$\begin{aligned} |G(x) - G(c)| &\leqq |G(x) - G(x_2)| + |G(x_2) - G(c)| \\ &\leqq \varepsilon(x - x_2) + \varepsilon(x_2 - c) = \varepsilon(x - c). \end{aligned}$$

This means that (4.3.19) actually holds in the interval $[x_1, x_3]$ which is larger than $[x_1, x_2]$. This contradicts the assumption that $[x_1, x_2]$ was the largest interval where (4.3.19) holds. It follows that we must have $x_2 = b$. In the same manner one shows that $x_1 = a$.

In this argument ε was arbitrary so that (4.3.19) holds for every choice of ε and for all x in $[a, b]$. This is possible if and only if $G(x) = G(c)$ for all x, that is, $G(x) = C$, a constant. This completes the proof.

We have now

THEOREM 4.3.10. *If $f(x) \in C[a, b]$, then (4.3.12) defines a primitive of $f(x)$ and every other primitive differs from this one by a constant.*

Proof. That (4.3.12) defines a primitive follows from Theorem 4.3.8. If there is another primitive, then their difference is a function whose derivative is 0 for all x in $[a, b]$ and by Theorem 4.3.9 such a function is a constant.

We come finally to the *fundamental theorem of the integral calculus.*

THEOREM 4.3.11. *If $f \in C[a, b]$ and if G is any primitive of f, then*

$$\int_a^b f(t)\,dt = G(b) - G(a). \tag{4.3.20}$$

Proof. If $F(x)$ is the function defined by (4.3.12), then there is a constant C such that $G(x) = F(x) + C$. It follows that

$$\begin{aligned} G(b) - G(a) &= [F(b) + C] - [F(a) + C] \\ &= F(b) - F(a) = F(b) \end{aligned}$$

since $F(a) = 0$. But $F(b)$ is the left member of (4.3.20) so the theorem is proved.

The following notation is commonly used:

$$G(x)\,\Big|_a^b = G(b) - G(a). \tag{4.3.21}$$

It is clear that the letter x on the left can be replaced by any other letter not already engaged.

As an application of Theorem 4.3.11 we note that

$$\int_0^1 t^n\,dt = \frac{t^{n+1}}{n+1}\Big|_0^1 = \frac{1}{n+1} \quad \text{for} \quad n > -1. \tag{4.3.22}$$

The result is correct as stated, but for $-1 < n < 0$ we are actually dealing with an improper integral, a topic that will be discussed later.

The student should note that *a definite integral is a number, while an indefinite integral is a function.*

EXERCISE 4.3

Compute the following integrals:

1. $\int_0^1 (t^2 - t + 1)\,dt.$ 2. $\int_0^1 (t^3 + 3t^4)\,dt.$ 3. $\int_5^6 x^3\,dx.$

4. $\int_{-1}^1 (t^3 - t)\,dt.$ 5. $\int_{-2}^2 (t^4 - 2t^2)\,dt.$ 6. $\int_0^1 t^{\frac{1}{2}}\,dt.$

7. $\int_1^8 t^{\frac{1}{3}}\,dt.$ 8. $\int_1^{10} u^{-2}\,du.$ 9. $\int_1^2 (1 + s)^{-3}\,ds.$

10. $\int_{-1}^1 (2 + u)^{-2}\,du.$

Find primitives of the following functions:

11. $\frac{1}{2}x^3 + 3x^2 - 1.$ 12. $x^{\frac{1}{3}} + x^{-\frac{1}{3}}.$ 13. $\dfrac{1}{x^3}.$ 14. $\dfrac{1}{(2x + 1)^2}.$

15. $\dfrac{2x}{(1 + x^2)^2}.$ 16. $x(x - 1)(x - 2).$ 17. $\dfrac{1}{x^2} + \dfrac{1}{(x - 1)^2}.$

18. $(x^2 + x + 1)^2.$

19.–26. Find the primitives in the preceding Problems 11–18 which are 0 for $x = 1$, that is, determine what constants should be added to the results already obtained.

27.–34. Choosing for $f(x)$ the functions of Problems 11–18, find the area between $y = f(x)$, the x-axis, and the lines $x = 2$, $x = 3$.

35. Find $\int_a^b (x - a)(b - x)\,dx.$ 36. Find $\int_a^b (x - a)^2(x - b)^2\,dx.$

37. If n is a positive integer, find the value of

$$\int_0^1 \frac{1 - x^n}{1 - x}\,dx.$$

38. If $f(t)$ is continuous, prove that the derivative with respect to x of $\int_x^b f(t)\,dt$ is $-f(x)$.

39. If $f(t)$ is continuous for $0 \leqq t$, find the derivative with respect to x of

$$\int_0^{x^2} f(t)\,dt.$$

[*Hint:* This is a composite function.]

40. Find the derivative with respect to x of

$$\int_x^{2x} \frac{dt}{t}.$$

Interpret the result.

41. If $f(x)$ is convex (see Definition 2.7.2), prove that

$$f[\tfrac{1}{2}(a+b)]\,(b-a) \leqq \int_a^b f(t)\,dt \leqq \tfrac{1}{2}[f(a)+f(b)]\,(b-a).$$

Draw a figure. The underlying procedure is known as *trapezoid integration.*

4.4. Rules of integration. Let us state some simple rules concerning primitives. They follow directly from the definition and the rules of differentiation.

1. If $F(x)$ is a differentiable function whose derivative is integrable, then

$$\int F'(x)\,dx = F(x) + C. \tag{4.4.1}$$

It suffices that $F'(x)$ is continuous.

2. If $f_1(x)$ and $f_2(x)$ have primitives, so does $f_1(x) + f_2(x)$ and

$$\int [f_1(x) + f_2(x)]\,dx = \int f_1(x)\,dx + \int f_2(x)\,dx. \tag{4.4.2}$$

3. If $f(x)$ has a primitive, and if C is a constant, then $Cf(x)$ has a primitive and

$$\int Cf(x)\,dx = C\int f(x)\,dx. \tag{4.4.3}$$

4. If $F(x)$ and $G(x)$ have integrable derivatives, then

$$\int [F(x)G'(x) + F'(x)G(x)]\,dx = F(x)G(x) + C. \tag{4.4.4}$$

We note that the integrand is actually integrable. Since it is the derivative of $F(x)G(x)$, the result follows from item 1.

To simplify the statement of the next rule we impose slightly more stringent conditions than are strictly necessary.

5. Let $G(x)$ be strictly monotone and have a continuous derivative in some interval (a, b). Let $F(u)$ be defined for u in the range of G. Let F have a continuous derivative. Then for values of x in (a, b)

$$\int F'[G(x)]G'(x)\,dx = F[G(x)] + C, \tag{4.4.5}$$

for the integrand is the derivative of $F[G(x)]$.

The art of integration consists essentially in a repeated use of these rules and a few basic primitives. Thus far we have had only one primitive, namely

$$\int x^n\,dx = \frac{x^{n+1}}{n+1} + C, \quad n \neq -1. \tag{4.4.6}$$

The excluded case $n = -1$ will be considered in the next section together with the inverse function. This will add two important primitives. Other additions will be made later.

Item 4 is the basis of the method of *integration by parts*. We can obviously rewrite (4.4.4) as

$$\int F(x)G'(x)\,dx = F(x)G(x) - \int G(x)F'(x)\,dx \tag{4.4.7}$$

or, equivalently, in terms of differentials,

$$\int F(x)\,dG(x) = F(x)G(x) - \int G(x)\,dF(x). \tag{4.4.8}$$

This is of course just an identity. It is of practical use if the integral on the right is simpler than the one on the left. The following is an example of the use of the method. Take

$$J = \int \frac{x^3\,dx}{(1 + x^2)^3}. \tag{4.4.9}$$

Here we note that

$$\frac{d}{dx}\frac{1}{(1 + x^2)^2} = -\frac{4x}{(1 + x^2)^3}.$$

Hence

$$\begin{aligned} J &= -\tfrac{1}{4}\int x^2 \frac{d}{dx}\left\{\frac{1}{(1 + x^2)^2}\right\} dx \\ &= -\frac{x^2}{4(1 + x^2)^2} + \tfrac{1}{4}\int \frac{2x}{(1 + x^2)^2}\,dx \\ &= -\tfrac{1}{4}\frac{x^2}{(1 + x^2)^2} - \tfrac{1}{4}\int \frac{d}{dx}\left\{\frac{1}{1 + x^2}\right\} dx \\ &= -\tfrac{1}{4}\left\{\frac{x^2}{(1 + x^2)^2} + \frac{1}{1 + x^2}\right\} + C = -\tfrac{1}{4}\frac{1 + 2x^2}{(1 + x^2)^2} + C. \end{aligned}$$

The student should verify the result by differentiation.

In using the method of integration by parts the student must keep in mind that the purpose is to replace a complicated integral by a simpler one more likely to be integrable directly. Usually the factors in the product can be chosen in several ways. One choice may be good and another may lead to something more complicated. In the present case (4.4.9), simplification means reducing the exponents of the powers. This requires integrating $(1 + x^2)^{-3}$ and differentiating x^3. Here there is a complication in that $(1 + x^2)^{-3}$ is not the derivative of a simple function. We can get around this difficulty, however, by "borrowing" a factor x since $2x(1 + x^2)^{-3}$ is the derivative of a rational function. This leaves a factor x^2 to be differentiated which is more advantageous than x^3. It should be noted that integrals are very sensitive to what looks like a minor change in the integrand. Thus the integral (4.4.9) is a rational function of x, but this is no longer true if we replace x^3 by either x^2 or x^4. Finally, one must keep in mind that while every continuous function has a primitive there is no guarantee that the primitive is an elementary function even when the integrand is elementary. We shall encounter many examples in the following.

Item 5 is the point of departure of another powerful, constantly used method of integration, namely *integration by substitution*. If in the left member of (4.4.5) we set

$$u = G(x), \tag{4.4.10}$$

then

$$\int F'(G(x))G'(x)\,dx = \int F'(u)\,du = F(u) + C = F[G(x)] + C. \tag{4.4.11}$$

The result of course agrees with (4.4.5) and on the face of it we do not seem to have accomplished much. It is again a question of simplification which makes it easier to see what is involved and how to proceed. In the present case it looks simple, but it should be remembered that in an actual problem the choice of F and G may be far from obvious and several different substitutions may suggest themselves. We shall use (4.4.9) also to illustrate this method.

This problem is suitable for demonstration purposes because of the large number of available substitutions. Besides trigonometric substitutions which we shall learn to use later, there are at least three rational substitutions that work. The most obvious one is

$$x^2 = t, \quad 2x\,dx = dt, \quad x^3\,dx = x^2x\,dx = \tfrac{1}{2}t\,dt,$$

so that

$$\begin{aligned} J &= \tfrac{1}{2}\int \frac{t\,dt}{(t+1)^3} = \tfrac{1}{2}\int \frac{(t+1)-1}{(t+1)^3}\,dt \\ &= \tfrac{1}{2}\int \frac{dt}{(t+1)^2} - \tfrac{1}{2}\int \frac{dt}{(t+1)^3} \\ &= -\tfrac{1}{2}\frac{1}{t+1} + \tfrac{1}{4}\frac{1}{(t+1)^2} + C \\ &= -\tfrac{1}{2}\frac{1}{x^2+1} + \tfrac{1}{4}\frac{1}{(x^2+1)^2} + C \end{aligned}$$

in agreement with our previous result.

It is somewhat more advantageous to take

$$1 + x^2 = u, \quad 2x\,dx = du, \quad x^3\,dx = \tfrac{1}{2}(u-1)\,du.$$

This gives

$$\begin{aligned} J &= \tfrac{1}{2}\int \frac{u-1}{u^3}\,du = \tfrac{1}{2}\int \frac{du}{u^2} - \tfrac{1}{2}\int \frac{du}{u^3} \\ &= -\frac{1}{2u} + \frac{1}{4u^2} + C \end{aligned}$$

which leads to the same result if we substitute $u = 1 + x^2$. The main advantage of this choice is that it avoids the artifice $t = (t+1) - 1$ used above, to bring the fraction $t/(t+1)^3$ into integrable form.

A third choice is given by

$$\frac{1}{1+x^2} = s, \quad \frac{2x\,dx}{(1+x^2)^2} = -\,ds, \quad x^2 = \frac{1}{s} - 1$$

so that

$$J = -\tfrac{1}{2}\int\left(\frac{1}{s} - 1\right)s\,ds = \tfrac{1}{2}\int(s-1)\,ds = \tfrac{1}{4}s^2 - \tfrac{1}{2}s + C.$$

Substituting the value of s in terms of x, we again get the same result.

It should be noted that the expression for the primitive may well depend upon the method used in obtaining it. Two such expressions must differ by a constant. Usually the constant is easily found, but occasionally deeper properties of the primitive are involved and the resulting relation implies a functional equation.

The student should note that the use of the differential in the notation for the integral is a great help when integration by substitution is to be used. It is part of the substitution to figure out the relation between the differentials of the old and the new variables of integration as was shown in the problem above.

So far we have dealt only with primitives, but what has been said applies also to definite integrals. Formula (4.4.1) now becomes

$$\int_a^b F'(x)\,dx = F(b) - F(a). \tag{4.4.12}$$

The formula for integration by parts gives

$$\int_a^b F(x)\,dG(x) = F(x)G(x)\Big|_a^b - \int_a^b G(x)\,dF(x). \tag{4.4.13}$$

In actual practice it may be advantageous to omit the limits of integration until the primitive of the integrand has been found and then substitute the limits.

Improper handling of the limits of integration when making a change of variables is a frequent source of error. If, as we assume, $G(x)$ is strictly monotone in (a, b), then we have

$$\int_a^b F'[G(x)]G'(x)\,dx = \int_{G(a)}^{G(b)} F'(u)\,du$$

$$= F(u)\Big|_{G(a)}^{G(b)} = F[G(x)]\Big|_a^b. \tag{4.4.14}$$

This illustrates that changing the limits of integration is part of the substitution. The value of the integral is not

$$F(b) - F(a)$$

for instance. There is another pitfall in working with substitutions in definite integrals. The substitution should really establish a one-to-one mapping of the interval $[a, b]$ onto an interval $[\alpha, \beta]$. If not, the resulting answer may have no bearing on the problem. This is illustrated by the following problem. The integral

$$\int_{-1}^{2} (1 + x^2)x^2\, dx$$

is readily computed by expanding the integrand and applying formula (4.4.6). The value is 9.6. Now a substitution enthusiast could possibly try to set

$$x^2 = t, \quad 2x\, dx = dt, \quad x = \sqrt{t}$$

and would note that $t = 1$ when $x = -1$, $t = 4$ when $x = 2$. This gives

$$\tfrac{1}{2}\int_{1}^{4} (1 + t)\sqrt{t}\, dt = \tfrac{7}{3} + \tfrac{31}{5} \neq 9.6.$$

The reason for the discrepancy is the following. When x goes from -1 to 0, t decreases from 1 to 0, moreover we must take the negative value of the square root since x is negative. When x goes from 0 to 1, t increases from 0 to 1 and now the square root must be given its positive determination. Both integrals give positive contributions which were completely lost when our enthusiast settled for the integral from 1 to 4. If account is taken of the neglected integrals, the correct result is obtained. The method is not recommended.

EXERCISE 4.4

Find primitives of the following functions using algebraic simplification and/or simple substitution. Verify your results by differentiation.

1. $(x - 1)^6$.
2. $(2 - x)^4$.
3. $\left(x + \dfrac{1}{x}\right)^2$.
4. $(3x - 4)^{-3}$.
5. $(1 + x)^{\frac{1}{2}}$.
6. $x(1 + x^2)^2$.
7. $x(1 + x^2)^{\frac{1}{2}}$.
8. $\dfrac{x}{\sqrt{1 + x^2}}$.
9. $\dfrac{x^2}{(1 - x^3)^{\frac{1}{2}}}$.
10. $x(1 + x)^{\frac{1}{2}}$.
11. $x^2(1 - x^3)^{\frac{1}{3}}$.
12. $\dfrac{x^3}{(1 - x^2)^{\frac{1}{2}}}$.

13.–15. Work Problems 6, 10, and 12 using partial integration.

Find primitives of the following functions using the method of substitution or otherwise.

16. $\dfrac{x^5}{(1 + x^6)^2}$.
17. $\dfrac{x}{(1 + x^2)^2}$.
18. $\dfrac{x - x^3}{(1 + x^2)^3}$.
19. $\dfrac{x^2}{(1 + x^3)^4}$.
20. $\dfrac{x + 2x^3}{x^6(1 + x^2)^3}$.
21. $\left(x + \dfrac{1}{x}\right)\left(1 - \dfrac{1}{x^2}\right)$.
22. Verify that

$$\int_{0}^{1} (x^2 - x)\, dx = -\tfrac{1}{6}.$$

What is the geometrical significance of the answer?

23. Compute

$$\int_0^{\sqrt{2}} (x^3 - x)\, dx$$

and discuss the geometric meaning of the answer.

Find primitives of the following functions by integration by parts, repeated if necessary.

24. $\dfrac{x}{(1 + x)^3}$. 25. $\dfrac{x^3}{(4 - x^2)^3}$. 26. $\dfrac{1 + 2x}{(1 + x + x^2)^3}$.

27. $\dfrac{x^3}{(1 + x^2)^{\frac{1}{2}}}$. 28. $\dfrac{x}{(5 + x)^{\frac{1}{3}}}$. 29. $\dfrac{x^5}{(5 + x^3)^{\frac{1}{3}}}$.

Evaluate the following definite integrals:

30. $\displaystyle\int_0^1 (1 - x)^{\frac{1}{2}}\, dx$. 31. $\displaystyle\int_0^1 x(1 - x)^{\frac{1}{2}}\, dx$.

32. $\displaystyle\int_{-1}^1 x(1 + x^2)^{\frac{1}{2}}\, dx$. 33. $\displaystyle\int_5^6 \frac{x\, dx}{\sqrt{1 + x^2}}$.

34. $\displaystyle\int_0^1 x(1 - x^2)^{\frac{1}{2}}\, dx$. 35. $\displaystyle\int_0^3 \frac{x^5\, dx}{\sqrt{1 + x^6}}$.

36. $\displaystyle\int_1^2 \frac{x^2\, dx}{\sqrt{1 + x^3}}$. 37. $\displaystyle\int_{-1}^1 x^2(1 - x^3)^{\frac{1}{3}}\, dx$.

38. $\displaystyle\int_1^2 \frac{t^2 - 1}{\sqrt{t}}\, dt$. 39. $\displaystyle\int_0^2 \frac{s^2\, ds}{\sqrt{s + 2}}$. 40. $\displaystyle\int_1^2 \frac{x^5\, dx}{(7 + x^3)^{\frac{1}{3}}}$.

4.5. The logarithm and the exponential function. It is time to consider the excluded case $n = -1$ in formula (4.4.6). We define

$$L(x) = \int_1^x \frac{ds}{s}, \quad 0 < x < \infty. \tag{4.5.1}$$

The integral is well defined for all positive values of x. The function $L(x)$ is negative for $0 < x < 1$ (cf. formula (4.2.25)), it is 0 for $x = 1$, and positive for $x > 1$. It is clear that $L(x)$ is strictly increasing. We shall see that

$$\lim_{x\to 0} L(x) = -\infty, \quad \lim_{x\to\infty} L(x) = +\infty, \tag{4.5.2}$$

so that the range of $L(x)$ is the set of all real numbers.

We have

$$L'(x) = \frac{1}{x}. \tag{4.5.3}$$

Thus $L(x)$ has derivatives of all orders and

$$L^{(n)}(x) = (-1)^{n-1}(n - 1)!\, x^{-n}. \tag{4.5.4}$$

The curve

$$y = L(x)$$

lies completely in the right half-plane and below the line $y = x$ since

$$L(x) \leqq x - 1 \tag{4.5.5}$$

by the mean value theorem. Since $L'(x)$ is positive and strictly decreasing to 0, the curve always lies below its tangent, that is, the curve is concave downward. See Figure 4.9 which also represents the inverse function

$$y = E(x)$$

where

$$\begin{aligned} E[L(x)] &= x, \quad x > 0, \\ L[E(x)] &= x, \quad \text{all } x. \end{aligned} \tag{4.5.6}$$

There will be much more about this function below.

The geometric interpretation of (4.5.1) for $x > 1$ is that $L(x)$ is the area under the equilateral hyperbola

$$st = 1$$

above the s-axis and between the lines $s = 1$ and $s = x$. See Figure 4.10. For $0 < x < 1$, $L(x)$ is the negative of the area below the curve, above the s-axis, and between the lines $s = x$ and $s = 1$.

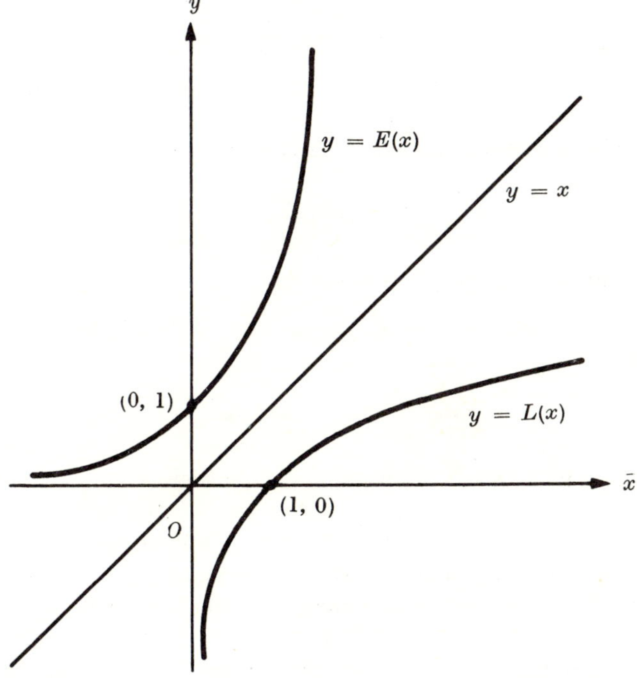

Figure 4.9

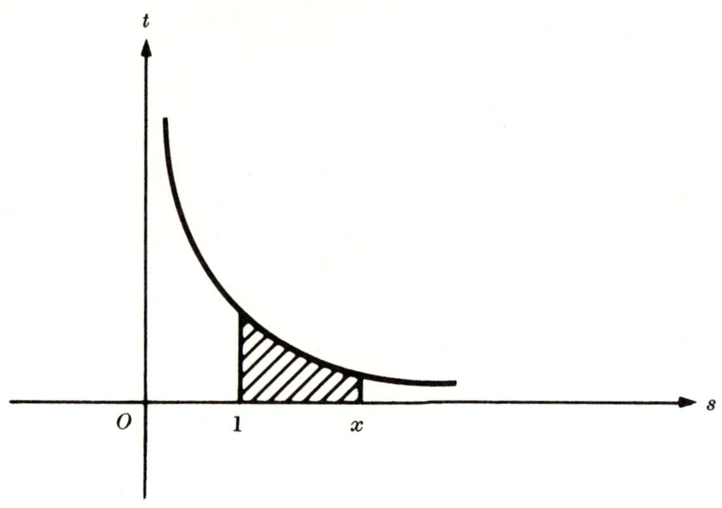

Figure 4.10

This area problem was considered by a Belgian Jesuit, Gregorius a Santo Vincentio, at the end of an otherwise fairly uninteresting geometrical work on conic sections and squaring of the circle, which appeared in 1647. Using Cavalieri's methods, Gregorius proved the relation

$$A(a, b) = A(ka, kb), \quad k > 0. \tag{4.5.7}$$

Here the left member represents the area between the hyperbola, the s-axis, and the lines $s = a$, $s = b$. The special case $a = 1$, $b = 2$, $k = 2$, figured in Problem 3 of Exercise 4.1. In particular we have

$$A(a, b) = A\left(1, \frac{b}{a}\right) = L\left(\frac{b}{a}\right). \tag{4.5.8}$$

Now by the additive property of areas

$$A(a, b) = A(a, c) + A(c, b), \quad a < c < b. \tag{4.5.9}$$

Hence, if we set

$$\frac{c}{a} = x_1, \quad \frac{b}{c} = x_2 \quad \text{so that} \quad \frac{b}{a} = x_1 x_2,$$

the additive property gives the functional equation

$$L(x_1 x_2) = L(x_1) + L(x_2). \tag{4.5.10}$$

This equation states the characteristic property of any system of *logarithms*. (The Greek word λόγος can also mean "proportion" while ἀριθμός means "number.")

At the beginning of the seventeenth century John Napier set himself the task of simplifying numerical calculations by replacing multiplication by addition, division by subtraction, and root extraction by division by an integer. In other words, he wanted a function having the property (4.5.10). He found that all such functions are constant multiples of a particular one, the *natural logarithm* of x which he computed and listed. That all continuous solutions of the functional equation

$$f(x_1x_2) = f(x_1) + f(x_2) \tag{4.5.11}$$

are indeed of the form $f(x) = CL(x)$ was proved two hundred years later by Cauchy. Gregorius realized that he was dealing with a logarithm, but not that $L(x)$ is the natural logarithm of x. We shall make the identification below.

Our first task is to prove formula (4.5.10) with the aid of (4.5.1). Note that we have not yet proved (4.5.7). For the sake of simplicity we restrict ourselves to the case $x_1 > 1$, $x_2 > 1$. We have then

$$L(x_1x_2) = A(1, x_1x_2) = A(1, x_1) + A(x_1, x_1x_2)$$
$$= L(x_1) + A(x_1, x_1x_2).$$

Now

$$A(x_1, x_1x_2) = \int_{x_1}^{x_1x_2} \frac{ds}{s},$$

and here we set

$$s = x_1t, \quad ds = x_1\,dt, \quad \frac{ds}{s} = \frac{dt}{t}.$$

For the limits of integration we note that $s = x_1$ gives $t = 1$ and $s = x_1x_2$ gives $t = x_2$. Hence

$$A(x_1, x_1x_2) = \int_1^{x_2} \frac{dt}{t} = A(1, x_2) = L(x_2).$$

Combining the results we see that (4.5.10) holds and we have also actually verified that (4.5.7) holds. The extension to the case where one or both of the variables are < 1 is merely a matter of examining the various possibilities and using the additive property of areas.

As a consequence of (4.5.10) we get

$$L\left(\frac{1}{x}\right) = -L(x). \tag{4.5.12}$$

We also note that for all $x > 0$

$$L(x^n) = nL(x), \tag{4.5.13}$$

where n is a positive integer to begin with. In view of (4.5.12) the formula also holds for negative integers. Next, let q be a positive integer and set $x = s^{1/q}$, $n = q$. Then

$$L(s) = qL(s^{1/q}) \quad \text{or} \quad L(s^{1/q}) = \frac{1}{q}\,L(s).$$

Here we set $s = t^p$, p an integer, and obtain

$$L(t^{p/q}) = \frac{1}{q}\,L(t^p) = \frac{p}{q}\,L(t).$$

Since p, q, and t are arbitrary, it follows that (4.5.13) is valid for any rational number n. The formula also holds for irrational numbers n; here the main difficulty lies in defining x^n when n is irrational. We need not face this obstacle at once.

Let us derive various consequences of these important formulas. The first application of (4.5.13) is to prove the limit relations in (4.5.2). Now $L(x)$ is strictly increasing, and so in order to prove that $L(x)$ becomes infinite with x, it is sufficient to show that $L(2^n)$ becomes infinite with n. This is obvious since

$$L(2^n) = nL(2).$$

This and (4.5.12) suffice to prove (4.5.2).

Next we identify $L(x)$ with the natural logarithm of x. To this end we note that

$$1 = L'(1) = \lim_{h\to 0} \frac{1}{h} [L(1 + h) - L(1)] = \lim_{h\to 0} \frac{1}{h} L(1 + h)$$

since $L(1) = 0$. Here it is immaterial how h goes to 0. We set

$$h = \frac{1}{n}, \quad n \text{ positive integer,}$$

and find that

$$1 = \lim_{n\to\infty} nL\left(1 + \frac{1}{n}\right) = \lim_{n\to\infty} L\left[\left(1 + \frac{1}{n}\right)^n\right] = L\left[\lim_{n\to\infty}\left(1 + \frac{1}{n}\right)^n\right] = L(e).$$

Here we have used (i) formula (4.5.13), (ii) the definition of e given in Section 1.7 by formula (1.7.15), and (iii) the fact that $L(x)$ is continuous. Thus

$$L(e) = 1. \tag{4.5.14}$$

Using (4.5.13) again we obtain

$$L(e^n) = n \tag{4.5.15}$$

for every rational number n.

The last step is to extend this relation to irrational numbers. Here it is decisive that $L(x)$ as a continuous, strictly increasing function has a continuous, strictly increasing inverse. Suppose that α is an irrational number and that $\{r_n\}$ is a strictly increasing sequence of rational numbers converging to α. Then

$$\{e^{r_n}\}$$

is also a bounded strictly increasing sequence and as such has a limit. We define

$$e^\alpha = \lim_{n\to\infty} e^{r_n}. \tag{4.5.16}$$

Since

$$L(e^{r_n}) = r_n,$$

the continuity gives

$$L(e^\alpha) = \alpha,$$

which is the desired relation.

Since the inverse function is uniquely defined, we have by (4.5.6)

$$E(x) = e^x \equiv \exp(x) \tag{4.5.17}$$

and, conversely, by definition,

$$L(x) = \log x. \tag{4.5.18}$$

Here and in the following, the symbol "log" always denotes the logarithm to the base e, *the natural logarithm*, unless otherwise stated. The notation $\ln x$ (sometimes log nat x) is often used, especially in the applications and in elementary texts. Formulas (4.5.6) now take the form

$$\begin{aligned} e^{\log x} &= x, \quad x > 0, \\ \log (e^x) &= x, \quad \text{all } x. \end{aligned} \tag{4.5.19}$$

Let us now consider some integrals involving the logarithm. If $0 < a < b$, we have

$$\int_a^b \frac{ds}{s} = \log b - \log a = \log \frac{b}{a}. \tag{4.5.20}$$

If a and b are both negative, there is the difficulty that $\log x$ is not defined for $x < 0$ and, in fact, cannot be defined as a real number for such values of x. On the other hand, both the first and the third members of (4.5.20) have a meaning since $b/a > 0$. Actually the third member gives the correct value of the integral as long as a and b have the same sign. If a and b are both negative, we use the substitution

$$s = -t, \quad ds = -dt$$

which gives

$$\int_a^b \frac{ds}{s} = \int_{-a}^{-b} \frac{dt}{t} = \log (-b) - \log (-a) = \log \frac{b}{a}$$

as asserted. Note that the integral (4.5.20) has no meaning if a and b have opposite signs.

This takes care of the definite integral of $1/x$. For the indefinite integral we have

$$\int \frac{dx}{x} = \log |x| + C \tag{4.5.21}$$

which is valid for negative as well as positive values of x.

Next we consider some integrals involving $\log x$ in the integrand. The simplest is

$$\int \log x \, dx.$$

This problem illustrates the old adage: When in doubt, integrate by parts! In formula (4.4.8) we take

$$F(x) = \log x, \quad G(x) = x$$

and obtain

$$\int \log x \, dx = x \log x - \int x \frac{1}{x} \, dx = x [\log x - 1] + C,$$

so that

$$\int \log x \, dx = x[\log x - 1] + C. \tag{4.5.22}$$

This is a special case of various other integrals of similar nature. Thus

$$\int x^m \log x \, dx = \frac{x^{m+1}}{m+1}\left\{\log x - \frac{1}{m+1}\right\} + C, \quad m \neq -1, \tag{4.5.23}$$

while

$$\int \frac{dx}{x \log x} = \log(|\log x|) + C. \tag{4.5.24}$$

In all these integrals x must be restricted to positive values.

We can generalize still further to integrals of the form

$$\int x^m (\log x)^n \, dx, \tag{4.5.25}$$

where m is any real number and n is a positive integer. If $m \neq -1$, we use repeated integration by parts to reduce the exponent of $\log x$ to 0. The result in this case is a polynomial in $\log x$ of degree n, multiplied by x^{m+1}. For $m = -1$, we have

$$\int (\log x)^n \frac{dx}{x} = \frac{(\log x)^{n+1}}{n+1} + C. \tag{4.5.26}$$

The restriction of n to positive integers is not absolutely necessary, but if n is a negative integer the reduction may lead to

$$\int \frac{dx}{\log x}, \tag{4.5.27}$$

where no further reduction is possible. The function

$$\operatorname{li}(x) = \int_2^x \frac{ds}{\log s}, \quad 1 < x, \tag{4.5.28}$$

is known as the *logarithm integral* of x. It is of considerable interest to mathematicians for if $\Pi(x)$ denotes the number of primes $\leqq x$, then

$$\lim_{x \to \infty} \frac{\Pi(x)}{\operatorname{li}(x)} = 1. \tag{4.5.29}$$

Within the range of the prime number tables

$$\operatorname{li}(x) - \Pi(x)$$

is positive, but it can be proved that the difference changes its sign infinitely often and that the first change occurs for an $x > 10^9$ but not exceeding $10^{10^{10^{34}}}$. The latter appears to be the largest number that ever served any definite purpose in mathematics.

Let us now turn to the inverse function of the logarithm which is known as the *exponential function*. We already know some of its properties. Thus, e^x is a strictly increasing positive function which tends to $+\infty$ as x goes to $+\infty$ and to 0 as x goes to $-\infty$. Since $\log x$ is differentiable, Theorem 3.5.1 asserts that e^x is also, and from

$$\log e^x = x$$

we get by differentiation

$$\frac{1}{e^x} D_x e^x = 1,$$

or

$$D_x e^x = e^x, \tag{4.5.30}$$

and hence also

$$D_x^n e = e^x \tag{4.5.31}$$

for all n. Thus the exponential function equals its own derivative and it may be shown that

$$y = e^x$$

is the unique solution of the differential equation

$$D_x y = y,$$

which has the value 1 for $x = 0$.

Let us now see how the equation (4.5.10) is reflected in properties of the exponential function. We set

$$u_1 = \log v_1, \quad u_2 = \log v_2$$

so that

$$v_1 = e^{u_1}, \quad v_2 = e^{u_2}.$$

We have then

$$e^{u_1+u_2} = e^{\log v_1 + \log v_2} = e^{\log(v_1 v_2)} = v_1 v_2 = e^{u_1} e^{u_2},$$

so that for all x_1 and x_2

$$e^{x_1+x_2} = e^{x_1} e^{x_2}. \tag{4.5.32}$$

Thus e^x is a solution of the functional equation

$$F(x_1 + x_2) = F(x_1)F(x_2). \tag{4.5.33}$$

In his *Cours d'Analyse* of 1828 Cauchy showed that every real continuous solution of this equation is of the form

$$F(x) = e^{ax}, \tag{4.5.34}$$

where a is any real constant.

We shall now take up for discussion some integrals involving the exponential function. We have obviously as a consequence of (4.5.30) the formulas

$$\int e^x \, dx = e^x + C, \tag{4.5.35}$$

$$\int_a^b e^s \, ds = e^b - e^a. \tag{4.5.36}$$

Integrals of the form

$$\int x^n e^x \, dx,$$

where n is a positive integer, can be found by integrating repeatedly by parts: e^x is integrated and the power is differentiated. The result is an expression of the form e^x times a polynomial in x of degree n. Thus, for instance,

$$\int x^3 e^x \, dx = e^x(x^3 - 3x^2 + 6x - 6) + C. \tag{4.5.37}$$

Negative integers require a change of tactics: we integrate the power and differentiate the exponential. This, however, leads to the integral

$$\int e^x \frac{dx}{x} = \int \frac{dt}{\log t} \quad \text{with} \quad t = e^x. \tag{4.5.38}$$

We leave these formulas and proceed to prove an inequality satisfied by e^x for $x > 0$. Since $e^0 = 1$ and e^x is strictly increasing, we have clearly $e^x > 1$ for $x > 0$. This inequality may be improved indefinitely with the aid of (4.5.36). In the latter relation we set $a = 0$, $b = x$ (> 0), and obtain

$$e^x = 1 + \int_0^x e^s \, ds > 1 + \int_0^x ds = 1 + x,$$

since $e^s > 1$. But if

$$e^s > 1 + s,$$

then again

$$e^x = 1 + \int_0^x e^s \, ds > 1 + \int_0^x (1 + s) \, ds$$

$$= 1 + \frac{x}{1!} + \frac{x^2}{2!}.$$

Thus we see that each improvement of the estimate leads automatically to a still better one. After n steps we have

$$e^x > 1 + \frac{x}{1!} + \frac{x^2}{2!} + \cdots + \frac{x^n}{n!}, \tag{4.5.39}$$

where n is any positive integer.

The underlying reason for this inequality will become apparent in Chapter 6 when we discuss the Maclaurin series for e^x. The right member of (4.5.39) is the nth partial sum of this series. We shall use the inequality for two special purposes. First we note the particular case $x = 1$ which gives lower bounds for e:

$$e > 1 + \frac{1}{1!} + \frac{1}{2!} + \cdots + \frac{1}{n!}. \tag{4.5.40}$$

Next we replace n by $n + 1$ in (4.5.3) and divide by x^n, obtaining

$$x^{-n}e^x > x^{-n} + x^{-n+1} + \cdots + \frac{1}{n!} + \frac{x}{(n+1)!} > \frac{x}{(n+1)!}.$$

This becomes infinite as x becomes infinite. Hence:

When x tends to $+\infty$, the exponential function e^x becomes infinite faster than any fixed positive power of x.

Since $e^{-x} = (e^x)^{-1}$, this also gives

$$\lim_{x \to -\infty} x^n e^x = 0 \tag{4.5.41}$$

for every fixed n.

The same limit relations give information concerning the rate of growth of the logarithm. We set $e^x = t$ and obtain

$$\lim_{t \to \infty} \frac{t}{(\log t)^n} = +\infty. \tag{4.5.42}$$

This is equivalent to saying:

The function $\log t$ *becomes infinite with* t *more slowly than any fixed power of* t *no matter how small the exponent is.*

EXERCISE 4.5

Find primitives of the following functions:

1. $x \log x$.
2. $(\log x)^2$.
3. $x(\log x)^2$.
4. $\dfrac{1}{x(\log x)^2}$.
5. $\dfrac{\log x}{x^2}$.
6. xe^x.
7. e^{2x}.
8. $(e^x + 1)^2 e^x$.
9. $(e^x + e^{-x})^2$.
10. $x^2 e^{-x}$.
11. xe^{x^2}.
12. $x^3 e^{-x^2}$.

13.–24. Find the first-order derivatives of the functions in the preceding problems.

Evaluate the following integrals:

25. $\displaystyle\int_1^2 s^2 \log s \, ds$.
26. $\displaystyle\int_1^e (\log s)^3 \, ds$.
27. $\displaystyle\int_2^e \frac{ds}{s(\log s)^3}$.
28. $\displaystyle\int_1^{1000} \frac{\log s}{s^2} \, ds$.
29. $\displaystyle\int_0^1 se^s \, ds$.
30. $\displaystyle\int_0^4 e^{-2s} \, ds$.
31. $\displaystyle\int_{-1}^1 (e^x - e^{-x}) \, dx$.
32. $\displaystyle\int_0^{10} se^{-s^2} \, ds$.
33. $\displaystyle\int_0^3 t^3 e^{-t} \, dt$.

34. Use the Bouniakovski-Schwarz inequality to show that

$$\int_a^b \frac{ds}{s} < \frac{b-a}{\sqrt{ab}}, \quad 0 < a < b.$$

35. Show that the primitive of $\log \log x$ involves the logarithm integral.
36. How large must x be in order that $\log \log x > 10$?
37. Find $\displaystyle\lim_{h \to 0} \frac{1}{h} (e^h - 1)$.
38. Show that if $F(x)$ is not identically zero and satisfies (4.5.33), then $F(0) = 1$.

4.6. Absolutely convergent improper integrals. Thus far we have restricted ourselves to finite intervals and functions bounded in such an interval. These restrictions are not necessary and will now be partly removed.

DEFINITION 4.6.1. *Let (a, b) be a finite or infinite interval. Let f be integrable over every closed finite subinterval of (a, b). If there exists a fixed finite number B such that*

$$\int_\alpha^\beta |f(s)|\, ds < B \tag{4.6.1}$$

for all $[\alpha, \beta] \subset (a, b)$, then f is said to be absolutely integrable over (a, b). If this is the case, then, by definition,

$$\int_a^b f(s)\, ds = \lim_{\substack{\alpha \to a \\ \beta \to b}} \int_\alpha^\beta f(s)\, ds. \tag{4.6.2}$$

We have to convince ourselves that this limit really exists. This is the object of

THEOREM 4.6.1. *If f is absolutely integrable, so are $|f|$, f^+, and f^-. Further*

$$\int_a^b f(s)\, ds = \int_a^b f^+(s)\, ds - \int_a^b f^-(s)\, ds, \tag{4.6.3}$$

$$\int_a^b |f(s)|\, ds = \int_a^b f^+(s)\, ds + \int_a^b f^-(s)\, ds, \tag{4.6.4}$$

where all integrals exist as finite limits.

Proof. It follows from the definition that $|f|$ is absolutely integrable when f is. We recall that

$$f^+(s) = \max\,[f(s), 0], \quad f^-(s) = \max\,[-f(s), 0].$$

From this it follows that

$$\int_\alpha^\beta f^+(s)\, ds \leqq \int_\alpha^\beta |f(s)|\, ds \leqq B,$$

$$\int_\alpha^\beta f^-(s)\, ds \leqq \int_\alpha^\beta |f(s)|\, ds \leqq B,$$

so that f^+ and f^- are absolutely integrable. Now, if g is a nonnegative function, absolutely integrable over (a, b), then

$$\int_\alpha^\beta |g(s)|\, ds = \int_\alpha^\beta g(s)\, ds$$

is a bounded, nondecreasing function of β and a bounded, nonincreasing function of α. It then has a finite limit as $\beta \to b$ and $\alpha \to a$. From this it follows that

$$\int_\alpha^\beta f^+(s)\, ds \quad \text{and} \quad \int_\alpha^\beta f^-(s)\, ds$$

tend to finite limits when α decreases to a and β increases to b. These limits are by definition the integrals

$$\int_a^b f^+(s)\,ds \quad \text{and} \quad \int_a^b f^-(s)\,ds.$$

Since

$$\int_\alpha^\beta f(s)\,ds = \int_\alpha^\beta f^+(s)\,ds - \int_\alpha^\beta f^-(s)\,ds,$$

it follows that the left member has a finite limit as $\alpha \to a$, $\beta \to b$, and that (4.6.3) holds. This shows that the limit in (4.6.2) really exists. Formula (4.6.4) is proved in the same manner.

DEFINITION 4.6.2. *$R^*(a, b)$ is the class of all functions absolutely integrable over (a, b).*

We have clearly (note that $R(a, b)$ is not defined for infinite intervals):

THEOREM 4.6.2. *$R(a, b) \subset R^*(a, b)$, if $b - a < \infty$.*

It is almost equally obvious that the analogue of Theorem 4.2.1 holds:

THEOREM 4.6.3. *The integral of a function in $R^*(a, b)$ is an additive function of intervals:*

$$\int_a^b f(s)\,ds = \int_a^c f(s)\,ds + \int_c^b f(s)\,ds. \tag{4.6.5}$$

The integral is also a linear function of the integrand: if f_1 and $f_2 \in R^*(a, b)$, and if C_1 and C_2 are real constants, then $C_1f_1 + C_2f_2 \in R^*(a, b)$ and

$$\int_a^b [C_1f_1(s) + C_2f_2(s)]\,ds = C_1\int_a^b f_1(s)\,ds + C_2\int_a^b f_2(s)\,ds. \tag{4.6.6}$$

Proof. If B is a bound for $\int_a^b |f(s)|\,ds$, then it is also a bound for the integrals over (α, c) and (c, β), whence it follows that f is absolutely integrable over (a, c) and (c, b). Since

$$\int_\alpha^\beta = \int_\alpha^c + \int_c^\beta,$$

formula (4.6.5) is obtained when $\alpha \to a$, $\beta \to b$. For the linearity, suppose that B_1 and B_2 are bounds for the integrals of the absolute values of f_1 and f_2, respectively. Then $|C_1|\,B_1 + |C_2|\,B_2$ is a bound for the integral of $|C_1f_1 + C_2f_2|$, whence it follows that $C_1f_1 + C_2f_2 \in R^*(a, b)$. Formula (4.6.6) holds if a and b are replaced by α and β, respectively, and hence also in the limit with a and b. This completes the proof.

It is customary in the Calculus to call (4.6.2) an *improper integral* if either the interval is infinite or the integrand is unbounded. There are, however, more general improper integrals for it could very well happen that the limit could exist in (4.6.2) without condition (4.6.1) being satisfied. This is why we emphasize the absolute integrability.

Let us now give some examples of unbounded functions which are absolutely integrable.

For $s > 0$, we have $s^{-\frac{1}{2}} > 0$ and

$$\int_\delta^1 s^{-\frac{1}{2}}\,ds = 2s^{\frac{1}{2}}\Big|_\delta^1 = 2(1 - \delta^{\frac{1}{2}}) < 2.$$

Thus $x^{-\frac{1}{2}}$ satisfies condition (4.6.1) in $(0, 1)$ and is consequently absolutely integrable and

$$\int_0^1 s^{-\frac{1}{2}}\,ds = 2.$$

This is a special case of the following proposition:

$$x^{-\sigma} \in R^*(0, 1) \quad \textit{if and only if } \sigma < 1. \tag{4.6.7}$$

To prove the sufficiency, suppose that $\sigma < 1$. Then

$$\int_\delta^1 s^{-\sigma}\,ds = \frac{s^{1-\sigma}}{1-\sigma}\Big|_\delta^1 < \frac{1}{1-\sigma}.$$

Thus condition (4.6.1) is satisfied, and the bound is the value of the integral over $(0, 1)$.

If $\sigma = 1$, then

$$\int_\delta^1 \frac{ds}{s} = \log\frac{1}{\delta},$$

and this tends to $+\infty$ when $\delta \to 0$. If $\sigma > 1$, similarly

$$\int_\delta^1 s^{-\sigma}\,ds = \frac{\delta^{1-\sigma} - 1}{\sigma - 1},$$

and this also becomes infinite when $\delta \to 0$.

For the interval $(1, \infty)$ the situation is reversed:

$$x^{-\sigma} \in R^*(1, \infty) \quad \textit{if and only if } \sigma > 1. \tag{4.6.8}$$

The proof follows directly from the three formulas listed above where we now have to replace the substitution $|_\delta^1$ by $|_1^\beta$. We note

$$\int_1^\infty s^{-\sigma}\,ds = \frac{1}{\sigma - 1}, \quad \sigma > 1. \tag{4.6.9}$$

In discussing more complicated integrals, the following theorem is very useful.

THEOREM 4.6.4. *Suppose that $f \in R(\alpha, \beta)$ whenever $[\alpha, \beta] \subset (a, b)$. Then a necessary and sufficient condition that f belongs to $R^*(a, b)$ is that there exists a function $g \in R^*(a, b)$ such that $g(s)$ is nonnegative and $|f(s)| \leq g(s)$ for all s in (a, b).*

Proof. The condition is clearly necessary for $|f(s)|$ is such a function. It is also sufficient for if B is a bound of $\int_\alpha^\beta g(s)\,ds$, then B is also a bound of $\int_\alpha^\beta |f(s)|\,ds$. Hence, $f \in R^*(a, b)$.

We shall give examples of the use of this theorem. It is required to show that the function $x^{-\frac{1}{2}}(1 + x)^{-1} \in R^*(0, \infty)$. We note first that

$$x^{-\frac{1}{2}}(1 + x)^{-1} < x^{-\frac{1}{2}}, \quad 0 < x < 1.$$

Using (4.6.7) and Theorem 4.6.4, we see that our function belongs to $R^*(0, 1)$. Further

$$x^{-\frac{1}{2}}(1 + x)^{-1} < x^{-\frac{3}{2}}, \quad 1 < x,$$

so that the function also belongs to $R^*(1, \infty)$ and hence to $R^*(0, \infty)$.

A second example is the following:

$$x^n e^{-x} \in R^*(1, \infty) \quad \textit{for every } n. \tag{4.6.10}$$

It is enough to prove this for n a positive integer. To this end we use (4.5.39) where we replace n by $n + 2$. Neglecting all but the last term on the right we obtain

$$e^x > \frac{x^{n+2}}{(n + 2)!},$$

or

$$(n + 2)!\, x^{-2} > x^n e^{-x}.$$

The left member belongs to $R^*(1, \infty)$ by (4.6.8), hence also the right member, as asserted. In (4.6.10) we can replace x by σx where σ is any positive constant. This gives

$$x^n e^{-\sigma x} \in R^*(1, \infty) \tag{4.6.11}$$

for any n and any positive σ.

As our last example we take:

$$x^{-1}(\log x)^{-\sigma} \in R^*(2, \infty) \tag{4.6.12}$$

if and only if $\sigma > 1$.

For

$$\int_2^\infty x^{-1}(\log x)^{-\sigma}\,dx = \int_{\log 2}^\infty s^{-\sigma}\,ds,$$

if we set $\log x = s$. The result then follows from (4.6.8). The lower limit 2 can be replaced by any number $a > 1$. We cannot allow $a = 1$, since $\log 1 = 0$ and the integrand becomes infinite too rapidly to be integrable when $\sigma > 1$.

In Theorem 4.6.4 we have a criterion for deciding if a given function belongs to $R^*(a, b)$. It is worth mentioning the negative side of this criterion which we formulate as a separate theorem.

THEOREM 4.6.5. *If $f \in R(\alpha, \beta)$ for every $[\alpha, \beta] \subset (a, b)$, and if there exist a function h such that $h(s) \geqq 0$, $|f(s)| \geqq h(s)$ for all s, $h(s)$ belongs to every $R[\alpha, \beta]$ but not to $R^*(a, b)$, then $f(s)$ does not belong to $R^*(a, b)$.*

As an example of the use of this theorem we go back to (4.6.7). We showed that x^{-1} is not in $R^*(0, 1)$. Since $x^{-\sigma} > x^{-1}$, if $1 < \sigma$, $0 < x < 1$, it follows that $x^{-\sigma}$ is not in $R^*(0, 1)$.

EXERCISE 4.6

Which of the following functions belong to $R^*(0, 1)$?

1. $x^{-\frac{1}{3}}(1 + x)^{-\frac{1}{3}}$. 2. $x^{-\frac{1}{3}}(1 - x)^{-\frac{1}{3}}$. 3. $\log x$.

4. $(\log x)^2$. 5. $[\log x]^{-1}$. 6. $x(\log x)^2$.

7. $x^{-1}(\log x)^2$. 8. $[x \log x]^{-\frac{1}{3}}$. 9. $x^{-1}[\log (ex)]^{-2}$.

10. $x^{-1}(\log x)^{-2}$. 11. $x^{-\frac{1}{2}}\left(\log \dfrac{1}{x}\right)^{-\frac{1}{2}}$. 12. $\log |\log x|$.

13. $\log \log \dfrac{e}{x}$. 14. $(\log |\log x|)^{-1}$.

Which of the following functions belong to $R^*(1, \infty)$?

15. $\dfrac{e^x}{x^2}$. 16. xe^{-x}. 17. e^{-3x}. 18. e^{-x^2}.

19. $\dfrac{x}{x^2 + 1}$. 20. $\left(\dfrac{x}{x^2 + 1}\right)^2$. 21. $e^x + e^{-x}$. 22. $(e^x + e^{-x})^{-1}$.

23. $(x^2 + 1)^{-\frac{3}{2}}$. 24. $\dfrac{e^x}{e^{3x} + 1}$. 25. $\dfrac{\log x}{x^2}$.

26. $[x(x^2 - 1)]^{-\frac{1}{2}}$. 27. $x(x^4 + 1)^{-\frac{1}{2}}$. 28. $\dfrac{1}{x(\log x)^2}$.

Evaluate the following integrals:

29. $\displaystyle\int_0^1 (1 - x)^{-\frac{1}{2}}\, dx$. 30. $\displaystyle\int_0^1 x \log x\, dx$. 31. $\displaystyle\int_0^1 x^{-\frac{1}{3}}(1 + x)\, dx$.

32. $\displaystyle\int_0^4 \frac{x\, dx}{\sqrt{4 - x}}$. 33. $\displaystyle\int_0^\infty x^3 e^{-x}\, dx$. 34. $\displaystyle\int_0^\infty xe^{-2x}\, dx$.

35. $\displaystyle\int_0^\infty e^{-\sqrt{x}} \frac{dx}{\sqrt{x}}$. 36. $\displaystyle\int_0^\infty xe^{-x^2}\, dx$. 37. $\displaystyle\int_0^\infty \frac{e^x\, dx}{(e^x + 1)^2}$.

38. $\displaystyle\int_0^\infty \frac{dx}{(e^x + e^{-x})^2}$. 39. $\displaystyle\int_0^\infty \frac{e^{2x}\, dx}{(e^{2x} + 1)^3}$.

40. Verify that

$$D_x \log [x + (x^2 + a)^{\frac{1}{2}}] = (x^2 + a)^{-\frac{1}{2}}$$

and use this to evaluate the integral

$$\int_0^2 (s^2 + 4)^{-\frac{1}{2}}\, ds.$$

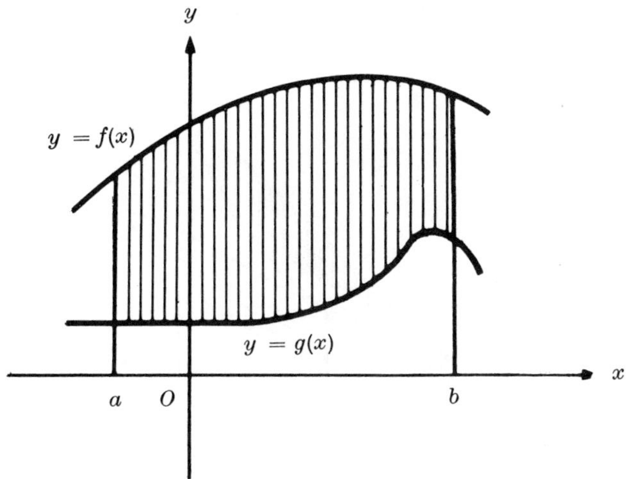

Figure 4.11

41. Verify that

$$\frac{2}{t(t+1)(t+2)} = \frac{1}{t} - \frac{2}{t+1} + \frac{1}{t+2},$$

and use this decomposition to evaluate

$$\int_1^\infty \frac{dt}{t(t+1)(t+2)}.$$

42. Evaluate $\int_0^\infty x^n e^{-x}\, dx$.

4.7. Some area problems. Table of integrals. The basic result concerning areas is Theorem 4.2.10 according to which the two-dimensional measure of the ordinate set

$$S: \{(x, y) \mid a < x < b, \quad 0 < y < f(x)\} \tag{4.7.1}$$

is

$$\mu(S) = \int_a^b f(s)\, ds. \tag{4.7.2}$$

This was proved for $f \in R(a, b)$. With the aid of the exhaustion principle and the definition of absolute integrability we can show that this formula is also valid for functions in $R^*(a, b)$.

If the lower boundary of the set S is not the x-axis cut a curve

$$y = g(x),$$

as in Figure 4.11, then S is the set

$$S: \{(x, y) \mid a < x < b, g(x) < y < f(x)\}, \tag{4.7.3}$$

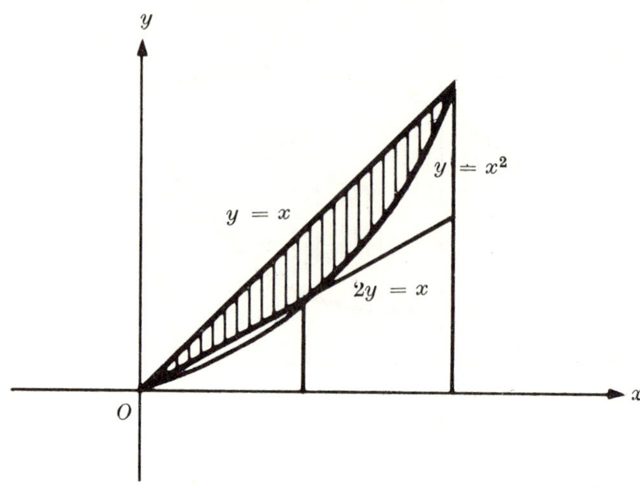

Figure 4.12

and the measure is

$$\mu(S) = \int_a^b [f(s) - g(s)]\, ds. \tag{4.7.4}$$

We shall illustrate by some examples.

Example 1. Find the area of the region bounded by the straight lines $y = \frac{1}{2}x$, $y = x$, and the parabola $y = x^2$. See Figure 4.12.

The lines intersect the parabola at the origin and at the points $(\frac{1}{2}, \frac{1}{4})$ and $(1, 1)$ respectively. The upper boundary is given by $y = x$, the lower by $y = \frac{1}{2}x$ for $0 \leqq x \leqq \frac{1}{2}$, and by $y = x^2$ for $\frac{1}{2} < x \leqq 1$. Hence

$$\begin{aligned}\mu(S) &= \int_0^{\frac{1}{2}} (s - \tfrac{1}{2}s)\, ds + \int_{\frac{1}{2}}^1 (s - s^2)\, ds \\ &= \tfrac{1}{4}s^2\Big|_0^{\frac{1}{2}} + (\tfrac{1}{2}s^2 - \tfrac{1}{3}s^3)\Big|_{\frac{1}{2}}^1 \\ &= \tfrac{1}{16} + \tfrac{1}{2} - \tfrac{1}{3} - \tfrac{1}{8} + \tfrac{1}{24} = \tfrac{7}{48}.\end{aligned}$$

Example 2. Find the area of the lens-shaped region bounded by the two parabolas $y^2 = 4ax$, $x^2 = 4ay$. See Figure 4.13.

The upper boundary is given by

$$f(x) = 2(ax)^{\frac{1}{2}},$$

the lower by

$$g(x) = \frac{x^2}{4a}.$$

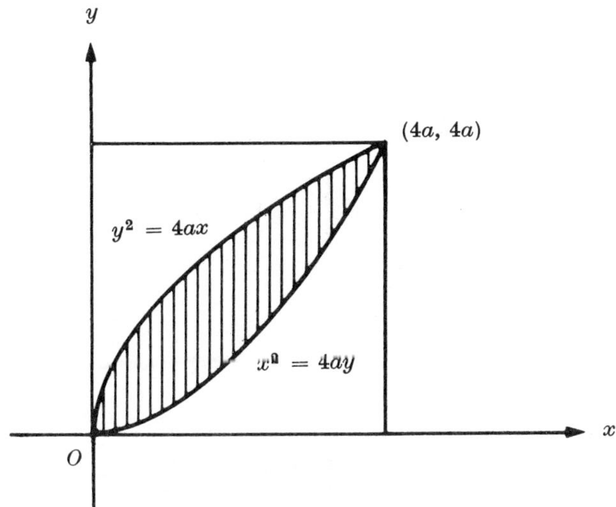

Figure 4.13

The two curves intersect at (0, 0) and $(4a, 4a)$. Hence the area is

$$\mu(S) = \int_0^{4a} \left\{ 2(ax)^{\frac{1}{2}} - \frac{x^2}{4a} \right\} dx$$
$$= \left\{ 2a^{\frac{1}{2}} \, \tfrac{2}{3} x^{\frac{3}{2}} - \frac{x^3}{12a} \right\}_{x=0}^{x=4a}$$
$$= a^2 \left[\tfrac{32}{3} - \tfrac{16}{3}\right] = \tfrac{16}{3} a^2.$$

This is one-third of the area of the square indicated in the figure. The result is in agreement with, and could have been obtained from (4.1.6).

Example 3. Find the area between the curves $xy = 1$ and $y(x^2 + 1) = x$ to the right of the line $x = 1$. See Figure 4.14.

The area extends to infinity and so there is a question of absolute integrability. Since

$$f(x) - g(x) = \frac{1}{x} - \frac{x}{x^2 + 1} = \frac{1}{x(x^2 + 1)} < \frac{1}{x^3},$$

condition (4.6.1) is satisfied and the integral exists as the limit of a bounded strictly increasing function of the upper limit β. We have

$$\int_1^\beta [f(x) - g(x)] \, dx = \int_1^\beta \frac{dx}{x} - \int_1^\beta \frac{x \, dx}{x^2 + 1}$$
$$= \log \beta - \tfrac{1}{2} \log (\beta^2 + 1) + \tfrac{1}{2} \log 2.$$

Now

$$\log \beta - \tfrac{1}{2} \log (\beta^2 + 1) = \log \frac{\beta}{\sqrt{\beta^2 + 1}} = \log \frac{1}{\sqrt{1 + \beta^{-2}}} \to \log 1 = 0,$$

so the desired area is $\frac{1}{2} \log 2$.

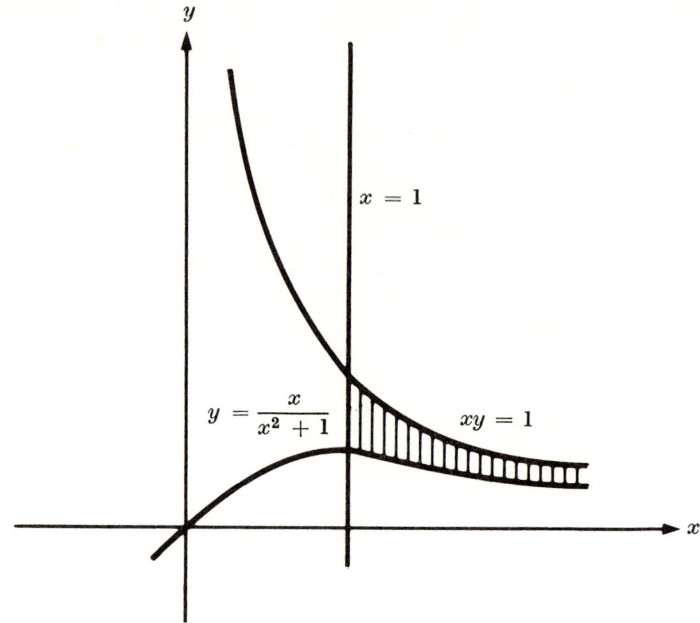

Figure 4.14

We have reduced the area problem to finding the measure of an *ordinate set* by integration. Sometimes it is easier to discuss the corresponding *abscissa set*. This is merely a question of interchanging the role of the variables. If the set is

$$S: \{(x, y) \mid F(y) < x < G(y), \quad c < y < d\}, \tag{4.7.5}$$

then

$$\mu(S) = \int_c^d [G(y) - F(y)]\, dy. \tag{4.7.6}$$

The following example illustrates the method.

Example 4. Find the area bounded by the curves $y = x^{\frac{1}{3}}$ and $y = x^{\frac{1}{2}}$.

Here we solve for x and obtain $x = y^3$ as the left boundary and $x = y^2$ as the right. Hence

$$\mu(S) = \int_0^1 (y^2 - y^3)\, dy = \left[\tfrac{1}{3}y^3 - \tfrac{1}{4}y^4\right]_{y=0}^{y=1} = \tfrac{1}{12}.$$

This does not exhaust the topic of area problems. In later chapters we shall give formulas suitable for the computation of areas where the boundary is given in terms of polar coordinates or parametrically. Before leaving the topic, however, we shall make some remarks concerning a class of limit problems which leads to area problems.

Let (a, b) be a finite interval and let $f(x)$ be a positive strictly monotonic function belonging to $R^*(a, b)$. Form

$$T_n = \frac{1}{n}(b - a)\sum_{k=1}^{n-1} f\left(a + \frac{k}{n}(b - a)\right). \tag{4.7.7}$$

Then

$$T_n < \int_a^b f(s)\,ds$$

and

$$\lim_{n\to\infty} T_n = \int_a^b f(s)\,ds. \tag{4.7.8}$$

See the proof of Theorem 4.2.9 for the case $f \in R(a, b)$. The extension to $R^*(a, b)$ is immediate.

Now the problem is the following. There is given a sequence of finite sums $\{T_n\}$. Is it possible to find a function $f(s)$ and an interval (a, b) such that T_n is expressible by a formula of type (4.7.7) and, if so, does the function satisfy formula (4.7.8)? We shall illustrate the question by a couple of examples.

Example 5. Given the sequence

$$T_n = \frac{1}{n+1} + \frac{1}{n+2} + \cdots + \frac{1}{2n}, \quad n = 1, 2, 3, \cdots, \tag{4.7.9}$$

prove that

$$\lim_{n\to\infty} T_n = \log 2. \tag{4.7.10}$$

This sequence was shown in Section 1.7, formula (1.7.12), to be strictly increasing and to have a limit. It belongs to the type considered here. We take

$$f(s) = \frac{1}{1+s}, \quad a = 0, \quad b = 1,$$

and note that

$$\frac{1}{n+k} = \frac{1}{n}\frac{1}{1+\dfrac{k}{n}} = \frac{1}{n} f\left(\frac{k}{n}\right).$$

Here $f(s) \in R(0, 1)$ and we have

$$\lim_{n\to\infty} T_n = \int_0^1 \frac{ds}{1+s} = \log 2,$$

as asserted.

Example 6. Given the sequence

$$T_n = \frac{1}{\sqrt{n}}\left\{\frac{1}{\sqrt{1}} + \frac{1}{\sqrt{2}} + \cdots + \frac{1}{\sqrt{n}}\right\}, \quad n = 1, 2, 3, \cdots, \tag{4.7.11}$$

prove that

$$\lim_{n\to\infty} T_n = \int_0^1 s^{-\frac{1}{2}}\,ds = 2. \tag{4.7.12}$$

It is simple to verify that

$$f(s) = s^{-\frac{1}{2}}, \quad a = 0, \quad b = 1,$$

and, since this is a function belonging to $R^*(0, 1)$, (4.7.8) holds and gives (4.7.12).

A brief list of primitives ends this section. All have figured before except (iii) which follows from the decomposition

$$\frac{1}{x^2 - a^2} = \frac{1}{2a}\left\{\frac{1}{x - a} - \frac{1}{x + a}\right\}.$$

Here and in the following a stands for a positive number while α denotes a real number.

(i) $\displaystyle\int x^n\,dx = \frac{x^{n+1}}{n+1} + C, \quad n \neq -1.$

(ii) $\displaystyle\int \frac{dx}{x - \alpha} = \log|x - \alpha| + C.$

(iii) $\displaystyle\int \frac{dx}{x^2 - a^2} = \frac{1}{2a}\log\left|\frac{x - a}{x + a}\right| + C.$

(iv) $\displaystyle\int \frac{x\,dx}{\sqrt{x^2 + \alpha}} = \sqrt{x^2 + \alpha} + C.$

(v) $\displaystyle\int \frac{dx}{\sqrt{x^2 + \alpha}} = \log|x + \sqrt{x^2 + \alpha}| + C.$

(vi) $\displaystyle\int e^{\alpha x}\,dx = \frac{1}{\alpha}e^{\alpha x} + C.$

Further formulas involving logarithms and exponential functions are to be found in Section 4.5. To these we add the recursion formula

(vii) $\displaystyle\int x^n e^{\alpha x}\,dx = \frac{1}{\alpha}x^n e^{\alpha x} - \frac{n}{\alpha}\int x^{n-1}e^{\alpha x}\,dx.$

Repeated applications of this formula with n replaced by $n - 1$, $n - 2$, etc., will reduce the left member of (vii) to formula (vi).

EXERCISE 4.7

1. Find by integration the area of the triangle bounded by the straight lines:

$$x - y + 1 = 0, \quad 5x + y - 7 = 0, \quad 2x + y - 13 = 0.$$

2. The cubic parabola $y = x^3$ has two tangents of slope 1. Each of them intersects the curve in another point. Find the area of the region bounded by the curve and one of the tangents.

3. Find the area of the region between the curves $y^2 = 2px$ and $27py^2 = 8(x - p)^3$. The curves intersect at two points of abscissa $4p$.

4. Verify that

$$\int (x^2 + \alpha)^{\frac{1}{2}}\,dx = \tfrac{1}{2}x(x^2 + \alpha)^{\frac{1}{2}} + \frac{\alpha}{2}\int (x^2 + \alpha)^{-\frac{1}{2}}\,dx.$$

5. Find the area of the region bounded by a branch of the hyperbola $b^2x^2 - a^2y^2 = a^2b^2$ and the corresponding latus rectum (vertical chord through a focus).

6. How should A and B be chosen so that

$$(x^2 + 1)^{\frac{1}{2}} - Ax - \frac{B}{x}$$

is integrable over the interval $(1, \infty)$? Evaluate the corresponding integral. Interpret the integral as an area.

7. The curve $y^2 = x^3 + 4x^2$ is symmetric with respect to the x-axis. It intersects itself at the origin. It forms a loop in the left half-plane with a vertical tangent at $(-4, 0)$. Find the area of the region bounded by this loop.

8. Find the area of the region bounded by the parabolas $y^2 = x$ and $y^2 = 2 - x$.

9. Find the area of the region in the first quadrant bounded by the curves $xy = 1$ and $y = 2 - x^2$.

Find the limit of T_n if T_n is given by:

10. $n^{-1-\alpha} \sum_{k=1}^{n} k^{\alpha}, \quad \alpha > 1.$ 11. $\sum_{k=1}^{3n} \frac{1}{n+k}.$ 12. $\sum_{k=2n}^{3n} \frac{k}{n^2} e^{k/n}.$

13. $\sum_{k=1}^{n-1} \frac{k}{k^2 + n^2}.$ 14. $\sum_{k=1}^{2n} \frac{1}{\sqrt{k^2 + n^2}}.$ 15. $\sum_{k=n}^{2n} \frac{n}{k(n+k)}.$

Evaluate the integrals:

16. $\int_{2a}^{\infty} \frac{ds}{s^2 - a^2}.$ 17. $\int_{a}^{\infty} \frac{ds}{s(s+a)}.$ 18. $\int_{0}^{\infty} (e^s + e^{-s})^{-\frac{1}{2}}\, ds.$

19. $\int_{1}^{\infty} \frac{ds}{e^s - e^{-s}}.$ 20. $\int_{\log 2}^{\infty} \frac{ds}{e^{2s} - 1}.$

MISCELLANEOUS PROBLEMS FOR CHAPTER 4

1. Find a primitive of $x^3(1 + x^2)^{-1}$.

2. Find a primitive of $(x^2 + x^3)^{-1}$.

3. Let $f(x)$ be a monotonic differentiable function in an interval (a, b) and let $x = g(y)$ be its inverse. Let $F(x)$ be a primitive of $f(x)$. Show that

$$G(y) \equiv yg(y) - F[g(y)]$$

is a primitive of $g(y)$.

4. Find a polynomial $P(x)$ of degree 5 such that $P(x) + 10$ is divisible by $(x + 2)^3$ while $P(x) - 10$ is divisible by $(x - 2)^3$. [*Hint*: What are the roots of $P'(x)$?]

5. Let $f(x) \in C[-\infty, \infty]$ and let

$$\lim_{x \to -\infty} f(x) = A, \quad \lim_{x \to +\infty} f(x) = B.$$

Find

$$\lim_{a\to\infty}\int_{-a}^{a}[f(x+1)-f(x)]\,dx.$$

6. Find

$$\lim_{x\to\infty}\log x\int_{x}^{2x}\frac{ds}{s\log s}.$$

[*Hint:* Use Theorem 4.3.6.]

7. What is the limit of $x^2 \log x$ when x decreases to 0? What is the value of the right-hand derivative at $x = 0$? Is there a point on the curve $y = x^2 \log x$ where the tangent is horizontal?

8. Prove that $\displaystyle\int_0^\infty \frac{\log s\,ds}{(1+s)^2} = 0.$

9. Evaluate $\displaystyle\int_0^1 [\log x]^4\,dx.$

10. Evaluate $\displaystyle\int_0^\infty (e^s+1)^{-\frac{1}{2}}\,ds.$

[*Hint:* A substitution reduces this integral to an application of formula (v) in the list of primitives.]

11. Set

$$f(a)=\int_{-\infty}^{\infty}\frac{ds}{(1+|s|)(1+|s-a|)}.$$

Find $f(0)$ and show that $f(a) \leqq f(0)$ for all a. [*Hint:* Try Theorem 4.3.7.]

12. Show that

$$\int_0^x t^4e^{-t^2}\,dt = C\int_0^x e^{-t^2}\,dt - \tfrac{1}{2}e^{-x^2}P(x),$$

where C is a constant and $P(x)$ a polynomial. Determine C and $P(x)$.

13. Determine the constants A and B such that

$$x(x^2+1)^{\frac{1}{2}} - Ax^2 - B$$

is integrable over $(0, \infty)$ and evaluate the integral.

14. Evaluate the integral

$$n\int_0^1 e^{-ns}\,ds$$

and use the result to prove that

$$\lim_{n\to\infty} n\int_0^1 f(s)e^{-ns}\,ds = f(0+0),$$

if $f(s) \in R(0, 1)$ and has a right-hand limit at $s = 0$.

15. Find $\lim P_n$ if

$$P_n = \left\{\left(1+\frac{1}{n}\right)\left(1+\frac{2}{n}\right)\cdots\left(1+\frac{n}{n}\right)\right\}^{1/n}.$$

[*Hint:* Consider $\log P_n$.]

16. If $F(x) \not\equiv 0$ is a solution of (4.5.33), then

$$F(s + h) - F(s) = [F(h) - 1]F(s).$$

Show that a solution which is continuous at $x = 0$ is continuous everywhere. Integrate the relation with respect to s from 0 to x, simplify the left member, and divide by h to obtain

$$\frac{1}{h}\int_x^{x+h} F(t)\,dt - \frac{1}{h}\int_0^h F(t)\,dt = \frac{1}{h}[F(h) - 1]\int_0^x F(t)\,dt.$$

Show that this implies the existence of a derivative at the origin and hence everywhere. Show that $F'(x) = F'(0)F(x)$.

17. Find the area between a parabola and one of its normals. Take $y = bx^2$, the normal at (x_0, y_0), and compare the result with (4.1.6).

Find the nth derivative of:

18. x^3e^x. 19. x^2e^{-x}. 20. $x \log x - x$. 21. $(\log x)^2$.

22. Examine

$$\lim_{x\to -0} e^{1/x}, \quad \lim_{x\to +0} e^{1/x}.$$

Set

$$f(x) = x\,\frac{e^{1/x} + 1}{e^{1/x} + 4}, \quad x \neq 0, \quad f(0) = 0.$$

Is f continuous at $x = 0$? Does it have a derivative there?

Determine how many derivatives the following functions have at $x = 0$.

23. $f(x) = \begin{cases} e^x, & x \leqq 0, \\ 1 + x + x^2, & 0 < x. \end{cases}$ 24. $f(x) = \begin{cases} e^{2x} + 1, & x \leqq 0, \\ 1 + 2x + 2x^2, & 0 < x. \end{cases}$

*25. By the first mean value theorem of the integral calculus

$$\int_0^x \frac{dt}{1+t} = \frac{x}{1 + x\theta(x)}, \quad x > -1,$$

where $0 < \theta(x) < 1$. Show that $\lim_{x\to 0} \theta(x) = \frac{1}{2}$.

26. Given that

$$a_n = \frac{1}{\sqrt{n}} + \frac{1}{\sqrt{n+1}} + \cdots + \frac{1}{\sqrt{n+n}},$$

find

$$\lim n^{-\frac{1}{2}}a_n.$$

Find the areas bounded by:

27. $y = 1 + x,\ y = 1 - x,\ y = x^2$.
28. $y = x,\ y = 2x,\ xy = 1$.
29. $x + y - 2 = 0,\ xy + x + y = 0$.
30. $x = 2y - y^2,\ xy + x + y = 0$.

5

Infinite Series

5.1. General concepts. The notions developed in this Chapter are closely related to the idea of a sequential limit, presented in Section 1.7, and the main tool used in our investigation is Cauchy's Principle of Convergence of Section 2.1.

Addition as defined in Section 1.3 is restricted to a finite number of summands and any extension to an infinite number must involve a limiting process of one kind or another. The resulting process may or may not possess the properties of associativity and commutativity pertaining to finite addition. In other words, we must start by making suitable definitions of symbols and processes, and we must adjust our definitions to the ends in view.

To define an infinite series, we need an infinite sequence $\{u_n\}$ whose elements are called the *terms* of the series. In the simplest case these terms are positive numbers, but we shall also allow real numbers, complex numbers, vectors, functions, in fact any objects for which we can define (finite) addition. With the sequence $\{u_n\}$ we define the series

$$u_0 + u_1 + u_2 + \cdots + u_n + \cdots, \tag{5.1.1}$$

which is usually written

$$\sum_{n=0}^{\infty} u_n. \tag{5.1.2}$$

This symbol is read "summation over all n from zero to infinity of u sub n." The summation sign is actually a capital Greek sigma.

This symbol does not have any intrinsic meaning, since our postulates for addition make no provision for infinitely many summands. In such a case, we have to define the meaning of the symbol, or, as we shall do here, define a procedure which, depending upon the outcome, will sometimes allow us to assign a definite meaning to the symbol.

For this purpose we introduce a second sequence $\{U_n\}$ whose elements are the so-called *partial sums* of the series:

$$U_n = u_0 + u_1 + u_2 + \cdots + u_n. \tag{5.1.3}$$

This expression is defined for every n. It is often written

$$U_n = \sum_{k=0}^{n} u_k, \tag{5.1.4}$$

which is read "summation over all k from zero to n of u sub k." The subscript k is a "dummy" in the sense that it can be replaced by any other letter which is not already engaged. Thus we have to avoid u and n since they are used for other purposes.

The student should distinguish carefully between the series, the terms of the series, and the partial sums of the series. In particular, if reference is made to the infinite series, (5.1.1) or (5.1.2) should be used, and it is wrong to use the right member of (5.1.3). In other words, the second set of interstitial points in (5.1.1) should not be forgotten. When (5.1.2) is used, the limits of summation may be omitted, that is, "$n = 0$" and "∞" if no ambiguity can arise. If the summation does not start with 0, but with some other integer, this must be noted and the symbol modified accordingly.

Once the sequence of partial sums $\{U_n\}$ is at our disposal, we can use it to assign a meaning to the series.

DEFINITION 5.1.1. *The series* $\sum_{n=0}^{\infty} u_n$ *is said to be convergent or divergent according as the sequence* $\{U_n\}$ *has a finite limit or not. In the former case, the limit is called the sum of the series, the series is said to converge to its sum, and the sum is the arithmetical meaning given to the symbol* (5.1.1).

This convention assigns a reasonable meaning to convergent series and leaves open the question what to do with the others.

DEFINITION 5.1.2. *A series* $\Sigma\, u_n$ *with real terms is said to be properly divergent if the sequence* $\{U_n\}$ *has a unique infinite limit. The series is then said to diverge to* $+\infty$ *or to* $-\infty$ *as the case may be.*

Infinite series which are divergent without being properly divergent occur frequently in the applications to mathematical physics. This alone makes it impossible to bar such series from further consideration and makes it imperative to construct other means of evaluating series. This need has led to elaborate theories of *summability* which complement the classical theory of convergence. Except for occasional remarks on summability, we restrict ourselves to the theory of convergence. The latter is largely the creation of Cauchy and of the Norwegian genius Niels Henrik Abel (1802–1829). The theory had reached a definitive form by the end of the nineteenth century.

Definition 5.1.1 refers the meaning of the series to Cauchy's Convergence Principle, Theorem 2.1.4. If $m < n$ we have

$$U_n - U_m = u_{m+1} + u_{m+2} + \cdots + u_n = \sum_{k=m+1}^{n} u_k. \tag{5.1.5}$$

The last member is read "summation over all k from $m + 1$ to n of u sub k." Here again k is a dummy which can be replaced by other letters, though not by m, n, or u. We set $n = m + p$, p positive integer, and can now formulate the convergence criterion as follows:

THEOREM 5.1.1. *The series* $\Sigma\, u_n$ *is convergent if and only if, given any* $\varepsilon > 0$, *there is an* $N(\varepsilon)$ *such that for* $m > N(\varepsilon)$

$$\left| \sum_{k=m+1}^{m+p} u_k \right| \leqq \varepsilon, \quad p = 1, 2, 3, \cdots. \tag{5.1.6}$$

Since the condition is necessary as well as sufficient, we see that in a convergent series

$$\lim_{n\to\infty} u_n = 0, \tag{5.1.7}$$

that is, the terms u_n form a null sequence. This condition is necessary, but far from sufficient as we shall see.

The case of positive terms is particularly simple and important. Here the sequence $\{U_n\}$ is strictly increasing and Theorem 1.7.1 leads to

THEOREM 5.1.2. *A series with positive terms is either convergent or properly divergent. It converges if and only if the partial sums form a bounded sequence.*

Convergent series with positive terms can be handled very much like finite sums: We can rearrange the terms, we can multiply and add such series just as finite sums, and the resulting series converge and have sums with the expected values. We shall take up such questions later. Some examples will now illustrate the definitions and the theorems.

Example 1. We take

$$u_n = \frac{1}{n(n+1)}, \quad n = 1, 2, 3, \cdots. \tag{5.1.8}$$

Here

$$U_n = \frac{1}{1\cdot 2} + \frac{1}{2\cdot 3} + \frac{1}{3\cdot 4} + \cdots + \frac{1}{n(n+1)} \tag{5.1.9}$$

coincides with the sequence defined by formula (1.7.8) which has the limit 1. It follows that the series

$$\sum_{n=1}^{\infty} \frac{1}{n(n+1)}$$

converges and has the sum 1, which we express by writing

$$\sum_{n=1}^{\infty} \frac{1}{n(n+1)} = 1. \tag{5.1.10}$$

Example 2. We take

$$u_n = n^{-\frac{1}{2}}, \quad n = 1, 2, 3, \cdots. \tag{5.1.11}$$

Here

$$U_n = 1 + 2^{-\frac{1}{2}} + 3^{-\frac{1}{2}} + \cdots + n^{-\frac{1}{2}} > n \cdot n^{-\frac{1}{2}} = n^{\frac{1}{2}},$$

for each of the n summands in U_n is greater than or equal to the last one $n^{-\frac{1}{2}}$. From this inequality it follows that

$$\sum_{n=1}^{\infty} n^{-\frac{1}{2}} = +\infty,$$

that is, the series is properly divergent. The terms of this series form a null sequence, and so condition (5.1.7) is not sufficient for convergence of an infinite series.

Example 3. The series

$$\sum_{n=1}^{\infty} (-1)^{n-1} \tag{5.1.12}$$

is an example of a divergent series which is not properly divergent. It is divergent since the terms do not tend to 0 when n becomes infinite. Here

$$U_{2k-1} = 1, \quad U_{2k} = 0, \quad k = 1, 2, 3, \cdots,$$

so the partial sums are bounded but fail to have a limit. It follows that the series is not properly divergent. The series has figured in the literature since 1696 when Jakob Bernoulli assigned the value $\frac{1}{2}$ to it for reasons explained in the next example.

Calculations with divergent series, even properly divergent ones, were popular during the eighteenth century. While masters like Euler normally did not get bizarre results in this manner, lesser mathematicians were less lucky. In the 1820's Abel and Cauchy found it necessary to exclude all such series to bring order in the threatening chaos. Once convergence theory was firmly established, it became desirable to reexamine the dismissed divergent series and to develop rules for safe handling. This development started in 1882 and is still going on.

Example 4. The *geometric series* (see Problem 29, Exercise 2.3)

$$\sum_{n=0}^{\infty} x^n, \quad x^0 \equiv 1, \tag{5.1.13}$$

is the simplest of all *power series*. Such series will be discussed in Section 5.4. It is also one of the simplest series whose terms are functions rather than numbers. In such a case we ask: *For what values of x is the series convergent, divergent, or properly divergent?* It is easy to find the answer. We set

$$U_n = U_n(x) = 1 + x + x^2 + \cdots + x^n.$$

If $x = 1$, the sum is $n + 1$, which becomes infinite with n. Hence the series diverges properly for $x = +1$. The same is true for any x with $x > 1$, for then

$$U_n(x) > U_n(1) = n + 1.$$

We note that

$$xU_n(x) = x + x^2 + x^3 + \cdots + x^n + x^{n+1}.$$

Subtracting this from $U_n(x)$, we obtain

$$(1 - x)U_n(x) = 1 - x^{n+1},$$

and

$$U_n(x) = \frac{1 - x^{n+1}}{1 - x}, \qquad x \neq 1. \tag{5.1.14}$$

For $|x| < 1$, the terms x^n form a null sequence, and

$$\lim_{n\to\infty} U_n(x) = \frac{1}{1 - x},$$

so that

$$\sum_{n=0}^{\infty} x^n = \frac{1}{1-x} \quad \text{for } |x| < 1. \tag{5.1.15}$$

Thus, the series is properly divergent for $1 \leqq x$, convergent for $-1 < x < 1$, and divergent for $x \leqq -1$. If $x = -1$, we obtain the series (5.1.12) of the preceding example and we see why Bernoulli and Euler wanted to assign it the value $\frac{1}{2}$. The series is divergent as we have seen but

$$\lim_{x \to -1+0} \sum_{n=0}^{\infty} x^n = \lim_{x \to -1+0} \frac{1}{1-x} = \tfrac{1}{2}.$$

With the same type of motivation Euler also claimed that

$$1 - 2 + 2^2 - 2^3 + \cdots = \tfrac{1}{3}.$$

Here

$$U_n(-2) = \tfrac{1}{3}[1 - (-2)^{n+1}],$$

which is an unbounded sequence with

$$\liminf U_n(-2) = -\infty, \quad \limsup U_n(-2) = +\infty.$$

There are various methods available for "summing" the series (5.1.13) for $x \leqq -1$ to the sum $(1 - x)^{-1}$ even though the series is divergent. This much may be said in vindication of Euler's speculations.

We have to give one more definition.

DEFINITION 5.1.3. *The series $\sum u_n$ is said to be absolutely convergent if the series $\sum a_n$ is convergent where $a_n = |u_n|$.*

It is clear that the concepts of convergence and absolute convergence coincide for series with positive terms (negative terms). For series with both negative and positive terms the concepts are distinct.

THEOREM 5.1.3. *An absolutely convergent series is convergent. There are convergent series which are not absolutely convergent.*

Proof. We have

$$\left| \sum_{k=n+1}^{n+p} u_k \right| \leqq \sum_{k=n+1}^{n+p} a_k.$$

If the series is absolutely convergent, we can choose $N(\varepsilon)$ so large that the right member is less than a preassigned ε for $n > N(\varepsilon)$ and $p = 1, 2, 3, \cdots$. This implies that the left member is less than ε for such n and p. Hence the original series is convergent.

To prove the second assertion we use a modification of Example 2. The series

$$\sum_{n=1}^{\infty} (-1)^{n-1} n^{-\frac{1}{2}} \tag{5.1.16}$$

is certainly not absolutely convergent since the series formed by the absolute values of the terms coincides with the divergent series of Example 2. On the other hand, grouping the terms we obtain

$$U_{2k+1} = 1 - (2^{-\frac{1}{2}} - 3^{-\frac{1}{2}}) - (4^{-\frac{1}{2}} - 5^{-\frac{1}{2}}) - \cdots - [(2k)^{-\frac{1}{2}} - (2k+1)^{-\frac{1}{2}}].$$

Since $(2k)^{-\frac{1}{2}} > (2k+1)^{-\frac{1}{2}}$, this shows that the partial sums of odd order form a strictly decreasing sequence. Further

$$U_{2k} = (1 - 2^{-\frac{1}{2}}) + (3^{-\frac{1}{2}} - 4^{-\frac{1}{2}}) + \cdots + [(2k-1)^{-\frac{1}{2}} - (2k)^{-\frac{1}{2}}],$$

so that the partial sums of even order form a strictly increasing sequence. Since

$$U_{2k+1} = U_{2k} + (2k+1)^{-\frac{1}{2}},$$

both sequences are bounded and hence convergent. From

$$\lim_{k\to\infty} U_{2k+1} = \lim_{k\to\infty} U_{2k},$$

we get that the series (5.1.16) converges. It consequently gives an example of a convergent series which is not absolutely convergent.

The discussion of the series (5.1.16) will be generalized in Section 5.3.

We shall define remainders of a convergent series. Suppose that the series (5.1.1) is convergent and its sum is U. We write

$$U = \sum_{k=0}^{\infty} u_k. \tag{5.1.17}$$

We can also write

$$U = U_n + R_n, \tag{5.1.18}$$

where

$$U_n = \sum_{k=0}^{n} u_k, \quad R_n = \sum_{k=n+1}^{\infty} u_k. \tag{5.1.19}$$

Here U_n of course is the nth partial sum. R_n is called *the remainder after n terms*. Since the original series is convergent by assumption, each of the remainders is a convergent infinite series whose sum R_n tends to zero as n becomes infinite. This means that given any $\varepsilon > 0$, we can find an N_ε such that

$$|R_n| \leqq \varepsilon \quad \text{if} \quad N_\varepsilon \leqq n. \tag{5.1.20}$$

EXERCISE 5.1

Suppose that u_n has one of the following values:

1. $\dfrac{1}{n(n+1)}$. 2. $\dfrac{1}{2n(2n+1)}$. 3. $n2^{-n}$. 4. $n^{-n}x^n$.

In each case consider the series $\sum_{n=1}^{\infty} u_n$ and find (i) the first five terms, (ii) their sum, and (iii) the tenth term.

5. Show that the series 4 converges for all x.

Prove the convergence and find the sum of:

6. $\sum_{n=1}^{\infty} \frac{1}{n(n+2)}$. 7. $\sum_{n=1}^{\infty} \frac{1}{n(n+1)(n+2)}$.

8. If

$$S_n(x) = 1 + 2x + 3x^2 + \cdots + (n+1)x^n,$$

prove that

$$(1-x)^2 S_n(x) = 1 - (n+2)x^{n+1} + (n+1)x^{n+2}.$$

Show that $\lim_{n\to\infty} S_n(x)$ exists as a finite number if and only if $|x| < 1$ and find the limit when it exists.

9. What does the preceding result imply for the power series

$$\sum_{n=1}^{\infty} (n+1)\, x^n?$$

Use the result also to prove the convergence of the series in Problem 3 and find its sum.

10. Prove that $\lim_{n\to\infty} \sum_{k=n+1}^{2n} k^{-\alpha} = 0$ if $1 < \alpha$.

11. If $u_n = n^{-\alpha}$, and if B_m is the sum of the block of terms u_n for which $2^m < n \leqq 2^{m+1}$, prove that

$$2^{-\alpha} 2^{(1-\alpha)m} < B_m < 2^{(1-\alpha)m}.$$

12. Use the preceding result to show that the series

$$\sum_{n=1}^{\infty} n^{-\alpha}$$

converges for $1 < \alpha$ and diverges properly for $\alpha \leqq 1$.

13. Use the method employed in the proof of Theorem 5.1.3 to show that the series

$$\sum_{n=1}^{\infty} (-1)^{n-1}\, n^{-\alpha}$$

converges for $\alpha > 0$. When is the series absolutely convergent?

14. If $a_n > 0$ and $\Sigma\, a_n$ converges, prove that

$$\Sigma\, (a_n a_{n+1})^{\frac{1}{2}}.$$

does also.

15. Under the same assumptions as in Problem 14, prove the convergence of

$$\Sigma \frac{2a_n a_{n+1}}{a_n + a_{n+1}}.$$

[In Problem 14 the nth term is the geometric mean of the nth and the $(n+1)$th terms of the given series. See (1.6.14). In Problem 15 the so-called *harmonic mean* is used instead.]

For what values of x are the following series convergent?

16. $\sum_{n=0}^{\infty} e^{-nx}$. 17. $\sum_{n=1}^{\infty} ne^{-nx}$. 18. $\sum_{n=1}^{\infty} n^{-2x}$.

Is either of the following series convergent?

19. $\sum_{n=1}^{\infty} (1 - e^{-1/n})$. 20. $\sum_{n=1}^{\infty} (1 - e^{-1/n})^2$.

21. Show the convergence and find the sum of

$$\sum_{n=1}^{\infty} [(n+1)n^{\frac{1}{2}} + n(n+1)^{\frac{1}{2}}]^{-1}.$$

22. Discuss the convergence of

$$(\sqrt{2} - 1)^{-1} - (\sqrt{2} + 1)^{-1} + (\sqrt{3} - 1)^{-1} - (\sqrt{3} + 1)^{-1} + \cdots.$$

23. If (u) is the distance from u to the nearest integer, show the convergence of the series

$$\sum_{n=1}^{\infty} (n^2 + n\pi)\left(\frac{1}{n^2 + n\pi}\right).$$

24. If $\alpha > 1$, prove the absolute convergence for all x of the *trigonometric series*

$$\sum_{n=1}^{\infty} n^{-\alpha} \sin nx.$$

25. Prove the absolute convergence for all x of the so-called *exponential series*

$$\sum_{n=0}^{\infty} \frac{x^n}{n!} \qquad (0! = 1).$$

Is either of the following series convergent?

26. $\sum_{n=1}^{\infty} \log\left(1 + \frac{1}{n}\right)$. 27. $\sum_{n=1}^{\infty} \left[\log\left(1 + \frac{1}{n}\right)\right]^2$.

5.2. Series with positive terms. Their basic character makes it suitable to begin with the elements of the theory of series with positive terms in a systematic discussion of infinite series. Series with positive terms are also involved in the theory of absolutely convergent series and the latter form the largest class of series with which we can operate freely.

Work on series with positive terms is based on comparison with other series of known behavior. The fundamental theorem is the following.

THEOREM 5.2.1. *A series with positive terms $\sum u_n$ is convergent if and only if there exists a convergent series $\sum c_n$ with $c_n > 0$ such that for all n*

$$u_n \leqq c_n. \tag{5.2.1}$$

It is divergent if and only if there is a divergent series $\Sigma\, d_n$ with $d_n > 0$ such that for all n

$$u_n \geqq d_n. \tag{5.2.2}$$

Proof. Suppose that (5.2.1) is satisfied. Then

$$U_n \leqq C_n < C,$$

where U_n is the nth partial sum of the given series, C_n the nth partial sum of the comparison series, and C is the sum of the latter. This shows that the given series has bounded partial sums and by Theorem 5.1.2 the series is convergent. This sufficient condition is clearly also necessary: if the series does converge we can take $c_n = u_n$.

Suppose that (5.2.2) is satisfied instead and let D_n denote the nth partial sum of the divergent comparison series. Then

$$U_n \geqq D_n \to \infty$$

as $n \to \infty$, so the given series is also divergent. Again, the condition is necessary as well as sufficient, for the choice $d_n = u_n$ is admissible.

Section 5.1 provides us with a couple of comparison series

$$\sum_{n=0}^{\infty} q^n \quad \text{and} \quad \sum_{n=1}^{\infty} n^{-\alpha}. \tag{5.2.3}$$

The former is a geometric series, convergent for $0 < q < 1$, and properly divergent for $1 \leqq q$. The second series is a special *Dirichlet series*; it converges for $1 < \alpha$ and is properly divergent for $\alpha \leqq 1$. The special case $\alpha = 1$ is known as the *harmonic series*. For the proof see Problems 11 and 12 of Exercise 5.1.

Comparison of series is frequently reduced to so-called *convergence criteria* of which there is an abundance. A few are quite useful and should be known to the student, but it is better to use one's head than to rely on criteria. It is important to develop a feeling for the order of magnitude of the terms of a series. An accurate appraisal may lead immediately to a decision of the convergence question. Even a rough estimate is valuable.

In a convergent series the terms form a null sequence. If the terms are positive, it is not necessary that the sequence be strictly decreasing. If the terms decrease sufficiently fast, then the series converges. But the existence of a subsequence of terms which go to zero very slowly as a function of the index n, need not prevent the series from converging. We shall see examples of this later. Convergence criteria usually work best for series with monotonic terms.

Comparison with a geometric series leads to several observations, theorems, and criteria.

THEOREM 5.2.2. *If $u_n \geqq 0$, and if there is a number q, $0 < q < 1$, such that $u_n \leqq q^n$ for all large values of n, then $\Sigma\, u_n$ is convergent.*

This is an immediate consequence of Theorem 5.2.1 and the convergence of the geometric series when the quotient $q < 1$. A special case of this result is furnished by the following.

THEOREM 5.2.3. *If $u_n \geqq 0$ and if*

$$\lim_{n\to\infty} (u_n)^{1/n} \equiv \rho \tag{5.2.4}$$

exists, then the series converges for $\rho < 1$. It diverges properly if $\rho > 1$, and no conclusion can be drawn if $\rho = 1$.

Proof. If $\varepsilon > 0$ is given, and $\rho \neq 0$, then (5.2.4) implies that

$$(\rho - \varepsilon)^n \leqq u_n \leqq (\rho + \varepsilon)^n$$

for all large values of n. If $\rho < 1$, then we can choose ε so small that $\rho + \varepsilon \equiv q < 1$. We can then apply Theorem 5.2.2 and see that the series is convergent. On the other hand, if $\rho > 1$, we can choose ε so small that $\rho - \varepsilon \geqq 1$. This means that $u_n > 1$ for all large values of n and so the series is certainly properly divergent. Finally, if $\rho = 1$, the following example shows that no conclusion can be drawn. We take

$$u_n = n^{-\alpha}, \quad 0 < \alpha.$$

Here

$$\frac{1}{n} \log u_n = -\alpha \frac{\log n}{n} \to 0$$

(see formula (4.5.42)). It follows that $\rho = 1$. The series converges if $\alpha > 1$ and diverges for $\alpha \leqq 1$. Thus $\rho = 1$ is compatible with convergence as well as with divergence of the series. This completes the proof.

The last theorem contains an unnecessary restriction and it is possible to give a formulation of the so-called *root test* which is equivalent to Theorem 5.2.2.

THEOREM 5.2.4. *Let $u_n \geqq 0$ and set*

$$\limsup (u_n)^{1/n} \equiv \lambda. \tag{5.2.5}$$

The series converges if $\lambda < 1$, diverges properly if $\lambda > 1$, and no conclusion can be drawn if $\lambda = 1$.

This is proved as in the preceding case. If $\lambda > 1$, we now have $u_n \geqq 1$ infinitely often instead of for all n, but the conclusion is the same.

We can make a comparison with a geometric series in a different manner using division instead of root extraction. This leads to one form or another of the so-called *ratio test*, a criterion much favored by students and examiners alike. When it works, it works very well, but the test is highly sensitive to trivial irregularities in the sequence of terms which may not affect the convergence at all. The simplest form of the test is the following theorem.

THEOREM 5.2.5. *Suppose that* $u_n > 0$ *and that*

$$\lim_{n=\infty} \frac{u_{n+1}}{u_n} \equiv \mu \tag{5.2.6}$$

exists. Then the series $\Sigma\, u_n$ *converges if* $\mu < 1$, *diverges if* $\mu > 1$, *and no conclusion can be drawn if* $\mu = 1$.

Proof. Suppose that $\mu < 1$. We can then find a number q, $\mu \leqq q < 1$, such that

$$\frac{u_{n+1}}{u_n} < q$$

for all large values of n, say for $n \geqq m$. This gives

$$u_{m+1} < qu_m, \quad u_{m+2} < qu_{m+1} < q^2u_m,$$

and generally

$$u_{m+k} < q^k u_m.$$

Hence the terms of the given series are ultimately dominated by the terms of a convergent geometric series so that $\Sigma\, u_n$ is convergent. If $\mu > 1$, the terms of the series do not decrease to zero which is a necessary condition for convergence. For the series with $u_n = n^{-\alpha}$ we have

$$\frac{u_{n+1}}{u_n} = \frac{(n+1)^{-\alpha}}{n^{-\alpha}} = \left(1 + \frac{1}{n}\right)^{-\alpha} \to 1$$

as $n \to \infty$ while the series converges provided $\alpha > 1$, but diverges if $\alpha \leqq 1$. For a more realistic form of the ratio test, see Problem 17 in the Exercise below.

We return to comparison theorems later, but this is the proper time for a discussion of two classes of infinite series which dominate the applications. We refer to power series of the form

$$\sum_{n=0}^{\infty} R(n)x^n, \tag{5.2.7}$$

and

$$\sum_{n=0}^{\infty} R(0)R(1) \cdots R(n)x^n, \tag{5.2.8}$$

where $R(t)$ is a rational function of t. It is of course understood that the non-negative integers belong to the domain of $R(t)$. For a study of these series we need the following theorem which is also of independent interest. For the notation we refer to Definition 3.1.1.

THEOREM 5.2.6. *Let*

$$R(t) = \frac{P(t)}{Q(t)},$$

where

$$P(t) = a_0t^j + a_1t^{j-1} + \cdots + a_j, \quad a_0 \neq 0,$$

$$Q(t) = b_0t^k + b_1t^{k-1} + \cdots + b_k, \quad b_0 \neq 0,$$

are polynomials. Then for $t \to \infty$

$$R(t) = \frac{a_0}{b_0} t^{j-k} \left\{1 + O\left(\frac{1}{t}\right)\right\}, \tag{5.2.9}$$

and

$$\frac{R(t+1)}{R(t)} = 1 + \frac{j-k}{t} + O\left(\frac{1}{t^2}\right). \tag{5.2.10}$$

In particular,

$$\lim_{t\to\infty} \frac{R(t+1)}{R(t)} = 1. \tag{5.2.11}$$

Proof. We have

$$P(t) = a_0t^j\left\{1 + \frac{a_1}{a_0t} + \cdots + \frac{a_j}{a_0t^j}\right\} = a_0t^j\left\{1 + O\left(\frac{1}{t}\right)\right\},$$

$$Q(t) = b_0t^k\left\{1 + O\left(\frac{1}{t}\right)\right\}.$$

Division gives (5.2.9). Similarly we have

$$\frac{R(t+1)}{R(t)} - 1 = \frac{P(t+1)Q(t) - P(t)Q(t+1)}{P(t)Q(t+1)}. \tag{5.2.12}$$

Here the denominator is a polynomial of degree $j + k$ and

$$P(t)Q(t+1) = a_0b_0t^{j+k}\left\{1 + O\left(\frac{1}{t}\right)\right\}.$$

The numerator is of degree $\leqq j + k - 1$ for the polynomials $P(t)Q(t + 1)$ and $P(t + 1)Q(t)$ have the same leading term $a_0b_0t^{j+k}$ which cancels. There are six terms of degree $j + k - 1$ of which four cancel, leaving only

$$(j - k)a_0b_0t^{j+k-1}.$$

If $j = k$, this term also drops out and the numerator is of degree $\leqq j + k - 2$. We disregard this possibility and obtain

$$P(t+1)Q(t) - P(t)Q(t+1) = (j-k)a_0b_0t^{j+k-1}\left\{1 + O\left(\frac{1}{t}\right)\right\}.$$

Division then gives (5.2.10).

The proof is finished except for some justification of the manipulations with symbols of the form $1 + O(1/t)$. Setting $1/t = h$ to simplify the notation, we have obviously used relations of the form

$$[1 + O(h)][1 + O(h)] \Rightarrow 1 + O(h), \tag{5.2.13}$$

$$[1 + O(h)]^{-1} \Rightarrow 1 + O(h), \tag{5.2.14}$$

which would seem to violate the rules of arithmetic.

These relations are puzzling until we start to analyze the meaning of the symbols. The first relation states that if

$$[1 + f_1(h)][1 + f_2(h)] \equiv 1 + f_3(h),$$

and if

$$|f_1(h)| \leqq M_1 |h|, \qquad |f_2(h)| \leqq M_2 |h|,$$

valid say for $|h| < 1$, then we can find an M_3 such that

$$|f_3(h)| \leqq M_3 |h|, \qquad |h| < 1.$$

We can choose $M_3 = M_1 + M_2 + M_1 M_2$. This justifies (5.2.13). We can obviously extend this to any finite number of factors.

Similarly (5.2.14) is an assertion that if

$$[1 + f_1(h)]^{-1} \equiv 1 + f_4(h),$$

and if $|f_1(h)| \leqq M_1 |h|$, then we can find an M_4 such that

$$|f_4(h)| \leqq M_4 |h|,$$

when $|h|$ is sufficiently small. Since

$$f_4(h) = -\frac{f_1(h)}{1 + f_1(h)},$$

we can take $M_4 = 2M_1$ provided $|h|$ is so small that $|f_1(h)| \leqq \frac{1}{2}$. Once we have a way of handling reciprocals, we can also take care of quotients and see that

$$\frac{1 + O(h)}{1 + O(h)} \Rightarrow 1 + O(h).$$

These rules were tacitly assumed in the proof of Theorem 5.2.6.

We now proceed to a discussion of the series (5.2.7) and (5.2.8). The results are sufficiently important to formulate as a theorem. The notation is that of the preceding theorem.

THEOREM 5.2.7. *Let $R(t)$ be a rational function defined for all non-negative integers and positive for large positive values of t. Then the series (5.2.7) is absolutely convergent for $-1 < x < 1$. The series converges for $x = 1$ if and only if $j < k - 1$. The series (5.2.8) converges for all x if $j < k$, it converges for $-b_0/a_0 < x < b_0/a_0$ if $j = k$, and for no value of $x \neq 0$ if $j > k$.*

Proof. We start with the series (5.2.7) and apply the ratio test, assuming $x > 0$. By (5.2.11)

$$\lim_{n\to\infty} \frac{R(n + 1)x^{n+1}}{R(n)x^n} = x,$$

so that the series converges for $0 \leqq x < 1$. The series also converges for $-1 < x < 0$ since it is absolutely convergent. For $x = 1$ we have by (5.2.9)

$$R(n) = \frac{a_0}{b_0} n^{j-k}\left\{1 + O\left(\frac{1}{n}\right)\right\},$$

and this shows that the series converges if and only if

$$\sum_{n=1}^{\infty} n^{j-k}$$

is convergent, that is, if and only if $j < k - 1$. In such a case we say that the two series are *equiconvergent*.

For the series (5.2.8) we have

$$\frac{u_{n+1}}{u_n} = R(n + 1)x,$$

where we take x to be positive. In view of (5.2.10) we can assert that the ratio tends to 0 if $j < k$, to $(a_0/b_0)x$ if $j = k$, and to $+\infty$ if $j > k$. This proves the assertion concerning these series and completes the proof.

We shall need the following

Lemma 5.2.1. *For any positive integers m and n, $m < n$,*

$$\log \frac{n+1}{m} < \frac{1}{m} + \frac{1}{m+1} + \cdots + \frac{1}{n} < \log \frac{n}{m-1}. \tag{5.2.15}$$

Proof. We use Theorem 4.3.3, the first mean value theorem of the integral calculus, to obtain

$$\frac{1}{k+1} < \int_k^{k+1} \frac{ds}{s} < \frac{1}{k}$$

for $k = 1, 2, 3, \cdots$, Addition of the right halves of these inequalities for $k = m, m + 1, \cdots, n$ gives the first half of (5.2.15) and the second is obtained in the same manner.

We shall now prove an extension of the ratio test which is often effective when the ratio test fails to give information. Actually it is based on the special Dirichlet series

$$\sum_{n=1}^{\infty} n^{-\sigma}$$

as comparison series.

Theorem 5.2.8. *If $u_n > 0$ and if*

$$\liminf_{n\to\infty} n \log \frac{u_{n+1}}{u_n} = \alpha, \ \limsup_{n\to\infty} n \log \frac{u_{n+1}}{u_n} = \beta, \tag{5.2.16}$$

where α and β may be finite or infinite, then the series $\Sigma\, u_n$ converges if $\beta < -1$, diverges if $\alpha > -1$, and no conclusion can be drawn if $\alpha \leqq -1 \leqq \beta$.

Proof. Suppose first that $\beta < -1$. We choose a δ, $\beta < \delta < -1$, and a corresponding integer m such that

$$k \log \frac{u_{k+1}}{u_k} < \delta \quad \text{for} \quad m \leqq k.$$

This gives

$$\frac{u_{k+1}}{u_k} < \exp\left\{\frac{\delta}{k}\right\},$$

where $\exp u = e^u$. We multiply these inequalities for $k = m, m + 1, \cdots, n$, simplify the left member and obtain, by Lemma 5.2.1,

$$\frac{u_{n+1}}{u_m} < \exp\left\{\delta \sum_{k=m}^{n} \frac{1}{k}\right\} < \exp\left\{\delta \log \frac{n}{m-1}\right\} = \left(\frac{n}{m-1}\right)^{\delta},$$

or

$$u_{n+1} < u_m(m-1)^{-\delta} n^{\delta}.$$

Since $\delta < -1$, the right member is the general term of a convergent series whence it follows that $\Sigma\, u_n$ is also convergent.

If, on the other hand, $\alpha > -1$, we find a $\gamma > -1$ and a corresponding integer m such that

$$k \log \frac{u_{k+1}}{u_k} > \gamma \quad \text{for} \quad k \geqq m.$$

This gives

$$\log \frac{u_{k+1}}{u_k} > \frac{\gamma}{k},$$

and hence, using the lemma as above,

$$u_n > u_m m^{-\gamma}(n+1)^{\gamma}.$$

Since the right member is the general term of a divergent series, the given series is also divergent.

The following example shows the limitations of the test. We take

$$u_n = 2^{(-1)^n} n^{-\sigma},$$

that is, we take the series $\sum_{n=1}^{\infty} n^{-\sigma}$ and multiply the terms by 2 if n is even or by $\frac{1}{2}$ if n is odd. This evidently does not affect the convergence properties: the two series are equiconvergent. But

$$n \log \frac{u_{n+1}}{u_n} = (-1)^{n+1} n \log 4 - \sigma n \log \left(1 + \frac{1}{n}\right).$$

Here the inferior limit is $-\infty$ and the superior one is $+\infty$ regardless of the value of σ. The trivial multipliers cause the malfunctioning of the test.

In order to apply Theorem 5.2.8 when u_n is a rational function of n we need

LEMMA 5.2.2. *We have*

$$\lim_{t \to \infty} t \log \left(1 + \frac{a}{t}\right) = a. \tag{5.2.17}$$

Proof. Suppose that $a > 0$. Then

$$t \log \left(1 + \frac{a}{t}\right) = t[\log (t + a) - \log t] = t \int_t^{t+a} \frac{ds}{s}.$$

By the mean value theorem the last member lies between

$$\frac{at}{t+a} \quad \text{and} \quad a.$$

This implies (5.2.17). We leave the case $a < 0$ to the reader.

We shall now illustrate the use of Theorem 5.2.8.

Example 1. We can base the discussion of the series

$$\sum_{n=0}^{\infty} R(n) \tag{5.2.18}$$

on Theorem 5.2.8. By (5.2.10)

$$n \log \frac{R(n+1)}{R(n)} = n \log \left\{1 + \frac{j-k}{n} + O\left(\frac{1}{n^2}\right)\right\}.$$

For large values of n

$$\frac{j-k-\varepsilon}{n} < \frac{j-k}{n} + O\left(\frac{1}{n^2}\right) < \frac{j-k+\varepsilon}{n}$$

and Lemma 5.2.2 gives

$$\lim_{n\to\infty} n \log \frac{R(n+1)}{R(n)} = j-k.$$

Hence we have convergence for $j < k-1$ and divergence for $j > k-1$. The case $j = k-1$, which leads to divergence, is left in doubt by this method.

Example 2. The series

$$\sum_{n=0}^{\infty} \frac{a(a+1)(a+2)\cdots(a+n-1)}{b(b+1)(b+2)\cdots(b+n-1)} \tag{5.2.19}$$

is of the type

$$u_n = R(0)R(1)\cdots R(n-1) \tag{5.2.20}$$

with

$$R(t) = \frac{a+t}{b+t}.$$

Since

$$t \log R(t) = t \log\left(1 + \frac{a}{t}\right) - t \log\left(1 + \frac{b}{t}\right) \to a-b,$$

we have convergence for $b > a+1$ and divergence for $b < a+1$. If $b = a+1$ the series reduces to

$$\sum \frac{a}{n+a}$$

which diverges as a harmonic series.

Example 3. The choice

$$R(t) = \frac{(\alpha + t)(\beta + t)}{(1 + t)(\gamma + t)} \tag{5.2.21}$$

leads to a power series of type (5.2.8), known as the *hypergeometric series*:

$$F(\alpha, \beta, \gamma; x) \equiv 1 + \sum_{n=1}^{\infty} \prod_{k=0}^{n-1} \frac{(\alpha + k)(\beta + k)}{(1 + k)(\gamma + k)} x^n. \tag{5.2.22}$$

The symbol that follows the summation sign is an oversize Greek capital pi and the symbol is read "product over all k from 0 to $n - 1$." Again the k is a dummy which can be replaced by any other letter not already committed. A number of functions commonly used in analysis are special cases or limiting cases of hypergeometric series. The latter was studied at length by the famous German astronomer, geodesist, mathematician, and physicist Carl Friedrich Gauss (1777–1855) in a paper which appeared in 1812 and it has been the object of much research up to our time. The power series converges for $|x| < 1$ as shown by Theorem 5.2.7. For $x = 1$ we can use Theorem 5.2.8 and Lemma 5.2.2. The latter shows that

$$\lim_{t \to \infty} t \log R(t) = \alpha + \beta - 1 - \gamma,$$

so the series converges for

$$\alpha + \beta < \gamma. \tag{5.2.23}$$

In a convergent series the terms form a null sequence but we know this need not be monotone. For the case of a series whose terms are positive and steadily decreasing, we have other criteria at our disposal. We shall give two of them, both due to Cauchy.

We start with Cauchy's *condensation test*, also accredited to V. Ermakov.

THEOREM 5.2.9. *If $u_n > 0$ and is steadily decreasing, then the series*

$$\sum_{n=0}^{\infty} u_n \quad \text{and} \quad \sum_{k=0}^{\infty} 2^k u_{2^k} \tag{5.2.24}$$

are equiconvergent.

Proof. We denote by B_k the sum of the block of terms u_n with

$$2^{k-1} < n \leqq 2^k.$$

There are 2^{k-1} terms in this block; each term is greater than or equal to the last term of the block and less than the last term of the preceding block. This gives the double inequality

$$2^{k-1} u_{2^k} < B_k < 2^k u_{2^{k-1}} \tag{5.2.25}$$

Now if the first series under (5.2.24) converges, so does ΣB_k, and hence also the second series under (5.2.24). Conversely, if the second series converges, so does ΣB_k, and hence also the first series.

We illustrate by giving two examples.

Example 4. The series

$$\sum_{n=2}^{\infty} n^{-1} (\log n)^{-\alpha} \tag{5.2.26}$$

converges if $\alpha > 1$ and diverges if $\alpha \leqq 1$.

For the terms are monotone and the "condensed series" is

$$(\log 2)^{-\alpha} \sum_{k=1}^{\infty} k^{-\alpha}.$$

Example 5. The series

$$\sum_{n=3}^{\infty} n^{-1}(\log n)^{-1}(\log \log n)^{-\alpha} \tag{5.2.27}$$

converges if $\alpha > 1$ and diverges if $\alpha \leqq 1$.

Here the condensed series is

$$(\log 2)^{-1} \sum_{2}^{\infty} k^{-1}(\log k + \log \log 2)^{-\alpha}$$

and $(\log k + \log \log 2)/(\log k)$ lies between $\frac{1}{3}$ and 1 for $2 \leqq k$. It follows that the condensed series is equiconvergent with (5.2.26) and the conclusion is immediate.

Other series on the so-called *logarithmic scale* can be handled in the same manner. These series can also be discussed using Cauchy's *integral test* stated in the next theorem. The latter has the advantage of giving information about the rapidity of convergence or divergence.

THEOREM 5.2.10. *Let $f(x)$ be a positive, steadily decreasing function defined for $0 \leqq x$. Set*

$$\delta_n \equiv f(0) + f(1) + \cdots + f(n-1) - \int_0^n f(x)\, dx. \tag{5.2.28}$$

Then $\{\delta_n\}$ is a steadily increasing sequence of positive numbers and $\lim_{n\to\infty} \delta_n < f(0)$. *In particular, the infinite series and the improper integral,*

$$\sum_{k=0}^{\infty} f(k) \quad \textit{and} \quad \int_0^{\infty} f(x)\, dx,$$

are equiconvergent.

Proof. Figure 5.1 shows a part of the graph of $y = f(x)$. The integral represents the area under the curve, above the x-axis and between the lines $x = 0$ and $x = n$. The finite sum on the other hand is the corresponding area for the upper function which has the value $f(k)$ for $k - 1 \leqq x < k$. The difference of the two areas is positive; it equals the sum of the areas of the shaded curvilinear triangles. This sum is clearly an increasing function of n. On the other hand, the kth triangle has an area less than

$$f(k-1) - f(k),$$

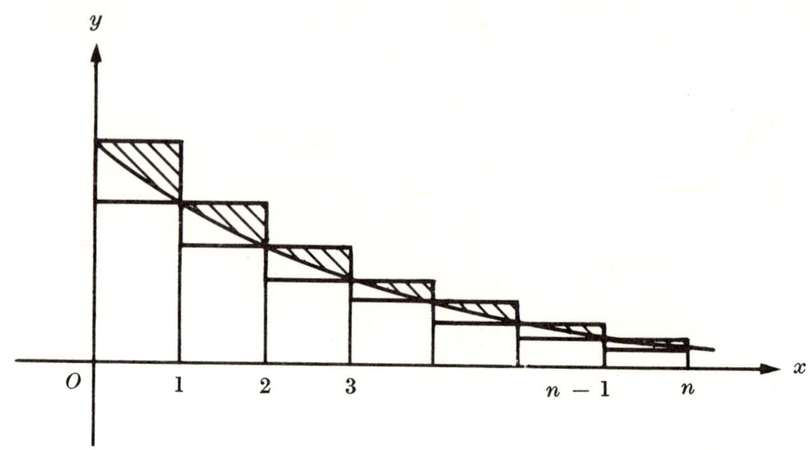

Figure 5.1

so that

$$\begin{aligned}\delta_n &< [f(0) - f(1)] + [f(1) - f(2)] + \cdots + [f(n-1) - f(n)] \\ &= f(0) - f(n) < f(0)\end{aligned}$$

as asserted. From the definition of δ_n and the existence of $\lim \delta_n$, it follows that

$$\lim_{n\to\infty} \sum_{k=0}^{n} f(k)$$

is finite if and only if

$$\lim_{n\to\infty} \int_0^n f(x)\,dx$$

is finite. This completes the proof.

Example 6. We shall reexamine Example 4 using the integral test. It is clear that the series (5.2.26) converges for $\alpha > 1$ because

$$\int_a^\infty x^{-1}(\log x)^{-\alpha}\,dx = (\alpha - 1)^{-1}(\log a)^{1-\alpha}.$$

On the other hand, the integral diverges for $\alpha \leqq 1$ so the series has the same property. We can use the inequalities proved in connection with Theorem 5.2.10 to estimate how fast the series converges or diverges, as the case may be. If $\alpha > 1$, the test shows

$$0 < \sum_{n=p}^{\infty} n^{-1}(\log n)^{-\alpha} - (\alpha - 1)^{-1}(\log p)^{1-\alpha} < p^{-1}(\log p)^{-\alpha}.$$

Suppose that $\alpha = 1.1$, and $p = 10^{10}$. Then, since $\log 10 = 2.30259$,

$$(\alpha - 1)^{-1}(\log p)^{1-\alpha} = 10\,(10 \log 10)^{-0.1} = 7.308,$$

that is, the remainder after ten billion terms exceeds 7 units. Even if we take $p = 10^{10^{10}}$, the corresponding partial sum is still more than 0.9 below the sum of the series. Using (5.2.27) we can get series which converge still slower.

On the other hand, if we take $\alpha = 1$, then we have a divergent series,

$$\int_3^p x^{-1}(\log x)^{-1}\,dx = \log\log p - \log\log 3,$$

so that

$$0 < \sum_3^{p-1} n^{-1}(\log n)^{-1} - \log\log p + \log\log 3 < 3^{-1}(\log 3)^{-1}.$$

If we take $p = 10^{10^{10}}$, we have

$$\log\log p = 10\log 10 + \log\log 10 < 24.$$

The partial sums do become infinite with p although exceedingly slowly. If we did not know in advance that the first series converges and the second diverges, numerical computation of partial sums would not enable us to tell the difference.

These considerations suggest rather strongly that, given any convergent series with positive terms, say $\Sigma\, a_n$, we can find a convergent series $\Sigma\, b_n$ such that $a_n/b_n \to 0$. This is true; that is, *for every slowly convergent series there is one which converges more slowly. A similar situation holds for divergent series.* See Problems 25 and 26 in the Exercise below.

EXERCISE 5.2

Discuss the convergence of the series $\Sigma\, u_n$ if u_n is given by:

1. $n^2\, 2^{-n}$.
2. $(1 + a^n)^{-1}, a > 0$.
3. $n!\, n^{-n}$.
4. $n^n e^{-n^2}$.
5. $\dfrac{(n!)^2}{(2n)!}$.
6. $(1 + 2n)^{-\alpha}$.
7. $n^{-\frac{1}{2}}(n + 1)^{-\frac{1}{2}}$.
8. $\dfrac{1}{n}[(n + 1)^{\frac{1}{2}} - n^{\frac{1}{2}}]$.
9. $n^{-\frac{1}{2}}(n^2 + 1)^{-\frac{1}{2}}$.
10. $(\log n)^{-a}$.
11. $(\log n)^{-\log n}$.
12. $(\log n)^{-\log\log n}$.
13. $(\log\log n)^{-\log n}$.
14. $n^{-1-(1/n)}$.
15. $\dfrac{1}{n} - \log\dfrac{n+1}{n}$.
16. If $0 < a < b < 1$, prove the convergence of

$$1 + a + b + a^2 + b^2 + a^3 + b^3 + \cdots.$$

Give the proof without use of convergence criteria and discuss what can be obtained from using various forms of the root test and of the ratio test.

17. Prove the following general form of the ratio test. Let $u_n > 0$, and set

$$\liminf_{n\to\infty}\frac{u_{n+1}}{u_n} = \alpha, \quad \limsup_{n\to\infty}\frac{u_{n+1}}{u_n} = \beta.$$

Then the series converges if $\beta < 1$, diverges if $1 < \alpha$, and no conclusion can be drawn if $\alpha \leqq 1 \leqq \beta$. Compare the preceding problem.

18. Prove that

$$\tfrac{1}{1} + \tfrac{1}{2} + \cdots + \frac{1}{n} - \log(n+1)$$

converges to a limit < 1. The limit is known as Euler's constant C and $C = 0.57721\,56649 \cdots$. In spite of much research it is not known whether C is rational or irrational.

19. For what values of α does the following series converge:

$$\sum_{n=1}^{\infty} \exp\left\{-\alpha\left[\tfrac{1}{1} + \tfrac{1}{2} + \tfrac{1}{3} + \cdots + \frac{1}{n}\right]\right\}.$$

20. If $\alpha > -1$, prove the absolute convergence for all x of

$$\sum_{k=1}^{\infty} \frac{(-1)^k x^{2k}}{k!\,(\alpha+1)(\alpha+2)\cdots(\alpha+k)}.$$

21. Prove that the following series converges for all x

$$\sum_{n=1}^{\infty} 2^{-n}\,\frac{(x^2+1^2)(x^2+2^2)\cdots(x^2+n^2)}{[(n+1)!]^2}.$$

22. Does the condensation test apply to Problem 14 above and, if so, with what result?

How should a be chosen in order that the series be convergent whose general term is given by

23. $n + \dfrac{1}{n} - (n^2+a)^{\frac{1}{2}}$.

24. $n^{-a}[(n+2)^{\frac{1}{2}} - (n-2)^{\frac{1}{2}}]$.

25. Let $a_n > 0$, let $\Sigma\, a_n$ be convergent, and set $\sum_{n+1}^{\infty} a_k = R_n$, $b_n = R_{n-1}^{\frac{1}{2}} - R_n^{\frac{1}{2}}$. Prove that (i) $a_n/b_n \to 0$; (ii) $\Sigma\, b_n$ is convergent.

26. Let $d_n > 0$, let $\Sigma\, d_n$ be divergent, and set $\sum_{1}^{n} d_k = D_n$, $e_n = D_n^{\frac{1}{2}} - D_{n-1}^{\frac{1}{2}}$. Prove that (i) $e_n/d_n \to 0$; (ii) $\Sigma\, e_n$ is divergent.

Prove the convergence of the series whose general term u_n equals

27. $\displaystyle\int_0^{1/n} x^{\frac{1}{3}}(1+x)\,dx$.

28. $\displaystyle\int_0^{1/n} \frac{x^{\frac{1}{2}}\,dx}{1+x^2}$.

29. $\displaystyle\int_0^{1/n} \frac{x^{\frac{2}{3}}\,dx}{1+x^3}$.

[*Hint:* Remember the mean value theorem!]

30. If $n!!$ denotes the product of the positive integers $\leqq n$ of the same parity as n, examine the convergence of

$$\sum_{n=1}^{\infty} \left[\frac{(2n-1)!!}{(2n)!!}\right]^2.$$

5.3. Nonabsolute convergence. Operations with series. The theory of infinite series would be simpler but also less interesting and useful, if we could restrict ourselves to series with positive terms or at least to absolutely convergent series. The mathematical description of physical phenomena of an oscillatory nature normally leads to nonabsolutely convergent series. For this and other reasons we have to study some simple classes of nonabsolutely convergent series, to develop methods useful in handling such series, and to become familiar with some of the peculiar phenomena associated with them. We start with a simple theorem, a generalization of a result proved for the series in formula (5.1.16).

THEOREM 5.3.1. *Let $\{a_n\}$ be a strictly decreasing null sequence. Then the series*

$$\sum_{n=0}^{\infty} (-1)^n a_n \tag{5.3.1}$$

is convergent but not necessarily absolutely convergent.

Proof. We use the same method as in the special case just mentioned. We form the partial sums and prove that the sums of even order form a strictly decreasing sequence of positive numbers, while the sums of odd order form an increasing sequence which is bounded above. Both sequences have limits which must coincide since $\lim a_n = 0$. First we note that

$$\begin{aligned} U_{2n} &= a_0 - a_1 + a_2 - a_3 + a_4 - \cdots - a_{2n-1} + a_{2n} \\ &= a_0 - (a_1 - a_2) - (a_3 - a_4) - \cdots - (a_{2n-1} - a_{2n}). \end{aligned}$$

Here each parentheses contains a positive number. This gives the following sequence of inequalities

$$a_0 = U_0 > U_2 > \cdots > U_{2n} > U_{2n-1}.$$

In the same manner we obtain

$$U_{2n+1} = (a_0 - a_1) + (a_2 - a_3) + \cdots + (a_{2n} - a_{2n+1})$$

which shows that

$$0 < U_1 < U_3 < \cdots < U_{2n-1} < U_{2n} < U_0.$$

These inequalities prove the asserted monotony properties and show that

$$0 < \lim U_{2n-1} \leqq \lim U_{2n} < U_0. \tag{5.3.2}$$

But we have

$$U_{2n-1} + a_{2n} = U_{2n}$$

and $\lim a_{2n} = 0$, whence it follows that the sign of equality holds in the middle of (5.3.2). This completes the proof.

If we denote the sum of the series (5.3.1) by U, we have the useful inequality

$$a_0 - a_1 < U < a_0. \tag{5.3.3}$$

We also note that the remainder of the series after the nth term satisfies a similar inequality, namely

$$a_{n+1} - a_{n+2} < (-1)^{n+1} \sum_{k=n+1}^{\infty} (-1)^k a_k < a_{n+1}. \tag{5.3.4}$$

In particular, the remainder has the same sign as its first term and is numerically less than the latter.

We illustrate by examples.

Example 1. The series

$$\sum_{n=0}^{\infty} (-1)^n (n+1)^{-\alpha} \tag{5.3.5}$$

converges for $\alpha > 0$. It is absolutely convergent if and only if $\alpha > 1$. The series (5.1.16) is the special case $\alpha = \frac{1}{2}$.

Example 2. The series

$$\frac{1}{\log\log 3} - \frac{1}{\log\log 4} + \cdots + \frac{1}{\log\log(2k-1)} - \frac{1}{\log\log 2k} + \cdots \tag{5.3.6}$$

is convergent since $\log\log n$ is increasing. A reference to the end of the preceding section shows that the series converges exceedingly slowly since

$$\log\log k < 24 \quad \text{for} \quad k < 10^{10^{10}}.$$

We note that if all negative terms are omitted in (5.3.6), then the remaining series of positive terms is properly divergent. Similarly the negative terms alone form a divergent series. The same holds in the case of (5.3.5) for $0 < \alpha \leqq 1$. We shall see later that this phenomenon occurs in every convergent series which is not absolutely convergent.

In the series (5.3.1) positive and negative terms alternate. We could also allow blocks of p positive terms to alternate with blocks of p negative terms. This gives a series of the type

$$\sum_{n=0}^{\infty} (-1)^{[n/p]} a_n, \tag{5.3.7}$$

where, we recall, $[n/p]$ is the largest integer $\leqq n/p$. Here the sign of the terms remains constant between consecutive multiples of p. If we assume that $\{a_n\}$ is a strictly decreasing null sequence, then the series (5.3.7) is convergent. This follows from Theorem 5.3.1 after some cogitation. We can set

$$b_k = a_{kp} + a_{kp+1} + \cdots + a_{(k+1)p-1}$$

and consider the series

$$\sum_{k=0}^{\infty} (-1)^k b_k. \tag{5.3.8}$$

Here we have

$$b_k > b_{k+1}, \qquad \lim b_k = 0,$$

so that (5.3.8) is convergent by Theorem 5.3.1. Now any partial sum of (5.3.8) is a partial sum of (5.3.7). The converse is not true, but a partial sum of (5.3.7), say the sum U_N with $N = np + q - 1$, $0 \leqq q < p$, differs from the partial sum V_{n-1} of (5.3.8) by the group of terms

$$\pm (a_{np} + a_{np+1} + \cdots + a_N).$$

The absolute value of the difference is less than b_n which tends to 0 when $n \to \infty$. It follows that the two series are equiconvergent and have the same sum.

In these two cases the positive and the negative blocks have the same length. Other patterns are possible, however. As an example we mention the series

$$\sum_{n=0}^{\infty} (-1)^{[\sqrt{n}]} a_n, \tag{5.3.9}$$

where the sign changes each time n passes through a square. Here the series converges if

$$\begin{aligned} &\text{(i)} \quad \{a_n\} \textit{ is a strictly decreasing null sequence,} \\ &\text{(ii)} \quad \textit{the series } \textstyle\sum a_{n^2} \textit{ is convergent.} \end{aligned} \tag{5.3.10}$$

We leave the proof to the student.

In the applications the sign changes are apt to occur much more at random than in the special series studied so far. Here almost the only general method that is available for the study of convergence is *summation by parts*, a worthy companion of the method of integration by parts which we found useful in the theory of integration. The method of summation by parts is due to Abel (1826). It applies to series of the form

$$\sum_{k=0}^{\infty} u_k v_k. \tag{5.3.11}$$

We introduce the following notation

$$U_n = \sum_{k=0}^{n} u_k, \quad V_n = \sum_{k=0}^{n} v_k. \tag{5.3.12}$$

We have then

$$u_n = U_n - U_{n-1}, \quad n = 0, 1, 2, \cdots,$$

where we define $U_{-1} = 0$. This gives

$$\sum_{k=m}^{n} u_k v_k = \sum_{k=m}^{n} (U_k - U_{k-1}) v_k.$$

We rearrange this expression to bring out the U's. Each individual U_p, $m \leqq p < n$, occurs twice in the sum on the right, multiplied by v_p in the first instance and by $-v_{p+1}$ in the second. On the other hand, U_{m-1} and U_n occur each once, the former multiplied by $-v_m$, the latter by v_n. It follows that

$$\sum_{k=m}^{n} u_k v_k = -v_m U_{m-1} + v_n U_n + \sum_{k=m}^{n-1} (v_k - v_{k+1}) U_k. \tag{5.3.13}$$

This formula should be compared with (4.4.13). Here U_k corresponds to $F(x)$, u_k to $dF(x)$, v_k to $G(x)$, and $v_{k+1} - v_k$ to $dG(x)$. The terms $-v_m U_{m-1}$ and $v_n U_n$ correspond to the integrated parts in (4.4.13).

It is clear that in (5.3.13) we can interchange the roles of the u's and the v's.

This formula is merely an identity, and becomes useful in the study of convergence if and only if we have some applicable information concerning the quantities in the right member. The next theorem, also due to Abel, shows one of the many uses of the formula.

THEOREM 5.3.2. *The series* (5.3.11) *converges if* (i) *the sequence* $\{U_n\}$ *is bounded,* (ii) $\lim v_n = 0$, *and* (iii) *the series* $\Sigma\, |v_n - v_{n+1}|$ *is convergent.*

Proof. Suppose that $|U_n| \leqq M$ for all n. Let $\varepsilon > 0$ be given and choose m so large that

$$|v_k| < \varepsilon \quad \text{for} \quad m \leqq k, \ \sum_{k=m}^{\infty} |v_k - v_{k+1}| < \varepsilon.$$

This is possible by virtue of our assumptions (ii) and (iii). Using (5.3.13) we then see that

$$\left| \sum_{k=m}^{n} u_k v_k \right| \leqq |v_m|\, |U_{m-1}| + |v_n|\, |U_n| + \sum_{k=m}^{n-1} |v_k - v_{k+1}|\, |U_k|$$
$$< M(\varepsilon + \varepsilon + \varepsilon) = 3M\varepsilon.$$

Since $\varepsilon > 0$ is arbitrary, the conditions of Theorem 5.1.1 are satisfied, and the series (5.3.11) is convergent. This completes the proof.

Two remarks are in order. Condition (i) is satisfied if the sequence $\{U_n\}$ converges to a finite limit. Conditions (ii) and (iii) are satisfied if, for instance, $\{v_n\}$ is a monotone null sequence.

Example 3. Theorem 5.3.1 is a direct consequence of Theorem 5.3.2. We take

$$u_n = (-1)^n, \quad v_n = a_n$$

so that

$$0 \leqq U_n = \tfrac{1}{2}[1 + (-1)^n] \leqq 1$$

and $\{v_n\}$ is a monotone null sequence. Hence the conditions of Theorem 5.3.2 are satisfied. Similarly the convergence of the series (5.3.7) follows from Theorem 5.3.2. We take

$$u_n = (-1)^{[n/p]}, \quad v_n = a_n.$$

In this case $0 \leqq U_n \leqq p$ and the conditions are satisfied.

Example 4. The power series

$$\sum_{n=0}^{\infty} b_n x^n \tag{5.3.14}$$

converges for $-1 \leqq x < 1$, if the sequence $\{b_n\}$ satisfied conditions (ii) and (iii). Here we set

$$u_n = x^n, \quad v_n = b_n,$$

and obtain [see (5.1.14)]

$$0 \leqq U_n = \frac{1 - x^{n+1}}{1 - x} < \frac{2}{1 - x}.$$

Hence the assertion follows from Theorem 5.3.2.

It is important to realize in what respects an absolutely convergent series differs from other series. Given two infinite series, we say that the second is a *subseries* of the first, if any two terms of the second series occur in the first series and in the same order.

THEOREM 5.3.3. *Every subseries of an absolutely convergent series is also absolutely convergent. A convergent series which is not absolutely convergent has subseries of all types: absolutely convergent, convergent but not absolutely convergent, divergent, properly divergent. In particular, the subseries of positive terms as well as the subseries of negative terms are properly divergent.*

Proof. Suppose that $\Sigma\, u_n$ is the given series, $\Sigma\, u_{n_p}$ the subseries. Here $\{n_p\}$ is a strictly increasing sequence of positive integers. If the given series is absolutely convergent and if $\varepsilon > 0$ is given, then we can find an $m = m(\varepsilon)$ such that

$$\sum_{k=m}^{\infty} |u_k| < \varepsilon.$$

Suppose that q is the least integer such that $n_q \geqq m$. Then

$$\sum_{p=q}^{\infty} |u_{n_p}| \leqq \sum_{k=m}^{\infty} |u_k| < \varepsilon,$$

that is, the subseries is also absolutely convergent.

Suppose next that the given series is convergent but not absolutely convergent. It must then have infinitely many positive terms as well as infinitely many negative ones. For a better survey, let us write

$$u_n^+ = \max\,(u_n, 0), \quad u_n^- = \max\,(-u_n, 0), \quad a_n = |u_n|,$$

$$P_n = \sum_{k=0}^{n} u_k^+, \quad Q_n = \sum_{k=0}^{n} u_k^-, \quad U_n = \sum_{k=0}^{n} u_k, \quad A_n = \sum_{k=0}^{n} |a_k|.$$

We have then

$$U_n = P_n - Q_n, \quad 2P_n = A_n + U_n,$$
$$A_n = P_n + Q_n, \quad 2Q_n = A_n - U_n.$$

Since U_n tends to a finite limit U, and A_n becomes infinite, we see that $P_n \to +\infty$, $Q_n \to +\infty$. This proves that the terms of fixed sign form properly divergent subseries.

Since $\lim U_n = U$ is finite, $\lim u_n = 0$. Hence we can select from the positive terms of the series a subseries, $\Sigma\, p_k$ say, such that each p_k is a u_n and

$$p_1 < 1, p_2 < 2^{-1}, \quad \cdots, \quad p_n < 2^{-n}, \cdots.$$

This series is convergent, even absolutely convergent, since the terms are positive.

Since the given series is convergent without being absolutely convergent and since any subseries obtained by omitting a finite number of terms has the same convergence properties, we have no lack of nonabsolutely convergent subseries.

To construct a subseries of the third type, divergent but not properly divergent, we proceed as follows. Our aim is to construct a subseries whose partial sums have preassigned inferior and superior limits S and T respectively. We assume $-\infty < S < T < +\infty$. The construction uses two facts: (1) that P_n and Q_n become infinite with n; (2) that $\lim u_n = 0$. Using the first fact we can add sufficiently many positive terms to get a sum $\geqq T$. We take just enough terms to achieve this goal so that one less term would give a sum $< T$. This is the first building block. The second block contains negative terms chosen from the negative terms with subscript greater than the subscript of the last term in the first block. We take just sufficiently many negative terms so that the sum of the first two blocks is $\leqq S$. We then form a third block, positive terms this time, and just sufficiently many so that the sum of the first three blocks is $\geqq T$. In this manner the alternating process continues. Since the terms tend to zero, we can be sure that beyond a certain point all partial sums lie between $S - \varepsilon$ and $T + \varepsilon$ where $\varepsilon > 0$ is a preassigned arbitrarily small number. Since there are infinitely many partial sums in each of the intervals $(S - \varepsilon, S + \varepsilon)$ and $(T - \varepsilon, T + \varepsilon)$, we see that S is the inferior limit and T the superior limit of the partial sums of the constructed subseries. Hence the latter is divergent without being properly divergent. This completes the proof.

The construction used in the last part of this proof can be generalized and used to prove an important theorem due to B. Riemann (1854, published in 1867). We say that a series $\sum v_k$ is a *rearrangement* of a series $\sum u_k$, if the two series have the same terms though in a different order.

THEOREM 5.3.4. *An absolutely convergent series can be rearranged in any way we please without affecting the absolute convergence or the sum of the series. A series which is convergent but not absolutely convergent can be rearranged so that the new series has a preassigned real number as its sum or so that the inferior and superior limits of its partial sums have preassigned finite of infinite values. In particular, the series may be rearranged so as to diverge to $+\infty$ or to $-\infty$.*

We omit the proof, but we recommend that the student try to adapt the argument above to this new situation.

It remains to discuss operations with infinite series. We can add convergent series to obtain a convergent series and we can multiply a convergent series by a constant. Thus

$$\sum u_n + \sum v_n = \sum (u_n + v_n), \tag{5.3.15}$$

$$C \sum u_n = \sum C u_n. \tag{5.3.16}$$

These operations conserve convergence and sums by virtue of Theorem 1.7.2 applied to the partial sums. Thus the sum of two convergent series has as its sum the sum of the two summands.

Multiplication of two series is a much more delicate problem. Actually the definition of the product series depends somewhat on the factor series. The most common definition applies to power series and gives the product of two power series as a power series. It is often referred to as the *Cauchy product*. It is defined by

$$\sum_{n=0}^{\infty} w_n = \sum_{n=0}^{\infty} \left\{ \sum_{k=0}^{n} u_k v_{n-k} \right\}, \tag{5.3.17}$$

if the factor series are

$$U = \sum_{j=0}^{\infty} u_j, \quad V = \sum_{k=0}^{\infty} v_k.$$

The series (5.3.17) is not necessarily convergent when U and V are convergent series, but if the formal product converges, it does converge to the sum UV. The situation is more favorable for absolutely convergent factor series.

THEOREM 5.3.5. *The Cauchy product series of two absolutely convergent series is also absolutely convergent and its sum is the product of the sums of the factor series.*

Proof. We start by proving the theorem for the case where both factors have nonnegative terms so that the same is true for the Cauchy product. Here absolute convergence and convergence coincide. Let

$$U_n = \sum_{k=0}^{n} u_k, \quad V_n = \sum_{k=0}^{n} v_k, \quad W_n = \sum_{m=0}^{n} \left\{ \sum_{k=0}^{m} u_k v_{m-k} \right\}, \tag{5.3.18}$$

where $u_j \geq 0$, $v_k \geq 0$. The product of U_n with V_n contains all products $u_j v_k$ with $0 \leq j, k \leq n$ and no other terms. In W_n we have only such products $u_j v_k$ for which $j + k \leq n$. This means that

$$W_n \leq U_n V_n < UV.$$

Since $W_n \leq W_{n+1}$ and the sequence $\{W_n\}$ is bounded, the series (5.3.17) is convergent and its sum W is $\leq UV$. On the other hand, W_{2n} contains all products $u_j v_k$ with $j + k \leq 2n$. In particular, it contains all products which occur in $U_n V_n$. It follows that

$$U_n V_n \leq W_{2n},$$

and

$$UV = \lim U_n V_n \leq \lim W_{2n} = W.$$

It follows that $UV = W$.

The general case where the factor series contain negative as well as positive terms can be reduced to the case just treated by elementary algebra. We set as above

$$u_n = u_n^+ - u_n^-, \; v_n = v_n^+ - v_n^-,$$

and obtain

$$W_n = W_n^{+,+} - W_n^{-,+} - W_n^{+,-} + W_n^{-,-},$$

where

$$W_n^{r,s} = \sum_{m=0}^{n} \left\{ \sum_{k=0}^{m} u_k^{\,r} \, v_{m-k}^{s} \right\}, \; r, s = +, -.$$

Since the four series

$$U^+ = \sum_{n=0}^{\infty} u_n^+, \; U^- = \sum_{n=0}^{\infty} u_n^-, \; V^+ = \sum_{n=0}^{\infty} v_n^+, \; V^- = \sum_{n=0}^{\infty} v_n^-$$

converge by assumption, we see that each of the four sequences $\{W_n^{r,s}\}$ converges to a finite limit $W^{r,s}$ and

$$W^{r,s} = U^r V^s.$$

It follows that the product series is absolutely convergent and its sum

$$\begin{aligned} W &= U^+V^+ - U^-V^+ - U^+V^- + U^-V^- \\ &= (U^+ - U^-)(V^+ - V^-) = UV. \end{aligned}$$

This completes the proof.

EXERCISE 5.3

1. Discuss the convergence of $\sum_{n=1}^{\infty} \frac{(-1)^n n}{n^2+1}$.

2. Prove the divergence of $\sum_{n=1}^{\infty} (-1)^{n-1} n^{-1/n}$.

3. [Addendum to Theorem 5.2.7.] If $R(t)$ is a rational function defined for all nonnegative integers and positive for large values of t, give necessary and sufficient conditions for the convergence of

$$\sum_{n=0}^{\infty} (-1)^n R(n).$$

4. Given a series $\sum_{n=0}^{\infty} (-1)^n u_n$ where $u_n > 0$. Suppose there exists an absolutely convergent series $\sum_{n=0}^{\infty} (-1)^n c_n$ such that if $a_n = u_n + c_n$, then the series $\sum_{n=0}^{\infty} (-1)^n a_n$ satisfies the conditions of Theorem 5.2.1. Prove the convergence of the original series.

Discuss convergence of the series:

5. $\sum_{n=1}^{\infty} (-1)^{n-1} \sum_{k=0}^{2n} \frac{1}{n^2+k}$. 6. $\sum_{n=1}^{\infty} (-1)^{n-1} \frac{(2n-1)!!}{(2n)!!}$.

7. $\sum_{n=1}^{\infty} \frac{(-1)^{n-1}}{x+n}$, $x > -1$.

8. Prove that the series $\sum_{n=0}^{\infty} (-1)^{[\sqrt{n}]}(n+1)^{-\frac{1}{2}}$ diverges to $-\infty$. [*Hint:* Prove that the sum of the terms between $n = (2p)^2$ and $(2p+2)^2 - 1$, limits included, is $-1/p + O(1/p^2)$.]

9. Suppose that $u_n = 1/n$, if n is not a perfect square and $u_n = -2/\sqrt{n}$ if n is a perfect square. Prove that the series converges. [*Hint:* Group each negative term with the positive terms following it.]

10. If $f_n(x) = (x - x_0)(x - x_1) \cdots (x - x_n)$, verify the identity

$$\frac{1}{x-a} = \frac{1}{f_0(x)} + \frac{f_0(a)}{f_1(x)} + \frac{f_1(a)}{f_2(x)} + \cdots + \frac{f_{n-1}(a)}{f_n(x)} + \frac{f_n(a)}{f_n(x)} \cdot \frac{1}{x-a}.$$

11. If $\{b_n\}$ satisfies the conditions (ii) and (iii) of Theorem 5.3.2 and if $a > 0$, show that the series

$$\frac{b_0}{x} + \frac{b_1 a}{x(x+1)} + \frac{b_2 a(a+1)}{x(x+1)(x+2)} + \cdots + \frac{b_n a(a+1)\cdots(a+n-1)}{x(x+1)\cdots(x+n-1)(x+n)} + \cdots$$

converges for $x > a$. This is known as a *factorial series*.

12. The series

$$S = \tfrac{1}{1} - \tfrac{1}{2} + \tfrac{1}{3} - \tfrac{1}{4} + \tfrac{1}{5} - \tfrac{1}{6} + \cdots + \frac{1}{2n-1} - \frac{1}{2n} + \cdots$$

is not absolutely convergent. Its sum is log 2. The series

$$T = \tfrac{1}{1} + \tfrac{1}{3} - \tfrac{1}{2} + \tfrac{1}{5} + \tfrac{1}{7} - \tfrac{1}{4} + \cdots + \frac{1}{4n-3} + \frac{1}{4n-1} - \frac{1}{2n} + \cdots$$

is a rearrangement of the first series. Prove (i) that it converges; (ii) that $T = \frac{3}{2}S$. [*Hint:* Consider the partial sums T_{3n} to prove convergence. Compare T_{3n} with S_{4n} and use Example 5 of Section 4.7.]

13. How could you construct a rearrangement of the series S that diverges to $+\infty$?

14. Given the convergent series

$$U = 1 - 2^{-\frac{1}{2}} + 3^{-\frac{1}{2}} - 4^{-\frac{1}{2}} + \cdots + (2n-1)^{-\frac{1}{2}} - (2n)^{-\frac{1}{2}} + \cdots,$$

form the rearrangement

$$1 + 3^{-\frac{1}{2}} - 2^{-\frac{1}{2}} + 5^{-\frac{1}{2}} + 7^{-\frac{1}{2}} - 4^{-\frac{1}{2}} + \cdots + (4n-3)^{-\frac{1}{2}} + (4n-1)^{-\frac{1}{2}} - (2n)^{-\frac{1}{2}} + \cdots,$$

and prove that it diverges to $+\infty$.

15. If U is the series of the preceding problem, the formal Cauchy product of U with itself is the series $\sum_{n=0}^{\infty} (-1)^n w_n$ where

$$w_n = \sum_{k=0}^{n} (k+1)^{-\frac{1}{2}}(n+1-k)^{-\frac{1}{2}}.$$

Show that w_n tends to a finite limit $\neq 0$, for instance, using formula (4.7.8). Does this prove the divergence of the product series?

16. If $\{u_n\}$ is a bounded sequence, prove the absolute convergence of $\sum_{n=0}^{\infty} u_n x^n$ for $|x| < 1$. Find also the Cauchy product of $\sum_{n=0}^{\infty} x^n$ with $\sum_{n=0}^{\infty} u_n x^n$. When is the product series absolutely convergent?

5.4. Series of functions. Power series. Uniform convergence. Earlier sections of this book contain many examples of infinite series of the form

$$\sum_{n=0}^{\infty} f_n(x), \tag{5.4.1}$$

where each $f_n(x)$ is a function of x and all these functions have a common domain of definition D. If the series converges in some subset S of D, we call S a region of convergence of the series. S is *the region of convergence* if the series converges in S and nowhere outside of S. The sum of the series defines a function $F(x)$ in S.

To illustrate, we consider the geometric series

$$\sum_{n=0}^{\infty} x^n. \tag{5.4.2}$$

Here D is R, the set of all real numbers. The region of convergence is the open interval $(-1, 1)$ and the sum function is $(1-x)^{-1}$. We note that this function is defined for all $x \neq 1$, but it gives the sum of the series (5.4.2) only for $-1 < x < 1$.

We return to (5.4.1). That this series converges in S implies and is implied by the statement that it converges for each x in S. This can be expressed in several different ways using the epsilon language. Thus, given an $\varepsilon > 0$ and an $x \in S$, there exists an integer $N = N(x, \varepsilon)$ such that for any $n > N$

$$\left| \sum_{k=n}^{\infty} f_k(x) \right| < \varepsilon. \tag{5.4.3}$$

For a fixed $\varepsilon > 0$ this defines a function $N(x, \varepsilon)$ on S to natural numbers. To make this assertion precise we take in each case the least integer that satisfies the requirements. We can scarcely expect such a function to be continuous. Can we assert that it is bounded on S? Simple examples show that it need not be. Thus for the geometric series condition (5.4.3) calls for the least integer N such that

$$|x|^{N+1}(1 - x)^{-1} < \varepsilon.$$

If $x = 0$, this is true for all $N > 0$, but the further away from 0 we get, the larger N must be chosen. Thus if $\varepsilon = 10^{-6}$, we can take $N = 152$ for $x = 0.9$ while $N = 1833$ is necessary when $x = 0.99$. On the other hand, the choice $N = 152$ will do not merely for $x = 0.9$ but for all x in the interval $(0, 0.9)$. Similarly $N = 1833$ will do for all x in $(0, 0.99)$. These considerations lead to

Definition 5.4.1. *The series* (5.4.1) *is uniformly convergent in S if given any $\varepsilon > 0$ there exists an $N = N(\varepsilon)$, independent of x such that for all $x \in S$ we have*

$$\left| \sum_{k=n}^{\infty} f_k(x) \right| < \varepsilon \quad \textit{for} \quad N(\varepsilon) < n. \tag{5.4.4}$$

It is clear that the geometric series is uniformly convergent in any interval $[-1 + \delta, 1 - \delta]$, $\delta > 0$, but not in $(-1, 1)$ as we saw above.

We shall return to general properties and implications of uniform convergence later. At this juncture we shall study a particular class of function series, namely *power series* which are of the form

$$\sum_{n=0}^{\infty} a_n x^n \tag{5.4.5}$$

where $\{a_n\}$ is a given sequence of real numbers, and x is a real variable. We ask what is the region of convergence, is the series uniformly convergent, and what are the properties of the sum when the series converges?

The region of convergence may possibly reduce to $x = 0$ and then the series is of no interest to us. The other extreme is where the series converges for all x. If neither case is present, the region of convergence is an interval which may be open, half-open, or closed. The endpoints of this interval are symmetrically located with respect to the origin and are determined by Theorem 5.2.4. We shall prove

THEOREM 5.4.1. *Define*

$$1/R = \limsup_{n\to\infty} |a_n|^{1/n}. \tag{5.4.6}$$

If the right member is infinite, we take $R = 0$. Then the series (5.4.5) converges only for $x = 0$. If the right member is 0, we take $R = \infty$. Now the power series is absolutely convergent for every finite x and the convergence is uniform with respect to x in any fixed bounded closed interval. If $0 < R < \infty$, the series converges absolutely for $-R < x < R$. It may converge at either or both endpoints of the interval. If it is absolutely convergent at one of these points, it has the same property at the other. Finally, the series converges uniformly in any closed interval contained in the interval of convergence.

Proof. Theorem 5.2.4 asserts that the series converges absolutely for a particular value of $x \neq 0$ if

$$\limsup_{n\to\infty} |a_n x^n|^{1/n} = |x| \limsup |a_n|^{1/n} < 1, \tag{5.4.7}$$

and it diverges if the expression is > 1. Hence, if $\limsup |a_n|^{1/n} = \infty$, the series diverges for each $x \neq 0$. On the other hand, if the superior limit is 0, then (5.4.7) is satisfied for every finite x and we have absolute convergence for all x.

Suppose that neither of these cases is present. Then (5.4.6) defines a finite positive number R and (5.4.7) shows that the power series is absolutely convergent for $|x| < R$ and diverges for every x with $|x| > R$. We recall from the proof of Theorem 5.2.4 that in the divergence case there exists a sequence of terms with absolute values tending to infinity. It follows that the only doubtful values of x are $x = \pm R$. Here anything can happen. If we choose for $\{a_n\}$ one of the sequences

$$\{1\}, \quad \left\{\frac{1}{n}\right\}, \quad \left\{\frac{(-1)^n}{n}\right\}, \quad \left\{\frac{1}{n^2}\right\},$$

we have $R = 1$ in all four cases, but the intervals of convergence are

$$(-1, 1), \quad [-1, 1), \quad (-1, 1], \quad [-1, 1],$$

respectively.

It remains to discuss uniform convergence. There are three different cases.

A. Suppose $0 < R < \infty$ and the series diverges for $x = \pm R$. Choose any R_1 and R_2 such that $0 < R_1 < R_2 < R$. Since

$$\limsup |a_n|^{1/n} = R^{-1},$$

we see that

$$|a_n|^{1/n} < R_2^{-1}$$

for all large values of n, say for $n > N$. Hence for $|x| \leqq R_1$, $N < n$,

$$\left| \sum_{k=n}^{\infty} a_k x^k \right| \leqq \sum_{k=n}^{\infty} \left(\frac{R_1}{R_2}\right)^k = \frac{R_1^{\,n}}{R_2^{n-1}(R_2 - R_1)}.$$

If n is sufficiently large, this is less than a preassigned ε. Hence condition (5.4.3) is satisfied and the series converges uniformly for $|x| \leqq R_1 < R$.

B. Suppose $R = \infty$. Here we can choose R_1 and R_2 arbitrarily, $0 < R_1 < R_2 < \infty$ and repeat the argument.

C. Suppose that R is finite and the series converges at one of the endpoints, say for $x = R$. In this case the previous argument shows uniform convergence in any interval $[-R_1, R_1]$ with $R_1 < R$. Actually we have uniform convergence in $[-R_1, R]$. This follows from a result due to Abel which we state as

THEOREM 5.4.2. *If $\Sigma\, u_n$ is convergent, then the power series*

$$\sum_{n=0}^{\infty} u_n x^n \tag{5.4.8}$$

is uniformly convergent with respect to x in $[0, 1]$.

Proof. The argument is based upon summation by parts, but uses remainders rather than partial sums, so the underlying formula differs from (5.3.13). We set

$$\sum_{k=n}^{\infty} u_k \equiv r_n, \tag{5.4.9}$$

so that

$$u_n = r_n - r_{n+1}. \tag{5.4.10}$$

We have then

$$\sum_{k=m}^{n} u_k x^k = \sum_{k=m}^{n} (r_k - r_{k+1}) x^k.$$

Here r_m and r_{n+1} occur only once, while every other r_k occurs twice. Picking out the coefficients, we get

$$\sum_{k=m}^{n} u_k x^k = r_m x^m - \sum_{k=m+1}^{n} r_k (x^{k-1} - x^k) - r_{n+1} x^n. \tag{5.4.11}$$

This is the analogue of (5.3.13). Let $\varepsilon > 0$ be given and choose $N(\varepsilon)$ so large that

$$|r_k| < \varepsilon, \quad N(\varepsilon) < k.$$

If $m > N(\varepsilon)$, formula (5.4.11) shows that

$$\left| \sum_{k=m}^{n} u_k x^k \right| < \varepsilon \left\{ x^m + \sum_{m+1}^{n} (x^{k-1} - x^k) + x^n \right\} = 2\varepsilon x^m \leqq 2\varepsilon.$$

This holds for all n, $n > m$, uniformly with respect to x in $[0, 1]$. We can then let $n \to \infty$ and see that condition (5.4.4) holds for all x in $[0, 1]$ so that (5.4.8) converges uniformly in this interval.

To obtain the result stated under C in the preceding theorem, we set

$$u_k = a_k R^k$$

and replace x by x/R.

The quantity R defined by (5.4.6) is known as the *radius of convergence* of the power series. Now "radius" suggests "circle of convergence" rather than "interval of convergence." This is intentional. In fact, a deeper study of power series calls for the use of complex numbers and in the complex case the region of convergence is a circular disk of radius R. For the time being we restrict ourselves to the real case.

We have defined R by formula (5.4.6) but this is not always the best and the quickest way of finding R. The following elementary theorem often provides a shortcut.

THEOREM 5.4.3. *R is $\leqq 1$ if the sequence $\{a_n\}$ is unbounded, $\geqq 1$ if it is bounded. In the first case $R = 1$ if there is a positive k such that the sequence $\{(n + 1)^{-k} a_n\}$ is bounded, in the second case $R = 1$ if $\{(n + 1)^k a_n\}$ is unbounded for some $k > 0$.*

Proof. The criterion uses the profound difference in the behavior of the terms of a power series for $|x| < R$, $|x| = R$, and $|x| > R$. In the present case we test the behavior for $x = 1$. If $R < 1$, then the sequence $\{a_n\}$ must be unbounded and contain a subsequence which becomes infinite faster than any fixed positive power of n. If $R > 1$, then a_n goes to zero faster than any fixed negative power of n. Hence if $\{a_n\}$ is unbounded then $R \leqq 1$, while $\{a_n\}$ bounded implies $R \geqq 1$. If we can convert the unbounded sequence $\{a_n\}$ to a bounded one by dividing by $(n + 1)^k$, then we must have $R = 1$. Similarly we have $R = 1$ if the bounded sequence $\{a_n\}$ becomes unbounded after multiplication by $(n + 1)^k$. Actually it suffices in the second case that $\{a_n\}$ is not a null sequence.

We shall give two examples involving somewhat unusual coefficients.

Example 1. We take $a_n = (n\pi)$, that is, the distance of $n\pi$ from the nearest integer. This is a bounded sequence and so $R \geqq 1$; it is not a null sequence and so $R = 1$. The latter assertion follows from the fact that if (u) is close to 0, then $(u + \pi)$ is close to $\pi - 3$, that is, away from 0.

Example 2. We take a_n to be the number of prime factors of n. This is a very irregular function of n, which equals 1 when n is a prime while $a_n = k$ if $n = 2^k$. Thus it is an unbounded function of n so that $R \leqq 1$. Since obviously a_n is small in comparison with n, in particular, $\lim a_n/n = 0$, we have $R = 1$.

To obtain more convenient criteria we have merely to go back to the discussion in Section 5.2. We derived formula (5.4.1) by applying the refined form of the root test. We can obviously determine R using the ratio test when it applies. In this manner we obtain

$$\frac{1}{R} = \lim_{n\to\infty} \frac{a_{n+1}}{a_n}, \tag{5.4.12}$$

whenever the limit exists. This test is convenient when the coefficients a_n are rational functions of n. Typical examples are furnished by the series (5.2.7) and (5.2.8) for which the pertinent results on convergence are listed in Theorem 5.2.7.

Example 3. We take $a_n = \log n$ for $n > 1$. Then

$$1 < \frac{a_{n+1}}{a_n} = \frac{\log (n+1)}{\log n} = \frac{\log n + \log \left(1 + \frac{1}{n}\right)}{\log n} < 1 + \frac{1}{n \log n} \to 1,$$

so that $R = 1$. We have used the inequality $\log (1 + u) < u$ for $0 < u$.

Let us now return to the discussion of uniform convergence without special reference to power series. We set

$$F_n(x) = \sum_{k=0}^{n} f_k(x), \qquad F(x) = \sum_{k=0}^{\infty} f_k(x) \tag{5.4.13}$$

for $x \in S$. According to formula (5.4.4) we have uniform convergence of the series (5.4.1) in S, if, given any $\varepsilon > 0$, there exists an $N(\varepsilon)$ such that

$$|F(x) - F_n(x)| < \varepsilon, \quad N(\varepsilon) < n, \tag{5.4.14}$$

for all x in S. By Cauchy's convergence principle this condition holds if and only if

$$|F_m(x) - F_n(x)| < \varepsilon, \quad N_0(\varepsilon) < m, n, \tag{5.4.15}$$

for all x in S.

Condition (5.4.14) admits of a simple geometric interpretation shown in Figure 5.2. Here S is an interval $[a, b]$, and $y = F(x)$ is shown by its graph. Further we have marked a strip of width 2ε centered on $y = F(x)$. If we have uniform convergence, the graphs of all partial sums $F_n(x)$ must fall in this strip as soon as n exceeds a certain $N(\varepsilon)$. One such graph is shown in the figure. If the convergence is nonuniform, the graph of $F_n(x)$ fails to be confined to the strip no matter how large the n. This does not prevent $F_n(x)$ from converging to $F(x)$ as we shall see in a moment.

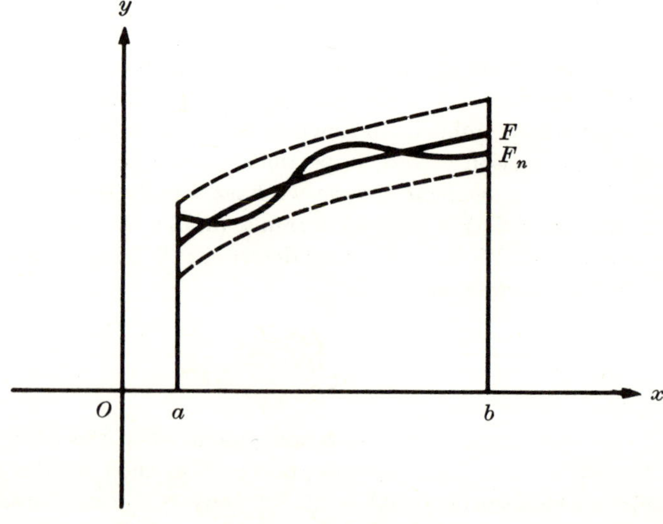

Figure 5.2

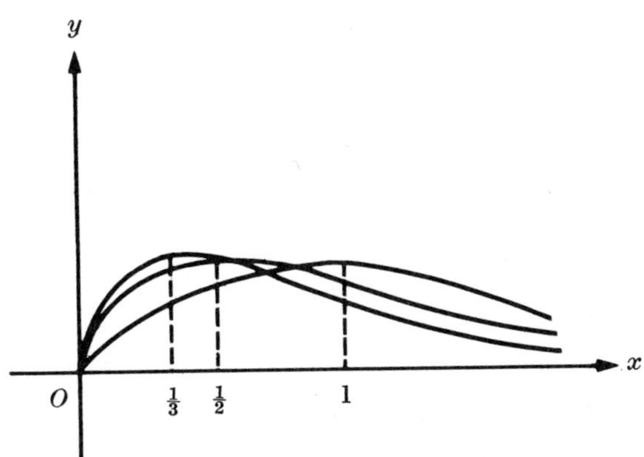

Figure 5.3

The question of uniform convergence does not always arise in connection with an infinite series. We may have a sequence of functions $\{F_n(x)\}$ converging to a limit $F(x)$ in a set S and we ask: does the sequence converge uniformly to its limit? We can answer this at once. Necessary and sufficient conditions are that (5.4.14) or, equivalently, (5.4.15) holds.

Actually there is no essential distinction between the two problems. Given the infinite series (5.4.1), we construct the sequence of partial sums (5.4.13) and translate condition (5.4.4) for uniform convergence of the series into condition (5.4.14) for the sequence of partial sums. Conversely, given a sequence $\{F_n(x)\}$, we can construct an infinite series whose partial sums are the given functions $F_n(x)$. We simply take

$$f_0(x) = F_0(x), \quad f_n(x) = F_n(x) - F_{n-1}(x), \tag{5.4.16}$$

so that the series becomes

$$F_0(x) + \sum_{n=1}^{\infty} [F_n(x) - F_{n-1}(x)]. \tag{5.4.17}$$

Example 4. See Figure 5.3. We take

$$F_n(x) = nxe^{-nx}, \qquad 0 \leqq x. \tag{5.4.18}$$

This sequence converges to the limit $F(x) \equiv 0$ for all x, $0 \leqq x$. The convergence is not uniform in any interval extending to the origin. The figure shows the first three functions $F_n(x)$ as well as the limit function. It is clear that no 2ε-strip about the x-axis can contain the graphs of $F_n(x)$ for all large n, because each graph exhibits a maximum of the same height.

To see what happens let us consider the function

$$F_1(x) = xe^{-x}, \quad 0 \leqq x.$$

Here $F_1(0) = 0$, $F_1(1) = 1/e$, and $F_1(x) \to 0$ as $x \to \infty$. Since

$$F_1'(x) = (1 - x)e^{-x},$$

the slope is positive for $0 \leqq x < 1$, zero for $x = 1$, and negative for $1 < x$. This means that the range of $F_1(x)$ is the interval $[0, 1/e]$ and $1/e$ is the maximum value of $F_1(x)$. Now

$$F_n(x) = F_1(nx)$$

has the same domain and the same range as $F_1(x)$. $F_n(x)$ has the same maximum value $1/e$ for $x = 1/n$. If $x \neq 0$, then $nx \to \infty$ with n so that

$$\lim_{n=\infty} F_n(x) = \lim_{n\to\infty} F_1(nx) = \lim_{t\to\infty} F_1(t) = 0.$$

On the other hand, $F_n(0) = 0$ for all n, so the limit relation holds for all x, $0 \leqq x$. The convergence of $F_n(x)$ to the limit zero is nonuniform in any interval extending to the origin, since each $F_n(x)$ with $n > 1/\delta$ assumes all values belonging to its range in the interval $[0, \delta]$. If we omit this interval and restrict ourselves to $[\delta, \infty]$, the convergence becomes uniform.

The sequence $\{F_n(x)\}$ converges *boundedly* but not uniformly to the limit zero. The emphasis in this statement is on "boundedly." This calls attention to the possibility of a sequence of functions to converge to a limit, zero for instance, without the sequence being uniformly bounded. An example is given by the sequence

$$G_n(x) = n^2xe^{-nx} = nF_n(x), \quad n = 1, 2, \cdots, \tag{5.4.19}$$

which converges to zero for $0 \leqq x$. $G_n(x)$ has the range $[0, n/e]$, and takes on its maximum value n/e at $x = 1/n$. While each function $G_n(x)$ is bounded, there is no common bound for all functions of the sequence. Naturally we do not have uniform convergence in $[0, \infty)$, but we still do in $[\delta, \infty)$ for any fixed positive δ.

EXERCISE 5.4

Find the radius of convergence of the power series $\Sigma\, a_n x^n$ when a_n is given by:

1. n. 2. $\dfrac{1}{n^2+1}$. 3. $2^n + 2^{-n}$. 4. n^2e^{-n}. 5. $\dfrac{(2n)!}{[n!]^2}$.

6. $\dfrac{(3n)!}{(n!)^3}$. 7. $(\log n)^{-n}$. 8. $3^{-\sqrt{n}}$.

9. $a^n + b^n \quad (a > 0, b > 0)$. 10. $(a^n + b^n)^{-1}$.

11. $\dfrac{(-1)^n}{n!}\left(\dfrac{n}{e}\right)^n$ 12. $0, n \neq k^2; \quad 2^{-k}, n = k^2, k = 1, 2, 3, \cdots$.

13. Find which of the power series in the preceding problems converge for $x = R$ or $-R$.

14. If the series $\Sigma\, a_n x^n$ and $\Sigma\, b_n x^n$ have R_1 and R_2 as their radii of convergence, what can be said concerning the corresponding radii for the two series:

$$\text{(i)} \quad \sum (a_n + b_n)x^n; \qquad \text{(ii)} \quad \sum a_n b_n x^n?$$

15. Let p_n be the nth prime ($p_1 = 2$), $\Pi(n)$ the number of primes $\leqq n$. Find the radii of convergence of the series:

$$\text{(i)} \sum_{n=2}^{\infty} \Pi(n)x^n; \qquad \text{(ii)} \sum_{2}^{\infty} \frac{x^n}{\Pi(n)}; \qquad \text{(iii)} \sum_{1}^{\infty} x^{p_n}.$$

16. Let $d(n)$ be the number of divisors of n including 1 and n. Let $D(n) = \sum_{k=1}^{n} d(k)$. Show that $2n - 1 \leqq D(n) \leqq \frac{1}{2}n(n+1)$. Find the radius of convergence of

$$\sum_{n=1}^{\infty} \frac{x^n}{D(n)},$$

and show that the series converges for $x = -R$. [It diverges for $x = R$, but the proof requires a profound study of $d(n)$.]

17. Show that the series in Problem 11, Exercise 5.3 converges uniformly with respect to x for $x \geqq a + \delta$, $\delta > 0$.

18. Show that the series

$$\sum_{n=1}^{\infty} \frac{1}{(x+n)(x+n+1)}$$

converges uniformly in x for $0 \leqq x$. What is its sum?

19. [Weierstrass's M-test.] Prove that the series (5.4.1) converges uniformly in S, if there exists a convergent series with positive terms, ΣM_n, such that $|f_n(x)| \leqq M_n$ for all n and all x in S.

Prove that these series converge uniformly for all x:

20. $\displaystyle\sum_{n=1}^{\infty} \frac{1}{n^2[1 + (nx)^2]}$. 21. $\displaystyle\sum_{n=1}^{\infty} \frac{x}{n[1 + (nx)^2]}$. 22. $\displaystyle\sum_{n=1}^{\infty} \frac{x^2}{1 + (nx)^2}$.

23. Show that the sequence

$$F_n(x) = \frac{nx}{1 + (nx)^2}$$

converges to zero for all x. Is the convergence uniform in any interval containing the origin? Prove bounded convergence.

24. Show that the series

$$\sum_{n=1}^{\infty} (1 - x)x^n$$

converges for $0 \leqq x \leqq 1$ and find its sum. Is the convergence uniform in $[1 - \delta, 1]$?

5.5. Implications of uniform convergence. We shall examine the relations between uniform convergence of a series and continuity, integrability, and differentiability of its sum.

THEOREM 5.5.1. *Let the functions $f_n(x) \in C[a, b]$ and let the series*

$$\sum_{n=0}^{\infty} f_n(x) \equiv F(x) \tag{5.5.1}$$

converge uniformly in $[a, b]$. Then $F(x) \in C[a, b]$.

Proof. For an arbitrary choice of n let us set

$$F(x) = F_n(x) + R_n(x), \tag{5.5.2}$$

where $F_n(x)$ is the nth partial sum and $R_n(x)$ is the remainder after n terms. Then $F_n(x)$ as the sum of a finite number of elements from $C[a, b]$ is also an element of $C[a, b]$, and $R_n(x)$ is uniformly small if n is large. Let $\varepsilon > 0$ be given and choose n so large that

$$|R_n(x)| \leqq \varepsilon$$

for all x in $[a, b]$. We can do this since the series is uniformly convergent. Then for any choice of x_1 and x_2 in $[a, b]$,

$$\begin{aligned} |F(x_1) - F(x_2)| &= |F_n(x_1) - F_n(x_2) + R_n(x_1) - R_n(x_2)| \\ &\leqq |F_n(x_1) - F_n(x_2)| + |R_n(x_1)| + |R_n(x_n)| \\ &\leqq |F_n(x_1) - F_n(x_2)| + 2\varepsilon \end{aligned}$$

by the choice of n. Since $F_n(x)$ is uniformly continuous in $[a, b]$, we can choose a $\delta = \delta(\varepsilon)$ such that

$$|F_n(x_1) - F_n(x_2)| \leqq \varepsilon \quad \text{whenever} \quad |x_1 - x_2| \leqq \delta.$$

Hence the same condition ensures that

$$|F(x_1) - F(x_2)| \leqq 3\varepsilon.$$

Since ε is arbitrary, this means that $F(x) \in C[a, b]$, and completes the proof.

We note that uniform convergence is a sufficient condition for the continuity of the sum function. It is not necessary as shown by the series (5.4.17) with $F_n(x)$ defined by either (5.4.18) or (5.4.19). Necessary and sufficient conditions for the continuity of the sum are known but are beyond our scope.

COROLLARY. *The sum of a power series is continuous in the interval of convergence.*

For if the interval is open, $-R < x < R$, we have uniform convergence in any closed subinterval, hence continuity of the sum in each such interval. This implies continuity in the open interval $(-R, R)$. Note that $F(x)$ need not be bounded in $(-R, R)$ and we have no assurance that $F(x)$ tends to any limit, finite or infinite, as x approaches R or $-R$. If the interval of convergence is $(-R, R]$, then we have uniform convergence and continuity of the sum in any interval of the form $[-R + \delta, R]$. In this case $F(x)$ tends to the finite limit $F(R)$ when $x \to R$.

THEOREM 5.5.2. *Under the assumptions of Theorem* 5.5.1 *and that* $[a, b]$ *is finite, we have*

$$\int_a^x F(s)\,ds = \sum_{n=0}^{\infty} \int_a^x f_n(s)\,ds,\ a \leqq x \leqq b. \tag{5.5.3}$$

Proof. We have seen that $F(s)$ is continuous. This means that all integrals involved in the formula have a sense, but we have to prove convergence of the series on the right and that its sum is the integral on the left. With the notation of (5.5.2) we have

$$\int_a^x F(x)\,ds = \int_a^x F_n(s)\,ds + \int_a^x R_n(s)\,ds.$$

Suppose that n is so large that $|R_n(s)| \leqq \varepsilon$ for $a \leqq s \leqq b$. Then by the mean value theorem

$$\left| \int_a^x F(s)\,ds - \int_a^x F_n(s)\,ds \right| = \left| \int_a^x R_n(s)\,ds \right| \leqq \varepsilon(b - a).$$

This shows that

$$\int_a^x F(s)\,ds = \lim_{n\to\infty} \int_a^x F_n(s)\,ds = \lim_{n\to\infty} \sum_{k=0}^{n} \int_a^x f_k(s)\,ds,$$

which is the assertion of (5.5.3).

The last theorem may be restated as follows:

A uniformly convergent series of continuous functions may be integrated term-by-term (*or termwise*) *over any finite interval and the sum of the integrals is the integral of the sum.*

Uniform convergence is a sufficient but not a necessary condition for termwise integrability. *Bounded convergence* will suffice if the terms and the sum are continuous.

COROLLARY. *For any* x *in the interval of convergence*

$$\int_0^x \left\{ \sum_{n=0}^{\infty} a_n s^n \right\} ds = \sum_{n=0}^{\infty} \frac{a_n}{n+1}\, x^{n+1}. \tag{5.5.4}$$

This follows from the uniform convergence and Theorem 5.5.2.
Let us apply this to the geometric series

$$\frac{1}{1-x} = \sum_{n=0}^{\infty} x^n.$$

Here

$$\int_0^x \frac{ds}{1-s} = \sum_{n=1}^{\infty} \frac{x^n}{n}, \quad -1 \leqq x < 1.$$

Now the left member is a logarithm so we have

$$\log \frac{1}{1-x} = \sum_{n=1}^{\infty} \frac{x^n}{n}, \quad -1 \leqq x < 1. \tag{5.5.5}$$

For $x = -1$ we get

$$\log 2 = \sum_{n=1}^{\infty} \frac{(-1)^{n-1}}{n}. \tag{5.5.6}$$

This series has figured in many examples, starting with formula (1.7.13). Here now is the sum.

Termwise differentiation can also be based on uniform convergence.

THEOREM 5.5.3. *Suppose that each function $f_n(x)$ has a continuous derivative in $[a, b]$. Suppose that* (i) $\Sigma f_n(a)$ *converges, and* (ii) $\Sigma f_n'(x)$ *converges uniformly in $[a, b]$. Then*

$$\sum_{k=0}^{\infty} f_k(x) \equiv F(x) \tag{5.5.7}$$

converges uniformly in $[a, b]$, $F(x)$ is differentiable, and

$$F'(x) = \sum_{k=0}^{\infty} f_k'(x). \tag{5.5.8}$$

Proof. Let us denote the right member of (5.5.8) by $G(x)$. By the preceding theorem the series can be integrated term-by-term and

$$\int_a^x G(s)\,ds = \sum_{k=0}^{\infty} \int_a^x f_k'(s)\,ds = \sum_{k=0}^{\infty} [f_k(x) - f_k(a)]$$

$$= \sum_{k=0}^{\infty} f_k(x) - \sum_{k=0}^{\infty} f_k(a) = F(x) - F(a). \tag{5.5.9}$$

Theorem 5.5.2 shows that the series in the third member converges. The second series in the fourth member is convergent by assumption whence it follows that the first series, which defines $F(x)$, must also converge. Actually it converges uniformly in x for if in the four series we extend the summation from a large integer n to infinity, all four series can be made small uniformly in x. This follows from the uniform convergence of the series in the right member of (5.5.8) and the convergence of the series defining $F(a)$. The first and the last members of (5.5.9) show that $G(s)$ is continuous so the integral is differentiable with respect to its upper limit and the derivative equals $G(x)$. Thus $F(x)$ is differentiable and $G(x)$ is its derivative. This proves (5.5.8) and completes the proof of the theorem.

The series (5.5.8) is often called the *derived series* of (5.5.7). The derived series of a power series is again a power series. As such it has an interval of convergence, not necessarily the same as that of the given power series. Actually the difference between the two intervals is at most two points as shown by the next proposition.

THEOREM 5.5.4. *A power series and its derived series have the same radius of convergence.*

Proof. Let the radii of the two series

$$F(x) \equiv \sum_{n=0}^{\infty} a_n x^n, \qquad F_1(x) \equiv \sum_{n=1}^{\infty} na_n\, x^{n-1}$$

be R and R_1 respectively. We have

$$R_1 \leqq R,$$

since

$$n\,|a_n| \geqq |a_n|.$$

On the other hand

$$\frac{1}{R_1} = \limsup_{n\to\infty} |na_n|^{1/n} \leq \limsup_{n\to\infty} n^{1/n} \limsup_{n\to\infty} |a_n|^{1/n} = 1 \cdot \frac{1}{R},$$

so that $R \leqq R_1$. Since the opposite inequality has already been proved, we have $R_1 = R$ as asserted.

THEOREM 5.5.5. *The sum $F(x)$ of a power series is a differentiable function of x for $|x| < R$ and*

$$F'(x) = \sum_{n=1}^{\infty} na_n x^{n-1}. \tag{5.5.10}$$

For the series in the right member converges uniformly in $|x| \leqq R(1 - \delta)$, $0 < \delta < 1$, so that Theorem 5.5.3 applies.

The next result is really a corollary of Theorem 5.5.5, but its fundamental importance makes it desirable to state it as

THEOREM 5.5.6. *The function defined by a convergent power series has derivatives of all orders and*

$$F^{(p)}(x) = \sum_{n=0}^{\infty} (n + p)(n + p - 1) \cdots (n + 1)a_{n+p}x^n. \tag{5.5.11}$$

All these series have the same radius of convergence.

Let us set $x = 0$ in (5.5.11). This gives

$$F^{(p)}(0) = p!\, a_p, \tag{5.5.12}$$

that is, the coefficients of the power series are expressible in terms of the derivatives of the sum of the series at $x = 0$. This gives

THEOREM 5.5.7. *If $F(x)$ is the sum of a power series convergent for $|x| < R$, then we have also*

$$F(x) = \sum_{n=0}^{\infty} \frac{F^{(n)}(0)}{n!}\, x^n. \tag{5.5.13}$$

A series of this type is known as a Maclaurin series, a special case of Taylor's series [after Colin Maclaurin (1698–1746) and Brook Taylor (1685–1731)]. The theorem asserts that *a power series is its own Maclaurin series*. Such series form the main topic of the next chapter.

EXERCISE 5.5

1. Check formula (5.5.12) for the logarithmic series (5.5.5).
2. Prove that

$$\tfrac{1}{2}\log\frac{1+x}{1-x} = \frac{x}{1} + \frac{x^3}{3} + \frac{x^5}{5} + \cdots, \quad -1 < x < 1.$$

3. Correct to ten decimal places $\log 1000 = 6.90775\,52790$. How would you choose x in the preceding formula in order to obtain $\log 1001$ with the same accuracy? How many terms are needed of the series?
4. Sum the series

$$\sum_{n=0}^{\infty} (n+p)(n+p-1)\cdots(n+1)x^n, \qquad -1 < x < 1, \qquad p \text{ positive integer.}$$

5. Sum the derived series of

$$\sum_{n=0}^{\infty} (-1)^n \frac{x^{2n+1}}{2n+1}$$

in closed form. When does the series converge?
6. Verify that the series

$$E(x) = \sum_{n=0}^{\infty} \frac{x^n}{n!} \qquad (0! = 1)$$

converges for all x and that $E(x)$ satisfies the differential equation

$$y' = y.$$

7. Verify that the series

$$C(x) = \sum_{n=0}^{\infty} (-1)^n \frac{x^{2n}}{(2n)!}, \qquad S(x) = \sum_{n=0}^{\infty} (-1)^n \frac{x^{2n+1}}{(2n+1)!}$$

converge for all x and satisfy the equations

$$S'(x) = C(x), \quad C'(x) = -S(x), \quad S''(x) + S(x) = 0, \quad C''(x) + C(x) = 0.$$

8. If $F_n(x)$ is defined by (5.4.18), compare

$$\int_0^{\infty} \{\lim F_n(s)\}\, ds \quad \text{with} \quad \lim \int_0^{\infty} F_n(s)\, ds.$$

9. Same question for the functions $G_n(x)$ defined by (5.4.19).

MISCELLANEOUS PROBLEMS FOR CHAPTER 5

1. If $\{a_n\}$ is a positive null sequence, show that the series

$$\sum_{n=1}^{\infty} \log(1 + a_n) \quad \text{and} \quad \sum_{n=1}^{\infty} a_n$$

are equiconvergent.

2. Under the same assumptions on $\{a_n\}$ show that the series

$$\sum_{n=1}^{\infty} \int_0^{a_n} \frac{dx}{1 + x^2} \quad \text{and} \quad \sum a_n$$

are equiconvergent.

3. If $\{b_n\}$ is a strictly increasing sequence tending to $+\infty$, what further conditions on the rate of growth of b_n will ensure that the following series be convergent:

$$\sum_{n=1}^{\infty} \int_{b_n}^{\infty} e^{-t^2}\, dt.$$

4. If u_n is the oscillation of $\log x$ in the interval $[a_n, b_n]$ where $0 < a_n < b_n < a_{n+1}$ and $a_n \to \infty$, what further conditions will ensure that $\Sigma\, u_n$ be convergent?

5. The equation

$$x^n + nx - 1 = 0$$

has a unique positive root x_n which tends to 0 when $n \to \infty$. For what values of α does the series

$$\sum_{n=1}^{\infty} x_n^{\alpha} \text{ converge?}$$

6. Find the radius of convergence of

$$\sum_{n=1}^{\infty} \left\{1 + \tfrac{1}{2} + \tfrac{1}{3} + \cdots + \frac{1}{n}\right\} x^n.$$

7. If α is a real number but not a positive integer, find the radius of convergence of

$$\sum_{n=1}^{\infty} \left\{\prod_{k=1}^{n} \left(1 - \frac{\alpha}{k}\right)^k\right\} x^n.$$

8. Under the assumptions of Problem 14, Exercise 5.4, find the radius of convergence of the series

$$\sum_{n=0}^{\infty} (a_0 b_n + a_1 b_{n-1} + \cdots + a_n b_0)\, x^n.$$

9. For what values of x is the so-called *Lambert series*

$$\sum_{n=1}^{\infty} \frac{1}{n} \frac{x^n}{1 - x^n}$$

uniformly convergent?

10. Let x be any positive number and define a sequence $\{x_n\}$ by the recursion formula

$$x_1 = (2 + x)^{\frac{1}{2}}, \quad x_{n+1} = (2 + x_n)^{\frac{1}{2}}, \quad n > 1.$$

Prove the convergence of the series

$$\sum_{n=1}^{\infty} (2 - x_n).$$

*11. Let p and q be positive real numbers, $1 \leqq p$, set $\alpha_n = [n^{1/p}]$ with $[u]$ equal to the largest integer $\leqq u$, and find what inequality must hold between p and q in order that the following series be convergent:

$$\sum_{n=1}^{\infty} (-1)^{\alpha_n} n^{-1/q}.$$

12. Prove that the series

$$\sum_{n-1}^{\infty} \frac{1}{n} \sin n \sin n^2$$

is convergent and moreover that its sum is positive and less than one. [*Hint:* The product of two sines is the difference of two cosines. Group terms!]

13. Let $u_n > 0$ and form

$$\liminf \frac{\log (1/u_n)}{\log n} = \alpha, \quad \limsup \frac{\log (1/u_n)}{\log n} = \beta.$$

Prove that $\Sigma\, u_n$ converges if $\alpha > 1$ and diverges if $\beta < 1$.

14. Let $c_k > 0$, $\Sigma\, c_k$ be convergent and set $R_n = \sum_{n+1}^{\infty} c_k$. Prove that the series

$$\sum_{k=1}^{\infty} c_k (R_{k-1})^{-\alpha}$$

converges if $0 < \alpha < 1$. What happens if $1 \leqq \alpha$? [*Hint:* Summation by parts as in the proof of Theorem 5.4.2 may be useful. An analogous result is found in Problem 25, Exercise 5.2.]

Find the radius of convergence R of the power series $\Sigma\, a_n x^n$, if a_n is given by:

15. $\dbinom{2n}{n}$.

16. $\left(1 + \dfrac{1}{n}\right)^{-n^2}$.

17. $n^{-1-1/n}$.

For what values of x are the following series convergent:

18. $\displaystyle\sum_{n=1}^{\infty} \frac{(-1)^n}{x^2 - n^2}$.

19. $\displaystyle\sum_{n=1}^{\infty} \frac{n^{2x}}{x^n + \log n}$.

20. $\displaystyle\sum_{n=1}^{\infty} (-1)^{n-1} \{e^{nx} - e^{(n-1)x}\}^{-1}$.

21. $\displaystyle\sum_{n=1}^{\infty} \frac{1}{n} \left(2\, \frac{x-3}{x+1}\right)^n$.

Sum the series:

22. $\sum_{n=0}^{\infty} \frac{(-1)^n 2^{-n}}{(n+1)(n+2)}$.

23. $\sum_{n=1}^{\infty} (-1)^{n-1} [e^n - e^{n-1}]^{-1}$.

24. The coefficients of the power series $\sum_{n=1}^{\infty} a_n x^n$ are positive integers, namely, the successive digits in the periodic decimal expansion of $\frac{4}{7}$. Find the radius of convergence of the series and find its sum when it is convergent.

Appendix B

Complex Analysis

B.1. Complex numbers. Algebra and Calculus have much to gain by the introduction of complex numbers. The student knows that a quadratic equation, if restricted to the real field, has either two roots which may coincide or none. Only in the larger field of complex numbers is it true that an algebraic equation of degree n has n roots. This situation has repercussions in the Calculus. Thus the integration of rational functions becomes more natural and systematic if we can use factorization of the denominator in linear factors which is available in the complex field. The behavior of power series is less of a mystery if we use complex variables. There are interesting relations between trigonometric functions and the exponential function which can be formulated only in terms of complex numbers, and so on. In this Appendix we shall develop enough complex analysis to serve our needs. We start with complex numbers.

We can introduce complex numbers as a field of number pairs (a, b) in which the arithmetic operations are suitably defined. This leads immediately to a geometric interpretation of complex numbers as vectors in the plane, the so-called *complex plane*. Addition of complex numbers is vector addition but multiplication has nothing to do with the traditional dot or cross products of vector analysis.

For the postulational approach we consider a set C of elements (a, b), where a and b are real numbers, a definition of equality and of two operations called addition and multiplication.

DEFINITION B.1.1. *We have* $(a, b) = (c, d)$ *if and only if* $a = c$, $b = d$. *We set*

$$(a_1, b_1) + (a_2, b_2) = (a_1 + a_2, b_1 + b_2), \tag{B.1.1}$$

$$(a_1, b_1) \cdot (a_2, b_2) = (a_1 a_2 - b_1 b_2, a_1 b_2 + a_2 b_1). \tag{B.1.2}$$

THEOREM B.1.1. *The set* C *is a field with* $(0, 0)$ *as zero element and* $(1, 0)$ *as unit element.*

Proof. We have to verify that the eleven field postulates of Section 1.3 are satisfied. Addition is defined for all elements of C and is clearly commutative and associative since addition of reals have these properties. This takes care of A_1 to A_3. To prove A_4 we note that

$$(a_1, b_1) + (a_3, b_3) = (a_2, b_2) + (a_3, b_3)$$

implies that

$$(a_1 + a_3, b_1 + b_3) = (a_2 + a_3, b_2 + b_3),$$

or

$$a_1 + a_3 = a_2 + a_3, b_1 + b_3 = b_2 + b_3.$$

By the cancellation property for reals we have

$$a_1 = a_2, b_1 = b_2 \quad \text{and} \quad (a_1, b_1) = (a_2, b_2)$$

as required.

Clearly we can solve

$$(a_1, b_1) + (x, y) = (a_2, b_2)$$

uniquely and the result is

$$(x, y) = (a_2 - a_1, b_2 - b_1).$$

This also shows that $(0, 0)$ acts as zero element. This completes the discussion of addition.

Multiplication is more laborious. Obviously M_1 and M_2 hold. For M_3 we note that

$$\begin{aligned}&(a_1a_2 - b_1b_2, a_1b_2 + a_2b_1) \cdot (a_3, b_3)\\ &= (a_1a_2a_3 - a_1b_2b_3 - a_2b_3b_1 - a_3b_1b_2, a_1a_2b_3 + a_2a_3b_1 + a_3a_1b_2 - b_1b_2b_3)\\ &= (a_1b_1) \cdot (a_2a_3 - b_2b_3, a_2b_3 + a_3b_2).\end{aligned}$$

Suppose that $(a_3, b_3) \neq (0, 0)$, and that

$$(a_1, b_1) \cdot (a_3, b_3) = (a_2, b_2) \cdot (a_3, b_3).$$

Then

$$a_1a_3 - b_1b_3 = a_2b_3 - b_2a_3,$$
$$a_1b_3 + a_3b_1 = a_2b_3 + a_3b_2.$$

or

$$(a_1 - a_2)a_3 - (b_1 - b_2)b_3 = 0,$$
$$(b_1 - b_2)a_3 + (a_1 - a_2)b_3 = 0.$$

Since $a_3{}^2 + b_3{}^2 \neq 0$, this requires that

$$a_1 - a_2 = 0, b_1 - b_2 = 0, \quad \text{and} \quad (a_1, b_1) = (a_2, b_2).$$

This proves M_4.

It is simple to verify that $(1, 0)$ acts as the unit element. It is also easy to verify M_5, for the equation

$$(a, b) \cdot (x, y) = (c, d), \quad a^2 + b^2 \neq 0,$$

is satisfied by

$$x = \frac{ac + bd}{a^2 + b^2}, \qquad y = \frac{ad - bc}{a^2 + b^2}.$$

Finally we have the distributive law. For

$$\begin{aligned}&(a_1, b_1) \cdot [(a_2, b_2) + (a_3, b_3)] = (a_1, b_1) \cdot (a_2 + a_3, b_2 + b_3)\\ &= (a_1a_2 + a_1a_3 - b_1b_2 - b_1b_3, a_1b_2 + a_1b_3 + a_2b_1 + a_3b_1)\\ &= (a_1a_2 - b_1b_2, a_1b_2 + a_2b_1) + (a_1a_3 - b_1b_3, a_1b_3 + a_3b_1)\\ &= (a_1, b_1) \cdot (a_2, b_2) + (a_1, b_1) \cdot (a_3, b_3)\end{aligned}$$

as asserted. This proves that C is a field, the *field of complex numbers*.

To make the identification of the symbols (a, b) which are the elements of C with the complex numbers familiar to the student, we need only a few steps. We start by defining *scalar multiplication* in C.

DEFINITION B.1.2. *If $(a, b) \in C$ and if c is a real number, then*

$$c(a, b) = (ca, cb). \tag{B.1.3}$$

The most important consequence of this convention is that

$$(a, b) = (a, 0) + (0, b)$$

can be written

$$(a, b) = a(1, 0) + b(0, 1). \tag{B.1.4}$$

Here $(1, 0)$ is the unit element of C, while the square of $(0, 1)$ is the negative of the unit element:

$$(0, 1) \cdot (0, 1) = (-1, 0) = -(1, 0). \tag{B.1.5}$$

To complete the identification we observe that the elements $\{(a, 0)\}$ form a closed set under addition and multiplication. In fact, the set contains the zero and the unit elements, $(0, 0)$ and $(1, 0)$, and

$$(a_1, 0) + (a_2, 0) = (a_1 + a_2, 0),$$
$$(a_1, 0) \cdot (a_2, 0) = (a_1 a_2, 0).$$

It follows that these elements form a subfield R_0 of C. This field can be ordered under the rule

$$(a_1, 0) < (a_2, 0) \quad \text{if and only if } a_1 < a_2.$$

The so-ordered field R_0 is *isomorphic* with the field R of real numbers. The correspondence

$$(a, 0) \leftrightarrow a$$

defines a one-to-one mapping which preserves addition, multiplication, and order. This is what we mean by an *isomorphism*. We do not distinguish between isomorphic fields, and identify R_0 with R.

As a consequence we write 1 for $(1, 0)$ and for $(0, 1)$ we introduce the symbol i so that

$$(1, 0) = 1, \quad (0, 1) = i. \tag{B.1.6}$$

With this notation (B.1.4) becomes

$$(a, b) = a + bi, \tag{B.1.7}$$

which completes the identification.

The real subfield R can be ordered in such a way that the order postulates of Section 1.4 are satisfied. This is not the case with the field C as a whole. Suppose on the contrary that such an order does exist and let $a + bi$ be a non-real complex number. Then we have either

$$a + bi > a - bi \quad \text{or} \quad a + bi < a - bi$$

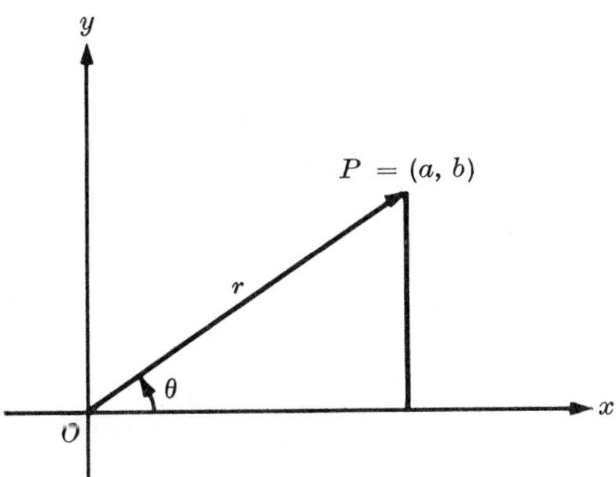

Figure B.1

by O_1. Using O_2 we see that either

$$2bi > 0 \quad \text{or} \quad 2bi < 0.$$

By O_3 this reduces to the alternatives

$$0 < i \quad \text{or} \quad 0 < -i.$$

Since by O_3 the square of a positive number is positive, both alternatives lead to the conclusion that

$$0 < -1$$

which is a contradiction. It follows that C cannot be ordered.

We come now to the geometric and vectorial representation of complex numbers. See Figure B.1. The point $P = (a, b)$ of the plane as well as the vector $\overrightarrow{OP}$ corresponds to the element $(a, b) = a + bi$. Either of these objects can be taken as the geometric representative of $a + bi$. Either representation gives a one-to-one correspondence between complex numbers and points or vectors. In particular, the unit element corresponds to the point $(1, 0)$ on the x-axis, and the element i corresponds to $(0, 1)$ on the y-axis.

In this connection we refer to the plane of representation as the *complex plane*; the x- and y-axes are known as the *real* and the *imaginary axes* respectively. Real numbers correspond to points on the real axis, while purely imaginary numbers (of the form bi) go into points on the imaginary axis.

The point P with Cartesian coordinates (a, b) has polar coordinates $[r, \theta]$. Here (see Section A.6)

$$a = r \cos \theta, \quad b = r \sin \theta \tag{B.1.8}$$

and

$$r = +\,(a^2 + b^2)^{\frac{1}{2}}. \tag{B.1.9}$$

We call r the *absolute value* and θ the *argument* of the complex number $a + ib$ and write

$$r = |a + ib|, \quad \theta = \arg\,(a + ib). \tag{B.1.10}$$

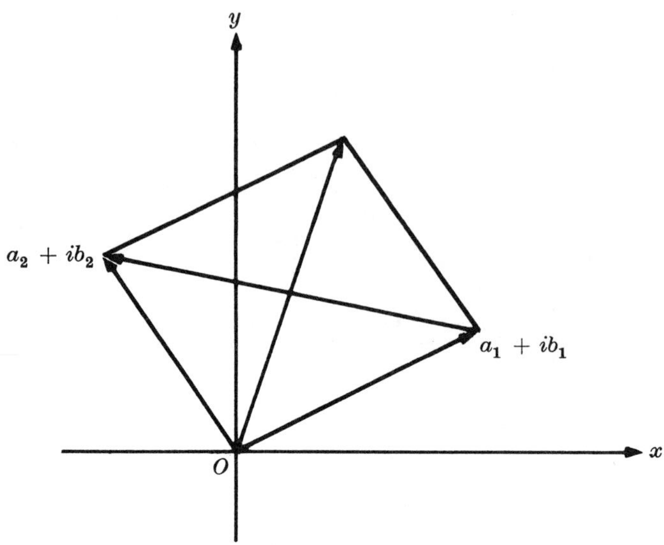

Figure B.2

Here r is a uniquely determined nonnegative real number which is zero if and only if $a + ib = 0$. The argument θ is determined up to an additive multiple of 2π when $r > 0$. One and only one of the admissible values of the argument lies in the half-open interval $[0, 2\pi)$ and is called the *principal determination* or value of the argument.

Geometrically speaking, *addition of complex numbers is vector addition.* See Figure B.2 which illustrates formula (B.1.1). The two given numbers $a_1 + ib_1$ and $a_2 + ib_2$ determine two sides of the parallelogram; their sum is the diagonal from the origin or, in the language of statics, *the sum is the resultant of the two given forces.* The other diagonal is the difference of the given numbers.

To interpret formula (B.1.2) geometrically, we use polar coordinates. It is convenient to denote complex numbers by a single letter. We shall write $x + iy = z$ and refer to x as the *real part* and y (not iy) as the *imaginary part.* Suppose then that the two factors are

$$z_1 = x_1 + iy_1 = r_1 (\cos \theta_1 + i \sin \theta_1),$$

$$z_2 = x_2 + iy_2 = r_2 (\cos \theta_2 + i \sin \theta_2).$$

By (B.1.2) the real part of the product equals

$$x_1x_2 - y_1y_2 = r_1r_2 (\cos \theta_1 \cos \theta_2 - \sin \theta_1 \sin \theta_2) = r_1r_2 \cos (\theta_1 + \theta_2),$$

while the imaginary part is

$$x_1y_2 + x_2y_1 = r_1r_2 (\cos \theta_1 \sin \theta_2 + \cos \theta_2 \sin \theta_1) = r_1r_2 \sin (\theta_1 + \theta_2).$$

This gives

$$z_1z_2 = r_1r_2[\cos (\theta_1 + \theta_2) + i \sin (\theta_1 + \theta_2)] \tag{B.1.11}$$

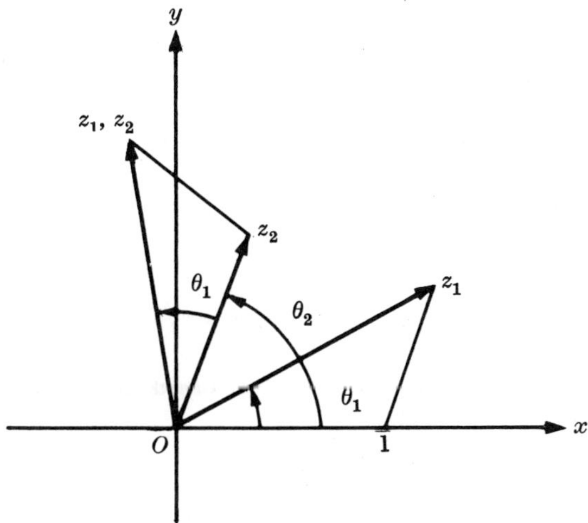

Figure B.3

so that

$$|z_1 z_2| = |z_1|\,|z_2|, \tag{B.1.12}$$

$$\arg(z_1 z_2) \equiv \arg z_1 + \arg z_2 \pmod{2\pi}. \tag{B.1.13}$$

The first relation states that:

The absolute value of a product is the product of the absolute values of the factors.

This extends to any finite number of factors.

(B.1.13) is not an equation but a particular type of an equivalence relation known as a *congruence*. The sign "$\equiv$" is here read "is congruent to" and what is inside the parentheses is read "modulo 2π." The meaning of the formula is the following. If for $\arg(z_1 z_2)$, $\arg z_1$, and $\arg z_2$ we take any of their admissible values, then the difference between the left and the right members is an integral multiple of 2π. Thus

$$(\sqrt{3} + 1) + i(\sqrt{3} - 1) = (1 - i)(1 + i\sqrt{3})$$

and the principal values of the three arguments are $\frac{\pi}{12}$, $\frac{7\pi}{4}$, and $\frac{\pi}{3}$ respectively. Here

$$\frac{\pi}{12} - \left(\frac{7\pi}{4} + \frac{\pi}{3}\right) = -2\pi.$$

Formula (B.1.13) suggests a construction of the product of two complex numbers, shown in Figure B.3. It involves two steps which are combined in the figure. The first is to construct a line segment whose length is the product of the lengths of two given line segments. This requires knowledge of the unit of length. Once it is known, the desired length is obtained from the proportion

$$1 : r_1 = r_2 : r_1 r_2.$$

The second step calls for the construction of an angle which is the sum of two given angles. In the figure the two triangles are similar and the desired results follow from the similarity.

Formula (B.1.13) generalizes in an obvious manner to the case of more than two factors. The following special case is of interest. We take n equal factors of absolute value 1 and argument θ. We then get

$$(\cos\theta + i\sin\theta)^n = \cos n\theta + i\sin n\theta, \tag{B.1.14}$$

which is known as the *theorem of de Moivre* after Abraham de Moivre (1667–1754).

The complex numbers $x + iy$ and $x - iy$ are said to be *conjugate*. The conjugate of z is denoted by $\bar{z}$. We observe that:

The conjugate of a sum is the sum of the conjugates of the summands. The conjugate of a product is the product of the conjugates of the factors.

We also note that

$$\arg\bar{z} \equiv -\arg z \pmod{2\pi}. \tag{B.1.15}$$

This gives

$$(\cos\theta + i\sin\theta)^{-1} = \cos\theta - i\sin\theta,$$

whence it follows that (B.1.14) is valid also if n is a negative integer.

EXERCISE B.1

Find the real and the imaginary parts of:

1. $(a + ib)^2$. 2. $(a + ib)^3$. 3. $(a - ib)^{-1}$. 4. $(2 + i\sqrt{2})^6$.

Find the absolute value and the argument of:

5. $(\sqrt{3} - i)^{14}$. 6. $\dfrac{1 + \cos\theta + i\sin\theta}{1 + \cos\theta - i\sin\theta}$.

7. How should θ be chosen so that the six points

$$z_p = (\cos\theta + i\sin\theta)^p, \quad p = 0, 1, 2, 3, 4, 5$$

form the vertices of a regular hexagon? Is the choice unique?

8. Find $z_0 + z_1 + \cdots + z_5$ if θ is so chosen.

9. Prove (B.1.14) by induction.

10. Show that $|\bar{z}| = |z|$ and $|z|^2 = z\bar{z}$.

11. If z_0 is a fixed point in the complex plane and r is a given positive number, find the locus of the points z such that

$$|z - z_0| = r.$$

12. Verify the properties of conjugates stated at the end of the section.

13. Find a number $x + iy$ whose square equals a given number $a + ib = r(\cos\theta + i\sin\theta)$. How many such numbers are there?

14. If a is fixed, $a > 0$, find the locus of the point z such that $|z^2 - a^2| = a^2$. Compare Exercise 3.6, Problem 8.

B.2. Complex variables. The geometric representation of complex numbers by points in a plane also defines a distance between such numbers as the Euclidean distance between their representative points:

$$d(z_1, z_2) = |z_1 - z_2|. \tag{B.2.1}$$

Once we have defined a distance between complex numbers, we can also define limits, cluster points, convergence, and so on.

Let S be a set of complex numbers. We say that S is *bounded* if

$$\sup_{z \in S} |z| < \infty.$$

This means that the corresponding point set in the complex plane is bounded, that is, there exists a circular disk which covers the set. Instead of a disk we can obviously use a rectangle or a square. An ε-neighborhood of z_0 is the set of points z such that

$$|z - z_0| < \varepsilon, \tag{B.2.2}$$

that is, an open circular disk with center at z_0 and radius ε. A point z_0 is a *cluster point* of the set S if every ε-neighborhood of z_0 contains infinitely many elements of S. The Bolzano-Weierstrass theorem is valid:

THEOREM B.2.1. *If S is an infinite bounded set of complex numbers, then S has at least one cluster point.*

Proof. We proceed as in the proof of Theorem 2.1.1. Since the set is bounded, it may be placed in a square

$$-a < x, y < a.$$

We divide the square into four equal subsquares. Since the set is infinite, at least one of the subsquares contains infinitely many elements of S in its interior or on the boundary. This subsquare is divided into four equal parts and one is chosen whose closure contains infinitely many elements of S. This procedure can be continued indefinitely and gives a sequence of nested closed squares

$$Q_0 \supset Q_1 \supset Q_2 \supset \cdots \supset Q_n \supset \cdots,$$

each containing infinitely many elements of S. Here Q_0 is the original square of side $2a$. The side of Q_n is $2^{1-n}a$. There is one and only one point z_0 common to all these squares and Q_n belongs to an ε-neighborhood of z_0 as soon as $2^{\frac{3}{2}-n}a < \varepsilon$. It follows that z_0 is a cluster point of S.

Let $\{z_n\}$ be a sequence of complex numbers. Then z_0 is a *cluster point* of the sequence if, given any $\varepsilon > 0$, we have

$$|z_n - z_0| < \varepsilon \tag{B.2.3}$$

for infinitely many values of n. If the inequality holds for all large values of n, no matter how small the ε, then z_0 is the *limit* of z_n and we write

$$z_0 = \lim_{n \to \infty} z_n. \tag{B.2.4}$$

We say that *the sequence converges to* z_0. A bounded sequence has at least one cluster point and Cauchy's Convergence Principle gives the condition that there be only one such point so the sequence has a limit. The proof of Theorem 2.1.4 is easily modified to apply to the complex case.

The student will recall that the theory of infinite series runs parallel to that of infinite sequences: any result in one of these theories carries over to the other. Let $\{w_n\}$ be a sequence of complex numbers, set

$$W_n = w_0 + w_1 + \cdots + w_n, \tag{B.2.5}$$

and consider the infinite series

$$\sum_{n=0}^{\infty} w_n. \tag{B.2.6}$$

This series converges, by definition, if and only if the sequence of its partial sums $\{W_n\}$ is convergent. Set

$$w_n = u_n + iv_n \tag{B.2.7}$$

where u_n and v_n are real numbers.

THEOREM B.2.2. *The series* (B.2.6) *converges if and only if the series*

$$\sum_{n=0}^{\infty} u_n \quad \textit{and} \quad \sum_{n=0}^{\infty} v_n \tag{B.2.8}$$

are convergent.

Proof. If U_n and V_n are the nth partial sums of the series (B.2.8), then

$$|W_{n+p} - W_n|^2 = (U_{n+p} - U_n)^2 + (V_{n+p} - V_n)_2.$$

It follows that the left member is small for large values of n if and only if each of the expressions in parentheses on the right is small. This implies the assertion of the theorem.

We say that the series (B.2.6) is absolutely convergent if

$$\sum_{n=0}^{\infty} |w_n|$$

converges. The proof of Theorem 5.1.3 obviously applies to the complex case, so we can assert that an absolutely convergent series is convergent. In Section 5.3 we noted without proof, that an absolutely convergent series with real terms can be rearranged *ad lib* without affecting the absolute convergence or the sum of the series. This also applies to series with complex terms. Similarly we can extend the proof of Theorem 5.3.5 to complex-valued series and obtain that the Cauchy product of two absolutely convergent series with complex terms is absolutely convergent. Again the sum of the product series is the product of the sums of the factors.

The most important special case is that of power series. Here we find that the series

$$\sum_{n=0}^{\infty} a_n z^n \tag{B.2.9}$$

is absolutely convergent for $|z| < R$ where

$$\frac{1}{R} = \limsup_{n\to\infty} |a_n|^{1/n}.$$

The proof of Theorem 5.4.1 applies to the complex-valued case without change. Here the power series converges absolutely in a circular disk of radius R in the complex plane. This explains the name "radius of convergence" introduced in Section 5.4.

Formula (B.2.9) defines a function of the complex variable z in the disk $|z| < R$. It is a very special class of functions, but in a certain sense the most important one. Let us briefly survey more general cases. The definitions of Section 2.2 apply to the case where X and Y are the space C, that is, the field of complex numbers with the Euclidean metric defined by formula (B.2.1). Thus f is a mapping of a set D onto a set S, both in the complex plane.

For the sake of simplicity we shall assume that D, the domain of definition of f, is a *convex domain*. By this we understand the following:

(1) if $z_0 \in D$, then there is an ε-neighborhood of z_0 which is in D, and

(2) if z_0 and z_1 are in D, then every point of the line segment joining z_0 and z_1 is also in D.

If D is a domain of this nature, if $z_0 \in D$, and if $f(z)$ is defined in D, we say that f is *continuous* at z_0 if, given any $\varepsilon > 0$, there exists a $\delta = \delta(\varepsilon)$ such that the disk $|z - z_0| < \delta$ is in D and for all such values of z

$$|f(z) - f(z_0)| < \varepsilon. \tag{B.2.10}$$

The function f is continuous in D, if it is continuous at all points of D. This convention is analogous to the definitions of Section 2.4. Just as in the real case, the functions which are continuous at a point of D or in D form algebras.

We say that f is *differentiable at* $z = z_0$, if there is a complex number α such that for any given $\varepsilon > 0$

$$\left|\frac{1}{h}[f(z_0 + h) - f(z_0)] - \alpha\right| < \varepsilon \tag{B.2.11}$$

as soon as $|h| < \delta(\varepsilon)$. We call α the *derivative* of $f(z)$ at $z = z_0$ and write $\alpha = f'(z_0)$. If $f(z)$ has a derivative at all points of D, then $f(z)$ is *differentiable* in D. In formula (B.2.11) h can be any complex number such that $|h| < \delta(\varepsilon)$. We can also write

$$\lim_{h\to 0} \frac{1}{h}[f(z_0 + h) - f(z_0)] = f'(z_0), \tag{B.2.12}$$

where the limit is independent of how h tends to zero.

The functions which are differentiable in a given domain form an algebra. The proof goes as in the real case. In particular, all the formulas of Sections 3.2 and 3.3 extend to the complex-valued case.

We now have no difficulty in finding differentiable functions of a complex variable. Clearly all polynomials have this property and so have rational functions, provided we exclude the zeros of the denominators. These functions actually have derivatives of all orders and not merely first-order derivatives. We shall give other classes of differentiable functions, but let us first give an example of a continuous function of a complex variable which is differentiable nowhere.

The absolute value of z is such a function. This is clear at the origin since

$$\frac{|h|}{h}$$

does not have a limit as $h \to 0$. There is a limit if h tends radially to zero, that is, arg h is kept constant. If this constant is γ, then the limit is

$$\cos \gamma - i \sin \gamma$$

so that it depends upon γ. At a point $z_0 = x_0 + iy_0 \neq 0$ the corresponding limit is instead

$$(x_0^2 + y_0^2)^{-\frac{1}{2}}(x_0 \cos \gamma + y_0 \sin \gamma)(\cos \gamma - i \sin \gamma). \tag{B.2.13}$$

We return to differentiable functions. Judging by Theorem 5.5.6 we would expect a function which is defined by a power series in a circular disk, to have derivatives of all orders. The proof for the real case depends upon four facts: (i) The derived power series has the same radius of convergence R as the given series. (ii) Any power series is uniformly convergent for $|x| \leqq R - \varepsilon$, $0 < \varepsilon$, ε fixed. (iii) A uniformly convergent series of continuous functions can be integrated termwise and the sum of the integrals is the integral of the sum. (iv) The definite integral of a continuous function is a differentiable function of its upper limit of integration. Here properties (i) and (ii) involve absolute convergence and are in no way affected if we replace real coefficients by complex ones and the real variable x by the complex variable z. Properties (iii) and (iv) are in a different category for they presuppose a process of integration of which we so far have no analogue. This will be the next item on our agenda.

DEFINITION B.2.1. *Let $f(z) = U(z) + iV(z)$ be continuous in the convex domain D and let a and b be points of D. Then*

$$\int_a^b f(z)\,dz \equiv (b - a)\left\{\int_0^1 U(a + t(b - a))\,dt + i\int_0^1 V(a + t(b - a))\,dt\right\}. \tag{B.2.14}$$

The two integrals on the right involve continuous functions of the real variable t taken over the real interval $[0, 1]$ so that the right member is well-defined. This notion of integration has all the properties listed in Section 4.2. In particular,

$$\left|\int_a^b f(z)\,dz\right| \leqq |b - a| \max |f(z)|, \tag{B.2.15}$$

where the maximum is taken with respect to z on the line segment joining a and b.

In the case of positive integral powers of z the integral has the expected value

$$\int_a^b z^n\,dz = \frac{1}{n+1}\{b^{n+1} - a^{n+1}\}. \tag{B.2.16}$$

For

$$(b-a)[a+t(b-a)]^n = \sum_{k=0}^{n} \binom{n}{k} a^{n-k}(b-a)^{k+1}t^k,$$

and we can clearly integrate termwise with respect to t without first separating reals and imaginaries. This gives

$$\frac{1}{n+1}\{[a+(b-a)]^{n+1} - a^{n+1}\},$$

which is just what we want.

THEOREM B.2.3. *Let the functions $f_n(z)$ be defined and continuous in the convex domain D. Let $a \in D$, $b \in D$, and let the series*

$$\sum_{n=0}^{\infty} f_n(z) \equiv F(z) \tag{B.2.17}$$

converge uniformly with respect to z on the line segment joining $z = a$ with $z = b$. Then $F(z)$ is a continuous function of z on the line segment and

$$\int_a^b F(z)\,dz = \sum_{n=0}^{\infty} \int_a^b f_n(z)\,dz. \tag{B.2.18}$$

Proof. The first assertion follows from Theorem 5.5.1. Indeed, let us set

$$z = a + t(b-a), \quad f_n(z) = u_n(z) + iv_n(z).$$

Then the uniform convergence of the series (B.2.17) implies and is implied by the uniform convergence of the two series

$$\sum_{n=0}^{\infty} u_n(a+t(b-a)) \equiv U(t) \quad \text{and} \quad \sum_{n=0}^{\infty} v_n(a+t(b-a)) \equiv V(t)$$

with respect to t in $[0, 1]$. By Theorem 5.51 $U(t)$ and $V(t)$ are continuous functions of t for $0 \leqq t \leqq 1$. Since

$$F(z) = U(t) + iV(t),$$

it follows that $F(z)$ is a continuous function of t in $[0, 1]$, that is, a continuous function of z on the line segment.

By Theorem 5.5.3

$$\int_0^1 U(t)\,dt = \sum_{n=0}^{\infty} \int_0^1 u_n(a+t(b-a))\,dt$$

and a similar equality with u replaced by v. Combining these equalities with Definition B.2.1, we obtain the desired result formula (B.2.18). This completes the proof.

COROLLARY 1. *If the power series* (B.2.9) *converges for* $|z| < R$ *and its sum is denoted by* $f(z)$, *then*

$$\int_0^z f(s)\,ds = \sum_{n=0}^{\infty} \frac{a_n}{n+1} z^{n+1}, \tag{B.2.19}$$

convergent for $|z| < R$.

COROLLARY 2. *If* a, b, c *are any three points in the disk* $|z| < R$, *then*

$$\int_a^b f(z)\,dz + \int_b^c f(z)\,dz + \int_c^a f(z)\,dz = 0. \tag{B.2.20}$$

Proof. From formula (B.2.16) it follows that

$$\int_a^b z^n\,dz + \int_b^c z^n\,dz + \int_c^a z^n\,dz = 0$$

for each n, that is, (B.2.20) holds for $f(z) = z^n$. If we multiply the last relation by a_n and sum for n from 0 to ∞, the result is still zero. But the three absolutely convergent series on the left equal the three integrals on the left in (B.2.20), so that the latter formula holds.

As a special case of (B.2.20) we obtain

$$\int_0^{z+h} f(s)\,ds - \int_0^z f(s)\,ds = \int_z^{z+h} f(s)\,ds, \tag{B.2.21}$$

valid for $|z| < R$, $|z+h| < R$.

We can now prove the analogue of Theorem 5.5.6.

THEOREM B.2.4. *If* $f(z)$ *is defined by* (B.2.6), *convergent for* $|z| < R$, *then* $f(z)$ *has derivatives of all orders and*

$$f^{(p)}(z) = \sum_{n=p}^{\infty} n(n-1)\cdots(n-p+1)\,a_n z^{n-p}, \quad |z| < R. \tag{B.2.22}$$

Proof. We know that each of the derived series converges and defines a continuous function for $|z| < R$. Denote the sum of the first derived series by $f_1(z)$ so that

$$f_1(z) = \sum_{n=1}^{\infty} n a_n z^{n-1}.$$

By Corollary 1

$$\int_0^z f_1(s)\,ds = \sum_{n=1}^{\infty} a_n z^n = f(z) - f(0).$$

We then use (B.2.21), where we replace $f(s)$ by $f_1(s)$, and obtain

$$f(z+h) - f(z) = \int_z^{z+h} f_1(s)\,ds.$$

Since $f_1(s)$ is continuous at $s = z$, the right member is of the form

$$hf_1(z) + o(h).$$

This shows that $f'(z)$ exists and

$$f'(z) = f_1(z) \quad \text{for} \quad |z| < R.$$

In the same way one proves that the pth derived series is the first derivative of the $(p - 1)$th derived series for $|z| < R$, whence it follows that $f(z)$ has derivatives of all orders and that (B.2.22) is valid. This completes the proof.

EXERCISE B.2

1. For what values of z is $\left|\dfrac{z-1}{z+1}\right| < 1$?

2. Given the set

$$a_{m,n} = \frac{1}{m} + \frac{i}{n}, \qquad m, n = 1, 2, 3, \cdots,$$

find the cluster points. Are they elements of the set?

3. Prove the absolute convergence of

$$\sum_{n=0}^{\infty} [n + 1 - in]^{-2}.$$

4. Prove (B.2.13).

5. Does the real part of z have a derivative with respect to z? Consider radial limits of the difference quotient.

6. Use (5.1.4) and Theorem 5.3.2 to prove that the series

$$L(z) \equiv \sum_{n=1}^{\infty} \frac{1}{n} z^n$$

converges uniformly with respect to z in the set of points z where $|z| \leqq 1$ and $|1 - z| > \delta > 0$. Draw a figure of the set.

7. Verify that

$$L(z) = \int_0^z \frac{ds}{1 - s}$$

for $|z| < 1$. For what values of z does the integral exist?

8. Verify that the imaginary part of this integral has the same sign as the imaginary part of z. [*Hint:* Set $s = zt$ and use the definition of the integral.]

9. Where does the series $\sum_{n=0}^{\infty} \left(\dfrac{z-i}{z+i}\right)^n$ converge? Converge uniformly? Sum the series.

MISCELLANEOUS PROBLEMS FOR APPENDIX B

1. Use de Moivre's theorem to find the cube roots of $+1$ and of -1.

2. Give a geometrical construction of $1/z$.

3. How does the function $f(z) = 1/z$ map the unit circle $|z| = 1$, the interior of the unit circle, the exterior? Are any points left invariant under this mapping?

4. If $P(z)$ is a polynomial in z with real coefficients, prove that $P(z)$ and $P(\bar{z})$ are conjugate complex numbers if z and $\bar{z}$ have this property.

5. Verify that if an algebraic equation with real coefficients has a complex root z_0 then $\bar{z}_0$ is also a root.

6. Plot the locus of the points z where the imaginary part of $z^3 - 3z + 1$ is zero.

7. Where on the unit circle does $|z^2 + 1|$ reach its maximum? What is the minimum value and where is it reached? How do we know that this function attains its least upper and greatest lower bounds on the unit circle? How do these bounds compare with the bounds for the disk $|z| \leqq 1$?

8. Is the function

$$A(z) \equiv \int_0^z \frac{ds}{1 + s^2}$$

bounded for $|z| < 1$?

9. Expand $A(z)$ in powers of z for $|z| < 1$.

10. Plot the image of $1 + z^2$ when z describes once the circle $|z| = 2$ in such a manner that the argument of z goes from 0 to 2π. How does the argument of $1 + z^2$ vary?

11. Define and study the function $\sqrt{z}$.

12. Show that $|z - 2| + |z| \geqq 4$ implies $|z - 1| \geqq \sqrt{3}$.

13. Compute

$$\prod_{k=1}^{n} \left(1 + \frac{2}{2 - z_k}\right),$$

where $z_1, z_2, \cdots, z_n$ are the roots of the equation $z^n - 1 = 0$.

14. Solve the equation

$$(1 - iz)^n + i(1 + iz)^n = 0.$$

15. If $z_k = \exp\left(\frac{k}{n} 2\pi i\right)$, verify that

$$\frac{n}{z^n - 1} = \frac{z_1}{z - z_1} + \frac{z_2}{z - z_2} + \cdots + \frac{z_n}{z - z_n}.$$

16. In Problem 6 the locus factors into a straight line and a hyperbola which intersect in two points. Determine these points and show how they are related to the function $z^3 - 3z + 1$.

17. Show that no root of the equation

$$z^3 - 3z + 1 = 0$$

can exceed 2 in absolute value.

18. Show that

$$||z_1| - |z_2|| \leqq |z_1 + z_2| \leqq |z_1| + |z_2|.$$

When does equality hold in either place? Interpret the inequalities geometrically.

6

Taylor's Series

6.1. Rolle's theorem. Mean value theorems. The basic problem of the present chapter is to express differences in terms of derivatives. This problem arose previously in Section 3.1. For a differentiable function of a real variable, by formula (3.2.2),

$$f(x+h) - f(x) = f'(x)h + o(h). \tag{6.1.1}$$

In Section 3.8 it was noted that the remainder is often $O(h^2)$. The question was raised when this holds and when greater precision is obtainable. Similar questions were posed for higher differences. At the opposite end of the scale we have formula (5.5.13) which gives an expansion of $f(x) - f(0)$ in powers of x when $f(x)$ is the sum of a convergent power series.

In Section 6.2 we shall generalize formula (6.1.1) to the case where $f(x)$ has derivatives of order $\leqq n$, and in Section 6.3 we consider the case where there are derivatives of all orders. As a preparation we consider further results involving only the existence of a first-order derivative. We start with the theorem of Rolle, a special case of which was announced in 1691 by Michel Rolle (1652–1719), a French proponent of Calculus.

THEOREM 6.1.1. *If $f \in C[a, b]$, if $f(a) = f(b) = 0$, and if $f(x)$ has a unique finite derivative for each x with $a < x < b$, then there exists at least one value c, $a < c < b$, such that $f'(c) = 0$.*

Proof. The theorem is trivial if $f(x) \equiv 0$. Suppose that $f(x) \not\equiv 0$ and that $f(x)$ takes on positive values in some subinterval. It then has a positive maximum value M and there exists at least one value c in (a, b) where $f(c) = M$. At $x = c$ the function $f(x)$ has a unique finite derivative. It is claimed that $f'(c) = 0$. If this is not so, consider the possibility that $f'(c) > 0$. Since

$$f(c+h) - f(c) = f'(c)h + o(h),$$

we conclude that $f(c+h) > f(c)$ for small positive values of h, and this contradicts the assumption that $f(c)$ is the maximum value of $f(x)$. In the same way we see that $f'(c)$ cannot be negative. This leaves $f'(c) = 0$ as the only possibility and proves the existence of a value c with the stated property. In the same manner we handle the case where $f(x) \leqq 0$ in (a, b). This completes the proof. See Figures 6.1 to 6.3. Figures 6.2 and 6.3 illustrate how the theorem breaks down when the function does not have a unique finite derivative everywhere in the interval.

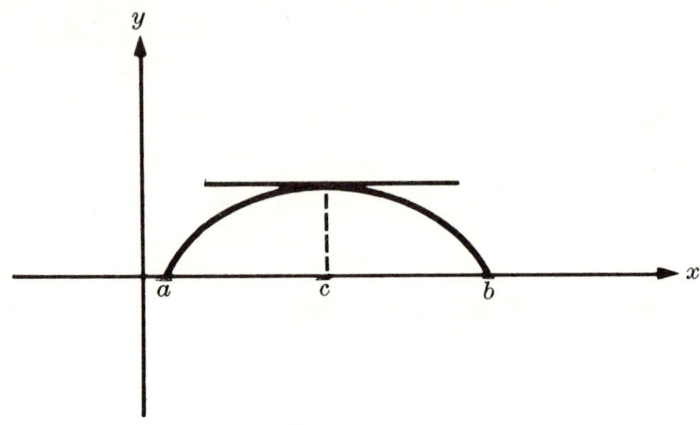

Figure 6.1

An immediate consequence of Rolle's theorem is *the Mean Value Theorem of the Differential Calculus*:

THEOREM 6.1.2. *If $f \in C[a, b]$, and if $f(x)$ has a unique finite derivative for all x with $a < x < c$, then there exists at least one value c such that*

$$f(b) - f(a) = f'(c)(b - a). \tag{6.1.2}$$

Proof. Geometrically speaking the theorem asserts that there exists at least one value c such that the curve tangent at $(c, f(c))$ is parallel to the chord joining $(a, f(a))$ with $(b, f(b))$. See Figure 6.4. For the proof we form the auxiliary function

$$F(x) = (b - a)[f(x) - f(a)] - (x - a)[f(b) - f(a)]. \tag{6.1.3}$$

This is an element of $C[a, b]$, and it has a unique finite derivative since $f(x)$ and x have this property. Finally $F(a) = F(b) = 0$. By Rolle's theorem there

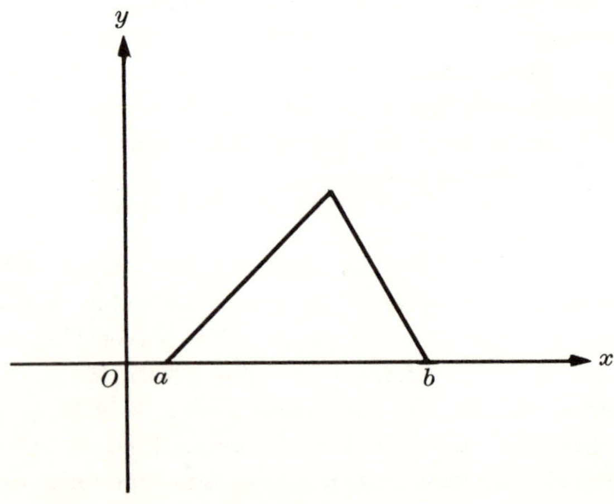

Figure 6.2

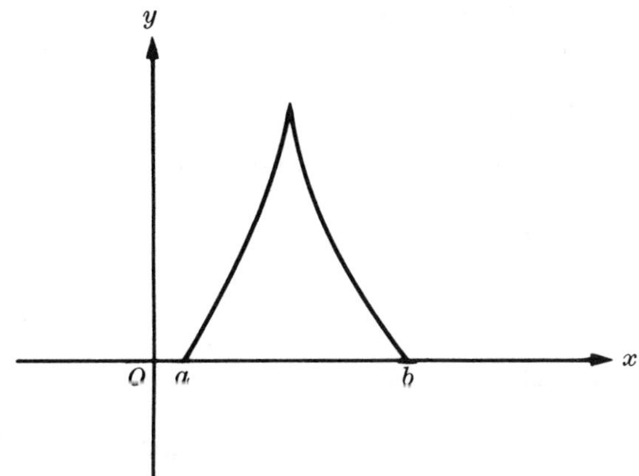

Figure 6.3

exists at least one c, $a < c < b$, such that $F'(c) = 0$. This gives

$$(b - a)f'(c) - [f(b) - f(a)] = 0,$$

which is the assertion of the theorem.

Let us compare this result with formula (6.1.1). If in the interval under consideration f has a unique finite derivative, then for each x and h such that

$$a < x, x + h < b$$

there exists a $\theta = \theta(x, h)$ such that

$$f(x + h) - f(x) = f'(x + \theta h)h, \quad 0 < \theta < 1, \tag{6.1.4}$$

by Theorem 6.1.2. Note that (6.1.1) and (6.1.4) are independent, neither implies the other, and that they are supplementary.

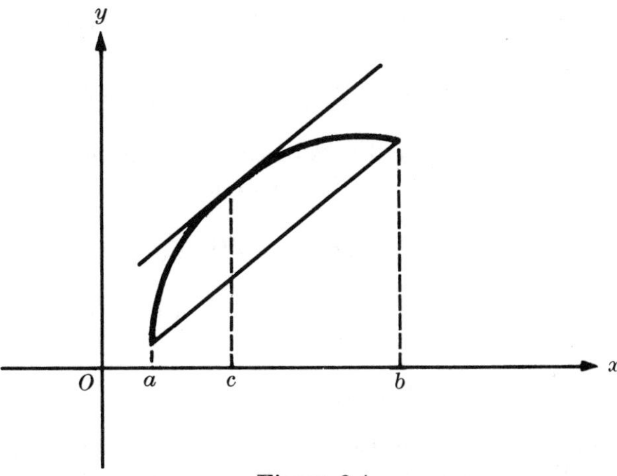

Figure 6.4

Among the consequences of Theorem 6.1.2 we list Theorem 4.3.9 which we restate as

THEOREM 6.1.3. *If $f(x) \in C[a, b]$ and is differentiable for all x in (a, b) and if $f'(x) = 0$ for all such x, then $f(x)$ is a constant.*

Proof. By Theorem 6.1.2 we have

$$f(x) - f(a) = (x - a)f'(a + \theta(x - a)) = 0$$

so that $f(x) = f(a)$ for all x in $[a, b]$.

COROLLARY. *Two functions in $C[a, b]$ differentiable for all x in (a, b) with the same derivative for all x, differ by a constant.*

The difference has the derivative identically equal to zero.

The mean value theorem was first formulated by Joseph Louis Lagrange (1736–1813), who, born in Turin, spent twenty years at the court of Frederick the Great, and died at Paris, one of the greatest mathematicians of his day. The theorem can be generalized as follows, known as *Cauchy's mean value theorem*:

THEOREM 6.1.4. *If $f(x)$ and $g(x)$ belong to $C[a, b]$, if $g(a) \neq g(b)$, and if both functions are differentiable in (a, b), then there exists at least one c with $a < c < b$ such that*

$$\frac{f(b) - f(a)}{g(b) - g(a)} = \frac{f'(c)}{g'(c)}. \tag{6.1.5}$$

Proof. Note first that this theorem is not obtainable from Theorem 6.1.1 by applying the latter to $f(x)$ and to $g(x)$ separately and dividing the results. This would give a right-hand side of the form $f'(c_1)/g'(c_2)$, and there is no reason for assuming that one can take $c_1 = c_2$. The proof of (6.1.5) is obtained by a slight modification of the proof of (6.1.2). We replace $F(x)$ by the auxiliary function

$$G(x) = [g(b) - g(a)][f(x) - f(a)] - [g(x) - g(a)][f(b) - f(a)] \tag{6.1.6}$$

which obviously satisfies the conditions of Rolle's theorem: it is a continuous differentiable function and $G(a) = G(b) = 0$. We have

$$G'(x) = [g(b) - g(a)]f'(x) - g'(x)[f(b) - f(a)]$$

and by Rolle's theorem there is at least one c such that $G'(c) = 0$. This gives (6.1.5).

EXERCISE 6.1

1. $P(x)$ is a polynomial with real coefficients. Prove that between two consecutive roots of the equation $P'(x) = 0$ there is at most one root of $P(x) = 0$.

2. Use the preceding result and Rolle's theorem to show that the equation

$$x^4 - 8x^3 + 22x^2 - 24x + 4 = 0$$

has exactly two real roots both positive.

3. Rolle's theorem applies to the function

$$(1 - x^2)^{\frac{1}{2}}(3 + x^2)^{-1}, \quad -1 \leqq x \leqq 1.$$

Find an admissible value of c.

4. If $f(x) = x$ for $0 \leqq x < 1$, and 0 for $x = 1$, then $f(x)$ is differentiable in (0, 1). Further $f(0) = f(1) = 0$, but $f'(x) \neq 0$ in (0, 1). Why does Rolle's theorem not apply?

5. Apply the mean value theorem to (i) $f(x) = (1 - x^2)^{\frac{1}{2}}$, $a = 0$, $b = 1$; and to (ii) $f(x) = x^3 - 3x$, $a = -1$, $b = 1$, and find appropriate values of c.

6. Suppose that $f(x)$ is differentiable in some neighborhood of $x = x_0$ and that $f'(x)$ has finite left- and right-hand limits as $x \to x_0$. Prove that these limits are equal and agree with $f'(x_0)$ so that $f'(x)$ is continuous at $x = x_0$.

7. Suppose that $f(x)$ is continuous together with its derivatives of order $\leqq n - 1$ in the closed interval $[a, b]$ and that $f^{(n)}(x)$ exists everywhere in the open interval (a, b). Suppose further that $f(x)$ has $n + 1$ distinct zeros in $[a, b]$. Prove that there is at least one c, $a < c < b$, such that $f^{(n)}(c) = 0$.

8. Suppose that $f(x)$ is continuous together with its derivatives of order $\leqq p + q$ in $[a, b]$ and that $f^{(p+q+1)}(x)$ exists everywhere in (a, b). Prove that the last-mentioned derivative is zero at least once in (a, b) if

$$f(a) = f'(a) = \cdots = f^{(p)}(a) = 0,$$
$$f(b) = f'(b) = \cdots = f^{(q)}(b) = 0.$$

9. The nth polynomial of Legendre is defined by

$$P_n(x) = \frac{1}{2^n n!} D^n[(x^2 - 1)^n].$$

Prove that it has n distinct real zeros which lie in the interval $(-1, 1)$.

10. Which is the most general function whose nth derivative is zero?

11. Prove that $e^{\alpha x}$ is a solution of

$$y' = \alpha y$$

and that a general solution is of the form $Ce^{\alpha x}$.

12. In formula (6.1.5) set

$$f(x) = \log (1 + x), g(x) = (1 + x)^{\frac{1}{2}}, a = 1, b = x > 1,$$

and find the limit of the quotient as x decreases to 1.

6.2. Taylor's theorem. We proceed to a generalization of formula (6.1.1) of the form

$$f(x + h) - f(x) = A_1 h + A_2 h^2 + \cdots + A_{n-1} h^{n-1} + R_n \qquad (6.2.1)$$

where $A_1, A_2, \cdots, A_{n-1}$ are functions of x, independent of h, and R_n is a function of x and h which, for fixed x, is $O(h^n)$ as $h \to 0$. In the applications the integer n is usually at our disposal and, if h and x are suitably restricted, the remainder $R_n \to 0$ as $n \to \infty$, so that an infinite power series in h is obtained. Such a series representation was discovered in 1712 by Brook Taylor (1685–1731), but the formula was disregarded until Lagrange called attention to it. Both he and Cauchy gave expressions for the remainder. In this section we consider the finite case and prove

THEOREM 6.2.1. *Suppose that f has derivatives of order $\leqq n$ all of which belong to $C[a, b]$. Then for $a \leqq x$, $x + h \leqq b$ we have*

$$f(x + h) = f(x) + \frac{h}{1!} f'(x) + \frac{h^2}{2!} f''(x) + \cdots + \frac{h^{n-1}}{(n-1)!} f^{(n-1)}(x) + R_n(x, h), \tag{6.2.2}$$

where

$$|R_n(x, h)| \leq \frac{|h|^n}{n!} \sup_{a \leqq t \leqq b} |f^{(n)}(t)|. \tag{6.2.3}$$

Remark. In view of the importance of this theorem we shall give two different proofs which lead to different expressions for the remainder. The first proof is based on repeated integration by parts and leads in a natural way to formula (6.2.2). The second proof is based on the mean value theorem and requires only the existence and boundedness of the nth derivative in (a, b) but not its continuity in $[a, b]$. It is shorter than the first proof but it employs an auxiliary function which requires much ingenuity plus an uncanny foresight to construct. It can be varied to give different forms of the remainder. Two such variants will be given making altogether three expressions for the remainder.

First proof. We start with the self-evident formula

$$f(x + h) - f(x) = \int_0^h f'(x + s)\, ds \equiv R_1(x, h). \tag{6.2.4}$$

Here we integrate by parts setting

$$u = f'(x + s), \qquad dv = ds,$$

$$du = f''(x + s)\, ds, \quad v = s - h.$$

The choice of $s - h$ as a primitive of 1 rather than s itself is decisive for this attack on the problem. We note that

$$\int_0^h f'(s + x)\, ds = \Big[(s - h)f'(x + s)\Big]_{s=0}^{s=h} - \int_0^h (s - h) f''(x + s)\, ds.$$

Here the integrated part vanishes in the upper limit and so we are left with

$$f(x + h) = f(x) + hf'(x) + \int_0^h (h - s) f''(x + s)\, ds. \tag{6.2.5}$$

This is formula (6.2.3) for $n = 2$ with

$$R_2(x, h) = \int_0^h (h - s) f''(x + s)\, ds.$$

It is clear how the argument continues: we have merely to integrate the remainder by parts again. It is important to note that we integrate the factor involving a power of $(h - s)$ and differentiate the other factor which involves a derivative of $f(x + s)$. Suppose that in this process we have arrived at the kth stage where $k \leqq n - 1$. We then have a formula ending with $R_k(x, h)$ where

$$R_k(x, h) = \frac{1}{(k-1)!} \int_0^h (h - s)^{k-1} f^{(k)}(x + s)\, ds.$$

Here we integrate by parts taking

$$u = f^{(k)}(x + s), \quad dv = \frac{1}{(k-1)!} (h - s)^{k-1}\, ds,$$

$$du = f^{(k+1)}(x + s)\, ds, \quad v = -\frac{1}{k!} (h - s)^k.$$

The integrated part vanishes in the upper limit, $s = h$, and we are left with

$$R_k(x, h) = \frac{h^k}{k!} f^{(k)}(x) + R_{k+1}(x, h).$$

After $n - 1$ integrations by parts, we obtain the desired formula (6.2.3) together with the expression

$$R_n(x, h) = \frac{1}{(n-1)!} \int_0^h (h - s)^{n-1} f^{(n)}(x + s)\, ds \tag{6.2.6}$$

for the remainder.

To estimate the remainder we can use formula (4.3.7) which, if $h > 0$, gives

$$\begin{aligned} |R_n(x, h)| &\leqq \sup_{0 \leqq s \leqq h} |f^{(n)}(x + s)| \frac{1}{(n-1)!} \int_0^h (h - s)^{n-1}\, ds \\ &\leqq \frac{h^n}{n!} \sup_{a \leqq t \leqq b} |f^{(n)}(t)|, \end{aligned}$$

and this is formula (6.2.4). If $h < 0$ instead, we merely replace h by $|h|$ in the final formula. This completes the first proof.

Second proof. To abbreviate, let us set $x + h = y$ and define the following function of s:

$$\begin{aligned} F(s) = f(s) &+ \frac{1}{1!}(y - s) f'(s) + \frac{1}{2!}(y - s)^2 f''(s) + \cdots \\ &+ \frac{1}{(n-1)!}(y - s)^{n-1} f^{(n-1)}(s). \end{aligned} \tag{6.2.7}$$

Here

$$F(x) = f(x) + \frac{h}{1!} f'(x) + \frac{h^2}{2!} f''(x) + \cdots + \frac{h^{n-1}}{(n-1)!} f^{(n-1)}(x),$$

$$F(y) = f(y) = f(x + h).$$

This shows that

$$R_n(x, h) = F(y) - F(x),$$

and the whole problem is reduced to obtaining a suitable expression for the difference on the right from which an estimate can be obtained. This suggests the mean value theorem of the differential calculus, either in its original form or in Cauchy's version. In either case we have to compute $F'(s)$. This function equals

$$f'(s) + \frac{1}{1!}(y - s)f''(s) + \cdots + \frac{1}{(n-2)!}(y - s)^{n-2}f^{(n-1)}(s)$$
$$+ \frac{1}{(n-1)!}(y - s)^{n-1}f^{(n)}(s)$$
$$- f'(s) - \frac{1}{1!}(y - s)f''(s) - \cdots - \frac{1}{(n-2)!}(y - s)^{n-2}f^{(n-1)}(s),$$

so that finally

$$F'(s) = \frac{1}{(n-1)!}(y - s)^{n-1}f^{(n)}(s).$$

We now use formula (6.1.2) where we replace f, a, b, c by F, x, $y = x + h$, $x + \theta h$, respectively. Since $y - c = (1 - \theta)h$, we find that

$$R_n(x, h) = \frac{1}{(n-1)!}(1 - \theta)^{n-1}h^n f^{(n)}(x + \theta h). \tag{6.2.8}$$

This form of the remainder is due to Cauchy. It should be kept in mind that θ is a function of x, h, and n and we have no idea of how θ varies. We know that $0 < \theta < 1$, but it may well happen that for fixed x and h the value of θ tends to 0 as $n \to \infty$.

With Cauchy's form of the mean value theorem we can get a variety of expressions for the remainder. If $F(s)$ is defined by (6.2.7), and $G(s)$ is a differentiable function with $G(x) \neq G(y)$, then Theorem 6.1.4 asserts that

$$R_n(x, h) = F(y) - F(x) = \frac{G(y) - G(x)}{G'(x + \theta h)} F'(x + \theta h). \tag{6.2.9}$$

A suitable choice of $G(s)$ is

$$G(s) = (y - s)^p, \tag{6.2.10}$$

where p is a positive integer. Here

$$G(x) = h^p, \quad G(y) = 0, \quad G'(s) = -p(y - s)^{p-1}.$$

If these expressions are substituted in (6.2.9), we obtain after simplification

$$R_n(x, h) = \frac{1}{p(n-1)!} h^n (1 - \theta)^{n-p} f^{(n)}(x + \theta h). \tag{6.2.11}$$

The choice $p = 1$ gives Cauchy's formula (6.2.8), while $p = n$ gives the remainder of Lagrange

$$R_n(x, h) = \frac{h^n}{n!} f^{(n)}(x + \theta h). \tag{6.2.12}$$

The estimate (6.2.3) follows immediately from this formula. This completes the second proof.

We shall illustrate these formulas by some special cases. We start with the function

$$f(x) = (1 + x)^\alpha, \tag{6.2.13}$$

where α is any fixed real number. If α is a positive integer, this function is defined for all values of x. If α is a negative integer, we omit the value $x = -1$, and for all other values of α we take the interval $(-1, \infty)$ as the domain of definition. As long as α is rational, $\alpha = p/q$, $q > 0$, we can define $(1 + x)^\alpha$ as the pth power of the positive qth root of the positive number $1 + x$, but if α is irrational a less direct procedure is required. The simplest is the following. We note that $\log (1 + x)$ is well-defined for $x > -1$. Hence we can set

$$(1 + x)^\alpha \equiv e^{\alpha \log (1 + x)}. \tag{6.2.14}$$

By the basic properties of the logarithm this definition agrees with the algebraic one, if α is rational, and it makes the power a continuous, differentiable function of α for fixed x.

We want the derivative with respect to x which is obtainable by the chain rule. We get

$$D_x e^{\alpha \log (1 + x)} = e^{\alpha \log (1 + x)} \frac{\alpha}{1 + x} = (1 + x)^\alpha \frac{\alpha}{1 + x} = \alpha(1 + x)^{\alpha - 1},$$

that is, the formula for the derivative of a power is valid also when the exponent is irrational. This gives

$$D_x^k (1 + x)^\alpha = \alpha(\alpha - 1)(\alpha - 2) \cdots (\alpha - k + 1)(1 + x)^{\alpha - k}. \tag{6.2.15}$$

We can now discuss

Example 1. If $f(x)$ is given by (6.2.13), then for $x = 0$ and $h = t$

$$(1 + t)^\alpha = 1 + \frac{\alpha}{1} t + \frac{\alpha(\alpha - 1)}{1 \cdot 2} t^2 + \cdots + \frac{\alpha(\alpha - 1) \cdots (\alpha - n + 2)}{1 \cdot 2 \cdots (n - 1)} t^{n-1} + R_n(0, t), \tag{6.2.16}$$

where $-1 < t < \infty$. For $0 < t$, the formula of Lagrange is preferable and gives

$$R_n(0, t) = \frac{\alpha(\alpha - 1) \cdots (\alpha - n + 1)}{1 \cdot 2 \cdots n} t^n (1 + \theta t)^{\alpha - n}, \tag{6.2.17}$$

while for $-1 < t < 0$ the formula of Cauchy gives the more advantageous expression

$$R_n(0, t) = \frac{\alpha(\alpha - 1) \cdots (\alpha - n + 1)}{1 \cdot 2 \cdots (n - 1)} t(1 + \theta t)^{\alpha-1}\left\{\frac{t(1 - \theta)}{1 + \theta t}\right\}^{n-1}. \quad (6.2.18)$$

Since

$$(1 + x + h)^\alpha = (1 + x)^\alpha\left\{1 + \frac{h}{1 + x}\right\}^\alpha,$$

the general case is easily reduced to the special case of (6.2.16) so there is no actual restriction involved in taking $x = 0$. We set $t = h/(1 + x)$ in (6.2.16) and multiply by $(1 + x)^\alpha$ in order to get the expansion for $(1 + x + h)^\alpha$.

Example 2. We take

$$f(x) = \log x, \quad 0 < x. \quad (6.2.19)$$

Here

$$f^{(k)}(x) = (-1)^{k-1}(k - 1)!\, x^{-k}, \quad k = 1, 2, 3, \cdots, \quad (6.2.20)$$

so that

$$\log (x + h) = \log x + \frac{h}{x} - \frac{h^2}{2x^2} + \frac{h^3}{3x^3} - \cdots + (-1)^n \frac{h^{n-1}}{(n - 1)x^{n-1}} + R_n(x, h). \quad (6.2.21)$$

The formula of Lagrange gives

$$R_n(x, h) = \frac{(-1)^{n-1}}{n} \left\{\frac{h}{x + \theta h}\right\}^n, \quad (6.2.22)$$

while the formula of Cauchy becomes

$$R_n(x, h) = \frac{(-1)^{n-1}}{x + \theta h} \left\{\frac{(1 - \theta)h}{x + \theta h}\right\}^{n-1} h. \quad (6.2.23)$$

Example 3. We take

$$f(x) = e^x \quad (6.2.24)$$

which equals its own derivative. This gives

$$e^x = 1 + \frac{x}{1!} + \frac{x^2}{2!} + \cdots + \frac{x^{n-1}}{(n - 1)!} + R_n(0, x), \quad (6.2.25)$$

where

$$R_n(0, x) = \frac{x^n}{n!} e^{\theta x}. \quad (6.2.26)$$

In particular, for $x = 1$

$$e = 1 + \frac{1}{1!} + \frac{1}{2!} + \cdots + \frac{1}{(n - 1)!} + R_n, \quad R_n < \frac{3}{n!}, \quad (6.2.27)$$

where we have used $e^\theta < e < 3$. From this expression e can be computed with any desired degree of accuracy.

EXERCISE 6.2

Estimate $R_2(0, h)$ if $h > 0$ and $f(x)$ equals:

1. $(1 + x)^{-2}$.
2. $(1 + x)^{\frac{1}{2}}$.
3. $(1 - x)^{-\frac{1}{2}}$.
4. $(1 + x)^{\frac{3}{2}}$.
5. $\log (1 - x)$.
6. $\log (1 - x^2)$.
7. If $x > 0$, $h > 0$, prove that

$$\sqrt{x+h} - \sqrt{x} < \frac{h}{2\sqrt{x}},$$

and find an upper bound for the difference of the two members.

8. Compare formulas (6.2.17) and (6.2.18) and explain why the first formula is better for $h > 0$ and the second better for $h < 0$.
9. Same question for formulas (6.2.22) and (6.2.23).
10. Use the identity

$$\frac{1}{1+s^2} = \frac{1}{2i}\left\{\frac{1}{s-i} - \frac{1}{s+i}\right\}$$

to find the nth derivative of the left member and evaluate it at $s = 0$.

11. If

$$A(x) = \int_0^x \frac{ds}{1+s^2},$$

expand in powers of x and estimate the remainder after the term involving x^{2n-1}.

12. Find log 1.01 with an error less than one unit in the ninth decimal place.
13. Compute e with an error less than one unit in the seventh decimal place.
14. Formula (6.2.27) shows that $e(n - 1)!$ differs from an integer by less than $3/n < 1$ as soon as $3 < n$. Use this to prove that e cannot be a rational number.

6.3. Convergence of remainder. Taylor's series. If $f(x)$ has derivatives of all orders, then formula (6.2.2) is valid for every n, so that

$$f(x+h) = f(x) + \frac{h}{1!}f'(x) + \frac{h^2}{2!}f''(x) + \cdots + \frac{h^{n-1}}{(n-1)!}f^{(n-1)}(x) + R_n(x, h),$$

with

$$R_n(x, h) = \frac{h^n}{n!}f^{(n)}(x + \theta_n h)$$

for every n. The usefulness of this representation depends largely upon the behavior of $R_n(x, h)$ as n becomes infinite. We take a fixed value of x, $x = c$, and restrict h so that c and $c + h$ lie in $[a, b]$. There are three essentially different possibilities.

(i) There is an interval $(-\delta, \delta)$ such that for $-\delta < h < \delta$ we have $\lim\limits_{n\to\infty} R_n(c, h) = 0$.

(ii) There is an interval $(-\delta, \delta)$ in which $\lim R_n(c, h)$ exists, but the limit is zero only if h is zero.

(iii) $R_n(c, h)$ does not tend to any limit for $h \neq 0$.

Only the first of these possibilities is favorable. In this case $f(c+h)$ is represented by an absolutely convergent power series in h for $|h| < \delta$, namely

$$f(c+h) = \sum_{n=0}^{n} \frac{f^{(n)}(c)}{n!} h^n. \tag{6.3.1}$$

This is known as *Taylor's series of the function f*. Since this series also converges for complex values of the variable h as long as $|h| < \delta$, it can serve as the definition of a differentiable function $f(z)$ of the complex variable z in the disk $|z - c| < \delta$ and this function coincides with $f(x)$ on the interval $(c - \delta, c + \delta)$. Moreover, it may be shown that there is at most one such function $f(z)$ which is differentiable in the disk $|z - c| < \delta$ and coincides with $f(x)$ on the real diameter of the disk. Thus the extension to the complex plane is unique whenever it exists.

In the second case the series in (6.3.1) is also convergent in $|h| < \delta$, but its sum is not $f(c + h)$. The series defines a differentiable function $F(z)$ of a complex variable $z = c + h$, but this function is not an extension of $f(x)$ to the complex plane since it does not coincide with $f(x)$ on the real axis. $F(z)$ has derivatives of all orders at $z = c$ and accidentally

$$F^{(n)}(c) = f^{(n)}(c)$$

for all n. On the left we have derivatives where z approaches c in an arbitrary manner; on the right only real approach is permitted.

In the third case the radius of convergence of the series in (6.3.1) equals zero so again the infinite series is useless as a representation of $f(x)$. We shall see that all three possibilities can occur.

Let us first examine the three examples discussed in the preceding section. We start with the exponential function

$$e^x = 1 + \frac{x}{1!} + \frac{x^2}{2!} + \cdots + \frac{x^{n-1}}{(n-1)!} + \frac{x^n}{n!} e^{\theta_n x}. \tag{6.3.2}$$

If x has any value $\leqq m$ in absolute value, the remainder does not exceed

$$r_n = \frac{m^n}{n!} e^m.$$

The sequence $\{r_n\}$ is strictly increasing at first and then decreases to zero:

$$r_1 < r_2 < r_3 < \cdots < r_{m-1} = r_m > r_{m+1} > \cdots.$$

Since

$$r_n = \frac{m}{n} r_{n-1},$$

and $m/n \to 0$ as $n \to \infty$, the sequence is strictly decreasing as soon as $n > m$. Thus it has a limit which must be zero. It follows that the remainder in (6.3.2) tends to zero uniformly with respect to x in any fixed interval $(-m, m)$. From this it follows that

$$e^x = 1 + \frac{x}{1!} + \frac{x^2}{2!} + \cdots + \frac{x^n}{n!} + \cdots, \tag{6.3.3}$$

where the series converges uniformly in any fixed finite interval.

Here we have taken $c = 0$. Since

$$f^{(n)}(c) = e^c$$

for all n, the discussion of the remainder also shows that

$$e^{c+h} = e^c\left\{1 + \frac{h}{1!} + \frac{h^2}{2!} + \cdots + \frac{h^n}{n!} + \cdots\right\} = e^c e^h, \tag{6.3.4}$$

that is, Taylor's series for the exponential function implies the addition theorem. Since the power series (6.3.3) converges for all x, we can define the exponential function for complex values by the *exponential series*

$$e^z = 1 + \frac{z}{1!} + \frac{z^2}{2!} + \cdots + \frac{z^n}{n!} + \cdots. \tag{6.3.5}$$

We consider next the binomial series. Using the notation of formula (2.3.20) we write

$$\frac{\alpha(\alpha - 1)\cdots(\alpha - k + 1)}{k!} = \binom{\alpha}{k}, \tag{6.3.6}$$

which is referred to as the kth *binomial coefficient* of order α. We then have

$$(1 + x)^\alpha = \sum_{k=0}^{n-1}\binom{\alpha}{k}x^k + R_n(0, x), \tag{6.3.7}$$

and want to prove that the remainder tends to zero as $n \to \infty$, provided $-1 < x < 1$. Hence we should have

$$(1 + x)^\alpha = \sum_{n=0}^{\infty}\binom{\alpha}{n}x^n \quad \text{for} \quad |x| < 1. \tag{6.3.8}$$

This infinite series is known as *the binomial series of order* α. Using the ratio test, we find that the radius of convergence of the series is 1. This explains the restriction of x to the interval $(-1, 1)$.

From the convergence of the power series it follows that

$$\lim_{n\to\infty}\binom{\alpha}{n}r^n = 0 \tag{6.3.9}$$

for any fixed r with $0 < r < 1$. Lagrange's form of the remainder, formula (6.2.17), gives

$$R_n(0, x) = \binom{\alpha}{n}x^n(1 + \theta x)^{\alpha-n}.$$

If $x > 0$ and $n > \alpha$, the last factor is < 1. Hence for $0 \leqq x \leqq r < 1$ the absolute value of the remainder is less than

$$\left|\binom{\alpha}{n}\right| r^n,$$

which tends to 0 by (6.3.9). The same conclusion is valid for $r = 1$ provided the binomial coefficient itself tends to zero, and this is the case if and only if $\alpha > -1$.

For negative values of x the Cauchy formula (6.2.18) gives a more advantageous expression for the remainder:

$$R_n(0, x) = n\binom{\alpha}{n} x(1 + \theta x)^{\alpha-1}\left\{\frac{x(1-\theta)}{1+\theta x}\right\}^{n-1}.$$

Suppose that $-1 < -r \leqq x \leqq 0$. Then

$$1 - r < 1 + \theta x \leqq 1,$$

so that

$$|x|\,(1 + \theta x)^{\alpha-1} \leqq r \max\,[1, (1-r)^{\alpha-1}].$$

Further

$$\left|\frac{x(1-\theta)}{1+\theta x}\right| \leqq r\frac{1-\theta}{1-\theta r} < r$$

and

$$n\binom{\alpha}{n} = \alpha\binom{\alpha-1}{n-1}.$$

Since

$$\binom{\alpha-1}{n-1} r^{n-1} \to 0,$$

the remainder approaches the limit zero also in the interval $(-1, 0)$. This discussion shows that formula (6.3.8) holds. If α is a positive integer, $\alpha = m$, then the series reduces to a polynomial of degree m; we have simply the binomial theorem and the formula is valid for all x. If $-1 < \alpha < 0$, the formula holds for $-1 < x \leqq 1$; it holds for $-1 \leqq x \leqq 1$ provided that $0 < \alpha$.

Finally we consider the logarithmic case. In formula (6.2.21) we replace x by a and h by $x - a$ obtaining

$$\log x = \log a + \frac{x-a}{a} - \frac{1}{2}\left(\frac{x-a}{a}\right)^2 + \frac{1}{3}\left(\frac{x-a}{a}\right)^3 - \cdots$$

$$+ \frac{(-1)^n}{n-1}\left(\frac{x-a}{a}\right)^{n-1} + R_n(a, x-a). \quad (6.3.10)$$

We want to prove that

$$\log\frac{x}{a} = \sum_{n=1}^{\infty} \frac{(-1)^{n-1}}{n}\left(\frac{x-a}{a}\right)^n \quad \text{for} \quad |x-a| < a. \qquad (6.3.11)$$

Actually this follows from formula (5.5.5), but here we want proof by a discussion of the remainder in (6.3.10). The series converges for $|x - a| < a$ whatever its sum. Further the series diverges for $x = 0$ since it reduces to the harmonic series. This shows that we are restricted to the interval $0 < x \leqq 2a$. For $a \leqq x \leqq 2a$ Lagrange's form of the remainder, formula (6.2.22), gives

$$R_n(a, x - a) = \frac{(-1)^{n-1}}{n}\left\{\frac{x - a}{(1 - \theta)a + \theta x}\right\}^n.$$

Here $(1 - \theta)a + \theta x \geqq (1 - \theta)a + \theta a = a$ so that

$$|R_n(a, x - a)| \leqq \frac{1}{n}\left(\frac{x - a}{a}\right)^n \leqq \frac{1}{n} \to 0 \quad \text{as} \quad n \to \infty.$$

For $0 < x < a$ Cauchy's formula (6.2.23) gives

$$R_n(a, x - a) = \frac{(-1)^{n-1}}{(1 - \theta)a + \theta x}\left\{\frac{(1 - \theta)(x - a)}{(1 - \theta)a + \theta x}\right\}^{n-1}.$$

Here $(1 - \theta)a + \theta x$ exceeds x as well as $(1 - \theta)a$ so that

$$|R_n(a, x - a)| < \frac{1}{x}\left(\frac{a - x}{a}\right)^{n-1} \to 0 \quad \text{as} \quad n \to \infty.$$

This proves (6.3.10).

After this discussion let us consider the general problem of proving the validity of the Taylor expansion for a given function of a real variable. Necessary and sufficient conditions can be read off from our first proof of Taylor's theorem which led to formula (6.2.6) for the remainder. The following theorem was proved by the German mathematician Alfred Pringsheim (1850–1941) in 1894.

THEOREM 6.3.1. *The function $f(x)$ defined in the interval $(c - r, c + r)$ is represented by the Taylor series*

$$f(x) = \sum_{n=0}^{\infty} \frac{f^{(n)}(c)}{n!}(x - c)^n \quad \textit{for} \quad |x - c| < r, \tag{6.3.12}$$

if and only if the following conditions hold: (i) *$f(x)$ has derivatives of all orders in $(c - r, c + r)$; and* (ii) *for each x in the interval*

$$\lim_{n\to\infty} \frac{1}{(n - 1)!}(x - t)^{n-1} f^{(n)}(t) = 0 \tag{6.3.13}$$

uniformly with respect to t for all t between c and x.

Proof. Referring to Section 6.2 we see that

$$f(x) = \sum_{k=0}^{n-1} \frac{f^{(k)}(c)}{k!}(x - c)^k + R_n(c, x - c), \tag{6.3.14}$$

where by formula (6.2.6)

$$R_n(c, x - c) = \frac{1}{(n - 1)!}\int_c^x (x - t)^{n-1} f^{(n)}(t)\, dt. \tag{6.3.15}$$

Condition (6.3.13) states that for any fixed x in $(c - r, c + r)$ the integrand in (6.3.15) converges to zero uniformly in t. Since the interval of integration is finite, the limit of the integral is the integral of the limit, that is, identically zero. Since $\lim R_n(c, x - c) = 0$, the power series in (6.3.12) converges and its sum is $f(x)$. This proves that the conditions are sufficient.

It is obvious that condition (i) is necessary. We omit the proof of the necessity of condition (ii) but refer to Problems 10–13 of Exercise 6.3 which give the three steps involved in the proof. This completes the proof of the theorem.

A Taylor series of the form

$$f(x) = \sum_{n=0}^{\infty} \frac{f^{(n)}(0)}{n!} x^n \tag{6.3.16}$$

is called a Maclaurin series [after Colin Maclaurin (1698–1746)].

This name is well-established by long usage though Taylor considered the special case $c = 0$ long before Maclaurin. The latter, a geometer of distinction, became professor of mathematics by competitive examination, at the University of Aberdeen at the early age of nineteen.

At the beginning of this section we stated three possibilities for the behavior of the remainder in Taylor's theorem. In case (i) the function has a convergent Taylor series. Theorem 6.3.1 states when this holds. Pringsheim has constructed some interesting examples showing that cases (ii) and (iii) may well occur. Two of his examples are found at the end of Exercise 6.3.

The standard example for case (ii) is the function

$$f(x) = e^{-x^{-2}}, \quad \text{with} \quad f(0) = 0, \tag{6.3.17}$$

at the origin. This function has derivatives of all orders, and each derivative tends to zero with x. Here x is restricted to be real. The complex-valued function

$$f(z) = e^{-z^{-2}}$$

is neither continuous at $z = 0$ nor can it have a derivative there.

We have

$$f'(x) = 2x^{-3}e^{-x^{-2}},$$

$$f''(x) = 2x^{-6}(2 - 3x^2)e^{-x^{-2}},$$

generally,

$$f^{(n)}(x) = x^{-3n} \sum_{k=0}^{n-1} a_{n,k} x^{2k} e^{-x^{-2}}, \tag{6.3.18}$$

where the coefficients $a_{n,k}$ are integers. This is easy to prove by induction. It was shown in Section 4.5 that the exponential function e^t becomes infinite faster than any fixed positive power of t, and formula (4.5.41) states that

$$\lim_{t \to +\infty} t^m e^{-t} = 0$$

for any fixed positive m. Setting $x^{-2} = t$, we see that

$$\lim_{x \to 0} x^{-3n} e^{-x^{-2}} = 0, \quad n = 1, 2, 3, \cdots. \tag{6.3.19}$$

This shows that for all n

$$\lim_{x \to 0} f^{(n)}(x) = 0.$$

At this point we raise the question of the existence of derivatives of $f(x)$ at $x = 0$, This is a different problem but it is solved using the same machinery. We recall that $f(0) = 0$ by definition which makes $f(x)$ continuous. We have

$$f'(0) = \lim_{h\to 0} \frac{1}{h}[f(h) - f(0)]$$

$$= \lim_{h\to 0} \frac{1}{h} f(h) = 0$$

by (6.3.19) which holds for all n, not merely for positive integers. In the same manner

$$f''(0) = \lim_{h\to 0} \frac{1}{h} [f'(h) - f'(0)]$$

$$= \lim_{h\to 0} 2h^{-4}e^{-h^{-2}} = 0,$$

again by (6.3.19). It is clear that we can prove the existence of derivatives of all orders at the origin and that for all n

$$f^{(n)}(0) = 0. \tag{6.3.20}$$

Taylor's theorem now gives

$$e^{-h^{-2}} = R_n(0, h) \tag{6.3.21}$$

for all n, so that

$$\lim_{n\to\infty} R_n(0, h) = e^{-h^{-2}},$$

that is, the limit is different from 0 for $h \neq 0$. The formal Maclaurin series is by (6.3.20)

$$\sum_{n=0}^{\infty} 0 \cdot x^n$$

which represents 0 and not the given function $f(x)$. This is a clear case of possibility (ii).

Case (iii) is represented by Problem 15 below. After we have studied differentiation of trigonometric functions in Section 7.1 we shall be able to prove that the function defined by

$$f(x) = \sum_{n=0}^{\infty} \frac{1}{n!} \sin(2^n x) \tag{6.3.22}$$

has derivatives of all orders and that

$$f^{(2n+1)}(0) = (-1)^n e^{2^{2n+1}}, \quad f^{(2n)}(0) = 0 \tag{6.3.23}$$

for all n. The corresponding formal Maclaurin series, based on these coefficients, diverges for all x different from zero.

EXERCISE 6.3

Evaluate the series:

1. $\displaystyle\sum_{n=0}^{\infty} \frac{1}{(2n+1)!}$.
2. $\displaystyle\sum_{n=0}^{\infty} \frac{2^n}{(2n)!}$.
3. $\displaystyle\sum_{n=0}^{\infty} \frac{n(n-1)}{n!}$.
4. $\displaystyle\sum_{n=0}^{\infty} \frac{n^2}{n!}$.

5. Suppose that x is fixed, $0 < x < 1$, prove that the series in (6.3.8) converges for all real or complex values of α. What is the situation for $x = 1$?

6. Prove that

$$\binom{\alpha}{0}\binom{\beta}{n} + \binom{\alpha}{1}\binom{\beta}{n-1} + \binom{\alpha}{2}\binom{\beta}{n-2} + \cdots + \binom{\alpha}{n}\binom{\beta}{0} = \binom{\alpha+\beta}{n}$$

for arbitrary real values of α and β. Is the relation also true for complex values?

7. Let $-1 < x < 1$ and define $(1 - x)^\alpha$ for complex values of α as the sum of the series (6.3.8). Prove that $|(1 + x)^\alpha| \equiv 1$, if α is purely imaginary. [*Hint:* Use $|z|^2 = z\bar{z}$.]

8. Prove that

$$\log \frac{x+1}{x-1} = \sum_{n=1}^{\infty} \frac{(-1)^{n-1}}{n} \left(\frac{2}{x-1}\right)^n \quad \text{for} \quad x \geqq 3.$$

9. In the proof of Theorem 6.3.1 it is stated that uniform convergence of the integrand implies that the limit of the integral is the integral of the limit. Show that this follows from the proof of Theorem 5.5.2 or give a direct proof.

10. If a power series $f(z) = \sum_{k=0}^{\infty} c_k z^k$ converges for $|z| < r$ and if $0 < \rho < r$, prove that there is a finite $M(\rho)$ such that

$$|c_k| \leqq M(\rho)\rho^{-k}$$

for all k. [*Hint:* Try $M(\rho) = \sum_{k=0}^{\infty} |c_k|\, \rho^k$.]

11. Under the same assumptions, prove that for $|z| < \rho$

$$|f^{(n)}(z)| \leqq \frac{\rho M(\rho) n!}{(\rho - |z|)^{n+1}}.$$

12. For x real, $0 \leqq |x| < \rho$, and t between 0 and x, prove that

$$\frac{1}{(n-1)!} |x - t|^{n-1} |f^{(n)}(t)| \leqq \frac{\rho M(\rho)}{(\rho - |t|)^2}\, n \left(\frac{|x - t|}{\rho - |t|}\right)^{n-1}.$$

13. Prove the necessity of condition (ii) of Theorem 6.3.1.

14. Let $a > 1$ be fixed and define

$$F(x) = \sum_{n=0}^{\infty} \frac{(-1)^n}{n!} \frac{1}{1 + a^n x}.$$

Show that this series and all its derived series converge absolutely and uniformly for all $x \geqq 0$. Show, in particular, that $F(x)$ has right-hand derivatives of all orders at $x = 0$ and

$$F^{(k)}(0) = (-1)^k k!\, e^{-a^k}.$$

Prove that the corresponding formal Maclaurin series converges for all x but does not represent $F(x)$. This is an example of case (ii). Note that $F(x)$ becomes infinite when x approaches any one of the points $x = -a^{-n}$, $n = 0, 1, 2, \cdots$, which cluster at $x = 0$.

15. Discuss

$$G(x) = \sum_{n=0}^{\infty} \frac{1}{n!} \frac{1}{1 + a^n x},$$

$a > 1$, in the same manner and show that the corresponding formal Maclaurin series diverges for all $x \neq 0$. This is an example of case (iii).

6.4. Miscellaneous questions. In this section we collect various results not too closely related but belonging to the general theme of this chapter.

I. Uniqueness

We prove that the representation furnished by Taylor's theorem is unique.

Theorem 6.4.1. *If $f(x)$ is defined and has derivatives of order $\leqq n$ for all x in the open interval (a, b), if for some c with $a < c < b$ and all small values of h we have*

$$f(c + h) = a_0 + a_1 h + a_2 h^2 + \cdots + a_{n-1} h^{n-1} + R_n(h), \qquad (6.4.1)$$

where the coefficients a_k are independent of h and $R_n(h) = O(h^n)$ as $h \to 0$, then

$$a_0 = f(c), \quad a_1 = \frac{f'(c)}{1!}, \cdots, a_{n-1} = \frac{f^{(n-1)}(c)}{(n-1)!}, \qquad R_n(h) = R_n(c, h). \quad (6.4.2)$$

Proof. Under the stated assumptions we have

$$f(c + h) = f(c) + \frac{f'(c)}{1!} h + \frac{f''(c)}{2!} h^2 + \cdots + \frac{f^{(n-1)}(c)}{(n-1)!} h^{n-1} + R_n(c, h), \quad (6.4.3)$$

so the assertion is that (6.4.1) must be identical with (6.4.3). If we equate the two representations of $f(c + h)$ and let $h \to 0$, we obtain $a_0 = f(c)$. Canceling a_0 and $f(c)$ and dividing the result by h we obtain

$$a_1 + a_2 h + \cdots + a_{n-1} h^{n-2} + h^{-1} R_n(h)$$
$$= f'(c) + \frac{f''(c)}{2!} h + \cdots + \frac{f^{(n-1)}(c)}{(n-1)!} h^{n-2} + h^{-1} R_n(c, h).$$

Letting h tend to zero we obtain $a_1 = f'(c)$. We repeat the process, cancel equal terms, divide by h, and let $h \to 0$. After n steps we are left with $h^{-n} R_n(h) = h^{-n} R_n(c, h)$ so that $R_n(h) = R_n(c, h)$ and the theorem is proved.

II. Finite Differences

The mean value theorem of the differential calculus associates the first difference of a function with its first derivative. This extends to differences and derivatives of higher order. We recall the notation introduced in Section

3.8. We have

$$\Delta f(x, h) = f(x + h) - f(x),$$
$$\Delta^2 f(x, h) = f(x + 2h) - 2f(x + h) + f(x),$$
$$\cdots\cdots\cdots\cdots\cdots\cdots \tag{6.4.4}$$
$$\Delta^n f(x, h) = \sum_{k=0}^{n} (-1)^{n-k} \binom{n}{k} f(x + kh),$$

and the recurrence relation

$$\Delta[\Delta^{k-1} f(x, h)] = \Delta^k f(x, h). \tag{6.4.5}$$

The mean value theorem now generalizes as follows:

THEOREM 6.4.2. *Suppose that $f(x)$ is defined and has derivatives of order $\leqq n$ in (a, b). If $a < c < b$, and if $n\,|h| < \min\,(c - a, b - c)$, then there exists an η_n, $0 < \eta_n < n$, such that*

$$\Delta^n f(c, h) = h^n f^{(n)}(c + \eta_n h). \tag{6.4.6}$$

If $f^{(n)}(x)$ is continuous at $x = c$, then

$$f^{(n)}(c) = \lim_{h\to 0} \frac{\Delta^n f(c, h)}{h^n}. \tag{6.4.7}$$

Proof. The mean value theorem is the case $n = 1$. The general case is obtained from the special one by alternately applying the difference operator and the mean value theorem. We start with

$$\Delta f(c, h) = hf'(c + \eta_1 h), \quad 0 < \eta_1 < 1.$$

We apply the operator Δ to both sides and obtain

$$\begin{aligned} \Delta^2 f(c, h) &= h\Delta f'(c + \eta_1 h, h) \\ &= h[f'(c + \eta_1 h + h) - f'(c + \eta_1 h)] \\ &= h^2 f''(c + \eta_2 h), \end{aligned}$$

where $\eta_1 < \eta_2 < \eta_1 + 1 < 2$, and the last step follows from an application of the mean value theorem to the function $f'(x + \eta_1 h)$. We then apply the difference operator once more and reduce the result with the aid of the mean value theorem. This can be repeated as long as the required derivatives exist. After n steps we have formula (6.4.6). If $f^{(n)}(x)$ is continuous at $x = c$, formula (6.4.7) follows immediately. This completes the proof and establishes the relations between differences and differentials of higher order stated at the end of Section 3.8.

III. TECHNIQUE OF TAYLOR EXPANSIONS

In principle we obtain formula (6.4.3) and the corresponding infinite series by computing the derivatives of $f(x)$ and evaluating at $x = c$. Often this is easier said than done and it pays to give some thought to less direct methods.

Two examples will illustrate ways and means which may lead to the desired result.

Example 1. Find the Maclaurin series of

$$f(x) = \log [x + (x^2 + 1)^{\frac{1}{2}}]. \tag{6.4.8}$$

This is a composite function and one could perhaps consider the expansions of logarithms occurring in Section 6.3. It is not hard to see, however, that this idea leads to impossible calculations. Now we have

$$f'(x) = (1 + x^2)^{-\frac{1}{2}}, \tag{6.4.9}$$

and from this we can obtain the higher derivatives by repeated differentiation. Again the calculations soon become unmanageable. If, however, we can expand $f'(x)$ in Maclaurin series, then we can also handle $f(x)$, for a power series can be integrated termwise in its circle of convergence by the Corollary of Theorem 5.5.2.

Using formula (6.3.8) with $\alpha = \frac{1}{2}$ and x replaced by x^2, we get

$$(1 + x^2)^{-\frac{1}{2}} = \sum_{n=0}^{\infty} \binom{-\frac{1}{2}}{n} x^{2n}, \quad |x| < 1,$$

and hence

$$\log [x + (x^2 + 1)^{\frac{1}{2}}] = \sum_{n=0}^{\infty} \binom{-\frac{1}{2}}{n} \frac{x^{2n+1}}{2n + 1}, \tag{6.4.10}$$

which is the required expansion.

Example 2. Find the first five terms of the Maclaurin series of

$$f(x) = e^{1/(1-x)}. \tag{6.4.11}$$

This is a composite function. We can find the first four derivatives without too much trouble, but the following argument is more instructive and gives more terms in less time. Since the exponential series (6.3.3) converges for all finite x, we can replace x by $(1 - x)^{-1}$ and obtain

$$f(x) = \sum_{n=0}^{\infty} \frac{1}{n!} (1 - x)^{-n}, \quad x \neq 1.$$

Now

$$(1 - x)^{-n} = 1 + \sum_{k=1}^{\infty} \frac{n(n + 1) \cdots (n + k - 1)}{k!} x^k \tag{6.4.12}$$

so that

$$f(x) = \sum_{n=0}^{\infty} \frac{1}{n!} + \sum_{n=0}^{\infty} \frac{1}{n!} \sum_{k=1}^{\infty} \frac{n(n + 1) \cdots (n + k - 1)}{k!} x^k$$

$$= \sum_{k=0}^{\infty} \frac{1}{k!} \left\{ \sum_{n=0}^{\infty} \frac{n(n + 1) \cdots (n + k - 1)}{n!} \right\} x^k. \tag{6.4.13}$$

Here we have interchanged the order of the two summations, one with respect to k and the other with respect to n. It may be shown that this is permitted for series with positive terms.

This method gives all coefficients at once but only as convergent infinite series. A slight variant of the method gives explicit expressions for the coefficients as finite sums. We note that

$$\frac{1}{1-x} = 1 + \frac{x}{1-x},$$

so that

$$e^{1/(1-x)} = e \cdot e^{x/(1-x)}.$$

This gives

$$\begin{aligned} e^{x/(1-x)} &= \sum_{n=0}^{\infty} \frac{1}{n!} x^n (1-x)^{-n} \\ &= \sum_{n=0}^{\infty} \frac{x^n}{n!} \left\{ \sum_{k=0}^{\infty} \frac{n(n+1)\cdots(n+k-1)}{k!} x^k \right\}. \end{aligned}$$

The presence of the factor x^n in the nth series ensures that any particular power of x, say x^k, occurs in exactly k places when $k > 0$. If we write out the result in a triangular array we get

$$\begin{array}{llllll} 1 + x + x^2 + & x^3 + & x^4 + & x^5 & + \cdots \\ \quad + \frac{1}{2}x^2 + & x^3 + & \frac{3}{2}x^4 + & 2x^5 & + \cdots \\ & + \frac{1}{6}x^3 + & \frac{1}{2}x^4 + & x^5 & + \cdots \\ & & + \frac{1}{24}x^4 + & \frac{1}{6}x^5 & + \cdots \\ & & & + \frac{1}{120}x^5 & + \cdots \\ & & & & + \cdots . \end{array}$$

We sum by columns and multiply by e to obtain the desired expansion

$$e^{1/(1-x)} = e + ex + \tfrac{3}{2}ex^2 + \tfrac{13}{6}ex^3 + \tfrac{73}{24}ex^4 + \cdots . \qquad (6.4.14)$$

This method is superior to repeated differentiation of $f(x)$ even though it does not give the general term of the series.

IV. The nth-order Derivative of a Composite Function

The problem of finding the nth-order derivative of a function is equivalent to finding the coefficient of $h^n/n!$ in the expansion of $f(x + h)$. This observation is useful for composite functions where the finding of an nth-order derivative is usually quite complicated. The following method is also laborious and complicated but makes systematic calculations possible and it gives information in advance of what terms can be expected. Suppose that we consider the composite function $F[f(x)]$ and have expansions

$$F(u + k) = \sum_{\alpha=0}^{p} A_\alpha k^\alpha, \qquad (6.4.15)$$

$$f(x + h) = \sum_{\beta=0}^{q} a_\beta h^\beta, \qquad (6.4.16)$$

where $A_p h^p$ and $a_q h^q$ are remainders and

$$A_\alpha = \frac{F^{(\alpha)}(u)}{\alpha!}, \quad a_\beta = \frac{f^{(\beta)}(x)}{\beta!}$$

for $\alpha < p$, $\beta < q$.

Here we take

$$u + k = f(x + h), \quad u = f(x)$$

and obtain

$$F[f(x + h)] = \sum_{\alpha=0}^{p} A_\alpha [\sum_{\beta=1}^{q} a_\beta h^\beta]^\alpha.$$

Suppose that $n < \min(p, q)$ and that we are looking for terms involving h^n. We note that such terms can arise for $\alpha = 1, 2, \cdots, n$ and nowhere else. This means that the coefficient of h^n is of the form

$$A_1 P_1 + A_2 P_2 + \cdots + A_n P_n,$$

where P_α is the coefficient of h^n in

$$[a_1 h + a_2 h^2 + \cdots + a_q h^q]^\alpha.$$

When this power is expanded, we obtain terms of the form

$$a_1^{\beta_1} a_2^{\beta_2} \cdots a_n^{\beta_n} h^{\beta_1 + 2\beta_2 + \cdots + n\beta_n}, \tag{6.4.17}$$

where the β's are positive integers or zero such that

$$\begin{aligned} \beta_1 + \beta_2 + \cdots + \beta_n &= \alpha, \\ \beta_1 + 2\beta_2 + \cdots + n\beta_n &= n. \end{aligned} \tag{6.4.18}$$

The last condition ensures that the exponent of h equals n. This gives the final expression

$$D^n F[f(x)] = \sum_{\alpha=0}^{n} F^{(\alpha)}[f(x)] \sum \frac{n!}{\beta_1! \cdots \beta_n!} \left[\frac{Df(x)}{1!}\right]^{\beta_1} \cdot \left[\frac{D^2 f(x)}{2!}\right]^{\beta_2} \cdots \left[\frac{D^n f(x)}{n!}\right]^{\beta_n}, \tag{6.4.19}$$

where the β's satisfy conditions (6.4.18).

V. Rational Functions in the Complex Domain

It is of importance to verify that Taylor's theorem and related expansions hold for rational functions of a complex variable. We need several results. The following is trivial.

THEOREM 6.4.3. *If $P(z)$ is a polynomial of degree n in the complex variable z, and if a is any complex number, then*

$$P(z) = P(a) + \frac{P'(a)}{1!}(z - a) + \frac{P''(a)}{2!}(z - a)^2 + \cdots + \frac{P^{(n)}(a)}{n!}(z - a)^n. \tag{6.4.20}$$

Proof. Clearly a polynomial of degree n can be written in one and only one way as a sum of powers of $z - a$

$$P(z) = A_0 + A_1(z - a) + A_2(z - a)^2 + \cdots + A_n(z - a)^n. \tag{6.4.21}$$

For $z = a$ this gives $A_0 = P(a)$. Differentiation and setting $z = a$ in the result gives $A_1 = P'(a)$. Repeating this procedure we see that (6.4.20) and (6.4.21) must be identical. Compare the proof of Theorem 6.4.1.

The following algebraic theorem is less obvious.

THEOREM 6.4.4. *If $Q(z)$ is a given polynomial of degree 2, then every polynomial $P(z)$ can be expressed in one and only one manner as*

$$P(z) = a_0 + b_0 z + (a_1 + b_1 z)Q(z) + (a_2 + b_2 z)[Q(z)]^2 + \cdots + (a_k + b_k z)[Q(z)]^k, \tag{6.4.22}$$

where the a's and b's are constants. If the degree n of $P(z)$ is even, then $2k = n$, $a_k \neq 0$, $b_k = 0$; if n is odd then $2k + 1 = n$ and $b_k \neq 0$.

Remark. This is the natural generalization of (6.4.21) but the coefficients do not have such simple expressions as in the linear case. We have similar representations in terms of polynomials of degree 3 or higher.

Proof. We use the so-called *division algorithm*. We divide $P(z)$ by $Q(z)$ and obtain a quotient $P_1(z)$ of degree $n - 2$ and a remainder $R_0(z)$ of degree one or zero. Next we divide $P_1(z)$ by $Q(z)$ and obtain a quotient $P_2(z)$ of degree $n - 4$ and a remainder $R_1(z)$ of degree $\leqq 1$. This procedure is repeated and after k steps the quotient is of degree $n - 2k$ and this is 0 or 1 so the process comes to an end. Let

$$R_0(z) = a_0 + b_0 z,\ R_1(z) = a_1 + b_1 z, \cdots,$$
$$R_{k-1}(z) = a_{k-1} + b_{k-1} z,\ P_k(z) = a_k + b_k z.$$

Expressing $P(z)$ in terms of $Q(z)$, the remainders, and the last quotient we get (6.4.22).

As an example we may take

$$P(z) = z^5 + 3z^2 + 1, \quad Q(z) = z^2 + 1$$

and obtain

$$z^5 + 3z^2 + 1 = z(z^2 + 1)^2 - (2z - 3)(z^2 + 1) + z - 2.$$

We have finally

THEOREM 6.4.5. *Let $P(z)$ and $Q(x)$ be polynomials without common divisor, let $Q(a) \neq 0$, and let n be any positive integer. Then the rational function $F(z) \equiv P(z)/Q(z)$ may be written*

$$F(z) = F(a) + \frac{F'(a)}{1!}(z - a) + \cdots + \frac{F^{(n-1)}(a)}{(n-1)!}(z - a)^{n-1} + (z - a)^n G_n(z), \tag{6.4.23}$$

where $G_n(z)$ is a rational function and $G_n(a)$ is finite.

Proof. Since $F(a)$ is finite, $F(z)$ has derivatives of all orders at $z = a$. In particular, the indicated derivatives exist. Let us now form

$$D(z) \equiv F(z) - F(a) - \frac{F'(a)}{1!}(z - a) - \cdots - \frac{F^{(n-1)}(a)}{(n-1)!}(z - a)^{n-1}.$$

This is also a rational function, in fact

$$D(z) = \frac{P_1(z)}{Q(z)},$$

where $P_1(z)$ is the polynomial

$$P_1(z) = P(z) - Q(z)\sum_{k=0}^{n-1} \frac{F^{(k)}(a)}{k!}(z - a)^k.$$

Since $Q(a) \neq 0$, $D(a)$ is finite and $D(z)$ has derivatives of all orders at $z = a$. A simple calculation shows that

$$D(a) = D'(a) = \cdots = D^{(n-1)}(a) = 0. \tag{6.4.24}$$

This implies that $P_1(z)$ is divisible by $(z - a)^n$, say $P_1(z) = (z - a)^n P_{n+1}(z)$. For $P_1(a)$ is clearly zero, and if we apply the formula for the kth derivative of a product to $P_1 \cdot Q^{-1}$ and evaluate the derivatives at $z = a$, taking (6.4.24) into account, we obtain successively

$$P_1(a) = 0,\ P_1'(a) = 0, \cdots, P_1^{(n-1)}(a) = 0.$$

This implies that $P_1(z)$ has $(z - a)^n$ as a factor. Hence we have

$$D(z) = (z - a)^n \frac{P_{n+1}(z)}{Q(z)} \equiv (z - a)^n G_n(z),$$

where $G_n(z)$ is rational and $G_n(a) \neq \infty$. This proves that an analogue of Taylor's theorem holds for rational functions in the complex domain.

EXERCISE 6.4

Find $\Delta^3 f(x, h)$ if $f(x)$ is:

1. x^2. 2. $\dfrac{1}{x}$. 3. e^x. 4. $\log x$. 5. x^3.

6. Prove that $\Delta^n f(x, h)$ is identically zero if $f(x)$ is a polynomial of degree $< n$.

7. Find the coefficient of x^n in the Maclaurin series of

$$\frac{\log(1 + x)}{1 + x}.$$

8. Is it possible to expand x in ascending (= increasing) powers of $y = \dfrac{x}{1 - x}$? If so, what is the expansion and for what values of x does it converge?

9. Expand x^2 in ascending powers of $\log(1 + x)$.

Find the first six terms of the Maclaurin series of:

10. $[x - \log(1 + x)]^2$ 11. $\exp[x(1 - x^2)^{-1}]$.

12. Show that the nth derivative of $F(e^x)$ is

$$\sum_{\alpha=1}^{n} \frac{P_\alpha}{\alpha!} e^{\alpha x} F^{(\alpha)}(e^x),$$

where

$$P_\alpha = \alpha^n - \alpha(\alpha - 1)^n + \frac{\alpha(\alpha - 1)}{1 \cdot 2} (\alpha - 2)^n - \cdots.$$

13. If $n = 4$, find all the solutions of (6.4.18) in nonnegative integers. Note that $\alpha = 1, 2, 3$, and 4.

14. Let $P(x)$ be a cubic polynomial with integral coefficients. Show that

$$P(x) = Ax(x - 1)(x - 2) + Bx(x - 1) + Cx + D,$$

where the coefficients are integers.

15. With $P(x)$ as in Problem 14, show that the sum of the series

$$P(0) + \frac{P(1)}{1!} + \frac{P(2)}{2!} + \cdots + \frac{P(n)}{n!} + \cdots$$

is an integral multiple of e.

16. Prove that (i) $zDz^n = nz^n$;

(ii) $zD\{zD(1 - z)^{-1}\} = \sum_{n=1}^{\infty} n^2 z^n$ for $|z| < 1$;

(iii) the sum of the series is $(z + z^2)(1 - z)^{-3}$.

6.5. The fundamental theorem of algebra. This theorem asserts that every algebraic equation has a root. This was taken for granted until the middle of the eighteenth century and the first proof, free from serious objections, was given by C. F. Gauss in his dissertation of 1799. Gauss gave three other proofs and Cauchy one. The proof given below uses the *principle of the minimum* and is due to Weierstrass.

Let $a_1, a_2, \cdots, a_n$ be arbitrary complex numbers and consider the polynomial

$$P(z) = z^n + a_1 z^{n-1} + a_2 z^{n-2} + \cdots + a_{n-1} z + a_n. \tag{6.5.1}$$

It is to be proved that there is at least one value of z, z_0 say, such that

$$P(z_0) = 0. \tag{6.5.2}$$

The basic idea is the following. The function $|P(z)|$ is a nonnegative continuous function of z. In a disk $|z| \leqq R$ this function must attain its maximum as well as its minimum and we can choose R so large that the minimum falls at an interior point because $|P(z)|$ is large everywhere on the boundary. At any point $z = a$ where $P(a) \neq 0$ it is possible to find nearby points where $|P(z)| < |P(a)|$. This implies that no such point can furnish the minimum value of $|P(z)|$, so that the minimum must be 0. Hence there is at least one point z_0 where (6.5.2) holds. We proceed to prove the various steps outlined here.

Set

$$\alpha = \max_{1 \leqq k \leqq n} |a_k|. \tag{6.5.3}$$

Then we have

LEMMA 6.5.1. *$P(z) \neq 0$ for any z with $|z| \geqq 1 + \alpha$.*

Proof. We set $|z| = r$ and note that

$$\begin{aligned}
|P(z)| &\geqq |z|^n - \sum_{k=1}^{n} |a_k|\, |z|^{n-k} \\
&\geqq r^n - \alpha(r^{n-1} + r^{n-2} + \cdots + r + 1) \\
&= r^n - \alpha \frac{r^n - 1}{r - 1} \\
&> r^n - \alpha \frac{r^n}{r - 1} \\
&= r^n \left\{1 - \frac{\alpha}{r - 1}\right\}.
\end{aligned}$$

The expression inside the braces is nonnegative if $r \geqq 1 + \alpha$ and this proves the lemma.

The same inequality shows that

$$|P(z)| \geqq \frac{r^n}{1 + \alpha} \quad \text{if} \quad r \geqq 2 + \alpha. \tag{6.5.4}$$

LEMMA 6.5.2. *The continuous function $|P(z)|$ reaches an absolute minimum at a point z_0 with $|z_0| < 2 + \alpha$.*

Proof. This is certainly true for $n = 1$ for then

$$P(z) = z + a_1$$

and the minimum value of $|P(z)|$ is zero. This value is reached at $z = -a_1$ and $|-a_1| = \alpha < 2 + \alpha$. Let us take $n > 1$ and consider the disk

$$D\colon [z \mid |z| \leqq 2 + \alpha].$$

By (6.5.4)

$$|P(z)| \geqq \frac{(2 + \alpha)^n}{1 + \alpha}$$

on the boundary of D. Since $n > 1$ this expression exceeds

$$1 + \alpha > |a_n|.$$

This says that all along the boundary of the disk the value of $|P(z)|$ exceeds its value at the center since $|P(0)| = |a_n|$. Now $|P(z)|$ has a minimum value in D

and the minimum cannot exceed $|P(0)| = |a_n|$. It follows that the minimum is reached at some point z_0 with $|z_0| < 2 + \alpha$ as asserted.

The crucial step of the proof is based on a theorem due to Weierstrass for analytic functions (= differentiable functions of a complex variable). Such a function can be expanded in a Taylor's series in $z - a$, if it is differentiable at $z = a$, and the existence of such a series is basic for the argument. We showed the existence of a trivial Taylor series for a polynomial (see Theorem 6.4.4). Hence we can formulate and prove Weierstrass's theorem for polynomials and this is the only case needed to complete the proof of the fundamental theorem.

THEOREM 6.5.1. *If $P(z)$ is a polynomial in z and if the complex number a is such that $P(a) \neq 0$, then there exist points in any neighborhood of a where $|P(z)| < |P(a)|$.*

Proof. We start with formula (6.4.21).

$$P(z) = A_0 + A_1(z - a) + \cdots + A_n(z - a)^n.$$

By assumption $A_0 = P(a) \neq 0$ and $A_n \neq 0$ if $P(z)$ is of exact degree n. Some or all of the intermediary coefficients may be zero. To clarify the ideas, suppose that

$$A_1 = A_2 = \cdots = A_{k-1} = 0, \quad A_k \neq 0.$$

We consider $P(z)$ as the sum of three groups of terms

$$A_0,\ A_k(z - a)^k,\ A_{k+1}(z - a)^{k+1} + \cdots + A_n(z - a)^n.$$

The first term is fixed; the other two groups we can vary by varying $z - a$. The idea of the proof is to choose $\arg(z - a)$ so that the vectors representing the first two terms point in opposite directions. If then $|z - a|$ is sufficiently small, the sum of these two vectors is a vector of the same direction as the first but of smaller magnitude. By diminishing $|z - a|$ if necessary, we can achieve that the vector representing the third term is small in comparison with the second so that the sum total is a vector shorter than the first vector. This is the outline; we now note the details.

Let us set

$$\arg A_0 \equiv \gamma_0, \quad \arg A_k \equiv \gamma_k, \quad \arg(z - a) \equiv \theta \pmod{2\pi}.$$

The vector $A_0 = P(a)$ is marked in Figure 6.6 as well as the other two vectors. The second vector points in the opposite direction to the first if

$$\pi + \gamma_0 \equiv \gamma_k + k\theta \pmod{2\pi},$$

or

$$\theta \equiv \frac{1}{k}(\pi + \gamma_0 - \gamma_k) \left(\operatorname{mod} \frac{2\pi}{k}\right). \tag{6.5.5}$$

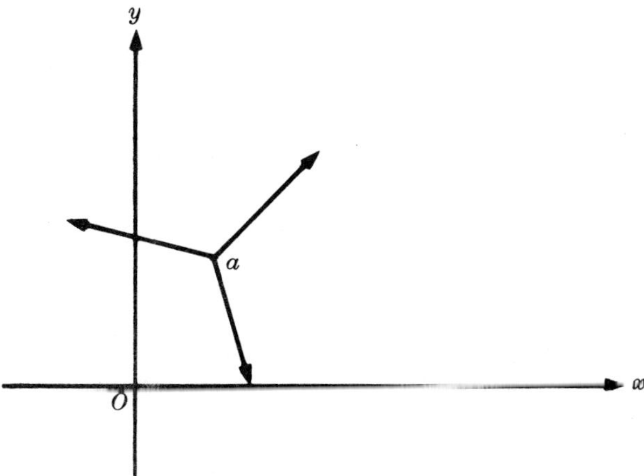

Figure 6.5

Thus there are k possible choices for θ and they make equal angles with each other. In Figure 6.5 $k = 3$.

Next we choose $|z - a|$ so that

$$|A_k|\,|z - a|^k < |A_0|, \quad |z - a| < 1.$$

The first condition requires that

$$\rho = |z - a| < \left|\frac{A_0}{A_\alpha}\right|^{1/k},$$

and it ensures that the second vector does not exceed the first in length. It remains to adjust ρ so that the second vector definitely dominates the third.

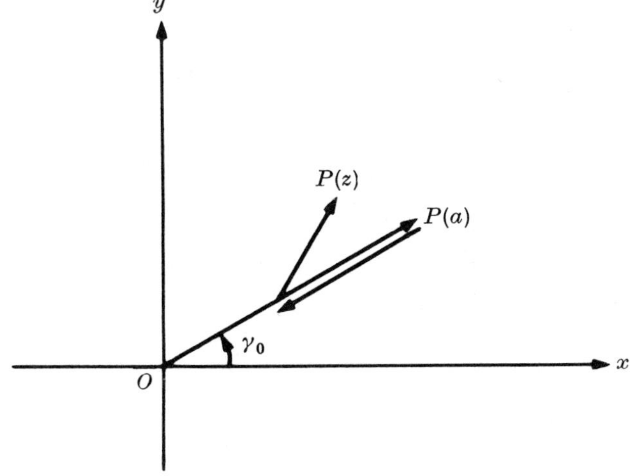

Figure 6.6

If $\rho < 1$, we have

$$|A_{k+1}(z-a)^{k+1} + \cdots + A_n(z-a)^n| < b\,\frac{\rho^{k+1}}{1-\rho}, \quad b = \max_{j>k} |A_j|.$$

Thus the second term is certainly at least twice as large as the third if

$$2b\,\frac{\rho^{k+1}}{1-\rho} < |A_k|\rho^k$$

or

$$\frac{\rho}{1-\rho} < \frac{|A_k|}{2b}$$

which gives finally

$$\rho = |z-a| \leqq \min\left\{\left|\frac{A_0}{A_k}\right|^{1/k}, \frac{|A_k|}{|A_k|+2b}\right\}, \quad b = \max_{j>k} |A_j|. \tag{6.5.6}$$

Suppose now that z is chosen subject to the conditions (6.5.5) and (6.5.6). The points z satisfying these conditions form a "k-pod" in the plane with center at $z = a$ (a "tripod" in Figure 6.5). For every z, $z \neq a$, belonging to the k-pod we have

$$\begin{aligned}|P(z)| &< |A_0| - |A_k|\,|z-a|^k + |A_{k+1}(z-a)^{k+1} + \cdots + A_n(z-a)^n| \\ &< |A_0| - \tfrac{1}{2}|A_k|\,|z-a|^k \\ &< |A_0| = |P(a)|.\end{aligned}$$

This proves the theorem.

It is also possible to find points z in any neighborhood of $z = a$ where $|P(z)| > |P(a)|$. It is easy to show that rotating the k-pod through an angle of π/k we obtain a k-pod on which the inequality $|P(z)| > |P(a)|$ is satisfied. Here we do not need the restriction $P(a) \neq 0$, but if $P(a) = 0$ it is clear that every point in a small neighborhood will give $|P(z)| > |P(a)|$.

We have now the fundamental theorem of algebra.

THEOREM 6.5.2. *Every polynomial in z has at least one zero, or equivalently every algebraic equation has at least one root.*

Proof. By Lemma 6.5.1 $P(z) \neq 0$ for $|z| \geqq 1 + \alpha$ and by Lemma 6.5.2 $|P(z)| > |P(0)|$ on the circle $|z| = 2 + \alpha$. It follows that the minimum of $|P(z)|$ in the disk D is reached in its interior. The minimum must be 0 by Theorem 6.5.1. Hence there is at least one point in D where $P(z) = 0$. This completes the proof.

Suppose that z_1 is a zero of $P(z)$. We have then

$$P(z) = P(z) - P(z_1) = \sum_{k=0}^{n} a_{n-k}(z^k - z_1{}^k) = (z - z_1)Q(z),$$

where $Q(z)$ is a polynomial of degree $n - 1$. If $n > 1$, we conclude that $Q(z)$

also has a zero, say $z = z_2$. Repeating the process we see that $P(z)$ can be factored

$$P(z) = A(z - z_1)(z - z_2) \cdots (z - z_n), \tag{6.5.7}$$

where the n complex numbers z_j are not necessarily distinct. A is the coefficient of the highest power in $P(z)$ and is 1, if $P(z)$ is given by (6.5.1).

If the coefficients in (6.5.1) are real numbers, the zeros of $P(z)$ fall into two groups. There is a set of real zeros $x_1, x_2, \cdots, x_k$, which contains at least one member, if n is odd, but may be void, if n is even. There is a group of pairwise conjugate complex zeros $\alpha_j + i\beta_j$ and $\alpha_j - i\beta_j$ which may be missing. Since

$$(x - \alpha - i\beta)(x - \alpha + i\beta) = (x - \alpha)^2 + \beta^2,$$

we see that $P(x)$ admits of a factorization in linear and quadratic factors with real coefficients:

$$P(x) = \prod_{\nu=1}^{k} (x - x_\nu) \prod_{j=1}^{p} [(x - \alpha_j)^2 + \beta_j^2], \quad n = k + 2p.$$

This is the factorization theorem stated in the title of the dissertation of Gauss.

EXERCISE 6.5

1. Show that the zeros of

$$P(z) = z^5 + 10z^2 + 1$$

lie in the annulus $r < |z| \leq R$, where r and R are the positive roots of

$$s^5 + 10s^2 = 1, \quad t^5 = 10t^2 + 1,$$

respectively. Show also that

$$11^{-\frac{1}{2}} < r < 10^{-\frac{1}{2}}, \quad 2.1 < R < 2.2.$$

2. Take this $P(z)$ and choose $a = -1$. Expand $P(z)$ in powers of $z + 1$. Determine k, θ, and ρ as in formulas (6.5.5) and (6.5.6) above.

3. Given the cubic equation

$$P(z) = z^3 + a_1z^2 + a_2z + a_3 = 0$$

with roots z_1, z_2, z_3, express a_1, a_2, a_3 in terms of the roots. Express

$$\frac{1}{z_1} + \frac{1}{z_2} + \frac{1}{z_3}$$

in terms of a_1, a_2, a_3.

4. Express the sum of the squares of the roots of the preceding equation.

5. How would you characterize the roots of the equation

$$\frac{1}{z - z_1} + \frac{1}{z - z_2} + \frac{1}{z - z_3} = 0.$$

6. Suppose that two roots of the cubic $P(z) = 0$ lie on the imaginary axis while the third lies in the right half-plane. It is assumed that the roots are not collinear. Form the real part of the left member in Problem 5 and show that

it is negative in the left half-plane and positive to the right of the vertical line through the third root of the cubic. Discuss also the imaginary part.

7. If the zeros of a cubic polynomial are not collinear, prove that the zeros of its derivative lie in the interior of the triangle formed by its own zeros. What happens if the zeros are collinear?

8. Assuming the vertices to be points in the complex plane, find the center of gravity of (i) a triangle; (ii) a quadrilateral.

9. What relations must hold between the coefficients of the polynomial

$$z^4 + a_1 z^3 + a_2 z^2 + a_3 z + a_4$$

in order that its zeros form the vertices of a parallelogram? Prove that the zeros of its derivative lie on a straight line. [*Hint:* Use a transformation $z = t + a$ to remove the third-order term and interpret geometrically.]

6.6. Partial fractions. We have seen that polynomials can be represented as sums of powers or by products of powers. There are similar representations for rational functions. Suppose $R(z)$ is the quotient of the two polynomials $P(z)$ and $Q(z)$ where

$$\begin{aligned} P(z) &= a_0 z^m + a_1 z^{m-1} + \cdots + a_m \\ &= a_0(z - z_1)(z - z_2) \cdots (z - z_m), \\ Q(z) &= b_0 z^n + b_1 z^{n-1} + \cdots + b_n \\ &= b_0(z - p_1)(z - p_2) \cdots (z - p_n), \end{aligned} \tag{6.6.1}$$

and $P(z)$ and $Q(z)$ have no roots in common. Then we have the two representations

$$R(z) = \frac{a_0 z^m + a_1 z^{m-1} + \cdots + a_m}{b_0 z^n + b_1 z^{n-1} + \cdots + b_n}, \tag{6.6.2}$$

$$R(z) = \frac{a_0(z - z_1)(z - z_2) \cdots (z - z_m)}{b_0(z - p_1)(z - p_2) \cdots (z - p_n)}. \tag{6.6.3}$$

The second representation exhibits the zeros $\{z_j\}$ and the poles $\{p_k\}$ of the rational function $R(z)$. Here a "zero" is a value of z for which $R(z) = 0$, while a "pole" is a value for which $R(z)$ becomes infinite or, equivalently, the rational function $1/R(z)$ has a zero. It is understood that $P(z)$ and $Q(z)$ may have multiple factors.

These representations are useful but do not meet all our requirements, in particular, they are of little use when it comes to integration of rational functions. Here we need more precise information about the behavior of the function in a neighborhood of a pole. Representation by so-called *partial fractions* meets these requirements. The following one is typical:

$$\frac{z^4 + z^2 + 3}{z^3 - z^2} = z + 1 - \frac{3}{z^2} - \frac{3}{z} + \frac{5}{z - 1}.$$

This representation shows that for large values of z

$$R(z) = z + 1 + o(1)$$

and that $R(z)$ becomes infinite when z tends to 0 or 1, so that

$$R(z) = -\frac{3}{z^2} - \frac{3}{z} + O(1), \quad z \to 0,$$

$$R(z) = \frac{5}{z-1} + O(1), \quad z \to 1.$$

The representation also gives

$$\tfrac{1}{2}z^2 + z + \frac{3}{z} - 3\log z + 5\log(z-1)$$

as a primitive of $R(z)$.

The representation of $R(z)$ by partial fractions involves certain basic fractions. Some of these may be improper like 1 and z above. The proper basic fractions are of the form

$$\frac{1}{(z-a)^k},$$

where a can be any complex number and k any positive integer. The problem is to determine what basic fractions occur in a given case and what are the corresponding constant multipliers. To answer the first question we need only know the location and multiplicity of the poles of $R(z)$, together with the behavior of $R(z)$ for large values of z. "The behavior at infinity" is the phrase used in analytic function theory for such a study and we shall adopt this usage.

We refer to the larger of the two numbers m and n as the *degree* of the rational function. The relation between m and n determines the behavior of $R(z)$ at infinity. We refer to formula (5.2.9) which gives

$$R(z) = z^{m-n}\left\{\frac{a_0}{b_0} + o(1)\right\}, z \to \infty. \tag{6.6.4}$$

If $m > n$ we see that $R(z)$ becomes infinite as $z \to \infty$ and, by definition $R(z)$ has a pole of order $m - n$ at $z = \infty$. If $m < n$, the function tends to the limit 0 and, by definition, $R(z)$ has a zero of order $n - m$ at infinity.

Improper basic fractions will appear if and only if $m \geqq n$. In this case we divide $P(z)$ by $Q(z)$ and obtain a quotient $K(z)$ of degree $m - n$ and a remainder $S(z)$ of degree $< n$ so that

$$R(z) = \frac{P(z)}{Q(z)} = K(z) + \frac{S(z)}{Q(z)}. \tag{6.6.5}$$

Here $K(z)$ is a polynomial and $S(z)/Q(z)$ is a proper fraction. $K(z)$ is a component of the partial fraction expansion of $R(z)$. The rest of the expansion comes from the proper fraction $S(z)/Q(z)$ and involves proper basic fractions $(z-a)^{-k}$ where a is a pole of $R(z)$ and k is less than or equal to the order of this pole which by definition is the same as the order of a as a zero of $Q(z)$.

In the special case exhibited above we have

$$P(z) = z^4 + z^2 + 3, Q(z) = z^3 - z^2, K(z) = z + 1, S(z) = 2z^2 + 3.$$

Let us consider the proper fraction

$$R_1(z) = \frac{2z^2 + 3}{z^3 - z^2}$$

with a view to finding a method which leads to the partial fraction expansion. This function becomes infinite when $z \to 0$ or 1. More precisely

$$z^2 R_1(z) = \frac{2z^2 + 3}{z - 1} \to -3 \quad \text{as} \quad z \to 0,$$

$$(z - 1)R_1(z) = \frac{2z^2 + 3}{z^2} \to 5 \quad \text{as} \quad z \to 1.$$

The first relation shows that $R_1(z)$ is approximately $-3/z^2$ as $z \to 0$. To find how good the approximation actually is, we take the difference of $R_1(z)$ and the approximate value $-3/z^2$ which is

$$\frac{2z^2 + 3}{z^3 - z^2} + \frac{3}{z^2} = \frac{2z + 3}{z(z - 1)}.$$

This still becomes infinite as $z \to 0$, but the difference has a pole of the first order at $z = 0$ while $R_1(z)$ has one of the second order. We have

$$z \frac{2z + 3}{z(z - 1)} = \frac{2z + 3}{z - 1} \to -3 \quad \text{as} \quad z \to 0.$$

Now

$$\frac{2z + 3}{z(z - 1)} + \frac{3}{z} = \frac{5}{z - 1},$$

so we have finally

$$\frac{2z^2 + 3}{z^2(z - 1)} = -\frac{3}{z^2} - \frac{3}{z} + \frac{5}{z - 1},$$

in agreement with the result stated.

This problem suggests a general reduction process that will always lead to the partial fraction expansion in n easy obvious steps, if n is the degree of Q and the fraction is proper so that $m < n$. It is not the quickest way as a rule and once we know what to expect, faster methods will be given.

Suppose that $R(z) = S(z)/Q(z)$ is a proper fraction of degree n and that $Q(z)$ has been factored into linear factors. Suppose that $z = a$ is a zero of $Q(z)$ of multiplicity α so that

$$R(z) = (z - a)^{-\alpha} R_1(z) = \frac{S(z)}{(z - a)^{\alpha} Q_1(z)}. \tag{6.6.6}$$

Here $R_1(z) = S(z)/Q_1(z)$ is a rational function which tends to the limit $R_1(a) \neq 0$ as $z \to a$. The degree of $R_1(z)$ is $\leq \max(n - 1, n - \alpha)$, since the degree of Q_1 is $n - \alpha$ and that of S is $\leq n - 1$. For small values of $|z - a|$ the function $R_1(a)(z - a)^{-\alpha}$ is a first approximation of $R(z)$ since $R_1(a)$ is a first approximation of $R_1(z)$. Theorem 6.4.5 could tell us how close the approximation is, but at this stage a purely algebraic argument is preferable. We form the difference between $R(z)$ and its first approximation

$$R(z) - R_1(a)(z - a)^{-\alpha} = [R_1(z) - R_1(a)](z - a)^{-\alpha}.$$

Here

$$R_1(z) - R_1(a) = \frac{S(z)}{Q_1(z)} - R_1(a) = \frac{S(z) - R_1(a)Q_1(z)}{Q_1(z)}$$

is a rational function which vanishes at $z = a$. This requires that the polynomial $S(z) - R_1(a)Q_1(z)$ is divisible by $z - a$. We denote the quotient by $S_1(z)$ and set $R_2(z) = S_1(z)/Q_1(z)$ so that

$$R_1(z) - R_1(a) = R_2(z)(z - a).$$

Here the degree of $S_1(z)$ is at most $n - 2$, and that of $Q_1(z)$ is $n - \alpha$ so the degree of $R_2(z)$ is $\leqq \max(n - 2, n - \alpha)$. Further, $R_2(z)$ has a finite limit $R_2(a)$ as $z \to a$; in fact

$$R_2(a) = \lim_{z \to a} \frac{R_1(z) - R_1(a)}{z - a} = R_1'(a).$$

This gives

$$R(z) - R_1(a)(z - a)^{-\alpha} = R_2(z)(z - a)^{-\alpha+1}. \tag{6.6.7}$$

If $\alpha > 1$, the right member still has a pole at $z = a$ but the order of the pole is at most $\alpha - 1$. Since the left member tends to zero as z becomes infinite, so does the right member, that is, the latter is a proper fraction. Now

$$R_2(z)(z - a)^{-\alpha+1} = \frac{S_1(z)}{Q_1(z)(z - a)^{\alpha-1}}.$$

Here the degree of the denominator is $n - \alpha + \alpha - 1 = n - 1$. This shows that by subtracting the first approximation $R_1(a)(z - a)^{-\alpha}$ from $R(z)$, we obtain a new proper fraction of degree $\leqq n - 1$ which has a pole of order at most $\alpha - 1$ at $z = a$. Thus the operation lowers both the degree of the rational function and the order of the pole at $z = a$ by at least one unit each.

This shows how the process is to be continued. If $R_2(a) \neq 0$, we can take $R_2(a)(z - a)^{-\alpha+1}$ as first approximation of the right member of (6.6.7). We form

$$R_2(z)(z - a)^{-\alpha+1} - R_2(a)(z - a)^{-\alpha+1} \equiv R_3(z)(z - a)^{-\alpha+2}.$$

Here $R_3(z)$ is a rational function which tends to a finite limit when $z \to a$ and the right member is a proper rational function of order $\leqq (n - 2)$. This holds also if $R_2(a) = 0$. After at most α steps we have

$$\begin{aligned} R(z) = A_0(z - a)^{-\alpha} + A_1(z - a)^{-\alpha+1} + \cdots, \\ + A_{\alpha-1}(z - a)^{-1} + R_{1,\alpha}(z). \end{aligned} \tag{6.6.8}$$

Here the A's are certain definite constants and $R_{1,\alpha}(z)$ is a rational function of degree $n - \alpha$ having a finite limit as $z \to a$. This means that we can shift the attention from the pole $z = a$ of $R(z)$ to one of the other poles, say one at $z = b$

of multiplicity β. We proceed in the same manner, and after $\leqq \beta$ steps we have

$$R_{1,\alpha}(z) = \sum_{j=0}^{\beta-1} B_j(z - b)^{-\beta+j} + R_{2,\beta}(z),$$

where $R_{2,\beta}(z)$ is a rational function of degree $n - \alpha - \beta$ and neither $z = a$ nor $z = b$ is a pole of this function. Since at each step of the reduction we lower the order of the pole under consideration as well as the degree of the remainder by the same positive integer, we get, after not more than n steps, a remainder of degree 0, that is, a constant. Since the remainder always has a zero at infinity, this constant must be 0. The reduction is now complete and if we add up the various basic fractions which were subtracted we obtain the desired representation of $R(z)$ by partial fractions. Thus we have proved

THEOREM 6.6.1. *If $R(z)$ is a rational function with (finite) poles at $z = a$, $b, \cdots$, of multiplicities $\alpha, \beta, \cdots$, respectively, and if $m \geqq n$, then there exists a polynomial $K(z)$ of degree $m - n$ and n constants*

$$A_0, A_1, \cdots, A_{\alpha-1}; B_0, B_1, \cdots, B_{\beta-1}; \cdots$$

such that

$$R(z) = K(z) + \sum_{j=0}^{\alpha-1} A_j(z - a)^{-\alpha+j} + \sum_{j=0}^{\beta-1} B_j(z - b)^{-\beta+j} + \cdots. \tag{6.6.9}$$

Here $K(z)$ is given by (6.6.5). *This term is missing if $m < n$. The coefficients A_j, $B_j, \cdots$ can be determined successively by the process given above.*

This process is not recommended for practical use, but it gives valuable information that can be used for a more rapid calculation of the coefficients. We note first that if $z = a$ is a pole of $R(z)$ of order α, then the coefficient of the highest negative power of $(z - a)$ in the expansion (6.6.9) is

$$A_0 = \lim_{z \to a} (z - a)^\alpha R(z). \tag{6.6.10}$$

Now such a formula holds for each of the poles and the limit process is quite trivial. We simply suppress the factor $(z - a)^\alpha$ in the denominator and substitute $z = a$ in what remains. This gives A_0. If all zeros of $Q(z)$ are simple, this observation gives all the coefficients.

Example 1. We take

$$R(z) = \frac{4z^2 - 3z + 1}{z(z - 1)(z - 2)}$$

which we set equal to

$$\frac{A}{z} + \frac{B}{z + 1} + \frac{C}{z - 2}$$

since the three poles are simple. Omitting one factor at a time in the denominator of $R(z)$, we obtain

$$R_1(z) = \frac{4z^2 - 3z + 1}{(z-1)(z-2)}, \quad R_2(z) = \frac{4z^2 - 3z + 1}{z(z-2)}, \quad R_3(z) = \frac{4z^2 - 3z + 1}{z(z-1)}.$$

Here we substitute $z = 0, 1, 2$, respectively, and obtain

$$A = R_1(0) = \tfrac{1}{2}, \quad B = R_2(1) = -2, \quad C = R_3(2) = \tfrac{11}{2},$$

so that

$$\frac{4z^2 - 3z + 1}{z(z-1)(z-2)} = \frac{\frac{1}{2}}{z} - \frac{2}{z-1} + \frac{\frac{11}{2}}{z-2}.$$

Example 2. We take once more

$$R(z) = \frac{2z^2 + 3}{z^2(z-1)} = \frac{A_0}{z^2} + \frac{A_1}{z} + \frac{B}{z-1}.$$

Here we multiply by z^2 or by $z - 1$ obtaining

$$R_1(z) = \frac{2z^2 + 3}{z - 1}, \quad R_2(z) = \frac{2z^2 + 3}{z^2}$$

and

$$A_0 = R_1(0) = -3, \quad B = R_2(1) = 5.$$

This leaves A_1 to be determined. We shall give a general method, but here we merely observe that

$$\frac{2z^2 + 3}{z^2(z-1)} = -\frac{3}{z^2} + \frac{A_1}{z} + \frac{5}{z-1}$$

is supposedly an identity in z. Hence the relation must hold for any specific choice of z. We take $z = -1$ and obtain

$$-\tfrac{5}{2} = -3 - A_1 - \tfrac{5}{2} \quad \text{or} \quad A_1 = -3$$

in agreement with our previous result.

In the case of multiple poles, Theorem 6.4.5 renders good service.

THEOREM 6.6.2. *If $z = a$ is a pole of $R(z)$ of multiplicity $\alpha > 1$, and if*

$$R_1(z) = (z - a)^\alpha R(z),$$

then the principal part of $R(z)$ at $z = a$ is

$$R_1(a)(z-a)^{-\alpha} + \frac{R_1'(a)}{1!}(z-a)^{-\alpha+1} + \frac{R_1''(a)}{2!}(z-a)^{-\alpha+2} + \cdots \tag{6.6.11}$$

$$+ \frac{R_1^{(\alpha-1)}(a)}{(\alpha-1)!}(z-a)^{-1}.$$

Remark. The *principal part* of $R(z)$ at $z = a$ is that portion of the partial fraction expansion which becomes infinite when $z \to a$. The theorem asserts that in formula (6.6.9)

$$A_j = \frac{R_1^{(j)}(a)}{j!}. \tag{6.6.12}$$

Proof. Since $R_1(z)$ is a rational function with a finite value at $z = a$, Theorem 6.4.5 applies and gives for $n = \alpha$

$$R_1(z) = R_1(a) + \frac{R_1'(a)}{1!}(z-a) + \frac{R_1''(a)}{2!}(z-a)^2 + \cdots$$
$$+ \frac{R_1^{(\alpha-1)}(a)}{(\alpha-1)!}(z-a)^{\alpha-1} + O[(z-a)^\alpha].$$

Here we multiply by $(z-a)^{-\alpha}$ to obtain $R(z)$. The result is (6.6.10) plus a remainder which is $O(1)$, that is, a rational function bounded at $z = a$. This proves the theorem.

We see now that in Example 2 we should have $A_1 = R_1'(0)$. Since

$$R_1'(z) = \frac{2z^2 - 4z - 3}{(z-1)^2} \to -3 \quad \text{as} \quad z \to 0,$$

we again see that $A_1 = -3$.

Using formulas (6.6.10) and (6.6.12) we can always determine the n constants entering in (6.6.9). There is still another method that can be used for this purpose. It is the *method of undetermined coefficients*. We know the existence of a representation of type (6.6.9) where the A's, B's, and so on are the undetermined coefficients. This relation is an identity in z and, hence, if we clear fractions, we get an identity between two polynomials, one of which is $P(z)$. Now two polynomials are identical if and only if equal powers have equal coefficients. On the left the coefficients are known numbers, on the right they are linear combinations of the unknown coefficients. This gives a system of linear equations for the unknowns from which the latter can be determined. If $m < n$ so that $K(z) \equiv 0$, there are n equations in n unknowns. If $m \geqq n$, we may introduce the coefficients of $K(z)$ as additional unknowns, $m - n + 1$ in number, and have then a system of $m + 1$ linear equations in $m + 1$ unknowns. We illustrate the method by an example.

Example 3. We take

$$\frac{z^3 - 3z^2 + 2z + 2}{z^2(z-1)^2} = \frac{A}{z^2} + \frac{B}{z} + \frac{C}{(z-1)^2} + \frac{D}{z-1}.$$

We clear fractions, expand, collect terms, and obtain finally

$$z^3 - 3z^2 + 2z + 2$$
$$= (B + D)z^3 + (A - 2B + C - D)z^2 + (-2A + B)z + A.$$

This gives the following system for the determination of the coefficients:

$$\begin{aligned} B + \quad\quad D &= 1, \\ A - 2B + C - D &= -3, \\ -2A + B \quad\quad &= 2, \\ A \quad\quad\quad &= 2. \end{aligned}$$

Here the fourth equation gives $A = 2$. Substituting this value in the third equation, we obtain $B = 6$. The first equation then gives $D = -5$. Finally the second equation gives $C = 2$. Hence

$$\frac{z^3 - 3z^2 + 2z + 2}{z^2(z-1)^2} = \frac{2}{z^2} + \frac{6}{z} + \frac{2}{(z-1)^2} - \frac{5}{z-1}.$$

This method is the most mechanical of all devices used to find the coefficients and it puts the least demand on understanding and ingenuity. For this reason it is often the only method presented in texts on the Calculus.

In this discussion we have so far made no distinction between real and complex roots. If $R(z)$ is real for real values of z but $Q(z)$ has complex zeros, we know that the latter come in pairs of conjugate numbers and we have a factorization of $Q(z)$ into real factors of the first and the second degree. Corresponding to this factorization of $Q(z)$ there is a representation of $R(z)$ by partial fractions where a pair of conjugate roots gives rise to basic fractions of the form

$$\frac{\alpha_k + \beta_k z}{[(z-a)^2 + b^2]^k}. \tag{6.6.13}$$

Here k can take on the values $1, 2, \cdots, \alpha$ and the coefficients are real. The existence of such an expansion is proved in

THEOREM 6.6.3. *If $R(z)$ is real when z is real, and if $a \pm bi$ are complex roots of $Q(z)$ of multiplicity α, then the corresponding principal parts are of the form*

$$\sum_{j=0}^{\alpha-1} [A_j(z - a - bi)^{-\alpha+j} + \bar{A}_j(z - a + bi)^{-\alpha+j}], \tag{6.6.14}$$

where $\bar{A}_j$ is the conjugate of A_j. This rational function also equals

$$\sum_{j=0}^{\alpha-1} \frac{c_j + d_j z}{[(z-a)^2 + b^2]^{\alpha-j}}, \tag{6.6.15}$$

where the coefficients are real.

Proof. The first assertion follows from Theorem 6.6.2. The two sets of coefficients involved are Taylor coefficients in the expansions of

$$\begin{aligned} R_1(z) &\equiv (z - a - bi)^\alpha R(z) \quad \text{at} \quad z = a + bi, \\ R_2(z) &\equiv (z - a + bi)^\alpha R(z) \quad \text{at} \quad z = a - bi, \end{aligned}$$

respectively. Since these functions assume conjugate imaginary values at conjugate imaginary points so that $\overline{R_1(z)} = R_2(\bar{z})$, we conclude that corresponding Taylor coefficients are also conjugate imaginary. This proves that $\bar{A}_j$ is indeed the conjugate of A_j as asserted. We denote by $H(z)$ the function defined by (6.6.14). We have

$$H(z) = \frac{G(z)}{[(z-a)^2 + b^2]^\alpha}, \tag{6.6.16}$$

where

$$G(z) = \sum_{j=0}^{\alpha-1} [A_j(z-a-bi)^j(z-a+bi)^\alpha + \bar{A}_j(z-a-bi)^\alpha(z-a+bi)^j] \tag{6.6.17}$$

is a polynomial in z of degree $2\alpha - 1$. If $z = x$ is real, then each bracket is the sum of two conjugate complex numbers and hence real. This means that $G(x)$ is real, so that $G(z)$ has real coefficients when written as a polynomial in z. We can now invoke Theorem 6.4.4 to obtain

$$G(z) = c_0 + d_0 z + (c_1 + d_1 z)[(z-a)^2 + b^2] + \cdots + (c_{\alpha-1} + d_{\alpha-1})[(z-a)^2 + b^2]^{\alpha-1}. \tag{6.6.18}$$

We recall that the coefficients of this expansion are obtained by repeated long divisions involving only real polynomials, if the original dividend and divisor are real. Such a process can only give real numbers. Since $G(x)$ and $(x-a)^2 + b^2$ are real polynomials, we conclude that the coefficients c_j and d_j are real. Combining (6.6.16) and (6.6.18) we get the desired formula (6.6.15).

If the A's are already known, we can get the corresponding c's and d's with the aid of Theorems 6.6.3 and 6.4.4. As long as $\alpha \leqq 2$, this procedure is not too laborious. We can of course also arrange the calculations to obtain the real partial fractions without the use of the complex ones. Thus if

$$R(z) = \frac{z^3 + 4z^2 - 4z - 1}{(z^2+1)^2}, \tag{6.6.19}$$

it is waste of time to work with (6.6.9). Since

$$z^3 + 4z^2 - 4z - 1 = (z+4)(z^2+1) - 5z - 5$$

by inspection, we have

$$R(z) = -\frac{5+5z}{(z^2+1)^2} + \frac{4+z}{z^2+1}$$

without further ado.

In more complicated cases it may be necessary to resort to the method of undetermined coefficients. This means that in (6.6.9) we first replace each group of terms of type (6.6.14) by the corresponding group (6.6.15) and then proceed to find the coefficients. If clearing fractions and collecting terms is laborious, it should be realized that the object is to find the coefficients and that equations joining the unknowns can be obtained by substituting particular values of z

just as well as by equating coefficients of like powers of z. If n coefficients are to be found, we need n conditions, and as long as these conditions are correctly derived from the original identity their use is legitimate.

EXERCISE 6.6

Obtain the expansions in partial fractions of the functions:

1. $\dfrac{2z+3}{(z-2)(z+5)}$.
2. $\dfrac{z}{(z+1)(z+2)(z+3)}$.
3. $\dfrac{z^2+1}{z(z^2-1)}$.
4. $\dfrac{z^2+3}{z^2-3z+2}$.
5. $\dfrac{z-1}{z^3+z^2}$.
6. $\dfrac{3z^2+2}{z^3-z^2}$.
7. $\dfrac{z^5-z^3-z^2}{z^2-1}$.
8. $\dfrac{z^2}{(z-1)^2(z+1)}$.
9. $\dfrac{z^6-z^2+1}{(z-1)^3}$.
10. $\dfrac{4}{(z^2-1)^2}$.
11. $\dfrac{z^4+z^3-5z^2+2}{z(z+1)(z-2)}$.
12. $\dfrac{2z-1}{(z+1)^2(z-2)^2}$.

13.–24. Find primitives of the functions in Problems 1–12.

25. If $P(x)$ is a polynomial of degree $< n$, find the coefficients in the partial fraction expansion of

$$P(x)(x-a)^{-n}.$$

26. Decompose $f'(x)/f(x)$ in partial fractions knowing the roots of the polynomial $f(x)$.

27. Decompose

$$\frac{n!}{z(z+1)(z+2)\cdots(z+n)}$$

into partial fractions and find the sum of the coefficients.

Decompose the following functions into real partial fractions:

28. $\dfrac{x^4+1}{x^3(x^2+1)}$.
29. $\dfrac{x^2+1}{(x^2+x+1)^2}$.
30. $\dfrac{3x^2-x+2}{x^3+2x^2+2x}$.
31. $\dfrac{x^3}{(x^2+2x+2)^2}$.
32. $\dfrac{2x^4}{x^4+x^2+1}$.
33. $\dfrac{x^4+2x^2+2}{(x^4-1)(x^2+1)}$.
34. $\dfrac{1}{x^4+4}$.
35. $\dfrac{x^5}{x^4+1}$.
36. $\dfrac{1}{x^3+1}$.
37. $\dfrac{x}{x^3-1}$.
38. $\dfrac{x^3}{x^3+8}$.

39. Decompose the function in Problem 35 into complex fractions and use this to find the real fractions.

40. Find constants a, b, c such that

$$\int\frac{dx}{(x^2+1)^2}=\frac{a+bx}{x^2+1}+c\int\frac{dx}{x^2+1}.$$

This is an adaptation of the method of undetermined coefficients to the problem of integrating a rational function and avoids the usual intermediary step of decomposing the integrand into partial fractions.

41. Show that a rational function of degree d assumes every value d times, if multiple roots are counted with proper multiplicity. What values can be assumed at infinity?

6.7. Indeterminate forms. Many problems in analysis lead to a discussion of limits of expressions which appear in an indeterminate form if the variable is replaced by its limit or if we interchange arithmetical operations and limiting processes when this is not justified. All derivatives belong to this type. Thus the difference quotient

$$\frac{f(x+h)-f(x)}{h}$$

tends to $f'(x)$ when the derivative exists. On the other hand, substituting 0 for h in numerator and denominator gives

$$\frac{0}{0}, \tag{6.7.1}$$

which has no meaning. We get the same result if we say that "the limit of a quotient is the quotient of the limits," thus forgetting the important proviso that the limit of the denominator must be different from 0 for this rule to apply.

Similarly

$$\frac{2z^2+3}{z^2(z-1)}-\frac{5}{z-1}$$

tends to the limit -6 when $z \to 1$, but if we are careless and say that "the limit of a difference is the difference of the limits," we obtain the meaningless form

$$\infty - \infty. \tag{6.7.2}$$

Other such indeterminate forms of frequent occurrence are

$$0 \cdot \infty, \quad \frac{\infty}{\infty}, \quad 1^{\infty}, \quad \infty^0, \quad 0^0. \tag{6.7.3}$$

In the present section we shall consider some general methods applicable to such questions. The seven forms listed fall into three groups. The first of these contains (6.7.1) and the first two forms under (6.7.3). The second group is made up of (6.7.2) and the third group are the remaining three forms under (6.7.3). Using logarithms we can reduce the forms in the third group to forms in the first. Arithmetical juggling of the type

$$0 \cdot \infty = 0 \cdot \frac{1}{0} = \frac{0}{0}$$

suggests that the three forms in the first group are not fundamentally distinct. Similar "considerations" show that the first and the second groups are related. Hence we are justified in concentrating on the form "zero over zero."

Here the basic tools are Taylor's series, Taylor's theorem, and Cauchy's mean value theorem. These tools presuppose differentiable functions, but it is clear that indeterminate forms can be generated by nondifferentiable functions. Such functions are excluded for the time being.

In the simplest case numerator and denominator of the fraction under consideration can be expanded in powers of a variable h. Here the constant terms of the two power series are zero, since for $h = 0$ we are supposed to get the form 0/0. Additional terms at the beginning of the series may vanish. In any case there is a lowest power of h actually present in the numerator and a lowest power of h present in the denominator. The true value of the indeterminate form is decided by these two terms as follows.

THEOREM 6.7.1. *Suppose that $f(x)$ and $g(x)$ can be expanded in Taylor's series at $x = a$ and that*

$$\begin{aligned} f(a+h) &= a_m h^m + a_{m+1} h^{m+1} + \cdots, \\ g(a+h) &= b_n h^n + b_{n+1} h^{n+1} + \cdots, \end{aligned} \tag{6.7.4}$$

where $a_m \neq 0$, $b_n \neq 0$. Then

$$\lim_{h \to 0} \frac{f(a+h)}{g(a+h)} = \begin{cases} 0 & \text{if } m > n, \\ \dfrac{a_m}{b_n} & \text{if } m = n, \\ \pm\infty & \text{if } m < n. \end{cases} \tag{6.7.5}$$

In the last case the sign of the infinite limit is opposite that of a_m/b_n if h increases to 0 and $m - n$ is odd; in all other cases it is the same as that of a_m/b_n.

Proof. We have clearly for small values of h

$$\frac{f(a+h)}{g(a+h)} = \frac{a_m}{b_n} h^{m-n}[1 + o(1)] \quad \text{as} \quad h \to 0, \tag{6.7.6}$$

and the various assertions follow from this formula.

The limit defined by formula (6.6.5) is called the "true value" of the indeterminate form. We shall illustrate this theorem by examples. These include applications to other forms and illustrate by what methods such forms can be handled.

Example 1. We take

$$a = 0, \quad f(x) = e^x - 1 - x, \quad g(x) = x + x^3.$$

Here

$$f(h) = \tfrac{1}{2}h^2 + \tfrac{1}{6}h^3 + \cdots, g(h) = h + h^3,$$

so that

$$m = 2, n = 1, a_2 = \tfrac{1}{2}, b_1 = 1,$$

$$\lim_{h \to 0} \frac{e^h - 1 - h}{h + h^3} = 0. \tag{6.7.7}$$

Example 2. We take

$$a = 0, \quad f(x) = e^x - 1 - x, \quad g(x) = \log(1 + x) - x.$$

Here $m = n = 2$, $a_2 = \frac{1}{2}$, $b = -\frac{1}{2}$, so that

$$\lim_{h \to 0} \frac{e^h - 1 - h}{\log(1 + h) - h} = -1. \tag{6.7.8}$$

Example 3. The following, a more complicated case, involves several different indeterminate forms. We set

$$G(x) = x\left\{\left(1 + \frac{1}{x}\right)^x - e\right\}, \tag{6.7.9}$$

and ask if $G(x)$ approaches a limit as $x \to +\infty$. The power is of the form 1^∞ but, as we shall see, it has the true value e so the brace has the true value 0 and $G(x)$ takes the form $\infty \cdot 0$ at infinity. To replace behavior at infinity by behavior at zero, we set $x = 1/h$. Further we set

$$F(h) \equiv (1 + h)^{1/h}. \tag{6.7.10}$$

It is known that

$$\lim_{n \to \infty} F\left(\frac{1}{n}\right) = \lim_{n \to \infty}\left(1 + \frac{1}{n}\right)^n = e,$$

and we want to show that

$$\lim_{h \to 0} F(h) = e. \tag{6.7.11}$$

More precisely we want to evaluate $F(h) - e$ and find $F'(0)$. For this purpose consider

$$\begin{aligned} \log F(h) &= \frac{1}{h}\log(1 + h) = \frac{1}{h}\left\{\frac{h}{1} - \frac{h^2}{2} + \frac{h^3}{3} - \cdots\right\} \\ &= 1 - \frac{h}{2} + \frac{h^2}{3} - \cdots, \quad |h| < 1. \end{aligned} \tag{6.7.12}$$

This shows first that $\log F(h)$ is a continuous differentiable function of h for $|h| < 1$. It follows that

$$F(h) = \exp[\log F(h)]$$

has the same property. Since

$$\lim_{h \to 0} \log F(h) = 1,$$

formula (6.7.11) must hold. Further, differentiation of (6.7.12) gives

$$\frac{F'(h)}{F(h)} = -\tfrac{1}{2} + \tfrac{2}{3}h - \cdots,$$

and here we can let $h \to 0$ obtaining

$$F'(0) = -\tfrac{1}{2}e.$$

Hence

$$\lim_{h\to 0} \frac{F(h) - e}{h} = -\tfrac{1}{2}e$$

and, finally,

$$\lim_{x\to\infty} G(x) = -\tfrac{1}{2}e. \tag{6.7.13}$$

Example 4. The following problem illustrates how to reduce a case of $\infty - \infty$ to $0/0$. It is required to prove that

$$\lim_{x\to 0} \left\{\frac{1}{e^x - 1} - \frac{1}{x}\right\} = -\tfrac{1}{2}. \tag{6.7.14}$$

This is equivalent to finding the limit of

$$\frac{1 + x - e^x}{x(e^x - 1)} \tag{6.7.15}$$

as $x \to 0$, and here we can apply Theorem 6.7.1. If

$$f(x) = x + 1 - e^x, \quad g(x) = x[e^x - 1],$$

we have

$$m = n = 2, \quad a_2 = -\tfrac{1}{2}, \quad b_2 = 1$$

and (6.7.14) follows.

The above theorem makes severe demands on the functions and we would like less restrictive conditions. This is the objective of the remaining theorems.

THEOREM 6.7.2. *Suppose that $f(x)$ and $g(x)$ are continuous functions defined in some neighborhood of $x = a$ and that $f(a) = g(a) = 0$. Suppose that they are differentiable at $x = a$ and that at least one of the derivatives is not zero. If $g'(a) \neq 0$, then*

$$\lim_{h\to 0} \frac{f(a + h)}{g(a + h)} = \frac{f'(a)}{g'(a)}. \tag{6.7.16}$$

If $g'(a) = 0$, the ratio $f(a + h)/g(a + h)$ may have (i) $+\infty$ *as two-sided limit,* (ii) $-\infty$ *as two-sided limit,* (iii) $+\infty$ *and* $-\infty$ *as one-sided limits, or* (iv) *no limit at all. In the last case every real number may be a cluster point of the ratio as $h \to 0$.*

Proof. The first part follows immediately from the definition of the derivative. If $g'(a) \neq 0$, then

$$\frac{f(a + h)}{g(a + h)} = \frac{f(a + h) - f(a)}{g(a + h) - g(a)} = \frac{[f(a + h) - f(a)/h}{[g(a + h) - g(a)]/h} \to \frac{f'(a)}{g'(a)}$$

as asserted. If $g'(a) = 0$, it is easy to get examples of any of the first three cases using Theorem 6.7.1 with $m = 1$, $n > 1$. We leave the construction to the student. An example of case (iv) is given below, Problem 22.

By demanding a little more of our functions, we can obtain different criteria. The most familiar of these is the *rule of l'Hospital,* called after Guillaume

François Antoine l'Hospital (1661–1704), who published a treatise *Analyse des infinitements petits*, Paris, 1696 where the rule was given. Actually it might be due to Johann Bernoulli, his teacher. We formulate the rule as follows.

THEOREM 6.7.3. *Let f and g be defined and differentiable with $g'(x) \neq 0$ in $0 < |x - a| < \delta$. Let $\lim_{x\to a} f(x) = \lim_{x\to a} g(x) = 0$. Then*

$$\lim_{x\to a} \frac{f(x)}{g(x)} \quad \textit{exists and equals} \quad \lim_{a\to x} \frac{f'(x)}{g'(x)}, \tag{6.7.17}$$

whenever the latter limit exists.

Proof. We set $f(a) = g(a) = 0$ so that f and g are defined and continuous in $|x - a| < \delta$. By Cauchy's mean value theorem [Theorem 6.1.4]

$$\frac{f(x)}{g(x)} = \frac{f(x) - f(a)}{g(x) - g(a)} = \frac{f'(\xi)}{g'(\xi)}, \tag{6.7.18}$$

where $\xi = \xi(x)$ lies between a and x. Hence, if the last member has a limit when $x \to a$, so does the first member and the theorem is proved.

We have formulated the theorem for two-sided approach of x to a, but it also holds for one-sided approach.

Example 5. Take

$$f(x) = e^x - 1, g(x) = x^2 + 2x, a = 0.$$

Here

$$f(0) = g(0) = 0, \quad f'(x) = e^x, \quad g'(x) = 2x + 2,$$

and

$$\lim_{x\to 0} \frac{e^x - 1}{2x + x^2} = \lim_{x\to 0} \frac{e^x}{2 + 2x} = \tfrac{1}{2}.$$

The limit does not exist necessarily if $f'(a) = g'(a) = 0$. But if $f''(x)$ and $g''(x)$ exist, and $g''(x) \neq 0$ for $0 < |x - a| < \delta$, then we can extend (6.7.18) by adding a fourth member

$$\frac{f''(\xi_1)}{g''(\xi_1)}$$

to the equality. Here $\xi_1 = \xi_1(x)$ lies between a and ξ. We now have

$$\lim_{x\to a} \frac{f(x)}{g(x)} = \lim_{x\to a} \frac{f''(x)}{g''(x)}, \tag{6.7.19}$$

provided the last limit exists. If need be, this procedure can be repeated as long as the necessary derivatives exist.

Let us apply this to Example 2 above. Here $a = 0$,

$$f(x) = e^x - 1 - x, f'(x) = e^x - 1, f''(x) = e^x,$$
$$f(0) = f'(0) = 0, \quad f''(0) = 1,$$

and

$$g(x) = \log(1 + x) - x, g'(x) = \frac{1}{1 + x} - 1, g''(x) = -\frac{1}{(1 + x)^2},$$

$$g(0) = g'(0) = 0, \quad g''(0) = -1.$$

This shows that (6.7.19) holds for this function and (6.7.8) is verified.

It is important to observe that the conclusion in (6.7.17) is one-sided: if the second limit exists, so does the first. The converse is false: it is possible for the first limit to exist while the second does not. Such a case will be given in the Miscellaneous Problems for Chapter 7. Actually formula (6.7.18) suggests that the first member cannot have a limit without the second limit in (6.7.17) existing. What does exist is the limit of the third member and we do not know how $\xi(x)$ tends to a as x tends to a. For instance, $\xi(x)$ need not be a continuous function of x.

In applying l'Hospital's rule some judgment is required. The rule should not be applied at all in evaluating something like

$$\lim_{x\to 0} \frac{x^2 - x^3}{x + 2x^2} = \lim_{x\to 0} \frac{x - x^2}{1 + 2x} = 0.$$

Here elementary algebra suffices. In general it pays to simplify the fractions before and after differentiation whenever possible. Suppose, for instance, that the problem is to find

$$\lim_{x\to 0} \frac{\log(1 + x^{\frac{1}{2}})}{x^{\frac{1}{2}} + x}. \tag{6.7.20}$$

Here the change of variable $x^{\frac{1}{2}} = t$ is recommended. The resulting fraction

$$\frac{\log(1 + t)}{t + t^2}$$

can be handled by series expansion or using l'Hospital's rule. Direct application of the latter to the original fraction leads to a complicated fraction of the form ∞/∞ at $x = 0$. Continued differentiation leads only to further complications. The quotient of the derivatives can be simplified, however; multiplication of numerator and denominator by $x^{\frac{1}{2}}$ leads to a fraction with a definite limit.

The rule of l'Hospital admits of various extensions. First, it is permitted to let a be infinite. Let $f(x)$ and $g(x)$ be defined and differentiable for large positive values of x and let $g'(x) \neq 0$ for such values. Further $f(x)$ and $g(x)$ shall tend to 0 as $x \to +\infty$. We can reduce the discussion to the previous case by setting $x = 1/t$. Then

$$\lim_{x\to\infty} \frac{f(x)}{g(x)} = \lim_{t\to 0} \frac{f(t^{-1})}{g(t^{-1})} = \lim_{t\to 0} \frac{-t^{-2}f'(t^{-1})}{-t^{-2}g'(t^{-1})}$$

$$= \lim_{t\to 0} \frac{f'(t^{-1})}{g'(t^{-1})} = \lim_{x\to\infty} \frac{f'(x)}{g'(x)},$$

provided the last limit exists. This proves the desired extension.

L'Hospital's rule is also valid for the case ∞/∞. If $f(x)$ and $g(x)$ become infinite as $x \to a$, then normally $f'(x)$ and $g'(x)$, supposed to exist, also become infinite. In this case it is essential to simplify the quotient of the first derivatives before attempting to differentiate again. Imprudent differentiation leads to a quotient where numerator and denominator become infinite at a faster rate than the original functions.

Example 6. Find

$$\lim_{x\to 0} x \log x = \lim_{x \to 0} \frac{\log x}{x^{-1}} \tag{6.7.21}$$

for approach from the right. The first expression is of the form $0 \cdot \infty$, the second ∞/∞. The quotient of the derivatives is

$$\frac{x^{-1}}{-x^{-2}} = -x \to 0,$$

the desired result. It is obvious that the quotient of the derivatives is of the form ∞/∞ and that the algebraic simplification puts an end to a process which otherwise could be continued indefinitely.

It is often possible and recommendable to replace a quotient of the form ∞/∞ by one of the form $0/0$. For this purpose it suffices to note that

$$f/g = g^{-1}/f^{-1}.$$

While correct, it does not always lead to the desired result. For this reason we shall prove that Theorem 6.7.2 holds also when $f(x) \to \infty$, $g(x) \to \infty$ as $x \to a$.

The proof is again based on a judicious use of Cauchy's mean value theorem. To fix the ideas, suppose that we are interested in one-sided approach from the right. Let $a < x < x_0$ where x_0 will be chosen later. The mean value theorem gives

$$\frac{f(x) - f(x_0)}{g(x) - g(x_0)} = \frac{f(x)}{g(x)} \cdot \frac{1 - f(x_0)[f(x)]^{-1}}{1 - g(x_0)[g(x)]^{-1}} = \frac{f'(\xi)}{g'(\xi)},$$

where $x < \xi < x_0$. Suppose that the last member has the limit Q as $\xi \to a$. We now choose x_0 so near to a that

$$\left|\frac{f'(\xi)}{g'(\xi)} - Q\right| < \varepsilon$$

for any ξ in (a, x_0). We fix x_0 and note that

$$\frac{f(x)}{g(x)} = \frac{f'(\xi)}{g'(\xi)} \cdot \frac{1 - g(x_0)[g(x)]^{-1}}{1 - f(x_0)[f(x)]^{-1}}$$

tends to the limit Q when x decreases to a for this is the limit of the first quotient on the right and the second factor tends to 1 since $f(x) \to \infty$, $g(x) \to \infty$.

A final remark concerning indeterminate forms of the third group. Suppose that

$$F(x) = [f(x)]^{g(x)}$$

produces an indeterminate form of the type 1^∞, ∞^0, or 0^0 as $x \to a$. Then

$$\log F(x) = g(x) \log f(x)$$

gives rise to a form of the type $0 \cdot \infty$ and this can be discussed by the methods developed above. If $\log F(x) \to \alpha$, then $F(x) \to e^\alpha$ by the continuity of the exponential function.

EXERCISE 6.7

Expressions of the form 0/0. Find the limit as $x \to 0$ of:

1. $\dfrac{(1+x)^{\frac{1}{2}} - 1}{x}$.

2. $\dfrac{(1 + x + x^2)^{\frac{1}{2}} - 1 - \frac{1}{2}x}{x^2 + x^3}$.

3. $\dfrac{(1+2x)^{\frac{1}{2}} - (1+3x)^{\frac{1}{3}}}{x^2}$.

4. $\dfrac{(1+x^2)^{\frac{1}{2}} - 1}{(1+x)^{\frac{2}{5}} - 1 - \frac{2}{5}x}$.

5. $\dfrac{\log(1+x+x^2) + \log(1-x+x^2)}{x(e^x - 1)}$.

6. $\dfrac{\exp[(1+x)^{\frac{1}{2}}] - e}{\log(1-x)}$.

Expressions of the form $\infty - \infty$. Find the limit as $x \to 0$ of:

7. $\dfrac{2}{x(x+1)(x+2)} - \dfrac{1}{x}$.

8. $(x - x^2)^{-\frac{1}{2}} - (x + x^2)^{-\frac{1}{2}}$.

9. $\dfrac{1}{x(1+x)} - \dfrac{\log(1+x)}{x^2}$.

10. $\dfrac{\pi x - 1}{2x^2} + \dfrac{\pi}{x(e^{2\pi x} - 1)}$.

Characterize the indeterminate forms and find the limits:

11. $\displaystyle\lim_{x\to\infty} \left(\frac{x^2-1}{x^2+1}\right)^3$.

12. $\displaystyle\lim_{x\to 0} (1 + ax^2)^{x^{-2}}$.

13. $\displaystyle\lim_{x\to 1} \frac{x^{\frac{1}{2}} - x^{\frac{1}{4}}}{x - 1}$.

14. $\displaystyle\lim_{x\to 1} \left\{\frac{x}{x-1} - \frac{1}{\log x}\right\}$.

15. $\displaystyle\lim_{x\to 0+} x^x$.

16. $\displaystyle\lim_{x\to\infty} x \log \frac{x-a}{x+a}$.

17. $\displaystyle\lim_{x\to\infty} \left(\frac{x+1}{x-1}\right)^x$.

18. $\displaystyle\lim_{x\to 0+} \left(\log \frac{1}{x}\right)^x$.

19. $\displaystyle\lim_{x\to 1} x^{1/(1-x)}$.

20. What relation must hold between a and b if

$$\lim_{x\to\infty} \{(x^6 + ax^4 + 1)^{\frac{1}{3}} - (x^4 + bx^2 + 1)^{\frac{1}{2}}\} = \tfrac{1}{6}?$$

21. If $0 < \alpha < \beta$ and if $f(x) = \exp(-x^{-2})$, find

$$\lim_{x\to 0} \frac{f(\alpha x)}{f(\beta x)}.$$

Note that all derivatives of f vanish at $x = 0$.

22. If (u) is the distance from u to the nearest integer, set

$$f(x) = (x - \tfrac{1}{4}) - \tfrac{1}{4}, \quad g(x) = x^2\left\{\left(\frac{1}{x}\right) - \tfrac{1}{4}\right\}, \quad g(0) = 0,$$

and consider the quotient f/g. What are its domain and range? Prove that $g'(0)$ exists and is 0 while $f'(0) = -1$. Show that this quotient illustrates case (iv) of Theorem 6.7.2 when $x \to 0$.

23. Discuss the limit in formula (6.7.20) using (i) the suggested change of variable followed by l'Hospital's rule and (ii) direct attack by the latter.

24. Determine a so that the following limit is finite and $\neq 0$:

$$\lim_{x\to\infty} x^a e^{-x^2} \int_0^x e^{t^2} dt.$$

MISCELLANEOUS PROBLEMS FOR CHAPTER 6

1. If $P(x)$ is a real polynomial with p real zeros, prove that $P(x) - P'(x)$ has at least p real zeros. [*Hint:* Consider the function $1 - P'(x)/P(x)$ near a zero of $P(x)$ and for large x.]

2. According to the mean value theorem

$$(1 + x)^{\frac{1}{2}} - 1 = \tfrac{1}{2}\frac{x}{[1 + x\theta(x)]^{\frac{1}{2}}},$$

where $0 < \theta(x) < 1$. Find the limit of $\theta(x)$ (i) as $x \to 0$; (ii) as $x \to \infty$. Prove that $\theta(x)$ is a continuous differentiable function of x for $-1 < x < \infty$.

3. Discuss the convergence of the remainder $R_n(0, x)$ as $n \to \infty$, using formula (6.2.6) with $f(x) = (1 + x)^\alpha$.

4. Expand $\log x$ in powers of $z = \dfrac{x-1}{x+1}$. For what values of x does the series converge?

5. Find the expansion of e^{2x-x^2} in Maclaurin's series including the terms of degree $\leqq 5$.

6. If $z_1, z_2, \cdots, z_n$ are distinct complex numbers, and

$$P(z) = (z - z_1)(z - z_2) \cdots (z - z_n),$$

find what rational function is represented by the partial fractions

$$\sum_{k=1}^{n} \frac{z_k^{n-1}}{P'(z_k)(z - z_k)}.$$

*7. If $a > 1$, let

$$\frac{x + a}{x^2 + 2ax + 1} = \sum_{n=0}^{\infty} a_n x^n.$$

Find the radius of convergence R of this series and show that for $|x| < R^2$

$$\sum_{n=0}^{\infty} a_n{}^2 x^n = \frac{a^2 - 3(a^2 - \frac{1}{2})x + x^2}{(1 - x)(1 - (4a^2 - 2)x + x^2]}.$$

8. Show that

$$\int_1^\infty \frac{dx}{x(x+1)(x+2)} = \tfrac{1}{2}\log \tfrac{4}{3}.$$

9. Find $\lim_{n\to\infty}\left(1+\frac{x}{n}\right)^n$.

10. Find $\lim_{x\to\infty} x^{\frac{3}{2}}[(x+1)^{\frac{1}{2}} - 2x^{\frac{1}{2}} + (x-1)^{\frac{1}{2}}]$.

11. Find $\lim_{x\to\infty} [(x^4+x^3)^{\frac{1}{4}} - (x^4-x^3)^{\frac{1}{4}}]$.

12. Find $\lim_{x\to 1-} (\log x) \sum_{n=0}^{\infty} \frac{x^n}{n+1}$.

13. Find $\lim_{n\to\infty} (1^{\frac{1}{2}} + 2^{\frac{1}{2}} + \cdots + n^{\frac{1}{2}})n^{-\frac{3}{2}}$.

14. Determine a number a such that

$$\lim_{n\to\infty} n^a \exp(-n^{\frac{1}{2}}) \sum_{k=1}^{n} \exp(k^{\frac{1}{2}})$$

exists as a positive finite quantity.

15. Let $f(x)$ and $g(x)$ be odd functions with continuous derivatives of order $\leqq 3$. Let $g'''(0) \neq 0$. Find

$$\lim_{x\to 0} \frac{f(x) - 2f(2x) + f(3x)}{g(x) - 2g(2x) + g(3x)}.$$

16. A function f is defined for $x < 1$ by

$$f(x) = 2\int_0^x [(1-u)^{-\frac{1}{2}} - 1]\frac{du}{u}.$$

Verify that the integrand is continuous and tends to a finite limit as $u \to 0$. Determine

$$\lim_{x\to 0} \frac{f(x) - x}{x^2}.$$

*17. The function $f(x)$ is supposed to be continuous for $0 \leqq x \leqq 1$. Prove that

$$\lim_{x\to 0} x\int_x^1 f(t)\frac{dt}{t^2} = f(0).$$

18. For $x > 0$ prove the inequality $\log x \leqq x - 1$. If $a_1, a_2, \cdots, a_n$ are positive numbers whose sum equals n, use the inequality to prove that $(a_1 a_2 \cdots a_n)^{1/n} \leqq 1$. With the aid of this prove the following generalization of (1.6.14) and determine when equality holds:

$$(b_1 b_2 \cdots b_n)^{1/n} \leqq \frac{1}{n}(b_1 + b_2 + \cdots + b_n), \quad 0 \leqq b_k.$$

19. Find $f^{(n)}(0)$, if $f(x) = (1+x^2)^{\frac{1}{3}}$.

20. Expand $f(x)$ in Maclaurin series, find the coefficient of x^n, and determine the radius of convergence of the series, if

$$f(x) = \frac{3 + 4x}{(1 - 2x)(1 + 3x)}.$$

21. Let $a > 0$. Show that if the inequality

$$(1 + x)^a > 1 + 3x$$

holds for all small positive values of x, then $a \geqq 3$.

22. Determine a polynomial $P(x)$ such that

$$\lim_{x \to \infty} \left\{ \int_1^x (t^6 + t^3)^{\frac{1}{2}}\, dt - P(x) \right\} \text{ exists.}$$

23. Determine

$$\lim_{x \to \infty} \frac{1}{\log x} \int_x^{2x} \frac{\log t}{1 + t}\, dt.$$

24. Find

$$\lim_{s \to +0} s \sum_{n=2}^{\infty} \frac{1}{n (\log n)^{s+1}}.$$

[*Hint:* Use Cauchy's Integral Test to estimate the series.]

25. If all the zeros of a polynomials $P(x)$ are real, show that the same holds for its derivative $P'(x)$. How about the higher derivatives?

26. The nth Laguerre polynomial is defined by

$$L_n(x) = e^x D^n(x^n e^{-x}).$$

Verify that this really is a polynomial in x of degree n and show that all its zeros are real positive numbers.

27. Expand

$$\log (x + 2) - 2 \log (x + 1) + \log x$$

in powers of $(x + 1)^{-1}$.

7

Some Transcendental Functions

7.1. Trigonometric functions. Analytic functions, that is, differentiable functions of a complex variable, are often subdivided into two classes: *algebraic functions* and *transcendental functions*. For the former, see item 4 under Section 2.3. Special cases such as polynomials, rational functions, and roots of such functions have figured prominently in preceding chapters. A transcendental function is a differentiable function of a complex variable which does not satisfy any algebraic equation with polynomial coefficients.

Actually we shall restrict ourselves in this chapter to a narrow class of such functions, namely the exponential function, the trigonometric functions, and their inverses. Historically these were among the first functions considered and they are used in all applications of mathematics. They also have an inner relationship the full extent of which cannot be appreciated without the use of complex variables. Thanks to these relations they have a number of properties in common. These will play an important role in this chapter. The direct functions are *periodic*, possess *addition theorems*, and satisfy simple *differential equations*. The inverse functions exhibit corresponding properties.

We start with trigonometric functions, in particular, $\sin x$ and $\cos x$ for real values of x. In Section A.6 we introduced radian measure and sines and cosines of arcs on the unit circle. The time has come to put this discussion on a firm foundation, and out first concern is to define the length of arc of a circle. We postpone the general problem of arc lengths until later. The starting point is

LEMMA 7.1.1. *Let ABC be a triangle, and let $P_0P_1P_2 \cdots P_n$ with $P_0 = A$, $P_n = C$, be a polygonal line joining A with C such that the points $P_0, P_2, \cdots P_{n-1}$ are interior to the triangle and $P_0P_1P_2 \cdots P_{n-1}P_nP_0$ is a convex polygon. Then*

$$\begin{aligned} l[A, C] &< l[P_0, P_1] + l[P_1, P_2] + \cdots + l[P_{n-1}, P_n] \\ &< l[A, B] + l[B, C]. \end{aligned} \tag{7.1.1}$$

Proof. Here and below $l[P, Q]$ denotes the length of the line segment $[P, Q]$. In Figure 7.1 $n = 4$ and we give the proof for this case. The extension to other cases is trivial once the principle of the proof is grasped. Here convexity of the pentagon $P_0P_1P_2P_3P_4P_0$ implies that if the sides $[P_0, P_1]$, $[P_1, P_2]$, $[P_2, P_3]$ are continued until they intersect side $[B, C]$ of the triangle, then the points of intersection Q_1, Q_2, Q_3 are encountered in this order when $[B, C]$ is described from B to C. See Section 3.4 for similar considerations.

Now the first half of (7.1.1) holds since the straight line is the shortest distance between the points A and C. For the second half, the sum of the lengths of

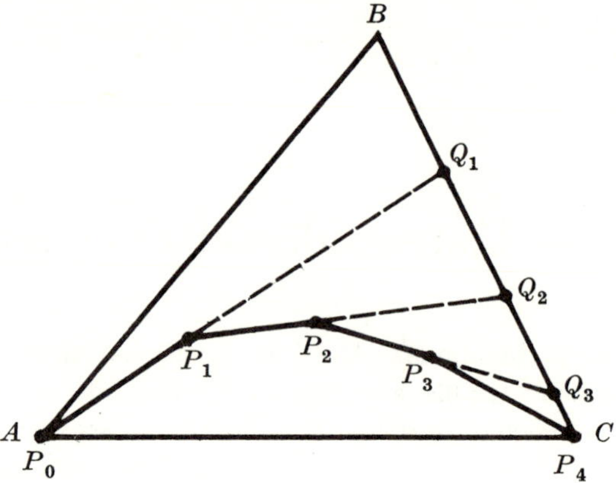

Figure 7.1

two sides of a triangle exceeds the length of the third; thus the following inequalities hold:

$$
\begin{aligned}
l[A, P_1] + l[P_1, Q_1] &< l[A, B] + l[B, Q_1],\\
l[P_1, P_2] + l[P_2, Q_2] &< l[P_1, Q_1] + l[Q_1, Q_2],\\
l[P_2, P_3] + l[P_3, Q_3] &< l[P_2, Q_2] + l[Q_2, Q_3],\\
l[P_3, C] &< l[P_3, Q_3] + l[Q_3, C].
\end{aligned}
$$

Adding these inequalities and noting that all terms $l[P_j, Q_j]$ cancel, we are left with

$$l[A, P_1] + l[P_1, P_2] + l[P_2, P_3] + l[P_3, C] < l[A, B] + l[B, C].$$

This is the second half of (7.1.1), and the lemma is proved.

The lemma applies in particular if the points P_j are located on a circular arc joining A and C and if the lines AB and BC are tangent to the circle. See Figure 7.2. It should be noted that the points P_j determine an order on the arc so that P_j lies between P_{j-1} and P_{j+1} for $j = 1, 2, \cdots, n-1$. Consider now the set L of all possible sums

$$S_n = \sum_{k=0}^{n-1} l[P_k, P_{k+1}], \tag{7.1.2}$$

where n is arbitrary as well as the points P subject to the stated order condition. As we have just observed

$$l[A, C] < S_n < l[A, B] + l[B, C] \tag{7.1.3}$$

for all $n > 1$. Since L is bounded it has a least upper bound.

DEFINITION 7.1.1. *The length $l(AC)$ of the arc AC is the least upper bound of the set L.*

THEOREM 7.1.1. *We have*

$$l[A, C] < l(AC) < l[A, B] + l[B, C]. \tag{7.1.4}$$

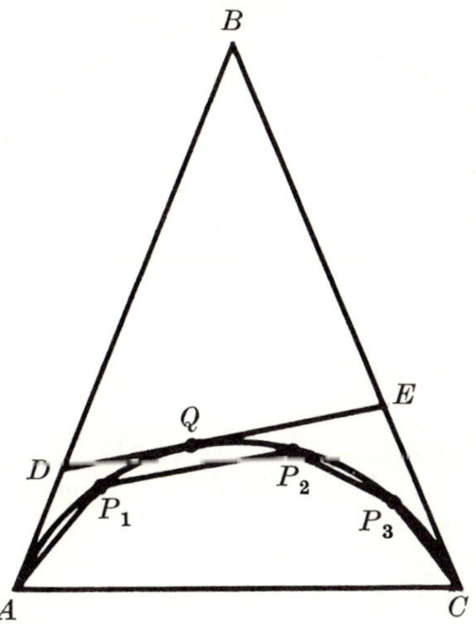

Figure 7.2

Proof. Here the first half of (7.1.4) follows immediately from the first half of (7.1.3) since $l(AC)$ is the least upper bound of L. But the second half of (7.1.3) implies only

$$l(AC) \leqq l[A, B] + l[B, C].$$

To clinch the argument we use the construction indicated in Figure 7.2. The line DQE is tangent to the circle at Q. We note first that

$$l(AC) = l(AQ) + l(QC). \tag{7.1.5}$$

This is almost self-evident. It can be proved by observing that the subset L^* of sums S_n having Q as a vertex has the same least upper bound as the whole set L. For a sum S_n not involving Q is less than a sum S_{n+1} involving the same points P_j and the point Q. Here $S_{n+1} \in L^*$ so that the least upper bound of L^* cannot be less than that of L. Hence it must be equal to the latter. Now a sum belonging to L^* can be split into two subsums, one involving vertices on the arc AQ, the other on the arc QC:

$$S_n = S_p' + S_q'' < l(AQ) + l(QC).$$

On the other hand we can find sums S_p' and S_q'' which differ from $l(AQ)$ and $l(QC)$ by less than ε. This gives

$$l(AQ) + l(QC) - 2\varepsilon < l(AC) \leqq l(AQ) + l(QC)$$

and this implies (7.1.5) since ε is arbitrary.

Since AD and DQ are tangent to the arc AQ, and QE and EC are tangent to the arc QC, we have, as observed above,

$$l(AQ) \leqq l[A, D] + l[D, Q], \; l(QC) \leqq l[Q, E] + l[E, C].$$

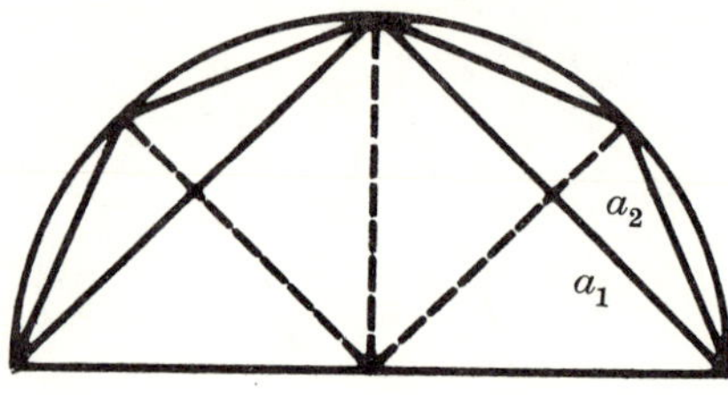

Figure 7.3

Adding the inequalities we get

$$\begin{aligned} l(AC) &\leqq l[A, D] + l[D, Q] + l[Q, E] + l[E, C] \\ &< l[A, D] + l[D, B] + l[B, E] + l[E, C] \\ &= l[A, B] + l[B, C]. \end{aligned}$$

This proves the second half of (7.1.4) and completes the proof.

For the purpose of actually computing the length of the arc, it is not necessary to consider the whole set L. Clearly the least upper bound of L can be determined as the limit of a suitably chosen strictly increasing sequence of sums. We shall illustrate this procedure presently for the case of the semicircle. The arc length has several important properties; the second of these was proved above, and the others we state without proof.

THEOREM 7.1.2. (i) *Congruent circular arcs have the same length.* (ii) *If B is a point on the arc AC, then $l(AB) + l(BC) = l(AC)$.* (iii) *If t is a number between* 0 *and 2π, then there exists an arc AC on the unit circle such that $l(A, C) = t$. Any two such arcs are congruent.*

With our present critical attitude toward elementary trigonometry we have to define the number π of the last theorem and indicate a procedure for determining the number as a limit.

DEFINITION 7.1.2. *π is the length of the semicircle of radius one.*

For the computation we use repeated bisection of the semicircle. This gives a sequence of polygonal lines with 2, 4, 8, 16, $\cdots$, 2^k, $\cdots$ equal sides. See Figure 7.3. If a_k denotes the length of the side at the kth stage, then

$$a_{k+1} = [2 - (4 - a_k{}^2)^{\frac{1}{2}}]^{\frac{1}{2}},\ a_1 = 2^{\frac{1}{2}}, \tag{7.1.6}$$

from which a_n can be computed by recurrence. More elegant formulas can be obtained by noting that

$$a_{k+1} = (2 - c_k)^{\frac{1}{2}},\ c_k = (2 + c_{k-1})^{\frac{1}{2}},\ c_1 = 2^{\frac{1}{2}}. \tag{7.1.7}$$

Thus c_k involves k superimposed square roots. It is easy to show that $\lim_{k\to\infty} c_k = 2$. This procedure gives

$$\pi = \lim_{k\to\infty} 2^k(2 - c_{k-1})^{\frac{1}{2}}. \tag{7.1.8}$$

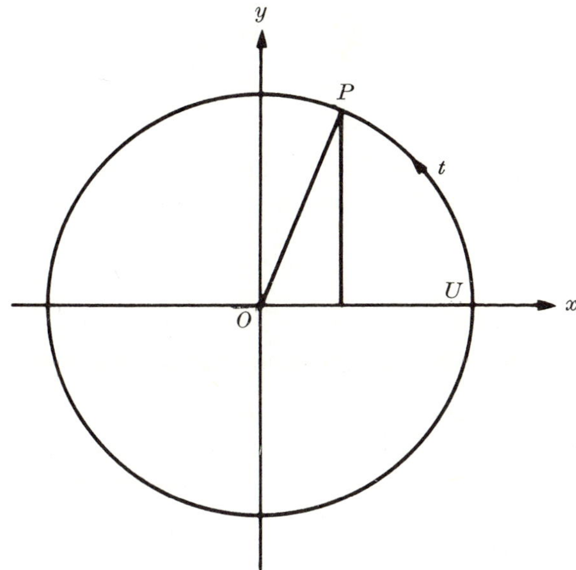

Figure 7.4

The method was invented by Archimedes who supplemented it by using circumscribed polygons. A Dutch mathematician Ludolph van Ceulen (1540–1610) used it to compute π to 35 places and, hence, for a long time π was known as "Ludolph's number." The procedure converges rather slowly and we shall later encounter infinite series for π which permit calculation to any number of places using a high-speed computer.

The preceding considerations make it possible to introduce radian measure as well as sines and cosines. In Figure 7.4 we have drawn the unit circle

$$x^2 + y^2 = 1$$

and marked the points $U(1, 0)$ and $P(x, y)$. The length of the arc UP is a number t with $0 < t < 2\pi$. We consider the coordinates of P as functions of the arc length t and set

$$\begin{aligned} x &= \cos t, \\ y &= \sin t, \end{aligned} \quad 0 \leqq t < 2\pi. \tag{7.1.9}$$

This definition is extended from the interval $[0, 2\pi)$ to all values of t by the convention

$$\cos (t + 2\pi) = \cos t, \quad \sin (t + 2\pi) = \sin t. \tag{7.1.10}$$

This makes the functions periodic of period 2π. Any integral multiple of 2π is also a period.

These functions are continuous for, with obvious notation,

$$\begin{aligned} |\cos t_1 - \cos t_2| &= |x_1 - x_2| \\ &= [(x_1 - x_2)^2 + (y_1 - y_2)^2]^{\frac{1}{2}} \\ &\leqq l[P_1, P_2] < l(P_1 P_2) \\ &= |t_1 - t_2| \end{aligned}$$

and the same inequality holds for the sine. Thus

$$\begin{aligned} |\cos t_1 - \cos t_2| &< |t_1 - t_2|, \\ |\sin t_1 - \sin t_2| &< |t_1 - t_2|. \end{aligned} \tag{7.1.11}$$

Since the functions satisfy a Lipschitz condition, they are continuous for all t.

From the interpretation of $\cos t$ and $\sin t$ as coordinates of a point on the unit circle it follows that

$$\cos^2 t + \sin^2 t = 1 \tag{7.1.12}$$

for all t. The unit circle is symmetric with respect to the origin. This shows that

$$\cos(t + \pi) = -\cos t, \ \sin(t + \pi) = -\sin t. \tag{7.1.13}$$

The unit circle is also symmetric with respect to the x-axis; that is, if (x, y) is on the circle, so is $(x, -y)$. If the former point corresponds to the arc t, counting from U to P, then the latter corresponds to $2\pi - t$ or equivalently to $-t$. This gives

$$\cos(-t) = \cos t, \ \sin(-t) = -\sin t. \tag{7.1.14}$$

It follows that the cosine is an even function of t and the sine is an odd function. The unit circle is also symmetric with respect to the y-axis and with respect to the line $y = x$, and so we obtain the identities

$$\cos(\pi - t) = -\cos t, \ \sin(\pi - t) = \sin t, \tag{7.1.15}$$

$$\cos\left(\frac{\pi}{2} - t\right) = \sin t, \ \sin\left(\frac{\pi}{2} - t\right) = \cos t. \tag{7.1.16}$$

The addition theorems are obtained by the following argument. In Figure 7.5 the arc UP is β, and the arcs P_1P and PP_2 are both α. Here $0 < \alpha, \beta < (\pi/2)$. From the triangle OQP_2 we see that the length of the chord P_1P_2 is $2 \sin \alpha$. We can also find the length of the chord from the distance formula in terms of coordinates of the points. This gives the identity

$$\begin{aligned} (2 \sin \alpha)^2 &= [\cos(\beta + \alpha) - \cos(\beta - \alpha)]^2 + [\sin(\beta + \alpha) - \sin(\beta - \alpha)]^2 \\ &= 2 - 2[\cos(\beta + \alpha)\cos(\beta - \alpha) + \sin(\beta + \alpha)\sin(\beta - \alpha)] \end{aligned}$$

or

$$\cos(\beta + \alpha)\cos(\beta - \alpha) + \sin(\beta + \alpha)\sin(\beta - \alpha) = 1 - 2\sin^2 \alpha.$$

In particular, if $\alpha = \beta$, then

$$\cos 2\alpha = 1 - 2\sin^2 \alpha,$$

and the preceding formula reduces to

$$\cos(\beta + \alpha)\cos(\beta - \alpha) + \sin(\beta + \alpha)\sin(\beta - \alpha) = \cos 2\alpha.$$

Here we set

$$\beta + \alpha = s, \ \beta - \alpha = t \text{ so that } 2\alpha = s - t,$$

and

$$\cos(s - t) = \cos s \cos t + \sin s \sin t. \tag{7.1.17}$$

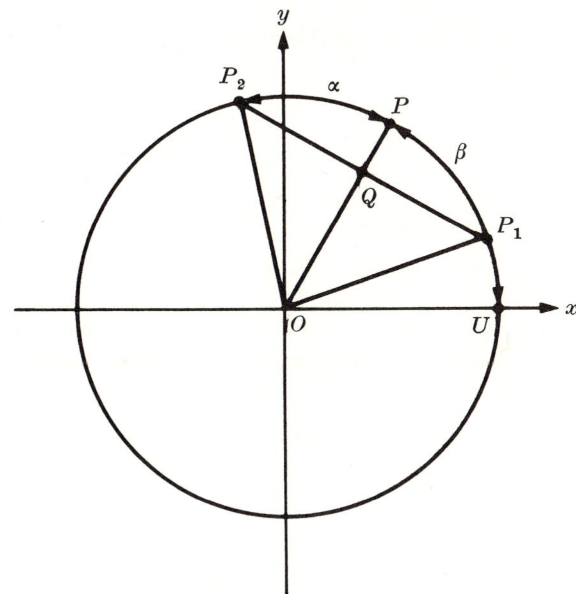

Figure 7.5

The limitations imposed on α, β give corresponding limitations for s and t, but formula (7.1.17) is true for all s and t. This may be seen as follows. First the formula is valid for $0 < s < \pi$, $-\frac{\pi}{2} < t < \frac{\pi}{2}$ by the given proof. By continuity we can replace the open intervals by closed intervals. Next, if s and t are arbitrary numbers, we can find two integers j and k such that

$$s = a + j\pi, \quad 0 \leqq a < \pi, \quad t = b + k\pi, \quad -\frac{\pi}{2} \leqq b < \frac{\pi}{2}.$$

Then

$$\cos (s - t) = \cos (a - b + (j - k)\pi) = (-1)^{j-k} \cos (a - b),$$

while the right member of (7.1.17) becomes

$$\cos (a + j\pi) \cos (b + k\pi) + \sin (a + j\pi) \sin (b + k\pi) \\ = (-1)^{j+k} [\cos a \cos b + \sin a \sin b] = (-1)^{j+k} \cos (a - b).$$

Since these two expressions are equal, formula (7.1.17) holds for all s and t.

Replacing t by $-t$ we get

$$\cos (s + t) = \cos s \cos t - \sin s \sin t, \tag{7.1.18}$$

which is the *addition theorem* for the cosine, also valid for all values of s and t. Replacing s by $\frac{\pi}{2} - s$ in (7.1.17) we get the addition theorem for the sine if we take into account formulas (7.1.16). After simplification one gets

$$\sin (s + t) = \sin s \cos t + \cos s \sin t. \tag{7.1.19}$$

In addition to the sine and the cosine, it is customary to consider four other functions, defined as follows:

$$\tan t = \frac{\sin t}{\cos t}, \quad \cot t = \frac{\cos t}{\sin t}, \tag{7.1.20}$$

$$\sec t = \frac{1}{\cos t}, \quad \operatorname{cosec} t = \frac{1}{\sin t}. \tag{7.1.21}$$

Here $\tan t$ and $\cot t$ have the period π, and the other functions have the period 2π. As quotients of continuous functions they are also continuous except at the points where the denominator in question is zero. These points are the multiples of π for $\cot t$ and $\operatorname{cosec} t$, the odd multiples of $\frac{\pi}{2}$ for the tangent and the secant.

Any one of the six functions can be expressed algebraically in terms of any other. If the expression involves a square root, the sign of the root may vary in different intervals. Thus

$$\sin t = \begin{cases} +(1 - \cos^2 t)^{\frac{1}{2}}, & 0 < t < \pi, \\ -(1 - \cos^2 t)^{\frac{1}{2}}, & \pi < t < 2\pi. \end{cases}$$

So far we have stayed essentially within the framework of trigonometry. The transition to Calculus is via

THEOREM 7.1.3. *For* $0 < t < \pi$

$$0 < \sin t < t < \tan t, \tag{7.1.22}$$

and

$$\lim_{h \to 0} \frac{\sin h}{h} = 1. \tag{7.1.23}$$

Proof. See Figure 7.6. Here RP and RQ are tangents to the circle, the chord PQ is $2 \sin t$, the arc PUQ has the length $2t$, and QR is $\tan t$ since $\wedge\, OQR$ is a right angle. The inequality then follows from Theorem 7.1.1. To prove the limit relation it is sufficient to consider positive values of h since the quotient is an even function of h. From (7.1.22) we obtain

$$\frac{\sin h}{h} < 1 < \frac{\tan h}{h} = \frac{\sin h}{h} \cdot \frac{1}{\cos h}.$$

This implies

$$\cos h < \frac{\sin h}{h} < 1,$$

and hence also (7.1.23), since $\cos h \to 1$ when $h \to 0$.

We can now verify that the six trigonometric functions considered here are differentiable for all t where they are defined and, in fact, possess derivatives of all orders. We start with $\sin t$. The addition theorem gives

$$\frac{1}{h}[\sin (t + h) - \sin t] = \sin t \frac{1}{h}[\cos h - 1] + \cos t \frac{\sin h}{h}.$$

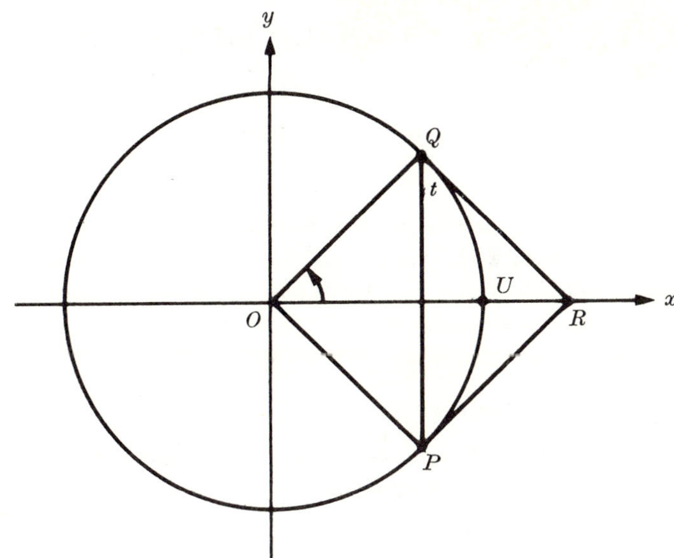

Figure 7.6

Since

$$\frac{1}{h}[1 - \cos h] = 2\,\frac{\sin^2 (h/2)}{h} = \frac{\sin (h/2)}{h/2} \sin (h/2)$$

is less than $\frac{1}{2}|h|$ in absolute value, Theorem 7.1.3 shows the existence of the limit of the difference quotient and gives

$$D_t \sin t = \cos t. \tag{7.1.24}$$

This formula implies that

$$D_t \cos t = D_t \sin \left(\frac{\pi}{2} - t\right) = -\cos \left(\frac{\pi}{2} - t\right) = -\sin t,$$

so that

$$D_t \cos t = -\sin t. \tag{7.1.25}$$

These relations show that $\sin t$ and $\cos t$ have derivatives of all orders which are periodic functions of the order. Thus the first four derivatives of $\sin t$ are

$$\cos t,\ -\sin t,\ -\cos t,\ \sin t.$$

These four derivatives form a cycle which is repeated indefinitely. It is easy to show that

$$D_t^n \sin t = \sin \left(t + n\frac{\pi}{2}\right),\ D_t^n \cos t = \cos \left(t + n\frac{\pi}{2}\right). \tag{7.1.26}$$

It follows from the proof that $\sin t$ and $\cos t$ satisfy the equation

$$D_t^2 y + y = 0. \tag{7.1.27}$$

This is called a *linear second-order differential equation*; "linear" because the left member is a linear function of y and of its derivatives, "second order" because this is the order of the highest derivative. If C_1 and C_2 are arbitrary constants, then

$$C_1 \cos t + C_2 \sin t \tag{7.1.28}$$

is a solution of the equation and it may be shown that every solution is of this form. An alternative form is

$$C \sin (t - \alpha), \tag{7.1.29}$$

where C and α are arbitrary constants. Every solution that is real for real values of t is of this form with C and α real.

The derivatives of the other four functions are obtained from the quotient formula. This gives after simplification

$$\begin{aligned} &D_t \tan t = \sec^2 t,\ D_t \cot t = -\operatorname{cosec}^2 t, \\ &D_t \sec t = \sec t \tan t,\ D_t \operatorname{cosec} t = -\operatorname{cosec} t \cot t. \end{aligned} \tag{7.1.30}$$

The derivatives of higher order can be found, but it is rather laborious and the expressions are complicated so they will be omitted here.

Since

$$\sec^2 t = 1 + \tan^2 t, \tag{7.1.31}$$

it is seen that $\tan t$ is a solution of

$$D_t y = 1 + y^2. \tag{7.1.32}$$

This is a *nonlinear first-order differential equation*. Its general solution is $\tan (t - \alpha)$, where α is an arbitrary constant. There are also two so-called *singular solutions*, not obtainable by special choice of α, namely the constants $+i$ and $-i$.

Since the trigonometric functions have derivatives of all orders, Taylor's theorem applies. In particular, the functions $\sin t$ and $\cos t$ can be expanded in Maclaurin's series convergent and valid for all t. This follows from formula (7.1.26) and Lagrange's form for the remainder, formula (6.2.12). Since the absolute value of the remainder is dominated by $|t|^n/n!$ and thus tends to 0 as $n \to \infty$ for every fixed t, it follows that the series converge and represent the functions. We get the coefficients from (7.1.26). Since $\sin t$ is odd, only terms of odd degree occur in its expansion while in the case of $\cos t$ only terms of even degree are present. We obtain

$$\sin t = \sum_{n=0}^{\infty} \frac{(-1)^n t^{2n+1}}{(2n+1)!} = t - \frac{t^3}{3!} + \frac{t^5}{5!} - \frac{t^7}{7!} + \cdots, \tag{7.1.33}$$

$$\cos t = \sum_{n=0}^{\infty} \frac{(-1)^n t^{2n}}{(2n)!} = 1 - \frac{t^2}{2!} + \frac{t^4}{4!} - \frac{t^6}{6!} + \cdots. \tag{7.1.34}$$

For $\tan t$ and $\sec t$ we can also obtain Maclaurin's series. These series converge and represent the functions for $|t| < \frac{\pi}{2}$. We give only a few terms of each series:

$$\tan t = t + \tfrac{1}{3}t^3 + \tfrac{2}{15}t^5 + \cdots, \tag{7.1.35}$$

$$\sec t = 1 + \tfrac{1}{2}t^2 + \tfrac{5}{24}t^4 + \cdots. \tag{7.1.36}$$

The coefficients are positive rational numbers. The easiest way of computing them is to use the method of undetermined coefficients and the identities

$$\cos t \tan t = \sin t, \ \cos t \sec t = 1.$$

Here we substitute the known power series for $\sin t$ and $\cos t$ and power series for $\tan t$ and $\sec t$ with undetermined coefficients. Multiplying the series and equating coefficients we obtain infinite systems of equations from which the desired coefficients can be determined successively.

The functions $\cot t$ and $\operatorname{cosec} t$ become infinite when $t \to 0$ and cannot be expanded in Maclaurin's series. On the other hand, $t \cot t$ and $t \operatorname{cosec} t$ admit of such expansions convergent for $|t| < \pi$.

The partial sums of the Maclaurin's series of $\sin t$ and $\cos t$ have the interesting property of "enveloping" the function in that for $t > 0$ and $n = 0, 1, 2, 3, \cdots$,

$$\sum_{k=0}^{2n+1} \frac{(-1)^k t^{2k+1}}{(2k+1)!} < \sin t < \sum_{k=0}^{2n} \frac{(-1)^k t^{2k+1}}{(2k+1)!}, \tag{7.1.37}$$

while for $t < 0$ the inequality is reversed. The corresponding inequalities for $\cos t$ hold for all t and are

$$\sum_{k=0}^{2n+1} \frac{(-1)^k t^{2k}}{(2k)!} < \cos t < \sum_{k=0}^{2n} \frac{(-1)^k t^{2k}}{(2k)!}. \tag{7.1.38}$$

For the proof see below.

Formulas (7.1.24), (7.1.25), and (7.1.30) enable us to add six more formulas to our list of primitives (see end of Section 4.7 for formulas (i)–(vii)):

(viii) $\displaystyle\int \sin x \, dx = -\cos x + C,$

(ix) $\displaystyle\int \cos x \, dx = \sin x + C,$

(x) $\displaystyle\int \sec^2 x \, dx = \tan x + C,$

(xi) $\displaystyle\int \operatorname{cosec}^2 x \, dx = -\cot x + C,$

(xii) $\displaystyle\int \sec x \tan x \, dx = \sec x + C,$

(xiii) $\displaystyle\int \operatorname{cosec} x \cot x \, dx = -\operatorname{cosec} x + C.$

We note that

$$\int_0^{2\pi} \sin t \, dt = \int_0^{2\pi} \cos t \, dt = 0,$$

that is, the integrals of sine and cosine over a period are zero. This does not extend to the other four trigonometric functions, since none of them is integrable over a period owing to the function becoming infinite in the interval of integration. The result listed is a special case of the following theorem which we state without proof.

THEOREM 7.1.4. *If $f(x)$ is a linear combination of sines and cosines of the form*

$$f(x) = \frac{a_0}{2} + a_1 \cos x + b_1 \sin x + \cdots + a_n \cos nx + b_n \sin nx, \quad (7.1.39)$$

then for every α

$$\frac{1}{\pi}\int_{\alpha}^{\alpha+2\pi} f(t)\, dt = a_0. \quad (7.1.40)$$

Such a function $f(x)$ is called a *trigonometric polynomial of degree* n (if $|a_n| + |b_n| \neq 0$). Expressions of this type are used to represent a mode of vibration involving a basic frequency and harmonic overtones up to and including the order n.

A discussion of trigonometric integrals will be given in Chapter 11, Vol. II. Here we restrict ourselves to some special cases.

Example 1. The integral

$$\int_0^{2\pi} \sin mx \sin nx\, dx = \begin{cases} \pi, & m = n,\ m \neq 0, \\ 0, & m \neq n, \end{cases} \quad (7.1.41)$$

where m and n are integers, is basic in the theory of trigonometric polynomials and series. It is evaluated using the formula

$$\sin s \sin t = \tfrac{1}{2}[\cos (s - t) - \cos (s + t)] \quad (7.1.42)$$

that is an immediate consequence of the addition theorem. Here we set $s = mx$, $t = nx$, and obtain

$$\sin mx \sin nx = \tfrac{1}{2}[\cos (m - n)x - \cos (m + n)x].$$

We can now apply Theorem 7.1.4 to get the desired result. The same result holds if both sines are replaced by cosines in (7.1.41). If only one sine is replaced by a cosine, the result is always 0.

Example 2. The integral

$$\int x^2 \sin x\, dx = (2 - x^2) \cos x + 2x \sin x + C \quad (7.1.43)$$

can be found by twofold integration by parts. One integrates the trigonometric function and differentiates the power, not vice versa.

Example 3. The integral

$$\int e^{-ax} \sin bx\, dx = -\frac{e^{-ax}}{a^2 + b^2}(a \sin bx + b \cos bx) + C \quad (7.1.44)$$

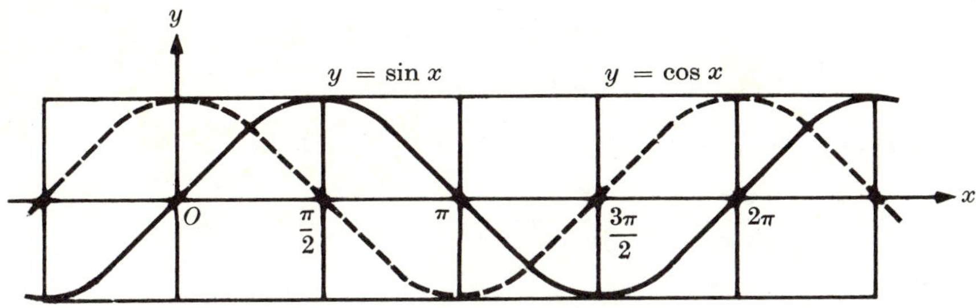

Figure 7.7

can also be obtained by twofold integration by parts, but we shall give a simpler method in the next section.

Example 4. Formulas (7.1.37) can be obtained by repeated integration of the basic inequality

$$\sin x < x,$$

which is valid for all positive x. This is the right half of (7.1.37) for $n = 0$. Here we replace x by t and integrate with respect to t from 0 to x (> 0). This gives

$$1 - \cos x < \tfrac{1}{2}x^2,$$

which is the first half of (7.1.38) for $n = 0$. Again replace x by t and integrate from 0 to x, obtaining

$$x - \sin x < \frac{1}{3!}\,x^3,$$

and this is equivalent to the first half of (7.1.37) for $n = 0$. This procedure can be repeated indefinitely and gives alternately one half or the other of the two formulas.

We end this discussion by giving the graphs in Figures 7.7 and 7.8 of the sine, the cosine, the tangent, and the cotangent. The vertical lines in the Figure 7.8 are asymptotes.

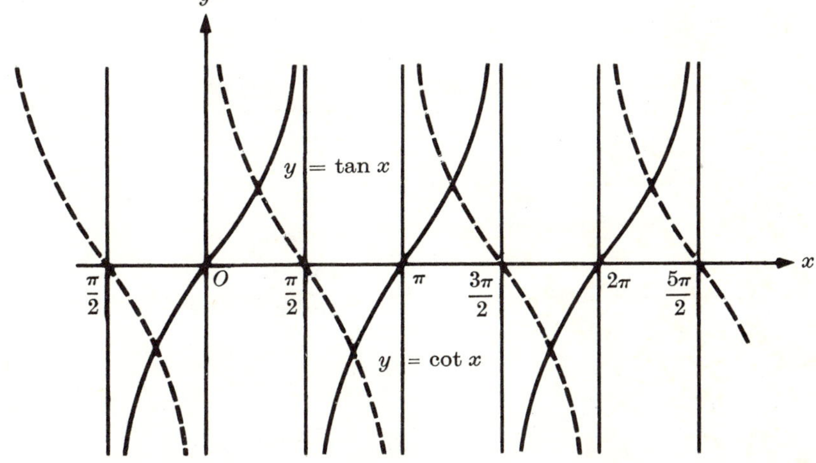

Figure 7.8

EXERCISE 7.1

Find the derivatives of the following functions:

1. $\sin x + \cos 2x$. 2. $\tan x \sec x$. 3. $\tan^2 x$.

4. $\dfrac{\tan x + \cot x}{\tan x - \cot x}$. 5. $\dfrac{2 \tan x}{1 - \tan^2 x}$. 6. $e^x \sin x$.

7. $(a^2 \cos^2 x + b^2 \sin^2 x)^{\frac{1}{2}}$. 8. $\sin (\log x)$. 9. $\log \sin x$.
10. $\log \tan x$. 11. $(\sin x)^{\frac{1}{3}}$. 12. $\exp (\sin x)$.
13. $\sin^5 x \cos x$. 14. $(\cos x)^{-2}$. 15. $\sin (e^x)$. 16. $1 + \tan^2 x$.

17. Show that if n is an odd integer, $n = 2k - 1$, then the nth derivative of $\tan x$ is a polynomial of degree k in the variable $\sec^2 x$ with coefficients that are integers. What happens if $n = 2k$? Use an induction argument.

18. If $-1 \leqq a \leqq +1$, discuss the distribution of the roots of the trigonometric equation $\sin x = a$. If $x = \alpha$ is the root of least absolute value, express the other roots in terms of α. For what values of a does the equation have double roots?

19. Verify that

$$\tfrac{1}{2} + \cos x + \cos 2x + \cdots + \cos nx = \frac{\sin (n + \frac{1}{2})x}{2 \sin \frac{1}{2}x}.$$

20. If $\Sigma\, a_n$ is absolutely convergent, verify that the cosine series

$$\frac{a_0}{2} + \sum_{n=1}^{\infty} a_n \cos nx$$

is absolutely and uniformly convergent for all x.

21. Use partial summation and Problem 19 to verify that the cosine series is uniformly convergent for $0 < \delta \leqq x \leqq 2\pi - \delta$ under the assumption that $\{a_n\}$ is a strictly decreasing null sequence. Note that the assumptions here and in Problem 20 are different and neither implies the other.

22. When does the series converge for $x = 0$?

23. If $f(x)$ is defined by (7.1.39), prove that

$$a_k = \frac{1}{\pi} \int_0^{2\pi} f(t) \cos kt\, dt,\ b_k = \frac{1}{\pi} \int_0^{2\pi} f(t) \sin kt\, dt.$$

24. Verify (7.1.43).

25. Verify (7.1.44).

Find the following integrals:

26. $\displaystyle\int_0^{\pi/4} \tan^2 x \sec^2 x\, dx$. 27. $\displaystyle\int \frac{\cos^3 x}{\sin^4 x}\, dx$.

28. $\displaystyle\int_0^{\pi/2} x^2 \cos x\, dx$. 29. $\displaystyle\int \cos \frac{x}{2} \cos \frac{x}{6}\, dx$.

30. $\displaystyle\int \cos^2 x \sin x\, dx$. 31. $\displaystyle\int \sin^3 x\, dx$.

32. $\int_{\pi/6}^{\pi/2} \cot x \, dx.$

33. $\int \sec x \, dx$ (set $t = \tan \frac{1}{2}x$).

34. $\int_0^{\pi/2} \cos^4 x \tan x \, dx.$

35. $\int_{-\pi}^{\pi} x^2 \sin^3 x \, dx.$

36. Use the observation that $\sin^2 (x/2) + \cos^2 (x/2) = 1$ to compute the integral

$$\int_0^{\pi} (1 + \sin x)^{\frac{1}{2}} \, dx.$$

37. If $P(x)$ is a polynomial of degree n, prove that

$$\int P(x) \cos x \, dx = \sin x \, [P(x) - P''(x) + P^{(4)}(x) - \cdots]$$
$$+ \cos x \, [P'(x) - P'''(x) + P^{(5)}(x) - \cdots].$$

How should the formula be changed if $\cos x$ is replaced by $\sin x$ on the left?

38. Find the addition theorem of the tangent.

39. Verify formulas (7.1.35) and (7.1.36) and find the next term in each of the expansions.

40. Can $\log \sin x$ be expanded in Maclaurin's series?

41. Find the first two terms different from zero in the Maclaurin's series of $\log \cos x$.

42. Sketch a graph of $y = \sec x$.

Find the limits as $x \to 0$ of:

43. $\dfrac{x \cos x - \sin x}{x^3}$.

44. $\cot x - \dfrac{1}{x}$.

45. $\operatorname{cosec}^2 x - \dfrac{1}{x^2}$.

46. Prove that the following series (a fixed positive number) converges but not absolutely:

$$\sum_{n=0}^{\infty} \sin [\pi(n^2 + a^2)^{\frac{1}{2}}].$$

7.2. Exponential function and logarithm. Both these functions were introduced in Section 4.5 and further properties were added later. In this section we shall review briefly the information obtained so far and then proceed to a discussion of the relationship between the exponential function and the trigonometric functions which goes via complex variables.

We first defined

$$\log x = \int_1^x \frac{dt}{t}, \; 0 < x, \tag{7.2.1}$$

with e^x as the inverse function satisfying

$$e^{\log x} = x, \; x > 0,$$
$$\log (e^x) = x, \text{ all } x. \tag{7.2.2}$$

Later it was found that

$$D_x e^x = e^x, \tag{7.2.3}$$

that is, e^x is a solution of the differential equation

$$D_x y = y \tag{7.2.4}$$

which is linear and of the first order and admits Ce^x as its general solution.

From (7.2.3) we also obtain the Taylor expansion

$$e^{x+h} = e^x\left\{1 + \frac{h}{1!} + \frac{h^2}{2!} + \cdots + \frac{h^n}{n!} + \cdots\right\}, \tag{7.2.5}$$

and, in particular, the Maclaurin series

$$e^x = 1 + \frac{x}{1!} + \frac{x^2}{2!} + \cdots + \frac{x^n}{n!} + \cdots. \tag{7.2.6}$$

This converges for all x.

As an example of an indeterminate form of the type 1^∞ we encountered (Problem 9, Miscellaneous Problems for Chapter 6)

$$e^x = \lim_{n\to\infty}\left(1 + \frac{x}{n}\right)^n. \tag{7.2.7}$$

Finally, from (7.2.5) we also obtain that e^x is a solution of the functional equation

$$f(x + h) = f(x)f(h). \tag{7.2.8}$$

This is the *addition theorem* of the exponential function.

In order to get further information we consider complex variables. Both (7.2.6) and (7.2.7) make sense for complex values of the variable. Let us work with the first formula. This means that

$$e^z = 1 + \frac{z}{1!} + \frac{z^2}{2!} + \cdots + \frac{z^n}{n!} + \cdots \tag{7.2.9}$$

by definition. This power series has derivatives of all orders and each derivative equals e^z. In particular, formulas (7.2.3) and (7.2.4) are valid for complex values. We do not have at our disposal a general theory of Taylor's series for complex values so we cannot prove (7.2.5) and (7.2.8) directly. But we can multiply two absolutely convergent power series and know that the Cauchy product series is absolutely convergent and its sum is the product of the sums of the factors (Theorem 5.3.5). This gives

$$e^{z_1}\cdot e^{z_2} = \sum_{n=0}^{\infty}\left\{\sum_{k=0}^{n}\frac{z_1^{\,k}z_2^{\,n-k}}{k!\,(n-k)!}\right\}.$$

Here

$$\frac{1}{k!\,(n-k)!} = \frac{1}{n!}\frac{n!}{k!\,(n-k)!} = \frac{1}{n!}\binom{n}{k}$$

and

$$\sum_{k=0}^{n}\binom{n}{k}z_1^{\,k}z_2^{\,n-k} = (z_1 + z_2)^n.$$

It follows that the sum of the product series is

$$\sum_{n=0}^{\infty} \frac{1}{n!}(z_1 + z_2)^n = e^{z_1+z_2}.$$

Hence

$$e^{z_1} \cdot e^{z_2} = e^{z_1+z_2}, \tag{7.2.10}$$

so the addition theorem holds also for complex values of the variables.

This implies that if $z = x + iy$, then

$$e^z = e^{x+iy} = e^x \cdot e^{iy}.$$

The first factor is well known; to evaluate the second we use (7.2.9). This gives

$$e^{iy} = 1 + \frac{iy}{1!} + \frac{(iy)^2}{2!} + \frac{(iy)^3}{3!} + \cdots + \frac{(iy)^n}{n!} + \cdots.$$

Separating reals and imaginaries we obtain

$$e^{iy} = 1 - \frac{y^2}{2!} + \frac{y^4}{4!} - \frac{y^6}{6!} + \cdots + \frac{(-1)^k y^{2k}}{(2k)!} + \cdots$$
$$+ i\left\{\frac{y}{1!} - \frac{y^3}{3!} + \frac{y^5}{5!} - \cdots + \frac{(-1)^k y^{2k+1}}{(2k+1)!} + \cdots\right\}.$$

Hence by (7.1.33) and (7.1.34)

$$e^{iy} = \cos y + i \sin y, \tag{7.2.11}$$

and finally

$$e^{x+iy} = e^x(\cos y + i \sin y). \tag{7.2.12}$$

These formulas, due to Leonard Euler 1740, have a number of important consequences. The most striking is perhaps the following:

THEOREM 7.2.1. *e^z is periodic with period $2\pi i$.*

This follows from the fact that $\cos y$ and $\sin y$ are periodic with the period 2π. We note that

$$e^{2\pi i} = \cos 2\pi + i \sin 2\pi = 1,$$

and

$$e^{z+2\pi i} = e^z \cdot e^{2\pi i} = e^z$$

as asserted.

Next we note the representation of a complex number by polar coordinates:

$$c = a + ib = r(\cos\theta + i \sin\theta),\ r = |c|,\ \theta = \arg c.$$

This can now be written

$$c = re^{i\theta}. \tag{7.2.13}$$

Now de Moivre's theorem (B.1.14)

$$(\cos\theta + i \sin\theta)^n = \cos n\theta + i \sin n\theta$$

becomes evident for

$$(\cos\theta + i \sin\theta)^n = (e^{i\theta})^n = e^{ni\theta} = \cos n\theta + i \sin n\theta$$

Further

$$e^{x+n\pi i} = (-1)^n e^x$$

so that the exponential function is real on the lines $y = n\pi$ in the complex plane, positive if n is even, negative if n is odd. Thus the exponential function can actually be negative though not for real values of z. Since $e^m \neq 0$ formula (7.2.12) shows that $e^z \neq 0$.

We have listed consequences of formula (7.2.12) for the exponential function; there are also important consequences for the trigonometric functions. From

$$e^{it} = \cos t + i \sin t,$$

$$e^{-it} = \cos t - i \sin t,$$

we get

$$\sin t = \frac{1}{2i}(e^{it} - e^{-it}), \ \cos t = \tfrac{1}{2}(e^{it} + e^{-it}), \tag{7.2.14}$$

and by division

$$\tan t = \frac{1}{i}\frac{e^{it} - e^{-it}}{e^{it} + e^{-it}}. \tag{7.2.15}$$

In other words the trigonometric functions are simple rational functions of

$$w = e^{it}.$$

Thus

$$\cos t = \frac{w^2 + 1}{2w}, \ \operatorname{cosec} t = \frac{2iw}{w^2 - 1}, \tag{7.2.16}$$

and so on.

Let us get some simple consequences of these formulas. If n is an integer, $n \neq 0$,

$$\int_0^t e^{nis}\, ds = \frac{1}{ni}(e^{nit} - 1). \tag{7.2.17}$$

In particular,

$$\int_0^\pi e^{nis}\, ds = \frac{1}{ni}[(-1)^n - 1], \int_0^{2\pi} e^{nis}\, ds = 0. \tag{7.2.18}$$

In working with trigonometric integrals one is often faced with a positive integral power of $\sin t$ or of $\cos t$. It is advantageous to express such a power as sums of $\sin kt$ or $\cos kt$. This can be done as follows. We have

$$(\sin t)^n = (2i)^{-n}[e^{it} - e^{-it}]^n = (2i)^{-n}\left(w - \frac{1}{w}\right)^n \tag{7.2.19}$$

with w defined as above. Here we expand using the binomial theorem and note that the binomial coefficients are symmetric with respect to the middle of the expansion. Collecting terms with the same coefficients and assuming n even,

$n = 2k$, we get

$$
\begin{aligned}
(\sin t)^{2k} &= (-1)^k 2^{-2k}\Big\{[w^{2k} + w^{-2k}] - \binom{2k}{1}[w^{2k-2} + w^{-2k+2}] + \cdots \\
&\quad + (-1)^{k-1}\binom{2k}{k-1}[w^2 + w^{-2}] + (-1)^k\binom{2k}{k}\Big\} \\
&= (-1)^k 2^{1-2k}\Big\{\cos 2kt - \binom{2k}{1}\cos(2k-2)t + \cdots \\
&\quad + (-1)^{k-1}\binom{2k}{k-1}\cos 2t + (-1)^k\tfrac{1}{2}\binom{2k}{k}\Big\}.
\end{aligned} \tag{7.2.20}
$$

If n is odd, there is no constant term and the expansion involves sines of odd multiples of t. Similar formulas hold for $(\cos t)^n$, but here the expansions always involve cosines of multiples of t, even multiples if n is even, odd if n is odd.

From these formulas we obtain, for instance,

$$\int_0^{\pi/2} (\sin t)^{2k}\, dt = \int_0^{\pi/2} (\cos t)^{2k}\, dt = 2^{-1-2k}\binom{2k}{k}\pi. \tag{7.2.21}$$

In this case we could tell in advance that only the constant term in (7.2.20) will give a contribution to the value of the integral since

$$\int_0^{\pi/2} \cos 2pt\, dt = 0$$

if p is any integer. Hence it is sufficient to find the constant term in the expansion and we can disregard all other terms. This speeds up the computation.

As another application of the same formulas we note that from the expansion

$$(\sin x)^5 = \tfrac{1}{16}[\sin 5x - 5\sin 3x + 10 \sin x],$$

it follows that

$$\int (\sin x)^5\, dx = -\tfrac{1}{16}[\tfrac{1}{5}\cos 5x - \tfrac{5}{3}\cos 3x + 10\cos x] + C.$$

The same method is of use in evaluating integrals of a product of a polynomial in x multiplied or divided by the square root of a quadratic polynomial. As an example we may consider

$$\int_0^a (a^2 - x^2)^{\frac{1}{2}}\, dx.$$

This is reduced to a trigonometric integral by the substitution

$$x = a \sin t,\ dx = a \cos t\, dt,$$

the limits $0, a$ become $0, \dfrac{\pi}{2}$ and the integral equals

$$a^2 \int_0^{\pi/2} \cos^2 t\, dt = \frac{\pi}{4} a^2.$$

We have

$$\int e^{-(a+bi)x}\,dx = -\frac{e^{-(a+bi)x}}{a+bi} + C,$$

where a and b are supposed to be real. By Euler's formula

$$e^{-(a+bi)x} = e^{-ax}(\cos bx - i\sin bx).$$

Further

$$\frac{1}{a+bi} = \frac{a-bi}{a^2+b^2},$$

and separation of reals and imaginaries gives

$$\int e^{-ax}\cos bx\,dx = -\frac{e^{-ax}}{a^2+b^2}(a\cos bx - b\sin bx) + C, \tag{7.2.22}$$

$$\int e^{-ax}\sin bx\,dx = -\frac{e^{-ax}}{a^2+b^2}(a\sin bx + b\cos bx) + C. \tag{7.2.23}$$

The second of these formulas figured as (7.1.44) in the preceding section. In particular we obtain

$$\int_0^\infty e^{-ax}\cos bx\,dx = \frac{a}{a^2+b^2}, \tag{7.2.24}$$

$$(a > 0)$$

$$\int_0^\infty e^{-ax}\sin bx\,dx = \frac{b}{a^2+b^2}. \tag{7.2.25}$$

Such integrals (as functions of a) are known as Laplace transforms after Pierre Simon Laplace (1749–1827) whose work on celestial mechanics, cosmogony, potential theory, and theory of probability is justly famous.

Let us consider briefly the logarithm for complex values. In analogy with the case when z is real positive we could use formula (7.2.1) as a definition of $\log z$. Now x is replaced by z which cannot be real and negative and we integrate along a straight line from 1 to z. However, it is better to use the first formula (7.2.2). This gives

A number w is a logarithm of the complex number $z \neq 0$, if

$$e^w = z. \tag{7.2.26}$$

Since the exponential function is periodic with period $2\pi i$ this means that, if w is a solution of the equation, every number of the form $w + 2n\pi i$ is also a solution, n being any integer. To solve the equation let us set

$$z = re^{i\theta},\ w = u + iv.$$

Then

$$e^u(\cos v + i\sin v) = re^{i\theta},$$

so that

$$u = \log r,\ v \equiv \theta \pmod{2\pi}$$

and

$$\log z = \log|z| + i\arg z. \tag{7.2.27}$$

Here the logarithm on the right is the ordinary real-valued natural logarithm. The symbol on the left is uniquely defined only when arg z is uniquely defined, for instance, as the principal value of the argument. In the following when we consider logarithms of complex numbers, we shall specify the meaning of the term whenever necessary for the understanding.

We say that the logarithm has the *additive period* $2\pi i$ corresponding to the period $2\pi i$ of the exponential function. We shall find similar additive periods when we study the inverse trigonometric functions.

EXERCISE 7.2

Evaluate the following integrals:

1. $\int_0^\pi (\cos x + \sin x)^2\, dx.$
2. $\int_0^\pi \sin^2 x \cos^2 x\, dx.$
3. $\int_0^{2\pi} |2 + 3e^{ix}|^2\, dx.$
4. $\int_0^1 (1 - x^2)^{\frac{3}{2}}\, dx.$
5. $\int_0^\pi \sin^2 x \cos^3 x\, dx.$
6. $\int_1^{e^\pi} \sin(\log x)\,\frac{dx}{x}.$
7. $\int_0^a \frac{dx}{(a^2 - x^2)^{\frac{1}{2}}}.$
8. $\int_0^\infty e^{-ax}\cos^2 x\, dx.$
9. $\int_0^\pi (1 + \cos x)^{\frac{1}{2}}\, dx.$
10. $\int_0^{\pi/2} \sin x \sin 2x \sin 3x\, dx.$
11. $\int_0^\pi (1 + \cos^2 x)^{\frac{1}{2}} \sin x\, dx.$
12. Prove that $|e^{ix} - 1| \leqq |x|$, x real.
13. Prove that

$$|1 + e^{ix} + e^{2ix} + \cdots + e^{nix}| \leqq \min(n + 1, \operatorname{cosec} \tfrac{1}{2}\delta) \text{ if } 0 < \delta \leqq x \leqq 2\pi - \delta.$$

14. Verify that

$$2^{-2k}\binom{2k}{k} = \frac{(2k - 1)!!}{(2k)!!},$$

where $m!!$ is the product of all positive integers not exceeding m and of the same parity as m. Thus $5!! = 1 \cdot 3 \cdot 5$, $6!! = 2 \cdot 4 \cdot 6$.

15. Prove that e^{2x} and e^{-2x} are solutions of the differential equation $D_x^{\,2}y = 4y$.

16. Prove that e^{it} and e^{-it} are solutions of (7.1.27). Can either of them be written in the form $C \sin(t - \alpha)$?

17. If $x > 1$ and $y = \log[x + (x^2 - 1)^{\frac{1}{2}}]$, prove that

$$x = \tfrac{1}{2}(e^y + e^{-y}).$$

18. Prove that for positive values of x

$$\left(1 + \frac{x}{n}\right)^n < e^x$$

by expanding both sides in powers of x and comparing coefficients.

19. What are the values of $\log(-1)$, $\log i$, $\log \frac{1}{2}(1 + i\sqrt{3})$?
20. If we define i^i as $\exp(i \log i)$, what values are then assigned to the power?

7.3. The hyperbolic functions. This name is given to the following functions:

$$\sinh x = \tfrac{1}{2}(e^x - e^{-x}), \quad \cosh x = \tfrac{1}{2}(e^x + e^{-x}),$$
$$\tanh x = \frac{\sinh x}{\cosh x}, \quad \coth x = \frac{\cosh x}{\sinh x}. \tag{7.3.1}$$

If time permits, pronounce the symbol "hyperbolic sine" rather than "cinch" and so on. Incidentally, British students call it "shine." There are at least three sets of notation used for these functions. The symbols used here are standard American and British notation. The very compact French notation is sh x, ch x, th x, and cth x. The Germans write 𝔖𝔦𝔫 x and so on.

The trigonometric functions are often called *circular functions* on account of their close relation to the *unit circle*. We recall that a parametric representation of the latter is

$$x = \cos t, \; y = \sin t,$$

where t is the arc of the circle from the point $(1, 0)$ to the point (x, y).

The functions (7.3.1) have similar close connections with the *equilateral hyperbola*

$$x^2 - y^2 = 1. \tag{7.3.2}$$

See Figure 7.9. A parametric representation of the right branch of the hyperbola is given by

$$x = \cosh t, \; y = \sinh t \tag{7.3.3}$$

for

$$\begin{aligned}\cosh^2 t - \sinh^2 t &= \tfrac{1}{4}(e^t + e^{-t})^2 - \tfrac{1}{4}(e^t - e^{-t})^2 \\ &= \tfrac{1}{4}[e^{2t} + 2 + e^{-2t} - e^{2t} + 2 - e^{-2t}] \\ &= 1\end{aligned}$$

as asserted.

Here the parameter t is not a length but an area. We shall show that

$$t = m(OP_1VPO) \tag{7.3.4}$$

that is, the area of the curvilinear triangle bounded by the two rays OP_1 and OP and the arc P_1VP of the hyperbola. Here

$$P = (\cosh t, \sinh t), \; P_1 = (\cosh t, -\sinh t).$$

We start by finding the area of the region bounded by the arc VP of the hyperbola and the line segments VX and XP. A point (u, v) lies on the hyperbola if

$$v = (u^2 - 1)^{\frac{1}{2}}.$$

It follows that the area of VPX is

$$A = \int_1^x (u^2 - 1)^{\frac{1}{2}}\, du.$$

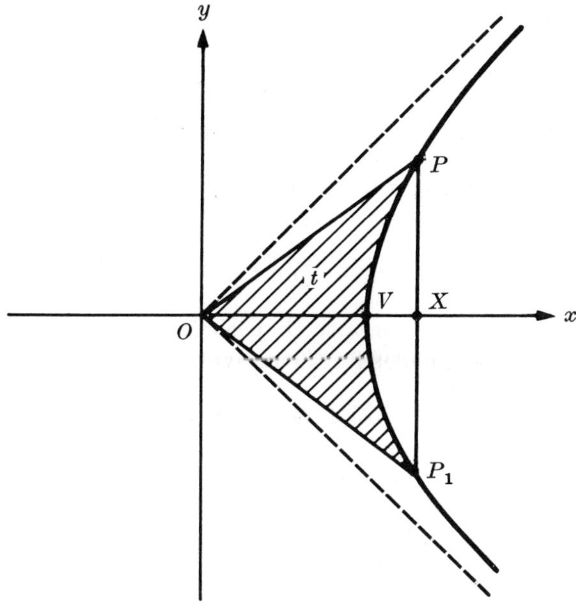

Figure 7.9

Here we substitute the parametric representation

$$u = \cosh s,\ v = \sinh s,\ du = \sinh s\, ds,\ (u^2 - 1)^{\frac{1}{2}} = \sinh s$$

and obtain

$$\begin{aligned} A &= \int_0^t \sinh^2 s\, ds \\ &= \tfrac{1}{4}\int_0^t [e^{2s} - 2 + e^{-2s}]\, ds \\ &= \tfrac{1}{8}[e^{2t} - e^{-2t}] - \tfrac{1}{2}t \\ &= \tfrac{1}{4}\sinh 2t - \tfrac{1}{2}t. \end{aligned}$$

On the other hand, the area of the triangle OXP is, by (7.3.7) below,

$$\tfrac{1}{2}xy = \tfrac{1}{2}\cosh t \sinh t = \tfrac{1}{4}\sinh 2t.$$

Subtracting the two areas we get $\frac{1}{2}t$ and this proves (7.3.4).

We get a closer analogy with the circular case by observing that the area of the circular sector $OPUQO$ in Figure 7.5 equals t when t is the length of the arc UQ.

In working with hyperbolic functions one should always keep in mind the defining relations (7.3.1). Any identity between hyperbolic functions must be a consequence of these formulas.

There is a "hyperbolic trigonometry" analogous to the circular kind. We list the basic formulas below. Some were already used in the proof of (7.3.4). We start with the *addition theorems*

$$\sinh (s + t) = \sinh s \cosh t + \cosh s \sinh t, \tag{7.3.5}$$

$$\cosh (s + t) = \cosh s \cosh t + \sinh s \sinh t. \tag{7.3.6}$$

Hence

$$\sinh 2s = 2 \sinh s \cosh s, \ \cosh 2s = \cosh^2 s + \sinh^2 s. \tag{7.3.7}$$

The basic identity is

$$\cosh^2 t - \sinh^2 t = 1, \tag{7.3.8}$$

used above in the parametrization of the hyperbola.

We also have obvious formulas for the derivatives

$$D_t \sinh t = \cosh t, \ D_t \cosh t = \sinh t. \tag{7.3.9}$$

They show that

$$C_1 \sinh t + C_2 \cosh t \tag{7.3.10}$$

is a solution (actually the general solution) of the second-order linear differential equation

$$D_t^2 y - y = 0. \tag{7.3.11}$$

Another consequence of formulas (7.3.9) is the Maclaurin series

$$\sinh t = t + \frac{1}{3!} t^3 + \frac{1}{5!} t^5 + \cdots + \frac{1}{(2k+1)!} t^{2k+1} + \cdots, \tag{7.3.12}$$

$$\cosh t = 1 + \frac{1}{2!} t^2 + \frac{1}{4!} t^4 + \cdots + \frac{1}{(2k)!} t^{2k} + \cdots. \tag{7.3.13}$$

These series can also be obtained from the first two formulas under (7.3.1), where we substitute the exponential series. It follows that the series converge for all t and represent the functions.

Obviously a close relationship exists between the circular and the hyperbolic functions. We pass from functions of one class to functions of the other essentially by replacing t by it and dividing the result by some power of i. The basic relation is obtained by defining the circular functions for complex values of the argument. This can be done either via Euler's formulas or via the Maclaurin series. The former alternative gives

$$\sin z = \frac{1}{2i} (e^{iz} - e^{-iz}), \ \cos z = \frac{1}{2} (e^{iz} + e^{-iz}). \tag{7.3.14}$$

If in these formulas we set $z = iy$, y real, then we see that

$$\sin (iy) = i \sinh y, \ \cos (iy) = \cosh y. \tag{7.3.15}$$

These relations underly all the above formulas and explain why the hyperbolic functions act so much like circular functions and at the same time are so strangely different.

We can add two formulas to our list of integrals:

$$\text{(xiv)} \quad \int \sinh t \, dt = \cosh t + C,$$

$$\text{(xv)} \quad \int \cosh t \, dt = \sinh t + C.$$

The hyperbolic functions are quite useful when it comes to evaluating integrals involving the product of a polynomial and the square root of a quadratic polynomial. We recall that circular functions could be used for similar purposes. Actually the two classes complete each other. Thus if the quadratic polynomial is of the form

$$a^2 - (x - c)^2, \quad \text{then} \quad x - c = a \sin t \tag{7.3.16}$$

is the best substitution, while

$$a^2 + (x - c)^2 \quad \text{calls for} \quad x - c = a \sinh t \tag{7.3.17}$$

and

$$-a^2 + (x - c)^2 \quad \text{requires} \quad x - c = a \cosh t. \tag{7.3.18}$$

The following example will illustrate the method.

Example 1. We take

$$I = \int_3^5 x^2(x^2 - 4x + 3)^{\frac{1}{2}} \, dx.$$

Here

$$x^2 - 4x + 3 = (x - 2)^2 - 1$$

is of type (7.3.18), so we set

$$x = 2 + \cosh t, \; dx = \sinh t \, dt, \; (x^2 - 4x + 3)^{\frac{1}{2}} = \sinh t.$$

If b is the positive root of the equation

$$\cosh b = 3 \quad \text{or} \quad e^{2b} - 6e^b + 1 = 0, \tag{7.3.19}$$

we get

$$\begin{aligned} I &= \int_0^b (2 + \cosh t)^2 \sinh^2 t \, dt \\ &= \int_0^b [4 \sinh^2 t + 4 \sinh^2 t \cosh t + \cosh^2 t \sinh^2 t] \, dt. \end{aligned}$$

The first two terms in the last member give

$$\sinh 2b - 2b + \tfrac{4}{3} \sinh^3 b,$$

while the third integral equals

$$\tfrac{1}{4} \int_0^b \sinh^2 2t \, dt = \tfrac{1}{8} \int_0^{2b} \sinh^2 s \, ds = \tfrac{1}{32} \sinh 4b - \tfrac{1}{16} b,$$

so that

$$I = \sinh 2b + \tfrac{4}{3} \sinh^3 b + \tfrac{1}{32} \sinh 4b - \tfrac{33}{16} b.$$

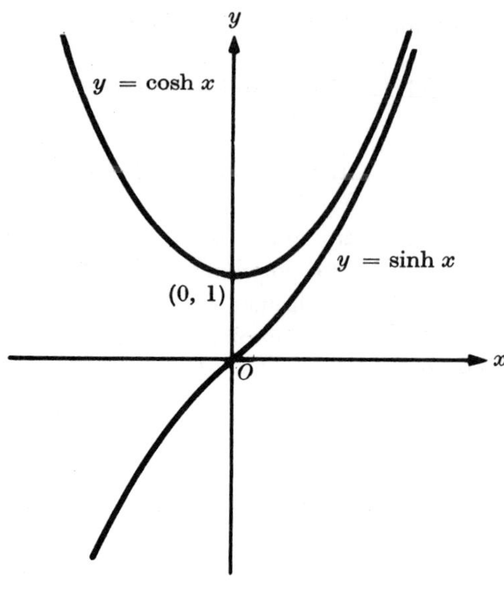

Figure 7.10

Now (7.3.19) is a quadratic in e^b and if $b > 0$, then $e^b > 1$. This gives

$$b = \log(3 + 2\sqrt{2}).$$

Knowing that $\cosh b = 3$, we can get $\sinh b$, $\sinh 2b$, and $\sinh 4b$ from formulas (7.3.7) and (7.3.8). We get successively

$$\sinh b = 2\sqrt{2},\ \sinh 2b = 12\sqrt{2},\ \cosh 2b = 17,\ \sinh 4b = 408\sqrt{2},$$

so that finally

$$I = \tfrac{553}{12}\sqrt{2} - \tfrac{33}{16}\log(3 + 2\sqrt{2}).$$

A sketch of the graphs of $\sinh x$ and $\cosh x$ is given in Figure 7.10. Here $\sinh x$ is odd and is strictly increasing, while $\cosh x$ is even and strictly decreasing for $x < 0$, strictly increasing for $x > 0$. The curve

$$y = \frac{c}{2}[e^{x/c} + e^{-x/c}] \tag{7.3.20}$$

is known as a *catenary* (L. *catena*, chain). A flexible, inelastic chain or cord of uniform line density, suspended at its two ends, will form an arc of a catenary under the influence of gravity. It is supposed that the thickness of the cord or chain can be disregarded.

Since $\sinh x$ is strictly increasing, it has a unique inverse which we can compute from the equation

$$e^{2x} - 2ye^x - 1 = 0. \tag{7.3.21}$$

This gives

$$x = \log[y + (y^2 + 1)^{\frac{1}{2}}]. \tag{7.3.22}$$

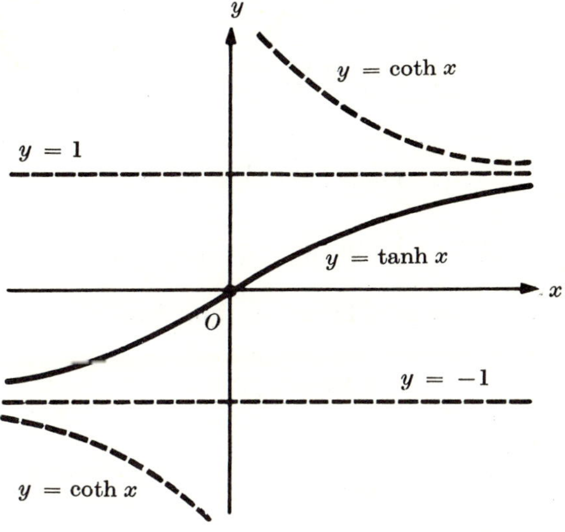

Figure 7.11

The hyperbolic tangent is an odd strictly increasing bounded function. This follows from the formula

$$\tanh x = \frac{\sinh x}{\cosh x} = \frac{e^x - e^{-x}}{e^x + e^{-x}} = 1 - \frac{2}{e^{2x} + 1}.$$

This shows that $\tanh x$ lies between -1 and $+1$ and, moreover, that

$$\lim_{x \to -\infty} \tanh x = -1, \quad \lim_{x \to +\infty} \tanh x = +1. \tag{7.3.23}$$

On the other hand, the hyperbolic cotangent is unbounded. It is an odd function which is greater than 1 for $x > 0$ and less than -1 if $x < 0$. We have

$$\lim_{x \to -\infty} \coth x = -1, \quad \lim_{x \to +\infty} \coth x = +1. \tag{7.3.24}$$

From formula (7.3.12) we conclude that

$$\lim_{x \to 0} \frac{\sinh x}{x} = 1 \quad \text{whence} \quad \lim_{x \to 0} x \coth x = 1. \tag{7.3.25}$$

The graphs of $\tanh x$ and $\coth x$ are sketched in Figure 7.11. Both curves have the lines $y = -1$ and $y = +1$ as horizontal asymptotes. Since $y = \tanh x$ is strictly increasing, there is a unique inverse given by

$$x = \tfrac{1}{2} \log \frac{1 + y}{1 - y}, \; -1 < y < 1. \tag{7.3.26}$$

The addition theorem of the hyperbolic tangent reads

$$\tanh (s + t) = \frac{\tanh s + \tanh t}{1 + \tanh s \tanh t}. \tag{7.3.27}$$

The derivative is given by

$$D_t \tanh t = \frac{4e^{2t}}{(e^{2t}+1)^2} = \frac{4}{(e^t+e^{-t})^2} = \frac{1}{\cosh^2 t} = \operatorname{sech}^2 t. \qquad (7.3.28)$$

We note also

$$\text{(xvi)} \qquad \int \tanh t\, dt = \log \cosh t + C,$$

$$\text{(xvii)} \qquad \int \coth t\, dt = \log |\sinh t| + C.$$

This completes our discussion of hyperbolic functions.

EXERCISE 7.3

Find the derivatives of:

1. $\sinh x \cosh 2x.$ 2. $\coth^2 x.$ 3. $\dfrac{\cosh x + \sinh x}{\cosh x - \sinh x}.$

4. $\tan x \tanh x.$ 5. $e^{-x} \sinh x.$ 6. $x^2 \sinh x.$

7. Find the area between the curves in Figure 7.10 to the right of the y-axis.

8. Find the area between the curve $y = \tanh x$ and the line $y = 1$ to the right of the y-axis.

Find the following integrals:

9. $\displaystyle\int \sinh^3 x\, dx.$ 10. $\displaystyle\int \tanh^3 x\, dx.$ 11. $\displaystyle\int_0^\infty \frac{\sinh^3 x}{\cosh^4 x}\, dx.$

12. Expand $\tanh x$ in Maclaurin series through the term in x^5.

13. Verify (7.3.25).

14. Express $\tanh 4x$ in terms of $\tanh x$.

15. Find the first three nonvanishing terms in the Maclaurin series of $x \coth x$.

16. Use the definition of $\cos z$ and $\sin z$ to find their derivatives for complex values of z.

17. Verify the addition theorems for $\sin z$ and $\cos z$ and prove that

$$\sin (x + iy) = \sin x \cosh y + i \cos x \sinh y.$$

18. What condition should a satisfy in order that the equation

$$\cosh x = a$$

have real roots? Find the roots in this case.

Find the limits as $x \to 0$ of:

19. $\dfrac{\sinh x - x}{x^3}.$ 20. $\dfrac{\cosh x - 1}{x^2}.$

21. $\coth x - \dfrac{1}{x}.$ 22. $\operatorname{cosech}^2 x - \dfrac{1}{x^2}.$

Find the limits as $x \to \infty$ of:

23. $\dfrac{\sinh x}{\cosh x + x}$. 24. $e^{-x} \sinh x$. 25. $x(1 - \tanh x)$.

26. Use Problem 17 to find all the (complex) roots of

$$\sin z = 2.$$

27. Find a differential equation satisfied by $\tanh x$.

7.4. The inverse trigonometric functions. The problem of solving trigonometric equations leads to the inverse trigonometric functions. The basic problem here is to solve equations like

$$\sin x = a, \cos y = b, \tan t = c, \tag{7.4.1}$$

where a, b, c are given real numbers. The reader knows that the first two equations have no (real) roots unless

$$-1 \leqq a \leqq +1, -1 \leqq b \leqq +1.$$

On the other hand, if these conditions are satisfied, then there are actually infinitely many roots. For the first equation this follows from the identities

$$\sin (x + 2\pi) = \sin x, \sin (\pi - x) = \sin x.$$

They imply that if x_0 is a root, so are

$$x_0 + 2k\pi \quad \text{and} \quad \pi - x_0 + 2k\pi, \tag{7.4.2}$$

where k is any integer. For the cosine and the tangent equations the root systems have different patterns:

$$\pm y_0 + 2k\pi \tag{7.4.3}$$

for the cosine equation, and

$$t_0 + k\pi \tag{7.4.4}$$

for the tangent case. Here (7.4.3) reflects the fact that $\cos y$ is an even function of period 2π, while (7.4.4) is implied by $\tan t$ having the period π.

The problem now is to find one root of each equation. We start with the sine case. The function

$$y = \sin x$$

is strictly increasing in the interval $\left[-\dfrac{\pi}{2}, +\dfrac{\pi}{2}\right]$ and goes from -1 to $+1$ in the interval. It follows that there is a unique inverse defined for $-1 \leqq y \leqq +1$ and strictly increasing. We denote this function by

$$x = \text{arc} \sin y. \tag{7.4.5}$$

Read arc sine y. It is usually denoted in American and British textbooks by

$$\sin^{-1} y.$$

In Section 2.8 we have stated our reasons for avoiding this type of notation.

From the basic properties of inverse functions we conclude that

$$\sin(\text{arc}\sin y) = y, \ -1 \leqq y \leqq 1, \tag{7.4.6}$$

$$\text{arc}\sin(\sin x) = x, \ -\frac{\pi}{2} \leqq x \leqq \frac{\pi}{2}. \tag{7.4.7}$$

Once we have defined the arc sine, then we can give all the solutions of the equation

$$\sin x = a, \ -1 \leqq a \leqq +1. \tag{7.4.8}$$

By (7.4.2) they are

$$\begin{aligned} &\text{arc}\sin a + 2k\pi, \\ &(\pi - \text{arc}\sin a) + 2k\pi, \end{aligned} \tag{7.4.9}$$

where k is any integer. The first of these formulas reflects the periodic character of the sine function. It is often said that the arc sine has the *additive period* 2π.

Since $\sin x$ is a differentiable function, its inverse has the same property, and by formula (3.5.1) the product of the derivatives equals one. Hence

$$(D_x \sin x)(D_y \text{ arc}\sin y) = 1,$$

so that

$$D_y \text{ arc}\sin y = \frac{1}{\cos x},$$

or

$$D_y \text{ arc}\sin y = +(1 - y^2)^{-\frac{1}{2}}, \tag{7.4.10}$$

since $\cos x > 0$ in the interval $\left(-\frac{\pi}{2}, \frac{\pi}{2}\right)$.

From this value of the derivative we can easily obtain the Maclaurin series of the arc sine. It is clear that there are derivatives of all orders, but repeated differentiation soon becomes unmanageable. But the derivative can be expanded in a binomial series, which is its Maclaurin series, and this series can be integrated term by term to obtain the Maclaurin series of the original function. Now

$$(1 - y^2)^{-\frac{1}{2}} = \sum_{n=0}^{\infty} (-1)^n \binom{-\frac{1}{2}}{n} y^{2n},$$

so that

$$\text{arc}\sin y = \sum_{n=0}^{\infty} (-1)^n \binom{-\frac{1}{2}}{n} \frac{y^{2n+1}}{2n+1}.$$

We can rewrite the coefficients of this series in various ways to bring out their properties. First, the coefficients are positive and form a strictly decreasing sequence, for

$$a_n \equiv (-1)^n \binom{-\frac{1}{2}}{n} \frac{1}{2n+1} = \frac{1 \cdot 3 \cdot 5 \cdots (2n-1)}{2 \cdot 4 \cdot 6 \cdots (2n)} \frac{1}{2n+1} = \frac{(2n-1)!!}{(2n)!!} \frac{1}{2n+1}$$

where $m!!$ is the product of all positive integers $\leqq m$ of the same parity as m.

Further

$$\frac{a_{n+1}}{a_n} = \frac{(2n+1)^2}{(2n+2)(2n+3)} < 1.$$

This gives

$$\begin{aligned}\arcsin y &= y + \tfrac{1}{6}y^3 + \tfrac{3}{40}y^5 + \cdots \\ &\quad + \frac{(2n-1)!!}{(2n)!!}\frac{1}{2n+1}\, y^{2n+1} + \cdots.\end{aligned} \tag{7.4.11}$$

Actually this is a hypergeometric series in the sense of Example 3 of Section 5.2. More precisely,

$$\arcsin y = yF(\tfrac{1}{2}, \tfrac{1}{2}, \tfrac{3}{2}; y^2). \tag{7.4.12}$$

We leave the verification to the student.

Now by formula (5.2.23) the hypergeometric series $F(\alpha, \beta, \gamma; 1)$ converges if and only if $\alpha + \beta < \gamma$. In our case $\alpha = \frac{1}{2}$, $\beta = \frac{1}{2}$, $\gamma = \frac{3}{2}$ so the condition is satisfied. It follows that the series (7.4.11) converges for $-1 \leqq y \leqq +1$. Since $\arcsin 1 = \dfrac{\pi}{2}$ we have a series representation for π, namely

$$\pi = 2F(\tfrac{1}{2}, \tfrac{1}{2}, \tfrac{3}{2}; 1).$$

This series converges too slowly to be of any value for computational purposes. Even a million terms would scarcely suffice to give π with three correct decimals. There are other possibilities, however. Thus the choice $y = \frac{1}{2}$, $\arcsin y = \dfrac{\pi}{6}$ gives

$$\pi = 3F(\tfrac{1}{2}, \tfrac{1}{2}, \tfrac{3}{2}; \tfrac{1}{4}), \tag{7.4.13}$$

and here we could get four decimals using only the first five or six terms. We shall encounter still more favorable expansions below in connection with the arc tangent.

We recall that the sine function has an addition theorem and we ask how this fact is reflected in properties of the arc sine. Suppose that

$$-\frac{\pi}{2} \leqq x_1, x_2 \leqq \frac{\pi}{2}, \sin x_1 = s, \sin x_2 = t.$$

Then

$$\begin{aligned}\sin(x_1 + x_2) &= \sin x_1 \cos x_2 + \sin x_1 \cos x_2 \\ &= s(1 - t^2)^{\frac{1}{2}} + t(1 - s^2)^{\frac{1}{2}},\end{aligned}$$

where the roots are taken to be positive. This shows that

$$\arcsin[\sin(x_1 + x_2)] = \arcsin[s(1 - t^2)^{\frac{1}{2}} + t(1 - s^2)^{\frac{1}{2}}].$$

If now

$$-\frac{\pi}{2} \leqq x_1 + x_2 \leqq \frac{\pi}{2},$$

then the left member of the preceding display equals $x_1 + x_2$, that is

$$\arcsin s + \arcsin t = \arcsin[s(1 - t^2)^{\frac{1}{2}} + t(1 - s^2)^{\frac{1}{2}}]. \tag{7.4.14}$$

It is possible for $x_1 + x_2$ to fall outside of the interval $\left[-\dfrac{\pi}{2}, \dfrac{\pi}{2}\right]$, namely either in the interval $\left(\dfrac{\pi}{2}, \pi\right]$ or in $\left[-\pi, -\dfrac{\pi}{2}\right)$. In the first case we replace the left member of (7.4.14) by

$$\pi - \text{arc sin } s - \text{arc sin } t,$$

and in the second case by

$$-\pi - \text{arc sin } s - \text{arc sin } t.$$

With this proviso we can regard (7.4.14) as the functional equation of the arc sine and as an inverse addition theorem.

We have a similar situation in the case of the logarithm. Since e^x satisfies

$$f(x_1 + x_2) = f(x_1) f(x_2),$$

the inverse function satisfies

$$\check{f}(s) + \check{f}(t) = \check{f}(st),$$

that is,

$$\log s + \log t = \log st. \tag{7.4.15}$$

There is no ambiguity as long as s and t are positive numbers and we take real logarithms. We do get into difficulties, if s and t are complex. If the logarithms are given their principal values corresponding to principal values of the arguments of s, t, and st, then (7.4.15) is not necessarily true, but the two members of the equation can differ at most by a multiple of $2\pi i$, the additive period of the logarithm.

We return for a moment to the consequences of (4.7.10). Upon integration this relation gives

$$\text{arc sin } y = \int_0^y (1 - t^2)^{-\frac{1}{2}}\, dt, \; -1 \leqq y \leqq +1. \tag{7.4.16}$$

In particular we have

$$\pi = 2\int_0^1 (1 - t^2)^{-\frac{1}{2}}\, dt. \tag{7.4.17}$$

We turn now to the inverse cosine. Since $\cos x$ is strictly decreasing from $+1$ to -1 in $[0, \pi]$, there is a unique inverse denoted by

$$x = \text{arc cos } y, \; -1 \leqq y \leqq +1. \tag{7.4.18}$$

Hence

$$\cos(\text{arc cos } y) = y, \; -1 \leqq y \leqq +1, \tag{7.4.19}$$

$$\text{arc cos}(\cos x) = x, \; 0 \leqq x \leqq \pi. \tag{7.4.20}$$

The arc cosine is strictly decreasing and

$$D_y \text{ arc cos } y = -(1 - y^2)^{-\frac{1}{2}}. \tag{7.4.21}$$

Actually we can reduce the discussion of the arc cosine to that of the arc sine, since the two functions are related by the identity

$$\text{arc cos } y = \frac{\pi}{2} - \text{arc sin } y. \tag{7.4.22}$$

The inverse tangent is more interesting. Since $\tan x$ is strictly increasing from $-\infty$ to $+\infty$ in the interval $\left(-\frac{\pi}{2}, +\frac{\pi}{2}\right)$, there is a unique inverse denoted by

$$x = \arc\tan y, \quad -\infty < y < +\infty, \tag{7.4.23}$$

such that

$$\tan(\arc\tan y) = y, \quad -\infty < y < +\infty, \tag{7.4.24}$$

$$\arc\tan(\tan x) = x, \quad -\frac{\pi}{2} < x < +\frac{\pi}{2}. \tag{7.4.25}$$

Since $\tan x$ is differentiable, we have

$$(D_x \tan x)(D_y \arc\tan y) = 1,$$

or

$$D_y \arc\tan y = \cos^2 x = \frac{\cos^2 x}{\cos^2 x + \sin^2 x} = \frac{1}{1 + \tan^2 x},$$

so that

$$D_y \arc\tan y = \frac{1}{1 + y^2}. \tag{7.4.26}$$

From this we get

$$\arc\tan y = \int_0^y \frac{dt}{1 + t^2}. \tag{7.4.27}$$

We can also get the Maclaurin series of the arctangent from the last two relations. We can expand $(1 + y^2)^{-1}$ in geometric series for $|y| < 1$, and this series can be integrated term by term to give the Maclaurin series of $\arc\tan y$ by formula (5.5.4). Hence

$$\arc\tan y = y - \tfrac{1}{3}y^3 + \tfrac{1}{5}y^5 - \cdots + \frac{(-1)^n}{2n + 1}\, y^{2n-1} + \cdots. \tag{7.4.28}$$

The radius of convergence of this series equals 1. This series is also hypergeometric. In fact

$$\arc\tan y = yF(\tfrac{1}{2}, 1, \tfrac{3}{2}; -y^2). \tag{7.4.29}$$

Here $\gamma - \alpha - \beta = \frac{3}{2} - \frac{1}{2} - 1 = 0$ so the series is not convergent for $y = \pm i$. For $y = 1$ we have an alternating series, convergent by Theorem 5.3.1. Since $\arc\tan 1 = \frac{\pi}{4}$ we have the following simple series

$$\frac{\pi}{4} = 1 - \tfrac{1}{3} + \tfrac{1}{5} - \cdots + \frac{(-1)^n}{2n + 1} + \cdots. \tag{7.4.30}$$

Leibniz seems to have found this formula in 1673. The general series (7.4.28) was known to James Gregory in 1671 but the discovery was not published until 1712. It is an elegant formula for π, but the series converges too slowly for computational purposes. We shall find more rapidly convergent expansions below.

We get them from the inverse addition theorem. Suppose that

$$-\frac{\pi}{2} < x_1, x_2 < \frac{\pi}{2},$$

and set

$$\tan x_1 = s, \tan x_2 = t..$$

Then by the addition theorem

$$\tan(x_1 + x_2) = \frac{\tan x_1 + \tan x_2}{1 - \tan x_1 \cdot \tan x_2} = \frac{s+t}{1-st}.$$

Hence

$$\text{arc tan}\,[\tan(x_1 + x_2)] = \text{arc tan}\,\frac{s+t}{1-st},$$

and if $-\frac{\pi}{2} < x_1 + x_2 < \frac{\pi}{2}$, formula (7.4.25) applies and gives

$$\text{arc tan}\, s + \text{arc tan}\, t = \text{arc tan}\,\frac{s+t}{1-st}. \qquad (7.4.31)$$

If $x_1 + x_2$ lies outside the interval, then the formula must be modified by adding or subtracting π, the additive period of the arc tangent.

We can regard (7.4.31) as a functional equation for the arc tangent or, equivalently, as the inverse addition theorem. The structure of the fraction in the right member is worthy of attention. We set

$$\frac{s+t}{1-st} = \frac{S_{21}}{S_{20} - S_{22}}, \quad \text{where} \quad S_{20} = 1, S_{21} = s + t, S_{22} = st.$$

S_{21} and S_{22} are the elementary symmetric functions of two variables in terms of which every rational symmetric functions on two letters can be expressed rationally.

Now formula (7.4.31) can be extended to three or more summands and the right member will always be the arc tangent of a quotient of sums of elementary symmetric functions. In the case of three summands the formula reads

$$\text{arc tan}\, r + \text{arc tan}\, s + \text{arc tan}\, t = \text{arc tan}\,\frac{S_{31} - S_{33}}{S_{30} - S_{32}}, \qquad (7.4.32)$$

where

$$S_{30} = 1, S_{31} = r + s + t, S_{32} = rs + st + tr, S_{33} = rst$$

are the elementary symmetric functions on three letters. This general pattern extends to any number of summands.

The resulting formulas provide us with a profusion of identities that can serve for the computation of π. It is necessary to choose the variables in such a way that the fraction on the right has the value one. It is also desirable that the variables should be given values that are small simple fractions. The following choices

$$\tfrac{1}{2}, \tfrac{1}{3}; \tfrac{1}{3}, \tfrac{1}{5}, \tfrac{1}{8}; \tfrac{1}{3}, \tfrac{1}{5}, \tfrac{1}{7}, \tfrac{1}{8}; \tfrac{1}{5}, \tfrac{1}{5}, \tfrac{1}{5}, \tfrac{1}{5}, -\tfrac{1}{239}$$

correspond to 2, 3, 4, and 5 summands, respectively. The last choice is particularly favorable. It gives

$$\frac{\pi}{4} = 4\,\text{arc tan}\,\tfrac{1}{5} - \text{arc tan}\,\tfrac{1}{239}. \qquad (7.4.33)$$

Here

$$\arctan\frac{1}{5} = \frac{1}{5} - \frac{1}{3}\cdot\frac{1}{5^3} + \frac{1}{5}\cdot\frac{1}{5^5} - \frac{1}{7}\,\frac{1}{5^7} + \frac{1}{9}\cdot\frac{1}{5^9} - \cdots$$

$$= 0.197\,395\,559\,851,$$

where the last two figures are in doubt. The five terms listed give

$$\begin{array}{rr} 0.200\,000\,000\,000 & -\;0.002\,666\,666\,667 \\ 0.000\,064\,000\,000 & 0.000\,001\,828\,571 \\ \cline{2-2} 0.000\,000\,056\,889 & 0.002\,668\,495\,238 \\ \cline{1-1} 0.200\,064\,056\,889 & \\ -0.002\,668\,495\,238 & \\ \cline{1-1} 0.197\,395\,561\,651 & \end{array}$$

This is a trifle too large. The remainder of the series is negative and very close to the sum of the first two neglected terms, that is, about

$$-1800 \times 10^{-12}.$$

This correction applied to the sum of the five terms gives the value given above. Further

$$\arctan\frac{1}{239} = \frac{1}{239} - \frac{1}{3}\,\frac{1}{239^3} + \cdots$$

$$\begin{array}{r} = 0.004\,184\,100\,418 \\ -\;0.000\,000\,024\,414 \\ \hline 0.004\,184\,076\,004. \end{array}$$

Here we can neglect the remainder of the series completely since it would contribute at most one unit in the thirteenth decimal place. From this we get

$$\pi = 16 \times 0.197\,395\,559\,851 - 4 \times 0.004\,184\,076\,004$$

$$= 3.141\,592\,653\,610.$$

The first nine decimals are correct but they should be followed by 590 if we want the error to be less than one unit in the twelfth place. It is clear that greater accuracy can be obtained by taking more terms into account. We shall not pursue this any further and restrict ourselves to giving the value of π to twenty decimals:

$$\pi = 3.14159\,26535\,89793\,23846.$$

We list two more primitives:

$$\text{(xviii)} \quad \int \frac{dx}{(a^2 - x^2)^{\frac{1}{2}}} = \arcsin\frac{x}{a} + C,$$

$$\text{(xix)} \quad \int \frac{dx}{a^2 + x^2} = \frac{1}{a}\arctan\frac{x}{a} + C.$$

It is of some interest to compare these formulas with (iii) and (v) at the end of Section 4.7. We have

$$\int \frac{dx}{x^2 - a^2} = \frac{1}{2a}\log\frac{a - x}{a + x} + C,$$

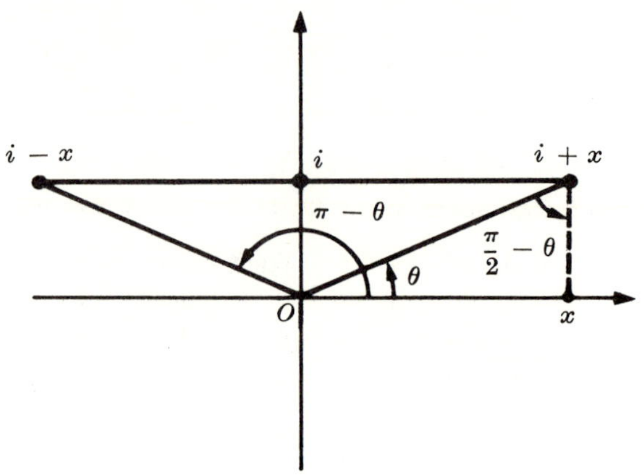

Figure 7.12

so that, replacing a by i we should have

$$\arctan x = \frac{1}{2i} \log \frac{i - x}{i + x}. \tag{7.4.34}$$

Similarly

$$\int \frac{dx}{(x^2 + a^2)^{\frac{1}{2}}} = \log [x + (x^2 + a^2)^{\frac{1}{2}}] + C$$

suggests that

$$\arcsin x = \frac{1}{i} \log [ix + (1 - x^2)^{\frac{1}{2}}]. \tag{7.4.35}$$

Actually both formulas are correct for a suitable choice of the logarithms. We show first that (7.4.34) is valid if the imaginary part of the logarithm is taken between $-\pi$ and π. In Figure 7.12 x is assumed to be positive. Then by the definition of the complex logarithm

$$\frac{1}{2i} \log \frac{i - x}{i + x} = \frac{1}{2i} \log \left| \frac{i - x}{i + x} \right| + \tfrac{1}{2} \arg \frac{i - x}{i + x}.$$

Here

$$\left| \frac{i - x}{i + x} \right| = 1 \quad \text{so} \quad \log \left| \frac{i - x}{i + x} \right| = 0.$$

The figure gives

$$\tfrac{1}{2} \arg \frac{i - x}{i + x} = \tfrac{1}{2}[(\pi - \theta) - \theta] = \frac{\pi}{2} - \theta,$$

and also

$$\tan \left(\frac{\pi}{2} - \theta \right) = x.$$

Since $0 < \dfrac{\pi}{2} - \theta < \dfrac{\pi}{2}$, we have

$$\frac{\pi}{2} - \theta = \arctan x,$$

and

$$\frac{1}{2i}\log\frac{i-x}{i+x} = \frac{\pi}{2} - \theta = \text{arc tan } x$$

as asserted. This is for $x > 0$. If we replace x by $-x$ in (7.4.34), both sides are changed into their negatives so the equation remains true.

In formula (7.4.35) we take $-1 < x < +1$ and set

$$x = \sin\theta,\ -\frac{\pi}{2} < \theta < \frac{\pi}{2},\ +(1-x^2)^{\frac{1}{2}} = \cos\theta.$$

Thus

$$\frac{1}{i}\log\,[ix + (1-x^2)^{\frac{1}{2}}] = \frac{1}{i}\log\,(i\sin\theta + \cos\theta)$$

$$= \frac{1}{i}\log\,(e^{i\theta}) = \frac{1}{i}\,i\theta = \theta = \text{arc sin } x.$$

This shows that the logarithm has to be chosen so that its imaginary part lies between $-\dfrac{\pi}{2}$ and $+\dfrac{\pi}{2}$. If this is done, the proof is valid.

EXERCISE 7.4

Differentiate the following functions:

1. $\text{arc sin } (1 - x^4)$.
2. $\text{arc cos } (1 - x^2)^{\frac{1}{2}}$.
3. $\text{arc tan}\left(\dfrac{1-x}{1+x}\right)^{\frac{1}{2}}$.
4. $\text{arc sin } (e^x)$.
5. $\text{arc tan } (\log x),\ x > 0$.
6. $e^{\text{arc sin } x}$.
7. $\text{arc sin}\left(\dfrac{x^2-a^2}{b^2-a^2}\right)^{\frac{1}{2}}$.
8. $\text{arc tan}\left(\dfrac{1-\cos x}{1+\cos x}\right)^{\frac{1}{2}}$.
9. $\cos\,(\text{arc sin } x)$.
10. $\cot\,(\text{arc tan } x)$.
11. Find the derivative of $\text{arc sin}\dfrac{x}{(1+x^2)^{\frac{1}{2}}}$ and examine the relation between this function and arc tan x.
12. Find the relation between $\text{arc tan}\dfrac{x}{(1-x^2)^{\frac{1}{2}}}$ and arc sin x if $-1 < x < 1$.
13. Prove that for $a > b > 0$

$$D_x\left\{(a^2-b^2)^{-\frac{1}{2}}\,\text{arc cos}\,\frac{b + a\cos x}{a + b\cos x}\right\} = \frac{1}{a + b\cos x}.$$

14. Prove that

$$D_x\left\{\log\frac{1 + x\sqrt{2} + x^2}{1 - x\sqrt{2} + x^2} + 2\,\text{arc tan}\,\frac{x\sqrt{2}}{1-x^2}\right\} = \frac{4\sqrt{2}}{1+x^4}.$$

15. If $y = \sin\,(b\,\text{arc sin } x)$, prove that

$$(1-x^2)y'' - xy' + b^2y = 0.$$

16. If n is a positive integer, prove that $\sin(n \arcsin x)$ is a polynomial in x of degree n. What expression for this polynomial is obtainable from Euler's formula plus (7.4.35)?

17. If $F_n(x)$ denotes the nth derivative of $F_0(x) = [\arcsin x]^2$, prove that

$$(1 - x^2)F_n''(x) - (2n + 1)xF_n'(x) - n^2F_n(x) = 0,\ n \geqq 1.$$

18. Expand $[\arcsin x]^2$ in Maclaurin's series. The coefficients may be obtained from the preceding problem or by forming the Cauchy product series of (7.4.11) with itself.

19. Verify (7.4.12) and (7.4.29).

20. Express $r^3 + s^3 + t^3$ in terms of the elementary symmetric functions S_{31}, S_{32}, S_{33}.

Evaluate the following integrals:

21. $\displaystyle\int_0^1 \arcsin x\, dx.$ 22. $\displaystyle\int_0^1 \arctan x\, dx.$ 23. $\displaystyle\int_1^\infty \arctan x \frac{dx}{x^2}.$

24. $\displaystyle\int_0^{\frac{1}{2}} x \arcsin x\, dx.$ 25. $\displaystyle\int_{-1}^1 [\arcsin x]^2\, dx.$

26. $\displaystyle\int_0^1 x \arctan x\, dx.$ 27. $\displaystyle\int_0^{\sqrt{3}} [\arctan x]^2\, x\, dx.$

28. $\displaystyle\int \frac{x^2\, dx}{1 + x^2}.$ 29. $\displaystyle\int \frac{x^2\, dx}{\sqrt{1 - x^2}}.$ 30. $\displaystyle\int \frac{dx}{x\sqrt{x^2 - 1}}.$ 31. $\displaystyle\int_1^\infty \frac{dx}{x\sqrt{x^2 + 1}}.$

Find the limits as $x \to 0$ of

32. $\arctan\left(\dfrac{\arcsin x}{x}\right).$ 33. $\dfrac{[\arcsin x]^2}{x \sin x}.$

34. $\dfrac{x}{\arctan x}.$ 35. $\dfrac{1}{x}\left\{\arctan\left(\dfrac{1 + x}{1 - x}\right) - \dfrac{\pi}{4}\right\}.$

36. Multiple choice question: The following verse (1) tests the student's knowledge of Spanish; (2) commemorates the siege of Syracuse; (3) gives π to thirty decimal places.

Que j'aime à faire apprendre un nombre utile aux sages!
Immortel Archimède, artiste ingénieur,
Que de ton jugement peut priser le valeur?
Pour moi, ton problème eut de pareils avantages.

7.5. On the role of the logarithm. Logarithms were originally introduced as an aid to computation and this is still their main function. In the present section we shall give two other applications of the logarithm: first to the theory of integration, and second to the technique of differentiation.

By this time the student should have a fair idea of what functions can appear as primitives of a rational function. The systematic discussion of the technique of integration is postponed until Chapter 11, Vol. II, and at this juncture we are not concerned with the best and speediest way of handling particular classes of integrals such as integrals of rational functions. What we want is a qualitative analysis of the problem rather than a quantitive one. The spirit of the investigation is shown by the following

THEOREM 7.5.1. *The integral of a rational function*

$$R(x) = \frac{P(x)}{Q(x)} \tag{7.5.1}$$

is the sum of two functions, a rational part $R_1(x)$ and a transcendental part $T(x)$,

$$\int R(x)\,dx = R_1(x) + T(x). \tag{7.5.2}$$

The poles of $R_1(x)$ are located at the multiple poles of $R(x)$: if $x = a$ is a k-fold zero of $Q(x)$ and hence a k-fold pole of $R(x)$, then $x = a$ is a pole of order $k - 1$ of $R_1(x)$. The point at infinity is a pole of $R_1(x)$ of order $m - n + 1$ if $m \geqq n$, where m is the degree of $P(x)$ and n is the degree of $Q(x)$. The rational part $R_1(x)$ is missing if and only if $m < n$ and all zeros of $Q(x)$ are simple. The transcendental part is of the form

$$T(x) = \Sigma\, \alpha_j \log (x - a_j), \tag{7.5.3}$$

where the summation extends over the distinct zeros of $Q(z)$.

Proof. This is a fairly straightforward consequence of Theorem 6.6.1 on partial fractions. Suppose $m > n$, and let the distinct zeros of $Q(x)$ be

$$a_1, a_2 \cdots, a_k,$$

of multiplicities

$$\nu_1, \nu_2, \cdots, \nu_k,$$

respectively, where

$$\nu_1 + \nu_2 \cdots + \nu_k = n.$$

Then

$$\frac{P(x)}{Q(x)} = \sum_{p=0}^{m-n} A_p x^p + \sum_{j=1}^{k} \frac{A_{j1}}{x - a_j} + \sum_{j=1}^{k} \left\{ \sum_{p=2}^{\nu_j} \frac{A_{jp}}{(x - a_j)^p} \right\}. \tag{7.5.4}$$

It follows that

$$\begin{aligned} \int R(x)\,dx &= C + \sum_{p=0}^{m-n} A_p \frac{x^{p+1}}{p+1} - \sum_{j=1}^{k} \left\{ \sum_{p=2}^{\nu_j} \frac{A_{jp}}{(p-1)(x - a_j)^{p-1}} \right\} \\ &\quad + \sum_{j=1}^{k} A_{j1} \log (x - a_j) \\ &\equiv R_1(x) + T(x). \end{aligned} \tag{7.5.5}$$

From these representations of $R_1(x)$ and $T(x)$ we can read off all the properties stated in the theorem. Note that A_{j1} is not zero, if $x = a_j$ is a simple zero of $Q(x)$.

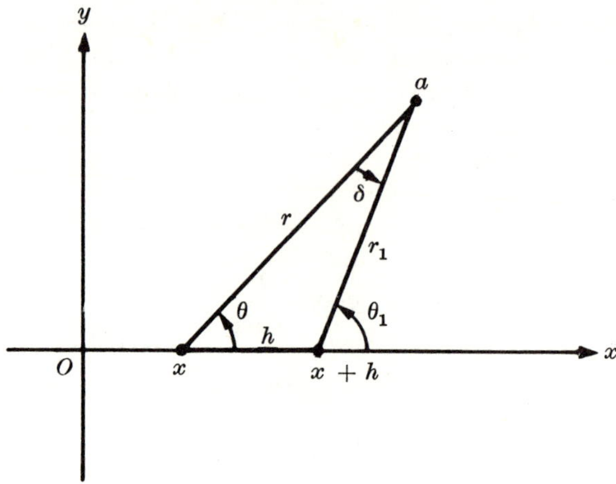

Figure 7.13

Some remarks are in order. In general, both the zeros a_j and the coefficients A_j and A_{jp} are complex numbers. Is it then true that

$$\int (x - a)^{-p}\, dx = (p - 1)^{-1}(x - a)^{-(p-1)},\ p > 1, \tag{7.5.6}$$

and

$$\int (x - a)^{-1}\, dx = \log (x - a) \tag{7.5.7}$$

up to an additive constant? The question is the following: Is the integrand equal to the derivative of the right-hand member? This is obvious in the first case regardless of the value of a, since

$$D_z(z - a)^n = n(z - a)^{n-1}$$

for all n. Here both a and z may be complex numbers. We are interested mainly in the case where $z = x$ is real. On the other hand, the formula

$$D_z \log (z - a) = \frac{1}{z - a} \tag{7.5.8}$$

is not proved in Section 7.2, and so it will be proved here that the formula holds at least for real z and complex a.

Set $a = \alpha + i\beta$, where we assume $\beta > 0$ to clarify the ideas. We have

$$\begin{aligned} \log (x - a) &= \log |x - a| + i \arg (x - a) \\ &= \log r + i\theta \end{aligned}$$

in the notation of Figure 7.13. Now

$$\log r = \tfrac{1}{2} \log [(x - \alpha)^2 + \beta^2],$$

the derivative of which is

$$\frac{x - \alpha}{(x - \alpha)^2 + \beta^2}.$$

The difference quotient for arg $(x - a)$, taken between 0 and π, is

$$\frac{\theta_h - \theta}{h} = \frac{\delta}{h},$$

where $\delta > 0$ when $h > 0$. Here we borrow a device from trigonometry and apply the law of sines to the triangle with vertices at $z = x, x + h$, and a. This gives

$$\frac{\sin \delta}{h} = \frac{\sin \theta}{r_1}.$$

Hence

$$\frac{\delta}{h} = \frac{\sin \delta}{h} \frac{\delta}{\sin \delta} = \frac{\sin \theta}{r_1} \frac{\delta}{\sin \delta}.$$

As $h \to 0$, $\delta \to 0$, $r_1 \to r$, $\delta/(\sin \delta) \to 1$, and

$$\lim \frac{\delta}{h} = \frac{\sin \theta}{r} = \frac{\beta}{r^2} = \frac{\beta}{(x - \alpha)^2 + \beta^2}.$$

It follows that

$$D_x \log (x - a) = D_x \log |x - a| + iD_x \arg (x - a)$$

$$= \frac{x - \alpha + i\beta}{(x - \alpha)^2 + \beta^2} = \frac{1}{x - \alpha - i\beta} = \frac{1}{x - a}$$

as asserted.

The second remark in connection with Theorem 7.5.1 is the following. Formula (xix) shows that arc tan x is a primitive of a rational function. Why is there no mention of arc tangents in the theorem? The answer is given by formula (7.4.34) which states that arc tan x is the difference of two logarithms and hence of the form (7.5.3). This does not reduce the importance of the function arc tan x, but at this juncture we want to emphasize that integration of rational functions basically leads to only one new transcendental function, namely the logarithm. If we are interested in expressing the integral of a real rational function in real form, then and only then do we need arc tangents.

There are other questions which arise in connection with the theorem. For instance, how do we know that the transcendental part is not really rational? This would imply that we could find a rational function $R_2(x) = P_2(x)/Q_2(x)$ such that

$$\sum_{j=1}^{k} \alpha_j \log (x - a_j) = \frac{P_2(x)}{Q_2(x)}, \tag{7.5.9}$$

where $P_2(x)$ and $Q_2(x)$ are polynomials without common factor. Differentiation gives

$$\sum_{j=1}^{k} \frac{\alpha_j}{x - a_j} = \frac{P_2'(x)Q_2(x) - P_2(x)Q_2'(x)}{[Q_2(x)]^2}. \tag{7.5.10}$$

We want to show first that $Q_2(x)$ must be a constant. If not, then there must be a zero b of $Q_2(x)$ and we have

$$Q_2(x) = (x - b)^p Q_3(x) \quad \text{with} \quad Q_3(b) \neq 0.$$

This gives

$$P_2'(x)Q_2(x) - P_2(x)Q_2'(x)$$
$$= (x - b)^{p-1}\{(x - b)[P_2'(x)Q_3(x) - P_2(x)Q_3'(x)] - p\,Q_3(x)P_2(x)\}.$$

This is divisible by $(x - b)^{p-1}$ but by no higher power of $x - b$ since $Q_3(b)P_2(b) \neq 0$. It follows that the right member of (7.5.10), when expressed in its lowest terms, has a factor $(x - b)^{p+1}$ in the denominator. On the other hand, the left member, when expressed as a rational function in its lowest terms, has no repeated factor in its denominator. Hence $p = 0$ and $Q_2(x)$ must be a constant. We may set $Q_2(x) \equiv 1$. Then

$$\sum_{j=1}^{k} \frac{\alpha_j}{x - a_j} = P_2(x).$$

Multiplying by $x - a_p$ and letting $x \to a_p$, we see that $\alpha_p = 0$. Since this holds for $p = 1, 2, \cdots, k$, we find that $T(x)$ must be identically zero. This is ordinarily not the case so we conclude that the "transcendental part" is truly transcendental.

It is interesting to note that the transcendental part ordinarily cannot be found unless we can effectively factor the denominator $Q(x)$ in (7.5.1). On the other hand, the rational part $R_1(x)$ can be determined by purely algebraic processes without knowledge of the zeros of $Q(x)$. This was proved independently by the Russian mathematician M. V. Ostrogradski [1801–1861] and the French mathematician Charles Hermite [1822–1901]. We refer to Section 11.2 for such questions.

Rational functions are special cases of algebraic functions and there are classes of algebraic functions such that the corresponding primitive is the sum of an algebraic function and logarithms of algebraic functions. Formulas (iv) and (v) in Section 4.7 and (xviii) together with (7.4.35) in the preceding section give examples of such integrals. They are all special cases of integrals of the form

$$\int R(x, y)\,dx \quad \text{where} \quad y^2 = P(x), \tag{7.5.11}$$

a polynomial in x, and $R(x, y)$ is a rational function of x and of y. Such integrals can be reduced to various normal forms with the aid of

LEMMA 7.5.1. *If $R(x, y)$ is a rational function of x and y and if $y^2 = P(x)$ is a polynomial in x, then there are three rational functions in x: R_0, R_1, R_2, such that*

$$R(x, y) = R_0(x) + yR_1(x) = R_0(x) + R_2(x)\frac{1}{y}, \tag{7.5.12}$$

where $R_2(x) = R_1(x)P(x)$.

Proof. This is only elementary algebra, but it may not strike the student as simple algebra. What is meant by saying that $R(x, y)$ is rational in y as well as in x? $R(x, y)$ is a rational function of y, if it is the quotient of two polynomials in y. Since $R(x, y)$ is also supposed to be rational in x, the coefficients of the powers of y in the expansion must be rational functions of x. Here we can clear fractions and see that $R(x, y)$ is the quotient of two polynomials in y whose coefficients are polynomials in x, that is, the quotient of two expressions of the form

$$P_0(x) + P_1(x)y + P_2(x)y^2 + \cdots + P_n(x)y^n, \tag{7.5.13}$$

where the P's are polynomials. But here $y^2 = P(x)$ is a polynomial in x, so every even power of y is a polynomial in x, and every odd power is the product of y and a polynomial in x. This means that (7.5.13) reduces to the form

$$P_0^*(x) + P_1^*(x)y,$$

where P_0^*, P_1^* are polynomials in x. Hence

$$R(x, y) = \frac{P_0^*(x) + P_1^*(x)y}{Q_0^*(x) + Q_1^*(x)y}.$$

Here we can rationalize the denominator by multiplying by

$$Q_0^*(x) - Q_1^*(x)y.$$

Now

$$\begin{aligned}(P_0^* + P_1^*y)(Q_0^* - Q_1^*y) &= P_0^*Q_0^* - P_1^*Q_1^*y^2 \\ &\quad + (P_1^*Q_0^* - P_0^*Q_1^*)y \\ &\equiv P_0 + Q_0y\end{aligned}$$

and

$$(Q_0^* + Q_1^*y)(Q_0^* - Q_1^*y) = (Q_0^*)^2 - (Q_1^*)^2y^2 \equiv D,$$

where P_0, Q_0, D are polynomials in x. Hence finally

$$R(x, y) = \frac{P_0(x)}{D(x)} + \frac{Q_0(x)}{D(x)}\,y. \tag{7.5.14}$$

Here is the first formula under (7.5.12) and the second is an immediate consequence of the first. This completes the proof.

Hence

$$\int R(x, y)\,dx = \int R_0(x)\,dx + \int R_2(x)\,\frac{dx}{y}. \tag{7.5.15}$$

The first integral on the right is of the form discussed in Theorem 7.5.1, the second one poses new problems.

By assumption $y^2 = P(x)$ is a polynomial in x which we may suppose to have no repeated factors. The degree of $P(x)$ is decisive for the nature of the integral. We shall show that if the degree does not exceed two, then the integral is the sum of two elements. The first is of the form: a rational function of x plus y times a rational function of x. The second, transcendental part, is the sum of logarithms of linear functions of y. In case $P(x)$ is of degree one, we can take

$$P(x) = x - a.$$

Here we introduce y as the new variable. This gives

$$x = a + y^2, \; dx = 2y\, dy$$

and

$$\int R_2(x) \frac{dx}{y} = 2 \int R_2(a + y^2)\, dy,$$

that is, the integral of a rational function of y. By Theorem 7.5.1 the result is of the form

$$S(y) + T(y),$$

where $S(y)$ is a rational function of y, that is,

$$S(y) = S_0(x) + yS_1(x),$$

while $S_0(x)$ and $S_1(x)$ are rational functions of x, and

$$T(y) = \sum \alpha_j \log (y - a_j).$$

This proves the assertion in the linear case.

Quadratic polynomials $P(x)$ are more interesting but also harder to treat. The problem is to rationalize the integrand by introducing a new variable t such that, on the one hand x and y are rational functions of t and on the other t is a rational function of x and y. Geometrically speaking the situation is as follows. The curve

$$y^2 = P(x) \tag{7.5.16}$$

is of the second degree and hence a conic. We want to represent the coordinates of a point (x, y) on this conic rationally in terms of a parameter t where conversely t should be a rational function of x and y. Rational representations for the conics will be found in Appendix A. It is clear that once we have the representation, we can introduce t as a new variable and obtain a rational integrand. The resulting integral can be worked out, using Theorem 7.5.1, and t can be eliminated from the result by expressing it in terms of x and y.

A systematic way of obtaining such representations is worthy of brief mention. We take a point (x_0, y_0) on the conic and lay a pencil of straight lines through it. See Figure 7.14. The line of slope t intersects the conic in two points, one fixed (x_0, y_0) and the other variable (x, y). Here x and y are quotients of quadratic polynomials in t and

$$t = \frac{y - y_0}{x - x_0}.$$

As an example we may take

$$y^2 = 1 - x^2$$

where the conic is the unit circle. Here we can use formula (A.3.21) and obtain

$$x = \frac{1 - t^2}{1 + t^2}, \; y = \frac{2t}{1 + t^2}, \; t = \frac{y}{x + 1}. \tag{7.5.17}$$

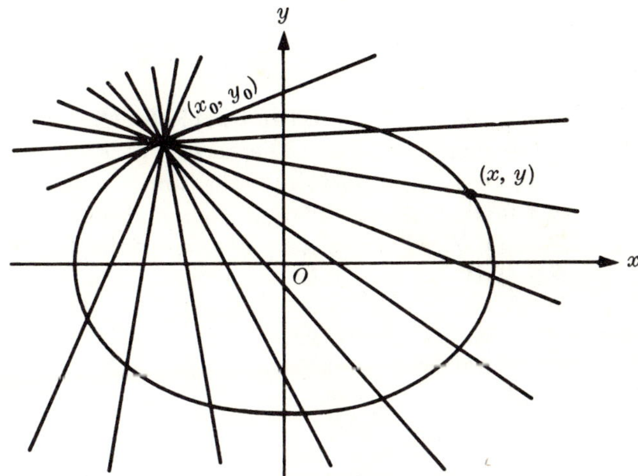

Figure 7.14

We can use this substitution to evaluate any integral of the form

$$\int R(x)\frac{dx}{y}, \; y = (1 - x^2)^{\frac{1}{2}}.$$

Since

$$dx = \frac{-4t\,dt}{(1+t^2)^2}, \; \frac{dx}{y} = -\frac{2\,dt}{1+t^2},$$

we get

$$\int R(x)\frac{dx}{y} = -2\int R\left(\frac{1-t^2}{1+t^2}\right)\frac{dt}{1+t^2}. \tag{7.5.18}$$

This is an integral of a rational function and can be handled as above. In the result we replace t by $y/(x+1)$.

If n, the degree of $P(x)$, exceeds two, the integration problem leads to new transcendental functions and the logarithm is no longer sufficient. If $n = 3$ or 4, we speak of an *elliptic integral* because this type of integral arises in finding the length of arc of an ellipse. The case where n is greater than 4 is known as *hyperelliptic*. Integrals of the form

$$\int R(x, y)\,dx, \quad \text{where} \quad P(x, y) = 0, \tag{7.5.19}$$

and $P(x, y)$ is a polynomial in x and y, are called *Abelian* after N. H. Abel who studied them in a memoir written in 1826 and published in 1841 long after his death. A study of such integrals is beyond the scope of this treatise, but special cases will be encountered intermittently. Thus all the integrals studied above are Abelian. With these remarks we end our discussion of applications of the logarithm to the theory of integration.

Our second topic for this section is *logarithmic differentiation*. This useful device enables us to find derivatives of complicated products and powers.

Suppose first that

$$F(x) = f_1(x)f_2(x)\cdots f_n(x), \tag{7.5.20}$$

where each $f_k(x)$ is defined as a positive differentiable function in an interval (a, b). Then $F(x)$ has the same properties. The derivative $F'(x)$ of $F(x)$ is of course obtained by differentiating one factor at a time, multiplying by the remaining factors, and adding the n different combinations. The following procedure is often time-saving and brings out properties of the derivative which may otherwise be overlooked. We note that

$$\log F(x) = \log f_1(x) + \log f_2(x) + \cdots + \log f_n(x).$$

Differentiation gives

$$\frac{F'(x)}{F(x)} = \frac{f_1'(x)}{f_1(x)} + \frac{f_2'(x)}{f_2(x)} + \cdots + \frac{f_n'(x)}{f_n(x)}. \tag{7.5.21}$$

This is the desired formula.

Example 1. If

$$F(x) = x(x-1)(x-2)(x-3),$$

then

$$\frac{F'(x)}{F(x)} = \frac{1}{x} + \frac{1}{x-1} + \frac{1}{x-2} + \frac{1}{x-3}.$$

This shows that, if δ is a small positive number, then the quotient is large positive for $x = \delta, 1+\delta, 2+\delta, 3+\delta$, and large negative for $x = -\delta, 1-\delta, 2-\delta, 3-\delta$. Hence $F'(x)$ changes sign once (and only once) in each of the intervals $(0, 1)$, $(1, 2)$, and $(2, 3)$.

Logarithmic differentiation is of more fundamental importance when powers are involved. Suppose that $f(x)$ and $g(x)$ are differentiable functions in the interval (a, b) and that $f(x)$ is positive. We want to consider

$$F(x) = [f(x)]^{g(x)}, \tag{7.5.22}$$

and our first question is what sense should be given to this symbol. We observe that the formal logarithm is well defined so we set

$$\log F(x) \equiv g(x) \log f(x), \tag{7.5.23}$$

and

$$F(x) = \exp[\log F(x)] = \exp[g(x)\log f(x)]. \tag{7.5.24}$$

This makes sense. The function $F(x)$ thus defined is obviously differentiable, and using the formula for differentiation of a composite function we obtain

$$D_x\{[f(x)]^{g(x)}\} = [f(x)]^{g(x)}\left\{g'(x)\log f(x) + g(x)\frac{f'(x)}{f(x)}\right\}. \tag{7.5.25}$$

Example 2. If

$$F(x) = x^x,$$

then the function is well defined for $x > 0$. Further

$$\lim_{x\to 0} x^x = \lim_{x\to 0} \exp[x \log x]$$
$$= \exp[\lim_{x\to 0} x \log x] = \exp 0 = 1.$$

Using (7.5.25) we obtain

$$D_x x^x = x^x(1 + \log x). \tag{7.5.26}$$

Example 3. We take

$$F(x) = (\tan x)^{\cos x}. \tag{7.5.27}$$

This function is well defined in the interval $(0, \pi/2)$. Here

$$\lim_{x\to 0} F(x) = 0, \quad \lim_{x\to\pi/2} F(x) = 1, \tag{7.5.28}$$

and

$$D_x\{(\tan x)^{\cos x}\} = (\tan x)^{\cos x}\left\{\frac{1}{\sin x} - \sin x \log \tan x\right\}. \tag{7.5.29}$$

EXERCISE 7.5

1. Use formula (7.5.17) to find $\displaystyle\int_0^1 \frac{dx}{(1 + x)(1 - x^2)^{\frac{1}{2}}}$.

2. Prove that $\displaystyle\int_{-1}^1 \left(\frac{1 - x}{1 + x}\right)^{\frac{1}{2}} dx = \pi$. The integral may be written in the form (7.5.18).

*3. Reduce

$$\int_0^1 \frac{(1 + x)^{\frac{1}{2}} - (1 - x)^{\frac{1}{2}}}{(1 + x)^{\frac{1}{2}} + (1 - x)^{\frac{1}{2}}}\, dx$$

to the form (7.5.18) and find the value of the integral.

4. Evaluate the integral

$$\int_0^4 \frac{dx}{y},$$

if (x, y) lies on the parabola $(y - x)^2 = 4x$ and $y = 8$ when $x = 4$.

5. Find a parametric representation of the hyperbola

$$x^2 - y^2 = 1$$

using the pencil $y = t(x + 1)$. Use this representation to work out the integral

$$\int_1^2 \frac{x\, dx}{y}.$$

Find for what values of x the following functions define mappings of reals into reals. Find also their first-order derivatives.

6. $(\log x)^x$. 7. $x^{\log x}$. 8. $(\tan x)^x$. 9. $(1 + \sin x)^{\cot x}$.

10. $\left(1 + \dfrac{1}{x}\right)^x$. 11. $(1 + \cot x)^x$. 12. $x^{\log \cos x}$. 13. $(\cos x)^{x^2}$.

14.–20. Find the limits of the preceding functions when x decreases to zero. Omit the function in Problem 6.

21. For what values of x is $x^x < e^x$?

7.6. Some differential equations. It was pointed out earlier in this chapter that the exponential function and its cognates satisfy linear differential equations with constant coefficients. Some of these functions and their differential equations are:

$$e^x, \quad y' - y = 0;$$

$$\sin x, \cos x, \quad y'' + y = 0;$$

$$\sinh x, \cosh x, \quad y'' - y = 0.$$

A linear differential equation with constant coefficients is by definition of the form

$$a_0 y^{(n)} + a_1 y^{(n-1)} + \cdots + a_{n-1} y' + a_n y = 0, \tag{7.6.1}$$

where the a's are constants. The properties of the solution are governed by the zeros of the associated polynomial

$$P(t) \equiv a_0 t^n + a_1 t^{n-1} + \cdots + a_{n-1} t + a_n. \tag{7.6.2}$$

Thus for the three listed equations the associated polynomials are

$$t - 1,\ t^2 + 1,\ t^2 - 1,$$

respectively. The zeros of these polynomials are

$$1;\ +i,\ -i;\ +1,\ -1.$$

In the first case e^x is a solution, in the second e^{ix} and e^{-ix} are solutions, while in the third e^x and e^{-x} are solutions.

In terms of the D_x notation for derivatives, we can rewrite (7.6.1) as

$$a_0 D_x{}^n y + a_1 D_x^{n-1} y + \cdots + a_{n-1} D_x y + a_0 y = 0, \tag{7.6.3}$$

or, symbolically,

$$P(D_x)[y] = 0. \tag{7.6.4}$$

We call $P(D_x)$ a *linear differential operator.*

We say that $y = f(x)$ *is a solution of the differential equation* (7.6.4) *in an interval* (a, b) *if for all* x *in* (a, b),

$$P(D_x)[f(x)] = 0. \tag{7.6.5}$$

It will be seen later that solutions of (7.6.4) are actually defined for all values of x.

Linearity of the operator implies and is implied by the following property:

THEOREM 7.6.1. *If $f_1(x)$ and $f_2(x)$ are solutions of the differential equation and if C_1 and C_2 are arbitrary constants, then*

$$C_1 f_1(x) + C_2 f_2(x) \tag{7.6.6}$$

is also a solution.

Proof. This is basically a consequence of Theorem 3.2.1 which asserts that

$$D_x[C_1f_1(x) + C_2f_2(x)] = C_1D_x[f_1(x)] + C_2D_x[f_2(x)].$$

From this it follows for all nonnegative integers k that

$$D_x^k[C_1f_1(x) + C_2f_2(x)] = C_1D_x^k[f_1(x)] + C_2D_x^k[f_2(x)].$$

If we write these relations for $k = 0, 1, 2, \cdots, n$, multiply the kth identity by a_{n-k}, and add, we get

$$P(D_x)[C_1f_1(x) + C_2f_2(x)] = C_1P(D_x)[f_1(x)] + C_2P(D_x)[f_2(x)] = 0.$$

This proves the assertion. In this argument $D_x^0f(x) = f(x)$ by definition.

We come now to the problem of finding solutions of a given differential equation (7.6.1). The situation holding in the three special cases above suggests trying solutions of the form

$$e^{rx},$$

where r is a constant to be determined. Since

$$D_x^ke^{rx} = r^ke^{rx}, \tag{7.6.7}$$

we see that

$$P(D_x)[e^{rx}] = P(r)e^{rx}. \tag{7.6.8}$$

It follows that e^{rx} is a solution if and only if r is a root of the algebraic equation

$$P(t) = 0, \tag{7.6.9}$$

usually called the *auxiliary equation.*

It is of degree n and has n roots

$$r_1, r_2, \cdots, r_n.$$

If these roots are distinct, then

$$C_1e^{r_1x} + C_2e^{r_2x} + \cdots + C_ne^{r_nx} \tag{7.6.10}$$

gives the general solution of the equation in a sense to be explained later. In any case the expression defines a solution.

If there are multiple roots, the differential equation has solutions which are not pure exponential functions or linear combinations of such functions. We shall prove

THEOREM 7.6.2. *If $t = \alpha$ is a root of* (7.6.9) *of multiplicity k, then*

$$(C_1 + C_2x + C_3x^2 + \cdots + C_kx^{k-1})e^{\alpha x} \tag{7.6.11}$$

is a solution of the differential equation.

Proof. The assumption that $t = \alpha$ is a k-fold root of (7.6.9) is equivalent to

$$P(t) = (t - \alpha)^kQ(t) \quad \text{where} \quad Q(\alpha) \neq 0. \tag{7.6.12}$$

From this factorization we see that

$$P(t),\ P'(t),\ P''(t),\ \cdots,\ P^{(k-1)}(t)$$

vanish for $t = \alpha$.

We now go back to (7.6.8). This is an identity in r as well as in x, and the right member is a differentiable function of r. Differentiation with respect to r gives a new identity

$$D_r P(D_x)[e^{rx}] = [xP(r) + P'(r)]e^{rx}. \tag{7.6.13}$$

Here the crucial observation is that

$$D_r\{P(D_x)[e^{rx}]\} = P(D_x)\{D_r e^{rx}\} \tag{7.6.14}$$

and generally

$$D_r^m\{P(D_x)[e^{rx}]\} = P(D_x)\{D_r^m e^{rx}\}. \tag{7.6.15}$$

This is a special case of a general theorem asserting the commutability of differential operators. Such a theorem will be proved in Chapter 14, Vol. II. At present it is sufficient to observe that (7.6.15) is true if it holds when $P(t)$ is a power of t, that is, if

$$D_r^m\{D_x^p e^{rx}\} = D_x^p\{D_r^m e^{rx}\} \tag{7.6.16}$$

for $m, p = 1, 2, 3, \cdots$. For $m = 1$ and p arbitrary the verification follows from the two identities

$$D_r[D_x^p e^{rx}] = D_r[r^p e^{rx}] = (r^p x + pr^{p-1})e^{rx},$$
$$D_x^p[D_r e^{rx}] = D_x^p[xe^{rx}] = (xr^p + pr^{p-1})e^{rx},$$

where the last step follows from the theorem of Leibniz, formula (3.7.11). Hence (7.6.16) is true for $m = 1$ and any value of p. This implies that (7.6.14) holds. Hence, by (7.6.13),

$$P(D_x)[xe^{rx}] = [xP(r) + P'(r)]e^{rx}. \tag{7.6.17}$$

If $k \geqq 2$ in Theorem 7.6.2, then the right member vanishes for $r = \alpha$, and it follows that

$$xe^{\alpha x}$$

is a solution of the differential equation.

Likewise it may be shown that

$$P(D_x)[x^2 e^{rx}] = [x^2 P(r) + 2xP'(r) + P''(r)]e^{rx}. \tag{7.6.18}$$

This vanishes for $r = \alpha$, if $k \geqq 3$, and shows that

$$x^2 e^{\alpha x}$$

is also a solution. The method applies for any value of k.

Suppose the auxiliary equation (7.6.9) has p, $p \leqq n$, distinct roots

$$r_1, r_2, \cdots, r_p,$$

of multiplicity

$$k_1, k_2, \cdots, k_p,$$

respectively, where

$$k_1 + k_2 + \cdots + k_p = n.$$

We can then assert that for any choice of the constants C_{jn} the function

$$y(x) \equiv \sum_{j=1}^{p} \left\{ \sum_{m=1}^{k_j} C_{jm} x^{m-1} \right\} e^{r_j x} \tag{7.6.19}$$

is a solution of (7.6.1).

This solution involves n arbitrary constants and is the general solution of (7.6.1) in the sense that to each solution $y(x)$ corresponds a unique set of constants C_{jm} giving the representation (7.6.19). Since the latter is unique, the solution which is identically zero has also a unique representation, all the C_{jm} being zero. Conversely, if the representation of zero is unique, then every solution has a unique representation.

THEOREM 7.6.3. *The function defined by* (7.6.19) *is identically zero if and only if every* C_{jm} *is zero.*

We do not prove this theorem, but we give some comments. The case $n = 2$ is practically trivial, for if $r_1 \neq r_2$, then

$$C_1 e^{r_1 x} + C_2 e^{r_2 x} = 0 \quad \text{with} \quad C_1 \neq 0$$

gives

$$e^{(r_1 - r_2)x} \equiv -C_2/C_1$$

which is absurd, and if $r_1 = r_2$, then

$$(C_1 + C_2 x)e^{r_1 x} \equiv 0 \quad \text{implies} \quad C_1 + C_2 x \equiv 0$$

or $C_1 = C_2 = 0$.

For $n > 2$ a more elaborate argument is required. We shall mention only one aspect, usually ignored. For a given set of coefficients C_{jm}, not all zero, one can always find directions in the complex plane, determined by the r_j's actually present, such that $y(z)$ becomes infinite when z tends to infinity in such a direction. Since the function is supposed to be identically zero, no such directions can exist, and this requires that every C_{jm} is zero.

Example 1. Take the function

$$C_1 e^{z} + C_2 e^{iz} + C_3 e^{-iz}. \tag{7.6.20}$$

It becomes infinite when $z = x \to +\infty$ unless $C_1 = 0$; it also becomes infinite if $z = iy$ and $y \to +\infty$ unless $C_3 = 0$, and when $y \to -\infty$ unless $C_2 = 0$.

DEFINITION 7.6.1. *A set of n functions $f_1(x), f_2(x), \cdots, f_n(x)$, is said to be linearly independent in an interval (a, b) if no linear combination of them can be identically zero, that is, if*

$$C_1 f_1(x) + C_2 f_2(x) + \cdots + C_n f_n(x) = 0 \tag{7.6.21}$$

for all x in (a, b) implies that $C_1 = C_2 = \cdots = C_n = 0$.

DEFINITION 7.6.2. *A set of n linearly independent solutions of a linear differential equation of order n is called a fundamental system.*

In this sense the n functions

$$x^{m-1}e^{r_j x};\ m = 1, 2, \cdots, k_j;\ j = 1, 2, \cdots, p; \tag{7.6.22}$$

form a fundamental system. Theorem 7.6.3 implies the existence of n linearly independent solutions of (7.6.1). This system is maximal in the sense that *any system of $n + 1$ solutions is necessarily linearly dependent.* To prove this we need

THEOREM 7.6.4. *If $f(x)$ is a solution of* (7.6.1), *then $f(x)$ has derivatives of all orders and $f(x)$ is represented by its formal Taylor series about any point c. The coefficients of the series are uniquely determined by the values of*

$$f(c), f'(c), \cdots, f^{(n-1)}(c). \tag{7.6.23}$$

If these n numbers are all zero, then $f(x)$ is identically zero.

Proof. These properties must be read off from the differential equation and not from formula (7.6.19) since we do not know *a priori* that every solution is of this form. Indeed, this is one of the facts that requires a proof.

Since $f(x)$ satisfies the differential equation in some interval (a, b), $a < c < b$, then

$$f^{(n)}(x) = b_1 f^{(n-1)}(x) + b_2 f^{(n-2)}(x) + \cdots + b_n f(x) \tag{7.6.24}$$

where

$$b_k = -\frac{a_k}{a_0}.$$

The right member of (7.6.24) is a differentiable function of x since $f(x), f'(x), \cdots, f^{(n-1)}(x)$ have this property. It follows that $f^{(n+1)}(x)$ also exists, and

$$f^{(n+1)}(x) = b_1 f^{(n)}(x) + b_2 f^{(n-1)}(x) + \cdots + b_n f'(x).$$

We can then repeat the argument and show that $f^{(n+2)}(x)$ exists and so on. In general, we have for any positive integer p

$$f^{(n+p)}(x) = b_1 f^{(n+p-1)}(x) + b_2 f^{(n+p-2)}(x) + \cdots + b_n f^{(p)}(x). \tag{7.6.25}$$

In particular, at $x = c$

$$f^{(n)}(c) = b_1 f^{(n-1)}(c) + b_2 f^{(n-2)}(c) + \cdots + b_n f(c),$$
$$f^{(n+1)}(c) = b_1 f^{(n)}(c) + b_2 f^{(n-1)}(c) + \cdots + b_n f'(c),$$

and so on. The first equation determines $f^{(n)}(c)$ in terms of the n numbers $f(c)$, $f'(c), \cdots, f^{(n-1)}(c)$. We can then get $f^{(n+1)}(c)$ from the second equation and step by step the values of all derivatives of $f(x)$ at $x = c$ in terms of the numbers (7.6.23). If the latter are all zero, every derivative vanishes at $x = c$. This does not imply that $f(x)$ is identically zero, however, since we do not know that $f(x)$ is actually represented by its Taylor series.

We prove this by Taylor's theorem. To show that the remainder tends to zero, we need some estimates of the derivatives. Let us consider some interval

$[\alpha, \beta]$, $a < \alpha < c < \beta < b$, and let M be the largest value assumed by any one of the n functions

$$|f(x)|, |f'(x)|, \cdots, |f^{(n-1)}(x)|$$

in the interval $[\alpha, \beta]$. Let

$$C = \max \{|f(c)|, |f'(c)|, \cdots, |f^{(n-1)}(c)|\},$$
$$B = |b_1| + |b_2| + \cdots + |b_n| + 1.$$

We then have for $\alpha \leqq x \leqq \beta$

$$|f^{(n)}(x)| \leqq |b_1|\,|f^{(n-1)}(x)| + |b_2|\,|f^{(n-2)}(x)| + \cdots + |b_n|\,|f(x)|$$
$$\leqq (|b_1| + |b_2| + \cdots + |b_n|)M < BM.$$

If $x = c$, we can replace M by C obtaining

$$|f^{(n)}(c)| < BC.$$

Next we have

$$|f^{(n+1)}(x)| \leqq |b_1|\,|f^{(n)}(x)| + |b_2|\,|f^{(n-1)}(x)| + \cdots + |b_n|\,|f'(x)|$$
$$< |b_1|\,BM + |b_2|\,M + \cdots + |b_n|\,M$$
$$< |b_1|\,BM + |b_2|\,BM + \cdots + |b_n|\,BM < B^2M$$

and

$$|f^{(n+1)}(c)| < B^2C.$$

By induction on p one proves that

$$|f^{(n+p)}(x)| < B^{p+1}M, \; |f^{(n+p)}(c)| < B^{p+1}C. \tag{7.6.26}$$

We can now use Taylor's theorem in the form given by (6.3.14) and (6.3.15) where we replace n by $n + p$. This gives

$$f(x) = \sum_{k=0}^{n+p-1} \frac{f^{(k)}(c)}{k!}(x - c)^k + \frac{1}{(n + p - 1)!}\int_c^x (x - t)^{n+p-1} f^{(n+p)}(t)\, dt$$

and for $\alpha \leqq x \leqq \beta$ the remainder does not exceed in absolute value

$$\frac{1}{(n + p)!} MB^{p+1}\, d^{p+1}, \; d = \max (c - \alpha, \beta - c).$$

Since this tends to zero when $p \to +\infty$, it follows that $f(x)$ is represented by its Taylor series for $\alpha \leqq x \leqq \beta$. Actually the series converges for all values of x as we see from the estimates of $|f^{(n+p)}(c)|$. It follows that the series satisfies the differential equation and every solution exists for all finite x. In particular we see that $f(x) \equiv 0$ if and only if

$$f(c) = f'(c) = \cdots = f^{(n-1)}(c) = 0.$$

This completes the proof of the theorem.

We are now ready to prove the final

THEOREM 7.6.5. *Every solution of* (7.6.1) *is of the form* (7.6.19) *for a suitable unique choice of the coefficients* C_{jm}.

Proof. Let us denote the n functions of the fundamental system by

$$y_1(x), y_2(x), \cdots, y_n(x).$$

Let $f(x)$ be a solution of the equation, and let us try to determine constants C_1, $C_2, \cdots, C_n$ such that

$$f(x) = C_1 y_1(x) + C_2 y_2(x) + \cdots + C_n y_n(x). \tag{7.6.27}$$

If such constants exist, then we have also

$$\begin{aligned} C_1 y_1(c) + C_2 y_2(c) + \cdots + C_n y_n(c) &= f(c), \\ C_1 y_1'(c) + C_2 y_2'(c) + \cdots + C_n y_n'(c) &= f'(c), \\ \cdots\cdots\cdots\cdots\cdots\cdots\cdots & \\ C_1 y_1^{(n-1)}(c) + C_2 y_2^{(n-1)}(c) + \cdots + C_n y_n^{(n-1)}(c) &= f^{(n-1)}(c). \end{aligned} \tag{7.6.28}$$

Conversely, if these equations hold, (7.6.27) also holds, for the two sides of the equation are solutions of (7.6.1) whose Taylor series have the same first n terms and consequently are identical by virtue of Theorem 7.6.4.

The system (7.6.28) either has a solution, or the equations are inconsistent and there is no solution. In the first case the solution must be unique. For if there were two sets of solutions, then the set of differences would satisfy the corresponding homogeneous system of equations. Thus there would exist numbers $\{H_k\}$, not all zero, such that

$$\sum_{k=1}^{n} H_k y_k^{(j)}(c) = 0, j = 0, 1, 2, \cdots, n-1. \tag{7.6.29}$$

This means that the solution

$$\sum_{k=1}^{n} H_k y_k(x)$$

vanishes for $x = c$ together with its derivatives of order $\leq n - 1$, and by Theorem 7.6.4 such a solution vanishes identically. This violates Theorem 7.6.3, since it would imply that the set $\{y_k(x)\}$ is not made up of linearly independent solutions. It follows that the solution of (7.6.28) must be unique if it exists. But if the equations (7.6.28) cannot be solved, then, by a theorem in linear algebra, the corresponding homogeneous system (7.6.29) has nontrivial solutions and we have just seen that this leads to contradictions. For the linear algebra, see Chapter 13, Volume II.

Hence the system (7.6.28) has a unique solution and this solution gives the desired representation (7.6.27). This completes the proof.

EXERCISE 7.6

Solve the following differential equations:

1. $y'' - 3y' + 2y = 0$.
2. $y'' - 2y' + y = 0$.
3. $y'' + 3y' = 0$.
4. $y''' - y'' + y' - y = 0$.
5. $y^{(4)} - y = 0$.
6. $y^{(4)} + y = 0$.
7. $y^{(4)} + y'' = 0$.
8. $y''' - 3y'' + 2y' = 0$.
9. $y''' - 4y'' + 5y' - 2y = 0$.
10. $y''' - 3y'' + 3y' - y = 0$.

11. Find the solution of equation 4 that satisfies the conditions $y(0) = 1$, $y'(0) = y''(0) = 0$.

12. If $\Sigma\, a_k x^k$ is a solution of equation 5 above, what relations must hold between the coefficients of the series?

13. If a linear differential equation with real constant coefficients has a solution of the form $u(x) + iv(x)$ where $u(x)$ and $v(x)$ are real, prove that $u(x)$ and $v(x)$ are also solutions. Illustrate by examples. See Problems 4 to 7 above.

14. What differential equation has (7.6.20) as its general solution?

15. Find a fundamental system of solutions of the equation $D_z{}^3 w - w = 0$. For each solution of the system find a ray from the origin of the complex plane on which the solution is unbounded.

16. A radioactive substance decomposes at a rate proportional to the amount present. What is the factor of proportionality if the half-life of the element is 1760 years, that is, the amount will be reduced to one-half of the present value in this time. Take $\log 2 = 0.693$.

17. An equation of the form

$$x^2 y'' + axy' + by = 0$$

is known as an *equidimensional* or *Euler equation*. Show that it has solutions of the form $C_1 x^{r_1} + C_2 x^{r_2}$ where r_1 and r_2 are the roots of a certain quadratic equation. If this equation has equal roots, the term x^{r_2} is to be replaced by $x^{r_1} \log x$. If r_1 and r_2 are conjugate complex numbers, give the solution in real form.

18. Verify the assertions in the preceding problem by introducing $t = \log x$ as a new variable. This gives a differential equation with constant coefficients.

19. Differential equations of the form

$$y'' + 4y = F(x)$$

arise in the theory of forced oscillations. Show that if $Y(x)$ is a particular solution, then the general solution is obtained by adding $Y(x)$ and the general solution of the homogeneous equation, where $F(x)$ is replaced by 0.

20. If $F(x) = e^{ax}$, $a \neq \pm 2i$, show that there is a particular solution of the form Ae^{ax}.

21. If $F(x) = x^2$, show that the equation is satisfied by a quadratic polynomial in x.

22. If $F(x) = \sin 2x$, show that the equation has a solution of the form $x(A \cos 2x + B \sin 2x)$. (This is essentially the excluded case of Problem 20. Here the free periods and those imposed by the perturbing force coincide, the resulting resonance causes unbounded oscillations.)

MISCELLANEOUS PROBLEMS FOR CHAPTER 7

1. If $\sum_{n=1}^{\infty} n^p\,|a_n|$ converges for every p, prove that the series

$$\sum_{n=0}^{\infty} a_n \cos nx \equiv F(x)$$

can be differentiated term-by-term as often as we please so that $F(x)$ has derivatives of all orders.

2. Verify formula (6.3.23).

3. If

$$x^2 D_x{}^2 y + x D_x y + y = 0,$$

prove that

$$x^2 D_x^{n+2} y + (2n + 1) x D_x^{n+1} y + (n^2 + 1) D_x{}^n y = 0.$$

4. Show that the expansion of $(1 + x)^{1+x}$ in powers of x starts with the terms $1 + x + x^2 + \frac{1}{2}x^3$.

5. Prove that for x real and $|x| > 2$,

$$\log \frac{(x-1)^2(x+2)}{(x+1)^2(x-2)} = 2 \sum_{n=0}^{\infty} \frac{1}{2n+1} \left(\frac{2}{x^3 - 3x}\right)^{2n+1}$$

6. The inverse function of the hyperbolic tangent, defined by (7.3.25), is sometimes denoted by arg tanh y. Verify that

$$\text{arg tanh } y = \sum_{n=0}^{\infty} \frac{y^{2n+1}}{2n+1} \quad \text{for} \quad |y| < 1.$$

7. Show that

$$\log 2 = 14 \text{ arg tanh } \tfrac{1}{31} + 10 \text{ arg tanh } \tfrac{1}{49} + 6 \text{ arg tanh } \tfrac{1}{161}.$$

Use the series of the preceding problem to calculate log 2 from this formula with an error less than one unit in the sixth decimal place. How many terms are needed? Discuss the remainder.

8. Prove that

$$\lim_{n\to\infty} n \sum_{k=0}^{n-1} \frac{1}{n^2 + k^2} = \int_0^1 \frac{dx}{1 + x^2} = \frac{\pi}{4}.$$

See Section 4.7.

9. If $a + b > 0$, prove that

$$\int_b^{\infty} \frac{dx}{(x + a)(x - b)^{\frac{1}{2}}} = \frac{\pi}{(a + b)^{\frac{1}{2}}}.$$

10. Verify that

$$D_x{}^n\left(\frac{1}{x^2 + 1}\right) = (-1)^n n! \frac{1}{2i}\left\{\frac{1}{(x - i)^{n+1}} - \frac{1}{(x + i)^{n+1}}\right\}$$

$$= (-1)^n n!\, (x^2 + 1)^{-(n+1)/2} \sin\left[(n + 1) \text{ arc tan } (1/x)\right].$$

[*Hint:* Set $x + i = re^{i\theta}$, $x - i = re^{-i\theta}$ with $-\dfrac{\pi}{2} < \theta < \dfrac{\pi}{2}$ and simplify.]

11. Prove that

$$\arctan(x\sqrt{2}+1) + \arctan(x\sqrt{2}-1) = \arctan\frac{x\sqrt{2}}{1-x^2}.$$

12. If $y(x-y)^2 = x$, prove that

$$\int \frac{dx}{x-3y} = \tfrac{1}{2}\log[(x-y)^2 - 1] + C.$$

[*Hint:* Set $x/y = t$.]

13. Prove that x, e^x, xe^{-x} are linearly independent in $(-a, a)$.

14. Determine the constants a and b so that the equation

$$y'' + ay' + by = xe^x$$

is satisfied by xe^x. Find the general solution of the resulting differential equation.

15. Let $f(x) = x \sin\dfrac{1}{x}$, $f(0) = 0$, $g(x) = |x|^{\frac{1}{2}} + x$. Show that

$$\lim_{x\to 0}\frac{f(x)}{g(x)} = 0 \quad \text{but} \quad \lim_{x\to 0}\frac{f'(x)}{g'(x)} \text{ does not exist.}$$

See comments to formula (6.7.17).

16. If $a > 0$, show that for all N and all a

$$\sum_{n=1}^{N} \frac{a}{a^2+n^2} < \frac{\pi}{2}.$$

Find the limits as $x \to 0$ of:

17. $\dfrac{\sin^2(4x)\tan(5x)}{x^2 \arctan(3x)}$.

18. $\dfrac{\sin(x^2) - \log(1+x^2)}{1-\cos(x^2)}$.

19. $\dfrac{\sqrt{1+x^2} - \cos x}{\log(1+x^2)}$.

20. $(x \sin x)^{-\tan^2 x}$.

21. Find $\lim \sin(\pi\sqrt{4n^2+n})$ as the integer $n \to \infty$.

22. Expand $\sin x$ in Taylor's series about $x = \dfrac{\pi}{4}$. Estimate the remainder after four terms.

23. Does the function $2\sin 3x + 10\cos 5x$ have a real zero?

24. Does the following integral converge?

$$\int_1^\infty \sin\frac{1}{x}\log\left(1+\frac{1}{x}\right)dx.$$

Evaluate the integrals:

25. $\displaystyle\int_0^1 x^{-1-\log x}\log x\,dx$.

26. $\displaystyle\int_1^2 \frac{dx}{x^2(x^2+2x+2)}$.

27. $\displaystyle\int_0^\infty e^{-x}(x-[x])\,dx$ where $[x]$ is the largest integer $\leqq x$.

8

Applications of the Differential Calculus

8.1. Tangents and normals. In this chapter we shall discuss applications of the differential calculus. These will include the use of the derivative in the local study of curves (global problems are discussed in the next chapter), questions of increase and decrease, rates of change, maxima and minima, and motion on a line or in a plane.

We start with tangents and normals. Given a differentiable function $f(x)$ defined in some interval (a, b), we saw in Section 3.1 that it is possible to assign a direction of the curve

$$y = f(x) \tag{8.1.1}$$

at every point (x_0, y_0) with $a < x_0 < b$. For we proved that the secant through the points $P = (x_0, y_0)$ and $Q = (x_1, y_1)$ tends to a limiting position as $x_1 \to x_0$. This limit we called the *tangent of the curve* at the point P and its equation is

$$y - y_0 = f'(x_0)(x - x_0). \tag{8.1.2}$$

In this connection the derivative is called the *slope of the curve* at $x = x_0$. Strictly speaking it is of course the slope of the tangent to the curve at the point, but the tangent may be regarded as a first approximation of the curve near the point. The student should note that the slope of the tangent at $x = x_0$ it $f'(x_0)$, not $f'(x)$, a common error. Always check the answer to tangent problems for linearity: the tangent is a straight line and its equation must be of the first degree in x and y.

Example 1. We take the ellipse

$$b^2x^2 + a^2y^2 = a^2b^2. \tag{8.1.3}$$

On the upper half of the ellipse

$$y = \frac{b}{a}(a^2 - x^2)^{\frac{1}{2}},$$

and

$$D_xy = -\frac{b}{a}x(a^2 - x^2)^{-\frac{1}{2}}.$$

At the point (x_0, y_0) the slope equals

$$-\frac{b}{a}x_0(a^2 - x_0{}^2)^{-\frac{1}{2}} = -\frac{b^2}{a^2}\frac{x_0}{y_0}.$$

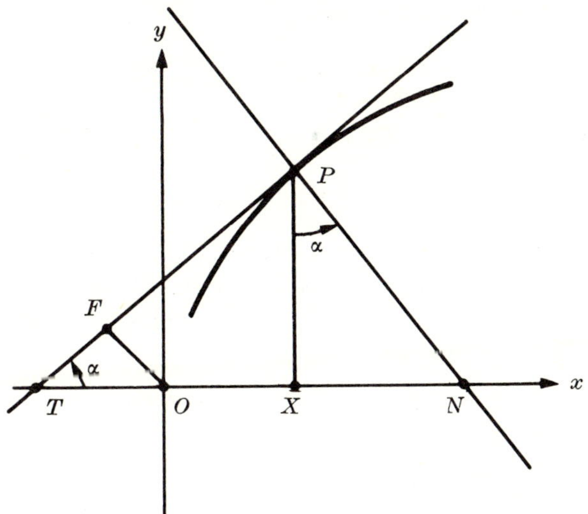

Figure 8.1

Thus the equation of the tangent is

$$y - y_0 = -\frac{b^2 x_0}{a^2 y_0}(x - x_0).$$

This simplifies to

$$b^2 x_0 x + a^2 y_0 y = a^2 b^2$$

in agreement with Formula (A.4.17).

Figure 8.1 shows an arc of a curve $y = f(x)$, the tangent at the point P, the line through P perpendicular to the tangent, and the intersections of these lines with the x-axis at the points T and N. The line perpendicular to the tangent is known as the *normal* at P. We have also marked the angle α which the tangent makes with the horizontal direction and the normal with the vertical.

The four line segments TP, TX, PN, and XN are examples of functions in the sense of Leibniz (see Section 2.3, item 8 under category III). They are known, respectively, as the *tangent*, *subtangent*, *normal*, and *subnormal* of the curve at P with respect to the chosen coordinate system. Note that the word "tangent" is used in three senses: the tangent of a curve, a line segment on this line from the point of contact to the intersection with the x-axis, and finally in the sense of trigonometry. Thus the trigonometric tangent of the angle between the curve tangent and the positive x-axis is the slope of the curve. The context will show which "tangent" is meant. Similarly the noun "normal" is used both for a straight line and for a segment on this line. The adjective "normal" is doing overtime in mathematics.

All these line segments can be expressed in terms of the local values of the function and of its derivative. The equation of the normal is

$$y - y_0 = -\frac{1}{f'(x_0)}(x - x_0), \tag{8.1.4}$$

while from the triangle PXN we get

$$PN = f(x_0)\{1 + [f(x_0)']^2\}^{\frac{1}{2}}. \tag{8.1.5}$$

This is positive if and only if $f(x_0) > 0$, that is, if the curve is above the x-axis. Similarly

$$XN = f(x_0)f'(x_0). \tag{8.1.6}$$

At present we shall attach no significance to the sign of these quantities. Lengths of the line segments are obtained by using absolute values.

From the same figure we read off expressions for the tangent and the subtangent:

$$TP = \frac{f(x_0)}{f'(x_0)}\{1 + [f'(x_0)]^2\}^{\frac{1}{2}}, \tag{8.1.7}$$

$$TX = \frac{f(x_0)}{f'(x_0)}. \tag{8.1.8}$$

Example 2. For the ellipse we find that the equation of the normal is

$$a^2y_0x - b^2x_0y = (a^2 - b^2)x_0y_0,$$

while the normal and subnormal are

$$PN = a^{-2}(a^4y_0^2 + b^4x_0^2)^{\frac{1}{2}},$$

$$XN = -\frac{b^2}{a^2}x_0.$$

For the tangent and subtangent we have, respectively,

$$TP = -\frac{y_0}{a^2x_0}(b^4x_0^2 + a^4y_0^2)^{\frac{1}{2}},$$

$$TX = -\frac{a^2y_0^2}{b^2x_0}.$$

In working with such formulas it is worthwhile to use what might be called "dimension control." All the line segments have the dimension of a length, since x_0, y_0, a, and b also are lengths, it is necessary that if the numerator has the dimension l^k then the denominator is of dimension l^{k-1}. Evidently all four formulas satisfy this condition.

We can draw two real tangents to an ellipse from a point exterior to the ellipse, but no tangent from a point in the interior. Normals present a totally different problem. There is a curve, known as the *evolute of the ellipse* (see Section 9.5), symmetric with respect to the coordinate axes, such that if the point is interior to this curve, four real normals can be drawn from the point to the ellipse while if the point is exterior, only two real normals exist. The evolute intersects the ellipse in four points if $a > \sqrt{2}b$ and it is then partly interior and partly exterior to the ellipse so the situation is not so simple as for the tangent problem. We have a similar situation for the hyperbola, but in the case of the parabola, there are either three normals or only one. We shall sketch the argument.

Example 3. We take the parabola with its axis coinciding with the positive y-axis

$$x^2 = 4ay. \tag{8.1.9}$$

A straight line through the point (x_0, y_0) of slope m

$$y - y_0 = m(x - x_0) \tag{8.1.10}$$

intersects the parabola in two points whose x-coordinates are

$$2am \pm 2[a^2m^2 + a(y_0 - mx_0)]^{\frac{1}{2}}.$$

The slope of the normal at the point (X, Y) on the parabola is

$$-\frac{2a}{X} = m,$$

if the straight line (8.1.10) is the normal of the parabola at the point of intersection. Substituting for X the x-coordinate of the point of intersection and simplifying we obtain a cubic equation for m

$$x_0m^3 + (2a - y_0)m^2 + a = 0. \tag{8.1.11}$$

This equation has three roots and one is always real. If the point (x_0, y_0) lies on the curve

$$4(y - 2a)^3 - 27ax^2 = 0, \tag{8.1.12}$$

then the equation (8.1.11) has two equal roots, namely

$$m = \tfrac{2}{3}\,\frac{y_0 - 2a}{x_0}.$$

The curve (8.1.12) is called a *semicubical parabola*, first studied by William Neil (1637–1670) it is often called a *Neil's parabola*. See Figure 8.2. It has a cusp with vertical tangent at the point $(0, 2a)$. In Section 9.5 we shall encounter it as the *evolute of the parabola*. It separates the plane into two parts. From any point above the semicubical parabola one can draw three real normals to the parabola; from a point below the curve only one can be drawn We saw above that on the semicubical parabola two of the normals coincide. We shall see later that *the evolute of a curve is the locus of the points of intersection of "nearby" normals* and also *the locus of the so-called center of curvature of the given curve.*

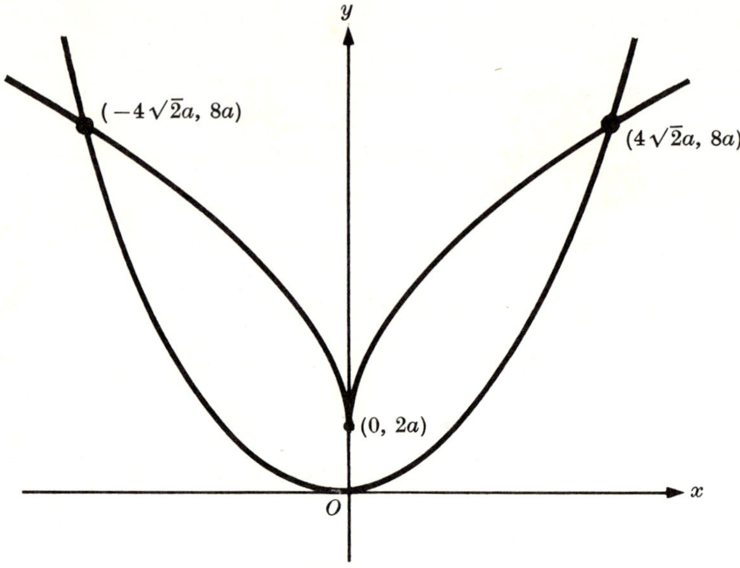

Figure 8.2

This section ends with a brief mention of another type of curve associated with the system of tangents of the given curve (8.1.1). From a fixed point, here taken as the origin, we drop a perpendicular to the curve tangent. The point of intersection F lies between the points P and T. See again Figure 8.1. As P varies, F describes a curve known as the *pedal* or the *foot point curve* of the given curve with respect to the fixed point. The coordinates of the foot point $F = (X, Y)$ are easily found in terms of the coordinates of the point of contact of the tangent $P = (x_0, y_0)$. The equation of the tangent is

$$xf'(x_0) - y - [x_0 f'(x_0) - f(x_0)] = 0.$$

The perpendicular from the origin has the equation

$$x + f'(x_0)y = 0.$$

Finding the intersection of the two lines we get

$$X = \frac{f'(x_0)[x_0 f'(x_0) - f(x_0)]}{1 + [f'(x_0)]^2}, \qquad Y = -\frac{x_0 f'(x_0) - f(x_0)}{1 + [f'(x_0)]^2}. \tag{8.1.13}$$

This is the parametric form of the equation of the pedal with x_0 as the parameter.

Example 4. For the ellipse (8.1.3) we get

$$X = \frac{a^2 b^4 x_0}{b^4 {x_0}^2 + a^4 {y_0}^2}, \quad Y = \frac{a^4 b^2 y_0}{b^4 {x_0}^2 + a^4 {y_0}^2}, \tag{8.1.14}$$

where

$$b^2x_0{}^2 + a^2y_0{}^2 = a^2b^2.$$

It is simple to verify that the equation of the pedal is

$$(X^2 + Y^2)^2 = a^2X^2 + b^2Y^2, \tag{8.1.15}$$

a curve known as *Booth's lemniscate*, after the British geometer, James Booth (1806–1878). It has an isolated point at the origin.

EXERCISE 8.1

1. Find the tangent, subtangent, normal, and subnormal of the hyperbola $b^2x^2 - a^2y^2 = a^2b^2$. Find also the equations of the tangent and the normal at a point (x_0, y_0) on the hyperbola.

2. Same question for the parabola $y^2 = 4ax$. Note that the subnormal has constant length.

3. Given the conic

$$y^2 = 2px + qx^2.$$

Show that the following gives a geometric construction of the normal at P. Let OP be the radius vector from the origin to the point P on the conic, let PQ be perpendicular to OP at P and intersect the x-axis at Q. Let N be the point on the x-axis such that $NQ = p$ (attention being paid to the sign of p). Then PN is the normal of the conic at P.

4. Prove that the curve $y = ae^{bx}$ has a constant subtangent.

5. Prove that any tangent to the semicubical parabola (8.1.12) is a normal to the parabola (8.1.9).

6. Find the equation of the tangent to the curve $y = x \log x + 1$ at the point (1, 1).

7. Find the equation of the tangent to the curve

$$2x^3 - x^2y^2 - 3x + y + 7 = 0$$

at the point (1, —2).

8. Find the equation of the normal to the curve

$$x^3 + y^2 + 2x - 6 = 0$$

at a point with ordinate $y = 3$.

9. For what values of a and b does the curve $y = \dfrac{ax}{1 + bx^2}$ intersect the x-axis under an angle of 45°?

10. Prove that the curves

$$x^2 - y^2 = a^2, \quad xy = b^2$$

intersect at right angles.

11. Verify equations (8.1.11) and (8.1.12).

12. Verify (8.1.14) and (8.1.15).

13. Show that the parabolas

$$y^2 + 4ax - 4a^2 = 0, \quad y^2 - 4bx - 4b^2 = 0$$

intersect at right angles.

14. Determine the tangent of the angle which the curves $x^2 + xy + y^2 = 7$ and $xy = 2$ make with each other at one of their intersections.

15. Find the equations of the tangents common to the curves $y^2 = 4x$ and $x^2 + y^2 = 2$.

16. Do the parabolas in Problem 13 have any normals in common?

17. What normal of the parabola $(x + y)^2 = x - y$ is vertical?

18. Find a normal of slope 1 to the ellipse $x^2 + 4y^2 = 4$.

19. Find the pedal of a hyperbola with respect to its center.

Find the pedal of a parabola with respect to:

20. The vertex.

21. The focus.

22. Take the ellipse $a^2x^2 + b^2y^2 = a^2b^2$ where the major axis falls along the y-axis. To each point $P = (x, y)$ on the ellipse order a point $Q = (X, Y)$ such that O, P, Q are collinear and $OP \times OQ = ab$. Show that Q describes (8.1.15). We say that the two curves transform into each other by an *inversion* (or a *reciprocation*) in the circle $x^2 + y^2 = ab$.

8.2. Rates of change. In Definition 3.1.3 we introduced a mode of measuring a rate of change. If the phenomenon under consideration is described by a differentiable function $f(x)$ of x, we defined the rate of change of $f(x)$ with respect to x at the instant $x = x_0$ as the value of the derivative at $x = x_0$, that is, as $f'(x_0)$. In the present section we shall elaborate on this and study simultaneous rates of interdependent quantities.

It is clear that "rate of change" is a relative notion and not an absolute one. We always study changes with respect to some other changing quantity, the time or the distance, for example. Often we cannot analyze what causes a change but even if ultimate causes are beyond our ken, it is helpful to formalize the observations and attempt an analytic description. Calculus provides us with means for evaluating the consequences of the description.

In a particularly simple instance the phenomenon may be described (actually or with sufficient accuracy for the purpose in mind) by a linear function:

$$y = mx + b. \tag{8.2.1}$$

In this case *y is said to vary uniformly with x*, and m is called the *rate of change of y with respect to x*. The numerical value of m is the number of units change in y per unit change in x, and the sign of m indicates whether there is increase or decrease. Conversely, if the ratio of the increment of y to the increment of x always equals m, then there exists a constant b such that (8.2.1) holds and y varies uniformly with respect to x.

If $f(x)$ is not linear, the difference quotient

$$\frac{1}{h}[f(x_0 + h) - f(x_0)]$$

is no longer a constant. The difference quotient gives an *average* or *mean rate* of change of $f(x)$ over the interval $[x_0, x_0 + h]$. In this case it is convenient to define the rate of change at $x = x_0$ as the limit of the difference quotient as $h \to 0$ whenever this limit exists. This convention is adopted in Definition 3.1.3.

We can interpret the mean value theorem of the differential calculus in this new terminology as stating that *if the mean rate of change of $f(x)$ in the interval $[a, b]$ equals m, then there exists at least one point c, $a < c < b$, where the rate of change equals m.*

Example 1. The area of a sphere is

$$A = 4\pi r^2,$$

where r is the length of the radius. The rate of change of the area with respect to the radius is

$$D_r A = 8\pi r$$

and has the dimension of length.

Example 2. The current c in an electric circuit varies inversely with the resistance R. To find the factor of proportionality it suffices to know the value of c for a particular value of R. Suppose, for instance, that the current is 4 amp when the resistance is 10 ohms. Then

$$c = \frac{40}{R}$$

and the rate of change of the current with respect to the resistance is

$$D_R c = -\frac{40}{R^2}.$$

When $R = 10$, this works out to be -0.4 amp/ohm^2.

We shall now consider interdependent rates and start with some typical cases.

Example 3. A moving particle P describes the unit circle in such a manner that its coordinates are differentiable functions of the time t, say

$$x = f(t), \quad y = g(t).$$

Then their rates of change are

$$D_t x = f'(t), \quad D_t y = g'(t),$$

respectively. But these rates are not independent of each other for

$$[f(t)]^2 + [g(t)]^2 \equiv 1, \tag{8.2.2}$$

so that

$$f(t)f'(t) + g(t)g'(t) \equiv 0. \tag{8.2.3}$$

Hence, knowing the position of P at the time $t = t_0$ and the rate of change of one of the coordinates, we can find the rate of the other. This does not require that $x = f(t)$ be an explicitly known function of t. All we need is that (8.2.2) holds and that $f(t)$ is differentiable. This implies that $g(t)$ is also differentiable and that (8.2.3) holds.

Example 4. A spherical balloon is expanding due to solar radiation. At the instant when its radius is 10 units of length, its area expands at the rate of 0.1 square units per second. Find the rates of change of the radius and of the volume at this instant. We have

$$A = 4\pi r^2, \quad V = \tfrac{4}{3}\pi r^3, \tag{8.2.4}$$

so that

$$D_tA = 8\pi r D_t r, \quad D_t V = 4\pi r^2 D_t r. \tag{8.2.5}$$

From the first equation we get

$$0.1 = 80\pi (D_t r)_{r=10}.$$

Hence

$$(D_t r)_{r=10} = \frac{1}{800\pi},$$

and from the second equation under (8.2.5)

$$(D_t V)_{r=10} = \frac{400\pi}{800\pi} = \tfrac{1}{2}.$$

If the problem had not called for the rate of change of r but only for that of V, we could have replaced (8.2.4) by the expression of the volume of the sphere in terms of its surface area, that is,

$$V = \frac{1}{6\sqrt{\pi}} A^{\frac{3}{2}}. \tag{8.2.6}$$

This gives

$$D_t V = \frac{1}{4\sqrt{\pi}} A^{\frac{1}{2}} D_t A.$$

When $r = 10$, $A = 400\pi$ and

$$(D_t V)_{r=10} = \frac{1}{4\sqrt{\pi}} \cdot 20\sqrt{\pi} \cdot \tfrac{1}{10} = \tfrac{1}{2}.$$

We can now formulate a general method for dealing with interdependent rates. We start by listing the relations that hold between the variable quantities. These quantities are supposed to be functions of a single variable, t say. We list only relations which hold for all values of t, restricted to some interval natural to the problem. This interval need not be known explicitly. We then differentiate all relations with respect to t remembering that all variable quantities are functions of t. The result will be one or more relations involving the variables and their rates of change. Here we substitute whatever numerical data that are given and solve for the required instantaneous rate or rates. It is very important to remember that only general relations are differentiated.

EXERCISE 8.2

1. The sides of a rectangle are $2x + 3$ and $3x + 4$. At what rate with respect to x is its area increasing when x increasing passes through the value 5?

2. A straight line is turning about the fixed point (1, 2). At a certain instant the y-intercept of the line equals 6 and is known to decrease at the rate of 2 units per second. Find the rate of change of the area of the triangle bounded by the line and its two intercepts at this instant. Interpret your result.

3. A point P moves on the equilateral hyperbola

$$x^2 - y^2 = 9$$

in such a manner that the radius vector (from O to P) sweeps out an area of two units per second. At what rate is the ordinate increasing when P is (5, 4)? [*Hint:* See Section 7.3.]

4. On the same hyperbola, let P move so that the ordinate increases by one unit per second. Find the rate of increase of x when P is at (5, 4).

5. A particle moves on the straight line $x + 2y = 2$ and its abscissa increases 4 units per second. How fast does the ordinate change?

6. P moves on the unit circle and at the time t

$$x = \frac{1 - t^2}{1 + t^2}, \quad y = \frac{2t}{1 + t^2}.$$

At what times is the rate of decrease of x equal to the rate of increase of y?

7. An isosceles triangle has the base $2a$ and opposite angle α which changes at the rate of 0.01 radian per second when $\alpha = \frac{1}{3}\pi$. At what rate is the area of the triangle then changing?

8. Suppose the fuel consumption of a ship is proportional to the cube of the speed. Let the factor of proportionality be 0.2 and find the rate of change in fuel consumption at the time when the speed is 20 knots and changes by 0.5 knot an hour.

9. In the special theory of relativity the mass of a particle moving at the velocity v is

$$m\left(1 - \frac{v^2}{c^2}\right)^{-\frac{1}{2}},$$

where m is the mass at rest and c is the velocity of light. If $v = \frac{1}{2}c$ and changes at $0.01c$ per second, find the rate of change of the moving mass.

10. *Boyle's law* for an ideal gas states that the product of volume and pressure is a constant when the temperature is kept fixed. Suppose that a given (ideal) gas mass expands isothermally and at a given instant $v = 100$ liters, $p = 0.25$ atm. while the rate of change of the volume is 0.1 liter/sec. Find the corresponding value of the rate of change of the pressure.

11. For an actual gas *van der Waal's equation*,

$$\left(p + \frac{a}{v^2}\right)(v - b) = RT$$

is supposed to hold. Here p is pressure, v volume, T absolute temperature, R

the universal gas constant, and a and b characteristic constants for the gas. In a certain case $a = 540$, $b = 1$, $T = 1000°\text{K}$, $v = 6$, and $D_T v = 0.03$. It is required to find $D_T p$ corresponding to these data. Take $R = 0.08207$.

12. For *adiabatic expansion* $pv^\gamma =$ constant where γ is characteristic for the gas. Suppose that $\gamma = \frac{5}{3}$, $p = 3$, $v = 64$, $D_t v = 3$, and find $D_t p$.

8.3. Rectilinear and planar motion. Suppose that a point P moves along a straight line under the influence of a force. Let $x = f(t)$ denote its position on the line, measured from a fixed point O, at the time t seconds after a certain epoch. Here $|x|$, measured in centimeters, denotes the distance of P from O and, as usual, x is positive to the right of O and negative to the left.

In the simplest case x is a linear function of t

$$x = mt + b. \tag{8.3.1}$$

In this case P describes equal line segments in equal time intervals, since the ratio

$$\frac{f(t+h) - f(t)}{h} = m,$$

independently of t. Here x changes at a uniform rate and so we speak of *uniform motion*. The constant m is called the *velocity* of the motion and is denoted by v. Conversely, if the velocity is a constant, then x is a linear function of t.

Newton's *First Law of Motion* states:

A material particle persists in its state of rest, or of motion in a straight line with constant velocity, except in so far as it is compelled to change that state by the action of a force upon it.

In other words, uniform motion can take place only in the absence of a force acting on the particle. This is sometimes called the "law of inertia."

If x is not a linear function of t, we conclude that a force is acting on the particle. The rate of change of x with respect to t is no longer a constant. We still refer to this rate of change as the *velocity of the motion*. We have then

$$v = D_t x = f'(t). \tag{8.3.2}$$

The absolute value of v is known as the *speed*. If v is positive, the particle is moving to the right, if negative, to the left.

In the present case v is also changing with t. Its rate of change is known as the *acceleration of the motion*. We write

$$\alpha = D_t v = D_t^2 x = f''(t). \tag{8.3.3}$$

If α is positive, the velocity is increasing, if α negative, decreasing. In cgs units, velocities are measured in centimeters per second while accelerations call for centimeters per second per second.

Example 1. If x is a quadratic polynomial in t

$$x = at^2 + bt + c, \tag{8.3.4}$$

we have

$$v = 2at + b, \quad \alpha = 2a, \tag{8.3.5}$$

that is, the acceleration is constant. A particular case is that of a particle moving vertically under gravity. Here

$$D_t^2 y = -g, \tag{8.3.6}$$

and hence

$$D_t y = -gt + v_0, \quad y = -\tfrac{1}{2}gt^2 + v_0 t + y_0, \tag{8.3.7}$$

where y_0 is the initial position of the particle, v_0 is its initial velocity, and g is approximately 980 cm/sec^2.

We know that where there is acceleration there is a force causing it. How this force acts is stated in Newton's *Second Law of Motion*:

"*Change of motion* [momentum] *is proportional to the impressed force, and takes place in the direction of the straight line in which that force is impressed.*

Here we can take this to mean that *mass times acceleration equals force*:

$$mD_t^2 x = F, \tag{8.3.8}$$

where $F = F(t)$ is the force acting on the particle at the time t.

Example 2. We assume that the particle P is attracted towards a fixed point O in the line of motion with a force varying as the distance from that point. Here the equation of motion becomes

$$D_t^2 x = -k^2 x, \tag{8.3.9}$$

where we have absorbed the mass in the factor of proportionality and $k > 0$. The sign in the right member is negative since the force is directed from P to O. Equation (8.3.9) is familiar to us and has the general solution

$$x = C \sin (kt + a). \tag{8.3.10}$$

If the position x_0 and velocity v_0 of the particle are given at $t = 0$, then a and C have to be determined from the equations

$$C \sin a = x_0, \quad Ck \cos a = v_0.$$

This gives

$$a = \arctan \frac{kx_0}{v_0}, \quad C^2 = x_0^2 + \left(\frac{v_0}{k}\right)^2,$$

and the sign of C agrees with that of v_0. The absolute value of C is known as the *amplitude* of the motion; a is the *initial phase*.

This type of motion is of great importance in Dynamics; it is called a *simple-harmonic motion* (oscillation, vibration). Special instances where this theory applies, at least in the first order of approximation, are: vertical oscillations of a weight which hangs from a fixed point by a helical spring, vertical oscillations of a ship, small oscillations of a pendulum, and so on. In each case we must neglect some feature such as the inertia of the spring or of the water, the difference between the sine of an arc and the arc itself, etc.

We turn now to motion in a plane. Here we need vector representation of position, velocity, acceleration, and force. The student should review the parts of Sections 1.5 and A.1 concerning vectors.

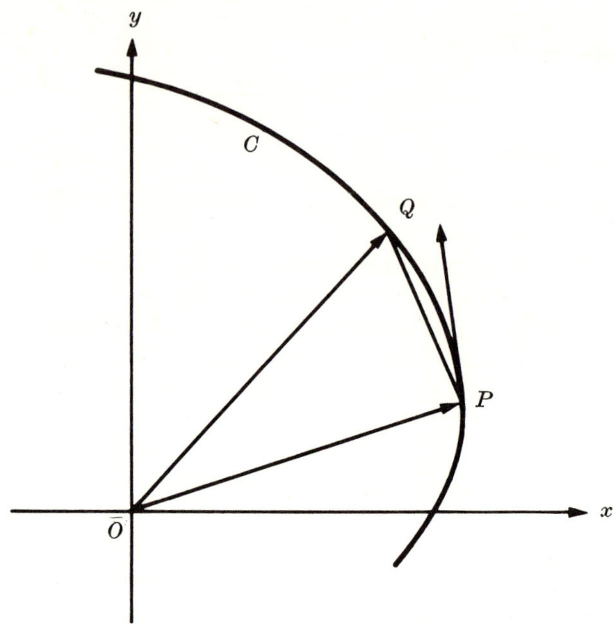

Figure 8.3

The position of the particle P is described by a *position vector*

$$\boldsymbol{x} = (x, y), \tag{8.3.11}$$

where the components x and y are supposed to be differentiable functions of t. The moving particle describes a path C (its *orbit*) given by the vector equation

$$\boldsymbol{x} = \boldsymbol{x}(t). \tag{8.3.12}$$

In Figure 8.3 we have drawn an arc of C and marked two position vectors $\boldsymbol{x}(t)$ and $\boldsymbol{x}(t + h)$ as well as their difference

$$\boldsymbol{x}(t + h) - \boldsymbol{x}(t)$$

which forms a chord of C. The figure suggests that C has a length. It will be shown in Appendix C that this is indeed the case and that actually the arc length is a differentiable function of t when the coordinates have this property.

When $h \to 0$, we see that

$$\lim_{h\to 0} \frac{1}{h}[\boldsymbol{x}(t + h) - \boldsymbol{x}(t)] \equiv \boldsymbol{x}'(t) \tag{8.3.13}$$

exists, for the difference quotient on the left equals

$$\left\{\frac{1}{h}[x(t + h) - x(t)],\quad \frac{1}{h}[y(t + h) - y(t)]\right\}$$

and each component has a limit, hence also the vector. This shows also that

$$\boldsymbol{x}'(t) = (x'(t), y'(t)), \tag{8.3.14}$$

that is, *the derivative of the vector is the vector formed by the derivatives of the components.*

We define the *velocity* of the particle at P to be

$$\boldsymbol{v}(t) \equiv \boldsymbol{x}'(t). \tag{8.3.15}$$

From the geometry of the situation we see that $\boldsymbol{v}(t)$ is a *tangent vector* at P and its length, the *speed*, equals

$$\|\boldsymbol{v}(t)\| = +\{[x'(t)]^2 + [y'(t)]^2\}^{\frac{1}{2}}. \tag{8.3.16}$$

Similarly we define the *acceleration* as the time derivative of $\boldsymbol{v}(t)$. Thus

$$\boldsymbol{\alpha}(t) = \boldsymbol{v}'(t) = \boldsymbol{x}''(t) = (x''(t), y''(t)), \tag{8.3.17}$$

provided $x(t)$ and $y(t)$ have second-order derivatives.

We may possibly interpret the Second Law of Motion as an assertion that position vectors corresponding to problems of Dynamics do possess second-order time derivatives. For the *momentum* is the vector $m\boldsymbol{v}(t)$ and in nonrelativistic mechanics m is a constant so the rate of change of the momentum is $m\boldsymbol{v}'(t) = m\boldsymbol{x}''(t)$ which is supposed to be proportional to the impressed force. This gives the equation of motion in vector form

$$m\boldsymbol{x}''(t) = \boldsymbol{F}(t), \tag{8.3.18}$$

where $\boldsymbol{F}(t)$ is the force acting on the particle at P.

Example 3. The motion of a projectile in a vertical plane is a simple case. We disregard air resistance and the curvature of the earth. The only force acting on the projectile is gravity which is directed vertically downward. The force then is

$$(0, -mg),$$

and the equations of motion for the components are

$$x''(t) = 0, \quad y''(t) = -g. \tag{8.3.19}$$

Hence

$$x(t) = v_x t + x_0, \quad y(t) = -\tfrac{1}{2}gt^2 + v_y t + y_0, \tag{8.3.20}$$

where (x_0, y_0) is the initial position and

$$\boldsymbol{v}_0 = (v_x, v_y)$$

the initial velocity. The path described by the projectile, if $v_x \neq 0$, is an arc of the parabola

$$y - y_0 = -\tfrac{1}{2}g\left(\frac{x - x_0}{v_x}\right)^2 + \frac{v_y}{v_x}(x - x_0). \tag{8.3.21}$$

For the sake of simplicity suppose that $x_0 = y_0 = 0$. The projectile will then hit zero level again at

$$x = \frac{1}{g}\, 2v_x v_y = \frac{1}{g}\, \|\boldsymbol{v}_0\|^2 \sin 2\theta, \tag{8.3.22}$$

where θ is the angle of elevation of the rifle. This shows that with a given initial velocity the range of the gun will reach a maximum when $\theta = 45°$.

This, we recall, is under the assumption that air resistance, wind, and other factors can be neglected.

Example 4. Let us consider finally the plane analogue of simple-harmonic motion. This is called *elliptic harmonic motion* since the orbits turn out to be ellipses. We have a particle P attracted by a force, directed towards a fixed point O, and of magnitude proportional to the distance from O to P. This gives the vector equation

$$\boldsymbol{x}''(t) = -k^2\boldsymbol{x}(t), \tag{8.3.23}$$

where k is a positive constant. The solution may be written

$$\boldsymbol{x}(t) = \boldsymbol{a} \cos kt + \boldsymbol{b} \sin kt. \tag{8.3.24}$$

Here $\boldsymbol{a}$ and $\boldsymbol{b}$ are arbitrary constant vectors, $\boldsymbol{a}$ is the position vector, and $k\,\boldsymbol{b}$ the initial velocity vector both for $t = 0$. If the two vectors $\boldsymbol{a}$ and $\boldsymbol{b}$ are collinear, the motion takes place on the same line; otherwise $\boldsymbol{x}(t)$ describes an ellipse.

EXERCISE 8.3

1. If $x = t^3 - 7x^2 + 8t$, describe the motion.

2. A ball rolling down an inclined plane travels $x = 100t^2$ centimeters in t seconds. Find v when $t = 4$; t and x when $v = 200$; and the acceleration α.

3. If x^2 is a quadratic polynomial in t which is not a perfect square, prove that the acceleration of x varies as x^{-3}.

4. Rectilinear motion can be described by a curve in the plane where the abscissa is $x(t)$ and the ordinate is $v(t)$. Show that the acceleration is given by the subnormal of this curve. Use the case of constant acceleration as an illustration.

5. Prove that a point cannot move so that its velocity is proportional to the distance that it has traveled from rest.

6. A particle moves so that $x = t^2$, $y = t^3$. Sketch the orbit (semicubical parabola). Determine the velocity and acceleration vectors. Plot the graph described by $\boldsymbol{v}(t)$. Such a velocity curve is called a *hodograph*.

7. If the equation of motion is

$$x = a \cos 2t, y = b \sin 2t,$$

prove that the point lies on an ellipse. Prove that the ellipse is described infinitely often in the counterclockwise sense. Show that the motion is periodic and determine the period.

8. Show that the hodograph of the motion in Problem 7 is also an ellipse. Sketch it.

9. Verify that (8.3.24) describes an ellipse. Consider only the case $\boldsymbol{a} = \alpha\boldsymbol{i}$, $\boldsymbol{b} = \gamma\boldsymbol{i} + \delta\boldsymbol{j}$, $\delta \neq 0$.

10. If P moves so that

$$x = \cos 2t, \quad y = 2 \sin t,$$

show that P describes an arc of a certain parabola and verify that its hodograph is the curve $x^2 = y^2(4 - y^2)$.

11. P moves along the straight line

$$x - 2y = 3$$

and $x = 4t^2$. Find $\mathbf{x}(t)$, $\mathbf{v}(t)$, $\boldsymbol{\alpha}(t)$, and sketch the hodograph.

12. P moves in a plane under an attractive force perpendicular to a fixed straight line and of magnitude proportional to the distance of P from the line. Show that P describes a sine curve. [*Hint:* Choose the line as x-axis.]

13. If the polar coordinates of P are $[r, \theta]$ and if $r = R$ is a constant, then P is said to rotate about 0. A *rotation* is completely determined if $\theta = \theta(t)$ is a known function of t. The quantities $\theta'(t)$ and $\theta''(t)$ are known as the *angular velocity* and the *angular acceleration* respectively. Express $\mathbf{v}(t)$ and $\boldsymbol{\alpha}(t)$ in terms of these quantities. Illustrate by considering the case where the angular velocity is a constant.

8.4. Geometric meaning of derivatives. In this section we shall be concerned with the applications of differential calculus to local curve discussion. This study really started in Section 3.1 but even for the first-order derivative there are geometrical aspects not yet sharply formulated. We need a discussion of the importance of the sign of the first derivative in all questions of increase or decrease. This will serve as an introduction to the *theory of maxima and minima.* In this study the sign of the second and, occasionally, of higher derivatives will also play a role, but the main use of the second derivative is in the discussion of *curvature*. We start with

THEOREM 8.4.1. *If $f(x)$ is differentiable in an interval (a, b), and if $f'(x)$ keeps a constant sign, then $f(x)$ is strictly monotone. More precisely, $f(x)$ is strictly increasing, if $f'(x) > 0$, strictly decreasing if $f'(x) < 0$.*

Proof. Suppose that $f'(x) > 0$ in (a, b). For any choice of α, β with $a \leqq \alpha < \beta \leqq b$ the mean value theorem gives

$$f(\beta) - f(\alpha) = (\beta - \alpha)f'(\gamma),\ \alpha < \gamma < \beta,$$

and hence

$$f(\alpha) < f(\beta) \quad \text{for} \quad \alpha < \beta.$$

This proves that $f(x)$ is strictly increasing when $f'(x) > 0$. Evidently the first inequality is reversed if $f'(x) < 0$.

Suppose that $f(x)$ has a continuous derivative in (a, b) and that $f'(x) = 0$ at a finite number of points $\{x_j\}$:

$$a < x_1 < x_2 < \cdots < x_k < b.$$

Then there are $k + 1$ subintervals in each of which $f'(x)$ keeps a constant sign so that the preceding discussion applies there. If the sign of $f'(x)$ is the same in two or more contiguous subintervals, they can be combined into a single interval in which the function is strictly monotone.

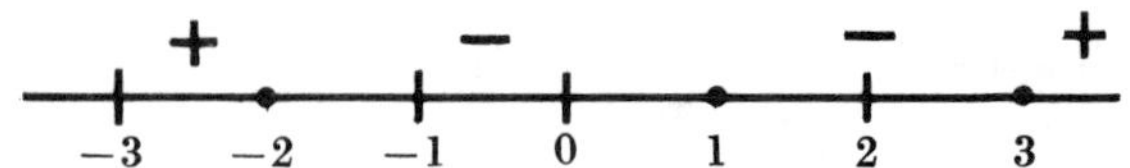

Figure 8.4

Example 1. Take

$$f(x) = 4x^5 - 15x^4 - 20x^3 + 110x^2 - 120x + 40 \tag{8.4.1}$$

with

$$\begin{aligned} f'(x) &= 20x^4 - 60x^3 - 60x^2 + 220x - 120 \\ &= 20(x^4 - 3x^3 - 3x^2 + 11x - 6) \\ &= 20(x + 2)(x - 1)^2(x - 3). \end{aligned}$$

The zeros of $f'(x)$ are $x = -2, 1, 3$. There is no change in sign of $f'(x)$ when x goes through $x = 1$. In Figure 8.4 we have marked these intervals and the sign of $f'(x)$ in the interval. Theorem 8.4.1 then shows that $f(x)$ increases in $(-\infty, -2)$, decreases in $(-2, 3)$, and then increases again in $(3, +\infty)$. This means that $f(-2) = 512$ is a *local maximum* and $f(3) = -113$ is a *local minimum* in the following sense. The first value is the largest value of $f(x)$ in some neighborhood of $x = -2$ and the second value is the smallest assumed by $f(x)$ in some neighborhood of $x = 3$. Actually 512 is the largest value assumed in $(-\infty, 3)$ and -113 is the least value assumed in $(-2, \infty)$.

Example 2. If $f(x) = \sin x$ then $f'(x) = \cos x$. Since $\cos x$ is positive in every interval of the form

$$(2n - \tfrac{1}{2})\pi < x < (2n + \tfrac{1}{2})\pi,$$

where n is any integer, and negative in the complementary (open) intervals, it follows that $\sin x$ is increasing in every interval of the first kind and decreasing in those of the second kind. There are local maxima at the points $(4k + 1)(\pi/2)$ and local minima at the points $(4k + 3)(\pi/2)$. In this case the maximum value, 1, is actually the absolute maximum of $\sin x$ for all x and similarly the local minimum value -1 is an absolute minimum. Compare Figure 7.7.

If $f(x)$ is a given differentiable function in (a, b), we find the *local* (also called *relative*) *maxima* and *minima* by differentiating and equating the derivative to zero. After the roots of the equation

$$f'(x) = 0 \tag{8.4.2}$$

have been found, they have to be examined for relevance to the problem.

DEFINITION 8.4.1. *If $x = x_0$ is a root where $f'(x)$ changes its sign from plus to minus as x increases, then $f(x_0)$ is called a relative maximum. If the sign goes from minus to plus instead, it is a relative minimum, and it is neither a maximum nor a minimum if $f'(x)$ does not change its sign.*

The values obtained in this manner need not give the largest or the least values assumed by $f(x)$ in (a, b). For a more detailed discussion of maxima and minima we refer to the next section.

Suppose that $f(x)$ has also a second derivative. We can then form an idea of how the graph of $y = f(x)$ is curved. If there are no rectilinear stretches and $f''(x)$ is continuous, we have arcs which are convex downward (= concave upward) separated by arcs which are convex upward (= concave downward). An arc of the first type lies below its chords and above its tangents while the situation is reversed for arcs of the second type. See Figure 8.5. In an interval corresponding to an arc of the first kind $f(x)$ is a convex function in the sense of Definition 2.7.2 while arcs of the second kind arise if $-f(x)$ is convex.

THEOREM 8.4.2. *If $f''(x)$ exists and keeps a constant sign in (a, b), then $f'(x)$ is monotone and $f(x)$ or $-f(x)$ is convex according as $f''(x) > 0$ or < 0 in (a, b).*

Proof. That $f'(x)$ is monotone follows from the preceding theorem. Suppose now that $f''(x) > 0$ in (a, b) and let us prove that the arc C of $y = f(x)$ lies above the tangent at the point $P = (x_0, f(x_0))$, $a < x_0 < b$, except for the point of contact. This follows from Taylor's theorem, more precisely from formula (6.2.2) where we set $x = x_0$, $n = 2$ and use (6.2.12) for the remainder. This gives

$$f(x_0 + h) = f(x_0) + hf'(x_0) + \tfrac{1}{2}h^2 f''(x_0 + \theta h), \tag{8.4.3}$$

where $0 < \theta < 1$. This formula does not presuppose that $f''(x)$ is continuous, merely that it exists everywhere in (a, b). Hence

$$f(x_0 + h) > f(x_0) + hf'(x_0). \tag{8.4.4}$$

The left member is the ordinate of the point on C whose abscissa is $x_0 + h$, and the right member is the ordinate of the corresponding point on the tangent at P. This proves that C lies above the tangent. The case $f''(x) < 0$ is treated in the same way. This completes the proof.

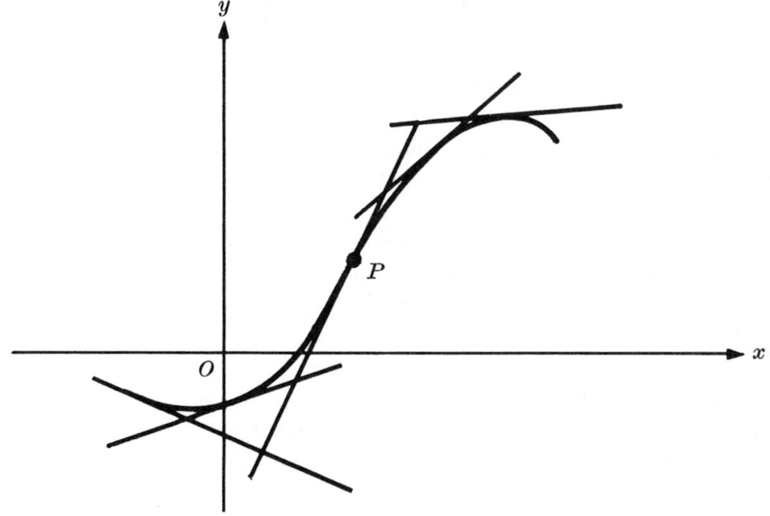

Figure 8.5

Suppose now that $f''(x)$ exists and is continuous in (a, b) and that $f''(x) = 0$ holds only at a finite number of points. Then (a, b) breaks up into a finite number of subintervals in each of which $f''(x)$ keeps a constant sign so that the preceding discussion applies. At this juncture we are interested in the points, if any, where $f''(x)$ changes its sign.

DEFINITION 8.4.2. *The point* $P = (x_0, f(x_0))$ *is a point of inflection of* C *if* $f''(x_0) = 0$ *and if*

$$f''(x_0 - h)f''(x_0 + h) < 0 \tag{8.4.5}$$

for all sufficiently small values of $h \neq 0$.

Such a point separates a convex arc from a concave one. Moreover, the *curve* C *crosses its tangent at a point of inflection*. For formula (8.4.3) shows that

$$f(x_0 + h) - f(x_0) - hf'(x_0)$$

changes its sign when h goes through zero since $f''(x)$ changes sign at $x = x_0$.

To get the points of inflection we differentiate $f(x)$ twice and equate $f''(x)$ to zero. The roots of

$$f''(x) = 0$$

do not necessarily satisfy (8.4.5). Only those roots which satisfy this condition give rise to points of inflection.

Example 3. We take $f(x)$ as in Example 1. Then

$$\begin{aligned} f''(x) &= 80x^3 - 180x^2 - 120x + 220 \\ &= 20(4x^3 - 9x^2 - 6x + 11) \\ &= 80(x - \rho)(x - 1)(x - \sigma), \end{aligned}$$

where

$$\left.\begin{matrix}\rho \\ \sigma\end{matrix}\right\} = \tfrac{1}{8}(5 \pm \sqrt{201}).$$

Each of the three roots gives a point of inflection.

Example 4. If $f(x) = \sin x$, then $f''(x) = -\sin x$ and every integral multiple of π gives a zero of $f(x)$ as well as a point of inflection.

Example 5. If $f(x) = x^n$ where n is a positive integer > 2, then $f''(x) = n(n - 1)x^{n-2}$ vanishes at $x = 0$. The second derivative changes sign at zero if and only if n is an odd integer. If n is odd there is a point of inflection at the origin otherwise an absolute minimum. The larger the n is, the flatter the curve at the origin.

The curve tangent at the point $P = (x_0, f(x_0))$ has an obvious extremal property: *Of all straight lines through* P, *the tangent gives the closest approximation to the curve in a small neighborhood of the point*. This is simply a restatement of the obvious relation

$$f(x) - f(x_0) - m(x - x_0) = o(x - x_0), \tag{8.4.6}$$

if and only if

$$m = f'(x_0).$$

Similarly we see that *the second Taylor polynomial*

$$f(x_0) + f'(x_0)(x - x_0) + \tfrac{1}{2}f''(x_0)(x - x_0)^2 \tag{8.4.7}$$

gives the best approximation to $f(x)$ of all second-order polynomials for

$$f(x) - A_0 - A_1(x - x_0) - A_2(x - x_0)^2 = o(x - x_0)^2, \tag{8.4.8}$$

if and only if

$$A_0 = f(x_0),\ A_1 = f'(x_0),\ A_2 = \tfrac{1}{2}f''(x_0).$$

We may regard this as the solution of the following problem: *Find the parabola with vertical axis which has the closest contact with C: $y = f(x)$, at the point P.* It is characterized uniquely by the fact that it goes through P where it has the same first- and second-order derivatives as $f(x)$.

For practical purposes it is often more important to find the *osculating circle* rather than the *osculating parabola*. This is the circle through P which has the same tangent as C at P and whose ordinate has the same second derivative with respect to the abscissa as C. The center of this circle must lie on the normal to C at P, and the circle is uniquely determined as soon as we know the coordinates of the center. See Figure 8.6.

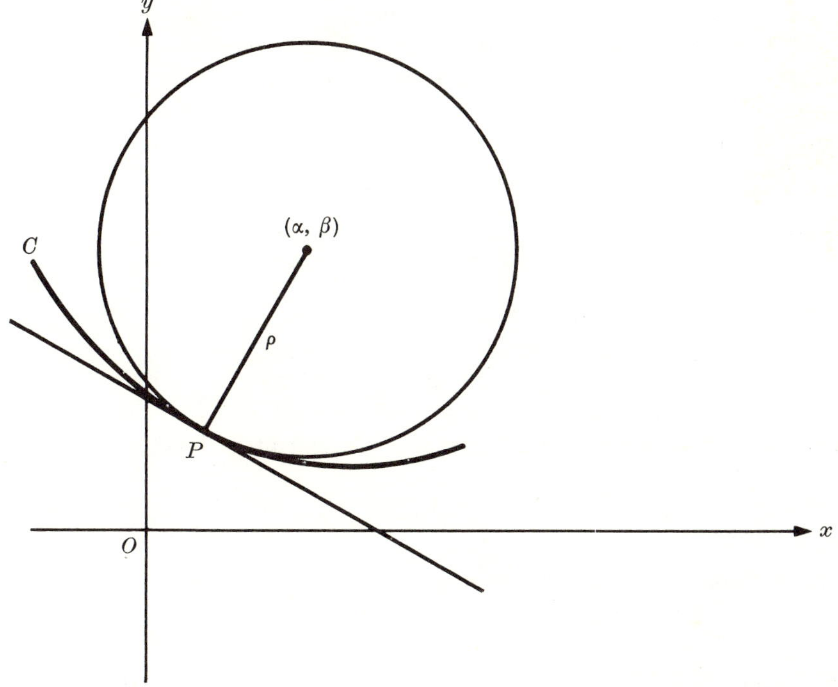

Figure 8.6

Suppose that the equation of the circle is

$$(x-\alpha)^2+(y-\beta)^2=\rho^2. \tag{8.4.9}$$

We differentiate this equation twice implicitly with respect to x and obtain

$$x-\alpha+(y-\beta)y'=0,$$

$$1+(y')^2+(y-\beta)y''=0.$$

In the last three equations we substitute

$$x=x_0,\quad y=f(x_0),\quad y'=f'(x_0),\quad y''=f''(x_0).$$

In this manner we take into account the conditions of the problem, namely that the circle is to pass through P where the first- and second-order derivatives of the ordinates of the two curves must agree. From the resulting equations we obtain successively β, α, ρ. The values are

$$\alpha=x_0-f'(x_0)\frac{1+[f'(x_0)]^2}{f''(x_0)}, \tag{8.4.10}$$

$$\beta=f(x_0)+\frac{1+[f'(x_0)]^2}{f''(x_0)}, \tag{8.4.11}$$

$$\rho=\pm\frac{\{1+[f'(x_0)]^2\}^{\frac{3}{2}}}{f''(x_0)}, \tag{8.4.12}$$

where the sign is that of $f''(x_0)$. These formulas determine the circle completely. It is known as the *circle of curvature*, (α, β) is the *center of curvature*, ρ is the *radius of curvature*, and $1/\rho$ is called the *curvature*. All these names are with reference to the point P on the given curve C: $y=f(x)$.

The radius of curvature of a circle is equal to its radius, and it is its own circle of curvature at each point.

Example 6. We take the parabola

$$x^2=4ay \tag{8.4.13}$$

considered in Section 8.1. Here

$$f(x_0)=\frac{x_0{}^2}{4a},\quad f'(x_0)=\frac{x_0}{2a},\quad f''(x_0)=\frac{1}{2a},$$

and formulas (8.4.10) to (8.4.12) give after simplification

$$\alpha=-\frac{x_0{}^2}{4a^2},\quad \beta=2a+\tfrac{3}{4}\frac{x_0{}^2}{a},\quad \rho=\frac{(4a^2+x_0{}^2)^{\frac{3}{2}}}{4a^2}. \tag{8.4.14}$$

If we eliminate x_0 between the first two equations, we obtain

$$4(\beta-2a)^3=27a\alpha^2 \tag{8.4.15}$$

as the equation of the locus of the center of curvature. This is equation (8.1.12), except for notation, that is, the locus is the evolute of the parabola.

Example 7. We now take the ellipse

$$b^2x^2 + a^2y^2 = a^2b^2. \tag{8.4.16}$$

Implicit differentiation gives

$$b^2x + a^2yy' = 0,$$

$$b^2 + a^2(y')^2 + a^2yy'' = 0.$$

Hence, at the point (x_0, y_0) on the ellipse the first- and second-order derivatives are given by

$$y_0' = -\frac{b^2x_0}{a^2y_0}, \quad y_0'' = -\frac{b^4}{a^2{y_0}^3}.$$

Using the equation of the ellipse and simplifying we obtain

$$\alpha = \frac{a^2 - b^2}{a^4}{x_0}^3, \quad \beta = -\frac{a^2 - b^2}{b^4}{y_0}^3, \quad \rho = \frac{(a^4{y_0}^2 + b^4{x_0}^2)^{\frac{3}{2}}}{a^4b^4}. \tag{8.4.17}$$

Elimination of x_0 and y_0 between the first two equations and the equation of the ellipse gives

$$(a\alpha)^{\frac{2}{3}} + (b\beta)^{\frac{2}{3}} = (a^2 - b^2)^{\frac{2}{3}} \tag{8.4.18}$$

as locus of the center of curvature. This is the evolute of the ellipse mentioned in Section 8.1. The four points on the axes are cusps.

EXERCISE 8.4

For the following functions determine (i) intervals of increase and decrease, (ii) maxima and minima, (iii) points of inflection, and (iv) intervals of convexity and concavity.

1. $x^3 - 3x + 2$. 2. $x^4 - 4x^2 + 6$. 3. $x^5 - 5x^3$.

4. $x^2(1 - x)^2$. 5. $x^3(x + 1)^3$. 6. $x(x + 1)(x + 2)$.

7. $x + \dfrac{1}{x}$. 8. $x^2 + \dfrac{1}{x^2}$. 9. $x^2 - \dfrac{1}{x}$. 10. xe^{-x}.

11. xe^{-x^2}. 12. $\dfrac{e^x - e^{-x}}{e^x + e^{-x}}$. 13. $x \log x$.

14. $\dfrac{1}{x}\log x$. 15. $\sin x + \cos x$. 16. $\sin^2 x$.

17. $\tan x$. 18. $\tan^2 x$.

Find the radius of curvature at an arbitrary point of the curve $y = f(x)$ if $f(x)$ equals:

19. x^3. 20. $(x^2 + 1)^2$. 21. $x + \dfrac{1}{x}$. 22. e^x.

23. $\cosh x$. 24. $\log x$. 25. $\log(1 + e^x)$. 26. $\arctan x$.

27. Find the radius of curvature of $y = \sin x$ and show that the minimum of the radius is 1. Is there a finite least upper bound?

28. Verify (8.4.14).

29. Verify (8.4.17) and (8.4.18).

30. Sketch the curve $y = \sin^2 x$. Determine maxima and minima, points of inflection, and radius of curvature.

31. What is the radius of curvature of

$$y = \cot x - \frac{1}{x}$$

at $x = 0$? Note that y, y', and y'' have finite limits as $x \to 0$.

32. What is the maximum value of the curvature of the hyperbola $xy = x^2 + 1$?

33. What is the limit of the curvature if x tends to infinity along the right branch of the hyperbola?

34. Is this curvature integrable over $(1, \infty)$? If so, evaluate the integral.

35. Is the curvature of $y = e^x$ integrable over $(0, \infty)$? If so, evaluate the integral.

Does the curvature tend to finite limits if x goes to $+\infty$ along the curve

36. $y = \log x$?

37. $y^2 = x$?

38. Find the center of curvature of the hyperbola

$$b^2x^2 - a^2y^2 = a^2b^2,$$

and find the locus of the center.

39. Show that for the parabola the radius of curvature is twice the length of the segment of the normal from the point in question to the directrix. Give a geometric construction of the center of curvature on the basis of this observation. Note that the center lies on the normal in the opposite direction to the segment just described.

40. One and only one circle passes through three nearby points (x_1, y_1), (x_0, y_0), (x_2, y_2) on the curve C: $y = f(x)$. Take $x_1 = x_0 - h$, $x_2 = x_0 + h$, and show that if $f(x)$ has continuous first- and second-order derivatives, then the circle tends to the circle of curvature when $h \to 0$. [*Hint:* The argument is essentially equivalent to the implicit differentiation used above.]

8.5. Maxima and minima. The general problem of maxima and minima is to determine extremal values of functions. There is a practical aspect to this question. In many problems of engineering and physics it is important to find how large or how small a certain quantity may be under stated conditions and still stay within safe bounds. Any construction of a machine or a building involves such problems, not to mention recent concern with radioactive fallout and reentry temperatures of space rockets. If the problem admits of a mathematical formulation, it becomes a problem of extremal values of one function or other. If this is a function of one variable defined in an interval (a, b), we have the following possibilities.

(i) The function $f(x)$ is differentiable in the interval (a, b), there is at least one value of x for which $f'(x) = 0$, and one of these values gives $f(x)$ a value which is the desired extremum.

(ii) Either there is no zero of $f'(x)$ or the relative maxima and minima, corresponding to the zeros of $f'(x)$, do not furnish the least upper or the greatest lower bound of $f(x)$ in (a, b).

(iii) The function is not differentiable and, hence, the extremal problem is not open to attack by the methods of calculus.

Case (iii) is outside the scope of this treatise but we shall discuss the other possibilities.

In the first two cases we proceed as indicated in the preceding section. We differentiate $f(x)$, equate the derivative to zero, and solve the equation

$$f'(x) = 0. \tag{8.5.1}$$

Suppose that this equation has solutions. Each solution must then be examined for its bearing on the problem. Essentially two distinct methods are available.

1. *Change of sign.* Suppose that $x = x_0$ is a root of (8.5.1) and let h be so small that x_0 is the only zero of $f'(x)$ in the interval $(x_0 - h, x_0 + h)$. If $f'(x_0 - h)$ and $f'(x_0 + h)$ have the same sign, then x_0 is eliminated from further consideration. It can give neither a maximum nor a minimum. We do have a relative extremum, if the two quantities have opposite signs, and $f(x_0)$ is

$$\begin{aligned} &\textit{a relative maximum, if } f'(x_0 - h) > 0 \quad \textit{and} \quad f'(x_0 + h) < 0, \\ &\textit{a relative minimum, if } f'(x_0 - h) < 0 \quad \textit{and} \quad f'(x_0 + h) > 0. \end{aligned} \tag{8.5.2}$$

2. *The second derivative.* This criterion is based upon Taylor's theorem. Suppose that $f'(x_0) = 0$ and $f''(x)$ is defined and continuous in some neighborhood of $x = x_0$. Then

$$f(x_0 + h) = f(x_0) + \tfrac{1}{2}h^2 f''(x_0 + \theta h). \tag{8.5.3}$$

Here there are two main alternatives (a) $f''(x_0) \neq 0$, (b) $f''(x_0) = 0$. In the first case we observe that since $f''(x)$ is supposed to be continuous, then for h sufficiently small $f''(x_0 + \theta h)$ and $f''(x_0)$ have the same sign. Here there are two subcases (a.1) $f''(x_0) > 0$ and (a.2) $f''(x_0) < 0$. In the first case

$$f(x_0 + h) > f(x_0) \quad \text{so } f(x_0) \textit{ is a relative minimum.} \tag{8.5.4}$$

In the second case

$$f(x_0 + h) < f(x_0) \quad \text{so } f(x_0) \textit{ is a relative maximum.} \tag{8.5.5}$$

In case (b) this argument breaks down and a more detailed examination is required.

If $f''(x)$ keeps a constant sign in $(x_0 - h, x_0 + h)$ and vanishes only at $x = x_0$, then either (8.5.4) or (8.5.5) holds and we have a relative extremum. On the other hand, if $f''(x)$ changes its sign when x goes through x_0, then we have a point of inflection rather than a relative extremum. These various possibilities can be separated by a discussion of the third, or if necessary, higher order derivatives, but the change of sign test is decisive right away while the higher derivatives may not be so.

In fact, there is still a third method, a "common sense method" which may be quite useful. Since common sense is hard to teach, this method can be learned only by experience. The idea is simple: If the problem obviously has a solution and if only one value of x can be found which can possibly give the extremum, then this is the desired value and testing can be dispensed with. The nature of the function might be such that a discussion of signs of derivatives is difficult to carry out. Here a "common sense argument" is often applicable.

Suppose now that we have found the roots of (8.5.1) and classified the roots as giving relative maxima or relative minima or points of inflection. If the problem calls only for relative maxima and minima, the solution has been found. If we are interested in least upper or greatest lower bounds, however, a supplementary discussion is needed. Suppose that there is only a finite number of zeros $\{x_j\}$ of $f'(x)$ and that

$$a < x_1 < x_2 < \cdots < x_n < b.$$

Suppose that p of these values give rise to relative maxima and q to relative minima where $p + q \leqq n$. Let M_0 be the largest of the p relative maxima and let m_0 be the least of the q relative minima. Then M_0 is the least upper bound of $f(x)$ in $[x_1, x_n]$ and m_0 is the greatest lower bound. In order to get the corresponding bounds for the interval (a, b) we must examine the terminal intervals (a, x_1) and $[x_n, b]$. In both of these intervals $f(x)$ is monotone so that

$$\lim_{x\to a} f(x) \equiv A, \quad \lim_{x\to b} f(x) \equiv B$$

exist as finite or infinite values. We have then

$$M \equiv \sup_{a<x<b} f(x) = \max(M_0, A, B), \tag{8.5.6}$$

$$m \equiv \inf_{a<x<b} f(x) = \min(m_0, A, B). \tag{8.5.7}$$

Here we have tacitly assumed that $f'(x)$ does have zeros in (a, b). If this is not the case, then $f(x)$ is monotone in (a, b) and its extremal values are the limits at the endpoints:

$$M = \max(A, B), \quad m = \min(A, B). \tag{8.5.8}$$

This gives a complete discussion of cases (i) and (ii) mentioned at the beginning of this section.

Maxima and minima problems are of three types: (1) the function whose extremal values are to be found is given explicitly; (2) the function is given implicitly; and (3) the function must be constructed from a verbal description of the problem.

In the first case we have nothing to add. In case (2) it may be possible to solve for the function and thus reduce the problem to one of the first type. Even if possible, it is frequently not desirable, for the resulting function may well lead to a complicated derivative. In such cases implicit differentiation is often the best choice both for the first and the second derivative.

No general rules can be laid down for case (3). It is necessary to make a choice of variables, to label all given quantities, and to express the conditions of the problem by a formula or formulas. If a geometrical figure is involved, its symmetries should be utilized in choosing coordinates. Even if there are no symmetries, one choice of axes may lead to simpler coordinates of important points and simpler expressions for possible distances than some other choice. All existing relations should be noted, and either all unknowns expressed in terms of one of them or else simultaneous implicit differentiation with respect to one unknown must be used.

We proceed to the discussion of some typical problems where various methods are illustrated.

Example 1. What is the largest value assumed by

$$f(x) = xe^{-x}? \tag{8.5.9}$$

This function is negative for $x < 0$ and positive for $x > 0$. Since it tends to zero when $x \to 0$ or $x \to \infty$, there exists obviously a maximum for some finite positive value of x. The derivative is

$$f'(x) = (1 - x)e^{-x},$$

which has $x = 1$ as its only zero. The corresponding value $f(1) = e^{-1}$ must then be the required maximum value. It is easy to confirm this with the usual methods. Thus $f'(x)$ goes from plus to minus as x passes through 1 and

$$f''(x) = (x - 2)e^{-x} \quad \text{gives} \quad f''(1) = -e^{-1} < 0.$$

Both tests show that $x = 1$ really gives the maximum value of $f(x)$.

Example 2. The function

$$f(x) = (x^2 - 1)^2 \tag{8.5.10}$$

is never negative and is zero for $x = \pm 1$. Thus the infimum of $f(x)$ is 0 and the supremum is obviously infinite. On the other hand

$$f'(x) = 4x(x^2 - 1), f''(x) = 4(3x^2 - 1),$$

so that $x = 0$ gives a relative maximum with $f(0) = 1$ while $x = \pm 1$ give relative as well as absolute minima.

Example 3. The function

$$f(x) = \tanh x \tag{8.5.11}$$

gives

$$f'(x) = \operatorname{sech}^2 x > 0;$$

thus $f(x)$ is strictly increasing and has neither a relative maximum nor a relative minimum. Here

$$\sup f(x) = \lim_{x \to +\infty} f(x) = 1, \quad \inf f(x) = \lim_{x \to -\infty} f(x) = -1.$$

Example 4. What point in the first quadrant on the leaf of Descartes

$$x^3 + y^3 = 3axy \tag{8.5.12}$$

is farthest from the x-axis?

We want to find the point with the largest ordinate y. In this case it is impractical to solve for y so we resort to implicit differentiation obtaining

$$x^2 + y^2D_xy = a(xD_xy + y), \tag{8.5.13}$$

or

$$D_xy = \frac{ay - x^2}{y^2 - ax}. \tag{8.5.14}$$

Equating the numerator to zero, we have to solve the two simultaneous equations

$$x^2 = ay, x^3 + y^3 = 3axy.$$

Elimination of y gives

$$x^3 + \frac{x^6}{a^3} = 3x^3.$$

Hence either

$$x^3 = 0, \quad \text{or} \quad x^3 = 2a^3.$$

The root $x = 0$ gives $y = 0$, that is, the lowest point on the curve in the first quadrant. The second alternative gives only one real value so we have

$$x = 2^{\frac{1}{3}}a, \quad y = 2^{\frac{2}{3}}a.$$

Since there is only one value that can possibly give a maximum, this is the required solution.

Here a test along the usual lines is still possible, but requires much more care. We can study the change in the sign of D_xy by the following consideration. A point (x, y) on the curve (8.5.12) such that $0 < x < 2^{\frac{1}{3}}a$ lies inside the parabola $x^2 = ay$; that is, for such a value $ay - x^2 > 0$. In a small neighborhood of the point $(2^{\frac{1}{3}}a, 2^{\frac{2}{3}}a)$, the denominator $y^2 - ax$ in (8.5.14) is positive, since it is positive at the point as well as continuous in both arguments. This makes $D_xy > 0$ at such a point. When x increases and passes through the critical value $2^{\frac{1}{3}}a$, then the point (x, y) crosses the parabola $ay = x^2$. Thus the numerator in (8.5.14) becomes negative while the denominator stays positive. This means that the derivative goes from plus to minus, and so we do have a maximum.

We can also use the second derivative. For this purpose we differentiate (8.5.13), once more obtaining

$$2x + 2y(D_xy)^2 + y^2D_x^{\,2}y = a(2D_xy + xD_x^{\,2}y).$$

Here we substitute

$$x = 2^{\frac{1}{3}}a, \quad y = 2^{\frac{2}{3}}a, \quad D_xy = 0,$$

and obtain

$$D_x^{\,2}y = -\frac{2}{a} < 0.$$

Thus both tests apply and confirm the common sense argument.

Example 5. Which of the tangents to the ellipse

$$x^2 + xy + y^2 = 3 \tag{8.5.15}$$

is the farthest from the origin?

See Section 3.6 where this ellipse and the corresponding implicit function were studied at great length. The equation of the tangent at the point (x_0, y_0) on the ellipse is by (3.6.7)

$$(2x_0 + y_0)x + (x_0 + 2y_0)y - 6 = 0.$$

In order to find the distance of this straight line from the origin we reduce it to normal form by dividing by the square root of the sum of the squares of the coefficients of x and y. The required distance is

$$6[(2x_0 + y_0)^2 + (x_0 + 2y_0)^2]^{-\frac{1}{2}}.$$

Since (x_0, y_0) is on the ellipse, this reduces to

$$6[15 + 3x_0y_0]^{-\frac{1}{2}}. \tag{8.5.16}$$

Our problem now is: How should (x_0, y_0) be chosen on the ellipse so that the expression in (8.5.16) becomes a maximum?

Here we make an observation which applies to much more general situations than the one at hand. Since the numerator 6 in (8.5.16) is a constant, the problem of maximizing the quotient is equivalent to minimizing the denominator. Moreover, $(15 + 3x_0y_0)^{\frac{1}{2}}$ will have its least value when $15 + 3x_0y_0$ has its least value and this will happen when x_0y_0 is algebraically as small as possible. We recall that the product x_0y_0 can be negative but is bounded below, since $15 + 3x_0y_0$ is necessarily positive when (x_0, y_0) lies on the ellipse. Thus our problem now has taken on the following form:

Given that (x, y) satisfies (8.5.15), find the least value assumed by

$$P = xy. \tag{8.5.17}$$

We now differentiate (8.5.15) and (8.5.17) with respect to x remembering that y is a function of x implicitly defined by the former relation. This gives

$$2x + xD_xy + y + 2yD_xy = 0,$$
$$D_xP = xD_xy + y.$$

Here we solve the first equation for D_xy and substitute in the second obtaining finally

$$D_xP = 2\frac{y^2 - x^2}{x + 2y}.$$

Equating this to zero, we see that the extremal values of P occur at the intersections of the lines

$$y = x \quad \text{and} \quad y = -x$$

with the ellipse. These points are $(1, 1)$ and $(-1, -1)$ for the first line and $(\sqrt{3}, -\sqrt{3})$, $(-\sqrt{3}, \sqrt{3})$ for the second. The corresponding values of P are $P = 1$ and $P = -3$ respectively. It is clear that $P = 1$ is a maximum and

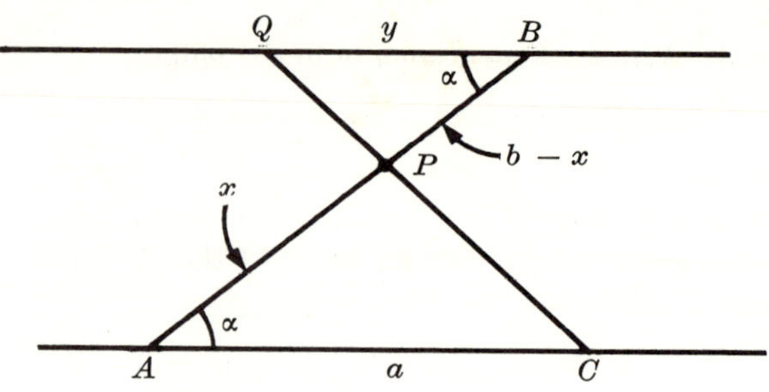

Figure 8.7

$P = -3$ is a minimum and it is the latter that is wanted. We conclude that the tangents at $(\sqrt{3}, -\sqrt{3})$ and $(-\sqrt{3}, \sqrt{3})$ are farthest away from the origin and by (8.5.16) both are at the distance $\sqrt{6}$. This is confirmed by Figure 3.9.

Example 6. Which rectangle of perimeter $4a$ has the largest area and what is the maximal area?

We gave Euclid's solution of this problem in Section 1.6. Using differential calculus we proceed as follows. Suppose that one of the sides of the rectangle is of length x where, obviously, $0 < x < 2a$. Then the adjacent sides are of length $2a - x$ and the area of the rectangle is

$$A = x(2a - x) = 2ax - x^2.$$

Since

$$D_xA = 2a - 2x, \quad D_x{}^2A = -2 < 0,$$

we see that the area is a maximum if $x = a$, that is, the square has the largest area of all rectangles of given perimeter. The maximum of the area is a^2.

Example 7. Two parallel lines are cut by a given line AB. See Figure 8.7. From the point C we draw a straight line intersecting AB. How should this line be chosen so that the sum of the areas of the two triangles ACP and QPB is a minimum?

This is known as the problem of (Vincenzo) Viviani (1622–1703), a pupil of Galileo. Suppose that the lengths of AC and AB are a and b respectively. Let the lengths of AP and QB be x and y. Since the triangles are similar we have

$$a:x = y:(b - x) \quad \text{or} \quad y = a\frac{b - x}{x}.$$

If α is the angle at A, the sum of the areas is

$$\tfrac{1}{2}ax \sin \alpha + \tfrac{1}{2}(b - x)y \sin \alpha.$$

We substitute the value of y and now the problem calls for the minimum value of the function

$$f(x) = x + \frac{(b - x)^2}{x}. \tag{8.5.18}$$

Here

$$f'(x) = 1 - \left(\frac{b - x}{x}\right)^2 - 2\,\frac{b - x}{x} = 2 - \frac{b^2}{x^2},$$

$$f''(x) = 2\,\frac{b^2}{x^3}.$$

It follows that

$$x = \frac{\sqrt{2}}{2}\,b.$$

gives the minimum area.

Example 8. Let ABC be an isosceles triangle where $AB = BC$. Let P be a point on the altitude from B. For what position of P is the sum of the distance of P from the vertices a minimum?

See Figure 8.8. We take $AC = 2a$, $OB = b$. Let $P = (0, y)$ where we assume $0 < y < b$. The function to be minimized is

$$F(y) = b - y + 2(y^2 + a^2)^{\frac{1}{2}}, \tag{8.5.19}$$

whose derivative is

$$F'(y) = -1 + 2y(y^2 + a^2)^{-\frac{1}{2}}.$$

The derivative goes from minus to plus when y increases through the value $y = \dfrac{\sqrt{3}}{3}\,a$. There are two possibilities.

(i) $a \leqq \sqrt{3}b$. In this case the critical value belongs to the interval $[0, b]$ and $y = \dfrac{\sqrt{3}}{3}\,a$ gives the minimum value of the sum of the distances. P is characterized by the line segments PA, PB, and PC making equal angles with each other, since the angle APO is 60° when $y = \dfrac{\sqrt{3}}{3}\,a$.

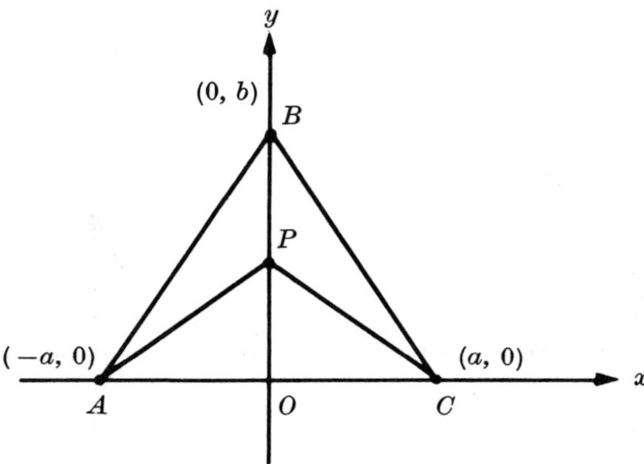

Figure 8.8

(ii) $a > \sqrt{3}b$. Now the critical value is outside the interval $[a, b]$ so $F'(y) < 0$ for $0 \leqq y \leqq b$. It follows that $F(y)$ decreases to $F(b)$ when y increases to b and the least sum of the distances is obtained if $P = B$, the minimum being $2(a^2 + b^2)^{\frac{1}{2}}$.

It may be objected that P should be allowed to lie on the altitude produced above B so that it is legitimate to consider values of y greater than b. This may be granted, but it is necessary to observe that if $y > b$, then the distance from P to B is $y - b$ and not $b - y$. Hence the expression for $F(y)$ changes and we have

$$F(y) = y - b + 2(y^2 + a^2)^{\frac{1}{2}}, \quad b < y, \tag{8.5.20}$$

and

$$F'(y) = 1 + 2y(y^2 + a^2)^{-\frac{1}{2}}$$

which is always positive. If we define $F(y)$ for $y \geqq 0$ by formula (8.5.19) for $0 \leqq y \leqq b$ and by (8.5.20) for $b < y$, and if $a > \sqrt{3}b$, then $F'(y) < 0$ in the first interval and > 0 in the second. Thus $y = b$ gives the minimum value, confirming our previous result. We note that $F'(y)$ changes sign when y passes through $y = b$, but $F'(b)$ does not exist and

$$\lim_{h \to 0} F'(b - h) \neq \lim_{h \to 0} F'(b + h).$$

This problem shows that if the function, whose extremal values are required, does not have a derivative everywhere, then the extremal values may well occur at points where the derivative does not exist. In such a case it is not

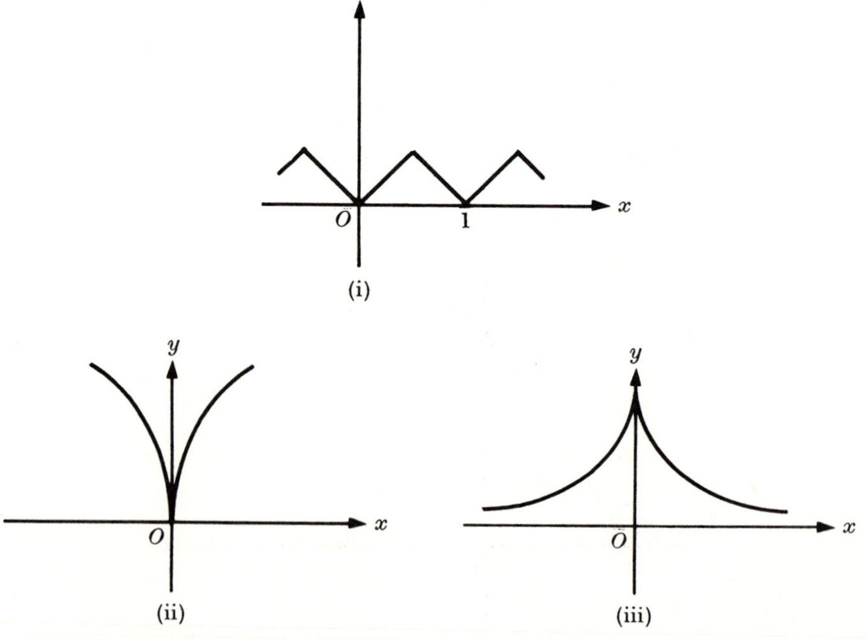

Figure 8.9

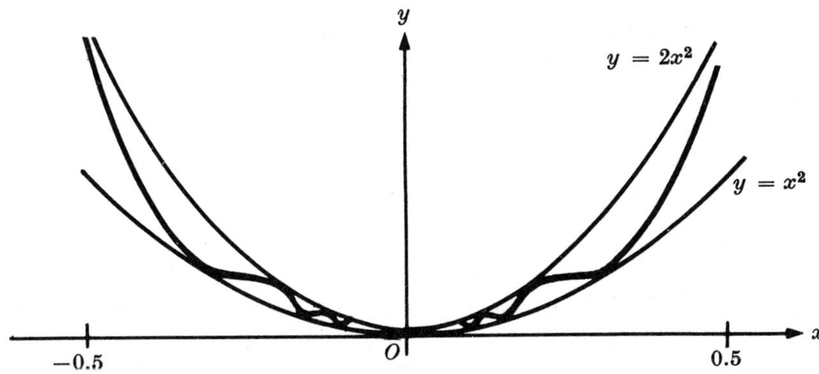

Figure 8.10

sufficient to look for the zeros of the derivative; the points of discontinuity must also be examined. Figure 8.9 shows some possibilities of this nature.

In case (i) we consider the function

$$f(x) = (x).$$

Here there are maxima at the points $x = n + \frac{1}{2}$, minima at $x = n$, and these are precisely the points where the derivative does not exist. In case (ii) we take

$$f(x) = x^{\frac{2}{3}}.$$

It is clear that $x = 0$ gives a minimum, but here $f'(-h) \to -\infty$, $f'(h) \to +\infty$ as h decreases to zero. Finally, in case (iii)

$$f(x) = \arctan(|x|^{-\frac{1}{2}}), \quad f(0) = \frac{\pi}{2}. \tag{8.5.21}$$

Here

$$f'(x) = -\frac{\operatorname{sgn} x}{2|x|^{\frac{1}{2}}(1 + |x|)},$$

where sgn x stands for the sign of x. There is a maximum at $x = 0$, and the derivative becomes infinite when x approaches 0.

Example 9. The following function shows another peculiarity:

$$f(x) = x^2\left[1 + \sin^2\frac{1}{x}\right], \quad x \neq 0, \quad f(0) = 0. \tag{8.5.22}$$

See Figure 8.10. This function has infinitely many relative maxima and minima which cluster at $x = 0$. Since

$$x^2 \leqq f(x) \leqq 2x^2,$$

we see that $x = 0, f(0) = 0$ is an absolute minimum. From

$$\frac{1}{h}[f(h) - f(0)] = h\left[1 + \sin^2\frac{1}{h}\right] \to 0$$

as $h \to 0$, it follows that $f'(0)$ exists and is 0. Further, for $x \neq 0$

$$f'(x) = 2x\left[1 + \sin^2 \frac{1}{x}\right] - \sin \frac{2}{x}, \tag{8.5.23}$$

so that $f'(x)$ exists for all x but is not continuous at $x = 0$. Further, $f'(x)$ changes its sign infinitely often in any neighborhood of the origin. Moreover

$$\liminf_{x \to 0} f'(x) = -1, \quad \limsup_{x \to 0} f'(x) = +1. \tag{8.5.24}$$

This shows that the criterion that $F(x)$ has a minimum at $x = a$ if $F'(x)$ exists in some neighborhood of $x = a$ where $F'(x)$ is continuous and changes from minus to plus is only a sufficient one. The same also applies of course to the criterion for a maximum. We have only to replace $f(x)$ by its negative in (8.5.22) to get an example where the criterion is violated.

EXERCISE 8.5

Find the relative maxima and minima of the following functions:

1. $x^4 - 8x^3 + 22x^2 - 24x + 12$.

2. $\dfrac{x}{1 + x^2}$. 3. $\dfrac{x^2 - x + 1}{x^2 + x - 1}$. 4. $\dfrac{(x - 3)^3}{(x - 2)^2}$.

5. $(a - x)^{\frac{2}{3}}(a + x)^{\frac{1}{3}}$, $a > 0$.

6. $\dfrac{x}{(a^2 + x^2)^{\frac{3}{2}}}$. 7. $\dfrac{\log x}{x^n}$, $n > 0$.

8. $x^m e^{-x^2}$. 9. $e^{-x} \sin x$. 10. $\cosh x - \cos x$.

Find the maxima and minima of y if y is defined implicitly by:

11. $xy^2 - yx^2 = 2a^3$, $a > 0$. 12. $y^4 + x^4 - 4xy + 2 = 0$.
13. $x^4 + y^4 - 4xy = 0$. 14. $y^2 - 2axy + x^2 - b^2 = 0$, $|a| < 1$.
15. $y^2 + 2yx^2 + 4x - 3 = 0$. 16. $\sin x + \sin y = 1$.
17. If $a_1, a_2, \cdots, a_n$ are given real numbers, where is the minimum of

$$(x - a_1)^2 + (x - a_2)^2 + \cdots + (x - a_n)^2.$$

18. If $a_1 < a_2 < a_3$, where does

$$|x - a_1| + |x - a_2| + |x - a_3|$$

reach its minimum and what is the minimum value?

19. Verify formula (8.5.23) and (8.5.24) and prove that $f(x)$ has infinitely many relative maxima and minima in any neighborhood of $x = 0$.

20. Give a similar discussion for

$$f(x) = x^2\left(1 - \cos \frac{1}{x}\right), \quad f(0) = 0.$$

21. The function

$$f(x) = x^2 \sin \frac{1}{x}, \quad f(0) = 0,$$

also has interesting extremal properties. Which? Does $x = 0$ give a relative extremum?

22. What is the largest value of

$$f(x) = \cot x - \frac{1}{x}, \quad f(0) = 0,$$

for $-\frac{1}{2}\pi \leqq x \leqq \frac{1}{2}\pi$? The least?

Find the maxima and minima of:

23. $\sin^4 x + \cos^4 x$.
24. $\cos 4x + 4 \sin 2x$.
25. $\sinh^4 x + \cosh^4 x$.
26. $\cosh 4x + 4 \sinh 2x$.
27. $e^{4x} - e^{-2x}$.

28. From a point on the hypotenuse of a right-angled triangle perpendiculars are drawn on the other two sides. When is the rectangle thus formed a maximum?

29. The coordinate axes are met at P and Q by a variable straight line through the point (a, b), $a > 0$, $b > 0$. Find the triangle OPQ of least area.

30. A rectangle is inscribed in a circle of radius R. Find the largest area of such a rectangle.

31. A rectangle is inscribed in an ellipse of semiaxes a and b. What is the maximum of the area?

32. Given a parabolic segment bounded by $y^2 = 4ax$ and $x = b$. What is the area of the largest rectangle that can be inscribed in such a segment?

33. A parallelogram is inscribed in a triangle ABC in such a way that one vertex coincides with A while the others fall one on each side of the triangle. When is the area of the parallelogram a maximum?

34. A trapezoid has three equal sides. If the area of the trapezoid is a maximum, show that one of its angles must be 60°.

35. If the perimeter of an isosceles triangle be given, show that the area of the triangle is greatest when it is equilateral.

36. From a rectangular piece of cardboard of dimensions a by b four equal squares are cut out, one at each corner, and the remaining crosslike piece is folded into an open box. What is the side of the squares when the volume of the box is a maximum?

Given a right circular cone of height h and radius of the base b, find:

37. The cylinder of greatest volume inscribed in the cone.
38. The cylinder of greatest curved surface.
39. The cylinder of greatest total surface.

Cylinders and right circular cones are inscribed in a sphere of radius R. Find:

40. The cylinder of greatest volume.
41. The cylinder of greatest curved surface.
42. The cone of greatest volume.

43. What is the maximum volume of a right circular cone of slant height a?

44. Given a right circular cone of height h, and a plane parallel to one of

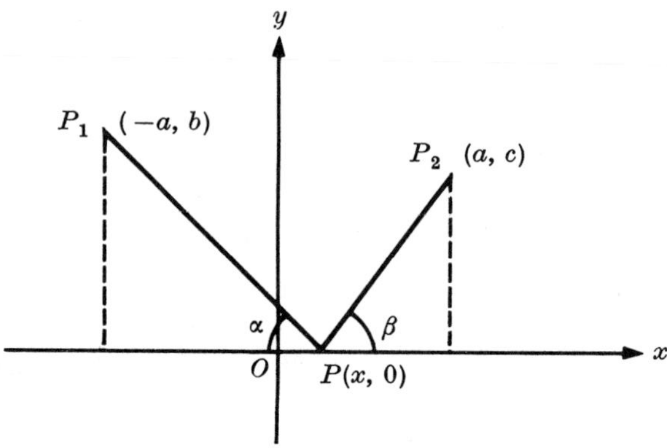

Figure 8.11

the generators of the cone, for what position of the plane is the area of its intersection with the cone a maximum?

45. The function

$$f(x) = \arctan(|x|^{-\alpha}), \quad x \neq 0, \quad f(0) = \tfrac{1}{2}\pi,$$

has a maximum at the origin for any $\alpha > 0$. Find $f'(x)$ and determine the limit of $f'(h)$, when h decreases to 0, as a function of α.

8.6. Maxima and minima continued. We shall treat some problems on maxima and minima important for the light they throw on other questions, mathematical or physical. The problems of Exercise 8.6 are accordingly more sophisticated than those of the preceding Exercise.

We start with two questions, aspects of the physical *Principle of Least Action*, which illustrate *Fermat's Principle* that light travels so that the travel time is a minimum.

Problem 1. Given a straight line in the plane and two points P_1 and P_2 in the plane to the same side of the line. Find a point P on the line such that the sum of the distances PP_1 and PP_2 is a minimum.

For the notation, see Figure 8.11. We take the line as the x-axis, and let

$$P = (x, 0), \quad P_1 = (-a, b), \quad P_2 = (a, c).$$

We have then to minimize

$$S = [(x + a)^2 + b^2]^{\frac{1}{2}} + [(x - a)^2 + c^2]^{\frac{1}{2}}. \tag{8.6.1}$$

Here

$$D_x S = \frac{x + a}{[(x + a)^2 + b^2]^{\frac{1}{2}}} - \frac{a - x}{[(x - a)^2 + c^2]^{\frac{1}{2}}}.$$

From the figure we get

$$\sin\alpha = \frac{x + a}{[(x + a)^2 + b^2]^{\frac{1}{2}}}, \quad \sin\beta = \frac{a - x}{[(x - a)^2 + c^2]^{\frac{1}{2}}},$$

so that the condition is

$$\sin\alpha = \sin\beta \quad \text{or} \quad \alpha = \beta, \tag{8.6.2}$$

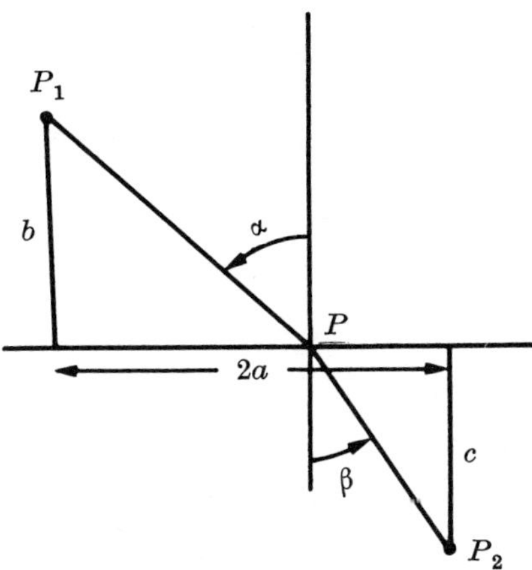

Figure 8.12

if D_xS is to be zero. Since there is only one possible solution and a minimum clearly exists, we conclude that the minimum occurs when P is so located on the line that the two angles α and β are equal.

The optical interpretation of this minimum problem is that the *law of reflection, the angle of incidence equals the angle of reflection,* gives the minimum travel time for the light from the source, to the reflector, to the observer.

The *law of refraction* discovered by Willebrord Snellius (1581–1626) in 1621 also corresponds to a minimum problem.

Problem 2. Assume two points P_1 and P_2 and a straight line separating them. A particle travels with constant velocity u from P_1 to a point P on the line and from P to P_2 with constant velocity v. How should P be chosen in order that the time it takes the particle to go from P_1 to P_2 will be as small as possible?

We arrange the discussion so that α, the angle of incidence, is taken as the independent variable. With the notation of Figure 8.12 we obtain

$$t = \frac{b}{u}\sec\alpha + \frac{c}{v}\sec\beta, \tag{8.6.3}$$

$$2a = b\tan\alpha + c\tan\beta, \tag{8.6.4}$$

where t, the time, is to be minimized. The variable angles satisfy the inequality

$$-\frac{\pi}{2} < \alpha, \beta < \frac{\pi}{2}.$$

We use (8.6.4) to define β as a function of α with domain and range as indicated. Differentiation of the two equations with respect to α gives

$$D_\alpha t = \frac{b}{u}\sin\alpha\sec^2\alpha + \frac{c}{v}\sin\beta\sec^2\beta\, D_\alpha\beta,$$

$$0 = b\sec^2\alpha + c\sec^2\beta\, D_\alpha\beta.$$

We eliminate $D_\alpha \beta$ between these two equations and obtain

$$D_\alpha t = b \sec^2 \alpha \left\{\frac{\sin \alpha}{u} - \frac{\sin \beta}{v}\right\}. \tag{8.6.5}$$

Since $\sec^2 \alpha$ is always positive, we see that $D_\alpha t = 0$ if and only if

$$\frac{\sin \alpha}{u} = \frac{\sin \beta}{v}, \tag{8.6.6}$$

and this is the *law of Snellius.*

Since $D_\alpha \beta < 0$, we conclude that β decreases when α increases. There is a unique pair of numbers α_0, β_0 in the interval $\left(0, \frac{\pi}{2}\right)$ which satisfies (8.6.6). If $\alpha < \alpha_0$, then $\beta_0 < \beta$ and $D_\alpha t < 0$. If on the other hand, $\alpha_0 < \alpha$, then $\beta < \beta_0$ and $D_\alpha t > 0$. Hence this choice of α and β minimizes t. It is fairly obvious that this is a true minimum for t becomes infinite when α tends to $\frac{\pi}{2}$ or $-\frac{\pi}{2}$.

Since rocketry and space travel have become of vital concern to everyone, it is perhaps not out of place to call attention to problems in maxima and minima involved here. We hear of the *apogee* and the *perigee* of a satellite, the points of the orbit farthest from and nearest to the earth. The maximum altitude and range of rockets are also news. Most of these problems are beyond Calculus proper, but the following, much simplified situation may give the student an idea. We follow the presentation of Richard Bellman in his *Adaptive Control Processes: A Guided Tour* (Princeton University Press, Princeton, N.J., 1961, pp. 46–47).

Problem 3. An object propelled straight up, is subjected only to gravity and the resistance of air. Let the equation of motion be

$$D_t^2 y = -g - H(D_t y), \tag{8.6.7}$$

where $H(u)$ is a given positive continuous function of u. It is supposed that the air resistance is a function of the velocity only and no account is taken of the fact that the density of the air rapidly decreases to zero as the object leaves the earth. The problem is to find the maximum altitude of the object as function of the initial velocity v.

Call this function $f(v)$. It could be determined by integrating equation (8.6.7) and solving the maximum problem for the solution $y(t)$ in the usual manner. This is suggested as an exercise for the case in which $H(u) = \gamma u$, where γ is a positive constant.

Bellman pointed out that a differential equation can be found for $f(u)$ from which the function can be found by straightforward integration. He argued that if δ is a short time interval, then, with an error which is $o(\delta)$, the maximum altitude $f(v)$ is the altitude gained over an initial time δ, plus the maximum altitude attained starting with a velocity $v - [g + H(v)]\delta$, this being the velocity of the object at the end of time δ. This gives

$$\begin{aligned} f(v) &= \delta v + f(v - [g + H(v)]\delta) + o(\delta) \\ &= \delta v + f(v) - [g + H(v)]\delta f'(v) + o(\delta), \end{aligned} \tag{8.6.8}$$

where we have used Taylor's theorem. Dividing by δ and passing to the limit we get

$$f'(v) = \frac{v}{g + H(v)} . \tag{8.6.9}$$

Since $f(0) = 0$ we have

$$f(v) = \int_0^v \frac{u\, du}{g + H(u)} . \tag{8.6.10}$$

Two particular cases are noteworthy. If $H(u) \equiv 0$, then

$$f(v) = \frac{v^2}{2g} .$$

This can also be read from (8.3.7). Secondly, if $H(u) = \gamma u$, then

$$f(v) = \frac{v}{\gamma} - \frac{g}{\gamma^2} \log\left(1 + \frac{\gamma}{g} v\right). \tag{8.6.11}$$

Explicit formulas for the maximum altitude can be found if

$$H(u) = \gamma u^\alpha,$$

where α is a positive rational number. A peculiar situation arises if $\alpha > 2$, for then the integral in the right member of (8.6.10) stays bounded regardless of the value of v. An atmosphere in which such a law of resistance is valid imposes an unattainable ceiling on all motion.

The remaining problems are of mathematical rather than physical interest.

Problem 4. If a and b are positive numbers, and if $0 \leqq x \leqq 1$, prove that

$$a^x b^{1-x} \leqq xa + (1 - x)b, \tag{8.6.12}$$

with strict inequality for $0 < x < 1$ unless $a = b$ when equality holds for all x.

To prove this we consider the auxiliary function

$$F(x; a, b) \equiv xa + (1 - x)b - a^x b^{1-x}. \tag{8.6.13}$$

This function is identically zero if $a = b$. Leaving this case to one side, we note that

$$F(0; a, b) = F(1; a, b) = 0.$$

Further

$$F'(x; a, b) = a - b - \left(\log \frac{a}{b}\right) a^x b^{1-x},$$

$$F''(x; a, b) = -\left(\log \frac{a}{b}\right)^2 a^x b^{1-x}.$$

Since $a \neq b$ by assumption, the second derivative is always negative. This means that $F(x; a, b)$ is concave downward. Since it is zero at $x = 0$ and $x = 1$, its graph lies above the x-axis for $0 < x < 1$ and

$$F(x; a, b) > 0, \quad 0 < x < 1.$$

Hence (8.6.12) holds for $0 < x < 1$, with strict inequality, and the problem is solved.

We use this result to prove *Hölder's inequality* (Otto Hölder, 1860–1937, professor at Leipzig).

Problem 5. If the two series

$$\sum_{n=1}^{\infty} |a_n|^{\alpha} \quad \text{and} \quad \sum_{n=1}^{\infty} |b_n|^{\beta} \tag{8.6.14}$$

converge, where

$$\frac{1}{\alpha} + \frac{1}{\beta} = 1, \tag{8.6.15}$$

prove that $\sum_{n=1}^{\infty} a_n b_n$ is absolutely convergent and

$$\sum_{n=1}^{\infty} |a_n b_n| \leqq \left\{\sum_{n=1}^{\infty} |a_n|^{\alpha}\right\}^{1/\alpha} \left\{\sum_{n=1}^{\infty} |b_n|^{\beta}\right\}^{1/\beta} \tag{8.6.16}$$

Equality holds if and only if there is a fixed positive μ such that $|b_n| = \mu |a_n|$ for all n.

To prove this we use (8.6.12) with

$$a = |a_n|^{\alpha},\ b = |b_n|^{\beta},\ x = \frac{1}{\alpha},\ 1 - x = \frac{1}{\beta}.$$

This gives

$$|a_n b_n| \leqq \frac{1}{\alpha} |a_n|^{\alpha} + \frac{1}{\beta} |b_n|^{\beta}.$$

Summing for n we get

$$\sum_{n=1}^{\infty} |a_n b_n| \leqq \frac{1}{\alpha} \sum_{n=1}^{\infty} |a_n|^{\alpha} + \frac{1}{\beta} \sum_{n=1}^{\infty} |b_n|^{\beta}. \tag{8.6.17}$$

This inequality, which is of some independent interest, shows that the series on the left is convergent.

To prove (8.6.16) let us write

$$\sum_{n=1}^{\infty} |a_n|^{\alpha} \equiv A,\ \sum_{n=1}^{\infty} |b_n|^{\beta} \equiv B,\ \sum_{n=1}^{\infty} |a_n b_n| \equiv C.$$

If we replace a_n and b_n by

$$t a_n \quad \text{and} \quad \frac{1}{t} b_n,$$

respectively, where t is any positive number, then C is unchanged, while A and B are replaced by

$$t^{\alpha} A \quad \text{and} \quad t^{-\beta} B,$$

respectively. This also modifies (8.6.17) which now becomes

$$C \leqq \frac{1}{\alpha} At^{\alpha} + \frac{1}{\beta} Bt^{-\beta} \equiv H(t). \tag{8.6.18}$$

This inequality is true for all values of $t > 0$. It holds in particular at the minimum of $H(t)$. There is such a point, for $H(t)$ becomes infinite when $t \to 0$ or ∞. Since

$$H'(t) = At^{\alpha-1} - Bt^{-\beta-1},$$

we see that the equation $H'(t) = 0$ has a unique solution

$$t = t_0 = \left(\frac{B}{A}\right)^{1/(\alpha+\beta)} = \left(\frac{B}{A}\right)^{1/(\alpha\beta)},$$

where we use (8.6.15) to replace $\alpha + \beta$ by $\alpha\beta$. From the structure of $H'(t)$ it follows that this function goes from minus to plus as t increasing passes through $t = t_0$. Thus $H(t)$ has a unique minimum and

$$H(t_0) = \frac{1}{\alpha} A\left(\frac{B}{A}\right)^{1/\beta} + \frac{1}{\beta} B\left(\frac{B}{A}\right)^{-1/\alpha}$$

$$= \frac{1}{\alpha} A^{1-(1/\beta)}B^{1/\beta} + \frac{1}{\beta} A^{1/\alpha}B^{1-(1/\alpha)}$$

$$= \left(\frac{1}{\alpha} + \frac{1}{\beta}\right) A^{1/\alpha}B^{1/\beta} = A^{1/\alpha}B^{1/\beta}.$$

Since

$$C \leqq H(t_0) = A^{1/\alpha}B^{1/\beta}, \tag{8.6.19}$$

we have proved (8.6.16).

For the sign of equality to hold in (8.6.16) we must have

$$C = H(t_0) = \frac{1}{\alpha} A\, t_0^{\alpha} + \frac{1}{\beta} t_0^{-\beta}.$$

On the other hand, for each n

$$|a_n b_n| \leqq \frac{1}{\alpha} |a_n|^{\alpha} t_0^{\alpha} + \frac{1}{\beta} |b_n|^{\beta} t_0^{-\beta},$$

and adding these inequalities we can get the preceding equality if and only if each inequality is actually an equality. By Problem 4 this requires that $|a_n|\, t_0 = |b_n|\, t_0^{-1}$ for each n or

$$|b_n| = t_0^2\, |a_n|,\ n = 1, 2, 3, \cdots.$$

Hence a multiplier of the stated nature exists and we have $\mu = t_0^2$. This completes the discussion.

Problem 6. Prove that for every x, $0 < x$, $x \neq 1$, and every n, $n > 0$,

$$\log x < n(x^{1/n} - 1). \tag{8.6.20}$$

We start with the case $n = 1$ and consider the function

$$f(x) = \log x - x + 1.$$

Here $f(1) = 0$ and

$$f'(x) = \frac{1}{x} - 1,$$

which is also zero for $x = 1$ and changes its sign from plus to minus as x increasing passes through 1. This shows that $f(x)$ has a maximum at $x = 1$, that is

$$f(x) < f(1) = 0$$

for $0 < x$, $x \neq 1$. This is (8.6.20) for $n = 1$.

Since this inequality is valid for all positive x, $x \neq 1$, we can replace x by $x^{1/n}$ without impairing the inequality. From

$$\frac{1}{n} \log x = \log x^{1/n} < x^{1/n} - 1$$

we get (8.6.20) by multiplying by n. This completes the proof.

Problem 7. Prove that

$$\log x = \lim_{n \to \infty} n(x^{1/n} - 1),\ x > 0. \tag{8.6.21}$$

We restrict ourselves to integral values of n and set

$$f_n(x) \equiv n(x^{1/n} - 1).$$

We start by proving that for all n and all x, $x > 0$, $x \neq 1$,

$$f_n(x) > f_{n+1}(x). \tag{8.6.22}$$

We use the same argument as in the proof of (8.6.20). Set

$$F_n(x) = f_n(x) - f_{n+1}(x) = nx^{1/n} - (n + 1)\, x^{1/(n+1)} + 1.$$

This function is zero for $x = 1$. Since

$$F_n'(x) = x^{\frac{1}{n} - 1} - x^{\frac{1}{n+1} - 1} = x^{-\frac{n}{n+1}}\left\{x^{\frac{1}{n(n+1)}} - 1\right\},$$

the derivative is also zero at $x = 1$ and changes its sign from minus to plus as x increasing passes through 1. It follows that $F_n(x)$ takes its least possible value 0 for $x = 1$. Hence $F_n(x)$ is positive everywhere else and (8.6.22) holds, that is,

$$\{f_n(x)\}$$

is a strictly decreasing sequence of functions bounded below by $\log x$. It follows that

$$\lim_{n \to \infty} f_n(x) \geqq \log x, \quad x > 0.$$

For $x > 1$ we can carry through the proof as follows. Consider

$$g(x) = \log x - (x - 1) + \tfrac{1}{2}(x - 1)^2.$$

Here $g(1) = 0$. Since

$$g'(x) = \frac{1}{x} - 1 + x - 1 = \frac{(x-1)^2}{x} > 0,$$

we see that $g(x)$ is always increasing and, since $g(1) = 0$, we have $g(x)$ positive for $x > 1$. This says that

$$\log x > (x - 1) - \tfrac{1}{2}(x - 1)^2 \quad \text{for} \quad x > 1. \tag{8.6.23}$$

Here we replace x by $x^{1/n}$ and multiply through by n. The result can be written

$$\log x > f_n(x) - \frac{1}{2n}[f_n(x)]^2 . \tag{8.6.24}$$

Since $\lim f_n(x)$ exists, we see that

$$\log x \geqq \lim f_n(x), \quad x > 1.$$

We have proved the opposite inequality. Hence (8.6.21) is valid for $x > 1$.

If $0 < x < 1$ then we observe that

$$n(x^{1/n} - 1) = x^{1/n}n(1 - x^{-1/n}).$$

Here the first factor on the right tends to 1 and the second to $-\log x^{-1}$ since $x^{-1} > 1$. This proves that the relation (8.6.16) is also valid for $0 < x < 1$ and completes the discussion.

EXERCISE 8.6

1. Discuss the maxima and minima of $C[f(x)]^2$ where C is a positive constant and $f(x)$ has a continuous derivative. The maxima, minima, and zeros of $f(x)$ are supposed to be known and to be finite in number.

2. At what point on the curve $y = \log x$ is the curvature a minimum?

3. What normal in the ellipse $b^2x^2 + a^2y^2 = a^2b^2$ is farthest away from the center?

4. What normal to the parabola $y^2 = 4ax$ forms the shortest chord with the curve?

5. Prove that $1 - x < e^{-x}$ for all $x \neq 0$.

6. The arithmetic, geometric, and harmonic means of two positive numbers a and b are

$$\tfrac{1}{2}(a + b), \sqrt{ab}, \quad \text{and} \quad \frac{2}{(1/a) + (1/b)},$$

respectively. Prove that for $0 < x < \pi/2$ all three means of $\sin x$ and $\tan x$ exceed x.

7. Sketch the curve $y = 1 - e^{-x} + \tfrac{1}{2}e^{-2x}$.
Determine relative extrema and points of inflection.

8. Prove that for $x < 1$, $x \neq 0$,

$$(1 - x)^{\frac{1}{2}} < 1 - \tfrac{1}{2}x,$$

and find the maximum of the difference between the two members for $-1 \leqq x < 0$.

9. Prove that

$$\frac{\sin x}{x} \geqq \frac{2}{\pi} \quad \text{for} \quad 0 < x \leqq \frac{\pi}{2},$$

with equality if and only if $x = \dfrac{\pi}{2}$.

10. If

$$f(x) = (x - 3)e^{2x} + 4xe^x + x + 3,$$

verify that $f(x)$, and all its derivatives are positive for x positive.

11. Prove that $(8a^2 + b^2)^2 > 16a^3b$ for $a > 0, b > 0$.

12. Choose the constants a, b, c in such a way that the fraction

$$\frac{ax + bx^2}{1 + cx}$$

differs as little as possible from $\log(1 + x)$ for small values of x. This done, find the maximum of the difference between the two functions for $0 \leqq x \leqq 1$.

13. For what value of a is

$$\int_0^1 [(1 + x^3)^{-\frac{1}{2}} - ax^2]^2\, dx$$

a minimum? It is not required to find the minimal value of the integral.

14. A function $y = f(x)$ is defined by

$$\left(y + \frac{a}{x^2}\right)(x - b) = c$$

where a, b, c are positive constants. Cf. Problem 11 of Exercise 8.2. Show that this function has neither a maximum nor a minimum in the interval (b, ∞) if $c > (8a)/(27b)$. [*Hint:* Express c in terms of x_0 if $f'(x_0) = 0$, $x_0 > b$, and determine between what limits c can vary.]

15. The differential equation

$$y'' + \gamma y' = -g \quad (y' = D_t y)$$

has a particular solution of the form at. Determine a. The general solution is of the form

$$C_1 + C_2 e^{-\gamma t} + at.$$

Verify this and determine C_1 and C_2 so that $y(0) = 0$, $y'(0) = v$. Find $\max y(t) \equiv f(v)$ and compare your result with (8.6.11).

16. Determine $f(v)$ for the case $H(u) = \gamma u^2$. This quadratic law of resistance appears to give fairly good agreement with observations for medium velocities v while the linear law works best for small velocities.

MISCELLANEOUS PROBLEMS FOR CHAPTER 8

1. For what values of x is

$$\exp\left(\frac{x - 1}{x}\right) \leqq x^2?$$

2. Determine maxima, minima, and inflection points on the curve

$$y^2 = x^5 - 80x.$$

3. Determine what points on the astroid

$$x^{\frac{2}{3}} + y^{\frac{2}{3}} = 1$$

are nearest to the origin and find the radius of curvature at one of these points.

4. Find the tangent to the astroid whose distance from the origin is a maximum.

5. The equations of motion of a point P are

$$x = a \cos t, \quad y = b \sin t.$$

Find the range of the speed.

6. Prove that $e^x < \dfrac{1}{1 - x}$ for $x < 1$.

7. Prove that $\arcsin x < \dfrac{x}{1 - x^2}$ for $0 < x < 1$.

8. Prove that for $u > 0$ and real values of v,

$$uv \leqq u \log u + e^{v-1},$$

and determine when equality holds.

9. Given two intersecting lines and a point P not on either of them. How should a third line be drawn through P such that the resulting triangle has the least possible area?

10. A beam of rectangular cross section is to be cut from a cylindrical log. What form of the cross section gives a beam of maximum strength if the strength is proportional to the width and to the square of the height of the cross section.

*11. Two spheres of radii r and R have the distance a between their centers where $a > r + R$. Find a point P on the line joining the centers from which the sum of the areas of the spherical zones visible at the point is a maximum. The area of a spherical zone is its height times the length of the great circle.

12. Prove Minkowski's inequality

$$\left\{\sum_{n=1}^{\infty} |a_n + b_n|^{\alpha}\right\}^{1/\alpha} \leqq \left\{\sum_{n=1}^{\infty} |a_n|^{\alpha}\right\}^{1/\alpha} + \left\{\sum_{n=1}^{\infty} |b_n|^{\alpha}\right\}^{1/\alpha}, \quad 1 \leqq \alpha.$$

The case $\alpha = 1$ is trivial. For $\alpha > 1$ note that for positive a and b

$$(a + b)^{\alpha} = a(a + b)^{\alpha-1} + b(a + b)^{\alpha-1}$$

and use Hölder's inequality.

13. Given the series

$$\sum_{n=1}^{\infty} a_n e^{\lambda_n z}$$

where a_n, λ_n, and z are complex numbers. Prove that if the series has a region of absolute convergence in the complex z-plane, then the region is convex.

In other words, if z_1 and z_2 are two points of the region, then all points on the line segment $[z_1, z_2]$ also belong.

14. If the λ_n's are positive numbers, prove that the region of absolute convergence is a left half-plane.

15. An arc of a curve is defined for $a < x < b$ by the equation $y = f(x)$ where $f(x)$ as well as its first two derivatives are continuous. A straight line intersects the arc in exactly three points with abscissas x_1, x_2, x_3, $a < x_1 < x_2 < x_3 < b$. Show that the curve has at least one point of inflection whose abscissa lies in the interval (x_1, x_2). Verify the theorem for the case of the curve $x^2y + y - 2 = 0$ and the straight line $x - 5y + 4 = 0$.

16. Determine if the function

$$f(x) = |x^2 - 1| + |x|, \ -1 < x \leqq 2,$$

has any relative extrema in this interval. Determine also the range of $f(x)$ in $(-1, 2]$.

17. A point P is given inside a parabola. Find the chord through P which cuts off the least area segment of the parabola.

18. At the point P on the catenary

$$y = \tfrac{1}{2}a(e^{x/a} - e^{-x/a}), \quad a > 0,$$

a tangent is drawn which intersects the x-axis at T. For what choice of P is PT a minimum?

19. Let

$$f(x) = \sin^2 x + \tan^2 x - 2x^2, \quad 0 \leqq x < \tfrac{1}{2}\pi.$$

Prove that $f(0) = f'(0) = f''(0) = 0$ and that $f''(x) > 0$ for $0 < x < \frac{1}{2}\pi$. Use this to prove that $f'(x)$ and $f(x)$ are positive strictly increasing functions in $(0, \frac{1}{2}\pi)$ and deduce the inequality

$$\{\tfrac{1}{2}(\sin^2 x + \tan^2 x)\}^{\frac{1}{2}} > x, \quad 0 < x < \tfrac{1}{2}\pi,$$

which is analogous to those of Problem 6, Exercise 8.6.

*20. Try to give a similar argument for

$$f(x) = \sin^3 x + \tan^3 x - 2x^3, \quad 0 < x < \tfrac{1}{2}\pi.$$

Here $f(0) = f'(0) = f''(0) = f'''(0) = 0$ and $f'''(x) > 0$.

21. If $f(x)$ is defined as in the preceding problem, find

$$\lim_{x\to 0} x^{-5} f(x).$$

22. Determine the extremal values of

$$y = \frac{\sin x}{1 - c^2 \cos^2 x},$$

where $c^2 < 1$. Note two cases according as $2c^2 > 1$ or < 1.

9

Plane Curves

9.1. Rational functions and their roots. This chapter is devoted to various aspects of curve tracing. We give a discussion of curves defined by equations of the form

$$y = R(x), \tag{9.1.1}$$

more generally

$$y^n = R(x), \tag{9.1.2}$$

where $R(x)$ is a rational function of x with real coefficients and n is a positive integer.

This is followed by remarks on more general algebraic curves: excluded territories, asymptotes, singularities, and so on. It is also necessary to study curves given in parametric form or in polar coordinates. The notion of arc length, its relation to curvature, and the theory of evolutes and involutes will demand our attention. Finally we discuss some special curves, algebraic as well as transcendental, which have figured prominently in the history of geometry and calculus.

We start with equation (9.1.1) and suppose that

$$R(x) = \frac{P(x)}{Q(x)}, \tag{9.1.3}$$

where $P(x)$ and $Q(x)$ are polynomials in x of degrees m and n, respectively. Both are supposed to have real coefficients. In the following discussion we need some of the results developed in Sections 6.5 and 6.6 concerning real factorization of polynomials and representation of rational functions by partial fractions. For most purposes we do not need the full representation theorem, only some of its consequences.

The object of the partial fraction theorem is to represent the rational function, up to an additive constant, as the sum of its principal parts. There is a principal part associated with each of the zeros of the denominator, $Q(x)$, and a principal part at infinity if $m > n$. By Theorem 6.6.2 the principal part at the k-fold zero $x = a$ of $Q(x)$ is of the form

$$\frac{A_0}{(x-a)^k} + \frac{A_1}{(x-a)^{k-1}} + \cdots + \frac{A_{k-1}}{x-a}. \tag{9.1.4}$$

It shows how $R(x)$ becomes infinite as $x \to a$. Similarly the principal part at infinity is a polynomial in x of degree $m - n$ which indicates how $R(x)$ becomes infinite with x.

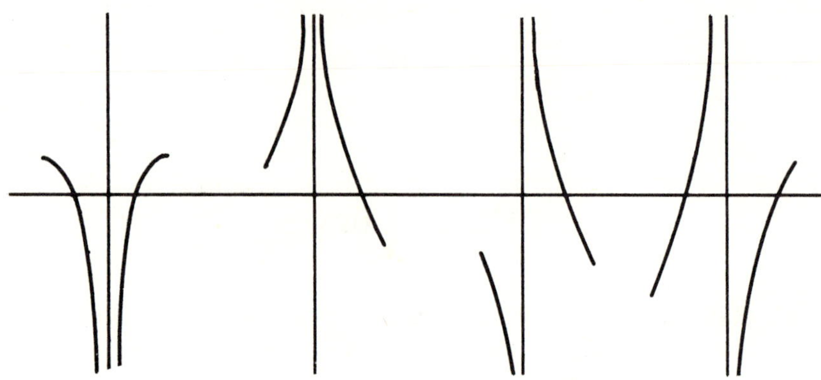

Figure 9.1

Not all this information is pertinent when it comes to plotting the graph of $y = R(x)$. The real zeros of $Q(x)$ show in a spectacular manner, since $R(x)$ becomes infinite when $x \to a$, $Q(a) = 0$, as shown by (9.1.4). Actually we do not need all of the principal part. It is enough to know k and the leading coefficient A_0.

Definition 9.1.1. *If $x = a$ is a real k-fold zero of $Q(x)$ where $k \geqq 1$, then the line $x = a$ is called a vertical asymptote of the curve $y = R(x)$.*

The curve approaches the asymptote when $x \to a$. This follows from

$$\lim_{x \to a} (x - a)^k R(x) = A_0. \tag{9.1.5}$$

The larger the k, the steeper is the approach of branches of the curve to the asymptote. In the following such branches will be referred to by the suggestive name "infinitary ends of the curve." For the approach there are four geometric possibilities, depicted in Figure 9.1. They correspond to the following analytical possibilities:

$$\begin{aligned}
&(1)\ \lim_{h \to 0} R(a - h) = -\infty,\ \lim_{h \to 0} R(a + h) = -\infty,\\
&(2)\ \lim_{h \to 0} R(a - h) = +\infty,\ \lim_{h \to 0} R(a + h) = +\infty,\\
&(3)\ \lim_{h \to 0} R(a - h) = -\infty,\ \lim_{h \to 0} R(a + h) = +\infty,\\
&(4)\ \lim_{h \to 0} R(a - h) = +\infty,\ \lim_{h \to 0} R(a + h) = -\infty.
\end{aligned} \tag{9.1.6}$$

Here cases (1) and (2) go with k even, (3) and (4) with k odd. (1) and (3) correspond to positive values of A_0, (2) and (4) to negative values.

On the other hand, the nonreal zeros of $Q(x)$ scarcely affect the graph. Here we can combine all the partial fractions corresponding to the complex zeros into a single fraction

$$\frac{P_1(x)}{Q_1(x)}. \tag{9.1.7}$$

The degree of $P_1(x)$ is at least one unit less than that of $Q_1(x)$. The latter has as zeros the complex zeros of $Q(x)$.

The principal part at infinity plus the additive constant coincides with the polynomial $K(x)$ of formula (6.6.5) obtained as a quotient when we divide $P(x)$ by $Q(x)$. If $m > n$, then $R(x)$ becomes infinite with x and if

$$K(x) = K_0 x^{m-n} + \cdots + K_{m-n}, \tag{9.1.8}$$

the sign of the infinite limit is

$$\operatorname{sgn}(K_0)\,[\operatorname{sgn} x]^{m-n}.$$

We have the same four possibilities for the sign of y for large values of $|x|$ as noted above for one-sided approach to a vertical asymptote.

Definition 9.1.2. *If* $m = n + 1$, *then*

$$y = K_0 x + K_1$$

is called a slanting asymptote, and if $m \leqq n$, *then*

$$y = K_0$$

is called a horizontal asymptote. We set $K_0 = 0$ *if* $m < n$.

As $x \to \infty$ the curve $y = R(x)$ approaches the curve

$$y = K(x). \tag{9.1.9}$$

If $m \leqq n + 1$, this is an assertion that the curve approaches its horizontal or slanting asymptote. In general we refer to (9.1.9) as the *asymptotic parabola*. It is a parabola in the strict sense if $m = n + 2$. The assertion concerning the approach follows from the relation

$$R(x) - K(x) = \frac{S(x)}{Q(x)}, \tag{9.1.10}$$

where the degree of the polynomial $S(x)$ is at most $n - 1$ so the right member goes to zero as $x \to \infty$.

It is important to note *how* the right member goes to zero. Since $S(x)$ and $Q(x)$ are polynomials, their quotient keeps a constant sign for large negative values of x as well as for large positive values. To clarify the ideas, suppose that $S(x)/Q(x)$ is negative for large positive values of x. This means that for such values

$$R(x) < K(x),$$

that is, the "positive end" of the curve (9.1.1) approaches the positive end of the asymptotic curve (9.1.9) *from below*. Again there are four possibilities for approach: from above at both ends, from below at both ends, from above at left end only, from below at left end only. If the number of sign changes of

$S(x)/Q(x)$ is even, one of the first two possibilities holds, if it is odd, then one of the second two choices holds.

We start the actual curve tracing by marking the infinitary ends of the curve in the figure. We can do this, at least roughly, knowing how the curve lies in relation to its asymptotes. Thus we obtain a certain number of loose ends, namely two for each asymptote (including asymptotic curve), and these must now be tied together. There are various things to keep in mind when this is done.

The first consideration is that the curve consists of a certain number of branches that begin and end at infinity, the number being one more than the number of vertical asymptotes. This means that each infinitary end has to be joined to another infinitary end. Moreover, for each x not a zero of the denominator, there is one and only one value of y, that is, each vertical line which is not an asymptote has one and only one point in common with the curve. Vertical asymptotes may be regarded as tangents at infinity; they have no finite intersections with the curve. On the other hand, the curve may intersect the asymptotic curve

$$y = K(x),$$

the x-axis if $K(x) \equiv 0$, namely at the points where $S(x) = 0$. This means in particular that slanting or horizontal asymptotes may have points in common with the curve. Vertical asymptotes separate the branches from each other and thus makes it possible to tell in advance the number of branches to be expected.

Returning to intersections of the curve with straight lines, we note that the number of such points is severely restricted. If the line has the equation

$$y = kx + b,$$

the intersections with $y = R(x)$ have abscissas which are roots of the equation

$$(kx + b)Q(x) - P(x) = 0, \tag{9.1.11}$$

the degree of which is normally max $(n + 1, m)$, that is, equal to the degree of the curve (9.1.1). This then is the upper limit to the number of intersections. The actual number may be smaller due to infinite or complex roots of (9.1.11). Keep this in mind when sketching curves: put a ruler on the figure, if it shows more intersections than the degree permits, the sketch should be revised.

The roots of $P(x) = 0$ give the intersections with the x-axis. They are at most m in number. If $Q(0) \neq 0$, the point on the y-axis has the ordinate $R(0)$. We get further information by discussing $R'(x)$ and $R''(x)$ with a view to finding maxima, minima, and points of inflection. These often cannot be found explicitly, but it may be possible to get a fair estimate of the number and location of these points. Plotting additional points on the graph may round out the picture, but the information obtained this way is usually less important than the qualitative observations obtained by following the general procedure outlined above.

We illustrate this procedure by a detailed discussion of special cases.

Example 1. We take

$$y = \frac{x^3 - 3x^2 + x + 2}{x^2 - 4x + 3}$$

$$= x + 1 + \frac{2x - 1}{(x - 1)(x - 3)} \tag{9.1.12}$$

$$= x + 1 - \tfrac{1}{2}\frac{1}{x - 1} + \tfrac{5}{2}\frac{1}{x - 3}.$$

Here

$$x = 1 \quad \text{and} \quad x = 3$$

are the vertical asymptotes and

$$y = x + 1$$

is a slanting asymptote. Further

$$R(1 - 0) = +\infty,\ R(1 + 0) = -\infty,$$

$$R(3 - 0) = -\infty,\ R(3 + 0) = +\infty,$$

$$y > x + 1 \quad \text{for} \quad x > 3,$$

$$y < x + 1 \quad \text{for} \quad x < \tfrac{1}{2}.$$

From this information we get a fairly good idea of the shape of the curve. There are three branches, one to the left of $x = 1$, one in the strip $1 < x < 3$, and one to the right of $x = 3$ above the slanting asymptote. There must be a point of inflection on the left branch, a maximum on the middle branch, and a minimum on the right branch. See Figure 9.2. These predictions from the geometry of the situation are confirmed by an examination of the function and its first two derivatives. First it is found that the curve intersects the x-axis at the points of abscissa

$$\tfrac{1}{2}(1 - \sqrt{5}),\ \tfrac{1}{2}(1 + \sqrt{5}),\ 2,$$

and the y-axis at $(0, \frac{2}{3})$. Since $R(x)$ is continuous and differentiable in the interval between the second and third zeros, Rolle's theorem asserts the existence of a zero of $R'(x)$ in this interval. Now

$$R'(x) = 1 + \tfrac{1}{2}\frac{1}{(x - 1)^2} - \tfrac{5}{2}\frac{1}{(x - 3)^2},\quad R''(x) = -\frac{1}{(x - 1)^3} + \frac{5}{(x - 3)^3}. \tag{9.1.13}$$

$R''(x)$ has a single real zero at

$$x = x_0 = -\tfrac{1}{2}(25^{\frac{1}{3}} + 5^{\frac{1}{3}} - 1) = -1.817.$$

This is the only real point of inflection. $R''(x)$ changes its sign three times, namely at the zero just found and at $x = +1$ and $+3$. From this we conclude that $y = R(x)$ is concave downward in the intervals $(-\infty, x_0)$ and $(1, 3)$ and concave upward in $(x_0, 1)$ and $(3, \infty)$. This confirms the existence of a maximum in $(1, 3)$ and a minimum in $(3, \infty)$. A little computation shows that the maximum

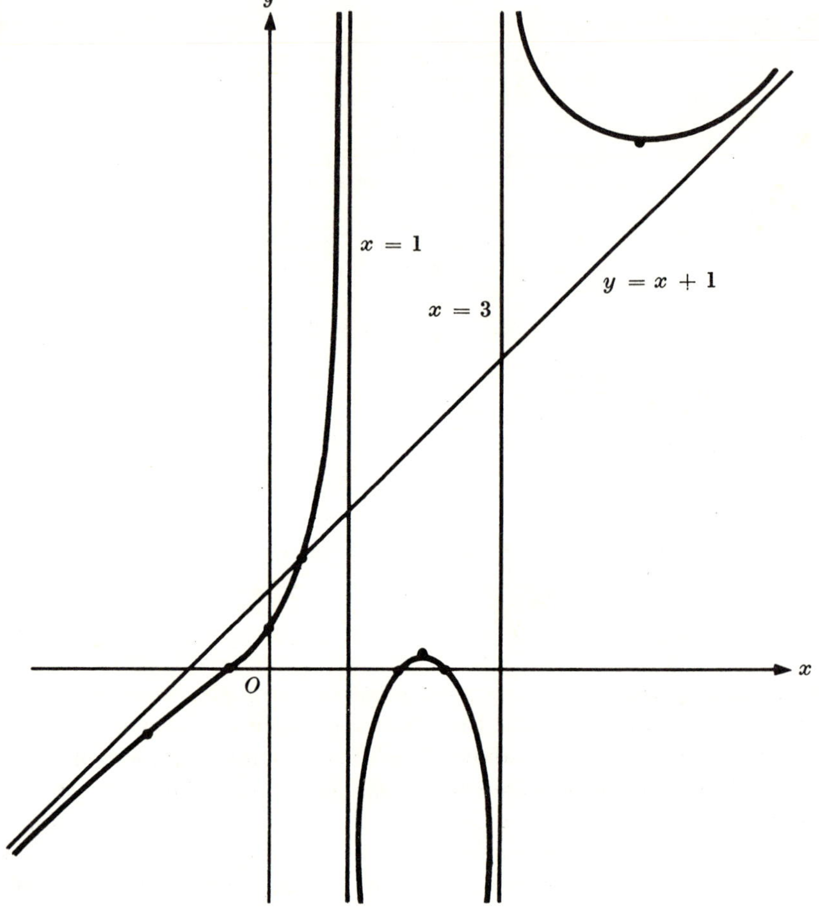

Figure 9.2

is close to $x = 1.81$, while the minimum lies between 4.5 and 4.6. We note that

$$\tfrac{5}{2}(x-3)^{-2} < 1 \quad \text{if} \quad x < 3 - \tfrac{1}{2}\sqrt{10}.$$

From this we conclude that for each $x < 1.81$, $x \neq 1$, $R'(x) > 0$, $R(x)$ is increasing in $(-\infty, 1)$ and in $(1, 1.81)$. It follows that $R'(x)$ has only two real zeros, both of which have been accounted for. This completes the discussion of (9.1.12).

Example 2. The curve defined by the equation

$$y = \frac{3x^2 - 4x + 1}{x^3 - 2x^2 + x - 2} = \frac{1}{x-2} + \frac{2x}{x^2+1} \tag{9.1.14}$$

exhibits only one vertical asymptote $x = 2$ with

$$R(2-0) = -\infty, \; R(2+0) = +\infty.$$

The x-axis is the horizontal asymptote and is intersected by the curve at $x = \frac{1}{3}$ and 1. There are two branches, one to the right of $x = 2$ and above the x-axis, and the other to the left of $x = 2$ and below the x-axis except for an arc between $x = \frac{1}{3}$ and $x = 1$. There is obviously a maximum between $\frac{1}{3}$ and 1 and there must also be a minimum somewhere on the left branch since it lies below the x-axis for $x < \frac{1}{3}$ and admits of the axis as an asymptote. The derivative is given by

$$R'(x) = -\frac{1}{(x-2)^2} - 2\,\frac{x^2-1}{(x^2+1)^2}\,. \tag{9.1.15}$$

The negative root of the equation $R'(x) = 0$ is about -0.9. This gives the desired minimum. The second derivative is

$$R''(x) = \frac{2}{(x-2)^3} + 4\,\frac{x^3-3x}{(x^2+1)^3}\,. \tag{9.1.16}$$

If is fairly easy to follow the changes in the sign of $R''(x)$. The first term changes its sign only once, namely at $x = 2$, but the second term changes its sign three times, namely at $x = 0$ and $\pm\sqrt{3}$. Noting the intervals where the signs agree, we see that $R''(x) > 0$ for $x > 2$ and < 0 for $x < -\sqrt{3}$ and in $0 < x < \sqrt{3}$. Actually $R''(x) < 0$ for $0 < x < 2$, for in the interval $(\sqrt{3}, 2)$ the first term in (9.1.16) clearly dominates. This leaves only the interval $(-\sqrt{3}, 0)$ in doubt. Here two and only two sign changes occur and, hence, two points of inflection. The curve is sketched in Figure 9.3.

We shall briefly discuss curves defined by equations of the form

$$y^2 = R(x), \tag{9.1.17}$$

where, as above, $R(x)$ is a rational function. Here a number of new features present themselves. Thus there are normally *excluded strips* where the curve does

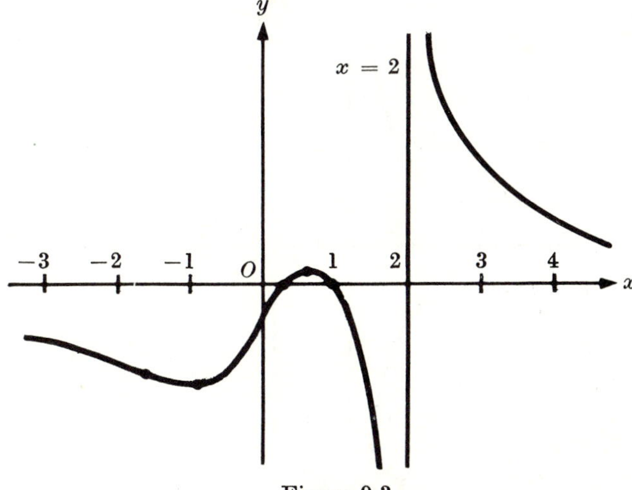

Figure 9.3

not exist. The curve is *symmetric with respect to the x-axis* and the points on the axis are either *singular* (double points, isolated points, or cusps) or feature *vertical tangents*. All three phenomena are exhibited by the following

Example 3. We take

$$y^2 = x^3 \frac{1 - x}{1 + x}. \tag{9.1.18}$$

Here the right member is positive for $x < -1$ and for $0 < x < 1$. $R(x) < 0$ for $-1 < x < 0$ and for $1 < x$. Thus there can be no points of the curve in

$$-1 \leqq x < 0 \quad \text{and in} \quad 1 < x.$$

These are excluded regions. We have

$$y = \pm x^{\frac{3}{2}} \left(\frac{1 - x}{1 + x}\right)^{\frac{1}{2}},$$

that is, branches symmetric with respect to the x-axis. The two branches in the left half-plane have no points in common. On the other hand, in the strip $0 \leqq x \leqq 1$ there is a single closed branch, for the two halves are connected at $x = 0$ and $x = 1$. At $x = 1$ the curve tangent is vertical, at the origin horizontal. The origin is a *cusp*. This is a singular point where the tangent vector reverses its direction. There is a single maximum and a single minimum on the oval both on the line

$$x = \tfrac{1}{3}(-1 + \sqrt{10}).$$

If the sign of the square root is changed, we obtain the abscissa of the maximum and minimum on the left branches. The line $x = -1$ is a vertical asymptote. Figure 9.4 gives a schematic drawing of the curve.

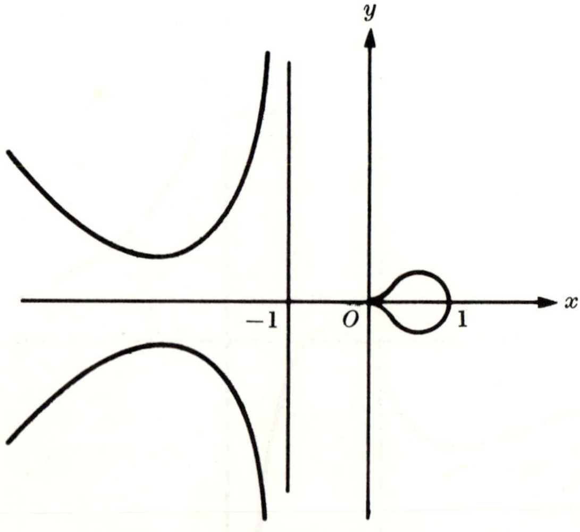

Figure 9.4

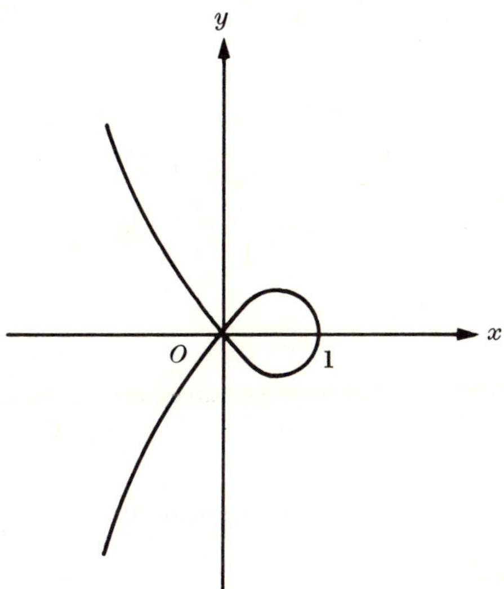

Figure 9.5

Example 4. The curve

$$y^2 = x^2(1 - x) \tag{9.1.19}$$

also has a singular point at the origin, but now we have a *double point* (also called a *node*) with real distinct tangents

$$y = x \quad \text{and} \quad y = -x.$$

The curve exists for $x \leq 1$; the rest of the plane is excluded. See Figure 9.5.

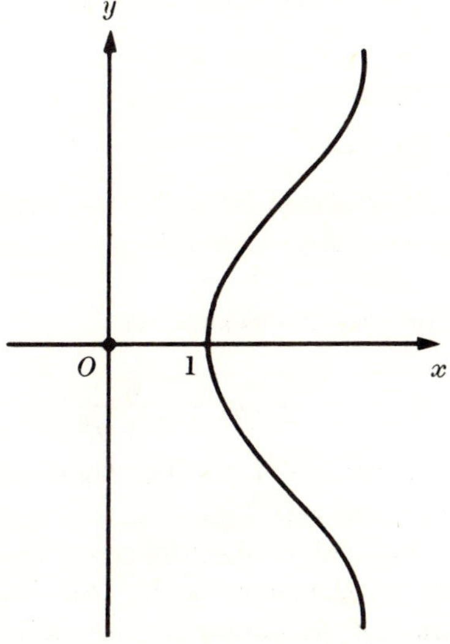

Figure 9.6

Example 5. The curve

$$y^2 = x^2(x-1) \tag{9.1.20}$$

has an *isolated point* at the origin. This may be regarded as a double point with distinct imaginary tangents. The rest of the curve lies in the half-plane $x \geqq 1$. See Figure 9.6.

Similar considerations apply to curves of the form

$$y^3 = R(x), \tag{9.1.21}$$

and more generally

$$y^n = R(x). \tag{9.1.22}$$

If n is even, there are excluded regions but not if n is odd. There are new possibilities of more complicated singularities. We refer to the problems below.

EXERCISE 9.1

Discuss $y = R(x)$ (asymptotes, maxima and minima, points of inflection) and sketch the curve when $R(x)$ is given by:

1. $\dfrac{x^2+1}{x^2-1}$. 2. $\dfrac{x^2-1}{x^2+1}$. 3. $\dfrac{x^3}{x^2-1}$.

4. $\dfrac{x^3}{x^2+1}$. 5. $\dfrac{x^3}{(x-2)^2}$. 6. $\dfrac{x^3-x}{x^2+1}$.

7. $\dfrac{x^4}{x^2-1}$. 8. $\dfrac{x^2+1}{x(x-1)(x-2)}$. 9. $\dfrac{2x^2+3}{x^2(x-1)}$.

10. $\dfrac{2}{x^2}+\dfrac{6}{x}+\dfrac{2}{(x-1)^2}-\dfrac{5}{x-1}$. 11. $\dfrac{4}{(x^2-1)^2}$. 12. $\dfrac{1}{x^3+1}$.

Discuss $y^2 = R(x)$ and sketch the curve when $R(x)$ is given by:

13. = 1. 14. = 2. 15. = 3.

16. = 4. 17. = 5. 18. = 6.

19. = 7. 20. = 8. 21. = 12.

Verify the existence of horizontal asymptotes in cases 13, 14, 20, and 21. There are slanting asymptotes in case 19 and asymptotic parabolas $y^2 = x + a$ in the other cases.

22. In the same diagram plot the two curves

$$y_1 = \frac{1}{x-2}, \quad y_2 = \frac{2x}{1+x^2}.$$

Adding ordinates one obtains the curve (9.1.14). Make fairly accurate drawings on coordinate paper and compare with Figure 9.2 and the discussion in the text.

23. Discuss $x^3 + y^3 = 1$ and show that the curve has a slanting asymptote. For the latter part of the problem one can use the factorization $x^3 + y^3 = (x+y)(x^2 - xy + y^2)$. Another possibility is to solve for y and to expand the cube root for large values of x. Find the points of inflection.

24. Discuss $x^3 + y^3 = x^2$ and show the existence of a slanting asymptote. Verify that the curve has a cusp at the origin with vertical tangent and that (1, 0) is an inflection point where the tangent is vertical.

25. Find the inflection points of the curve (9.1.20).

26. Find the maxima and minima of the curve (9.2.19) and find the area of the loop.

27. Discuss the curve defined by

$$y^2 = x^2 \frac{1 - x}{1 + x}$$

and find the area of the loop.

9.2. Remarks on algebraic curves. Let $P(x, y)$ be a polynomial in x and y with real coefficients, say

$$P(x, y) = \sum a_{jk} x^j y^k. \tag{9.2.1}$$

The degree of the polynomial is

$$n = \max (j + k). \tag{9.2.2}$$

We assume that $P(x, y)$ cannot be factored into polynomial factors. The set of points (x, y) in the plane such that

$$P(x, y) = 0 \tag{9.2.3}$$

forms an *algebraic curve C of order n*.

All curves considered in the preceding section are algebraic curves. Further examples are

$$\begin{gathered} b^2x^2 + a^2y^2 = a^2b^2, \quad x^3 + y^3 = 3axy, \\ (x^2 + y^2)y^2 = a^2x^2, \; (x^2 + y^2)^2 = a^2(x^2 - y^2), \end{gathered} \tag{9.2.4}$$

known as an *ellipse*, a *folium of Descartes*, a *kappa curve*, and a *lemniscate*, respectively. Their orders are 2, 3, 4, and 4.

An algebraic curve is not always given to us in the form (9.2.3). Thus the curve known as a *regular astroid* is usually given by one of the equations

$$x^{\frac{2}{3}} + y^{\frac{2}{3}} = a^{\frac{2}{3}}. \tag{9.2.5}$$

or

$$x = a \cos^3 t, \; y = a \sin^3 t, \tag{9.2.6}$$

rather than by

$$(x^2 + y^2 - a^2)^3 + 27a^2x^2y^2 = 0. \tag{9.2.7}$$

All three equations are equivalent and the third shows that the order of the curve is six.

The order of an algebraic curve puts a limit to the number of intersections of the curve with a straight line. The maximum number is n but some intersections may be nonreal or infinite. The student can visualize the situation by considering the intersections of a hyperbola with a straight line where there are 2, 1, or 0 real intersections depending upon the location of the line. An asymptote has two infinite intersections with the curve, a line parallel to an asymptote one finite and one infinite intersection. Finally, a line between the two branches has complex intersections. The order is invariant under coordinate transformations of the form

$$x = a + c_{11}s + c_{12}t,$$

$$y = b + c_{21}s + c_{22}t.$$

Another important invariant is the *class* of the curve. This is the *maximum number of tangents* that can be drawn to the curve from any point. Again it is understood that tangents may be imaginary and that the point of tangency may be infinite. For the four curves listed under (9.2.4) the class is 2, 4, 6, 6, respectively. It is not always greater than or equal to the order. Thus it is 4 for the astroid. The maximum value of the class number is $n(n - 1)$; it is lowered by the presence of singularities.

The classification of second-order curves, the conics, is given in Appendix A. We found three different types: ellipses, hyperbolas, and parabolas. They have totally different shapes, but all stem from a cone. More precisely, given any two conics, we can find a right circular cone such that two plane intersections of the cone are congruent to the given conics. No matter how different the conics may look, after they have been placed on the same cone, they look alike when viewed from the vertex of the cone. We say that any two conics are *projectively equivalent*. In this sense there is only one basic conic, say a circle, from which any other conic may be obtained by a suitable projection.

The notion of projective equivalence may be extended to curves of higher order. Without giving a strict definition, let us say that two curves are projectively equivalent if it is possible by rigid motion in three-space to place them in positions where they appear to coincide when viewed from a suitable vantage point.

This may be used for classifying curves of higher order than the second. Already for the curves of degree three, the so-called cubics, it is found that there are infinitely many curves which are projectively inequivalent. This makes it necessary to introduce a cruder criterion of equivalence.

It may be shown that any real cubic is projectively equivalent to a curve whose equation has the form

$$y^2 = x^3 + px + q,$$

where the constant

$$J \equiv (27q^2 + 4p^3)p^{-3}$$

has the same value for all cubics belonging to the same projective equivalence class. Here p^3J can be < 0, $= 0$, or > 0, according as the equation

$$x^3 + px + q = 0$$

has three real distinct roots, a multiple root, or only one real root. These cases give cubics of entirely different shape. In the second case the cubic has a singular point so we distinguish between three subcases according as this point is a node, an isolated point, or a cusp. This gives a total of five basic classes of cubics.

A classification of cubics was given by Isaac Newton in his *Enumeratio linearum tertii ordinis* published in 1704 as an Appendix to his *Opticks*. He found 78 different types of cubics, but they fall into five subdivisions corresponding to the five cases mentioned above. Newton's five basic cubics are the following so-called "divergent parabolas":

$$\text{(i)}\quad y^2 = (x - a)(x - b)(x - c), \qquad a < b < c,$$

$$\text{(ii)}\quad y^2 = (x - a)[x - \alpha)^2 + \beta^2], \qquad a,\ \alpha,\ \beta \text{ real},$$

$$\text{(iii)}\quad y^2 = (x - a)(x - b)^2, \qquad a < b,$$

$$\text{(iv)}\quad y^2 = (x - a)^2(x - b), \qquad a < b,$$

$$\text{(v)}\quad y^2 = (x - a)^3, \qquad a \text{ real}.$$

We have already met three of these curves. The last one is a Neil's parabola and has a cusp at $(a, 0)$. This lowers the class to 3. The curve was encountered in Chapter 8, see formulas (8.1.12) and (8.4.15), where however the axes are interchanged. Type (iv) has an isolated point at $(a, 0)$. See Figure 9.6 where the case $a = 0$, $b = 1$ is depicted. Type (iii) on the other hand has a node at $(b, 0)$. See Figure 9.5 for an equivalent curve. In both cases the class is 4. Finally, cases (i) and (ii) are cubics without singularities and hence of class 6. Type (i) has two branches one of which is a closed oval, type (ii) has a single branch. See Figures 9.7 and 9.8.

We note that no cubic can have more than one singular point which is either a node, an isolated point, or a cusp. The possible number of singularities increases with the order. For a curve of order n the number of singular points is at most

$$\tfrac{1}{2}(n - 1)(n - 2).$$

Thus if $n = 4$ the maximum number is 3. Any combination of d nodes and c cusps

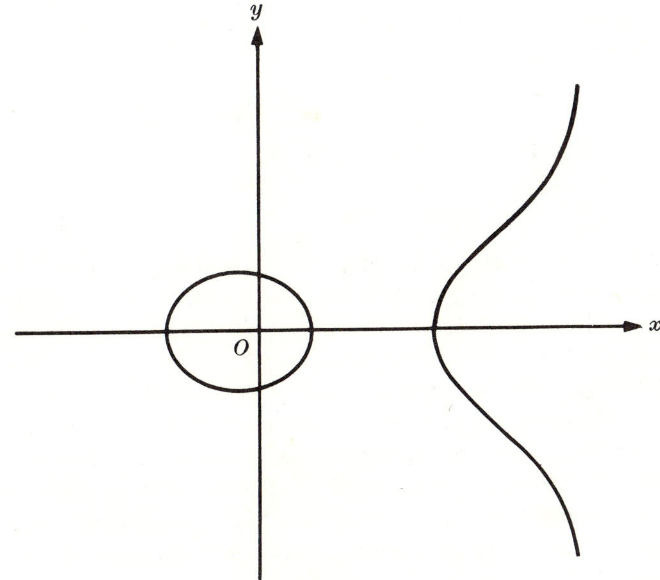

Figure 9.7

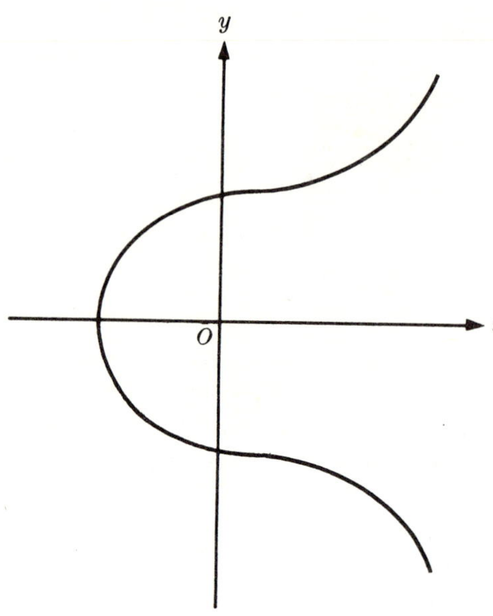

Figure 9.8

is permissible as long as c, d and $c + d$ lie between 0 and 3. This gives a classification of quartics into ten sets according to the number and character of the singular points. Each set naturally contains a considerable number of different types of quartics. We give no further details, since the methods of the Calculus are only secondary for these questions.

We turn instead to the discussion of singular points where the Calculus is of greater use. Suppose the curve goes through the origin so that the constant term in the expansion (9.2.1) is missing. Thus

$$P(x, y) = a_{10}x + a_{01}y + a_{20}x^2 + 2a_{11}xy + a_{02}y^2 + \cdots.$$

Suppose to begin with that $|a_{10}| + |a_{01}| \neq 0$. Then

$$a_{10}x + a_{01}y = 0$$

is tangent to the curve at the origin and there is a single branch of the curve passing through the origin. If $a_{01} \neq 0$, the ordinate y can be expanded in a series of integral ascending powers of x. This and similar results will be proved in Chapter 15, Vol. II.

If $a_{10} \neq 0$ but $a_{01} = 0$, the roles of x and y are interchanged; we can now expand x in integral ascending powers of y, and this series starts with a power of y greater than the first. Here the tangent at the origin is vertical and either the origin is a point of inflection or gives an extremum of x as a function of y depending upon the parity of the lowest power of y present in the series.

If $a_{10} = a_{01} = 0$, the situation is more complicated. We inspect the terms of the second order

$$a_{20}x^2 + 2a_{11}xy + a_{02}y^2, \tag{9.2.8}$$

and consider the discriminant

$$D = a_{11}{}^2 - a_{20}a_{02}.$$

We assume that the second-order terms are not entirely missing. There are several possibilities.

If $D > 0$, then the quadratic polynomial (9.2.8) can be factored into real factors

$$a_{20}x^2 + 2a_{11}xy + a_{02}y^2 = (\alpha x + \beta y)(\gamma x + \delta y).$$

There are now two distinct branches going through the origin having the lines

$$\alpha x + \beta y = 0 \quad \text{and} \quad \gamma x + \delta y = 0$$

as their tangents. This is a double point with real distinct tangents, also called a node.

An example is given by

$$x^2 - y^2 - x^3 = 0. \tag{9.2.9}$$

See Figure 9.5. Here the quadratic terms give

$$x^2 - y^2 = (y - x)(y + x)$$

so the two branches intersect at right angles at the origin. Since

$$y = \pm x(1 - x)^{\frac{1}{2}} = \pm[x - \tfrac{1}{2}x^2 + \cdots],$$

the statements concerning the branches and their tangents are easy to verify.

Let us take next the case $D < 0$. Here the quadratic terms can also be factored, but the factors are no longer real. In this case the origin is an isolated point. An example is

$$x^2 + y^2 - x^3 = 0. \tag{9.2.10}$$

See Figure 9.6.

The remaining case $D = 0$ contains a number of possibilities. We restrict ourselves to a few. The quadratic terms now form a perfect square

$$a_{20}x^2 + 2a_{11}xy + a_{02}y^2 = (\alpha x + \beta y)^2$$

and the line

$$\alpha x + \beta y = 0$$

is in some sense a double tangent at the origin.

We may have a cusp at the origin. This is the case for the semicubical (= Neil's) parabola

$$y^2 = x^3,$$

where the x-axis is the double tangent. We recall that at a cusp the tangent vector changes its direction by 180°. Problem 24 of the preceding Exercise

$$x^2 - x^3 - y^3 = 0$$

also has a cusp at the origin but now with a vertical tangent.

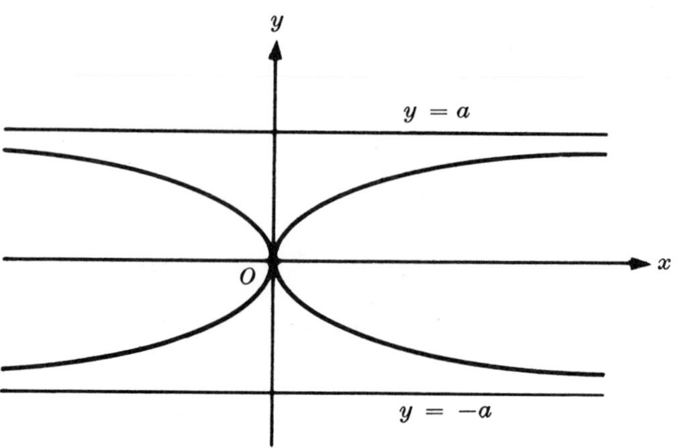

Figure 9.9

The *kappa curve*, the third item under (9.2.4), offers a different possibility. The curve is sketched in Figure 9.9. Two branches of the curve are tangent to each other, the y-axis being the double tangent. From the equation

$$(x^2 + y^2)y^2 = a^2x^2,\ a > 0,$$

one obtains series expansions of the form

$$x = \pm \left\{\frac{1}{a}\,y^2 + \frac{1}{2a^2}\,y^4 + \cdots\right\}$$

for the two branches. Since

$$y^2 = a^2\,\frac{x^2}{x^2 + y^2} < a^2,$$

the whole curve is contained in the strip $-a < y < a$. The bounding lines $y = a$ and $y = -a$ are asymptotes. The singularity of the kappa curve at the origin is called a *tacnode*. It is equivalent to two double points. There is a third double point at infinity. The kappa curve was known to Isaac Barrow in 1670. It is of some importance in machine construction. Finally its equation in polar coordinates has the simple form

$$r = a \cot \theta. \tag{9.2.11}$$

The curve

$$(y - x^2)^2 = x^5, \tag{9.2.12}$$

where y^2 is the term of lowest order, has a peculiar singularity at the origin which the algebraic geometers count as a combination of a cusp and a double point. Since

$$y = x^2 \pm x^{\frac{5}{2}},$$

the curve exists only for $x \geqq 0$. There are two branches in the right half-plane which stop at the origin where they are tangent to each other, to the x-axis, and to the parabola

$$y = x^2.$$

One branch lies above the parabola, the other below. Actually the latter branch lies in the first quadrant for $0 < x < 1$ so that both branches lie above their common tangent at the origin. This is different from an ordinary cusp where the two branches are separated by the cusp tangent. The lower branch exhibits a maximum and a point of inflection. Note also that the radius of curvature is $\neq 0$ at $x = 0$ on both branches. In the case of an ordinary cusp the radius would be zero.

An infinite variety of singular points are possible if $D = 0$. The three cases listed above are the simplest ones. There are totally different possibilities available if the second-order terms are missing, but we shall not pursue this question.

The following example throws some light on the technique of excluded regions which is very helpful in curve tracing. It also adds to the discussion of asymptotes. We take the cubic

$$y^2(x + 1) - x^3 = 0. \tag{9.2.13}$$

Here we note first that if x is negative but $x + 1$ is positive, the equation cannot hold, that is, the strip

$$-1 < x < 0$$

is excluded territory.

We can rewrite the equation as

$$x(y^2 - x^2) + y^2 = 0. \tag{9.2.14}$$

This factorization shows that the following regions are excluded:

(1) $x > 0,\ y - x > 0,\ y + x > 0,$
(2) $x > 0,\ y - x < 0,\ y + x < 0,$
(3) $x < 0,\ y - x > 0,\ y + x < 0,$
(4) $x < 0,\ y - x < 0,\ y + x > 0.$

See Figure 9.10 and Problem 5 below.

From

$$y^2 = \frac{x^3}{1 + x} \tag{9.2.15}$$

it follows that the origin is an ordinary cusp with the x-axis as the tangent.

There are three asymptotes. Thus

$$x + 1 = 0$$

is obviously a vertical asymptote by virtue of (9.2.15). There are two slanting asymptotes $y - x + \frac{1}{2} = 0$ and $y + x - \frac{1}{2} = 0$. These can be obtained from (9.2.15) by noting that

$$y = \pm x\left(\frac{x}{1 + x}\right)^{\frac{1}{2}} = \pm x\left(1 + \frac{1}{x}\right)^{-\frac{1}{2}} = \pm x\left\{1 - \frac{1}{2x} + O\left(\frac{1}{x^2}\right)\right\}$$

$$= \pm\left\{x - \tfrac{1}{2} + O\left(\frac{1}{x}\right)\right\}.$$

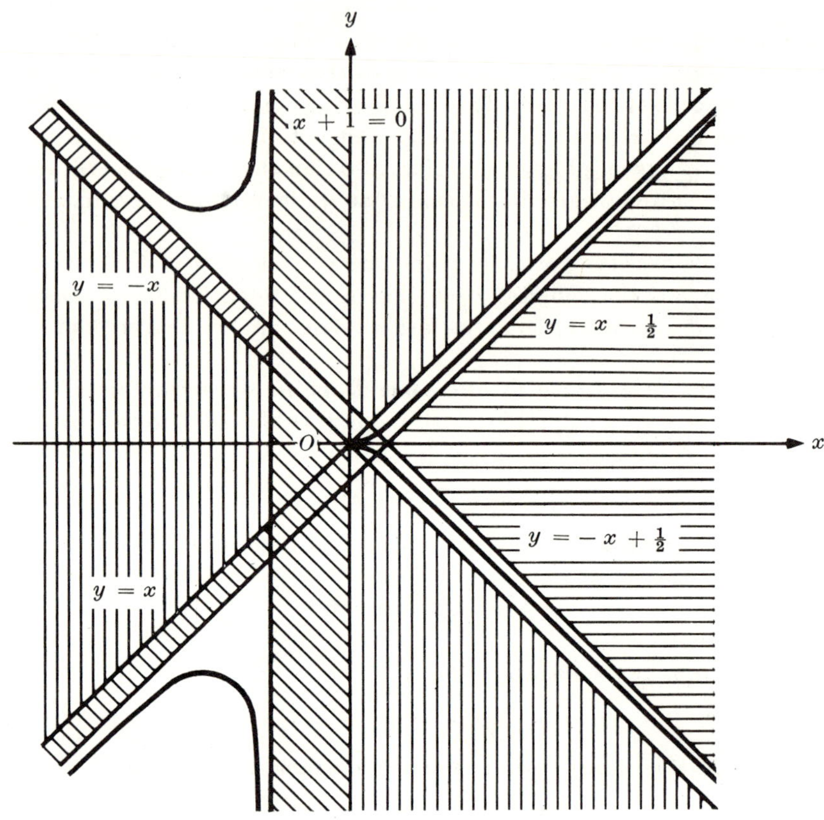

Figure 9.10

This shows that the difference between the ordinates on the line and on the corresponding branch of the curve is $O\left(\frac{1}{x}\right)$ and hence tends to 0 as $x \to \infty$.

There is another neat way of finding the asymptotes which is connected with (9.2.14). We note that each asymptote is parallel to one of the three lines obtained by equating the third degree terms $x(y - x)(y + x)$ to zero and factoring. This suggests as a general procedure to take the aggregate of terms of highest order and equate to zero and factor. Complex factors lead to nonreal asymptotes, and multiple factors may lead to asymptotic parabolas rather than to rectilinear asymptotes. Even if we get only real distinct factors we still have the problem of finding the constant terms in the equations of the hypothetical asymptotes. Here it turns out that decomposing the fraction whose numerator is the terms of degree $(n - 1)$ and the denominator those of degree n into partial fractions gives the desired constants as coefficients of the partial fractions. We illustrate this in the case of (9.2.14).

Here the terms of degree three is $x(x - y)(x + y)$ and y^2 is the only quadratic term. This gives the quotient

$$\frac{y^2}{x(x - y)(x + y)} \equiv 1.$$

We take x as the variable, y as a parameter, and decompose into partial fractions as usual, obtaining

$$-\frac{1}{x}+\frac{\frac{1}{2}}{x-y}+\frac{\frac{1}{2}}{x+y}=1. \tag{9.2.16}$$

This is another form of the equation of the curve, but if we equate each fraction separately to 1 we obtain successively the equations of the three asymptotes! Why?

It is enough to give the reason for one of the lines, say the second one, and it is enough to examine the situation in the first quadrant. We know, for instance, from (9.2.15), that for large positive values of x there is a branch on which y takes large positive values. If we follow this branch, we see that the first and the third terms of (9.2.16) tend to zero so that the middle term has a limit which is obviously 1. Hence on this branch

$$\lim_{x\to\infty}(y-x)=-\tfrac{1}{2} \quad \text{or} \quad \lim_{x\to\infty}(y-x+\tfrac{1}{2})=0.$$

This shows that the branch has the line

$$y-x+\tfrac{1}{2}=0$$

as its asymptote. A variant of this argument shows that the other lines are also asymptotes.

Equation (9.2.13) and its variants are all special cases of a class of equations which give a general method of generating cubics and, by extension, curves of higher order. We use L and Q as generic notation for linear and quadratic polynomials in x and y with subscripts to distinguish between different polynomials.

Suppose L, L_1, L_2, L_3 are four distinct linear polynomials and form

$$L_1L_2L_3-\alpha L^k=0, \tag{9.2.17}$$

where α is any real number and k stands for 0, 1, 2, or 3. Substituting their expressions in x and y for the four polynomials we see that (9.2.17) defines a cubic. Suppose that no two of the four lines

$$L=0,\ L_1=0,\ L_2=0,\ L_3=0 \tag{9.2.18}$$

are parallel and no three are concurrent. Then the lines divide the plane usually into eleven domains, triangular or quadrilateral. In each of these domains the signs of L, L_1, L_2, L_3 are constant, for a straight line $L_0=0$ divides the plane into two half-planes in which L_0 keeps a constant sign. This sign changes when we cross the line. We assign to the domain in question a signature, plus or minus, according as

$$(\operatorname{sgn} L_1)(\operatorname{sgn} L_2)(\operatorname{sgn} L_3)$$

and

$$(\operatorname{sgn}\alpha)(\operatorname{sgn} L)^k$$

are the same or not. The domains of negative signature form excluded domains for the cubic. For the case $k=1$ or 3 the pattern is of the type shown in Figure 9.11 which also show how the cubic can wind intricately through the domains

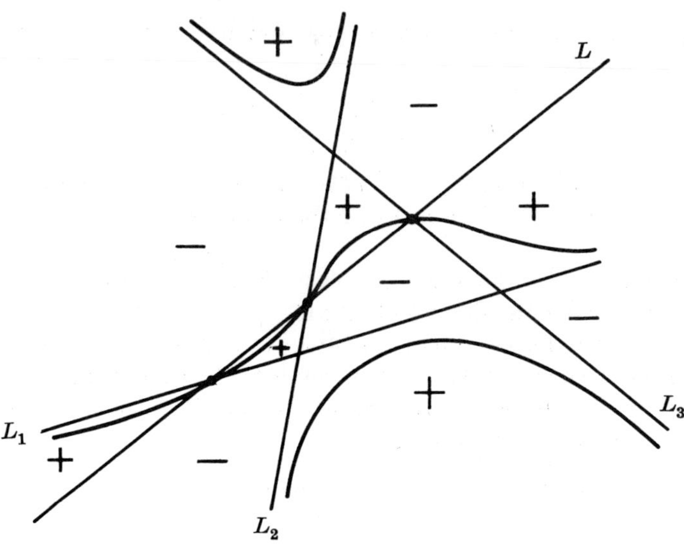

Figure 9.11

of positive signature. It goes through the points of intersection of the first line in (9.2.18) with the other three lines but cannot pass through the other points of intersection of the lines. If $k = 1$, the last three lines are asymptotes of the curve.

Another possible mode of generation of cubics involves a conic and two straight lines:

$$L_1Q - \alpha L^k = 0, \quad k = 0,\ 1,\ 2, \text{ or } 3. \tag{9.2.19}$$

See Figure 9.12 where we have marked signatures of the domains formed by the two lines and the conic. The cubic is shown threading its way through the domains of positive signature. It goes through the intersections of L with L_1 and Q but it is not allowed to pass through the intersections of L_1 with Q. If $k = 1$ the line L_1 is an asymptote. In the figure shown, Q is of the elliptic type

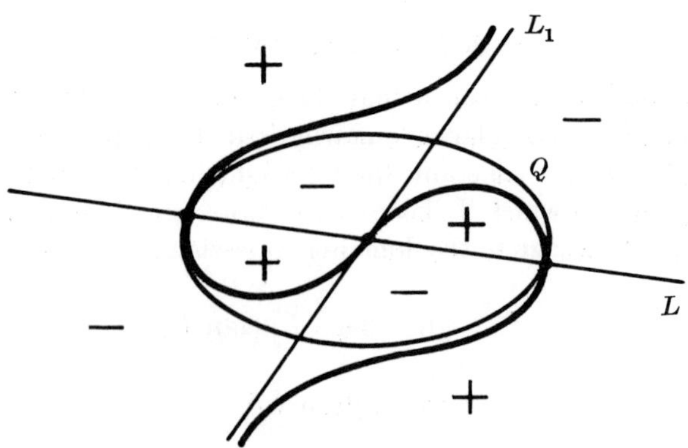

Figure 9.12

and there is only one asymptote. If Q is hyperbolic, the asymptotes of the hyperbola are asymptotes of the cubic as well, and in the parabolic case, the parabola plays the role of asymptotic curve of the cubic, in both cases provided $k = 0$ or 1. Some special cases figure in the problems below.

EXERCISE 9.2

1. Prove that on the branch of the kappa curve in the first quadrant, y is strictly increasing from 0 to a.

2. What relations must hold between a, α, and β for the following cubic to have real maxima and minima:

$$y^2 = (x - a)[(x - \alpha)^2 + \beta^2]?$$

3. Find maxima, minima, and points of inflection of

$$y^2 = x^3 - x.$$

4. Sketch the curve defined by (9.2.12). Find the maximum, the point of inflection, and the radius of curvature at $x = 0$.

5. Show that (9.2.13) can also be written

$$(x + 1)(y - x + \tfrac{1}{2})(y + x - \tfrac{1}{2}) = \tfrac{1}{4}(3x - 1),$$

and use this factorization to get further excluded regions (see Figure 9.10) and to obtain the asymptotes.

6. Let $L_1 = 0$, $L_2 = 0$, $L_3 = 0$ be three distinct lines through the origin and let α, β, γ be real numbers. Prove that

$$\frac{\alpha}{L_1} + \frac{\beta}{L_2} + \frac{\gamma}{L_3} = 1$$

is the equation of a cubic with the asymptotes

$$L_1 = \alpha,\ L_2 = \beta,\ L_3 = \gamma.$$

7. Sketch the curve

$$y(x^2 + y^2 - 1) = \alpha$$

for a positive value of α. How does the shape of the curve change when α decreases towards 0?

8. Same question for the curve

$$y(x^2 + y^2 - 1) = \alpha x.$$

Discuss the following curves where $\alpha > 0$. Determine excluded regions, common points, and asymptotes:

9. $y(x^2 - y^2 - 1) = \alpha$.

10. $y(x^2 - y^2 - 1) = \alpha x$.

11. $(x + y)(xy - 1) = \alpha$.

12. $(x + y)(xy - 1) = \alpha(x - y)$.

13. $(x - y)(y^2 - 4x) = \alpha x$.

14. $(x + 1)(y^2 - 4x) = \alpha y$.

15. $[(x + 1)^2 + y^2 - 3][(x - 1)^2 + y^2 - 3] = \alpha y$.

9.3. Parametric equations. Polar coordinates. An algebraic curve C is the locus of points (x, y) such that

$$P(x, y) = 0, \tag{9.3.1}$$

where $P(x, y)$ is a polynomial in x and y. Usually there are several values of y corresponding to a given value of x and vice versa. This means that we have only local representations of y as a function of x or of x as a function of y. The local character of the representation and the necessity of switching from functions of x to functions of y and back is frequently felt as a handicap and has led to much search for global representations or at least for representations where x and y are on the same footing. This is the background and justification for the use of parametric representations.

DEFINITION 9.3.1. *We say that*

$$x = x(t), \quad y = y(t), \quad a < t < b, \tag{9.3.2}$$

is a parametric representation of an arc of C *if* (i) $x(t)$ *and* $y(t)$ *are defined and continuous in* (a, b), (ii) *for every such* t

$$P[x(t), y(t)] = 0, \tag{9.3.3}$$

and (iii) *for every point* (x, y) *on the arc there is at least one value of* t *for which* (9.3.2) *holds.*

Note that we do not require that to each point (x, y) there be a unique value of t. We have seen that an algebraic curve can intersect itself a finite number of times and a parametric representation of an algebraic curve must allow for multiple points.

Moreover we do not need to restrict ourselves to algebraic curves. There is nothing in the definition about algebraic curves except the references to (9.3.1) where $P(x, y)$ is supposed to be a polynomial. If we drop this convention and think of $P(x, y)$ as a function of two variables defined in some domain in the (x, y)-plane, the definition still makes sense and will be used as such in the following.

More generally, we can consider two continuous functions x and y defined for $a < t < b$. Then equations (9.3.2) give a parametric representation of a curve C. Here we do not know *a priori* if there is a corresponding function $P(x, y)$. In simple cases we can find such a function by eliminating t between the two equations. Actually, if we assume merely that $x(t)$ and $y(t)$ are continuous, then the set of points (x, y) defined by (9.3.2) may be a geometrical object which has little to do with what is normally considered a curve. For our purposes it is sufficient to consider functions $x(t)$ and $y(t)$ having piecewise continuous first- and second-order derivatives. This ensures the existence of arc length and curvature as we shall see in the next section and will enable us to use calculus. This convention will be used in the following unless otherwise stated.

Let us first recall some parametric representations already used. We started in Section A.2 with the straight line for which two parametric forms were given. We restate formula (A.2.21)

$$x = x_0 + t, \quad y = y_0 + mt, \quad -\infty < t < \infty. \tag{9.3.4}$$

Parametric equations were also given for the conics. Thus for the circle with center at the origin and radius r we gave the rational representation

$$x = r\frac{1 - t^2}{1 + t^2}, \quad y = r\frac{2t}{1 + t^2}, \quad \infty < t < \infty, \tag{9.3.5}$$

and the transcendental representation

$$x = r \cos \theta, \quad y = r \sin \theta, 0 \leqq \theta < 2\pi. \tag{9.3.6}$$

If here we replace the first "r" by an "a" and the second one by a "b" in both formulas, we obtain representations for the ellipse of semiaxes a and b. The point $(-r, 0)$ plays a somewhat exceptional role in the first representation: it is obtained as the limit of (x, y) when $t \to \infty$.

For the hyperbola we have three representations, a rational one

$$x = a\frac{t^2 + 1}{t^2 - 1}, \quad y = b\frac{2t}{t^2 - 1}, \quad t \neq -1, +1, \infty, \tag{9.3.7}$$

and two transcendental representations

$$x = a \sec \theta, \quad y = b \tan \theta, \quad 0 \leqq \theta < 2\pi, \quad \theta \neq \frac{\pi}{2}, \quad -\frac{\pi}{2}, \tag{9.3.8}$$

$$x = a \cosh s, \quad y = b \sinh s, \quad -\infty < s < \infty. \tag{9.3.9}$$

The finite exceptional values give the infinite points. If $t \to \infty$, then $(x, y) \to (a, 0)$. The third representation gives only the right branch; we get a representation of the left branch by inserting minus signs in front of a and b.

Besides the conics we have given parametric representations for the folium of Descartes

$$x = a\frac{3t}{1 + t^3}, \quad y = a\frac{3t^2}{1 + t^3}, \quad t \neq -1, \infty, \tag{9.3.10}$$

and the lemniscate

$$x = a\frac{1 - t^4}{(1 + t^2)^2 + 4t^2}, \quad y = a\frac{2t(1 - t^2)}{(1 + t^2)^2 + 4t^2}, \quad -\infty < t < \infty. \tag{9.3.11}$$

For the folium we note that the double point $(0, 0)$ corresponds to $t = 0$ and $t \to \infty$. The excluded value $t = -1$ gives the infinite ends of the curve. For the lemniscate the double point at $(0, 0)$ corresponds to $t = -1, +1$. The point $(-a, 0)$ is the limit of (x, y) when $t \to \infty$. To these representations we add that of the kappa curve of the preceding section

$$x = a\frac{(1-t^2)^2}{2t(1+t^2)}, \quad y = a\frac{1-t^2}{1+t^2}, \quad t \neq 0, \infty. \tag{9.3.12}$$

Here the excluded values correspond to the infinite ends.

A curve which admits of a parametric representation in terms of rational functions is said to be rational. Algebraic geometers also use the term *unicursal.* Conics are rational. The folium of Descartes, the kappa curve, and the lemniscate are rational. A cubic is rational if and only if it has a singular point, node or cusp. A quartic is rational if it has three singular points, the maximum number, and so on. We make no attempt to prove this. We merely observe that the curves of order three and four for which rational representations were found above have the maximum number of permissible singularities provided we take into account imaginary singularities at infinity in the case of the two quartics.

We leave such questions and take up the problem of studying curves defined by parametric equations. We recall that the functions $x(t)$ and $y(t)$ are supposed to be continuous together with their first two derivatives.

Our first question concerns slopes and tangents. Suppose that (x_0, y_0) is a point on our curve and that

$$x_0 = x(t_0), \quad y_0 = y(t_0).$$

If $x'(t_0) \neq 0$, Cauchy's mean value theorem gives

$$\frac{y-y_0}{x-x_0} = \frac{y(t)-y(t_0)}{x(t)-x(t_0)} = \frac{y'(t+\theta(t-t_0))}{x'(t+\theta(t-t_0))}, \quad 0 < \theta < 1.$$

For small values of $|t - t_0|$ the denominators are different from zero, since $x'(t)$ is continuous and $x'(t_0) \neq 0$. It follows that the limit of the difference quotient is

$$\frac{y'(t_0)}{x'(t_0)}.$$

Hence we have proved that

$$D_x y = \frac{D_t y}{D_t x} = \frac{y'(t)}{x'(t)}, \tag{9.3.13}$$

wherever $x'(t) \neq 0$.

This means that the tangent at (x_0, y_0) can be written

$$y'(t_0)[x - x(t_0)] - x'(t_0)[y - y(t_0)] = 0. \tag{9.3.14}$$

In this form the equation is also valid at the points where $x'(t_0) = 0$ but $y'(t_0) \neq 0$. Such a tangent is vertical and its equation is

$$x = x(t_0).$$

Similarly the normal to the curve at (x_0, y_0) becomes

$$x'(t_0)[x - x(t_0)] + y'(t_0)[y - y(t_0)] = 0. \tag{9.3.15}$$

We can also find the line segments introduced in Section 8.1 under the names of "tangent," "subtangent," "normal," and "subnormal." It suffices to introduce the new expression for the slope in the formulas of Section 8.1. After simplification we obtain the following expressions

$$\text{tangent } TP = \frac{y(t_0)}{y'(t_0)} \{[x'(t_0)]^2 + [y'(t_0)]^2\}^{\frac{1}{2}}, \tag{9.3.16}$$

$$\text{subtangent } TX = \frac{y(t_0)x'(t_0)}{y'(t_0)}, \tag{9.3.17}$$

$$\text{normal } PN = \frac{y(t_0)}{x'(t_0)} \{[x'(t_0)]^2 + [y'(t_0)]^2\}^{\frac{1}{2}}, \tag{9.3.18}$$

$$\text{subnormal } XN = \frac{y(t_0)y'(t_0)}{x'(t_0)}. \tag{9.3.19}$$

Next we examine for points of inflection. Here we need second-order derivatives. Since

$$D_x^2y = D_x[D_xy] = D_x\left\{\frac{D_ty}{D_tx}\right\} = D_t\left\{\frac{D_ty}{D_tx}\right\}\frac{1}{D_tx},$$

we obtain

$$D_x^2y = \frac{D_tx\, D_t^2y - D_ty\, D_t^2x}{[D_tx]^3} = \frac{x'(t)y''(t) - y'(t)x''(t)}{[x'(t)]^3}, \tag{9.3.20}$$

which is the desired representation.

As $P = (x, y)$ moves along the curve C, the curve tangent keeps turning. See Figure 9.13. Suppose that the tangent makes the angle $\varphi = \varphi(P)$ with the positive horizontal direction. We are interested to know how φ changes with t.

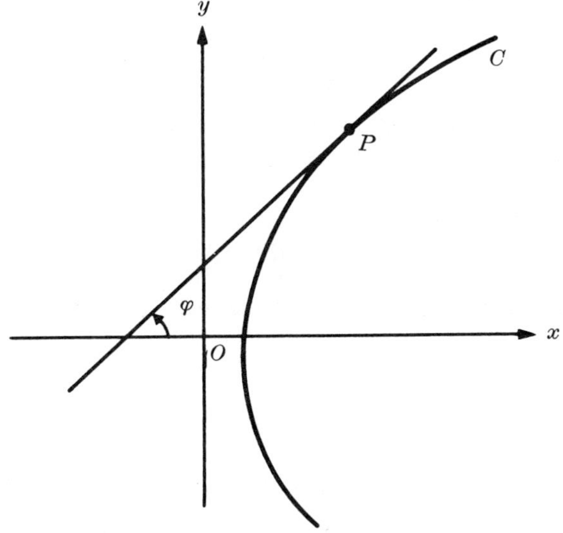

Figure 9.13

We have just seen that

$$D_t[D_x y] = \frac{x'(t)y''(t) - y'(t)x''(t)}{[x'(t)]^2}. \tag{9.3.21}$$

Since φ can differ from the arctangent of $D_x y$ only by a multiple of π, the two functions must have the same derivative with respect to t

$$D_t\varphi = D_t \arctan (D_x y) = \frac{D_t(D_x y)}{1 + (D_x y)^2},$$

and this simplifies to

$$D_t\varphi = \frac{x'(t)y''(t) - y'(t)x''(t)}{[x'(t)]^2 + [y'(t)]^2}. \tag{9.3.22}$$

It follows that

$$\operatorname{sgn} (D_t\varphi) = \operatorname{sgn} [x'(t)y''(t) - y'(t)x''(t)],$$

that is, φ is an increasing or a decreasing function of t according as

$$x'(t)y''(t) - y'(t)x''(t) \tag{9.3.23}$$

is positive or negative and the points of inflection of the curve correspond to those values of t for which (9.3.23) changes its sign.

The same expression naturally determines the sign of the curvature. For the center (α, β) and the radius ρ of the circle of curvature we find the expressions

$$\alpha = x(t_0) - y'(t_0)\frac{[x'(t_0)]^2 + [y'(t_0)]^2}{x'(t_0)y''(t_0) - y'(t_0)x''(t_0)}, \tag{9.3.24}$$

$$\beta = y(t_0) + x'(t_0)\frac{[x'(t_0)]^2 + [y'(t_0)]^2}{x'(t_0)y''(t_0) - y'(t_0)x''(t_0)}, \tag{9.3.25}$$

$$\rho = \pm\frac{\{[x'(t_0)]^2 + [y'(t_0)]^2\}^{\frac{3}{2}}}{x'(t_0)y''(t_0) - y'(t_0)x''(t_0)}. \tag{9.3.26}$$

For the proof we go back to formulas (8.4.10)–(8.4.12) and replace $D_x y$ and $D_x{}^2 y$ by the expressions (9.3.13) and (9.3.20).

Example 1. We take

$$x(t) = t^2, \quad y(t) = t^4 - t^5. \tag{9.3.27}$$

This is a parametric representation of the lower branch of the curve defined by (9.2.12). Here

$$x'(t) = 2t, \quad y'(t) = 4t^3 - 5t^4,$$

$$x''(t) = 2, \quad y''(t) = 12t^2 - 20t^3,$$

$$x'(t)y''(t) - y'(t)x''(t) = 2t^3(8 - 15t).$$

It follows that the maximum on the curve corresponds to $t = \frac{4}{5}$, the point of inflection to $t = \frac{8}{15}$. The singular point at the origin corresponds to $t = 0$ and is an absolute minimum for x and a relative minimum for y. The radius of curvature is

$$\frac{\{4 + t^4(4 - 5t)^2\}^{\frac{3}{2}}}{16 - 30t}$$

and this reduces to $\frac{1}{2}$ for $t = 0$.

We could simplify the formulas for the circle of curvature with the aid of (9.3.22). To complete this we need an expression for the denominator in (9.3.22) and we shall see in the next section that we are actually dealing with $[D_t s]^2$ where s is the arc length.

The rest of this section will be devoted to curves in polar coordinates. The basic relations are

$$x = r \cos \theta, \quad y = r \sin \theta. \tag{9.3.28}$$

See Section A.6. These formulas enable us to pass from an equation in Cartesian coordinates to one in polar coordinates.

It is sometimes worthwhile making this transformation for the purpose of finding if the curve is bounded or not. Thus if we did not know the result in advance, we could tell from the equations

$$r^2(b^2 \cos^2 \theta + a^2 \sin^2 \theta) = a^2b^2, \tag{9.3.29}$$

$$r^2(b^2 \cos^2 \theta - a^2 \sin^2 \theta) = a^2b^2, \tag{9.3.30}$$

that the ellipse is bounded, $b \leqq r \leqq a$, and the hyperbola is not.

The problem does not present itself for cubics, for every such curve is unbounded. If its equation is of degree 3 in y, then the assertion follows from the fact that an algebraic equation of degree 3 and real coefficients has at least one real root so that at least one value of y corresponds to every value of x. The same argument applies if the equation is of degree 3 in x. If the equation is at most quadratic in either variable, then a discussion of the discriminant will lead to the desired conclusion. The same result holds for any algebraic curve of odd degree.

When it comes to curves of even degree, the situation is different. Thus the kappa curve is unbounded, for its equation in polar coordinates is

$$r^2 = a^2 \cot^2 \theta, \tag{9.3.31}$$

and the cotangent is unbounded. On the other hand the lemniscate has the equation

$$r^2 = a^2 \cos 2\theta, \tag{9.3.32}$$

so r can never exceed a. The equation also shows that the curve is confined to the angles

$$|\theta| < \frac{\pi}{4}, \quad |\theta - \pi| < \frac{\pi}{4}.$$

We return to formula (9.3.28) where we consider r as a function of θ so that the equation of the curve under consideration is

$$r = r(\theta). \tag{9.3.33}$$

We can then consider (9.3.28) as a parametric representation of the curve with θ as the parameter. Denoting derivatives with respect to θ by primes we get

$$\begin{aligned} x' &= r'\cos\theta - r\sin\theta, \quad y' = r'\sin\theta + r\cos\theta, \\ x'' &= r''\cos\theta - 2r'\sin\theta - r\cos\theta, \quad y'' = r''\sin\theta + 2r'\cos\theta - r\sin\theta, \end{aligned} \tag{9.3.34}$$

and hence

$$\begin{aligned} (x')^2 + (y')^2 &= r^2 + (r')^2, \\ x'y'' - y'x'' &= r^2 + 2(r')^2 - rr''. \end{aligned} \tag{9.3.35}$$

This gives

$$\begin{aligned} D_x y &= \frac{y'}{x'} = \frac{r'\sin\theta + r\cos\theta}{r'\cos\theta - r\sin\theta}, \\ D_x{}^2 y &= \frac{x'y'' - y'x''}{(x')^3} = \frac{r^2 + 2(r')^2 - rr''}{(r'\cos\theta - r\sin\theta)^3}. \end{aligned} \tag{9.3.36}$$

We also get the radius of curvature

$$\rho = \pm\frac{[r^2 + (r')^2]^{\frac{3}{2}}}{r^2 + 2(r')^2 - rr''}. \tag{9.3.37}$$

To illustrate the use of these formulas we consider some special cases.

Example 2. We take the so-called *logarithmic spiral*

$$r = e^{\theta}. \tag{9.3.38}$$

Here r is a positive strictly increasing function of θ which tends to $+\infty$ with θ and to 0 when $\theta \to -\infty$. We have

$$D_x y = \frac{\cos\theta + \sin\theta}{\cos\theta - \sin\theta} = \frac{1 + \tan\theta}{1 - \tan\theta} = \tan\left(\theta + \frac{\pi}{4}\right).$$

Since $D_x y = \tan\varphi$, where φ is the angle that the curve tangent makes with the positive x-axis ($=$ the polar axis), the curve tangent to the logarithmic spiral makes the constant angle $\pi/4$ with the radius vector. Computation shows that $D_t(D_x y) > 0$ for all θ that was to be expected. Finally

$$\rho = \sqrt{2}e^{\theta} = \sqrt{2}r.$$

This means that the center of curvature (α, β), corresponding to the point $[r, \theta]$, has polar coordinates $\left[r, \theta + \frac{\pi}{2}\right]$ and describes a congruent spiral.

Example 3. We take the lemniscate (9.3.32). Here

$$rr' = -a^2\sin 2\theta \quad \text{so that} \quad \frac{r'}{r} = -\tan 2\theta.$$

It follows that

$$D_x y = \frac{r'}{r}\,\frac{\sin\theta + r\cos\theta}{\cos\theta - r\sin\theta} = \frac{1 + \dfrac{r'}{r}\tan\theta}{\dfrac{r'}{r} - \tan\theta}$$

$$= \frac{1 - \tan 2\theta \tan\theta}{-\tan 2\theta - \tan\theta} = -\cot 3\theta.$$

This says that

$$\tan\left(\varphi - \frac{\pi}{2}\right) = \tan 3\theta.$$

But $\varphi - \dfrac{\pi}{2}$ is the angle which the normal makes with the axis. Hence this angle is three times the polar angle. From

$$\frac{r'}{r} = -\tan 2\theta$$

we get first

$$r^2 + (r')^2 = r^2(1 + \tan^2 2\theta) = r^2 \sec^2 2\theta$$

$$= a^2 \sec 2\theta = \frac{a^4}{r^2}.$$

Secondly, by differentiation,

$$D_\theta\left(\frac{r'}{r}\right) = -2\sec^2 2\theta = -2\,\frac{a^4}{r^4},$$

or

$$\frac{r''}{r} - \left(\frac{r'}{r}\right)^2 = -2\,\frac{a^4}{r^4}.$$

This gives

$$r^2 + 2(r')^2 - rr'' = r^2 + (r')^2 - r^2\left\{\frac{r''}{r} - \left(\frac{r'}{r}\right)^2\right\}$$

$$= 3\,\frac{a^4}{r^2}.$$

Hence

$$\rho = \frac{(a^4 r^{-2})^{\frac{3}{2}}}{3a^4 r^{-2}} = \frac{a^2}{3r}.$$

Thus ρ has its least value $\frac{1}{3}a$ for $r = a$ and is strictly increasing to $+\infty$ as $r \to 0$.

EXERCISE 9.3

Find $D_x y$ and $D_x{}^2 y$ if x and y are given parametrically by:

1. $x = t^2, \quad y = t^3.$

2. $x = 1 + \dfrac{1}{t}, \quad y = t^2.$

3. $x = \dfrac{1 - t^2}{1 + t^2}, \quad y = \dfrac{2t}{1 + t^2}.$ 4. $x = \dfrac{3at}{1 + t^3}, \quad y = \dfrac{3at^2}{1 + t^3}.$

5. $x = a \cos^3 t, \quad y = a \sin^3 t.$ 6. $\begin{cases} x = b(n \cos t - \cos nt). \\ y = b(n \sin t - \sin nt). \end{cases}$

7. $x = \dfrac{t^2}{t - 1}, \quad y = \dfrac{t}{t^2 - 1}.$

8. Show that the radius of curvature of the astroid in No. 5 is $3(axy)^{\frac{1}{3}}$. Show that the centers of curvature form another astroid where a is replaced by $2a$ and t by $t + \dfrac{\pi}{4}$.

9. Show that the radius of curvature in No. 3 equals 1.

10. Show that the two branches through the origin of the folium in No. 4 have the same radius of curvature there and compute this value.

*11. Find the asymptotes and sketch the curve in No. 7.

12. Find the maximum distance from the origin of the points on the curve in No. 6, if $n = 4$.

13. Show that the curve $x^4 - y^4 = 4x^2y$ is unbounded. Find the asymptotes and sketch the curve.

14. If a cubic has the equation

$$xy^2 = Ax^2 + Bxy + Cy^2 + Dx + Ey + F,$$

prove that the curve is unbounded.

15. Show that the quartic

$$(x^2 + xy + y^2)^2 = ay^3 \quad (a > 0)$$

is bounded and lies in the upper half-plane except for the origin.

16. Show that the angle between the curve tangent and the radius vector is given by

$$\tan \psi = \frac{r}{r'}.$$

17. Given a curve in polar coordinates $r = r(\theta)$. Let OP be the radius vector and draw a line l through O perpendicular to OP. Consider the curve tangent and curve normal at P and let their intersections with l be T and N respectively. The line segments PT, PN, TO, and ON are known as the (polar) *tangent*, *normal*, *subtangent*, and *subnormal*, respectively. Show that

$$OT = \frac{r^2}{r'}, \quad ON = r', \quad PT = \frac{r}{r'}[r^2 + (r')^2]^{\frac{1}{2}}, \quad PN = [r^2 + (r')^2]^{\frac{1}{2}}.$$

18. The spiral of Archimedes has the equation $r = a\theta$. Find OT, ON, PT, and PN.

19. For the cardioid

$$r = 2a(1 - \cos \theta)$$

prove that $\psi = \frac{1}{2}\theta$ and that

$$PN = r \operatorname{cosec} \psi, \quad PT = r \sec \psi, \quad OT = r \tan \psi, \quad ON = 2a \sin \theta.$$

20. Find the radius of curvature of the cardioid.

9.4. Arc length and curvature. In discussing curves we have taken for granted that finite arcs have a length and only for the circle has length been defined. We shall discuss such questions here for the case of curves given by parametric equations involving functions $x(t)$ and $y(t)$ with continuous derivatives. The general case will be discussed in Appendix C.

We proceed as in Section 7.1. Suppose that an arc C is defined by

$$x = x(t), \quad y = y(t), \quad 0 \leqq t \leqq 1, \tag{9.4.1}$$

where, to start with, $x(t)$ and $y(t)$ are merely assumed to be continuous. The parametric representation defines an order on C where

$$P' = [x(t'), y(t')] \quad \text{precedes} \quad P'' = [x(t''), y(t'')]$$

if $t' < t''$. Take a partition π of the interval $[0, 1]$ by points

$$0 = t_0 < t_1 < t_2 < \cdots < t_n = 1,$$

and let the corresponding points on C be

$$P_0, P_1, P_2, \cdots, P_n \quad \text{where} \quad P_k = [x(t_k), y(t_k)].$$

These points form an ordered set on C. Consider the line segments joining consecutive points and form

$$S_\pi = \sum_{k=0}^{n-1} l[P_k, P_{k+1}], \tag{9.4.2}$$

the length of the "inscribed" polygonal line corresponding to the partition π. It is a positive number. Consider the set L of all such numbers S_π corresponding to all possible partitions π.

Definition 9.4.1. *The arc C is said to be rectifiable if and only if the set L is bounded and, if so, the least upper bound of L is called the length of C.*

If the functions $x(t)$ and $y(t)$ are merely continuous, C is normally not rectifiable. We get rectifiable arcs by imposing stronger restrictions on the functions $x(t)$ and $y(t)$. We shall show that the existence of continuous derivatives ensures rectifiability.

Theorem 9.4.1. *If $x'(t)$ and $y'(t)$ are continuous in $[0, 1]$, then C is rectifiable and its length is given by*

$$l(C) \equiv \int_0^1 \{[x'(t)]^2 + [y'(t)]^2\}^{\frac{1}{2}}\, dt. \tag{9.4.3}$$

Proof. For any partition π we have

$$S_\pi = \sum_{k=0}^{n-1} \{[x(t_{k+1}) - x(t_k)]^2 + [y(t_{k+1}) - y(t_k)]^2\}^{\frac{1}{2}}. \tag{9.4.4}$$

Now by the mean value theorem

$$x(t_{k+1}) - x(t_k) = x'(\sigma_k)(t_{k+1} - t_k),$$
$$y(t_{k+1}) - y(t_k) = y'(\tau_k)(t_{k+1} - t_k),$$

where $t_k < \sigma_k, \tau_k < t_{k+1}$. Hence

$$S_\pi = \sum_{k=0}^{n-1} \{[x'(\sigma_k)]^2 + [y'(\tau_k)]^2\}^{\frac{1}{2}} (t_{k+1} - t_k). \tag{9.4.5}$$

Since $x'(t)$ and $y'(t)$ are continuous in the closed interval $[0, 1]$, they are bounded functions. Suppose that

$$|x'(t)| \leqq B, \quad |y'(t)| \leqq B$$

for all t in $[0, 1]$. Then for each k

$$[x'(\sigma_k)]^2 + [y'(\tau_k)]^2 \leqq 2B^2,$$

so that

$$S_\pi \leqq \sqrt{2}B \sum_{k=0}^{n-1} (t_{k+1} - t_k) = \sqrt{2}B.$$

Hence $L = \{S_\pi\}$ is a bounded set and its least upper bound does not exceed $\sqrt{2}B$. It follows that C is rectifiable and that its length cannot exceed $\sqrt{2}B$. Actually this limit is the best of its kind for if

$$x(t) = Bt, \quad y(t) = Bt,$$

then C is a line segment joining the origin with the point (B, B) and $l(C) = \sqrt{2}B$.

To prove (9.4.3) we make four observations. First, the functions $x'(t)$ and $y'(t)$ are continuous in the closed interval $[0, 1]$ and hence uniformly continuous (see Section 2.6). It follows that, if $\varepsilon > 0$ is prescribed, we can find a δ such that in any subinterval of $[0, 1]$ of length not exceeding δ the oscillations of $x'(t)$ and of $y'(t)$ do not exceed ε. We use this to estimate the error made in S_π by replacing σ_k and τ_k by t_k when the partition π is so fine that no subinterval exceeds δ in length. Thus we are interested in

$$D \equiv \sum_{k=0}^{n-1} |\{[x'(\sigma_k)]^2 + [y'(\tau_k)]^2\}^{\frac{1}{2}} - \{[x'(t_k)]^2 + [y'(t_k)]^2\}^{\frac{1}{2}}| (t_{k+1} - t_k).$$

Here we make our second observation. We have

$$|(x_1^2 + y_1^2)^{\frac{1}{2}} - (x_2^2 + y_2^2)^{\frac{1}{2}}| \leqq \{(x_1 - x_2)^2 + (y_1 - y_2)^2\}^{\frac{1}{2}}, \tag{9.4.6}$$

for if $P_1 = (x_1, y_1)$, $P_2 = (x_2, y_2)$, then this is the triangle inequality applied to OP_1P_2:

$$|d(O, P_1) - d(O, P_2)| \leqq d(P_1, P_2).$$

Now in the expression for D for each k

$$|x'(\sigma_k) - x'(t_k)| \leqq \varepsilon, \quad |y'(\tau_k) - y'(t_k)| \leqq \varepsilon$$

since $\sigma_k - t_k < \delta$, $\tau_k - t_k < \delta$. Hence by (9.4.6)

$$D \leqq \sqrt{2}\varepsilon \sum_{k=0}^{n-1} (t_{k+1} - t_k) = \sqrt{2}\varepsilon.$$

If the partition π is so fine that the maximum length of the subintervals does not exceed δ, then S_π differs from

$$\sum_{k=0}^{n-0} \{[x'(t_k)]^2 + [y'(t_k)]^2\}^{\frac{1}{2}} (t_{k+1} - t_k) \tag{9.4.7}$$

by an amount not exceeding $\sqrt{2}\varepsilon$ in absolute value.

Our third observation is that the integral in (9.4.3) exists since the integrand is continuous and by the definition of the definite integral the two numbers in (9.4.3) and in (9.4.7) differ arbitrarily little as soon as δ is sufficiently small. From this it follows that all sums S_π corresponding to fine partitions are close to each other and to the value of the integral.

To clinch the argument we observe finally that if we start out with a partition π and subdivide it by intercalating points to obtain a partition π^*, then

$$S_\pi \leqq S_{\pi^*}.$$

For if the interval $[t_k, t_{k+1}]$ corresponding to the points P_k, P_{k+1} is subdivided and Q is the intercalated point on the arc between P_k and P_{k+1}, then

$$l[P_k, P_{k+1}] \leqq l[P_k, Q] + l[Q, P_{k+1}].$$

From this we conclude that no sum S_π can exceed the value of the integral. If such a sum should exist, then any sum obtained from it by refining the partition must have the same property. This is impossible since all sums corresponding to fine partitions lie close to the value of the integral. It follows that the integral is the least upper bound of all sums S_π and that the length of the arc C is given by (9.4.3). This completes the proof.

It is obvious that if the arc C is given by

$$x = x(t), \quad y = y(t), \quad a \leqq t \leqq b, \tag{9.4.8}$$

then (9.4.3) is to be replaced by

$$l(C) = \int_a^b \{[x'(t)]^2 + [y'(t)]^2\}^{\frac{1}{2}} \, dt. \tag{9.4.9}$$

Here it is tacitly assumed that $b - a$ is finite. If the interval is infinite, the integral may not exist as a finite number. This requires a special investigation. Some instances will be encountered below.

If, in particular, C is given by

$$y = f(x), \quad a \leqq x \leqq b, \tag{9.4.10}$$

where $f'(x)$ exists and is continuous and the interval is finite, then we can take $x = t$, $y = f(t)$ in (9.4.8) and obtain

$$l(C) = \int_a^b \{1 + [f'(x)]^2\}^{\frac{1}{2}} \, dx. \tag{9.4.11}$$

The assumption that $x'(t)$ and $y'(t)$ are continuous in $[a, b]$ is unnecessarily restrictive. It suffices that the interval breaks up into a finite number of subintervals such that $x'(t)$ and $y'(t)$ are continuous and bounded in each of the open subintervals.

If C is given in polar coordinates by

$$r = r(\theta), \quad \theta_1 \leqq \theta \leqq \theta_2, \tag{9.4.12}$$

we can take θ as the parameter in (9.4.8) so that

$$x = r(\theta)\cos\theta, \quad y = r(\theta)\sin\theta.$$

By formula (9.3.25)

$$[x'(\theta)]^2 + [y'(\theta)]^2 = [r(\theta)]^2 + [r'(\theta)]^2,$$

so that

$$l(C) = \int_{\theta_1}^{\theta_2} \{[r(\theta)]^2 + [r'(\theta)]^2\}^{\frac{1}{2}}\, d\theta. \tag{9.4.13}$$

We shall illustrate these formulas by some examples. We note that there are comparatively few curves which lead to integrals that can be handled by calculus. We shall give some examples where the work can be carried through and also instances where integrals are obtained which are not elementary functions.

Example. 1. We take the parabola

$$y^2 = 4ax \quad \text{or} \quad x = at^2,\ y = 2at.$$

We restrict t to the interval $[0, T]$. The length of the corresponding arc is then

$$\int_0^T \{(2at)^2 + (2a)^2\}^{\frac{1}{2}}\, dt = 2a\int_0^T (1 + t^2)^{\frac{1}{2}}\, dt, \tag{9.4.14}$$

and this gives

$$S(T) = aT(1 + T^2)^{\frac{1}{2}} + a\log[T + (T^2 + 1)^{\frac{1}{2}}]. \tag{9.4.15}$$

Compare Section 7.3 where similar integrals are discussed.

Example 2. For the ellipse

$$b^2x^2 + a^2y^2 = a^2b^2 \quad \text{or} \quad x = a\cos t,\ y = b\sin t$$

we find that the length of the curve is

$$4\int_0^{\pi/2} (a^2\sin^2 t + b^2\cos^2 t)^{\frac{1}{2}}\, dt. \tag{9.4.16}$$

This is called an *elliptic integral* and cannot be expressed in finite closed form. The integral can be evaluated by expanding the integrand in a suitable series and integrating term by term. The calculation is postponed until Chapter 11, Vol. II, but we list the result:

$$4aF\left(\tfrac{1}{2}, -\tfrac{1}{2}, 1; \frac{a^2 - b^2}{a^2}\right), \tag{9.4.17}$$

where, as usual, the symbol F stands for a hypergeometric series. See formula (5.2.22).

Example 3. The hyperbola

$$b^2x^2 - a^2y^2 = a^2b^2 \quad \text{or} \quad x = a \cosh t,\ y = b \sinh t$$

also leads to an elliptic integral. The arc corresponding to the t-interval $(0, T)$ has the length

$$s(T) = \int_0^T (a^2 \sinh^2 t + b^2 \cosh^2 t)^{\frac{1}{2}}\, dt. \tag{9.4.18}$$

It becomes infinite with T.

Example 4. The astroid

$$x = a \cos^3 t, \quad y = a \sin^3 t$$

leads to simple results. Here

$$[x'(t)]^2 + [y'(t)]^2 = 9a^2 \sin^2 t \cos^2 t,$$

that is, a perfect square. If $0 < T \leqq \frac{\pi}{2}$,

$$s(T) = 3a \int_0^T \sin t \cos t\, dt = \tfrac{3}{2} a \sin^2 T. \tag{9.4.19}$$

Here $T = \frac{\pi}{2}$ gives one-fourth of the total length. Thus the total length of the astroid is $6a$, a little less than the circumference of the circumscribed circle. Formula (9.4.19) is not valid for $\frac{\pi}{2} < T$. A correct general formula is obtainable by inserting absolute values in the integrand to ensure that we get the positive square root.

Example 5. We take finally the logarithmic spiral

$$r = ae^{c\theta}, \quad a > 0, \quad c > 0.$$

Here

$$r^2 + (r')^2 = a^2(1 + c^2)e^{2c\theta}$$

so that the length of arc from $\theta = 0$ to $\theta = \omega$ is

$$s(\omega) = \frac{a}{c}(1 + c^2)^{\frac{1}{2}}\,(e^{c\omega} - 1). \tag{9.4.20}$$

This becomes infinite when $\omega \to +\infty$. On the other hand, if $\omega \to -\infty$, $s(\omega)$ tends to a finite limit. This means that the end of the spiral, which winds around the origin as an asymptotic point, is of finite length

$$\frac{a}{c}(1 + c^2)^{\frac{1}{2}},$$

if we count the arc from $\theta = 0$.

Suppose now that the arc C is given by equation (9.4.10). For $a \leqq t \leqq b$ the integral

$$\int_a^t \{[x'(u)]^2 + [y'(u)]^2\}^{\frac{1}{2}}\, du \equiv s(t) \tag{9.4.21}$$

gives the length of the subarc of C defined by

$$x = x(u), \quad y = y(u), \quad a \leqq u \leqq t.$$

This function $s(t)$ is strictly increasing with t and since the integrand is continuous, $s(t)$ is also a differentiable function of t and

$$s'(t) = \{[x'(t)]^2 + [y'(t)]^2\}^{\frac{1}{2}}, \tag{9.4.22}$$

or

$$[s'(t)]^2 = [x'(t)]^2 + [y'(t)]^2. \tag{9.4.23}$$

We encountered the right member of this expression in the preceding section. We can now associate it with the arc length and reformulate our previous results. We start with formulas (9.3.22) and (9.3.26) and refer to Figure 9.13. Let $\varphi = \varphi(t)$ be the angle which the tangent makes with the x-axis at the point $[x(t), y(t)]$. We have seen that

$$\varphi'(t) = \frac{x'(t)y''(t) - y'(t)x''(t)}{[x'(t)]^2 + [y'(t)]^2}, \tag{9.4.24}$$

while the curvature, the reciprocal of the radius of curvature $\rho(t)$, is given by

$$K(t) = \frac{x'(t)y''(t) - y'(t)x''(t)}{\{[x'(t)]^2 + [y'(t)]^2\}^{\frac{3}{2}}}. \tag{9.4.25}$$

Combining these formulas with (9.4.23) we obtain the following expressions for the curvature and its reciprocal

$$K(t) = \frac{\varphi'(t)}{s'(t)} = \frac{d\varphi}{ds}, \tag{9.4.26}$$

$$\rho(t) = \frac{ds}{d\varphi}, \tag{9.4.27}$$

where we imagine the arc length expressed as a function of the inclination φ of the curve tangent.

For some purposes the arc length s is the most advantageous parameter for the representation of a rectifiable curve. This follows from (9.4.23) which shows that if t is replaced by s, then

$$[x'(s)]^2 + [y'(s)]^2 \equiv 1, \tag{9.4.28}$$

while

$$\varphi'(s) = K(s) = x'(s)y''(s) - y'(s)x''(s). \tag{9.4.29}$$

For the center of curvature we also obtain the simple formulas

$$\alpha = x(s) - \frac{dy}{d\varphi}, \qquad \beta = y(s) + \frac{dx}{d\varphi}. \tag{9.4.30}$$

In this connection a general remark concerning the representation of curves by equations is in order. The use of Cartesian coordinates for the study of geometrically defined curves was an enormous step forward and led to a development of geometry far beyond the dreams of the Greek masters. Nevertheless it may be said that Cartesian coordinates are artificial and the corresponding equations are not invariant under rigid motions of the plane. This leads to the question of what constitutes *natural coordinates* and *natural equations* for a curve. *The relation holding between the radius of curvature and the arc length is such an equation*. It is independent of the orientation of the curve.

A "natural geometry" based upon the use of s and ρ as coordinates was developed at great length by the Italian mathematician Ernesto Cesàro (1859–1906). Do not expect that what we regard as simple curves will also have simple equations in natural coordinates or vice versa. To illustrate the concepts we shall derive the natural equations of the logarithmic spiral and of the astroid. The first is linear, the second quadratic.

For the logarithmic spiral

$$r = ae^{c\theta}$$

we found above that

$$s(\omega) = \frac{a}{c}(1 + c^2)^{\frac{1}{2}}(e^{c\omega} - 1).$$

A simple computation gives

$$\rho(\omega) = a(1 + c^2)^{\frac{1}{2}} e^{c\omega},$$

so that elimination of the term $e^{c\omega}$ gives

$$\rho = cs + a(1 + c^2)^{\frac{1}{2}}. \tag{9.4.31}$$

We can suppress the additive term $a(1 + c^2)^{\frac{1}{2}}$. This amounts to counting arc length from the origin instead of from $[0, 1]$.

Similarly, for the astroid we get the arc length from (9.4.19) and ρ from Problem 6, Exercise 9.3. We simplify the equation by counting arc length from the point with $t = \dfrac{\pi}{4}$ rather than from $t = 0$. For $0 < T < \dfrac{\pi}{2}$ we have then

$$\rho(T) = 3\,a \sin T \cos T,$$

$$s(T) = \tfrac{3}{4}a(2 \sin^2 T - 1).$$

This may be rewritten

$$\rho(T) = \tfrac{3}{2}a \sin 2T,\ s(T) = -\tfrac{3}{4}a \cos 2T.$$

Squaring and adding, we get

$$\tfrac{1}{4}\rho^2 + s^2 = \tfrac{9}{16}a^2. \tag{9.4.32}$$

EXERCISE 9.4

Find the lengths of the arcs defined in Problems 1–10.

1. Parabola, $y = \sqrt{x}$, $0 \leqq x \leqq 1$.
2. Semicubical parabola, $y^2 = x^3$, $0 \leqq x \leqq 44$.
3. $y = \frac{1}{3}(x - 3a)\left(\dfrac{x}{a}\right)^{\frac{1}{2}}$, $0 \leqq x \leqq \omega < \infty$.
4. Catenary, $y = \cosh x$, $0 \leqq x \leqq \omega$.

5. $x = a\cos^5 t$, $y = a\sin^5 t$, $0 \leqq t \leqq \dfrac{\pi}{2}$. In the resulting integral substitute successively $\sin^2 t = u$, $u = \frac{1}{2} + \frac{1}{6}\sqrt{3}s$.

6. $y = \log x$, $\sqrt{3} \leqq x \leqq \sqrt{8}$. In the resulting integral set $x = 1/s$ and integrate by parts.

7. $y = 1 - \log\cos x$, $0 \leqq x \leqq \dfrac{\pi}{4}$. In the resulting integral substitute successively $x = 2t$, $\tan t = s$.

8. Cardioid, $r = a(1 - \cos\theta)$, $0 \leqq \theta \leqq 2\pi$.

9. Spiral of Archimedes, $r = c\theta$, $0 \leqq \theta \leqq \omega < \infty$.

10. Hyperbolic spiral $r\theta = c$, $c \leqq \theta \leqq \omega$.

11. Show that the radius of curvature of the cardioid is $\rho = \frac{4}{3}a\sin\frac{1}{2}\theta$ and that its natural equation is

$$9\rho^2 + s^2 = 16a^2,$$

where s is counted from the point where $\theta = \pi$.

12. Show that the natural equation of the semicubical parabola may be written

$$18\rho = (27s + 8)\{(27s + 8)^{\frac{2}{3}} - 4\}^{\frac{1}{2}},$$

where s is counted from the cusp.

13. A curve C: $z = z(t)$, $0 \leqq t \leqq 1$, is traced in the complex plane. If $z(t)$ is a differentiable function of t, express $l(C)$ as an integral.

14. If $s(T)$ is given by (9.4.18), prove that

$$s(T) - (a^2 + b^2)^{\frac{1}{2}}\sinh T$$

tends to a finite limit as $T \to \infty$. [*Hint:* Express the difference as an integral.]

9.5. Evolute and involute. We have already encountered the concept of an *evolute* in Section 8.1. Such a curve occurred in the discussion of the number of normals that can be drawn from a point to a conic. We found that the plane separates into two domains. From a point in one of these four normals could be drawn to the conic (three in the case of a parabola) while from a point in the complementary domain only two normals (one for the parabola) could be drawn. The two domains are separated by a curve which we called the evolute of the conic, and in Section 8.4 the same curve figured as the locus of center of curvature. The latter is the more natural characterization.

DEFINITION 9.5.1. *If C is defined by*

$$x = x(t), \quad y = y(t), \quad a \leqq t \leqq b, \tag{9.5.1}$$

where $x(t)$ and $y(t)$ have continuous first- and second-order derivatives, then the evolute E of C is the locus of the center of curvature and is defined parametrically by

$$\begin{aligned} \alpha = \alpha(t) &\equiv x(t) - y'(t)\,\frac{[x'(t)]^2 + [y'(t)]^2}{x'(t)y''(t) - x''(t)y'(t)}, \\ \beta = \beta(t) &\equiv y(t) + x'(t)\,\frac{[x'(t)]^2 + [y'(t)]^2}{x'(t)y''(t) - x''(t)y'(t)}. \end{aligned} \tag{9.5.2}$$

Conversely C is known as an involute of E.

The relation between evolute and involute is not one-to-one. If an evolute exists, it is unique, but a curve has infinitely many involutes.

THEOREM 9.5.1. *If $x(t)$ and $y(t)$ have continuous third-order derivatives, the normal of C at the point P is tangent to the evolute at the corresponding center of curvature.*

Proof. Since $x(t)$ and $y(t)$ have continuous third-order derivatives, we see from (9.5.2) that $\alpha(t)$ and $\beta(t)$ have continuous first-order derivatives at all points where

$$x'(t)y''(t) - x''(t)y'(t) \neq 0,$$

that is, at all nonsingular points where the curvature is not zero. In this connection we consider a point singular if

$$[x'(t)]^2 + [y'(t)]^2 = 0.$$

Consider now the equation of the circle of curvature at $P = [x(t), y(t)]$. It is

$$[X - \alpha(t)]^2 + [Y - \beta(t)]^2 = [\rho(t)]^2. \tag{9.5.3}$$

The equation of the tangent to the circle at P is then

$$[X - \alpha(t)]x'(t) + [Y - \beta(t)]y'(t) = 0.$$

In particular, this equation holds at the point P of tangency, so we have

$$[x(t) - \alpha(t)]x'(t) + [y(t) - \beta(t)]y'(t) = 0. \tag{9.5.4}$$

If P and hence t vary, this is a differentiable function of t which is identically zero. The same then is true for its derivative with respect to t, that is

$$\begin{aligned} &[x'(t)]^2 + [y'(t)]^2 - \alpha'(t)x'(t) - \beta'(t)y'(t) \\ &+ [x(t) - \alpha(t)]x''(t) + [y(t) - \beta(t)]y''(t) = 0. \end{aligned} \tag{9.5.5}$$

Here most terms cancel leaving only

$$\alpha'(t)x'(t) + \beta'(t)y'(t) = 0. \tag{9.5.6}$$

To see that this is indeed the case we have only to form

$$[x(t) - \alpha(t)]x''(t) + [y(t) - \beta(t)]y''(t).$$

Using (9.5.2) we see that this expression reduces to

$$-[x'(t)]^2 - [y'(t)]^2$$

and this cancels the first group of terms in (9.5.5).

Now the tangent to E at $Q = [\alpha(t), \beta(t)]$ has the slope $\beta'(t)/\alpha'(t)$ and the normal to C at $P = [x(t), y(t)]$ has the slope $-x'(t)/y'(t)$. By (9.5.6) the two lines are parallel and since they have the point Q in common, they are identical. This completes the proof.

THEOREM 9.5.2. *Under the same assumptions the length of an arc of the evolute E equals the difference between the lengths of the radii of curvature of C which are tangent to the arc at its endpoints, provided $\rho(t)$ is strictly monotone.*

Proof. We return to (9.5.3). This relation holds also at P, that is, for $X = x(t)$ and $Y = y(t)$. Hence

$$[x(t) - \alpha(t)]^2 + [y(t) - \beta(t)]^2 = [\rho(t)]^2. \tag{9.5.7}$$

This is an identity in t and all five functions involved are differentiable since $x'''(t)$ and $y'''(t)$ exist. Taking the derivative with respect to t we get

$$[x(t) - \alpha(t)][x'(t) - \alpha'(t)] + [y(t) - \beta(t)][y'(t) - \beta'(t)] = \rho(t)\rho'(t).$$

We subtract this relation from (9.5.4) obtaining

$$[x(t) - \alpha(t)]\alpha'(t) + [y(t) - \beta(t)]\beta'(t) = -\rho(t)\rho'(t). \tag{9.5.8}$$

On the other hand, eliminating $x'(t)$ and $y'(t)$ between equations (9.5.4) and (9.5.6) we get

$$\frac{x(t) - \alpha(t)}{\alpha'(t)} = \frac{y(t) - \beta(t)}{\beta'(t)}.$$

A well-known lemma in the theory of proportions asserts that if

$$\frac{A}{B} = \frac{C}{D},$$

then the common value of the fraction also equals

$$\frac{AB + CD}{B^2 + D^2} \quad \text{and} \quad \left\{\frac{A^2 + C^2}{B^2 + D^2}\right\}^{\frac{1}{2}}.$$

Here we take

$$A = x(t) - \alpha(t), C = y(t) - \beta(t),$$
$$B = \alpha'(t), D = \beta'(t),$$

and obtain

$$\frac{[x(t) - \alpha(t)]\alpha'(t) + [y(t) - \beta(t)]\beta'(t)}{[\alpha'(t)]^2 + [\beta'(t)]^2} = \left\{\frac{[x(t) - \alpha(t)]^2 + [y(t) - \beta(t)]^2}{[\alpha'(t)]^2 + [\beta'(t)]^2}\right\}^{\frac{1}{2}}.$$

Since $\alpha'(t)$ and $\beta'(t)$ are continuous, the evolute is rectifiable. We denote its arc length by $S(t)$ so that

$$[S'(t)]^2 = [\alpha'(t)]^2 + [\beta'(t)]^2. \tag{9.5.9}$$

Combining this with (9.5.7) and (9.5.8) we get

$$\pm \frac{\rho(t)\rho'(t)}{[S'(t)]^2} = \pm \frac{\rho(t)}{S'(t)},$$

whence

$$S'(t) = \pm \rho'(t).$$

Here the ambiguous sign corresponds to the ambiguity in the definition of $S(t)$. If we count $S(t)$ in the direction of increasing $\rho(t)$, then

$$S'(t) = \rho'(t) \tag{9.5.10}$$

and

$$S(t_2) - S(t_1) = \rho(t_2) - \rho(t_1), \tag{9.5.11}$$

if the arc on E corresponds to the parameter interval $[t_1, t_2]$. In stating (9.5.10) we have tacitly assumed that $\rho(t)$ is monotone. If $\rho(t)$ has an extremum on the arc, then (9.5.10) is not true in the whole interval and (9.5.11) does not follow. This completes the proof.

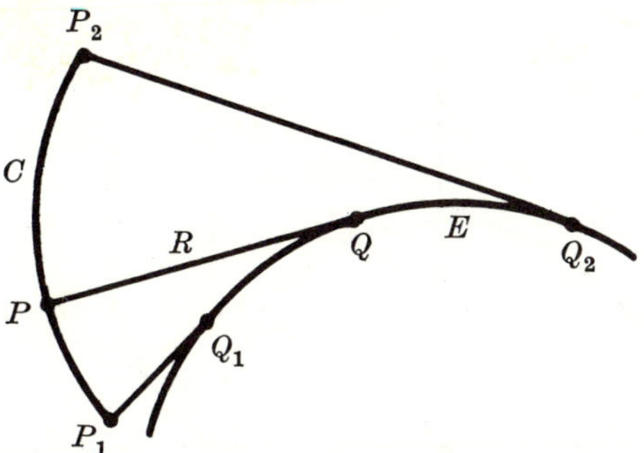

Figure 9.14

Figure 9.14 suggests a thread construction of involutes to a given curve E. Let us imagine a perfectly flexible, nonelastic thread of thickness zero. Let this thread be wrapped along E where one end is firmly attached at Q_2 while the free end is held at P_1. We suppose that the given arc of E has a continuously turning tangent which always turns in the same direction when the arc is described from Q_1 to Q_2. The thread is held taut along E and the free end is rectilinear and tangential to E at Q_1. We now start unwinding the thread, holding it taut during the process. The free end then describes an arc C. It is clear from Theorem 9.5.2 and from this construction that the relation between C and E is that between an involute and its evolute. We also see why C is not the only involute of E. We have merely to change the total length of the thread to get another involute of E.

Suppose now that E is given by an equation and let us find the equations of its involutes. Let

$$E\colon \alpha = \alpha(t), \quad \beta = \beta(t), \quad a \leqq t \leqq b, \tag{9.5.12}$$

where $\alpha(t)$ and $\beta(t)$ are differentiable functions of t. Let Q be a point between Q_1 and Q_2 and let $S(t)$ be the length of the arc Q_1Q on E. We have

$$S(t) = \int_a^t \{[\alpha'(u)]^2 + [\beta'(u)]^2\}^{\frac{1}{2}}\, du. \tag{9.5.13}$$

by formula (9.4.21).

Let φ be the angle which the curve tangent to E at Q makes with the α-axis. Then the coordinates of the point P on the involute are

$$x(t) = \alpha(t) - R(t)\cos\varphi, \quad y(t) = \beta(t) - R(t)\sin\varphi, \tag{9.5.14}$$

where $R(t)$ is the radius of curvature of the involute at P. Here we have

$$\cos\varphi = \frac{\alpha'(t)}{\{[\alpha'(t)]^2 + [\beta'(t)]^2\}^{\frac{1}{2}}}, \quad \sin\varphi = \frac{\beta'(t)}{\{[\alpha'(t)]^2 + [\beta'(t)]^2\}^{\frac{1}{2}}},$$

where the square root is taken to be positive. Further

$$R(t) = S(t) + C$$

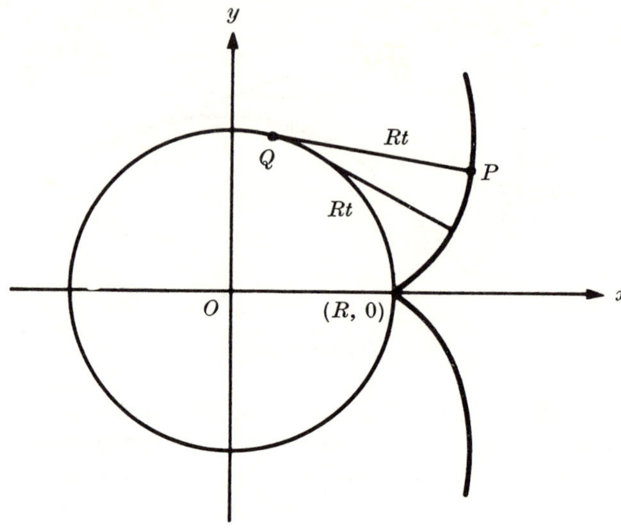

Figure 9.15

by Theorem 9.5.2. Combining these formulas we obtain

$$x(t) = \alpha(t) - \alpha'(t)\frac{S(t) + C}{S'(t)},$$
$$y(t) = \beta(t) - \beta'(t)\frac{S(t) + C}{S'(t)}. \tag{9.5.15}$$

In deriving these formulas we have tacitly assumed $R(t) > 0$ along the arc. If $R(t) < 0$ instead, we should change the minus signs to plus in (9.5.14), but we have also to replace $S(t) + C$ by its negative, so in the end formulas (9.5.15) remain unchanged. Note that the formulas contain an arbitrary constant C. This is because there are infinitely many involutes, one corresponding to each value of C.

Example 1. We shall derive the equation of the involute of the circle. We take

$$\alpha = R\cos t, \quad \beta = R\sin t, \quad 0 \leqq t < 2\pi.$$

Here

$$S'(t) = R, \quad S(t) = R\int_0^t du = Rt.$$

We take $C = 0$ and obtain

$$x(t) = R(\cos t + t\sin t),$$
$$y(t) = R(\sin t - t\cos t). \tag{9.5.16}$$

A small portion of the curve is sketched in Figure 9.15. The point $t = 0$, $x = R$, $y = 0$, is a cusp since

$$x - R = \tfrac{1}{2}Rt^2 + \cdots, \quad y = \tfrac{1}{3}Rt^3 + \cdots.$$

The curve describes ever-widening loops about the origin as t goes to $+\infty$ or $-\infty$.

We can characterize this curve as a *roll curve*, also called a *roulette*. Suppose we take a straight line tangent to the circle, and let it roll without gliding on the circle. Suppose that we follow the movement of the point on the line which at the time $t = 0$ is in contact with the circle at the position $(R, 0)$. This point describes the particular involute given above. The other involutes correspond to other choices of the point on the line.

The involute of the circle has a simple natural equation which is easily derived. We have $\rho(t) = Rt$ by the figure and by Theorem 9.5.2. Since

$$x'(t) = Rt \cos t, \; y'(t) = Rt \sin t,$$

we have

$$[s'(t)]^2 = R^2t^2, \; s(t) = \tfrac{1}{2}Rt^2,$$

if arc length is counted from the cusp. Elimination of t gives the desired equation

$$\rho^2 = 2Rs. \tag{9.5.17}$$

EXERCISE 9.5

1. Find the involute of the parabola $x = at^2$, $y = 2at$ which has a cusp at the origin. Show that the curve lies in the left half-plane.

2. Find the involute of the catenary $y = \cosh x$ which has a cusp at $(0, 1)$. This is the so called *tractrix*. Show that the curve lies in the strip $0 < y \leqq 1$ and has the x-axis as an asymptote.

3. The angle between the curve tangent and the polar axis is

$$\varphi = \frac{\pi}{2} + \theta - \text{arc tan}\,\frac{r'}{r}.$$

Verify that the center of curvature is

$$\alpha = r \cos \theta - \frac{(r \sin \theta)'}{\varphi'}, \quad \beta = r \sin \theta + \frac{(r \cos \theta)'}{\varphi'},$$

where r and φ are functions of θ and the primes denote differentiation with respect to θ.

4. Show that the evolute of a logarithmic spiral $r = ae^{c\theta}$ is another logarithmic spiral obtained by replacing θ by $\theta - \dfrac{\pi}{2}$.

5. Prove that the radius of curvature ρ_1 of the evolute of a given curve C is

$$\rho_1 = \rho \frac{d\rho}{ds},$$

where ρ is the radius of curvature and s the arc length of C. Continuous third-order derivatives of $x(t)$ and $y(t)$ are supposed to exist.

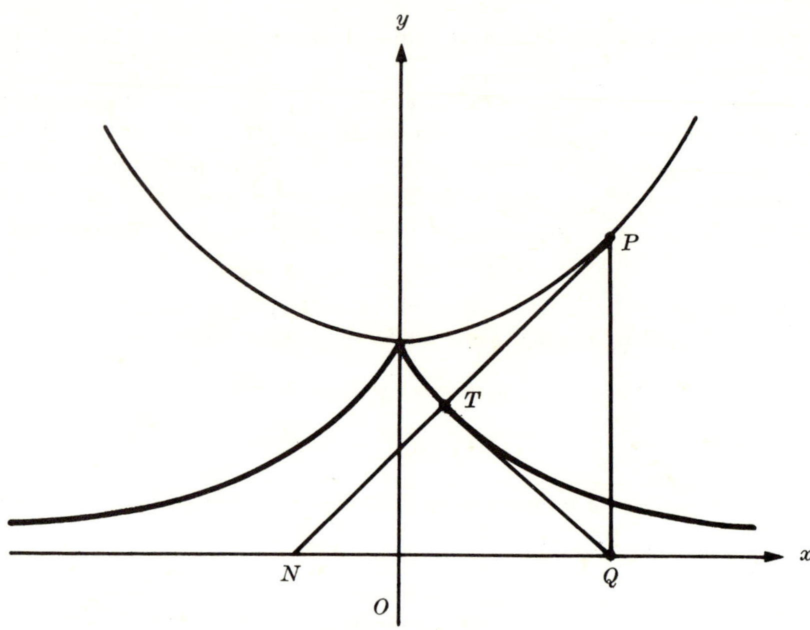

Figure 9.16

9.6. Further special curves. We end this chapter by a discussion of other roll curves. We get a roll curve by tracing the orbit of a point P in a fixed position relative to a curve C when the latter rolls without gliding on a fixed curve Γ. Such curves are of importance in mechanical engineering where they are used to change one type of motion into another. There is an infinite variety of such curves and all we can do is to consider a few samples.

A relatively simple case is where C is a straight line. We have seen that the problem is to find the involutes of Γ. In text and problems we have considered cases where Γ is a circle, a parabola, and a catenary. We shall reexamine the last case because the corresponding curve, the *tractrix* of the straight line, has other interesting properties. See Figure 9.16.

Suppose Q and T are two points in the plane, a fixed distance a from each other. Assume that Q and T are connected by an inelastic taut string. Q moves along a straight line and drags T with it. What is the orbit of T?

It is of course assumed that T is not on the fixed line at the start. The velocity vector at T is directed along TQ and the length of TQ is fixed, equal to a. This means that the orbit of T is a curve such that the segment of the curve tangent intercepted by the fixed line is of constant length a. If we take the fixed line as x-axis, then in the terminology of Section 8.1 we are looking for a curve whose "tangent" is constant. Compare formula (8.1.7). We shall verify that the tractrix has this property.

We start out with the evolute, the catenary, in the form

$$\alpha = t, \quad \beta = a \cosh \frac{t}{a}. \tag{9.6.1}$$

A simple calculation gives the involute, the tractrix, as

$$x = t - a \tanh \frac{t}{a}, \quad y = a \operatorname{sech} \frac{t}{a}. \tag{9.6.2}$$

When the curve is in parametric form, the length of the tangent is

$$TQ = \frac{y(t)}{y'(t)} \{[x'(t)]^2 + [y'(t)]^2\}^{\frac{1}{2}}. \tag{9.6.3}$$

Here

$$x'(t) = 1 - \operatorname{sech}^2 \frac{t}{a}, \quad y'(t) = \operatorname{sech} \frac{t}{a} \tanh \frac{t}{a},$$

so that

$$[x'(t)]^2 + [y'(t)]^2 = 1 - 2 \operatorname{sech}^2 \frac{t}{a} + \operatorname{sech}^4 \frac{t}{a} + \operatorname{sech}^2 \frac{t}{a} \tanh^2 \frac{t}{a}.$$

Using the basic identity

$$\cosh^2 u - \sinh^2 u = 1$$

repeatedly, we find

$$\{[x'(t)]^2 + [y'(t)]^2\}^{\frac{1}{2}} = \tanh \frac{t}{a}, \tag{9.6.4}$$

and substituting in (9.6.3) we see that TQ is indeed equal to a.

The representation of the tractrix by (9.6.2) is convenient for the computation of other lengths associated with the curve. Thus we find for the length of arc and the radius of curvature

$$s(t) = a \log \cosh \frac{t}{a}, \quad \rho(t) = a \sinh \frac{t}{a}. \tag{9.6.5}$$

If so desired, we can eliminate t between the two equations (9.6.2) and obtain

$$x = a \log \frac{a - (a^2 - y^2)^{\frac{1}{2}}}{y} - (a^2 - y^2)^{\frac{1}{2}}. \tag{9.6.6}$$

Since the most useful object that rolls is a wheel, it is natural to consider curves generated by a rolling circle. The simplest case is where the circle rolls on a straight line and the point P is on the circumference. See Figures 9.17 and 9.18. This, the simplest of all roulettes, figures quite early in the history of science. It was considered by the learned Cardinal Cusanus [Nicolaus de Cusa] in 1451. It was studied by Galileo Galilei (1564–1642) around 1599 to whom the name *cycloid* [Gr. κύκλος circle, ειδος form] appears to be due. Descartes studied it, G. P. de Roberval found the area, rediscovered by Galileo's pupil Evangelista Torricelli of barometric fame, Sir Christopher Wren found the length between two consecutive cusps. This list of famous men who have worked on the cycloid could be continued. See further below.

Let R be the radius of the circle, take the line on which the circle rolls as x-axis, and choose the origin as the point of contact of the circle with the line at $t = 0$. The circle is supposed to roll with uniform angular velocity so that at the time t the point of contact is located at $Q = (Rt, 0)$. The point P that was at the origin at $t = 0$ is now at $[x(t), y(t)]$, say, and the arc QP equals $OQ = Rt$. Let

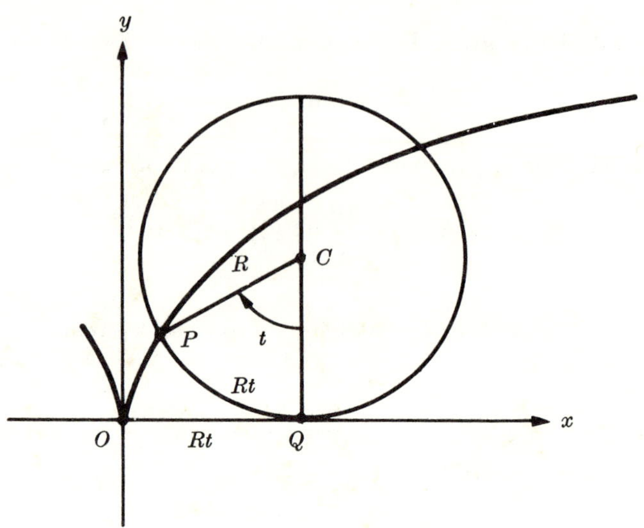

Figure 9.17

C be the center of the circle. Then the vector $\overrightarrow{CP}$ has the components $-R \sin t$ and $-R \cos t$, and this gives

$$x = R(t - \sin t), \quad y = R(1 - \cos t). \tag{9.6.7}$$

The curve consists of infinitely many congruent arches confined to the strip $0 \leqq y \leqq 2R$. The points on the x-axis are cusps with vertical tangents. They correspond to $t = 2n\pi$. For $t = (2n + 1)\pi$ the ordinate reaches its maximum value $2R$.

We can eliminate t between the equations (9.6.7) obtaining

$$x = R \arccos \frac{R - y}{R} \mp (2Ry - y^2)^{\frac{1}{2}}. \tag{9.6.8}$$

This shows that the curve is nonalgebraic, but the parametric equations are more convenient for the study. Using these we find that the length of one of the

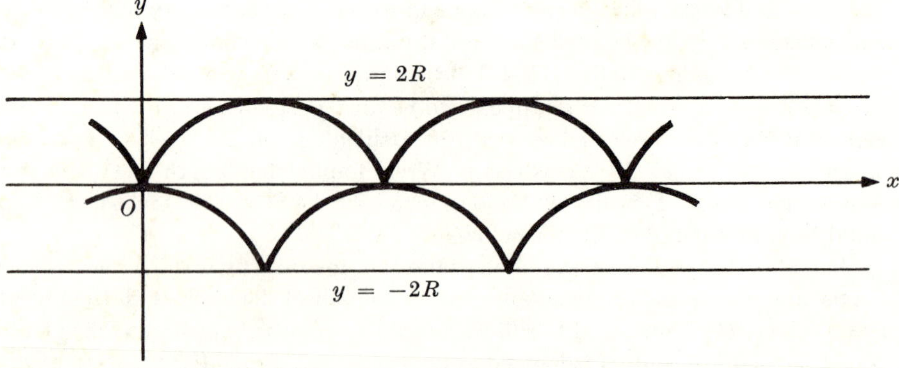

Figure 9.18

arches is $8R$ and the area between an arch and the x-axis is $3\pi R^2$, that is, three times the area of the rolling circle. Galileo had tried to prove this by weighing a model.

The arc length from the highest point of an arch to a point $[x(t), y(t)]$ on the same arch is

$$s(t) = 4R \cos \tfrac{1}{2}t. \tag{9.6.9}$$

Since the radius of curvature is

$$\rho(t) = -4R \sin \tfrac{1}{2}t, \tag{9.6.10}$$

the natural equation of such an arch is

$$\rho^2 + s^2 = 16R^2. \tag{9.6.11}$$

The cycloid arises in several kinematic problems involving descent of a particle under gravity along a curve in a vertical plane. In 1673 Christiaan Huygens showed that if the time of descent is independent of the point of release, then the curve is an arch of an inverted cycloid. This is called the *tautochrone* property (Gr. *το αυτο*, the same, *χρονος* time). For the condition that the time of descent shall be a minimum, Johann Bernoulli and Leibniz in 1694 showed that the solution is still given by a cycloid. This is the *brachysto-chrone* property (Gr. *βραχιστος* shortest).

The evolute of a cycloid is another cycloid, see Figure 9.18,

$$x = R(t + \sin t), \quad y = -R(1 - \cos t). \tag{9.6.12}$$

Huygens used this property to construct a pendulum for securing constancy of the rate of a clock in spite of variations in the amplitude of oscillation. The *cycloidal pendulum* is a clever device even if the practical solution of the problem has followed other lines.

Thus far we have considered only the locus of a point on the circumference of the rolling disk, a speck on the rim of the wheel so to speak. But one may just as well consider the motion of a point P on a spoke of the wheel or on the continuation of a spoke. If the distance of P from the center is h, the argument used above gives the equations

$$x = Rt - h \sin t, \quad y = R - h \cos t. \tag{9.6.13}$$

If $0 < h < R$, this is an *elongated cycloid*, if $R < h$ a *foreshortened cycloid*. The former has no singular points, the latter infinitely many equidistant double points.

Another interesting class of roulettes is obtained by letting a circle Γ roll on a fixed circle Γ_0, either externally or internally. Let R be the radius of Γ_0, r the radius of Γ, and h the distance of P from the center of Γ. If Γ rolls on the exterior of Γ_0, we obtain *epicycloids* (ordinary, elongated, or foreshortened according as $h = r$, $h < r$ or $h > r$). The equations are

$$\begin{aligned} x &= (R + r) \cos t - h \cos \left(\frac{R + r}{r} t\right), \\ y &= (R + r) \sin t - h \sin \left(\frac{R + r}{r} t\right). \end{aligned} \tag{9.6.14}$$

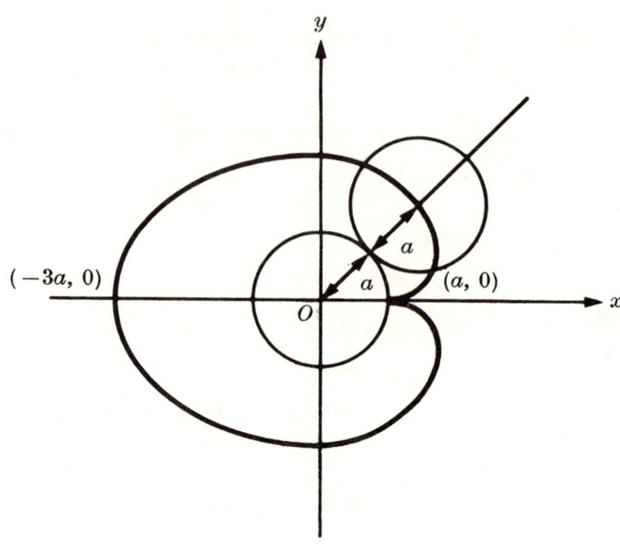

Figure 9.19

We get these equations by noting that the vector $\overrightarrow{OP}$ is the sum of the vectors $\overrightarrow{OC}$ and $\overrightarrow{CP}$ where C is the center of the rolling circle. The curve is closed and finite in length if and only if R/r is a rational number. In this case the four trigonometric functions have a common period so that $[x(t), y(t)]$ ultimately comes back to the starting point. If R/r is not merely rational but an integer, the common period is 2π, so P comes back to the starting point after the center C has made one revolution about the origin. Otherwise several revolutions are needed. If R/r is irrational, P never returns to the starting point. It may be shown, however, that P comes infinitely often into any preassigned neighborhood of the starting point. In fact, if Q is any point whose distance from the origin lies between $\max(0, R + r - h)$ and $R + r + h$, and if N_Q is any neighborhood of Q, then P passes infinitely often through N_Q. Thus the behavior of the epicycloid is totally different according as R/r is rational or irrational. In the rational case, assuming $h = r$, the points of the epicycloid which lie on the circle Γ_0 are cusps with tangents passing through the origin. If R/r is not an integer, there are also ordinary double points. In the irrational case cusps are dense on Γ_0 and double points are dense in the annulus

$$R^2 < x^2 + y^2 < (R + 2r)^2 \quad (h = r).$$

The cardioid is the simplest of all epicycloids. It corresponds to $R = r = h = a$. See Figure 9.19.

If Γ rolls on the interior of Γ_0 instead, we obtain *hypocycloids*. Formally this corresponds to replacing r and h by $-r$ and $-h$ in (9.6.14). This gives

$$\begin{aligned} x &= (R - r)\cos t + h \cos\left(\frac{R - r}{r} t\right), \\ y &= (R - r)\sin t - h \sin\left(\frac{R - r}{r} t\right). \end{aligned} \tag{9.6.15}$$

There are various interesting special cases. The choice $R = 2a$, $r = h = a$ gives

$$x = 2a \cos t, y = 0.$$

Thus the point P describes a rectilinear harmonic motion when a circle rolls on the inside of a circle of twice its own radius and P is on the circumference of the smaller circle. This can be used to construct gears that will change a rotary motion into a rectilinear motion.

The case $R = 3a$, $h = r = a$ gives

$$x = 2a \cos t + a \cos 2t, y = 2a \sin t - a \sin 2t. \tag{9.6.16}$$

This curve has three cusps for $t = 0, \frac{2}{3}\pi, \frac{4}{3}\pi$, located on the circle of radius $3a$ and center at the origin. The cusp tangents pass through the origin. It is possible to eliminate t between the parametric equations. The result is a quartic

$$(x^2 + y^2 + 12ax + 9a^2)^2 - 4a(2x + 3a)^3 = 0.$$

Jakob Steiner (1796–1863), Swiss-born German mathematician, hailed as the greatest geometer since Euclid, discovered this curve, known as "Steiner's curve," in 1856. He did not know it was a hypocycloid, and studied it as an algebraic curve of order four and class three. According to the observations made in Section 9.3 such a curve must be rational. We get a rational parametric representation for the

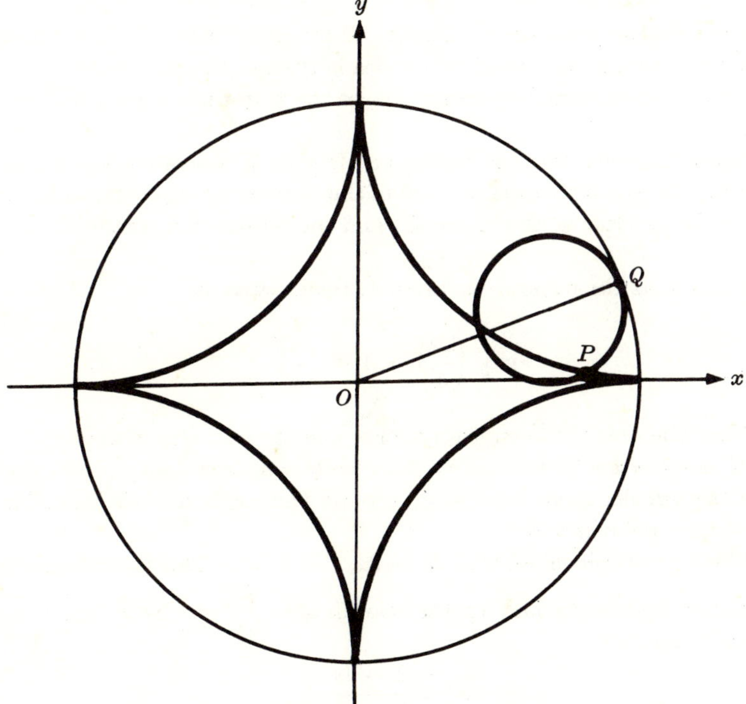

Figure 9.20

curve using (9.6.16), where we set

$$\cos t = \frac{1 - u^2}{1 + u^2}, \sin t = \frac{2u}{1 + u^2}.$$

Compare the two parametric representations of the circle. The result of this substitution is

$$x = a\frac{3 - 6u^2 - u^4}{(1 + u^2)^2}, y = 8a\frac{u^3}{(1 + u^2)^2}. \tag{9.6.17}$$

Finally we take $R = 4a$, $h = r = a$, and use well-known trigonometric identities to obtain

$$x = 4a \cos^3 t, \quad y = 4a \sin^3 t. \tag{9.6.18}$$

This is the *astroid* we have encountered in several places. See Figure 9.20.

EXERCISE 9.6

1. Verify that the length of an arch of the cycloid is $8R$ and that the area under the arch is $3\pi R^2$.
2. Verify formulas (9.6.9)–(9.6.12).
3. Show that the foreshortened cycloid

$$x = R(t - 2 \sin t), \quad y = R(1 - 2 \cos t)$$

must have a double point on the y-axis. [*Hint:* Show that $t = 2 \sin t$ has three and only three real roots one of which lies between 1.89 and 1.90.]

4. Give a parametric representation of the tractrix using the arc length as parameter.
5. Show that the tangent to the tractrix at T intersects the x-axis at a point whose abscissa is the same as that of the corresponding center of curvature. Show that the product of the "normal" and the radius of curvature at T equals a^2.
6. Show that the natural equation of the tractrix is

$$\exp\left(\frac{2s}{a}\right) = 1 + \left(\frac{\rho}{a}\right)^2.$$

7. Find the area between the tractrix and the coordinate axes.
8. If in equation (9.6.14) $R = 3a$, $r = h = a$, how many times does the center of the smaller circle have to go around the larger circle before P is back at the starting point ($t = 0$)?
9. Sketch the epicycloid with $R = 2a$, $r = h = a$. This is a rational curve. Find rational representations of the coordinates. $\Big[$*Hint:* Note that $x + iy = a(3e^{it} - e^{3it})$ and set

$$e^{it} = \frac{1 - is}{1 + is}.\Big]$$

10. The cardioid is stated in the text as

$$x = 2a \cos t - a \cos 2t, \quad y = 2a \sin t - a \sin 2t$$

and previously in terms of polar coordinates as

$$r = a(1 - \cos \theta).$$

Show that these two equations are consistent if $\theta = t$ and r is the distance from $(1, 0)$ to (x, y).

11. Prove formulas (9.6.14) and (9.6.15).

12. Sketch Steiner's curve and show that its length is $16a$.

13. Find the area of the astroid.

14. The astroid is a rational sextic. Express the coordinates rationally in terms of a parameter.

MISCELLANEOUS PROBLEMS FOR CHAPTER 9

1. Discuss $x(x^2 - y^2) + ay^2 = 0$.

2. Discuss the "conchoid of Nicomedes"

$$x = \pm \frac{a + y}{y} (a^2 - y^2)^{\frac{1}{2}}.$$

Find the corresponding equation in polar coordinates.

3. Sketch the "trisectrix of Maclaurin"

$$2x(x^2 + y^2) = a(y^2 - 3x^2).$$

Find the tangents at the origin and the vertical asymptote.

4. Another trisectrix, this one named for G. de Longchamps, is given by

$$r \cos 3\theta = a.$$

Find the Cartesian equation and the three asymptotes. A trisectrix is a curve that may be used to trisect an angle.

5. Sketch the "cross curve"

$$x = a \sec t, \quad y = b \operatorname{cosec} t.$$

Find the asymptotes and the Cartesian equation. The curve is a rational quartic. Find rational representations of the coordinates.

6. Sketch the "windmill"

$$r = 2a \cot 2\theta,$$

find its equation in Cartesian coordinates, and verify that the four asymptotes are $x = \pm a$, $y = \pm a$.

7. The curve

$$r = a \cos \theta + b$$

is known as a *limaçon* (= conch) *of Pascal* [not the famous philosopher and mathematician Blaise Pascal but his father Étienne]. Sketch the curve, find its equation in Cartesian coordinates. Show that the origin is a double point, a cusp, or an isolated point according as $b < a$, $b = a$, or $b > a$. For $b = a$, the limaçon becomes a cardioid.

8. Prove that the limaçon is the pedal (= foot point curve) of a circle. More precisely, take a circle C, a fixed point F, draw a tangent to the circle, drop the perpendicular from F to the tangent, and let P be the foot point. Then P describes a limaçon whose singular point is at F. The singular point is a double point, a cusp, or an isolated point according as F is outside, on, or inside C.

9. In the complex plane a rectifiable curve C is defined by

$$z = f(s), \quad 0 < s < L,$$

where L is the arc length. If $f(s)$ also has continuous second-order derivative, show that the curvature $K(s)$ is the imaginary part of $\overline{f'(s)}f''(s)$ where the bar indicates the complex conjugate.

10. The equation

$$z = \int_0^t \exp(iu^2)\, du$$

defines a double spiral in the complex plane known under various names; one is the *clothoid* (Gr. κλώϑω I spin). Find the length of arc, counting from the origin, the radius of curvature, and the natural equation. The curve has two asymptotic points, symmetrically located with respect to the origin. Try to sketch the curve, knowing that one of the asymptotic points is $\lim_{t\to\infty} z(t) = \frac{1}{2}(1 + i)\sqrt{\pi}$. The integral plays a role in the refraction of light.

11. A curve C given by its natural equation

$$F(\rho, s) = 0,$$

rolls without gliding on the x-axis in the x,y-plane. The center of curvature P of the point Q in contact with the x-axis traces a curve Γ in the plane. What is the equation of Γ? Γ is known as the *Mannheim curve* of C after the French geometer Amédée Mannheim (1831–1906) who created *kinematic geometry*.

12. Assume a closed curve C: $x = x(s)$, $y = y(s)$, which does not intersect itself. Here s is arc length and $x(s)$, $y(s)$ have continuous first- and second-order derivatives with respect to s. On the exterior normal to C at the point P we mark a point Q at the fixed distance a from P. As P varies, Q describes a parallel curve C_a to C. Find the equation of C_a.

13. If $S(s)$ is arc length on C_a, show that

$$S'(s) = 1 + aK(s),$$

where $K(s)$ is the curvature of C at P. Using (9.4.26) or otherwise, show that the length of C_a exceeds that of C by $2\pi a$.

14. If $K^*(s)$ is the curvature of C_a, show that

$$K^*(s) = K(s)[1 + aK(s)]^{-1}.$$

15. Sketch the curve

$$(2x - 3)y = e^{-x^2},$$

determine the asymptotes, and discuss extremal values.

16. Sketch the curve

$$3y = x^2(6 - x^2).$$

From a point P on the y-axis, tangents are drawn to the curve. How does the number of tangents vary with the position of P on the y-axis?

17. Sketch the curve

$$x = t^2 + 1, \quad y = t^2 + \frac{2}{t}, \quad t \neq 0,$$

and determine asymptotes and extrema.

Discuss the following three curves:

18. $x^4 = 4(x^2 - 2)y$.

19. $yx^2 = x^3 + 5x^2 - 4$.

20. $(y - 2x)^2(x^2 - 3) = 1$.

21. Construct the curve $r = \cos^2 \theta$, $0 \leqq \theta \leqq 2\pi$, and find its radius of curvature.

22. Construct the curve $r = \sin^2 \frac{1}{2}\theta$, $0 \leqq \theta \leqq 2\pi$, and find its length.

23. As $|\theta| \to \infty$ on the spiral of Archimedes, $r = a\theta$, show that the difference between the radius vector and the radius of curvature tends to zero.

24. Show also that the center of curvature describes a spiral which approaches indefinitely the circle $r = a$ as $|\theta| \to \infty$.

25. Let $\{C_n\}$ be the centers of curvature of the spiral of Archimedes at its points of intersection with the polar axis, $\theta = 0$. Show that the points C_n lie on a certain ellipse. Find the equation of the latter.

26. Determine the minimum value of the curvature on the ellipse $x = a \cos t$, $y = b \sin t$.

27. Show that at the point where the curvature is a minimum the ellipse and its circle of curvature have third-order contact, that is, also third-order derivatives coincide.

28. Determine the parabola $y = ax^2 + bx + c$ which goes through the point $(\frac{1}{2}\pi, 1)$, where it has second-order contact with $y = \sin x$, so that the curves have the same tangent and the same curvature.

What order of contact do the curves $y = f(x)$, $y = g(x)$ have at the origin, that is, how far do the derivatives coincide, if f and g are given by:

29. $x - x^2$, $\sin x$.

30. $2x^3$, $\sin^3 x + \tan^3 x$.

Appendix C

Some Basic Concepts

C.1. Functions of bounded variation. The importance of algebraic structure has been emphasized in this book. For each new class of functions introduced, the question was raised of how the elements behave under arithmetic operations. We have shown that rational functions form an algebra and repeated use has been made of this fact. Our discussion of limits and continuity is greatly simplified by the fact that the functions which have limits at a point form an algebra. This implies that continuous functions as well as differentiable functions form algebras. Periodic functions with the same period form an algebra, and so do bounded integrable functions.

This brings up the question of the algebraic structure of the functions of Calculus. This is an ill-defined set, but is customarily restricted to functions satisfying the *conditions of Dirichlet*, in other words, functions with only a finite number of relative maxima and minima in a given interval. Suppose we take a fixed interval (a, b) and consider all functions $f(x)$ continuous with a continuous first-order derivative such that $f(x)$ has only a finite number of relative maxima and minima in (a, b). This class D is well-defined and its members would seem acceptable from the point of view of elementary Calculus. What about the structure of D? Does it contain sums and products of its elements? The answer is *no* as is shown by the following example.

Consider the interval $(0, 1)$ and the three functions

$$f(x) = x - \tfrac{1}{3}x^2 \sin\frac{1}{x}, \quad g(x) = -x, \quad h(x) = \frac{1}{x}. \tag{C.1.1}$$

Since the interval is open, all three functions and their derivatives of all orders are continuous. It is clear that $g(x)$ and $h(x)$ are strictly decreasing, but less obvious that $f(x)$ is strictly increasing. We have

$$f'(x) = 1 - \tfrac{2}{3}x\sin\frac{1}{x} + \tfrac{1}{3}\cos\frac{1}{x}.$$

Since the sine and the cosine lie between -1 and $+1$, we see that

$$f'(x) \geqq 1 - \tfrac{2}{3}x - \tfrac{1}{3} > 0 \quad \text{for} \quad 0 < x < 1.$$

Hence $f(x)$ is strictly increasing. On the other hand,

$$f(x) + g(x) = -\tfrac{1}{3}x^2 \sin\frac{1}{x}$$

has infinitely many relative maxima and minima in (0, 1) and the same is true for

$$f(x)\,h(x) = 1 - \tfrac{1}{3}x \sin \frac{1}{x}.$$

Here f, g, h are in D, but neither $f + g$ nor fg is in D.

The class D was introduced by Dirichlet for study of trigonometric series. He showed that if $f \in D$ and is periodic, then $f(x)$ can be represented by a convergent trigonometric series. But then obviously, if $f \in D$, $g \in D$ and have the same period, then $f + g$ can also be represented by a convergent trigonometric series, the sum of the two series of f and of g. For this conclusion it is immaterial if $f + g \in D$. This shows that the class D is not a natural one to consider in this problem.

The appropriate class, *the functions of bounded variation*, was introduced by the French mathematician Camille Jordan (1838–1922) in 1887 in his *Cours d'Analyse*. These functions form a natural generalization of the class D and also permit us to characterize rectifiable curves in a simple manner. Further, these functions lead to an important generalization of the Riemann integral due to Thomas Jan Stieltjes [1856–1894], a Dutch-born astronomer who later became a French mathematician.

DEFINITION C.1.1. *Let $f(x)$ be defined in $[a, b]$. Let π be a partition of $[a, b]$ by points $\{t_k\}$,*

$$a = t_0 < t_1 < t_2 < \cdots < t_n = b,$$

and consider the sum

$$S_\pi[f] \equiv \sum |f(t_k) - f(t_{k-1})|. \tag{C.1.2}$$

If there is a finite number M such that

$$S_\pi[f] \leq M \tag{C.1.3}$$

for all possible partitions, then f is said to be of bounded variation in $[a, b]$, written $f \in BV[a, b]$. The least value of M, that is, the supremum of the set $\{S_\pi[f]\}$, is called the total variation of f in $[a, b]$ and is denoted by $V_a{}^b[f]$.

THEOREM C.1.1. *Any function f which is bounded and monotone in $[a, b]$ belongs to $BV[a, b]$.*

Proof. Suppose, for instance, that f is nondecreasing in $[a, b]$. Then

$$|f(t_k) - f(t_{k-1})| = f(t_k) - f(t_{k-1})$$

and

$$S_\pi[f] = \sum_{k=1}^{n} [f(t_k) - f(t_{k-1})].$$

This sum telescopes into

$$f(b) - f(a).$$

It follows that (C.1.3) holds and

$$V_a{}^b[f] = f(b) - f(a), \tag{C.1.4}$$

which is finite by assumption. Similarly we show that

$$V_a^b[f] = f(a) - f(b),$$

if f is nonincreasing in $[a, b]$.

THEOREM C.1.2. *If f and g belong to $BV[a, b]$, and A and B are arbitrary constants, then $Af + Bg \in BV[a, b]$ and*

$$V_a^b[Af + Bg] \leqq |A|\, V_a^b[f] + |B|\, V_a^b[g]. \tag{C.1.5}$$

Proof. We have

$$\begin{aligned} &|Af(t_k) + Bg(t_k) - Af(t_{k-1}) - Bg(t_{k-1})| \\ &\quad = |A[f(t_k) - f(t_{k-1})] + B[g(t_k) - g(t_{k-1})]| \\ &\quad \leqq |A|\, |f(t_k) - f(t_{k-1})| + |B|\, |g(t_k) - g(t_{k-1})|. \end{aligned}$$

Summing for k we get

$$\begin{aligned} S_\pi[Af + Bg] &\leqq |A|\, S_\pi[f] + |B|\, S_\pi[g] \\ &\leqq |A|\, V_a^b[f] + |B|\, V_a^b[g]. \end{aligned}$$

This proves the assertion and the inequality (C.1.5).

This theorem shows that any finite linear combination of elements of $BV[a, b]$ is also an element of this class. Since a power of x belongs to $BV[a, b]$ as long as $b - a$ is finite, we see that any polynomial in x is in $BV[a, b]$.

Oscillatory functions should also be considered. We observe that the total variation of $\sin x$ in $[0, 2\pi]$ is 4. For $\sin x$ increases to a maximum in $[0, \frac{1}{2}\pi]$, then decreases to a minimum at $x = \frac{3}{2}\pi$, and then gets back to the value 0. If we project this part of the sine curve on the y-axis, we see that $\sin x$ describes the interval $[-1, +1]$ twice, as x goes from 0 to 2π. The same result holds for $\cos x$. On the other hand,

$$V_0^{2\pi}[\sin 2x] = 8, \quad V_0^{2\pi}[\cos 5x] = 20,$$

in general

$$V_0^{2\pi}[\sin nx] = V_0^{2\pi}[\cos nx] = 4n, \tag{C.1.6}$$

because each of the functions $\sin nx$ and $\cos nx$ describes $2n$ times the range $[-1, +1]$.

We conclude that a function $f(x)$ with infinitely many extrema runs the risk of not being of bounded variation. There are functions with infinitely many extrema, which are of bounded variation, but there are others which are not. Thus the two functions

$$\begin{aligned} f(x) &= x \sin \frac{1}{x}, \quad 0 < x \leqq 1, \quad f(0) = 0, \\ g(x) &= x^2 \sin \frac{1}{x}, \quad 0 < x \leqq 1, \quad g(0) = 0, \end{aligned} \tag{C.1.7}$$

have infinitely many extrema. The second is of bounded variation while the first is not.

To prove this we note that

$$f\left(\frac{2}{(2k+1)\pi}\right) = \frac{2(-1)^k}{(2k+1)\pi},$$

so that

$$\left|f\left(\frac{2}{(2k+1)\pi}\right) - f\left(\frac{2}{(2k+3)\pi}\right)\right| = \frac{2}{\pi}\left\{\frac{1}{2k+1} + \frac{1}{2k+3}\right\} > \frac{2}{\pi}\frac{1}{k+1}.$$

Since the series

$$\sum \frac{1}{k+1}$$

is divergent, f cannot be of bounded variation in $[0, 1]$. To show that $g \in BV[0, 1]$, it suffices to show that the sum of the absolute values of the ordinates at the extrema converges, for the total variation is at most twice this sum. Now the extrema occur at the roots of the transcendental equation

$$\tan \frac{1}{x} = \frac{2}{x}. \tag{C.1.8}$$

If we draw the curves

$$v = \tan u, \quad v = 2u,$$

we see that they intersect once in every interval

$$(k\pi, (k + \tfrac{1}{2})\pi), \quad k = 0, 1, 2, \cdots.$$

We are not interested in the negative zeros. There is one and only one extremum of $g(x)$ in each interval

$$\left(\frac{2}{(2k+1)\pi}, \frac{1}{k\pi}\right), \quad k = 1, 2, 3, \cdots.$$

It follows that the absolute value of the ordinate at the kth extremum cannot exceed

$$\frac{1}{(k\pi)^2},$$

and this is the general term of a convergent series. Hence $g \in BV(a, b]$.

To complete our study of $BV[a, b]$ we need

THEOREM C.1.3. *If f and g belong to $BV[a, b]$, so does fg and*

$$V_a^b[fg] \leq M[f]\, V_a^b[g] + M[g] V_a^b[f] \tag{C.1.9}$$

where $M[f]$ and $M[g]$ stand for the supremum of $|f(x)|$ and $|g(x)|$ in $[a, b]$.

Proof. We have

$$\begin{aligned}|f(t_k)\,g(t_k)-f(t_{k-1})\,g(t_{k-1})| &= |[f(t_k)-f(t_{k-1})]\,g(t_k)+f(t_{k-1})\,[g(t_k)-g(t_{k-1})]|\\ &\leqq |f(t_k)-f(t_{k-1})|\,|g(t_k)|+|f(t_{k-1})|\,|g(t_k)-g(t_{k-1})|\\ &\leqq M[g]\,|f(t_k)-f(t_{k-1})|+M[f]\,|g(t_k)-g(t_{k-1})|.\end{aligned}$$

Summing for k we get

$$\begin{aligned}S_\pi[fg] &\leqq M[g]S_\pi[f]+M[f]S_\pi[g]\\ &\leqq M[g]V_a^{\,b}[f]+M[f]V_a^{\,b}[g].\end{aligned}$$

This implies (C.1.9) and proves the theorem.

Corollary. *$BV[a, b]$ is an algebra.*

To this result we add

Theorem C.1.4. *If $f \in BV[a, b]$ and if f is bounded away from 0, so that $\inf |f(x)| \equiv m[f] > 0$, then $1/f \in BV[a, b]$ and*

$$V_a^{\,b}\left[\frac{1}{f}\right] \leqq \{m[f]\}^{-2}V_a^{\,b}[f]. \tag{C.1.10}$$

Proof. We have

$$\begin{aligned}\left|\frac{1}{f(t_k)}-\frac{1}{f(t_{k-1})}\right| &= \frac{|f(t_k)-f(t_{k-1})|}{|f(t_k)|\,|f(t_{k-1})|}\\ &\leqq \{m[f]\}^{-2}\,|f(t_k)-f(t_{k-1})|,\end{aligned}$$

whence (C.1.10) follows.

Combining this result with the preceding theorem, we see that the quotient of two elements of $BV[a, b]$ belongs to the class, provided the denominator is bounded away from 0. Note that it is not sufficient that the denominator be different from 0, for a function in $BV[a, b]$ need not be continuous and 0 may be a cluster point of the range of f without actually being assumed.

If $f \in BV[a, b]$, and if $a \leqq \alpha < \beta \leqq b$, then the restriction of f to the interval $[\alpha, \beta]$ belongs to $BV[\alpha, \beta]$. This means that $V_\alpha^{\,\beta}[f]$ is well-defined and does not exceed $V_a^{\,b}[f]$. The variation $V_\alpha^{\,\beta}[f]$ is an additive function of intervals:

Theorem C.1.5. *If $a \leqq \alpha < \beta < \gamma \leqq b$, and if $f \in BV[a, b]$, then*

$$V_\alpha^{\,\gamma}[f] = V_\alpha^{\,\beta}[f] + V_\beta^{\,\gamma}[f]. \tag{C.1.11}$$

Proof. Consider any partition π of the interval $[\alpha, \gamma]$ and let $S_\pi'[f]$ be the corresponding sum. The prime serves as a reminder that we are concerned with the interval $[\alpha, \gamma]$ rather than with $[a, b]$. We may suppose that β is a partition point of π, for if this is not the case at the outset, we can insert β and obtain a new sum, greater than or equal to $S_\pi'[f]$. The modified sum consists of two parts, one sum involving the partition of $[\alpha, \beta]$ and the other the partition of $[\beta, \gamma]$. The first sum is dominated by $V_\alpha^{\,\beta}[f]$, the second by $V_\beta^{\,\gamma}[f]$.

This implies that

$$V_\alpha^\gamma[f] \leqq V_\alpha^\beta[f] + V_\beta^\gamma[f].$$

On the other hand, if the partition is fine enough, the first sum exceeds $V_\alpha^\beta[f] - \varepsilon$ and the second exceeds $V_\beta^\gamma[f] - \varepsilon$ where ε is a preassigned positive number. This shows that

$$V_\alpha^\gamma[f] \geqq V_\alpha^\beta[f] + V_\beta^\gamma[f] - 2\varepsilon.$$

Since $\varepsilon > 0$ is arbitrary, these two inequalities imply (C.1.11).

It is obvious that we have a corresponding equality for any subdivision of $[\alpha, \gamma]$ into intervals. We note the inequality

$$V_\alpha^\beta[f] \geqq |f(\beta) - f(\alpha)| \tag{C.1.12}$$

where equality holds if and only if f is monotone in the interval.

Suppose now that $a \leqq t \leqq b$ and set

$$V(t; f) \equiv V_a^t[f]. \tag{C.1.13}$$

This is a nonnegative nondecreasing bounded function of t in $[a, b]$. As such it is obviously an element of $BV[a, b]$. Suppose that $a \leqq \alpha < \beta \leqq b$. Then, if f is real,

$$V(\beta; f) - V(\alpha; f) = V_\alpha^\beta[f] \geqq |f(\beta) - f(\alpha)| \geqq f(\beta) - f(\alpha). \tag{C.1.14}$$

This implies that

$$V(\beta; f) - f(\beta) \geqq V(\alpha; f) - f(\alpha). \tag{C.1.15}$$

Thus the function

$$V(x; f) - f(x)$$

is not decreasing in $[a, b]$.

Since

$$f(x) = V(x; f) - [V(x; f) - f(x)], \tag{C.1.16}$$

we see that $f(x)$ is the difference of two monotonic functions. Thus we have proved

THEOREM C.1.6. *A real-valued function of bounded variation is the difference of two nondecreasing bounded functions.*

Formula (C.1.16) gives such a decomposition; it is evidently not unique. We can add any nondecreasing function to each of the components. If, as is perfectly feasible, f is not real-valued, we note that the real and the imaginary parts of f are also elements of $BV[a, b]$ and for each of them we get decompositions into monotone components. This means a decomposition for f of the form

$$f = f_1 + if_2 - f_3 - if_4, \tag{C.1.17}$$

where each of the four functions f_k is real and nondecreasing.

It remains to say something about continuity of functions of bounded variation. In view of the decomposition theorem we can restrict ourselves to nondecreasing functions. This problem was discussed in Section 2.7 where it was shown that a monotone function has right- and left-hand limits everywhere. If $f(x)$ is nondecreasing,

$$f(x+0) - f(x-0)$$

is never negative and the difference is positive for at most a countable set of values of x. This means that we can enumerate these values and order them for instance according to decreasing values of the oscillation at the point, that is, the difference $f(x_0+0) - f(x_0-0)$. It follows that a function of bounded variation has finite right- and left-hand limits everywhere in (a, b), a right-hand limit at $x = a$ and a left-hand limit at $x = b$. There is a point of discontinuity at any point $x = x_0$ where the three numbers

$$f(x_0-0), \quad f(x_0), \quad f(x_0+0)$$

do not coincide. Again the set of such points can be enumerated.

EXERCISE C.1

The following functions belong to $BV[0, 1]$. Find the total variation.

1. $\sin x$. 2. $x^2 - x^3$. 3. $x^4 - x^2$. 4. $\sin(4x)$. 5. xe^{-2x}.

6. If $[x]$ and (x) have the meaning defined in Section 2.2, find the total variation of $[4x]$ and $(4x)$ in $[0, 1]$.

7. If $f(x) = x\left[\dfrac{1}{x}\right]$, $0 < x \leqq 1$, $f(0) = 1$, does f belong to $BV[0, 1]$?

8. If $f'(x)$ exists and is bounded in $[a, b]$, show that $f \in BV[a, b]$.

9. We recall that f satisfies a Lipschitz condition of order 1 in $[a, b]$, if there is a fixed number M such that for all x_1, x_2 in $[a, b]$ we have

$$|f(x_1) - f(x_2)| \leqq M\,|x_1 - x_2|.$$

Show that this implies $f \in BV[a, b]$.

10. If $f \in BV[a, b]$, and if f is complex-valued, prove that the real and the imaginary parts of f belong to $BV[a, b]$.

11. What modifications if any are required in Definition C.1.1 for it to apply to the interval $[0, \infty]$?

Which of the following functions belong to $BV[0, \infty]$?

12. $\dfrac{1}{1+x^2}$. 13. $\dfrac{1}{1+x}$. 14. $\dfrac{\sin x}{1+x}$. 15. $e^{-x}\sin x$. 16. $\dfrac{[x]}{1+x}$.

17. If $f_n(x) \in BV[a, b]$ for $n = 1, 2, 3, \cdots$, and if $\Sigma\,|f_n(a)|$ and $\Sigma\, V_a^b[f_n]$ converge, prove that the series $\Sigma f_n(x)$ converges absolutely and uniformly in $[a, b]$ and that its sum belongs to $BV[a, b]$.

18. Does $f(x) \equiv \sum_{n=1}^{\infty} n^{-3}\cos nx$ belong to $BV[0, 2\pi]$?

C.2. Rectifiable curves. Functions of bounded variation have immediate applications to the existence of arc length. See Definition 9.4.1.

Suppose that an arc C is given by the parametric equations

$$x = x(t),\, y = y(t),\, a \leqq t \leqq b, \tag{C.2.1}$$

where $x(t)$ and $y(t)$ are continuous. The parametrization defines an order on C in which the point $P_1 = [x(t_1), y(t_1)]$ precedes the point $P_2 = [x(t_2), y(t_2)]$ if $t_1 < t_2$. We have seen that C is rectifiable if and only if the lengths of the inscribed polygonal lines form a bounded set. Here only such polygonal lines are considered whose vertices occur in the order prescribed by the parametrization. Analytically this means that we consider partitions π and corresponding sums

$$S_\pi = \sum_{k=0}^{n} \{[x(t_k) - x(t_{k-1})]^2 + [y(t_k) - y(t_{k-1})]^2\}^{\frac{1}{2}}\,. \tag{C.2.2}$$

C is rectifiable if the set of all such sums is bounded and the least upper bound of the set $\{S_\pi\}$ is the length of C.

This situation is familiar from the preceding section and we can get still closer analogy by setting

$$z(t) = x(t) + iy(t) \tag{C.2.3}$$

and considering

$$C\colon\ z = z(t) \tag{C.2.4}$$

as an arc in the complex plane where the same metric holds as in the Euclidean plane. Then (C.2.2) becomes

$$S_\pi = S_\pi[z] = \sum_{k=1}^{n} |z(t_k) - z(t_{k-1})|. \tag{C.2.5}$$

This leads to

THEOREM C.2.1. *The arc C defined by* (C.2.4) *is rectifiable if and only if $z(t)$ is a function of bounded variation. An equivalent condition is that $x(t)$ and $y(t)$ are of bounded variation. The length of C is the total variation of $z(t)$.*

To state the equivalent condition we used the fact (see Problem 10 of the preceding Exercise) that the real and the imaginary parts of a complex-valued function are of bounded variation if and only if the function itself has this property.

We may compare this theorem with Theorem 9.4.1 which asserts that the length of C is given by

$$l(C) = \int_a^b \{[x'(t)]^2 + [y'(t)]^2\}^{\frac{1}{2}}\, dt = \int_a^b |z'(t)|\, dt, \tag{C.2.6}$$

provided the derivatives are continuous. Let us first look at the case in which both formulas apply. We see that

$$V_a^b[z] = \int_a^b |z'(t)|\, dt, \tag{C.2.7}$$

if $z'(t)$ is continuous. More generally, if $f \in BV\,[a, b]$ and has a continuous derivative, then

$$V_a^b[f] = \int_a^b |f'(t)|\, dt. \tag{C.2.8}$$

Observe, however, that Theorem C.2.1 applies to a more general situation than does Theorem 9.4.1. Formula (C.2.7) is valid under more general assumptions than continuity of $z'(t)$, but it is not valid for an arbitrary continuous function of bounded variation.

The difficulties are twofold, the existence of the derivative and the integrability of the derivative. If $f(x) \in BV[a, b]$, then $f'(x)$ exists in general, that is, $f'(x)$ exists if we disregard a relatively sparse set of exceptional points. More precisely, the exceptional set can be covered by overlapping intervals of total length as small as we please. Such a function $f'(x)$ possesses a generalized integral, but it is not always true that

$$\int_\alpha^\beta f'(x)\,dx = f(\beta) - f(\alpha).$$

For this reason formulas (C.2.6)–(C.2.8) may fail to hold.

This is obvious for step functions. Thus if, for instance, $f(x) = [x]$, $0 \leqq x \leqq 2$, then $f'(x)$ exists except for $x = 0, 1, 2$, and its value is zero. Here

$$0 = \int_0^2 f'(x)\,dx \neq f(2) - f(0) = 2.$$

It is of greater interest that such discrepancies may actually occur for continuous functions of bounded variation, but this we shall not prove.

Theorem C.2.1 is due to Camille Jordan. We note the following properties of arc length.

THEOREM C.2.2. *If a rectifiable arc is partitioned into pieces, then each piece is rectifiable and the sum of their lengths equals the length of the whole arc.*

This follows from general properties of the variation as applied to the function $z(t)$.

THEOREM C.2.3. *The length of arc $V(t; z)$ is a continuous strictly increasing function in $[a, b]$.*

Proof. We already know that $V(t; z)$ is nondecreasing. From this it follows that $V(t - 0; z)$ and $V(t + 0; z)$ exist for each t with $a < t < b$ and that

$$V(t - 0; z) \leqq V(t; z) \leqq V(t + 0; z).$$

If the inequality holds in one of these relations, say in the second one, this means that $z(t + 0) \neq z(t)$. This is impossible, however, since $z(t)$ is continuous. In the same way one concludes that $V(t - 0; z) = V(t; z)$. Thus $V(t; z)$ must be continuous as asserted. If $V(t_1; z) = V(t_2; z)$ for some $t_1 < t_2$, then the nondecreasing function $V(t; z)$ must keep a constant value in $[t_1, t_2]$. This in turn is possible if and only if $z(t)$ keeps a constant value in $[t_1, t_2]$, and is obviously not a good choice of parameter for the curve C. We do not restrict the generality by excluding such freaks. If $z(t)$ is not allowed to be constant in any interval, $V(t; z)$ cannot have intervals of constancy and $V(z; t)$ is strictly increasing. This completes the proof.

Since $s = V(t; z)$ is a continuous strictly increasing function of t, the theorem on inverse functions assures that t is a continuous strictly increasing function of s in the interval $[0, L]$, where L is the length of C. If this function is $t(s)$, then

$$z = z[t(s)] \equiv Z(s) \tag{C.2.9}$$

is a parametric representation of C in terms of arc length. This proves

THEOREM C.2.4. *A rectifiable curve C can always be represented parametrically by*

$$z = Z(s), \quad 0 \leq s \leq L, \tag{C.2.10}$$

where $Z(s)$ is a continuous function of bounded variation.

EXERCISE C.2

1. Consider the total variation of the function $[x]$ in the intervals $[1 - h, 1 + h]$, $[1 - h, 1]$, and $[1, 1 + h]$ and find the limits of the variations as h decreases to 0.
2. Prove that if $f \in BV[a, b]$ and if f is continuous at $x = x_0$, so is $V[x; f]$.
3. If $f(x) \in BV[a, b]$, and if $x = x(t)$ is a continuous strictly increasing function which maps the interval $[\alpha, \beta]$ into $[a, b]$, prove that $f[x(t)]$ is of bounded variation in $[\alpha, \beta]$ and that its total variation in $[\alpha, \beta]$ is the same as the total variation of $f(x)$ in $[a, b]$.

The object of the following discussion is to construct a nondecreasing continuous nonconstant function $f(x)$ whose derivative is zero "almost everywhere" and to determine the length of the corresponding curve $y = f(x)$. Take the interval $[0, 1]$ and mark the "middle third" $[\frac{1}{3}, \frac{2}{3}]$ where $f(x)$ is given the constant value $\frac{1}{2}$. In each of the remaining intervals mark the middle third and set $f(x) = \frac{1}{4}$ in $[\frac{1}{9}, \frac{2}{9}]$ and to $\frac{3}{4}$ in $[\frac{7}{9}, \frac{8}{9}]$. Continue indefinitely. At the nth stage there are 2^{n-1} intervals to be marked. The right endpoint of such an interval is

$$\frac{a_1}{3} + \frac{a_2}{3^2} + \cdots + \frac{a_n}{3^n}, \quad a_k = 0 \quad \text{or} \quad 2,$$

and its length is 3^{-n}. In this interval the function is given the value

$$\frac{A_1}{2} + \frac{A_2}{2^2} + \cdots + \frac{A_n}{2^n}, \quad A_k = \tfrac{1}{2}a_k,$$

where $n = 1, 2, 3, \cdots$.

4. Prove that the sum of the lengths of all marked intervals is 1 and that every x in $[0, 1]$ either belongs to such an interval or is the cluster point of points belonging to such intervals.
5. Show that at any point x_0 not in one of the marked intervals, $f(x_0)$ can be defined uniquely by continuity, that is, if $x_n \to x_0$ and $f(x_n)$ is defined, then $f(x_0) = \lim f(x_n)$ and this is independent of what sequence $\{x_n\}$ is used.
6. Prove that $f(x)$ is continuous and nondecreasing.

7. At the nth stage of the construction $f(x)$ is defined in $2^n - 1$ intervals. Draw a polygonal line from $(0, 0)$ to $(1, 1)$ consisting of the $2^n - 1$ horizontal segments from the curve $y = f(x)$ and 2^n slanting line segments connecting consecutive horizontal pieces with each other, the first piece with $(0, 0)$ and the last with $(1, 1)$. Find the length of this line and show that it converges to 2 when $n \to \infty$. Here $l(C) = 2$ but $1 + [f'(x)]^2 = 1$ for almost all x. This example is due to H. Lebesgue.

C.3. Nondifferentiable functions. Formula (C.1.7) defines a function $f(x)$ which is not of bounded variation in $[0, 1]$ or in any interval $[0, \delta]$. On the other hand it is obviously of bounded variation in any interval $[\delta, 1/\delta]$, for in such an interval it has only a finite number of extrema. In this case the origin is a cluster point of extrema and the oscillations are insufficiently damped in the neighborhood of the origin. We want to exhibit an example of a function which fails to be of bounded variation in any interval no matter how small. We shall construct such a function and it will also be a continuous function without a derivative anywhere.

The point of departure is Problem 17 of Exercise C.1. From this we conclude that a trigonometric series

$$f(x) = \tfrac{1}{2}a_0 + \sum_{n=1}^{\infty} (a_n \cos nx + b_n \sin nx) \tag{C.3.1}$$

defines a function of bounded variation, if

$$\sum_{n=1}^{\infty} n\,(|a_n| + |b_n|) \tag{C.3.2}$$

is convergent. This is a sufficient, but not a very sharp condition so from a violation thereof we cannot conclude that $f(x)$ is of unbounded variation. But if the violation is sufficiently flagrant, unbounded variation may follow. This is indeed the case as we shall illustrate with the aid of the series

$$F(x) \equiv \sum_{k=1}^{\infty} a^k \sin(b^k \pi x), \tag{C.3.3}$$

where b is a positive integer > 1, $0 < a < 1$, and $ab > 1$. More stringent conditions will be imposed as they are needed.

We note that this series converges absolutely and uniformly for all x since

$$\sum a^k$$

converges. Hence $F(x)$ is a continuous function. Further $n = b^k$, $b_n = a^k$ and

$$\sum n\,|b_n| = \sum (ab)^k$$

diverges badly so condition (C.3.2) is strongly violated. The function $F(x)$ is obviously periodic and

$$F\left(x + \frac{1}{b}\right) = -F(x), \quad F\left(x + \frac{2}{b}\right) = F(x). \tag{C.3.4}$$

We note further that the remainder of the series after n terms is expressible in terms of $F(b^n x)$. In particular we have

$$F(x) = a \sin(b\pi x) + aF(bx). \tag{C.3.5}$$

This is a functional equation which restricts the properties of $F(x)$ severely and which forces the function to behave in the same way everywhere.

In particular, we shall show that (C.3.5) requires that $F(x)$ is either of bounded variation in any finite interval or else in no interval. For suppose that $F(x)$ is of bounded variation in some interval $[\alpha, \beta]$. Since $a \sin(b\pi x)$ is of bounded variation in every finite interval, $aF(bx)$ and hence also $F(bx)$ must be of bounded variation in $[\alpha, \beta]$. This requires that $F(x)$ is of bounded variation in two intervals $[\alpha, \beta]$ and $[b\alpha, b\beta]$. We can repeat the argument and see that $F(x)$ must be of bounded variation in any interval of the form

$$[b^m\alpha, b^m\beta],$$

where m is any positive integer. If m is sufficiently large, the length of the interval exceeds the period $2/b$ and this shows that $F(x)$ must be of bounded variation in any finite interval since $F(x)$ is periodic. Thus we have proved:

THEOREM C.3.1. *The function $F(x)$ defined by* (C.3.3) *is either of bounded variation in every finite interval or in no interval.*

We shall show that the second alternative holds at least if a and b satisfy suitable restrictions.

THEOREM C.3.2. *There is a constant $C \geqq 1$ such that $F(x)$ is nowhere of bounded variation if $ab > C$.*

Remark. The calculations below show that we can take $C = 1 + \frac{2}{3}\sqrt{3}$. There is no reason to assume that this is the best possible value.

Proof. We have to show that there is a sequence of partitions π_n of $[0, 2/b]$ such that the corresponding sums $S_{\pi_n}[F]$ form an unbounded sequence provided ab exceeds some fixed number C. It is convenient to take the partition points in the nth partition π_n at the multiples of b^{-n}, since for such a choice of x

$$F(b^n x) = 0,$$

and this means that the remainder of the series after $n - 1$ terms is 0. We have then

$$\begin{aligned}
\sum_j |F(jb^{-n}) - F((j-1)b^{-n})| &= \sum_j \left| \sum_{k=1}^{n-1} a^k[\sin(b^{k-n}j\pi) - \sin(b^{k-n}(j-1)\pi)] \right| \\
&\geqq a^{n-1} \sum_j |\sin(b^{-1}j\pi) - \sin(b^{-1}(j-1)\pi)| \qquad \text{(C.3.6)} \\
&\quad - \sum_{k=1}^{n-2} a^k \sum_j |\sin(b^{k-1}j\pi) - \sin(b^{k-n}(j-1)\pi)|.
\end{aligned}$$

The first term in the last member can be found accurately. If b is an even integer, the first $b/2$ differences are positive by themselves without the absolute value signs and add up to the largest sine present, namely $\sin \frac{\pi}{2} = 1$. The following $b/2$ differences are negative, the sum of the absolute values add up to 1, and so on. All told this sum contributes

$$4a^{n-1}b^{n-2}. \tag{C.3.7}$$

If b is odd, there is a difference, $\sin \frac{\pi}{2}$ is not reached, the largest sine present is $\sin \left(\frac{b-1}{2b}\,\pi\right)$ and this factor appears in (C.3.7). Let us write

$$S(b) = \begin{cases} 1 & , \quad b \text{ even}, \\ \sin \left(\dfrac{b-1}{2b}\,\pi\right), & b \text{ odd}. \end{cases} \tag{C.3.8}$$

Then the first term in the last member of (C.3.6) amounts to

$$4a^{n-1}b^{n-2}S(b). \tag{C.3.9}$$

Since the total variation of $\sin (b^k \pi x)$ in the interval $(0, 2/b)$ equals $4b^{k-1}$, we obtain finally

$$\begin{aligned} \sum_j |F(jb^{-1}) - F((j-1)b^{-1})| &> 4a^{n-1}b^{n-2}S(b) - 4\sum_{k=1}^{n-2} a^k b^{k-1} \\ &= 4a\left\{(ab)^{n-2}\, S(b) - \frac{(ab)^{n-2} - 1}{ab - 1}\right\} \\ &= \frac{4a}{ab-1}\,\{S(b)(ab)^{n-1} - [S(b) + 1](ab)^{n-2} + 1\}. \end{aligned} \tag{C.3.10}$$

If now

$$ab \geqq 1 + \frac{1}{S(b)} + \delta, \tag{C.3.11}$$

the brace exceeds

$$\delta(ab)^{n-2} + 1$$

and this becomes infinite with n.

Since $S(b)$ has its least value $\sin \frac{\pi}{3} = \frac{1}{2}\sqrt{3}$ for $b = 3$, it is seen that

$$C = 1 + \tfrac{2}{3}\sqrt{3} \tag{C.3.12}$$

is an admissible value. This completes the proof.

It was stated above that a function of bounded variation has a derivative for "almost all" values. Conversely, a function whose derivative exists and is bounded is automatically of bounded variation. Comparison with Theorem C.3.1 indicates that the existence of a derivative of $F(x)$ is highly doubtful and if a derivative should exist, it cannot be bounded in any interval. We shall prove directly that $F(x)$ does not have a derivative provided ab satisfies a more stringent condition. The proof is parallel to that of Theorem C.3.1 and goes back to Karl Weierstrass.

THEOREM C.3.3. *$F(x)$ is not differentiable for any value of x if $ab > 1 + \frac{3}{2}\pi$ and $b \geqq 5$.*

Proof. We take any point x in the interval $(0, 2/b)$. We assume that

$$x = \sum_{p=1}^{\infty} c_p b^{-p}, \tag{C.3.13}$$

where $c_1 = 0$ or 1 and $c_p = 0, 1, 2, \cdots, b - 1$ for $p > 1$. Such a representation of x can be obtained by the method used in Section 1.4 for the special case $b = 10$. We give x an increment h of the special form

$$h = 2kb^{-m}, \tag{C.3.14}$$

where k and m are integers, $m \geqq 1$, and $1 - b \leqq k \leqq b - 1$.

We have now

$$F(x + h) - F(x) = \sum_{n=1}^{\infty} a^n\{\sin [b^n\pi(x + h)] - \sin (b^n\pi x)\}.$$

Here we note that

$$\sin [b^n\pi(x + h)] - \sin (b^n\pi x) = 0 \quad \text{for} \quad n \geqq m,$$

owing to the special choice of h. On the other hand, for $n = m - 1$,

$$\sin [b^{m-1}\pi(x + h)] - \sin (b^{m-1}\pi x) = 2 \sin \frac{k\pi}{b} \cos \left[b^{m-1}\,\pi x + \frac{k\pi}{b}\right],$$

where we have used the familiar trigonometric identity

$$\sin B - \sin A = 2 \sin \tfrac{1}{2}(B - A) \cos \tfrac{1}{2}(B + A)$$

with

$$B = b^{m-1}\pi(x + h), \quad A = b^{m-1}\pi x.$$

Finally, for $n < m - 1$, the mean value theorem gives

$$\sin [b^n\pi(x + h)] - \sin (b^n\pi x) = b^n\pi h \cos [b^n\pi(x + \theta_n h)].$$

It follows that

$$\frac{1}{h}[F(x + h) - F(x)] = \sum_{n=1}^{m-2} (ab)^n\pi \cos [b^n\pi(x + \theta_n h)]$$

$$+ (ab)^{m-1}\pi \frac{\sin (k\pi/b)}{(k\pi/b)} \cos \left[b^{m-1}\pi x + \frac{k\pi}{b}\right] \tag{C.3.15}$$

$$\equiv S_{m-2} + T_{m-1},$$

with obvious notation. It is clear that

$$|S_{m-2}| \leqq \pi ab \frac{(ab)^{m-2} - 1}{ab - 1}. \tag{C.3.16}$$

Next we note that

$$b^{m-1}x + \frac{k}{b} = [b^{m-2}c_1 + b^{m-3}c_2 + \cdots + c_{m-1}] + \frac{c_m + k}{b} + \frac{c_{m+1}}{b^2} + \cdots .$$

The expression in brackets is an integer, N_m, that depends upon x and m only. It follows that

$$T_{m-1} = (-1)^{N_m}\, \pi \frac{\sin (k\pi/b)}{(k\pi/b)} \cos \left\{\pi\left(\frac{c_m + k}{b} + \frac{c_{m+1}}{b^2} + \cdots\right)\right\} \cdot (ab)^{m-1}. \quad \text{(C.3.17)}$$

Now the integer k is at our disposal. We can use it to make sure that the multiplier of $(ab)^{m-1}$ in T_{m-1} is bounded away from zero and has a given sign, plus or minus. If this is done, and if ab is sufficiently large, then T_{m-1} will dominate S_{m-2} and the sequence

$$\{S_{m-2} + T_{m-1}\} \quad \text{(C.3.18)}$$

is unbounded. This excludes the existence of a finite derivative, and requires that one or more of the following possibilities hold:

$$\limsup \frac{1}{h}[F(x+h) - F(x)] = +\infty, \quad \liminf \frac{1}{h}[F(x+h) - F(x)] = -\infty,$$

$$\limsup \frac{1}{h}[F(x) - F(x-h)] = +\infty, \quad \liminf \frac{1}{h}[F(x) - F(x-h)] = -\infty$$

where $h \to 0$ through positive values. In particular, if there is a subsequence of (C.3.18) corresponding to positive values of k for which T_{m-1} goes to $-\infty$, then the second of the four possibilities holds. This does not exclude the validity of the others.

Suppose now that for the values of m and k under consideration we have

$$|T_{m-1}| \geqq A\pi(ab)^{m-1} \quad \text{(C.3.19)}$$

for a fixed positive A and consider the inequality

$$P\frac{P^{m-2} - 1}{P - 1} < AP^{m-1}, \quad P = ab,$$

or

$$\frac{P}{P-1}\{AP^{m-1} - (1 + A)P^{m-2} + 1\} > 0. \quad \text{(C.3.20)}$$

This is certainly satisfied for $m \geqq 2$ if

$$P \geqq 1 + \frac{1}{A} + \delta, \delta > 0. \quad \text{(C.3.21)}$$

For such values of P the left member of (C.3.20) exceeds

$$\frac{P^{m-1}}{P-1}[AP - (1 + A)] > \frac{\delta P}{P-1}P^{m-1} > \delta P^{m-1},$$

which goes to infinity with m.

Since, for $P = ab$,

$$|S_{m-2} + T_{m-1}| \geqq |T_{m-1}| - |S_{m-2}|$$
$$> \pi A P^{m-1} - \pi P \frac{P^{m-2} - 1}{P - 1} > \delta \pi P^{m-1},$$

we see that

$$ab > 1 + \frac{1}{A}$$

implies

$$\limsup_{|h| \to \infty} \left| \frac{1}{h} [F(x + h) - F(x)] \right| = +\infty, \tag{C.3.22}$$

that is, the nonexistence of the derivative of $F(x)$.

It remains to find an admissible value for A. Here we find it convenient to assume that $b \geqq 5$. We note first that

$$\frac{\sin u}{u}$$

is decreasing in $(0, \pi)$. Hence if $|k| \leq \frac{3}{4}b$, we have

$$1 > \frac{\sin (k\pi/b)}{(k\pi/b)} \geqq \frac{2\sqrt{2}}{3\pi}.$$

If $b \geqq 5$ we still have sufficient freedom in the choice of k. We note that it is now possible to choose k within these limits in such a manner that

$$\left| \cos \pi \left\{ \frac{c_m + k}{b} + \frac{c_{m+1}}{b^2} + \cdots \right\} \right| \geqq \cos \frac{\pi}{b}.$$

To illustrate the procedure, let us take $b = 5$. Then the range for k is $-3 \leqq k \leqq 3$, $k \neq 0$. A suitable choice for k is

$$-1, -1, -2, -3, 1$$

according as c_m equals

$$0, 1, 2, 3, 4,$$

respectively. We note that

$$0 \leqq \frac{c_{m+1}}{b^2} + \cdots \leqq (b - 1) \sum_{n=2}^{\infty} \frac{1}{b^n} = \frac{1}{b}.$$

It follows that in the five cases considered

$$\pi \left\{ \frac{c_m + k}{b} + \frac{c_{m+1}}{b^2} + \cdots \right\}, \, b = 5,$$

lies in the following intervals

$$\left(-\frac{\pi}{5}, 0\right), \quad \left(0, \frac{\pi}{5}\right), \quad \left(0, \frac{\pi}{5}\right), \quad \left(0, \frac{\pi}{5}\right), \quad \left(\pi, \frac{6\pi}{5}\right),$$

respectively. In each case the cosine is at least equal to

$$\cos\frac{\pi}{5} = \tfrac{1}{4}(1 + \sqrt{5})$$

in absolute value. Since $\cos(\pi/b) \geqq \cos(\pi/5)$ for $b \geqq 5$, we conclude that

$$A = \frac{2\sqrt{2}}{3\pi}\cos\frac{\pi}{5} \tag{C.3.23}$$

is an admissible value, and since

$$1 + \frac{1}{A} = 1 + \tfrac{3}{4}\pi\sqrt{2}(\sqrt{5} - 1) < 1 + \tfrac{3}{2}\pi, \tag{C.3.24}$$

the theorem is proved.

Let us add a few remarks concerning the significance of this example. Mathematicians of the eighteenth century made no distinction between continuous and differentiable functions. In 1806 the French mathematical physicist A. M. Ampère attempted to prove that every continuous function is differentiable. While his contemporaries realized that the proof was wrong, there was little doubt as to the correctness of the assertion. Hence it came as a profound shock to the mathematical world when examples of nondifferentiable functions became known in the 1870's. The example of Weierstrass

$$\sum_{n=0}^{\infty} a^n \cos(b^n \pi x), \quad ab > 1 + \tfrac{3}{2}\pi, \tag{C.3.25}$$

is supposed to go back to 1861, though the first reference to it in print dates from 1874 in a paper by one of his pupils, Paul du Bois-Reymond. By this time the French geometer Gaston Darboux had independently arrived at an example of a nondifferentiable function. An example due to Bernard Bolzano, probably dating from 1830, came to light some ninety years later. In the mean time a number of examples have been found based on analytical, arithmetical, and geometrical considerations.

Note that the curve

$$y = F(x), \tag{C.3.26}$$

where $F(x)$ is nondifferentiable, has many strange features. It has nowhere a tangent, that is nowhere a direction, its length between any two points is infinite, there are infinitely many oscillations between any two points of the curve, we can plot points, but we can not speak of the shape of the curve, and no graphical representation can be given.

EXERCISE C.3

1. If sines are replaced by cosines in (C.3.3), are Theorems C.3.1 and C.3.2 true for the resulting function?

2. Verify (C.3.13). The representation is ordinarily unique but there exists an infinite set of numbers admitting two distinct representations. Find these numbers. [*Hint:* The case $b = 10$ is well-known to the student. Generalize!]

3. Any number in [0, 1] can be represented in the form

$$x = \sum_{n=1}^{\infty} \frac{c_n}{n!}, \quad c_n = 0, 1, 2, \cdots, n-1.$$

Verify! When is the representation nonunique? [*Hint:* Examine the case where $c_n = n - 1$ for $n \geqq m$.]

4. Discuss nondifferentiability of

$$G(x) = \sum_{n=1}^{\infty} \frac{c^n}{n!} \sin (n!\, \pi x)$$

for large values of c using the method developed in the text. Take $h = 2k/(m!)$ where k is an integer, $1 \leqq |k| \leqq m - 1$, and use No. 3.

5. Carry through the discussion of T_{m-1} for the case $b = 6$.

6. Take $b = 5$ and $c_p = 1$ for all p. Determine lim sup and lim inf of the difference quotient for one-sided approach, right and left. Note that $(-1)^{N_m} = (-1)^{m-1}$.

*7. If, as usual, (x) is the distance to the nearest integer, if b is an integer, $0 < a < 1$, $ab > 2$, prove the nondifferentiability of

$$\sum_{n=0}^{\infty} a^n (b^n x) \equiv K(x).$$

MISCELLANEOUS PROBLEMS FOR APPENDIX C

1. If $z(t) = x(t) + iy(t) \in BV[a, b]$, prove that

$$V_a^b[z] \leqq V_a^b[x] + V_a^b(y).$$

Can equality hold?

2. Given that $f(x)$ is continuous in $[a, b]$ and has a continuous derivative in $[a, b)$, prove that $f \in BV[a, b]$ if $\int_a^b |f'(t)|\, dt$ exists.

3. Is

$$z(t) = \int_0^t \frac{e^{iu^2}\, du}{u + 1}$$

of bounded variation in $[0, \infty]$? If not, how fast does $V_0^t[z]$ grow with t?

4. Find the length of arc from 0 to t for the curve defined in the preceding problem.

5. If $f(x) = (1 - x)^{\delta + i}$ is defined as $\exp [(i + \delta) \log (1 - x)]$ for $0 < x < 1$ and is extended by continuity at the endpoints, show that $f \in BV[0, 1]$. Here $\delta > 0$ and the logarithm is real.

6. Prove that $x^{-2} \sin^2 x \in BV[1, \infty]$.

7. Find the curvature of $z = z(s)$, where $z(s)$ is a complex-valued, twice differentiable function of the arc length s.

8. Is the spiral $z = \dfrac{1}{\theta} e^{i\theta}$, $0 < \theta < \infty$, of finite length?

9. Find a functional equation satisfied by $K(x)$ as defined by Problem 7 of Exercise C.3. Use this equation to prove that $K(x)$ is either of bounded variation in every interval or in no interval.

10. The absolute value of the function $F(x)$ of (C.3.3) is dominated by the smallest of the expressions

$$\frac{a}{1-a}, \frac{a^2}{1-a} + \pi abx, \frac{a^3}{1-a} + \pi ab(1 + ab)x, \cdots,$$

$$\frac{a^n}{1-a} + \pi ab[1 + ab + (ab) + \cdots + (ab)^{n-2}]x, \cdots.$$

Verify this. The first expression gives the best bound for $x \geqq (\pi b)^{-1}$ when the second expression takes over. In what interval does the nth expression give the lowest bound?

11. Find the area between $y = F(x)$, $y = \dfrac{a}{1-a}$, $x = 0$, and $x = 1/b$. Here $F(x)$ is given by (C.3.3).

12. Find $\int_0^1 K(x)\,dx$, where $K(x)$ is defined as in Problem 7 of Exercise C.3.

13. What is the total variation of $\sin(\log x)$ for $\exp(-100\pi) < x < 1$?

14. Find a differential equation of the form

$$x^2y'' + axy' + by = 0,$$

a and b constants, satisfied by $\sin(\log x)$ and $\cos(\log x)$.

15. Prove that the series

$$\sum_{n=1}^{\infty} \sin(n!\, \pi x)$$

converges trivially if x is a rational number. Prove also that it converges for $x = e$ and for $2/e$. [Riemann]

16. Riemann also claimed that the series converges for $x = \sin 1$ and $\cos 1$. Express these numbers as infinite series and verify the assertion.

17. If $\cos \dfrac{\pi}{5} = \alpha$, $\sin \dfrac{\pi}{5} = \beta$, then

$$(\alpha + i\beta)^5 = -1.$$

Obtain from this a biquadratic equation for α and determine its value.

18. Show that $x \sin(\log x)$ is of bounded variation in $(0, 1)$. [*Hint:* Compare Problem 2 above.]

19. In the preceding problem the factor x is evidently sufficient to dampen the oscillations of $\sin(\log x)$. Would the same hold if x is replaced by x^α, $0 < \alpha < 1$?

MISCELLANEOUS PROBLEMS FOR VOLUME I

Find the derivatives of the following functions:

1. $\left(\dfrac{x-1}{x+1}\right)^{\frac{3}{2}}$. 2. $[e^{3x}-1]^{\frac{1}{3}}$. 3. $\log \dfrac{x^2-1}{x^2+1}$.

4. $\cos^3 x \sin^2 x$. 5. $\sin(\log x)$. 6. $\sin \dfrac{2\tan x}{1-\tan^2 x}$.

7. $\arcsin \dfrac{2\tan x}{1+\tan^2 x}$. 8. $\log \cosh x$. 9. $\arctan e^x$.

10. $e^{\arcsin x}$ 11. $\log \dfrac{1-\sin x}{1+\sin x}$. 12. $\log(1+e^{2x})$.

13. A mapping on reals to reals is defined by

$$y = \log[(x^2+1)^{\frac{1}{2}} - x].$$

Find the domain and the range. Is the correspondence one-to-one?

14. Between what limits must a stay if

$$\tan x - a\tan^2 x$$

is to be monotone in $[-\frac{1}{4}\pi, \frac{1}{4}\pi]$?

15. How large must a be chosen in order that

$$a^x - 1 - x$$

be nonnegative for all x?

16. What point of the straight line $2x + 3y = 6$ is nearest to the point (3, 2)? What is the equation of the circle with center at (3, 2) and tangent to the line? At what point does this circle have vertical tangents?

17. Find the equation of the pencil of circles through the point (a, b) and of slope b/a at this point. Does any circle of the pencil pass through the origin? Note $a \neq 0$, $b \neq 0$.

18. Find the coordinates x_0 and y_0 of the center of the conic

$$x^2 + 2xy - y^2 + x - y = 0.$$

Lay a pencil of straight lines through (x_0, y_0) and determine which of these lines are asymptotes.

19. A family of parallel straight lines intersects the ellipse $b^2x^2 + a^2y^2 = a^2b^2$. What is the locus of the midpoint of the chord?

20. Given the conic

$$Ax^2 + 2Bxy + Cy^2 + 2Dx + 2Ey + F = 0, \quad C > 0.$$

Solve the equation for y and determine the nature of the conic by discussing the quadratic polynomial in x under the radical.

21. Prove that there are one or two excluded intervals for the value of the slope of the normal of a given hyperbola. What is the geometric significance of this? The center may be taken at the origin, but the axes need not coincide with the coordinate axes.

22. Show that the two hyperbolas $x^2 - y^2 = a^2$ and $xy = \frac{1}{2}a^2$ have exactly two real normals in common. Find their equations.

23. Given that the cubic

$$C(z) \equiv z^3 - az^2 + bz - c = 0$$

has the three roots α, β, γ. Construct a cubic with roots α^2, β^2, γ^2.

24. Show that the roots of the three equations

$$C(z) = 0, \quad C'(z) = 0, \quad C''(z) = 0,$$

regarded as points in the complex plane, have the same center of gravity.

25. Determine the cubic with roots $\alpha + d$, $\beta + d$, $\gamma + d$.

26. Assuming $c \neq 0$, determine the cubic with roots $1/\alpha$, $1/\beta$, $1/\gamma$.

27. Determine a so that the equation

$$x^4 - 4x^2 + 4ax - 1 = 0$$

has a double root and find all roots for these values of a.

28. Prove that

$$\tfrac{1}{2} + \cos x + \cos 2x + \cdots + \cos nx = \frac{\sin (n + \frac{1}{2})x}{2 \sin \frac{1}{2}x}.$$

A sequence $\{x_n\}$ is defined by recurrence, taking $x_1 = 1$ and $n \geqq 1$. Prove the existence and find the value of $\lim x_n$, if the recurrence relation is:

29. $x_{n+1} = \frac{1}{2}x_n + \frac{1}{3}(x_n{}^2 + 1)^{\frac{1}{2}}$.

30. $x_{n+1} = \log (1 + x_n{}^2)$.

31. $x_{n+1} = \frac{1}{2}(x_n{}^2 - 1)$.

32. $x_{n+1} = \frac{1}{2}\left(x_n + \dfrac{a}{x_n}\right), \quad a > 0.$

33. $x_{n+1} = \frac{1}{3}\left(2x_n + \dfrac{a}{x_n{}^2}\right), \quad a > 0.$

34. Two sequences are defined by

$$a_{n+1} = \tfrac{1}{2}(a_n + b_n), \quad b_{n+1} = (a_n b_n),^{\frac{1}{2}} \quad n \geqq 1,$$

where $0 < a_1 = a < b_1 = b$. Prove the existence of $\lim a_n = \lim b_n$. The limit is called the *arithmetic-geometric mean* of a and b (Gauss). It may be convenient to set $a = b \cos \theta$ and to use formulas for half the angle.

Which of the following series $\Sigma\, u_n$ converge if, for $n > 1$, u_n is given by:

35. $\dfrac{(-1)^{n-1}}{n + (-1)^n}$.

36. $\dfrac{1}{3n} + \dfrac{1}{3n + 1} - \dfrac{2}{3n + 2}$.

37. $\sin \pi\left(n + \dfrac{1}{n}\right)$.

38. $\log \left(1 + \dfrac{1}{n}\right) - \dfrac{1}{n}$.

39. $\exp \left\{-\dfrac{\alpha}{b} - \dfrac{\alpha}{b + 1} - \dfrac{\alpha}{b + 2} - \cdots - \dfrac{\alpha}{b + n}\right\}, \quad b > 0.$

Find the radius of convergence of the power series $\Sigma\, a_n z^n$, if for $n > 1$ the coefficients are given by:

40. $1 + a^n$.   41. $\dfrac{1}{1 + a^n}$, $a > 0$.   42. $\dfrac{(n!)^4}{(4n)!}$.

43. $\displaystyle\int_0^1 (t \log t)^n \, dt$.   44. $\displaystyle\frac{1}{n}\int_2^n \frac{dt}{\log t}$.

The coefficients of a power series $\Sigma\, a_n z^n$ are given by recurrence relations valid for $n \geqq 1$ with $a_1 = 1$. Find the radius of convergence of the series if:

45. $a_{n+1} = \log(1 + a_n)$.   46. $a_{n+1} = \arctan a_n$.

47. $a_{n+1} = a_n[\frac{1}{3}(1 + a_n)]^{\frac{1}{3}}$ 48. $a_{n+1} = a_n \log(1 + a_n)$.

49. $(n + 2)a_{n+2} - (n + 1)a_{n+1} - a_n = 0$, $a_2 = 2$.

50. Given the power series $f(z) = \sum_{n=0}^{\infty} a_n z^n$ where

$$a_0 = 0, \quad a_1 = 1, \quad a_2 = 0, \quad a_3 = 2, \quad a_{n+4} = a_{n+2} - a_n$$

for $n = 0, 1, 2, \cdots$. Find the function represented by the series. Find a_n as a function of n and determine the radius of convergence. [*Hint:* Show that $(z^4 - z^2 + 1) f(z)$ is a polynomial.]

51. Let r_1, r_2, r_3 be the roots of the cubic

$$x^3 + px + q = 0,$$

where p and q are given real numbers. Form the power series

$$F(z) \equiv \sum_{n=0}^{\infty} a_n z^n \quad \text{where} \quad a_n = r_1{}^n + r_2{}^n + r_3{}^n$$

for all n. Show that there is a relation

$$Aa_n + Ba_{n+1} + Ca_{n+2} + Da_{n+3} = 0$$

valid for all n. Determine $F(z)$ and find what conditions must be satisfied by p and q in order that the radius of convergence of the series be equal to 1.

Find the limits when $n \to \infty$ of:

52. $\sin \dfrac{a}{n^2} + \sin \dfrac{3a}{n^2} + \cdots + \sin \dfrac{(2n - 1)a}{n^2}$.

53. $\displaystyle\sum_{k=1}^{n} \left\{\left(1 + \frac{k^2}{n^3}\right)^{\frac{1}{3}} - 1\right\}$.   54. $n^2[a^{1/n} - a^{1/(n+1)}]$, $a > 0$.

55. $\{\frac{1}{2}[a^{1/n} + b^{1/n}]\}^n$, $a \geqq b > 0$.

56. $\dfrac{a^n}{(1 + a)(1 + a^2) \cdots (1 + a^n)}$, $a > 0$.

Find the limits when $x \to 0$ of:

57. $\dfrac{\sin mx}{\sin nx}$.   58. $\dfrac{1}{x}(a^x - b^x)$.   59. $\dfrac{\arcsin x - x}{\sin^3 x}$.

60. $\dfrac{\sin(a + x) - \sin(a - x)}{\cos(a + x) - \cos(a - x)}$.   61. $\left(\dfrac{\tan x}{x}\right)^{1/x^2}$.

62. Choose a so that the following function has a finite limit as $x \to 0$ and find the limit:

$$x^{-3}[(1 + x)^{1/x} - e + \frac{e}{2}x + ax^2].$$

63. If $0 < a < b$, find the sum of the series

$$\sum_{n=0}^{\infty} (a^{-n} + b^{-n})z^n.$$

64. Assuming that

$$(1 - z - z^2)^{-1} = a_0 + a_1 z + a_2 z^2 + \cdots$$

for small values of z, show that

$$a_0 = 1, \quad a_1 = 1, \quad a_2 = 2, \quad a_3 = 3, \quad a_4 = 5,$$

and generally that

$$a_{n+2} = a_{n+1} + a_n, \quad n \geqq 0.$$

These coefficients are known as the numbers of Fibonacci (Leonardo da Pisa, the son of Bonaccio, about 1175–1250, the first renaissance mathematician.)

65. If w and $-w^{-1}$ are the two roots of

$$z^2 + z - 1 = 0,$$

prove that

$$a_n = \frac{w^{-n-1} + (-1)^n w^{n+1}}{w + w^{-1}}.$$

What is the radius of convergence of the series?

66. Verify that for $|z| < 1$

$$\log(1 - 2z\cos\alpha + z^2) = -2\sum_{n=1}^{\infty} \frac{\cos n\alpha}{n} z^n.$$

Evaluate the following intervals:

67. $\displaystyle\int_2^3 \frac{x^4 + x + 1\, dx}{x^3 - 2x^2 + x}$.

68. $\displaystyle\int_0^1 \frac{dx}{(2 - x)(1 - x)^{\frac{1}{2}}}$.

69. $\displaystyle\int_0^{\infty} (1 + x^2)^{-\frac{3}{2}}\, dx$.

70. $\displaystyle\int_1^{\infty} (1 + x^2)^{-\frac{1}{3}} \frac{dx}{x}$.

71. $\displaystyle\int_1^{\infty} (1 + x^5 + x^{10})^{-\frac{1}{2}} \frac{dx}{x}$.

72. $\displaystyle\int_0^{\pi} (\cos x)^4\, dx$.

Verify that:

73. $\displaystyle\int_{-1}^1 \frac{dx}{(a - x)(1 - x^2)^{\frac{1}{2}}} = \frac{\pi}{(a^2 - 1)^{\frac{1}{2}}}, \quad a > 1.$

74. $\displaystyle\int_0^{2\pi} \frac{dt}{a + b\cos t} = \frac{2\pi}{(a^2 - b^2)^{\frac{1}{2}}}, \quad |b| < a.$

75. $\displaystyle\int_0^\pi \frac{\cos (4n + 1)x}{\cos x}\, dx = \pi$ [use $\cos x = \frac{1}{2}(e^{ix} + e^{-ix})$].

76. $\displaystyle\int_0^{\pi/2} (\tan x)^{\frac{1}{2}}\, dx = \int_0^{\pi/2} (\tan x)^{-\frac{1}{2}}\, dx = \frac{\sqrt{2}}{2}\,\pi.$

77. $\displaystyle\int_0^1 x^{-x}\, dx = \sum_{n=1}^{\infty} n^{-n}.$

78. $\displaystyle\lim_{a\to\infty} \int_0^\infty \frac{dx}{1 + ax^2} = 0.$

79. Determine the numbers p and q so that the function

$$F(x) = \int_x^\infty (t^2 + pt + q)e^{-t}\, dt$$

and its first derivative vanish for $x = 1$. Is the series $\Sigma\, F(n)$ convergent?

80. If $f(t)$ is a positive continuous function which decreases to zero, show that

$$\sum_{n=0}^{\infty} \int_{n\pi}^{(n+1)\pi} f(t) \sin t\, dt$$

is a convergent series. Give an example to show that the series need not be absolutely convergent.

81. Show that

$$\lim_{a\to\infty} \int_0^a \sin (t^2)\, dt$$

exists as a finite number.

82. Verify that

$$\frac{x}{\pi} \int_{-\infty}^{\infty} \frac{dt}{t^2 + x^2} = 1 \quad \text{for all} \quad x > 0,$$

and use this to show that

$$\lim_{x\to 0} \frac{x}{\pi} \int_{-\infty}^{\infty} \frac{f(t)\, dt}{t^2 + x^2} = f(0),$$

if $f(x)$ is a bounded continuous function. Express $f(a)$ as a similar limit if $a \neq 0$.

83. If $f(x)$ is a positive continuous function in a finite interval $[a, b]$, prove that

$$\lim_{p\to\infty} \left\{ \int_a^b [f(x)]^p\, dx \right\}^{1/p} = \max [f(x)].$$

84. Plot the curve

$$y^2 = \frac{x^4 + x^2 + 1}{x^2(x + 1)^2}.$$

Determine the asymptotes.

85. (i) Plot the curve (a *strophoid*)

$$(2a - x)y^2 = x(x - a)^2.$$

(ii) Express the curve parametrically by

$$x = \frac{2at^2}{t^2 + 1}, \quad y = \frac{at(t^2 - 1)}{t^2 + 1}.$$

(iii) Under what conditions on t_1, t_2, t_3 are the corresponding three points located on a straight line?

86. Find the locus of the point from which two perpendicular tangents can be drawn to a given hyperbola $b^2x^2 - a^2y^2 = a^2b^2$.

87. Show that the locus of the foot point of the perpendicular from the origin to the normal of the hyperbola has the equation

$$(X^2 + Y^2)^2(a^2Y^2 - b^2X^2) = (a^2 + b^2)^2X^2Y^2.$$

88. Find the length of the evolute of the ellipse. See equations (8.4.17) and (8.4.18).

*89. Show that the evolute of the lemniscate $(x^2 + y^2)^2 = a^2(x^2 - y^2)$ has the equation

$$9(\alpha^{\frac{2}{3}} + \beta^{\frac{2}{3}})^2(\alpha^{\frac{2}{3}} - \beta^{\frac{2}{3}}) = 4a^2.$$

90. Find the evolute of $y^2 = x^3$.

91. PP_1 is a short arc of length σ on the curve C. On the tangent to C at P is marked a point Q whose distance from P is also σ. Let λ be the distance of Q from P_1. Prove that the ratio σ^2/λ tends to the limit 2ρ when $\sigma \to 0$. Here ρ is the radius of curvature of C at P and is finite by assumption. [*Hint:* Place P at the origin, let the tangent be the x-axis, use parametric representation of C in terms of arc length s, and remember that $[x'(s)]^2 + [y'(s)]^2 = 1$.]

92. On the bisector of the angle A $(= 2\alpha)$ in the triangle ABC a point P is chosen. Find the relation holding between AP, AB, AC, and α when the difference between the angles APB and APC is a maximum or a minimum.

93. Find the triangle of largest area with a given base a and perimeter $2p$.

94. Which normal to the parabola $y^2 = 4ax$ forms the shortest chord?

95. Which is the longest chord that can be drawn from one of the endpoints of the minor axis of an ellipse? Note that the character of the extremum changes when the eccentricity passes through the critical value $\frac{1}{2}\sqrt{2}$.

96. Through a fixed point P interior to an ellipse a chord is drawn such that the two resulting elliptic segments have, one the largest and the other the smallest area. Prove that P bisects the cord.

97. Verify that the point $P = (\frac{1}{2}\sqrt{2}a, \frac{1}{2}\sqrt{2}b)$ on the ellipse $b^2x^2 + a^2y^2 = a^2b^2$ has the following extremal properties:

(i) The tangent at P forms the least angle with the radius vector.

(ii) The normal at P forms the largest triangle with the coordinate axes.

98. A vessel with vertical walls of height h stands on a horizontal plane. It is filled with water which squirts out of a tiny opening in the wall with the initial horizontal velocity $(2gx)^{\frac{1}{2}}$ where x is depth of the opening below the water level

(law of Torricelli). Where should the opening be placed so that the initial jet hits the plane at the largest possible distance from the vessel?

99. Find the angle at the vertex of a right circular cone of least volume which circumscribes a given sphere.

100. Discuss the maxima and minima of

$$f(x) = x^3 + px + q \quad (p, q \text{ real}),$$

and use the result to show that the cubic equation

$$x^3 + px + q = 0$$

has three real and distinct roots if and only if

$$4p^3 + 27q^2 < 0.$$

101. Show by induction that

$$\frac{1 \cdot 3 \cdot 5 \cdots (2n-1)}{2 \cdot 4 \cdot 6 \cdots 2n} < \frac{1}{(3n+1)^{\frac{1}{2}}} \quad \text{for} \quad n = 1, 2, 3, \cdots.$$

102. Show that the following equation has one and only one real root and determine the root:

$$\text{arc tan } x - x + \tfrac{1}{3}x^3 = 0.$$

103. Show that the functional equation

$$f\left(\frac{2x}{1+x^2}\right) = (1 + x^2) f(x)$$

is satisfied by

$$f(x) = 1 + \tfrac{1}{3}x^2 + \tfrac{1}{5}x^4 + \tfrac{1}{7}x^6 + \cdots, |x| < 1.$$

104. Show that

$$\tan x + 2 \log \cos x > 0 \quad \text{for} \quad 0 < x < \tfrac{1}{2}\pi.$$

105. Show that $\tan x < x + x^3$ for $0 < x < \frac{1}{3}\pi$.

106. Show that $x < \frac{1}{3} \tan x + \frac{2}{3} \sin x$ for $0 < x < \frac{1}{2}\pi$.

107. Show that $\log x - \text{arc tan } x$ is strictly increasing for $0 < x$.

108. Let f be continuous and $f(x) < 1$ for $0 \leqq x \leqq 1$. Show that the equation

$$2x - \int_0^x f(t)\, dt = 1$$

has one and only one solution in this interval.

109. Show that the function $g(x) = f(x)/x$ is strictly increasing for $x > 0$, if $f'(x)$ exists and is strictly increasing for $x \geqq 0$ and $f(0) = 0$.

110. Let f be a differentiable function and let a fixed number K exist such that $|f'(x)| \leqq K < 1$ for all x. Let x_0 be given and define a sequence $\{x_n\}$ by the formula

$$x_n = f(x_{n-1}), \quad n = 1, 2, 3, \cdots.$$

Show that $a = \lim\limits_{n\to\infty} x_n$ exists, that a satisfies the equation $f(a) = a$, and that a is independent of the choice of x_0.

111. If f has a continuous derivative, if $f'(x) \leqq K < 1$ for all $x \geqq 0$, and if $f(0) > 0$, show that the equation $f(x) = x$ has a single positive root.

112. Let h be a positive function such that $h(x) \to +\infty$ with x. Show that

$$\liminf_{x \to +\infty} \frac{g(x)}{h(x)} > 1$$

if and only if there is a constant $k > 1$ such that

$$g(x) - kh(x) \to +\infty \quad \text{when} \quad x \to +\infty.$$

113. Let f be continuous for $0 \leqq x < \infty$ and satisfy

$$\lim_{x \to \infty} \frac{f(x)}{x} = A.$$

Show that

$$\lim_{u \to 0} u^2 \int_0^\infty e^{-ut} f(t)\, dt = A.$$

114. Show that the function

$$f(x) = \int_1^x e^{-(xt)^2} \frac{dt}{t}$$

has a maximum for a positive x. Find this x.

115. It is known that the sequence $\{f_n(x)\}$ converges to zero in $[0, 1]$ and that $|f_n'(x)| \leqq M$ for all n and all x in $[0, 1]$. Show that the convergence is uniform.

116. Given that a function $f(x)$ is continuous but not uniformly continuous in an interval (a, b). Show that this implies the existence of a number $k > 0$ and two sequences $\{x_n\}$ and $\{y_n\}$ of points in (a, b) such that $\lim (x_n - y_n) = 0$ and

$$|f(x_n) - f(y_n)| > k \quad \text{for all } n.$$

117. Examine the existence of

$$\lim_{n \to \infty} n^{-\frac{1}{2}} e^{-n} \int_n^{n+1} x^{\frac{1}{2}} e^x \, dx$$

and find its value if it exists.

118. It is known that the series

$$\sum a_n \quad \text{and} \quad \sum |a_n|^5$$

converge. Prove the convergence of $\Sigma\,(2 \sin a_n + \tan a_n)$, if no a_n is an odd multiple of $\frac{1}{2}\pi$.

119. The function f and its first- and second-order derivatives are continuous in $[-1, +1]$. Further $f(0) = 1, f'(0) = 0, f''(0) = 2$. Does the series $\sum \left\{ f\left(\frac{1}{n}\right) - 1 \right\}$ converge?

120. Set

$$f_n(x) = x \frac{x^n - 1}{x^n + 1}.$$

Show that $\lim f_n(x)$ exists for all x in $[0, 1]$. Is the convergence uniform?

121. For what real values of x is the series $\sum_1^\infty \left(x + \frac{1}{\sqrt{n}}\right)^n$ convergent?

122. Find $\lim_{n\to\infty} \int_0^{\pi/2} (1 + \tan^n x)^{-1}\, dx$.

123. Find $\lim_{n\to\infty} n \int_0^\infty \frac{e^{-x}\, dx}{(x + n)^2}$.

124. The sequence $\{f_n(x)\}$ is defined in the interval $[0, 1]$ by $f_0(x) = 1$, $f_n(x) = \{xf_{n-1}(x)\}^{\frac{1}{2}}$, $n > 0$. Show that $\lim_{n\to\infty} f_n(x)$ exists and that the convergence to the limit is uniform.

125. Show that there is a number a such that

$$\lim_{n\to\infty} \left(an^2 - n \cot \frac{1}{n}\right)$$

exists. Find the limit.

126. Find $\lim_{n\to\infty} \left(1 + \sin \frac{2}{2n + 1}\right)^{(n+1)^{\frac{1}{2}}}$

127. Determine the two values of the integer p for which

$$\lim_{x\to\infty} \frac{e^x - \sin x - p}{x^p \sin x}$$

exists and is $\neq 0$.

128. If $f_n(x) = ne^{-nx} \sin nx$, and a is a fixed positive number, compute

$$\lim_{n\to\infty} \int_0^a f_n(x)\, dx \quad \text{and} \quad \int_0^a \lim_{n\to\infty} f_n(x)\, dx.$$

129. A function f is defined so that

$$f(x) = \begin{cases} \frac{1}{2}x & \text{for } x < 0, \\ P(x) & \text{for } 0 \leqq x \leqq 1, \\ 1 & \text{for } 1 < x, \end{cases}$$

where $P(x)$ is a third degree polynomial. How should $P(x)$ be chosen so that $f(x)$ and $f'(x)$ are continuous for all x?

130. Let f be defined by

$$f(x) = \frac{1}{x}|x^2 - 2x| + |1 - x|^{\frac{3}{2}}, \quad x \neq 0, \quad f(0) = 2.$$

For what values of x is f continuous, differentiable?

131. Find the minimum value of the radius of curvature of the catenary

$$y = \frac{a}{2}(e^{x/a} + e^{-x/a}).$$

132. For what positive value of a and b does

$$\int_0^\infty x^{-b}(1 - \cos 2x)^a \sin x\, dx$$

converge?

133. Compute $\int_0^4 (4x + 4\sqrt{x} + 2)^{-\frac{1}{2}}\, dx$.

Abbreviations

abs., absolute, -ly
appr., approximate, -ly
arb., arbitrary, -ily
asymp., asymptote, -ic
bd., bounded
coeff., coefficient
cond., condition
cont., continuous
coord., coordinate
cvg., cvgce., cvgt., converge, -ence, -ent
decr., decrease, -ing
def., define, -ing, -ition
dvg., dvgce., dvgt., diverge, -ence, -ent
elmt., element
excl., excluded
fcn., function
g.l.b., greatest lower bound
iff, if and only if
imag., imaginary
ineq., inequality
infl., inflection
incr., increase, -ing
l.u.b., least upper bound
max., maximum, -a
min., minimum, -a
nasc, necessary and sufficient condition
neg., negative
No., no., number
par., parallel
pargram., parallelogram
polyn., polynomial
pos., positive
pt., point
rat., rational
rel., relative, -ly
resp., respectively
sat., satisfy, -ied, -ies
sq., square
tgt., tangent (line or verb)
Th., th., theorem
unif., uniform, -ly
vert., vertical
w.r.t., with respect to

Solutions to Odd-Numbered Problems

CHAPTER 1

Exercise 1.2. **1.** $A \cup B = \{1, 2, 3, \cdots, 9, 10\}$, $A \cap B = \varnothing$. **3.** $A \bigcirc B = \{n \mid n = 4k\}$. **7.** Use induction: $3^{2n+3} + 2^{n+3} = 9(3^{2n+1} + 2^{n+2}) - 7 \cdot 2^{n+2}$. **9.** $\{a, b, c, d\}$, $\{a, b, c\}$, $\{a, b, d\}$, $\{a, c, d\}$, $\{b, c, d\}$, $\{a, b\}$, $\{a, c\}$, $\{a, d\}$, $\{b, c\}$, $\{b, d\}$, $\{c, d\}$, $\{a\}$, $\{b\}$, $\{c\}$, $\{d\}$, $\varnothing$. **11.** Intersections of the preceding sets with $\{a, b, c\}$ are resp. $\{a, b, c\}$, $\{a, b, c\}$, $\{a, b\}$, $\{a, c\}$, $\{b, c\}$, $\{a, b\}$, $\{a, c\}$, $\{a\}$, $\{b, c\}$, $\{b\}$, $\{c\}$, $\{a\}$, $\{b\}$, $\{c\}$, $\varnothing$, $\varnothing$. **13.** $A \cup B \cup C$ is the set of elmts. $\in$ of at least one of the sets A, B, C. $A \cap B \cap C$ is the set of elmts. $\in$ of all three. **17.** T takes n into $n + 1$ and $n + 1$ is the image of n only, hence T is (1, 1). T is not onto since 1 is not the image of any $n \in N$. **19.** -27. Six mappings are onto and (1, 1), namely those with the images $\{a, b, c\}$, $\{b, c, a\}$, $\{c, a, b\}$, $\{a, c, b\}$, $\{b, a, c\}$, $\{c, b, a\}$. In 18 mappings one elmt. occurs twice as image, in 3 mappings one elmt. occurs thrice. **21.** 10. **23.** $S_2(n + 1) = S_2(n) + (n + 1)^2$. Simplify!

Exercise 1.3. **1.** $mq = np$. **3.** If $3 = p/q$, p, q rel. prime, then $3q^2 = p^2$ and $p = 3m$, $q^2 = 3m^2$ or $q = 3n$, contradiction. **5.** $x^4 - 10x^2 + 1 = 0$, roots $\pm\sqrt{2} \pm\sqrt{3}$. **7.** $n^2x^2 - 2mnx + m^2 - 2p^2 = 0$, m, n, p arb. integers. **15.** A_0 is zero, A_1 is one, $A_2 = -A_1$, $A_2A_2 = A_1$.

Exercise 1.4. **1.** 1.41, 3.14. **3.** If $10^{p/q} = 2$, then $10^p = 2^q$. The left is divisible by 5, the right is not. Contradiction. 0.30103 is rat. but is only an appr. value of log 2. **5.** Sums and products of nos. $r + s\sqrt{2}$ have the same form. $Q(\sqrt{2})$ contains 0 and 1 and they are zero and unit elmts. of $Q(\sqrt{2})$. $r + s\sqrt{2}$ has a reciprocal in $Q(\sqrt{2})$ if $r^2 + s^2 \neq 0$. **7.** If α sat. $a_0x^n + a_1x^{n-1} + \cdots + a_n = 0$, then $\sqrt{\alpha}$ sat. $a_0t^{2n} + a_1t^{2n-2} + \cdots + a_n = 0$ with integral coeffs. **9.** Let the equation def. α be $P(x^2) + xQ(x^2) = 0$. Then α^2 sat. $[P(t)]^2 - t[Q(t)]^2 = 0$. **11.** P_1. If $a \neq 0$, then either a or $-a$ is pos. P_2. If a and b are pos., so are $a + b$ and ab. **13.** If $A_0 < A_1$, then $A_0 + A_1 < A_1 + A_1 = A_0$ or $A_1 < A_0$, absurd. Same for $A_1 < A_0$. **15.** $\sup S = \frac{13}{36}$, $\inf S = 0$. **17.** $\sqrt{3}, -\sqrt{3}$. **19.** $\frac{1}{3}, \frac{2}{9}$. **21.** Let S be the set of rat. nos. r with $r^2 + r < 1$. The roots are $\inf S$ and $\sup S$. **23.** E.g. the set of nos. $r + s\sqrt{2}$ with sq. <3. **25.** Let $A = \{s \mid s \leqq 0\} \cup \{t \mid 0 < t^3 < 2\}$, $B = \{u \mid 2 < u^3\}$, s, t, u in Q.

Exercise 1.5. **1.** 3, 3, $5\sqrt{2} - 6$, $\pi - 3$, $10 - 3\pi$, $5\sqrt{2} - \pi - 1$. **3.** $|x - a| + |x - b| = |a - b|$, if x lies between a and b, if not either $|x - a| > |a - b|$ or $|x - b| > |a - b|$. **5.** The property of being par. is reflexive, symmetric and transitive. **9.** $(0, \frac{1}{2})$, $(0, \frac{3}{2})$. **13.** Length of the diagonal. **15.** $\sqrt{3}$ times the side. **23.** $ac + bd = 0$.

Exercise 1.6. **1.** If x is age, then $A \cap B = \{x \mid 25 \leqq x < 50\}$. $A \cup B$ is all persons. **3.** $S_1 \cup S_2 = \{x \mid 0 < x < 4\}$, $S_1 \cup S_2 \cup S_3 = \{x \mid -1 < x < 4\}$, $S_1 \cap S_2 = \{x \mid 1 < x \leqq 2\}$, $S_1 \cap S_2 \cap S_3 = \varnothing$. The third ineq. is incompatible with the first as well as the second. **5.** $(x^2 + y^2)^{\frac{1}{2}}$ if $x \geqq 0$, $|y|$ if $x < 0$. **7.** $a > 1$. **9.** The interior of the sq. with vertices $(1, 0)$, $(0, 1)$, $(-1, 0)$, $(0, -1)$. **11.** Yes, since $2x^2 + 3xy + y^2 = (\frac{3}{2}x + y)^2 - (\frac{1}{2}x)^2$. **13.** Use (1.6.7). **15.** Either $x < -1 - \sqrt{2}$ or $x > 1$. **17.** Use binomial th. **19.** See Fig. 1.15. **21.** If $a > c$, $b > c$, then $ab > c^2$. If $ab > c^2$, then $a \leqq c$, $b \leqq c$ is impossible. **23.** $x_1 x_2 \cdots x_n(1 - x_1)(1 - x_2) \cdots (1 - x_n) = [x_1(1 - x_1)][x_2(1 - x_2)] \cdots [x_n(1 - x_n)] \leqq 4^{-n}$ by **22**. Then use **21**.

Exercise 1.7. **1.** $\frac{1}{11} = 0.0909 \cdots$. **3.** True. **5.** $a_n = (n + 1/n)^{-1} \to 0$. **7.** $a_n = (2^n + 2^{-n})^{-1} \to 0$. **9.** $A_n = (n + 1)^{\frac{1}{2}} - 1$ strictly incr. to ∞. **11.** $q_n = (1 - n^{-2})(3 + n^{-2})^{-1} \to \frac{1}{3}$. **13.** $q_n = [1 + (1 + n^{-2})^{\frac{1}{2}}]^{-1} \to 1$. **17.** $a^{1/n}$ decr. to 1 if $a > 1$, is always 1 if $a = 1$, and incr. to 1 if $0 < a < 1$. Note that $a^{1/n} < a^{1/(n+1)}$ iff $a^{n+1} < a^n$ or $a < 1$ etc. **21.** 0. **23.** ∞. **25.** If $|a_n - a| < \varepsilon$ for all large n, then this is also true when n runs through the subsequence. **27.** If $a_{n-1} < a_n$ for some n, then $2a_{n-1} = (a_n)^2 > (a_{n-1})^2$ or $a_{n-1} < 2$. Hence $(a_n)^2 < 2a_n$ or $a_n < 2$ and $(a_{n+1})^2 = 2a_n < 2^2$ or $a_{n+1} < 2$ and $2a_n = (a_{n+1})^2 < 2a_{n+1}$ so that $a_n < a_{n+1}$. Hence $\{a_n\}$ is strictly incr. and bd. by 2. Thus $\lim a_n \equiv a$ exists and $a \leqq 2$. Since $(a_n)^2 = 2a_{n-1}$, $(\lim a_n)^2 = 2 \lim a_{n-1}$ and $a^2 = 2a$, $a = 2$. **29.** 1. **33.** $n!/k! = n(n - 1) \cdots (n - k + 1)$.

Miscellaneous problems for Chapter 1

1. If $x = m$, $y = n$, with m, n integers, then $2m + 4n$ is even while 11 is odd. **3.** There is an n such that $2^n \leqq p < 2^{n+1}$. Thus $a_1 = 1$. There is an m such that $2^m \leqq p - 2^n < 2^{m+1}$. This gives $a_m = 1$ and $a_k = 0$ for $m < k < n$, etc. $1000 = 2^9 + 2^8 + 2^7 + 2^6 + 2^5 + 2^3 = 1111101000$. **5.** If $\sqrt{2} + \sqrt{3} = r$, then $5 + 2\sqrt{6} = r^2$ or $\sqrt{6} = \frac{1}{2}(r^2 - 5)$, a rat. no. Contradiction. **7.** If C_1, C_2 are convex, if P_1, P_2 are pts. in their intersection, then the line segment joining P_1 and P_2 lies in both sets, hence in $C_1 \cap C_2$ and the latter is convex. **9.** See Exercise 1.6, No. **22**, and (1.6.4) $a = x$, $b = 1 - x$. **11.** Since $(a_{n+1})^2 - (a_n)^2 = 3(a_n - a_{n-1})$, the sequence is monotone and incr. since $a_2 - a_1 = \sqrt{7} - 1 > 0$. From $(a_n)^2 = 4 + 3a_{n-1} < 4 + 3a_n$, follows that $a_n < 4$, the pos. root of $x^2 = 4 + 3x$ and $\lim a_n = 4$. **15.** Since every $a_n > 1$ and $a_n - a_{n-1} = \dfrac{1}{a_{n-1}} - \dfrac{1}{a_{n-2}} = -\dfrac{a_{n-1} - a_{n-2}}{a_{n-1}a_{n-2}}$, the sign of $a_n - a_{n-1} = (-1)^{n-1}$. Similarly $\operatorname{sgn}(a_{2k} - a_{2k-2}) = -1$, $\operatorname{sgn}(a_{2k+1} - a_{2k-1}) = +1$, so $\{a_{2k}\}$ is decr., $\{a_{2k+1}\}$ incr. and $1 < \lim a_{2k+1} \leqq \lim a_{2k} < a_2$. For $n > 4$, $a_{n-1}a_{n-2} > (a_3)^2$ so that $a_n - a_{n-1} \to 0$ and $\lim a_n \equiv a$ exists. Here $a = \frac{1}{2}(1 + \sqrt{5})$, the pos. root of $x^2 = x + 1$. **17.** Since $a + b > 2\sqrt{ab}$, $\dfrac{2ab}{a + b} < \dfrac{ab}{\sqrt{ab}} = \sqrt{ab}$ unless $a = b$. **19.** If

$x \geqq 2$, then $|x - 2| + |x| = 2(x - 1)$ and this is $\geqq 4$ iff $|x - 1| \geqq 2$. For $0 < x < 2$, $|x - 2| + |x| = 2 < 4$. For $x < 0$, $|x - 2| + |x| = 2(1 - x) \geqq 4$ iff $|x - 1| \geqq 2$. **21.** True for $n = 3$ since $15 < 27$. If true for $n = k$, then $(2k + 1)!! = (2k - 1)!!\,(2k + 1) < k^k(2k + 1) = (k + 1)^{k+1}\left(\frac{k}{k+1}\right)^k \frac{2k+1}{k+1}$. Here $\left(\frac{k}{k+1}\right)^k \leqq \left(\frac{3}{4}\right)^3$ by the def. of e and $\frac{2k+1}{k+1} \leqq \frac{7}{4}$ if $k \geqq 3$. Now $\frac{7}{4}\left(\frac{3}{4}\right)^3 = \frac{189}{256} < 1$ so the ineq. holds for $n = k + 1$ and hence for all $n \geqq 3$. **23.** No. Place one vertex at the origin and assume that the coords. of the other vertices are integers m, n, p, q with $m^2 + n^2 = p^2 + q^2 = (m - p)^2 + (n - q)^2$. Note, if a triangle with rat. coords. exists we can enlarge it to obtain integral coords. Then

$$m^2 + n^2 = p^2 + q^2 = 2(mp + nq).$$

Hence m, n have the same parity and so do p, q since the sums of their sq. are even. If m, n are odd, then $2(p + q - 1)$ must be divisible by 4. This requires that p and q have different parities. Since this is impossible, m, n, p, q are all even. This makes $m^2 + n^2$ and $p^2 + q^2$ multiples of 4. The same type of argument shows that m, n, p, q must be multiples of 4, 8, 16, $\cdots$. After a finite number of steps $m = 2^k m_1$, $n = 2^k n_1$, $p = 2^k p_1$, $q = 2^k q_1$, where m_1, n_1, p_1, q_1 are odd integers with ${m_1}^2 + {n_1}^2 = {p_1}^2 + {q_1}^2 = 2(m_1p_1 + n_1q_1)$. Since this system has no solution in odd integers, no such triangle can exist.

APPENDIX A

Exercise A.1. **3.** $(1, -3)$, $(-3, -1)$, $(-4, -2)$, $(-3, 2)$, $(4, -5)$, $(5, -4)$. **5.** $(-1, -3)$, $(3, -1)$, $(4, -2)$, $(3, 2)$, $(-4, -5)$, $(-5, -4)$. **7.** $\sqrt{10}$, $\sqrt{20}$, $\sqrt{41}$. **9.** $y = x$. **11.** 2, $2\sqrt{5}$, $\sqrt{26}$, $(-3.5, 1.5)$, $(-1, 2)$, $(-1.5, 2.5)$. **13.** $\pm(\boldsymbol{i} - \boldsymbol{j})$, $\pm(4\boldsymbol{i} + 2\boldsymbol{j})$, $\pm(5\boldsymbol{i} + \boldsymbol{j})$. **15.** 0, $3\sqrt{2}$, $3\sqrt{2}$. **17.** $x_1x_2 + y_1y_2 = 0$.

Exercise A.2. **1.** (a) 4, 3, $-\frac{3}{4}$; (b) -2, 2, 1; (c) $-\frac{13}{5}$, $\frac{13}{6}$, $\frac{5}{6}$; (d) 1, 1, -1; (e) 3, 4, $-\frac{4}{3}$; (f) -2, 4, 2. **3.** -3, $-\frac{5}{4}$, -1, $-\frac{5}{7}$, $-\frac{2}{3}$, $-\frac{5}{8}$, $-\frac{1}{2}$, 0, $\frac{1}{4}$, ∞. **5.** (b) and (d). **7.** -5.5, 11. **9.** $23x - 36y + 49 = 0$, $31x - 31y + 44 = 0$. **11.** $2x + 3y - 8 = 0$. **13.** $x - 2y - 3 = 0$. **17.** $5x + 4y - 20 = 0$.

Exercise A.3. **1.** (a) $(-\frac{1}{2}, -\frac{1}{2})$, $\frac{1}{2}\sqrt{2}$; (b) $(\frac{3}{2}, -2)$, $\frac{3}{2}$; (c) pt $(1, -1)$; (d) $(-\frac{5}{2}, 0)$, $\frac{1}{2}\sqrt{21}$; (e) $(0, \frac{11}{2})$, $\frac{9}{2}$; (f) $(5, 4)$, 3. **3.** $x^2 - 2ax + y^2 = 0$. **5.** (a) $x = -\frac{1}{2} \pm \frac{1}{2}\sqrt{2}$; (b) $x = 0, 3$; (d) $x = -\frac{5}{2} \pm \frac{1}{4}\sqrt{21}$; (e) $x = \pm\frac{9}{2}$; (f) $x = 2, 8$. **7.** $a = 1 \pm \sqrt{10}$. **9.** $(2, 0)$, $(\frac{6}{5}, -\frac{8}{5})$. **13.** All circles are tgt. to the y-axis at the origin. **15.** $(x + \frac{3}{4})^2 + (y - \frac{3}{2})^2 = \frac{125}{16}$. **17.** $x^2 - 10x + y^2 = 0$, $4(x^2 + y^2) - 5x + 30y = 0$, $x^2 + 5x + y^2 - 5y = 0$, $2(x^2 + y^2) - 5(x - y) = 15$.

Exercise A.4.1. **1.** Only as a limiting case. Set $de = a$, a fixed, and let $e \to 0$. The focus stays at $(a, 0)$ but the directrix recedes to ∞. The locus problem becomes meaningless. **3.** $(x - y + 1)^2 = 8(x + y - 1)$, $(0, 1)$. **5.** $y - 2x$ equals $-2, -3, -1, 0, 5, \frac{11}{4}$ resp. **7.** $x + y = 6$. **9.** The line is $y = 2a/m$. **11.** $y^2 = -8x$. **13.** $x^2 = -8y$. **15.** $y^2 = -6x$. **17.** $y^2 = 4x$. **19.** $y = \frac{2}{9}x^2$. **21.** $y^2 = b(2x - b)$. **23.** $(x - y)^2 = -6(x + y)$. **25.** $(x + y - 9)^2 = -2(x - y - \frac{3}{2})$.

Exercise A.4.2. **1.** (a) 2, 1; $(\pm\sqrt{3}, 0)$, $x = \pm\frac{4}{3}\sqrt{3}$; $e = \frac{1}{2}\sqrt{3}$. (b) 2, $\frac{4}{3}$; $(\pm\frac{2}{3}\sqrt{5}, 0)$; $x = \pm\frac{6}{5}\sqrt{5}$; $\frac{1}{3}\sqrt{5}$. (c) $\sqrt{5}$, 1; $(\pm 2, 0)$; $x = 2.5$; $\frac{2}{5}\sqrt{5}$. (d) $\sqrt{6}$, 2; $(0, \pm\sqrt{2})$; $y = \pm\sqrt{3}$; $\frac{1}{3}\sqrt{3}$. (e) 2, 1; $(0, \pm\sqrt{3})$; $y = \pm\frac{4}{3}\sqrt{3}$; $\frac{1}{2}\sqrt{3}$. (f) 2, $\frac{2}{3}\sqrt{3}$; $(\pm\frac{2}{3}\sqrt{6}, 0)$; $x = \pm\sqrt{6}$; $\frac{1}{3}\sqrt{6}$. **3.** $7x^2 - 2xy + 7y^2 - 8(x + y) = 0$, $\frac{4}{3}\sqrt{2}$. **5.** (a) $y \mp \frac{1}{17}\sqrt{17} = 2(x \pm \frac{8}{17}\sqrt{17})$, (b) $y \mp \frac{2}{15}\sqrt{10} = 2(x \pm \frac{3}{5}\sqrt{10})$, (c) $y \mp \frac{1}{11}\sqrt{11} = 2(x \pm \frac{10}{11}\sqrt{11})$, (d) $y \mp \frac{3}{11}\sqrt{22} = 2(x \pm \frac{4}{11}\sqrt{22})$, (e) $y \mp \frac{1}{2}\sqrt{2} = 2(x \pm \sqrt{2})$, (f) $y \mp \frac{2}{39}\sqrt{39} = 2(x \pm \frac{4}{13}\sqrt{39})$. **7.** $y - \frac{1}{2} = -\frac{1}{2}\sqrt{3}(x - \sqrt{3})$. **9.** The coords. of the midpt. are $\left(-\dfrac{ma^2}{b^2 + a^2m^2}\,q, \dfrac{b^2}{b^2 + a^2m^2}\,q\right)$. **13.** $4x^2 + 3y^2 = 48$. **15.** $9x^2 + 5y^2 = 80$. **17.** $3x^2 + 2y^2 = 11$. **19.** $(9 + \sqrt{17})x^2 + (1 + \sqrt{17})y^2 = \frac{1}{2}(26 + 10\sqrt{17})$. **21.** $4x^2 - xy + 4y^2 = 144$.

Exercise A.4.3. **1.** (a) $a = 2$, $b = 1$; $(\pm\sqrt{5}, 0)$; $x = \pm\frac{4}{5}\sqrt{5}$; $y = \pm\frac{1}{2}x$. (b) $a = 1$, $b = 2$; $(\pm\sqrt{5}, 0)$; $x = \pm\frac{1}{5}\sqrt{5}$; $y = \pm 2x$. (c) $a = 2$, $b = \sqrt{6}$; $(\pm\sqrt{10}, 0)$; $x = \pm\frac{8}{5}\sqrt{10}$; $y = \pm\frac{1}{2}\sqrt{6}x$. (d) $a = \sqrt{5}$, $b = 1$; $(\pm\sqrt{6}, 0)$; $x = \pm\frac{5}{6}\sqrt{6}$; $y = \pm\frac{1}{5}\sqrt{5}x$. (e) $a = 2$, $b = 1$; $(0, \pm\sqrt{5})$; $y = \pm\frac{4}{5}\sqrt{5}$; $y = \pm 2x$. (f) $a = 1, b = \sqrt{5}$; $(0, \pm\sqrt{6})$; $y = \pm\frac{5}{6}\sqrt{6}$; $y = \pm\frac{1}{5}\sqrt{5}x$. **3.** $(1, 1), (-1, -1)$; $(\sqrt{2}, \sqrt{2})$, $(-\sqrt{2}, -\sqrt{2})$; $a = \sqrt{2}$, $b = \sqrt{2}$. **5.** (a) $y \mp \frac{1}{3}\sqrt{3} = x \mp \frac{4}{3}\sqrt{3}$; (b) no; (c) no; (d) $y \mp \frac{1}{2} = x \mp \frac{5}{2}$; (e) $y \mp \frac{4}{3}\sqrt{3} = x \mp \frac{1}{3}\sqrt{3}$; (f) no. **9.** Coords. of midpt. $\left(\dfrac{a^2mq}{b^2 - a^2m^2}, \dfrac{b^2q}{b^2 - a^2m^2}\right)$. Slope $b^2/(a^2m)$. **11.** $9x^2 - 16y^2 = 144$. **13.** $x^2 - y^2 = 8$. **15.** $20x^2 - 9y^2 = 36$. **17.** $6x^2 - y^2 = 2$. **19.** $14(x^2 + y^2) - 25(xy + x + y - 1) = 0$. **21.** $x^2 + y^2 + 4xy = 6$. **23.** Right branch $-\frac{1}{2}\pi < \theta < \frac{1}{2}\pi$, left $\frac{1}{2}\pi < \theta < \frac{3}{2}\pi$.

Exercise A.5. **1.** Real hyperbola. **3.** $(0, 0)$. **5.** Straight lines $y = 1, y = -x$. **7.** Imag. par. lines. **9.** $\tan 2\alpha = 4$, $0 < 2\alpha < \frac{1}{2}\pi$, gives $\frac{1}{2}(\sqrt{17} - 3)s^2 - \frac{1}{2}(\sqrt{17} + 3)t^2 = 1$. **11.** $s = \frac{1}{2}\sqrt{2}(x + y)$, $t = \frac{1}{2}\sqrt{2}(x - y)$, $s^2 + 3t^2 = 2$. **13.** $s = \frac{1}{2}\sqrt{2}(x - y)$, $t = \frac{1}{2}\sqrt{2}(x + y)$, $s^2 + 3t^2 = 6$. **15.** $\tan 2\alpha = 2$, $\pi < 2\alpha < \frac{3}{2}\pi$, gives $3(\sqrt{5} - 1)s^2 + 3(\sqrt{5} + 1)t^2 = 10\sqrt{5}$. The acute angle gives the ellipse with major axis along the t-axis. **17.** $s = \frac{1}{2}\sqrt{2}(x + y + \frac{1}{4})$, $t = \frac{1}{2}\sqrt{2}(x - y - \frac{1}{4})$, $t^2 = \frac{1}{2}\sqrt{2}s$. **19.** $y = \frac{1}{5}x, y = -x$.

Exercise A.6. **7.** $(x^2 + y^2)^{\frac{3}{2}} = x^2$. **9.** $(x^2 + y^2)^{\frac{3}{2}} = y$. **11.** Ellipse.

Miscellaneous problems for Appendix A

1. (a) $(\frac{1}{2} \pm \frac{1}{2}\sqrt{7}, -\frac{1}{2} \pm \frac{1}{2}\sqrt{7})$; (b) $(\pm\sqrt{10}, \pm\frac{1}{2}\sqrt{2})$. 2 and 4 pts. resp. **3.** $\pm\sqrt{7}y + 3x = 8, \pm\sqrt{15}y + x = 8$. **5.** $x^2 - xy + 9y^2 = 9$. **7.** $x^2 + 3y^2 = 4$. **9.** The coords. of P_5 determine α uniquely. **11.** If a conic passes through the four pts. P_1 to P_4 and also through a fifth P_5, not on any line $L_{jk} = 0$, then there is a conic in the pencil with the same properties. Since there is only one conic on five pts., the given conic must belong to the pencil. **13.** Hyperbola if the lines intersect, otherwise a pair of par. lines. **15.** $Cx^2 + (A + B)xy - (A - B)ay - Ca^2 = 0$. Hyperbola if $A + B \neq 0$, reducing to intersecting straight lines if $C = 0$. Parabola if $A + B = 0$, $C \neq 0$. The line $y = 0$ if $A + B = C = 0$. **17.** Parabola $y^2 = 2mx + m^2$. **19.** $y = 2x - 1, y = \frac{2}{3}x - \frac{1}{9}$. **21.** $(x - y)^2 = 6x + 10y + 7$, $x + y + 2 = 0$. **23.** $y = 4 + 6/(x - 3)$. This conic exists for all x, $x \neq 3$, and all y, $y \neq 4$, hence a hyperbola. The lines $x = 3$, $y = 4$ are asymps. **25.** $y = \frac{1}{2}[-x \pm 3^{\frac{1}{2}}(4 - x^2)^{\frac{1}{2}}]$, real for $-2 \leqq x \leqq 2$. This conic must be an ellipse since restricted to a vertical strip of finite width.

CHAPTER 2

Exercise 2.1. **1.** 2, 1, 0, 1. **3.** $a_n > a_{n+1}$, $\lim a_n = 0$. **5.** True. **7.** If $|a_m - a_n| \leqq 1$ for $m, n \geqq N$, then $|a_n|$ has the finite bound max $\{|a_1|, |a_2|, \cdots, |a_N|, |a_{N+1}| + 1\}$. **9.** If $\bar{S}$ is finite or void, then $\bar{S}$ is closed by def. If $\bar{S}$ is infinite but $\bar{S}' = \varnothing \subset \bar{S}$, so $\bar{S}$ is closed. If a is a cluster pt. of $\bar{S}$, then there are infinitely many pts. of $\bar{S}$ in any neighborhood of a, hence also infinitely many pts. of S, i.e. $a \in \bar{S}$ and $\bar{S}$ is closed. **11.** $0 < a_{m,n} < (n^2 + 2n + 1)^{\frac{1}{2}} - n = 1$. Let $m(n)$ be the largest integer $\leqq 2\alpha n$. Then

$$\frac{2n\alpha - 1}{2n + 1} < a_{m(n),n} < \frac{2n\alpha}{2n} = \alpha$$

and the first member cvgs. to α. **13.** $1/e$.

Exercise 2.2.

1.	Domain	$f(0)$	$f(-\frac{1}{2})$	$f(-1)$
(a)	$\{x \mid x \neq 1\}$	1	$\frac{1}{3}$	0
(b)	all x	0	$\frac{3}{8}$	0
(c)	$\{x \mid x \leqq 1\}$	1	$\frac{1}{2}\sqrt{6}$	$\sqrt{2}$
(d)	all x	0	-2	-1
(e)	$\{x \mid -1 \leqq x \leqq 1\}$	3	$\frac{1}{3}(8 + \sqrt{3} + 2\sqrt{2} + 2\sqrt{6})$	$\frac{1}{2}(2 + \sqrt{2})$

3. All bd. except (a). **5.** $f(x) = 2x$ in $(0, \frac{1}{2})$, $2(1 - x)$ in $(\frac{1}{2}, 1)$. $g(x) = \frac{1}{2}f(x) = \frac{1}{2}[1 - |2x - 1|]$ in $(0, 1)$. **9.** $3x^2 + 3x + 1$. **11.** (i) $R[f] = [y \mid 3 \leqq y \leqq 4]$, (ii) $R[f] = [y \mid 7 < y < 28]$. **13.** $f(x) = x(\frac{1}{2}a - x)$, $D[f] = [x \mid 0 < x < \frac{1}{2}a]$, $R[f] = [y \mid 0 < y \leqq \frac{1}{16}a^2]$. **15.** $f \circ g = (x^3)^2 + 1 = x^6 + 1, g \circ f = (x^2 + 1)^3 = x^6 + 3x^4 + 3x^2 + 1$. **17.** $f \circ g = (1 - x^2)^{\frac{1}{2}}$, $-1 \leqq x \leqq 1$. $g \circ f = 1 - x$, all x. **19.** No. If $f > 0$, $-f < 0$ and not in the set. **21.** $R[f] = [z \mid -5 \leqq z]$. $f = 0$ on the ellipse $x^2 + 4y^2 = 5$. $f(0, \frac{3}{2}) = 4$, $f(1, 2) = 12$, $f(2, 1) = 3$. **23.** $f(x, y) = [(|x| - 1)^2 + y^2]^{\frac{1}{2}}$. $f = 2$ is the circular arcs $(x + 1)^2 + y^2 = 4$, $x \leqq 0$, and $(x - 1)^2 + y^2 = 4$, $x > 0$. **25.** A surface in R_3; paraboloid of revolution.

Exercise 2.3. **1.** $x = \dfrac{1+a}{1-a}$, $a \neq 1$. $1 - \dfrac{x-1}{x+1} = \dfrac{2}{x+1} = 0.000002$ if $x = 999{,}999$. **3.** $D[f] = [x \mid x \neq -1]$, $R[f] = [y \mid y \neq 1]$. **5.** $D[f] =$ all x, $R[f] = [y \mid -1 < y \leqq 1]$. **7.** $D[f] = R[f] = R$. **9.** $D[f] =$ all x, $R[f] = [y \mid y \leqq \frac{1}{4}]$ **11.** $D[f] = [x \mid -1 < x < 1]$, $R[f] = [y \mid 2 \leqq y]$. **13.** $D[f] = [x \mid 0 \leqq x \leqq 1]$, $R[f] = [y \mid 1 \leqq y \leqq \sqrt{2}]$. **15.** $D[f] \cap D[g]$ is the set where both f and g exist, i.e., all x except the (real) zeros of their denominators. In this set $f + g$, fg, αf and βg exist by Def. 2.2.6.—No. If $\alpha \in R$ then $f_\alpha(x) \equiv (x - \alpha)^{-1}$ is def. for $x \neq \alpha$ and every real no. is a pt. of non-existence for some fcn. f_α. **17.** $(1 - x^2)y^2 - 2 = 0$; $u^2 - (1 + x^2) = 0$; $v^4 - 2v^2 + (1 - 2x)^2 = 0$. **19.** $f = 0$ for $x \leqq 0$, $f = 2x$ for $0 < x$. **27.** Yes. **29.** $\lim S_n(x) = (1 - x)^{-1}$ for $-1 < x < 1$. **31.** $OT(x) = a^2 |x|^{-1}$. $D[f] = [x \mid 0 < |x| < a]$. $R[f] = [f \mid a < f]$.

Exercise 2.4. **1.** 3. **3.** $\frac{7}{8}$. **5.** 4. **7.** $-\frac{3}{5}$. **9.** 0. **11.** 0. **13.** $C + D + E$. **15.** C^2. **17.** $(C + D)^2$. **19.** f and $|f|$ have the same domain. If $\lim f(x) = C$, then $\lim |f(x)| = |C|$. **21.** Limit 0 since $[x + 1] = 0$, $-1 \leqq x < 0$, and $[x] = 0$, $0 \leqq x < 1$. **23.** Here $f(-x) = f(x)$ and $1 - x < x\left[\dfrac{1}{x}\right] \leqq 1$ for $x > 0$, so the limit is 1. **29.** $-1, +1$. **31.** $-1, +1$. **33.** 0, 1.

Exercise 2.5. **1.** Cont. in $[0, 1]$ iff $f(0) = 3$, $f(1) = 2$. **3.** Cont. in $[0, 1]$ iff $f(0) = 1$. **5.** Cont. in $[0, 1]$ iff $f(0) = f(1) = 0$. **7.** Cont. in $D[f] = [x \mid 0 \leqq x, x \neq 1]$, becomes infinite when $x \to 1$. **9.** Cont. in $D[f] = [x \mid 0 \leqq x, x \neq 4]$, becomes infinite when $x \to 4$. **11.** $x^{m/n} = (x^{1/n})^m$. The mth power of a cont. fcn. is cont. **13.** Cont. for $x \neq -1$, becomes infinite when $x \to -1$. **15.** Discont. for $x = n$, n any integer. **17.** $f(0) = 1$, $f(x) = 0$, $x \neq 0$. **19.** If $f(c) = c > 0$, there is an interval containing $x = c$ where $|f(x) - C| < C$ and hence $f(x) > C - C = 0$. Let (α, β) be the largest interval, containing c, where $f > 0$. If $a < \alpha$, then $f(\alpha) \not> 0$ since otherwise $f > 0$ in a larger interval. Similarly, $f(\alpha) < 0$ is impossible since this implies $f < 0$ in some neighborhood of $x = \alpha$. Hence $f(\alpha) = 0$. Same argument applies for $x = \beta$. **21.** $2\delta(3c^2 + \delta^2)$. **23.** Since $|f(x_1)g(x_1) - f(x_2)g(x_2)|$

$$\begin{aligned} &= |f(x_1)[g(x_1) - g(x_2)] + g(x_2)[f(x_1) - f(x_2)]| \\ &\leqq |f(x_1)|\, |g(x_1) - g(x_2)| + |g(x_2)|\, |f(x_1) - f(x_2)| \\ &\leqq \|f\|\, \omega(g, I) + \|g\|\, \omega(f, I) \end{aligned}$$

and x_1, x_2 can be so chosen that the first member differs arb. little from $\omega(fg, I)$, the ineq. follows. **25.** $[x \mid (n - \frac{1}{4})^{-1} < x < (n + \frac{1}{4})^{-1}, n = 0, \pm 1, \pm 2, \cdots]$. **27.** f is cont. at (x_0, y_0) if, given any $\varepsilon > 0$, there is a δ such that $|x - x_0| < \delta$, $|y - y_0| < \delta$ implies $|f(x, y) - f(x_0, y_0)| < \varepsilon$.

Exercise 2.6. **1.** If f is cont. in (a, b), then f is cont. in $[a + \delta, b - \delta]$ and the range $R[f, \delta]$ of f, restricted to this interval, is a closed interval. Here $R[f, \delta_2] \supseteqq R[f, \delta_1]$ if $\delta_1 > \delta_2$. Hence $R[f] = \lim R[f, \delta]$ is an interval. **3.** No, the inequalities permit f to keep a constant sign in $[a, b]$. **5.** $[x_1] = -2$, $[x_2] = 0$, $[x_3] = 1$.

7. $x^2 - (x - \delta)^2 = 2x\delta - \delta^2 \leqq 2\delta - \delta^2$ in $[0, 1]$. Similarly $\sup[x^n - (x - \delta)^n] = 1 - (1 - \delta)^n$. **9.** Follows from $|f(x) - f(c)| \geqq ||f(x)| - |f(c)||$.

Exercise 2.7. **1.** Strictly decr. for $x < -1$, strictly incr. for $x > -1$. **3.** Strictly decr. for $-1 < x < 0$ and for $0 < x < 1$, strictly incr. for $x < -1$ and for $1 < x$. **5.** Decr. for $0 < x < 1$, incr. for $-1 < x < 0$. **7.** If $0 < x_1 < x_2 < a$, then $-x_2 < -x_1$ and $f(-x_2) - f(-x_1) = -[f(x_2) - f(x_1)]$. Thus, f is monotone in $(-a, 0)$ and varies in the same sense as in $(0, a)$. It need not be monotone in $(-a, a)$ as shown by $f(x) = 1/x$. If f is cont. at $x = 0$, then $f(0) = 0$ since $f(0) = \lim f(-h) = -\lim f(h) = -f(0)$. **9.** $1/x$ and $-1/(1 - x)$ are decr. in $(0, 1)$, hence also their sum f. $f(h) \to +\infty$, $f(1 - h) \to -\infty$ when h decr. to 0. Hence $R[f] = R$ and, since f is monotone, it takes every real value exactly once in $(0, 1)$. **11.** $[\alpha x_1 + (1 - \alpha)x_2]^2 = \alpha^2 x_1^2 + 2\alpha(1 - \alpha)x_1x_2 + (1 - \alpha)^2 x_2^2 \leqq \alpha^2 x_1^2 + \alpha(1 - \alpha)(x_1^2 + x_2^2) + (1 - \alpha)^2 x_2^2 = \alpha x_1^2 + (1 - \alpha)x_2^2$. **13.** (i) and (ii) non-incr.

Exercise 2.8. **1.** $y^{\frac{1}{2}}$, $0 \leqq y \leqq 1$. **3.** $2 - y^{\frac{1}{4}}$, $1 \leqq y \leqq 16$. **5.** $\frac{1}{2}\{y - 1 + [(y - 1)(y + 3)]^{\frac{1}{2}}\}$, $1 \leqq y \leqq \frac{3}{2}$. **7.** $(-\infty, 0]$. In Nos. **2** and **4** no other inverses; in No. **3** $2 + y^{\frac{1}{4}}$, $2 \leqq y < \infty$; in No. **5** $\frac{1}{2}\{y - 1 - [(y - 1)(y + 3)]^{\frac{1}{2}}\}$, $-\infty < y \leqq -3$, $1 \leqq y < \infty$.

9.

$$f = \begin{cases} \frac{2}{3}(x - 3), & 3 \leqq x < 6, \\ -\frac{2}{3}(x - 12), & 6 \leqq x \leqq 9, \\ \frac{2}{3}(x - 3), & 9 < x < 12, \\ -\frac{2}{3}(x - 24), & 12 \leqq x \leqq 15. \end{cases} \qquad \hat{f} = \begin{cases} \frac{3}{2}y + 3, & 0 \leqq y < 2, \\ -\frac{3}{2}y + 12, & 2 \leqq y \leqq 4, \\ \frac{3}{2}y + 3, & 4 < y < 6, \\ -\frac{3}{2}y + 24, & 6 \leqq y \leqq 8. \end{cases}$$

11. $\hat{f} = y$ if y is rat., otherwise $-y$. **13.** No, $f = x$, $-1 \leqq x \leqq 1$, has an inverse, its sq. does not. **15.** $\hat{f} = -5y - y^5$ is strictly decr. for all y and $D[\hat{f}] = R[\hat{f}] = R$. Hence it has a unique inverse with same domain and range.

Miscellaneous problems for Chapter 2

1. If f_n is strictly incr., so are $f_n + x$ and its sq. root f_{n+1}. This holds for $n = 1$, hence for all n. Suppose $f_{n-1}(x) < f_n(x)$ for some n and x. Then $[f_{n+1}(x)]^2 - [f_n(x)]^2 = f_n(x) - f_{n-1}(x) > 0$ and since $f_n(x) > 0$ for all n and all $x > 0$, $f_{n+1}(x) > f_n(x)$. Since $f_1(x) < f_2(x)$ for $x \geqq 0$, the sequence $\{f_n(x)\}$ is incr. for each fixed pos. x. If $f_n(x) < 1 + \sqrt{x}$, then $f_{n+1}(x) < (x + 1 + \sqrt{x})^{\frac{1}{2}} < (x + 2\sqrt{x} + 1)^{\frac{1}{2}} = 1 + \sqrt{x}$. The ineq. is true for $n = 1$, hence for all n. Thus $\lim f_n(x) \equiv f(x)$ exists and $f(x) = [x + f(x)]^{\frac{1}{2}}$ so $f(x) = \frac{1}{2} + (\frac{1}{4} + x)^{\frac{1}{2}}$, an incr. fcn. **3.** $\lim S_n(x) = (1 - x)^{-2}$ for $|x| < 1$. The fcn. $(n + 2)x^{n+1} - (n + 1)x^{n+2}$ is unbd. for $|x| > 1$. **5.** $f = \frac{1}{2}x - 1 + |x - \frac{1}{2}| + \frac{1}{2}|x - 3|$. **7.** Obvious from (2.7.7). **9.** f^+ is cont. at all pts. where $f^+(x) = f(x) > 0$. The pts. where $f^+(x) = 0$ form closed intervals in the interior of which f^+ is cont. At the end pts. $f^+(x) = f(x) = 0$ and both f and f^+ are cont. there. **11.** Convex in $(-\infty, 0)$, concave in $(0, \infty)$. **13.** Strictly decr. in each of the intervals $(-\infty, 0)$, $(0, 1)$, $(1, \infty)$ but not in $(-1, 0) \cup (0, 1)$. **15.** Inverse $x = [y + 2 - (y^2 + 4)^{\frac{1}{2}}]/(2y)$.

Range $(-\infty, 0) \cup (1, \infty)$. **17.** $|f(x + h_1 + h_2) - f(x)| \leqq |f(x + h_1 + h_2) - f(x + h_2)| + |f(x + h_2) - f(x)|$ for any choice of x, h_1, h_2 in $D[f]$. If $|h_1| \leqq \delta_1$, $|h_2| \leqq \delta_2$, the right member $\leqq \mu(f, \delta_1) + \mu(f, \delta_2)$, while there are values of x, h_1, h_2 such that the left member equals $\mu(f, \delta_1 + \delta_2)$. **19.** If $\mu(f, \delta) \equiv 0$, then $f(x + h) - f(x) = 0$ for all x, $x + h$ in $D[f]$ and f is a constant. **21.** a_n is incr. and bd. by $2a + 1$. Note that $(2a + 1)^{\frac{1}{2}} < a + 1$ for $0 < a$. $\lim a_n = a + \frac{1}{2} + (a^2 + a)^{\frac{1}{2}}$. **23.** $a = 0, b = -2$. **25.** f is cont. everywhere in $(-1, 2)$. The pts. $x = 0$ and $x = 1$ are not discont. **27.** $\check{f} = (y^2 + 4)/(2y)$. $D[\check{f}] = R[\check{f}] = (-2, 2)$. **29.** $\check{f} = [(\frac{1}{4} + y)^{\frac{1}{2}} - \frac{1}{2}]^4$, $D[\check{f}] = R[\check{f}] = [0, \infty)$.

CHAPTER 3

Exercise 3.1. **5.** 0. **7.** $6x + 1$. **9.** g. **11.** $3x^2 + 6x$. **13.** $-x^{-2}$. **15.** $2x + x^{-2}$. **17.** $2x^{-3}$. **19.** $3x^2 - 3 - x^{-2}$. **21.** $A = \frac{1}{2}d^2$, $A' = d$. **23.** $V' = 4\pi r^2 = A$, $A' = 8\pi r$. **25.** $V' = 3s^2$, $A' = 12s$. **27.** $s(0) = b$, $v(0) = a$. **29.** $v = s' = 3t^2$, $a = v' = 6t$. **31.** $y - 20 = 24(x - 2)$. **33.** $y - \frac{1}{2} = \frac{1}{4}(x - 2)$. **35.** $y - \frac{7}{2} = \frac{17}{4}(x - 2)$. **37.** $y - \frac{15}{4} = \frac{9}{2}(x - 2)$. **39.** $y - 243 = 405(x - 2)$. **45.** No vert. tgt. **47.** $S_n = \frac{1}{3}n^3 + \frac{1}{3}n^2 + \frac{1}{6}n = \frac{1}{3}n^3 + O(n^2)$. **49.** $O(x^2) \Rightarrow o(x)$ if $x \to 0$, $o(x) \Rightarrow O(x^2)$ if $x \to \infty$. **51.** $y'(0) = \lim_{h \to 0} h\{(1/h) - h\} = 0$; $y = 0$ once in each interval $(2(2n + 1)^{-1}, n^{-1})$ and $((n + 1)^{-1}, 2(2n + 1)^{-1})$, $n = 2, 3, \cdots$.

Exercise 3.2. **1.** -3. **3.** $10x^4 - 40x^3 + 15x^2$. **5.** $3[(x + 1)^2 - (x - 1)^2] = 12x$. **7.** $-x^{-2}$. **9.** $2x + x^{-2}$. **11.** $2x - 3x^{-2}$. **13.** $2x - x^{-2}$. **15.** $\dfrac{(ad - bc)2x}{(cx^2 + d)^2}$ **17.** $\dfrac{-12x^2}{(x^3 - 3)^2}$. **19.** $nx^{n-1} - nx^{-n-1}$. **21.** $-(x^{-2} + 2x^{-3} + 3x^{-4} + 4x^{-5})$. **23.** $\frac{1}{3}x^{-\frac{2}{3}} + \frac{2}{3}x^{-\frac{1}{3}}$. **25.** $\dfrac{1 - \sqrt{x}}{2\sqrt{x}(1 + \sqrt{x})^3}$. **27.** $-\frac{3}{2}\dfrac{(\sqrt{x} + 1)^2}{(\sqrt{x})^5}$. **29.** $(2x^{\frac{1}{2}} + 1)^{-2}(x^{-\frac{4}{5}} + x^{-\frac{2}{3}} + 2x^{-\frac{3}{10}} + 2x^{-\frac{1}{6}} - 3x^{-\frac{7}{6}} - 5x^{-\frac{13}{10}})$. **31.** $-\{\frac{1}{2}x^{-\frac{3}{2}} + \frac{1}{3}x^{-\frac{4}{3}} + \frac{1}{4}x^{-\frac{5}{4}}\}$. **33.** $-x(x^2 + 1)^{-\frac{3}{2}}$. **35.** $-x^2(x^2 + 1)^{-\frac{4}{3}}$. **41.** $P'(x) = k(x - a)^{k-1}Q(x) + (x - a)^kQ'(x) = (x - a)^{k-1}[kQ(x) + (x - a)Q'(x)]$. **43.** $c = 0$, double root $x = 1$; $c = \frac{4}{27}$, double root $x = \frac{1}{3}$. **45.** $y - \frac{16}{3} = 10(x - 1)$, $y - 72 = 10(x + 9)$. **47.** $f'(x) = 2x$ for $x \geqq 0$, 0 for $x < 0$.

Exercise 3.3. **1.** $\left(1 - \dfrac{1}{x}\right)\dfrac{1}{2\sqrt{x}}$. **3.** $(1 + \sqrt{x})^{-2}x^{-\frac{1}{2}}$. **5.** $\dfrac{ad - bc}{(c\sqrt{x} + d)^2}\dfrac{1}{2\sqrt{x}}$. **7.** $\frac{1}{2}(x + 1)^{-\frac{1}{2}}$. **9.** $\frac{1}{2}\left(\dfrac{x - 1}{x + 1}\right)^{-\frac{1}{2}}\dfrac{2}{(x + 1)^2} = (x - 1)^{-\frac{1}{2}}(x + 1)^{-\frac{3}{2}}$. **11.** $-\frac{1}{2}[x + (x^2 + 1)^{\frac{1}{2}}]^{-\frac{1}{2}}(x^2 + 1)^{-\frac{1}{2}}$. **13.** $na(ax + b)^{n-1}$. **15.** $-3x(1 + x^2)^{-\frac{5}{2}}$. **17.** $(1 + 2x^2)(1 + x^2)^{-\frac{1}{2}}$. **19.** $\frac{4}{3}(2x + 1)^{-\frac{1}{3}}$. **21.** $-x(a^2 - x^2)^{-\frac{1}{2}}$. **23.** $-12x^{-7} + 7x^{-8}$. **25.** $-(x^2 - 1)^{-\frac{3}{4}}(6x^2 - 2x + 3)^{-\frac{5}{3}}(5x^3 - \frac{1}{3}x^2 - \frac{19}{2}x + \frac{4}{3})$. **27.** $\frac{1}{5}(s^{\frac{1}{3}} + 1)^{-\frac{2}{5}}s^{-\frac{2}{3}}$. **29.** $u^{18}(1 - u)^{-2}(19 - 18u)$. **31.** $(w - 1)^2(w + 5)(w + 1)^{-3}$. **33.** $(2, 1)$. **35.** $(1, 2)$, $(-1, 2)$. **37.** $(1, 0)$, $(-5, -\frac{27}{2})$.

Exercise 3.4. **1.** Left-hand derivative -2, right-hand $+2$.

Exercise 3.5. **1.** At any pt. where $f'(x) = 0$. **3.** If $y = 1/x$, $y' = -1/x^2$ and $xy' + y = 0$. **5.** $f_1(x)f_1'(x) = f_2(x)f_2'(x) = -x$.

Exercise 3.6. **1.** Implicit differentiation gives the slope at (x_0, y_0) as $-b^2x_0/(a^2y_0)$ so the tgt. is $y - y_0 = -[b^2x_0/a^2y_0)](x - x_0)$. Simplify! **3.** The slope is $2a/y$ and the tgt. is $y_0y = 2a(x - x_0)$. **5.** If $t \to 0$ then $x \to 0$, $y \to 0$ and $y/x \to 0$ so there is a branch through the origin with slope 0. If $t \to \infty$, then $x = \frac{3a}{t^2}(1 + o(1)) \to 0$, $y = \frac{3a}{t}(1 + o(1)) \to 0$ and $y/x = t \to \infty$ so there is a branch through the origin with vert. tgt. **7.** If $x + y = s$, then it sat. $st^2 - (s + 3a)t + s = 0$. The roots are real iff $-a \leqq s \leqq 3a$. **9.** Substitute $y = tx$ and solve for x. When $t \to 1$, $x \to 0$, $y \to 0$, $y/x \to 1$. When $t \to -1$, $x \to 0$, $y \to 0$, $y/x \to -1$. Two branches through the origin, with slopes $+1$ and -1. The left loop is obtained by changing the sign of the first sq. root. **11.** $D_xy = -(2x + y - 1)/(x + 2y - 1)$. **13.** $D_xy = -(3x^2 + 2xy + y^2)/(x^2 + 2xy + 3y^2 + 1)$. **15.** $D_xy = (5x^4 - y)/(x - 5y^4)$. **17.** $(0, 1)$, $(\frac{2}{3}, -\frac{1}{3})$. **19.** Nowhere.

Exercise 3.7. **1.** 7!. **3.** 8.6!. **5.** $-6(x - 1)^{-4}$. **7.** $\frac{40}{81}x^{-\frac{7}{3}}$. **9.** $8c(ad - bc)x(2cx^2 - d)(cx^2 + d)^{-4}$. **11.** $\frac{1}{2}(-1)^n n!\,[(x - 1)^{-n-1} - (x + 1)^{-n-1}]$. **13.** $-(1 - x^2)^{-\frac{3}{2}}$. **15.** If $f'' > 0$, f' is incr. and the curve lies above its tgt. **17.** $9x^6f'''(x^3) + 54x^3f''(x^3) + 6f'(x^3)$. **19.** $\frac{1}{4}x^{-1}f''(x^{\frac{1}{2}}) - \frac{1}{4}x^{-\frac{3}{2}}f'(x^{\frac{1}{2}})$. **21.** $(1 + 2x)^3f'''(1 + x + x^2) + 2(4x + 3)f''(1 + x + x^2)$. **23.** $y = x^{\frac{3}{2}}$ or $-x^{\frac{3}{2}}$. **25.** $D_x^2y = \pm\frac{3}{4}x^{-\frac{1}{2}} = 3/(4t)$.

Exercise 3.8.

	Δf	δf	$\Delta f - \delta f$	$\frac{1}{2}f''(a)h^2$
1.	0.0301	0.03	0.0001	0.0001
3.	0.100601	0.1	0.000601	0.0006
5.	0.012461 . .	0.0125	-0.000039 . .	-0.000039 . .

7. 10.05 with appr. error -0.000125. **9.** 1.02 with appr. error $+0.0003$. **11.** 2.0125 with appr. error -0.000156. In Nos. **13–19** $f(x)$ has the following values: **13.** x^2. **15.** $x^{\frac{1}{2}}$. **17.** x^{-2}. **19.** $x^{\frac{1}{3}}$.

23. $$D = \frac{1 + h - (1 + \frac{1}{2}h)^2}{(1 + h)^{\frac{1}{2}} + 1 + \frac{1}{2}h} = -\frac{1}{4}\frac{h^2}{(1 + h)^{\frac{1}{2}} + 1 + \frac{1}{2}h} > -\frac{1}{8}h^2.$$

25. $$D = \frac{3h^2 + 2h^3}{(1 + h)^2} < 5h^2.$$

27. $$D = -\frac{1}{3}\frac{h^2(1 + \frac{1}{9}h)}{(1 + h)^{\frac{2}{3}} + (1 + h)^{\frac{1}{3}}(1 + \frac{1}{3}h)^{\frac{1}{3}} + (1 + \frac{1}{3}h)^{\frac{2}{3}}}$$
$$> -\tfrac{1}{9}h^2(1 + \tfrac{1}{3}h)^{\frac{1}{3}} > -\tfrac{1}{9}(\tfrac{4}{3})^{\frac{1}{3}}h^2 > -\tfrac{4}{27}h^2,$$

since $1 + \frac{1}{9}h < 1 + \frac{1}{3}h < 1 + h$. **29.** The following estimates hold: **21.** No change. **22.** $2h^2$. **23.** $-\frac{1}{5}h^2$. **24.** $\frac{4}{5}h^2$. **25.** $12h^2$. **26.** $\frac{3}{2}h^2$. **27.** $-\frac{4}{25}h^2$. **31.** $\frac{1}{3}ax^{-\frac{2}{3}}$. **33.** 0.033 . . . **35.** 0.10769. **39.** If $f(x) = ax^2 + 2bx + c$, $f''(x) = 2a$ and $\Delta^2f(x, h) = 2ah^2$. **41.** $f(x + h) = f(x) + \frac{h}{1!}f'(x) + \frac{h^2}{2!}f''(x) + \frac{h^3}{3!}f'''(x)$.

Miscellaneous problems for Chapter 3

1. $y - x + 2 = 0$. **3.** $y - 3x + 7 = 0$. **5.** $y - x = \sqrt{5}$ or $-\frac{3}{5}\sqrt{5}$. **7.** $(1 - x^2)^{-\frac{3}{2}}$. **9.** $6x(5x^3 + 2)$. **11.** $24x(5x^2 + 3)$. **13.** $\dfrac{g(h)}{h} = h\left(\dfrac{1}{h}\right) \to 0$ with h. **15.** $\frac{1}{4}x^4 + x$. **17.** $\frac{1}{2}x^2 + 1/x$. **19.** $\frac{2}{3}x^{\frac{3}{2}}$. **21.** $\frac{3}{4}x^{\frac{4}{3}}$. **23.** $-1/x$. **25.** No. **27.** No. **29.** $-x^{-3} - (x - 1)^{-1} - \frac{1}{2}x^2$. **31.** $f'(x)$ exists and is the sum of three neg. fcns. for $x \neq -3, -1, 0$. Hence $f(x)$ is cont. and strictly decr. in each of the intervals $(-\infty, -3)$, $(-3, -1)$, $(-1, 0)$, $(0, \infty)$. Note that $f(-\infty) = 0$, $f(-3 - 0) = -\infty$, $f(-3 + 0) = +\infty$, $f(-1 - 0) = -\infty$, $f(-1 + 0) = +\infty$, $f(-0) = -\infty$, $f(+0) = +\infty$, $f(+\infty) = 0$, and use Th. 2.6.1. **33.** $[n + k/(2n)]^2 = n^2 + k + k^2/(4n^2) > n^2 + k$. The difference

$$n + \frac{k}{2n} - (n^2 + k)^{\frac{1}{2}} = \frac{k^2/(4n^2)}{n + k/(2n) + (n^2 + k)^{\frac{1}{2}}} < \frac{k^2}{8n^3}.$$

35. $f'(x)$ exists by assumption and is non-decr. by Th. 3.4.2. Since $f''(x) = \lim \dfrac{1}{h}[f'(x + h) - f'(x)]$ exists, it is $\geqq 0$. **37.** If $f'(x) = f(x)$, then the right member is differentiable, hence also the left, so that $f''(x)$ exists and equals $f'(x) = f(x)$. Induction! **39.** Here $(1 - x^2)D_x^2f = 2xD_xf - n(n + 1)f$ by assumption. The right member has a derivative, hence also the left. Thus D_x^3f exists and $(1 - x^2)D_x^3f = 4xD_x^2f - [n(n + 1) - 2]D_xf$. Induction! **41.** If $P(x) = x(x - 1)(x - 2)(x - 3)$, then

$$P'(x) = (x - 1)(x - 2)(x - 3) + x(x - 2)(x - 3) + x(x - 1)(x - 2).$$

Here $P'(0) = -6$, $P'(1) = 2$, $P'(2) = -2$, $P'(3) = +6$. By Lemma 2.6.1 $P'(x)$ has a zero in each of the intervals $(0, 1)$, $(1, 2)$, $(2, 3)$. No other zeros exist since $P'(x)$ is of degree 3. **43.** $2a^3xy(ax - y^2)^{-3}$.

CHAPTER 4

Exercise 4.1. **1.** $\frac{1}{2}hb$. Here $\mu(C_n) = hb\dfrac{n(n + 1)}{2n^2} \to \dfrac{1}{2}hb$. **3.** If each interval is divided into n equal parts, corresponding rectangles have the same area. $A(1, 2) = A(k, 2k)$ for all k. **5.** $\frac{1}{4}(b^4 - a^4)$. **7.** $\frac{1}{3}b^3 + \frac{1}{2}b^2$.

Exercise 4.2. **3.** $\frac{3}{2}$. **5.** $\frac{1}{4}$. **7.** $\frac{13}{18}$.

Exercise 4.3. **1.** $\frac{5}{6}$. **3.** $\frac{1}{4} \cdot 671$. **5.** $\frac{32}{15}$. **7.** $\frac{45}{4}$. **9.** $\frac{1}{12}$. **11.** $\frac{1}{6}x^4 + x^3 - x$. **13.** $-\frac{1}{2}x^{-2}$. **15.** $-(1 + x^2)^{-1}$. **17.** $-\dfrac{1}{x} - \dfrac{1}{x - 1}$. **19.** Add $-\dfrac{1}{6}$. **21.** Add $\frac{1}{2}$. **23.** Add $\frac{1}{2}$. **25.** The primitive does not exist for $x = 1$. **27.** $\frac{173}{6}$. **29.** $\frac{5}{72}$. **31.** $\frac{1}{10}$. **33.** $\frac{2}{3}$. **35.** $\frac{1}{6}(b - a)^3$. **37.** $1 + \dfrac{1}{2} + \dfrac{1}{3} + \cdots + \dfrac{1}{n}$. **39.** $2xf(x^2)$. **41.** We may suppose $f(t) > 0$ and so large that the tgt. to the curve $y = f(x)$ intersects the lines $x = a$ and $x = b$ at pts. P and Q of pos. ordinate. Then the area under the curve lies between those of the trapezoids $ABQPA$ and $ABDCA$.

Exercise 4.4. Constants of integration are omitted.

1. $\frac{1}{7}(x-1)^7$. **3.** $\frac{1}{3}x^3+2x-1/x$. **5.** $\frac{2}{3}(1+x)^{\frac{3}{2}}$. **7.** $\frac{1}{3}(1+x^2)^{\frac{3}{2}}$. **9.** $\frac{2}{9}(1+x^3)^{\frac{3}{2}}$. **11.** $-\frac{1}{4}(1-x^3)^{\frac{4}{3}}$. **13.** $\frac{1}{6}(1+x^2)^3$. **15.** $-\frac{1}{2}(2+x^2)(1-x^2)^{\frac{1}{2}}$. **17.** $-\frac{1}{2}(1+x^2)^{-1}$. **19.** $-\frac{1}{9}(1+x^2)^{-3}$. **21.** $\frac{1}{2}\left(x+\dfrac{1}{x}\right)^2$. **23.** The curve lies below the x-axis for $0<x<1$, above for $1<x\leq\sqrt{2}$. The two areas are the same but $\int_0^1 (x^3-x)\,dx$ is the neg. of the area between the curve and the x-axis. **25.** $\frac{1}{2}(x^2-2)(4-x^2)^{-2}$. **27.** $\frac{1}{3}(x^2-2)(1+x^2)^{\frac{3}{2}}$. **29.** $\frac{1}{10}(2x^3-15)(x^3+5)^{\frac{2}{3}}$. **31.** $\frac{4}{15}$. **33.** $\sqrt{37}-\sqrt{26}$. **35.** $\frac{1}{3}(\sqrt{730}-1)$. **37.** $2^{-\frac{2}{3}}$. **39.** $\frac{1}{15}(112-64\sqrt{2})$.

Exercise 4.5. Constants of integration are omitted in Nos. **1–9**.

1. $\frac{1}{4}x^2(2\log x-1)$. **3.** $\frac{1}{4}x^2[2(\log x)^2-2\log x+1]$. **5.** $-\dfrac{1}{x}(\log x+1)$. **7.** e^{2x}. **9.** $\frac{1}{2}(e^{2x}+e^{-2x})+2x$. **11.** e^{x^2}. **13.** $1+\log x$. **15.** $(\log x)^2+2\log x$. **17.** $x^{-3}(1-2\log x)$. **19.** $2e^{2x}$. **21.** $2(e^{2x}-e^{-2x})$. **23.** $(2x^2+1)e^{x^2}$. **25.** $\frac{8}{3}\log 2-\frac{7}{9}$. **27.** $\frac{1}{2}[(\log 2)^{-2}-1]$. **29.** 1. **31.** 0. **33.** $6-78e^{-3}$. **35.** $\int\log\log x\,dx=x\log\log x-\int(\log x)^{-1}\,dx$. **37.** $F(0)=[F(0)]^2$ implies either $F(0)=0$ or $F(0)=1$. $F(x)=F(0)F(x)$ shows that if $F(0)=0$, then $F(x)\equiv 0$. If this is excl. then $F(0)=1$.

Exercise 4.6. Nos. **1, 3, 9, 11, 13** belong to $R^*(0,1)$; **5, 7** do not. Nos. **15, 17, 23, 25, 27** belong to $R^*(1,\infty)$; **19, 21** do not. **29.** 1. **31.** $\frac{19}{6}$. **33.** 6. **35.** 2. **37.** 1. **39.** $\frac{1}{4}$. **41.** $\displaystyle\int_1^b\frac{dt}{t(t+1)(t+2)}=\log b-2[\log(b+1)-\log 2]+\log(b+2)-\log 3=\log\frac{4}{3}+\log\frac{b(b+2)}{(b+1)^2}\to\log\frac{4}{3}$ when $b\to\infty$.

Exercise 4.7 **1.** $\displaystyle\int_{-2}^1[(13-2x)-(7-5x)]\,dx+\int_1^4[(13-2x)-(x+1)]\,dx=27$. **3.** $2(2\sqrt{2}+\frac{34}{15})p^2$. **5.** $\dfrac{b^2}{a}(a^2+b^2)^{\frac{1}{2}}-ab\log\{[b+(a^2+b^2)^{\frac{1}{2}}]b^{-1}\}$. **7.** $\frac{256}{15}$. **9.** $\frac{4}{3}(1-a)+\log a$ where $a=\frac{1}{2}(\sqrt{5}-1)$. **11.** $\lim T_n=\displaystyle\int_0^3\frac{dx}{1+x}=\log 4$. **13.** $\lim T_n=\displaystyle\int_0^1\frac{x\,dx}{1+x^2}=\frac{1}{2}\log 2$. **15.** $\lim T_n=\displaystyle\int_1^2\frac{dx}{x(1+x)}=\log\frac{x}{x+1}\Big]_1^2=\log\frac{4}{3}$. **17.** $\log 2$. **19.** $\displaystyle\int_1^\infty\frac{e^s\,ds}{e^{2s}-1}=\int_e^\infty\frac{dt}{t^2-1}=\frac{1}{2}\lim_{b\to\infty}\log\frac{t-1}{t+1}\Big]_e^b=\frac{1}{2}\log\frac{e+1}{e-1}$.

Miscellaneous problems for Chapter 4

1. $\frac{1}{2}[1+x^2-\log(1+x^2)]$. **3.** $D_y[yg(y)-F(g(y))]=g(y)+yg'(y)-F'(g(y))g'(y)=x+yg'(y)-f(g(y))g'(y)=x+yg'(y)-yg'(y)=x$. But if

$G(y) = \int_{y_0}^{y} g(t)\,dt$, then $G'(y) = g(y) = x$. Hence the fcn. is a primitive of $g(y)$. **5.** $\int_{-a}^{a} [f(x+1) - f(x)]\,dx = \int_{a}^{a+1} f(s)\,ds - \int_{-a}^{-a+1} f(s)\,ds \to A - B$ when $a \to \infty$. **7.** $\lim_{x\to 0} x^2 \log x = 0$. $\lim_{h\to 0} \frac{1}{h}[h^2 \log h - 0] = \lim_{h\to 0} h \log h = 0$. The derivative of $x^2 \log x$ is zero for $x = 0$ and $x = e^{-\frac{1}{2}}$. The tgt. is horizontal at these pts. **9.** 24. **11.** $f(0) = 2\int_0^{\infty} (1+s)^{-2}\,ds = 2$. By Th. 4.3.7

$$[f(a)]^2 \leqq \int_{-\infty}^{\infty} (1+|s|)^{-2}\,ds \int_{-\infty}^{\infty} (1+|s-a|)^{-2}\,ds = f(0)\cdot f(0).$$

13. $A = 1$, $B = \frac{1}{2}$, integral $= -\frac{1}{3}$. Integrate the difference between 0 and b and note that

$$\tfrac{1}{3}\{b^2+1)^{\frac{3}{2}} - (Ab^3 + 3Bb)\} = \frac{1}{3}\frac{(1-A^2)b^6 + (3-6AB)b^4 + (1-9B^2)b^2 + 1}{(b^2+1)^{\frac{3}{2}} + Ab^3 + 3Bb}$$

The denominator is $O(b^3)$ when $b \to \infty$. The numerator is $O(b^6)$ if $A \neq 1$, $O(b^4)$ if $A = 1$ but $B \neq \frac{1}{2}$, $O(b^2)$ if $A = 1$, $B = \frac{1}{2}$. With this choice the right member is $O(b^{-1})$ and goes to 0 when $b \to \infty$. The lower limit gives the value of the integral. **15.** $\lim P_n = 4/e$. Note

$$\log P_n = \frac{1}{n}\sum_1^n \log\left[1 + \frac{k}{n}\right] \to \int_0^1 \log(1+x)\,dx = \log 4 - 1.$$

17. $\frac{1}{6}b(x_1 - x_0)^3$ where $x_1 = -x_0 - (2b^2x_0)^{-1}$ is the abscissa of the other intersection of the normal with the parabola. The answer agrees with (4.1.6), if we replace $4a$ by $1/b$ and interchange x and y. Difficult algebra! **19.** $(-1)^n[x^2 - 2nx + n(n-1)]\,e^{-x}$. **21.** Set $D_x{}^n[\log x]^2 = 2(-1)^n x^{-n}(A_n - B_n \log x)$. Then $A_1 = 0$, $B_1 = 1$, $A_{n+1} = nA_n + B_n$, $B_{n+1} = nB_n$. This gives

$$A_n = (n-1)!\left\{\frac{1}{1} + \frac{1}{2} + \frac{1}{3} + \cdots + \frac{1}{n-1}\right\},\ B_n = (n-1)!,\ n > 1.$$

23. Only one derivative $f'(0) = 1$. Since $f''(-0) = 1$, $f''(+0) = 2$, $f''(0)$ does not exist. **25.** The integral equals $\log(1+x)$ so $\theta(x) = \dfrac{x - \log(1+x)}{x\log(1+x)}$ and $\lim \theta(x) = \frac{1}{2}$. For

$$\log(1+x) = \int_0^x \frac{dt}{1+t} = x - \int_0^x \frac{t\,dt}{1+t} = x - \tfrac{1}{2}x^2 + \int_0^x \frac{t^2\,dt}{1+t},$$

so that if $x > 0$

$$x - \tfrac{1}{2}x^2 + \tfrac{1}{3}x^3 > \log(1+x) > x - \tfrac{1}{2}x^2$$

and

$$\tfrac{1}{2} - \tfrac{1}{3}x < \theta(x) < \frac{1}{2-x}.$$

A similar inequality holds for $x < 0$. Hence the limit is $\frac{1}{2}$. See Section 6.7. **27.** $\frac{1}{6}(5\sqrt{5} - 7)$. **29.** $4\sqrt{3} - 2\log(2+\sqrt{3})$.

CHAPTER 5

Exercise 5.1. **1.** First five terms $\frac{1}{2}, \frac{1}{6}, \frac{1}{12}, \frac{1}{20}, \frac{1}{30}$; their sum $\frac{5}{6}$; tenth term $\frac{1}{110}$. **3.** First five terms $\frac{1}{2}, \frac{1}{2}, \frac{3}{8}, \frac{1}{4}, \frac{5}{32}$; their sum $\frac{57}{32}$; tenth term $\frac{5}{512}$. **5.** If $m > 2\,|x|$ and $k \geqq m$, then $\left|\frac{x}{k}\right|^k < 2^{-k}$ and for $n > m$, $\sum_n^{n+p} \left|\frac{x}{k}\right|^k < \sum_n^{n+p} 2^{-k} < 2^{-n+1} \to 0$ as $n \to \infty$, so the series is abs. cvgt. for all x. **7.** $\frac{1}{n(n+1)(n+2)} = \frac{1}{n} - \frac{2}{n+1} + \frac{1}{n+2}$ so $\sum_1^N \frac{1}{n(n+1)(n+2)} = \frac{1}{1} - \frac{2}{2} + \frac{1}{3} - \frac{1}{N+1} + \frac{1}{N+2} \to \frac{1}{3}$ as $N \to \infty$. Hence the series cvgs. to the sum $\frac{1}{3}$. **9.** The series cvgs. for $|x| < 1$, its sum is $(1 - x)^{-2}$. If $x = \frac{1}{2}$, then $\sum_0^\infty (n + 1)2^{-n} = 4$. Hence the series in No. **2** cvgs., its sum is 2. **11.** The least term in the block is $2^{-\alpha(m+1)}$, the largest $< 2^{-\alpha m}$, the no. of terms 2^m. Hence B_m lies between $2^{-\alpha}\, 2^{(1-\alpha)m}$ and $2^{(1-\alpha)m}$ as asserted. **13.** $\{U_{2k+1}\}$ is strictly decr., $\{U_{2k}\}$ strictly incr. and $U_{2k+1} = U_{2k} + (2k + 1)^{-\alpha}$ so $\lim U_{2k+1}$ and $\lim U_{2k}$ exist and are equal. Abs. cvgt. for $\alpha > 1$. **15.** Cvgt. since $0 < \frac{2a_n a_{n+1}}{a_n + a_{n+1}} \leqq 2a_n$. **17.** Cvgt. for $x > 0$. Cf. No. **9**. **19.** Dvgt., for $e > \left(1 + \frac{1}{n}\right)^n$, $e^{1/n} > 1 + \frac{1}{n}$ and $1 - e^{-1/n} > 1 - \frac{1}{1 + 1/n} = \frac{1}{n+1}$.
21. Cvgt., sum 1. Note $u_n < n^{-\frac{3}{2}}$ and

$$\frac{1}{(n+1)n^{\frac{1}{2}} + n(n+1)^{\frac{1}{2}}} = \frac{(n+1)n^{\frac{1}{2}} - n(n+1)^{\frac{1}{2}}}{(n+1)^2 n - n^2(n+1)} = \frac{1}{n^{\frac{1}{2}}} - \frac{1}{(n+1)^{\frac{1}{2}}}.$$

23. Cvgt., sum $<\frac{1}{2}$. Note $(n^2 + \pi n) < \frac{1}{2}, \left(\frac{1}{n^2 + \pi n}\right) < \frac{1}{n(n+1)}$. **25.** Note $(2\,|x|)^n < n!$ for all large n and proceed as in No. **5**. **27.** Dvgt. Since $x - \frac{1}{2}x^2 < \log(1 + x) < x$ for $0 < x$, $\log\left(1 + \frac{1}{n}\right) > \frac{1}{2n}$.

Exercise 5.2. **1.** Cvgt. **3.** Cvgt. $\frac{u_{n+1}}{u_n} = \left(\frac{n}{n+1}\right)^n \to \frac{1}{e} < 1$. **5.** Cvgt. **7.** Dvgt. **9.** Cvgt. **11.** Cvgt. **13.** Cvgt. **15.** Cvgt. **17.** If $\liminf \frac{u_{n+1}}{u_n} = \alpha > 1$, then $u_{n+1} > u_n$ for all large n and the series dvgs. If $\limsup \frac{u_{n+1}}{u_n} = \beta < 1$ and $\beta < \gamma < 1$, then $u_{n+1} < \gamma u_n$ for all large n and the series cvgs. In No. **16** $\liminf \frac{u_{n+1}}{u_n} = 0$, $\limsup \frac{u_{n+1}}{u_n} = \infty$. The series cvgs. if $b < 1$, dvgs. if $b \geqq 1$. **19.** $\alpha > 1$. Note $1 + \frac{1}{2} + \cdots + \frac{1}{n} = \log(n + 1) + O(1)$ so the series is equicvgt. with $\sum n^{-\alpha}$. **21.** Cvgt. $\frac{u_{n+1}}{u_n} \to \frac{1}{2}$. **23.** $a = 2$. **25.** Cvgt. $\sum_1^N b_n = R_0{}^{\frac{1}{2}} - R_N{}^{\frac{1}{2}} \to R_0{}^{\frac{1}{2}}$. Here $\frac{a_n}{b_n} = \frac{R_{n-1} - R_n}{R_{n-1}^{\frac{1}{2}} - R_n{}^{\frac{1}{2}}} = R_{n-1}^{\frac{1}{2}} + R_n{}^{\frac{1}{2}} \to 0$ as $n \to \infty$.

27. Cvgt. $\int_0^{1/n} x^{\frac{1}{3}}(1 + x)\,dx < 2\int_0^{1/n} x^{\frac{1}{3}}\,dx = \frac{3}{2}n^{-\frac{4}{3}}$. **29.** Cvgt. $\int_0^{1/n} \frac{x^{\frac{2}{3}}\,dx}{1 + x^2} < \int_0^{1/n} x^{\frac{2}{3}}\,dx = \frac{3}{5}n^{-\frac{5}{3}}$.

Exercise 5.3. **1.** Cvgt. by Th. 5.3.1. **3.** $k > j$. **5.** Cvgt. Here

$$B_n \equiv \sum_{k=0}^{2n} \frac{1}{n^2 + k} = \frac{2}{n} + \frac{\theta_n}{n^2}, \quad |\theta_n| \leq 3$$

since

$$\frac{2}{n} - \frac{3}{n^2 + 2n} = \frac{2n + 1}{n^2 + 2n} < B_n < \frac{2n + 1}{n^2} = \frac{2}{n} + \frac{1}{n^2}.$$

Thus

$$\textstyle\sum_1^\infty (-1)^{n-1}B_n = 2\sum_1^\infty \frac{(-1)^{n-1}}{n} + \sum_1^\infty \frac{\theta_n}{n^2}.$$

The first series cvgs. by Th. 5.3.1, the second is abs. cvgt. **7.** Cvgt. **9.** Set

$$\begin{aligned} A_k &= -\frac{2}{k} + \frac{1}{k^2 + 1} + \frac{1}{k^2 + 2} + \cdots + \frac{1}{k^2 + 2k} \\ &= -\left\{\left(\frac{1}{k^2} - \frac{1}{k^2 + 1}\right) + \left(\frac{1}{k^2} - \frac{1}{k^2 + 2}\right) + \cdots + \left(\frac{1}{k^2} - \frac{1}{k^2 + 2k}\right)\right\} \\ &= -\left\{\frac{1}{k^2(k^2 + 1)} + \frac{2}{k^2(k^2 + 2)} + \cdots + \frac{2k}{k^2(k^2 + 2k)}\right\} \\ &> -\tfrac{1}{2}2k(2k - 1)\frac{1}{k^2(k^2 + 1)} > -\frac{2}{k^2}. \end{aligned}$$

Hence $\Sigma\,|A_k|$ cvgs. A partial sum U_n of $\Sigma\,u_m$ differs from a partial sum of $\Sigma\,A_k$ at most by a group of terms whose sum tends to 0 as $n \to \infty$. Thus $\Sigma\,u_m$ and $\Sigma\,A_k$ are equicvgt. $\Sigma\,u_m$ is not abs. cvgt. **11.** Take $f_n(x) = x(x + 1)\cdots(x + n - 1)$ in No. **10**. Then for $x > a$ and all n,

$$\frac{1}{x} + \textstyle\sum_1^{n-1} \frac{a(a + 1)\cdots(a + k - 1)}{x(x + 1)\cdots(x + k - 1)(x + k)} < \frac{1}{x - a}.$$

By Th. 5.3.2 the series cvgs. for $x > a$. **13.** Such rearrangement is obtained, e.g., if blocks B_k are formed containing 2^{k-1} pos. terms and a single neg. one,

$$B_k = (2^k - 1)^{-1} + (2^k + 1)^{-1} + \cdots + (2^{k+1} - 3)^{-1} - (2k)^{-1}.$$

Here the pos. terms add up to at least $\frac{1}{4}$ so that

$$\textstyle\sum_1^n B_k > \frac{1}{4}(N - 2\log N) \to \infty \quad \text{as } N \to \infty.$$

15. The product series is dvgt. since $\lim w_n = \int_0^1 [s(1 - s)]^{-\frac{1}{2}}\,ds \neq 0$. The series $\Sigma\,(-1)^n w_n$ is obviously dvgt. since the terms do not go to 0. If the terms of the product series are taken to be the individual products $(-1)^k(-1)^{n-k}u_k u_{n-k}$, arranged according to incr. values of n and k, the series is still dvgt. but now because the lim sup and the lim inf of the partial sums differ by $\lim w_n$.

Exercise 5.4. **1.** 1. **3.** $\frac{1}{2}$. **5.** $\frac{1}{4}$. **7.** ∞. **9.** $\min\left(\frac{1}{a}, \frac{1}{b}\right)$. **11.** The series cvgs. at $x = R$ in Nos. **2, 8, 11, 12,** at $x = -R$ in Nos. **2, 5, 6, 8, 12.** **17.** By formula (5.3.13) the series is unif. cvgt., if the partial sums $U_n(x)$ are unif. bd. The ineq. in No. **11,** Exercise 5.3 shows that this holds for $x \geqq a + \delta$. **19.** If $|f_n(x)| \leqq M_n$ and $\Sigma M_n < \infty$, then, given $\varepsilon > 0$, there exists an N such that for all p, $\sum_N^{N+p} |f_k(x)| \leqq \sum_N^{N+p} M_k < \varepsilon$. Hence the series cvgs. abs. and unif. **21.** Since $u(1 + u^2)^{-1} \leqq \frac{1}{2}$ for all u, the M-test applies with $M_n = (2n^2)^{-1}$. **23.** No, for $x = \pm \frac{1}{n}$, $F_n(x)$ takes on the extreme values $\pm\frac{1}{2}$ Since $F_n(x) \to 0$ for each fixed x, we have bd. but not unif. cvgce.

Exercise 5.5 **1.** If $F(x) = -\log(1 - x)$, then $F^{(n)}(x) = (n - 1)!\,(1 - x)^{-n}$. **3.** $x = 1/(2001)$. Only one term needed since the remainder is $< 4 \cdot 10^{-11}$. We get $\log 1001 = 6.9087547793$. **5.** The derived series is $\sum_{n=0}^{\infty} (-1)^n x^{2n} = (1 + x^2)^{-1}$ for $|x| < 1$. **7.** The series cvg. for all x as the exponential series. Termwise differentiation is permitted and gives the desired relations. **9.** $\int_0^\infty G_n(x)\,dx = \int_0^\infty te^{-t}\,dt = 1$ for all n while $\int_0^\infty \lim_{n\to\infty} G_n(x)\,dx = 0$.

Miscellaneous problems for Chapter 5

1. Equicvgce., $x - \frac{1}{2}x^2 < \log(1 + x) < x$. **3.** For $a > \frac{1}{2}$ we have

$$\int_a^\infty e^{-t^2}\,dt < \int_a^\infty 2te^{-t^2}\,dt = e^{-a^2}.$$

Thus the series cvgs. if $\sum_{n=0}^{\infty} \exp(-b_n^2)$ cvgs. and this holds, if, e.g., $b_n > c(\log n)^{\frac{1}{2}}$, c fixed > 1. A nasc is that $\Sigma\, b_n^{-1} \exp(-b_n^2)$ cvgs. **5.** Since $(x_n)^n + nx_n = 1$, $x_n < 1/n$ and the series cvgs. for $\alpha > 1$. More precisely, $x_n = n^{-1}(1 + o(1))$ since $(x_n)^n \to 0$. Hence the series dvgs. for $\alpha \leqq 1$. **7.** $R = e^\alpha$. Use the ratio test and note $\left(1 - \frac{\alpha}{n}\right)^n \to e^{-\alpha}$. **9.** $|x| \leqq 1 - \delta$, $0 < \delta$. The terms become infinite as $x \to \pm 1$. **11.** The series is abs. cvgt. for $q < 1$. Suppose $1 \leqq q$ and set

$$B_k = \sum n^{-1/q},\ k \leqq n^{1/p} < k + 1.$$

The given series cvgs. iff $\Sigma\,(-1)^k B_k$ cvgs. which requires $\lim B_k = 0$. Use summation by parts with $u_n = (-1)^{\alpha_n}$, $v_n = n^{-1/q}$. $\{U_n\}$ is not bd. Suppose n is chosen so that $[n^{1/p}] = m$, $[(n + 1)^{1/p}] = m + 1$. Then sgn $U_m = (-1)^m$ and $|U_n|$ is less than the no. of terms in B_m, i.e.,

$$< (m + 1)^p - m^p = p\int_m^{m+1} s^{p-1}\,ds < p(m + 1)^{p-1}.$$

Moreover, $|U_n| < p(m + 1)^{p-1}$ for all n with $m - 1 \leqq n^{1/p} \leqq m$. Let $\sum_{m,p} w_n$ denote a summation of w_n where $(m - 1)^p \leqq n < m^p$. Then

$$\sum_n |U_n|\,|v_n - v_{n+1}| = \sum_{m=1}^{\infty} \sum_{m,p} |U_n|\,|v_n - v_{n+1}| < p \sum_{m=1}^{\infty} (m + 1)^{p-1} \sum_{m,p} \{n^{-1/q} - (n + 1)^{-1/q}\}.$$

If $a < b$ are pos. integers

$$\sum_{a \leqq n < b} \{n^{-1/q} - (n + 1)^{-1/q}\} = a^{-1/q} - b^{-1/q} = \frac{1}{q} \int_a^b s^{-1-1/q}\, ds < \frac{1}{q}(b - a)a^{-1-1/q}.$$

We set $a = [(m - 1)^p]$, $b = [m^p]$. Then

$$b - a < m^p - (m - 1)^p + 2 < pm^{p-1} + 2 < (p + 2)(m + 1)^{p-1}$$

and $a^{-1-1/q} = [(m - 1)^p]^{-1-1/q}$. For $m > 2$ the last expression does not exceed $(m + 1)^{-p(1+1/q)}$ multiplied by a constant depending only on p and q. Hence

$$\sum |U_n|\,|v_n - v_{n+1}| < C(p, q) \sum (m + 1)^{p-1+p-1-p-p/q} = C(p, q) \sum (m + 1)^{p(1-1/q)-2}$$

This series cvgs. if $p - p/q < 1$ or $p < q/(q - 1)$. This cond. also gives $U_n v_n \to 0$ and is sufficient for cvgce. of the given series. Cond. also necessary since $B_m > Cm^{p-p/q-1}$ which does not tend to 0 unless $p < q/(q - 1)$. **13.** If $\alpha > 1$, there is a $\delta > 0$ such that $\alpha - \delta > 1$ and for all large n

$$\frac{\log (1/u_n)}{\log n} > \alpha - \delta \quad \text{or} \quad u_n < n^{-(\alpha-\delta)}.$$

Hence the series cvgs. If $\beta < 1$, $u_n > 1/n$ for all large n and the series dvgs. **15.** $R = \frac{1}{4}$. **17.** $R = 1$. **19.** $x > 1$ or $x < -\frac{1}{2}$. **21.** $\frac{5}{3} \leqq x < 7$. **23.** $e/(e^2 - 1)$.

APPENDIX B

Exercise B.1. **1.** $a^2 - b^2$, $2ab$. **3.** $a(a^2 + b^2)^{-1}$, $b(a^2 + b^2)^{-1}$. **5.** 16384, $\frac{\pi}{3}$. **7.** $\theta = k \cdot \frac{\pi}{3}$, $k = 1$ or 5. **11.** Circle with center z_0, radius r. **13.** $r^{\frac{1}{2}}[\cos (\frac{1}{2}\theta + k\pi) + i \sin (\frac{1}{2}\theta + k\pi)]$, $k = 0$ or 1.

Exercise B.2. **1.** For z in the left half-plane. **3.** $w_n = (2n^2 + 2n + 1)^{-1}$. **5.** No. $(x + r \cos \alpha - x)/(re^{i\alpha}) = \cos \alpha e^{-i\alpha}$ depends on α. **7.** Expand $(1 - s)^{-1}$ in geometric series and integrate termwise. The integral exists for all z not on the pos. real axis beyond $z = 1$. **9.** Cvgt. in the upper half-plane, unif. in any circular disk $|z - i| < a\,|z + i|$, $0 < a < 1$. Sum $\frac{1}{2}(1 - iz)$.

Miscellaneous problems for Appendix B

1. $\sqrt[3]{1} = 1, -\frac{1}{2} \pm i\frac{1}{2}\sqrt{3}$, $\sqrt[3]{-1} = -1, \frac{1}{2} \pm i\frac{1}{2}\sqrt{3}$. **3.** The unit circle maps onto itself, the interior onto the exterior and vice versa. $z = \pm 1$ invariant. **5.** If $P(z)$ is a polyn. with real coeffs., then $P(x + iy) = U(x, y) + iV(x, y)$ and $U(x, -y) = U(x, y)$, $V(x, -y) = -V(x, y)$. Hence, if $P(x_0 + iy_0) = 0$, then

$U(x_0, y_0) = V(x_0, y_0) = 0$ and also $U(x_0, -y_0) = V(x_0, -y_0) = 0$ and $P(x_0 - iy_0) = 0$ as asserted. **7.** Max. value 2 at $z = \pm 1$, min. value 0 at $z = \pm i$. We have $|e^{i\theta} + 1| = (2 + 2\cos\theta)^{\frac{1}{2}}$, a cont. fcn. of θ which attains its l.u.b. and g.l.b. For $|z| \leqq 1$, $0 \leqq |1 + z^2| \leqq 1 + |z|^2 \leqq 2$ so the bounds for the disk are the same as for the circumference. **9.** $\sum_{n=0}^{\infty} (-1)^n \dfrac{z^{2n} + 1}{2n + 1}$. **11.** Cf. No. **13,** Exercise B.1. The sq. root is uniquely determined if we take $k = 0$ and restrict θ so that $0 \leqq \theta < 2\pi$. **13.** The product is $\prod_{k=1}^{n} \dfrac{4 - z_k}{2 - z_k} = \dfrac{4^n - 1}{2^n - 1} = 2^n + 1$ since $\prod_{k=1}^{n} (z - z_k) = z^n - 1$. **15.** Assume $\dfrac{n}{z^n - 1} = \sum_{k=1}^{n} \dfrac{A_k}{z - z_k}$ and multiply both sides by $z^n - 1$. Then $n = \sum_{k=1}^{n} A_k \dfrac{z^n - 1}{z - z_k}$. Let $z \to z_m$. Then $\dfrac{z^n - 1}{z - z_m} \to nz_m^{n-1}$ while $\dfrac{z^n - 1}{z - z_k} \to 0$ if $k \neq m$. Hence $n = nA_m z_m^{n-1}$ or $A_m = z_m$. **17.** If $z = re^{i\theta}$ is a root of $z^3 = 3z - 1$, then $r^3 \leqq 3r + 1$ and this cannot hold for any $r \geqq 2$.

CHAPTER 6

Exercise 6.1. **1.** If $P(x)$ had two zeros, there would be a zero of $P'(x)$ between them. **3.** $c = 0$. **5.** (i) $c = \frac{1}{2}\sqrt{2}$, (ii) $c = \pm\frac{1}{3}\sqrt{3}$. **7.** If $f(x)$ has $n + 1$ distinct zeros in $[a, b]$, then by Rolle's th. $f'(x)$ has at least n distinct zeros there, $f''(x)$ has $n - 1, \cdots, f^{(n-1)}(x)$ has two and $f^{(n)}(x)$ at least one zero. **9.** $(x^2 - 1)^n$ has n-tuple zeros at $x = \pm 1$. Its first derivative has $(n - 1)$-tuple zeros there and also a zero at $x = 0$. Its second derivative has $(n - 2)$-tuple zeros at $x = \pm 1$ and, by Rolle's th., a zero in each of the intervals $(-1, 0)$ and $(0, 1)$. Each time we differentiate, two zeros are lost at $x = \pm 1$ and an additional zero appears in $(-1, 1)$. After n differentiations there are no zeros at the endpts. but at least n zeros in $(-1, 1)$. Since $P_n(x)$ is a polyn. of degree n, there are exactly n zeros in $(-1, 1)$ and none elsewhere. **11.** $D_x Ce^{\alpha x} = C\alpha e^{\alpha x}$.

Exercise 6.2. **1.** $R_2(0, h) = 3h^2(1 + \theta h)^{-4} < 3h^2$. **3.** By Cauchy's formula $R_2(0, h) = \frac{3}{4}(1 - \theta)(1 - \theta h)^{-\frac{5}{2}}h^2$. **5.** By Cauchy's formula $R_2(0, h) = (1 - \theta) \times (1 - \theta h)^{-2}h^2$. **7.** By Lagrange's formula $R_1(x, h) = \frac{1}{2}(x + \theta h)^{-\frac{1}{2}}h < \frac{1}{2}x^{-\frac{1}{2}}h$. The difference in question is $R_2(x, h) = -\frac{1}{8}(x + \theta h)^{-\frac{3}{2}}h^2 > -\frac{1}{8}x^{-\frac{3}{2}}h^2$. **9.** If $x \geqq h > 0$, Lagrange's formula gives a remainder $R_n(x, h)$ which is at most $O(1/n)$ as $n \to \infty$. The possibility that $\theta = \theta(x, h, n) \to 0$ as $n \to \infty$ does not affect this conclusion. Cauchy's formula gives same result for $x > h$, but is not conclusive for $x = h$ since θ may go to 0. For $h < 0$ the situation is reversed: Lagrange's formula gives $R_n(x, h) \to 0$ only for $x > 2|h|$, while Cauchy's formula works for $x > |h|$ since $x + \theta h = x - \theta|h| > (1 - \theta)|h|$ if $x > |h|$, $R_n(x, h)$ does not go to 0 for $x = |h|$. **11.** $A(x) = \sum_0^\infty (-1)^k \dfrac{x^{2k+1}}{2k + 1}$, $R_{2n}(0, x) = (-1)^n \displaystyle\int_0^x \frac{s^{2n}\,ds}{1 + s^2}$. $|R_{2n}(0, x)| \leqq \dfrac{|x|^{2n+1}}{2n + 1}$. **13.** Take $n = 11$ in (6.2.27), carry computation to 8 decimal places. R_{11} lies between $2 \cdot 10^{-8}$ and $3 \cdot 10^{-8}$.

Exercise 6.3. **1.** $\frac{1}{2}(e - e^{-1})$. **3.** e. **5.** Since $\left|\binom{\alpha}{n}\right| \leqq |\alpha|\,(|\alpha| + 1) \cdots (|\alpha| + n - 1)/(n!) = (-1)^n\binom{-|\alpha|}{n}$, we have $\sum_0^\infty \left|\binom{\alpha}{n}\right| |x|^n \leqq \sum_0^\infty \binom{-|\alpha|}{n}(-|x|)^n = (1 - |x|)^{-|\alpha|}$ for $|x| < 1$. Thus the series cvgs. for all α if $0 < x < 1$. For $x = 1$ use Lemma 5.2.2 and find cvgce. for $\Re(\alpha) > 0$, abs. cvgt. for $\Re(\alpha) > 1$. **7.** Set $\alpha = i\gamma$, $\beta = -i\gamma$ in No. **6**. Then

$$\binom{i\gamma}{0}\binom{-i\gamma}{n} + \cdots + \binom{i\gamma}{n}\binom{-i\gamma}{0} = \binom{0}{n} = \begin{cases} 1, & n = 0, \\ 0, & n > 0. \end{cases}$$

The left member is the coeff. of x^n in the product series

$$\sum_0^\infty \binom{i\gamma}{k} x^k \cdot \sum_0^\infty \binom{-i\gamma}{m} x^m = (1 + x)^{i\gamma}(1 + x)^{-i\gamma},$$

so the series reduces to 1 identically. **11.** For $|z| < \rho$ we have

$$\begin{aligned} |f^{(n)}(z)| &\leqq \sum_{k=n}^{\infty} k(k-1)\cdots(k-n+1)\,|c_k|\,|z|^{k-n} \\ &\leqq M(\rho)n! \sum_{m=0}^{\infty} \frac{(m+1)(m+2)\cdots(m+n)}{n!}\,\rho^{-m-n}\,|z|^m \\ &= M(\rho)n!\,\rho^{-n}\left(1 - \frac{|z|}{\rho}\right)^{-n-1} = \frac{\rho M(\rho)n!}{(\rho - |z|)^{n+1}}. \end{aligned}$$

13. If (6.3.12) holds, then $f(z) \equiv \sum_0^\infty \frac{f^{(n)}(c)}{n!}(z - c)^n$ sat. the cond. of Nos. **10–12**. The ineq. in No. **12** implies

$$\frac{1}{(n-1)!}\,|x - t|^{n-1}\,|f^{(n-1)}(t)| \leqq \frac{\rho M(\rho)}{(\rho - |t - c|)^2}\, n\left(\frac{|x - t|}{\rho - |t - c|}\right)^{n-1}.$$

For t between c and x, the largest value of the last fraction is for $t = c$ and equals $|x - c|/\rho$. This is a fixed no. < 1 when x is fixed and $|x - c| < \rho$. Hence the limit in (6.3.13) holds unif. in t so the cond. is necessary. **15.** Termwise differentiation gives $G^{(k)}(0) = (-1)^k k! \exp(a^k)$. It is clear that the series

$$\sum_{k=0}^{\infty} \frac{G^{(k)}(0)}{k!} x^k = \sum_{k=0}^{\infty} (-1)^k \exp(a^k) x^k$$

dvgs. for all $x \neq 0$.

Exercise 6.4. **1.** 0. **3.** $e^x(e^h - 1)^3$. **5.** $6h^3$. **7.** Form the Cauchy product of the Maclaurin series of $\log(1 + x)$ and $(1 + x)^{-1}$. The coeff. of x^n in the product is $(-1)^{n-1}\left\{1 + \frac{1}{2} + \frac{1}{3} + \cdots + \frac{1}{n}\right\}$. **9.** Set $\log(1 + x) = t$, then $x^2 = (e^t - 1)^2 = e^{2t} - 2e^t + 1$. Hence

$$x^2 = \sum_{n=2}^{\infty} \frac{2^n - 2}{n!} [\log(1 + x)]^n, \quad x \neq -1.$$

11. $1 + x + \frac{1}{2}x^2 + \frac{7}{6}x^3 + \frac{25}{24}x^4 + \frac{121}{120}x^5 + \cdots$. **13.** Five solutions: 0, 0, 0, 1; 1, 0, 1, 0; 0, 2, 0, 0; 2, 1, 0, 0; 4, 0, 0, 0. **15.** $\sum_{n=0}^{\infty} P(n)/(n!) = (A + B + C + D)e$.

Exercise 6.5. **1.** If z is a root, then $|z|^5 = |10z^2 + 1|$ which is at most $10|z|^2 + 1$ and at least $|10|z|^2 - 1|$. This gives $r < |z| < R$ where $r^5 + 10r^2 = 1$, $R^5 = 10R^2 + 1$. Evidently $10r^2 < 1 < 11r^2$ which gives the first stated ineq. For the second use a table of powers. **3.** $-a_1 = z_1 + z_2 + z_3$, $a_2 = z_1z_2 + z_2z_3 + z_3z_1$, $-a_3 = z_1z_2z_3$, $\frac{1}{z_1} + \frac{1}{z_2} + \frac{1}{z_3} = -\frac{a_2}{a_3}$. **5.** The roots of $P'(z) = 0$. **7.** It follows from No. **6** that the roots of $P'(z) = 0$ lie in the strip bd. by two par. lines, one through two zeros of $P(z)$, the other through the third. This holds for any numbering of the zeros of $P(z)$, so the zeros of $P'(z)$ must lie interior to the triangle with vertices at z_1, z_2, z_3. If the latter are collinear, the zeros of $P'(z)$ lie on the same line. **9.** If $P(z) = z^4 + a_1z^3 + a_2z^2 + a_3z + a_4$, then

$$P(t + a) = P(a) + P'(a)t + \tfrac{1}{2}P''(a)t^2 + \tfrac{1}{6}P'''(a)t^3 + t^4.$$

The term in t^3 is missing if $P'''(a) = 0$, that is, if $4a + a_1 = 0$ or $a = -\frac{1}{4}a_1$. With this choice of a, the sum of the roots of $P(t + a) = 0$ is zero, that is, $t = 0$ is the center of gravity of the four zeros t_1, t_2, t_3, t_4. If the original zeros z_1, z_2, z_3, z_4 form a pargram. in the z-plane, so do the zeros t_1, t_2, t_3, t_4 in the t-plane. The center of the latter pargram. is at $t = 0$, so the roots are of the form $t_1, t_2, -t_1, -t_2$. This requires that the transformed equation is biquadratic. Thus $P'(a) = 0$, whence $a_1^3 - 4a_1a_2 + 8a_3 = 0$. Conversely, if this cond. holds, the roots form the vertices of a pargram.

Exercise 6.6. **1.** $\frac{1}{z-2} + \frac{1}{z+5}$. **3.** $\frac{1}{z+1} - \frac{1}{z} + \frac{1}{z-1}$.

5. $-\frac{2}{z+1} - \frac{1}{z^2} + \frac{2}{z}$. **7.** $z^3 - 1 - \frac{1}{2(z+1)} + \frac{1}{2(z-1)}$.

9. $z^3 + 3z^2 + 6z + 10 + \frac{14}{z-1} + \frac{4}{(z-1)^2} + \frac{1}{(z-1)^3}$.

11. $z + 2 - \frac{7}{3(z+1)} + \frac{1}{z} - \frac{1}{3(z-2)}$. **13.** $\log(z - 2)(z + 5)$.

15. $\log\frac{z^2-1}{z}$. **17.** $\frac{1}{z} + 2\log\frac{z}{z+1}$. **19.** $\frac{1}{4}z^4 - z + 2\log\frac{z-1}{z+1}$.

21. $\frac{1}{4}z^4 + z^3 + 3z^2 + 10z + 14\log(z-1) - \frac{4}{z-1} - \frac{1}{2}\frac{1}{(z-1)^2}$.

23. $\frac{1}{2}z^2 + 2z - \frac{1}{3}\log\frac{(z+1)^7(z-2)}{z^3}$. **25.** Expand $P(x)$ in powers of $x - a$ and divide by $(x - a)^n$. Thus if $P(x)$ is of degree $k < n$

$$P(x)(x-a)^{-n} = P(a)(x-a)^{-n} + \frac{1}{1!}P'(a)(x-a)^{-n+1} + \cdots + \frac{1}{k!}P^{(k)}(a)(x-a)^{-n+k}.$$

27. The fcn. is $\Delta^n f(z, 1)$, where $f(z) = 1/z$, that is,

$$\frac{1}{z} - \binom{n}{1}\frac{1}{z+1} + \binom{n}{2}\frac{1}{z+2} - \cdots + (-1)^n\frac{1}{z+n}.$$

The sum of the coeffs. is $(1-1)^n = 0$. **29.** $\dfrac{-x}{(x^2+x+1)^2} + \dfrac{1}{x^2+x+1}$.

31. $\dfrac{2x-4}{(x^2+2x+2)^2} + \dfrac{x-2}{x^2+2x+2}$.

33. $\dfrac{5}{8}\dfrac{1}{x-1} - \dfrac{5}{8}\dfrac{1}{x+1} - \dfrac{1}{2}\dfrac{1}{(x^2+1)^2} - \dfrac{1}{4}\dfrac{1}{x^2+1}$.

35. $x - \frac{1}{4}\sqrt{2}\left\{\dfrac{1}{x^2-\sqrt{2}x+1} - \dfrac{1}{x^2+\sqrt{2}x+1}\right\}$.

37. $\dfrac{1}{3}\dfrac{1}{x-1} - \dfrac{1}{3}\dfrac{x-1}{x^2+x+1}$.

39. $\dfrac{z}{z^4+1} = \dfrac{i}{4}\left\{-\dfrac{1}{z-\omega} + \dfrac{1}{z-\omega^3} - \dfrac{1}{z-\omega^5} + \dfrac{1}{z-\omega^7}\right\}$, $\omega = \exp(\frac{1}{4}\pi i)$.

Note $\omega^2 = i$, $\omega^4 = -1$. **41.** If $R(z) = \dfrac{P(z)}{Q(z)}$ where $\deg P = p$, $\deg Q = q$, $d = \max(p, q)$, then $R(z) = a$, a finite, is equivalent to $P(z) - aQ(z) = 0$. This equation is normally of degree d and has d real or complex roots, i.e., the value a is assumed d times if each root is counted with its proper multiplicity. No exceptional values if $p > q$, but if $p \leqq q$, the value $b = \lim_{z\to\infty} R(z)$ is assumed at ∞ and the infinite roots must be included to account for the d roots of the equation $R(z) = b$. If $p \leqq q$, $R(z) = \infty$ is equivalent to $Q(z) = 0$ and this equation has $q = d$ roots. If $p > q$, the value ∞ is assumed $p - q$ times at ∞ and q times at the roots of $Q(z) = 0$ or $p - q + q = p = d$ times in all.

Exercise 6.7. **1.** $\frac{1}{2}$. **3.** $\frac{1}{2}$. **5.** 1. **7.** $-\frac{3}{2}$. **9.** $-\frac{1}{2}$. **11.** $\dfrac{\infty}{\infty}$, 1. **13.** $\frac{0}{0}$, $\frac{1}{4}$. **15.** 0^0, 1. **17.** 1^∞, e^2. **19.** 1^∞, e^{-1}. **21.** 0.

23. $\lim\limits_{t\to 0} \dfrac{\log(1+t)}{t+t^2} = \lim\limits_{t\to 0} \dfrac{\dfrac{1}{1+t}}{1+2t} = 1.$

$$\frac{D_x \log(1+x^{\frac{1}{2}})}{D_x[x^{\frac{1}{2}}+x]} = \frac{(1+x^{\frac{1}{2}})^{-1}(2x^{\frac{1}{2}})^{-1}}{(2x^{\frac{1}{2}})^{-1}+1} = \frac{1}{(1+x^{\frac{1}{2}})(1+2x^{\frac{1}{2}})} \to 1 \quad \text{as } x \to 0.$$

Miscellaneous problems for Chapter 6

1. If $P(x) = \prod_1^p(x-x_j)Q(x)$ where $Q(x) = \Pi[(x-\alpha_k)^2 + \beta_k{}^2]$,

$$R(x) \equiv 1 - \frac{P'(x)}{P(x)} = 1 - \sum_1^p \frac{1}{x-x_j} - \frac{Q'(x)}{Q(x)}.$$

Here $R(x_j + \delta) < 0$, $R(x_{j+1} - \delta) > 0$, so $P(x) - P'(x)$ has at least one zero in each of the $p - 1$ intervals (x_j, x_{j+1}), $j = 1, 2, \cdots, p-1$, $x_j < x_{j+1}$. Since $R(x_p + \delta) < 0$ and $R(x) \to 1$ as $x \to \infty$, there is another zero in (x_p, ∞). This gives at least p real zeros. **3.** $R_n(0, x) = n\binom{\alpha}{n}\int_0^x \left(\frac{x - s}{1 + s}\right)^{n-1} (1 + s)^{\alpha-1}\, ds.$ The largest value of $\left|\frac{x - s}{1 + s}\right|$ for $-1 < x$ is assumed for $s = 0$. Hence

$$|R_n(0, x)| \leqq |\alpha| \left|\binom{\alpha - 1}{n - 1}\right| |x|^{n-1} \left|\int_0^x (1 + s)^{\alpha-1}\, ds\right|$$
$$= \left|\binom{\alpha - 1}{n - 1}\right| |x|^{n-1}\, |(1 + x)^\alpha - 1|.$$

This goes to 0 as $n \to \infty$, provided $|x| < 1$. **5.** $1 + 2x + x^2 - \frac{2}{3}x^3 - \frac{5}{6}x^4 - \frac{11}{15}x^5 + \cdots$. **7.** Let $\alpha = (a^2 - 1)^{\frac{1}{2}} - a$, $\beta = -(a^2 - 1)^{\frac{1}{2}} - a$. Then $\alpha + \beta = -2a$, $\alpha\beta = 1$ and $(x - \alpha)(x - \beta) = x^2 + 2ax + 1$. This gives

$$\frac{x + a}{x^2 + 2ax + 1} = \frac{1}{2}\left\{\frac{1}{x - \alpha} + \frac{1}{x - \beta}\right\} = -\tfrac{1}{2}\textstyle\sum_0^\infty (\alpha^{-n-1} + \beta^{-n-1})x^n.$$

Hence $a_n = -\frac{1}{2}(\alpha^{-n-1} + \beta^{-n-1})$ and $R = |\alpha|$. Further

$$\textstyle\sum_0^\infty a_n{}^2 x^n = \frac{1}{4}\sum_0^\infty (\alpha^{-2n-2} + 2(\alpha\beta)^{-n-1} + \beta^{-2n-2})x^n$$
$$= \frac{1}{4}\left\{\frac{1}{\alpha^2 - x} + \frac{2}{1 - x} + \frac{1}{\beta^2 - x}\right\}$$

for $|x| < \alpha^2$. This reduces to the given expression. **9.** e^x. **11.** $\frac{1}{2}$. Factor out x and expand in powers of $1/x$. Only two terms matter. **13.** $\int_0^1 x^{\frac{1}{2}}\, dx = \frac{2}{3}$.

15. $\dfrac{f'''(0)}{g'''(0)}$. Note that $f(0) = f''(0) = g(0) = g''(0) = 0$. **17.** Note that

$$x\int_x^1 t^{-2}\, dt = 1 - x \to 1 \quad \text{as } x \to 0.$$

Hence

$$x\int_x^1 f(t)t^{-2}\, dt - f(0) = x\int_x^1 [f(t) - f(0)]t^{-2}\, dt - xf(0).$$

Since $f(t)$ is cont. at $t = 0$, given an $\varepsilon > 0$ there is a δ such that $|f(t) - f(0)| < \varepsilon$ when $0 \leqq t \leqq \delta$. In the rest of the interval $[0, 1]$, $|f(t) - f(0)| \leqq 2M$ where $M = \max |f(t)|$ in $[0, 1]$. Then for $0 < x < \delta$

$$x\int_x^1 |f(t) - f(0)|\, t^{-2}\, dt \leqq 2Mx\int_\delta^1 t^{-2}\, dt + \varepsilon x\int_x^\delta t^{-2}\, dt$$
$$= 2Mx\left(\frac{1}{\delta} - 1\right) + \varepsilon x\left(\frac{1}{x} - \frac{1}{\delta}\right) \to \varepsilon \quad \text{as } x \to 0.$$

Here ε is arb. small. Hence

$$\lim_{x\to 0}\left\{x\int_x^1 [f(t) - f(0)]t^{-2}\, dt + x\right\} = 0.$$

Assertion is proved. **19.** $f^{(2k+1)}(0) = 0$, $f^{(2k)}(0) = \binom{\frac{1}{3}}{k}(2k)!$. **21.** By Taylor's th. $(1 + x)^a = 1 + ax + \frac{1}{2}a(a - 1)(1 + \theta x)^{a-2}x^2$. If this exceeds $1 + 3x$ for all small $x > 0$, then

$$a - 3 > \tfrac{1}{2}a(a - 1)(1 + \theta x)^{a-2}x$$

for all such x, requiring that $a \geqq 3$. **23.** $\log 2$. **25.** Suppose that $P(x)$ has only real zeros, $x_1 < x_2 < \cdots < x_p$, of multiplicities $m_1, m_2, \cdots, m_p$, $m_1 + m_2 + \cdots + m_p = n$, the degree of $P(x)$. By Rolle's th. $P'(x)$ has at least one zero in each of the intervals (x_j, x_{j+1}), giving at least $p - 1$ real zeros. In addition, $P'(x)$ has a zero of multiplicity $m_j - 1$ at each of the pts. x_j. This gives at least $\sum_{j=1}^{p} (m_j - 1) + p - 1 = n - 1$ real zeros. Thus all zeros of $P'(x)$ are real and all derivatives have only real zeros. **27.** $\log (x + 2) - 2 \log (x + 1) + \log x = \log \dfrac{x(x+2)}{(x+1)^2} = \log \{1 - (x + 1)^{-2}\} = -\sum_{n=1}^{\infty} \dfrac{1}{n} (x + 1)^{-2n}$.

CHAPTER 7

Exercise 7.1. **1.** $\cos x - 2 \sin 2x$. **3.** $2 \tan x \sec^2 x$. **5.** $\dfrac{2 \sec^4 x}{(1 - \tan^2 x)^2} = 2 \sec^2 2x$. **7.** $(a^2 \cos^2 x + b^2 \sin^2 x)^{-\frac{1}{2}}(b^2 - a^2) \sin 2x$. **9.** $\cot x$. **11.** $\frac{1}{3}(\sin x)^{-\frac{2}{3}} \cos x$. **13.** $5 \sin^4 x - 6 \sin^6 x$. **15.** $e^x \cos (e^x)$. **17.** True for $n = 1$. Suppose true for $n = 2k - 1$:

$$D_x^{(2k-1)} \tan x = \sum_{j=1}^{k} A_{k,j} \sec^{2j} x,\ A_{k,j} \text{ integers.}$$

$$D_x^{(2k)} \tan x = \{\textstyle\sum_1^k 2jA_{k,j} \sec^{2j} x\} \tan x,$$

Then

$$D_x^{(2k+1)} \tan x = \{\textstyle\sum_1^k (2j)^2 A_{k,j} \sec^{2j} x\} \tan^2 x + \{\textstyle\sum_1^k 2jA_{k,j} \sec^{2j} x\} \sec^2 x.$$

Since $\tan^2 x = \sec^2 x - 1$,

$$D_x^{(2k+1)} \tan x = \sum_{j=1}^{k+1} A_{k+1,j} \sec^{2j} x$$

and the coeffs. are integers. This proves the assertion and shows that $D_x^{2k} \tan x$ is a polyn. of degree k in $\sec^2 x$ multiplied by $\tan x$. **19.** $2 \sin \frac{1}{2}x \cos kx = \sin (k + \frac{1}{2})x - \sin (k - \frac{1}{2})x$. Sum for k from 1 to n, add $\sin \frac{1}{2}x$, and divide result by $2 \sin \frac{1}{2}x$. **21.** Use summation by parts. $\Sigma |a_n - a_{n+1}|$ is cvgt., $a_n \to 0$, and $\left|\dfrac{\sin (n + \frac{1}{2})x}{2 \sin \frac{1}{2}x}\right| \leqq \dfrac{1}{2} \operatorname{cosec} \left(\dfrac{1}{2} \delta\right)$ for $0 < \delta \leqq x \leqq 2\pi - \delta$. Hence the series cvgs. unif. in this interval. It need not be abs. cvgt. **23.** Multiply by $\cos kx$ or $\sin kx$ and integrate termwise using (7.1.41). **25.**

$$\int e^{-ax} \cos bx\, dx = -\frac{1}{a} e^{-ax} \cos bx + \frac{b}{a} \int e^{-ax} \sin bx\, dx,$$

$$\int e^{-ax} \sin bx\, dx = -\frac{1}{a} e^{-ax} \sin bx - \frac{b}{a} \int e^{-ax} \cos bx\, dx.$$

27. $\operatorname{cosec} x - \frac{1}{3} \operatorname{cosec}^3 x$. **29.** $\frac{3}{4} \sin (\frac{2}{3}x) + \frac{3}{2} \sin (\frac{1}{3}x)$. **31.** $-\cos x + \frac{1}{3} \cos^3 x$.

33. $\log \tan(\frac{1}{2}x + \frac{1}{4}\pi) = \log \dfrac{1 + \sin x}{\cos x} = \log(\sec x + \tan x)$. **35.** 0. **37.** Integrate by parts.

$$\int P(x) \sin x \, dx = -\cos x [P(x) - P''(x) + P^{(4)}(x) - \cdots] + \sin x [P'(x) - P'''(x) + P^{(5)}(x) - \cdots].$$

39. $\frac{17}{305}t^7$ in (7.1.35); $\frac{61}{720}t^6$ in (7.1.36). **41.** $\log \cos x = -\frac{1}{2}x^2 - \frac{1}{12}x^4 - \cdots$. **43.** $-\frac{1}{8}$. **45.** $\frac{1}{3}$.

Exercise 7.2. **1.** π. **3.** 26π. **5.** 0. **7.** $\frac{1}{2}\pi$. **9.** $\sqrt{2}$. **11.** $\sqrt{2} + \log(1 + \sqrt{2})$. **13.** $\left|\sum_{k=0}^{n} e^{kix}\right| = \left|\dfrac{1 - e^{(n+1)ix}}{1 - e^{ix}}\right| < \operatorname{cosec} \dfrac{1}{2}\delta < \dfrac{\pi}{\delta}$ if $0 < \delta \leqq x \leqq 2\pi - \delta$. The last ineq. comes from $\sin t > \dfrac{2}{\pi}t$, valid for $0 < t < \frac{1}{2}\pi$. **15.** Differentiate! **17.** If $x = \frac{1}{2}(e^y + e^{-y})$, then $(e^y)^2 - 2xe^y + 1 = 0$ or $e^y = x \pm (x^2 - 1)^{\frac{1}{2}}$. **19.** $(2k + 1)\pi i$, $(2k + \frac{1}{2})\pi i$, $(2k + \frac{1}{3})\pi i$ resp. where k is any integer.

Exercise 7.3. **1.** $\cosh x \cosh 2x + 2 \sinh x \sinh 2x$. **3.** $2e^{2x}$. **5.** e^{-2x}. **7.** 1. **9.** $\frac{1}{3}\cosh^3 x - \cosh x$. **11.** $\frac{2}{3}$. **13.** $\dfrac{\sinh x}{x} = 1 + \frac{1}{6}x^2 + \cdots$. **15.** $x \coth x = 1 + \frac{1}{3}x^2 - \frac{1}{45}x^4 + \cdots$. **17.** Since $e^{i(z_1+z_2)} = e^{iz_1}e^{iz_2} = (\cos z_1 + i \sin z_1) \times (\cos z_2 + i \sin z_2)$,
we have

$$\cos(z_1 + z_2) + i \sin(z_1 + z_2) = \cos z_1 \cos z_2 - \sin z_1 \sin z_2 + i[\sin z_1 \cos z_2 + \sin z_1 \cos z_2].$$

We replace i by $-i$ and add and subtract the two resulting relations. This gives the addition ths. Hence

$$\sin(x + iy) = \sin x \cosh y + i \cos x \sinh y.$$

19. $\frac{1}{6}$. **21.** 0. **23.** 1. **25.** 0. **27.** $y' = 1 - y^2$.

Exercise 7.4. **1.** $-4x(2 - x^4)^{-\frac{1}{2}}$. **3.** $-\frac{1}{2}(1 - x^2)^{-\frac{1}{2}}$. **5.** $x^{-1}(1 + \log^2 x)^{-1}$. **7.** $x[(b^2 - x^2)(x^2 - a^2)]^{-\frac{1}{2}}$. **9.** $x(1 - x^2)^{-\frac{1}{2}}$.
11. $D_x \arcsin \dfrac{x}{(1 + x^2)^{\frac{1}{2}}} = \dfrac{1}{1 + x^2} = D_x \arctan x$. The fcns. are identical since they agree at $x = 0$. **13.** Differentiate! **15.** $y' = b \cos(b \arcsin x)(1 - x^2)^{-\frac{1}{2}}$.

$$y'' = -b^2 \sin(b \arcsin x)(1 - x^2)^{-1} + b \cos(b \arcsin x)x(1 - x^2)^{-\frac{3}{2}}.$$

Eliminate! **17.** Verify formula for $n = 1$ and use induction.

$$F_1(x) = 2(1 - x^2)^{-\frac{1}{2}} \arcsin x,$$
$$F_1'(x) = 2(1 - x^2)^{-1} + 2x(1 - x^2)^{-\frac{3}{2}} \arcsin x,$$
$$F_1''(x) = 6x(1 - x^2)^{-2} + 2(1 - x^2)^{-\frac{5}{2}}(1 + 2x^2) \arcsin x.$$

Hence true for $n = 1$. **19.** 1. **21.** $\frac{1}{2}\pi + 1$. **23.** $\frac{1}{4}\pi + \frac{1}{2}\log 2$. **25.** $\frac{1}{2}\pi^2 - 4$. **27.** $\frac{2}{9}\pi^2 - \frac{1}{3}\sqrt{3}\pi + \log 2$. **29.** $\frac{1}{2}[\text{arc}\sin x - x(1 - x^2)^{\frac{1}{2}}]$. **31.** $\log(1 + \sqrt{2})$. **33.** 1. **35.** 1.

Exercise 7.5. **1.** 1. **3.** (7.5.17) gives $4\int_0^1 \frac{(1-t)\,dt}{(1+t)(1+t^2)^2} = \frac{1}{2}\pi + \log 2 - 1$. **5.** $x = \frac{1+t^2}{1-t^2}$, $y = \frac{2t}{1-t^2}$. $\int_1^2 \frac{x\,dx}{y} = 2\int_0^{3^{-\frac{1}{2}}} \frac{(1+t^2)\,dt}{(1-t^2)^2} = \sqrt{3}$. **7.** $x > 0$. Derivative $2x^{\log x - 1}\log x$. **9.** All x which are not integral multiples of π. Derivative

$$(1 + \sin x)^{\cot x}\left\{\frac{\cos^2 x}{\sin x\,(1 + \sin x)} - \operatorname{cosec}^2 x \log(1 + \sin x)\right\}.$$

11. $n\pi < x < (n + \frac{3}{4})\pi$, n any integer. Derivative

$$(1 + \cot x)^x\left\{\log(1 + \cot x) - \frac{x}{\sin x(\sin x - \cos x)}\right\}.$$

13. $(2n - \frac{1}{2})\pi < x < (2n + \frac{1}{2})\pi$, n any integer. Derivative

$$(\cos x)^{x^2}x(-x\tan x + 2\log\cos x).$$

15. 1. **17.** 1. **19.** 1. **21.** $0 < x < e$.

Exercise 7.6. **1.** $C_1e^x + C_2e^{2x}$. **3.** $C_1 + C_2e^{-3x}$. **5.** $C_1e^x + C_2e^{-x} + C_3\cos x + C_4\sin x$. **7.** $C_1 + C_2x + C_3\cos x + C_4\sin x$. **9.** $(C_1 + C_2x)e^x + C_3e^{2x}$. **11.** $\frac{1}{2}(e^x + \cos x + \sin x)$. **13.** If the equation is $P(D)[y] = 0$ and if $y = u + iv$, then $P(D)[u + iv] = P(D)[u] + iP(D)[v] = 0$. $P(D)[u]$ and $P(D)[v]$ are real-valued fcns. since P has real coeffs. It follows that $P(D)[u] = 0$ and $P(D)[v] = 0$, i.e., $u(x)$ and $v(x)$ are solutions. $e^{ix} = \cos x + i\sin x$ is a solution of No. **5** and so are $\cos x$ and $\sin x$. **15.** Fundamental system e^z, $e^{\omega z}$, $e^{\omega^2 z}$, where $\omega = -\frac{1}{2} + i\frac{1}{2}\sqrt{3}$, $\omega^3 = 1$. The first solution is unbd. on the real axis when $z = x \to +\infty$, the second on $\arg z = -\frac{2}{3}\pi = -\arg\omega$, the third on $\arg z = \frac{2}{3}\pi$. **17.** r_1 and r_2 are the roots of $r(r-1) + ar + b = 0$. If $r_1 = \alpha + i\beta$, $r_2 = \alpha - i\beta$, then

$$C_1x^\alpha\cos(\beta\log x) + C_2x^\alpha\sin(\beta\log x)$$

is a solution. **19.** Set $y(x) = Y(x) + u(x)$ and substitute in the equation. This gives $u'' + 4u = 0$. **21.** $\frac{1}{4}x^2 - \frac{1}{8}$.

Miscellaneous problems for Chapter 7

1. The p times derived series is $\Sigma\, n^p a_n \cos\left(nx + p\frac{\pi}{2}\right)$ and this cvgs. abs. and unif. Hence its sum is $F^{(p)}(x)$. **3.** Use induction on n. **5.** $\frac{(x-1)^2(x+2)}{(x+1)^2(x-2)} = \frac{x^3 - 3x + 2}{x^3 - 3x - 2}$. Use $\frac{1}{2}\log\frac{1+u}{1-u} = \sum_0^\infty \frac{u^{2n+1}}{2n+1}$ with $u = 2(x^3 - 3x)^{-1}$. Here

$|u| < 1$ if $|x|\,|x^2 - 3| > 2$ and this holds iff $|x| > 2$. Hence the expansion cvgs. for $|x| > 2$. **7.** Follows from the identity

$$\left(\frac{1 + \frac{1}{31}}{1 - \frac{1}{31}}\right)^7 \left(\frac{1 + \frac{1}{49}}{1 - \frac{1}{49}}\right)^5 \left(\frac{1 + \frac{1}{161}}{1 - \frac{1}{161}}\right)^3 = \left(\frac{32}{30}\right)^7 \left(\frac{50}{48}\right)^5 \left(\frac{162}{160}\right)^3 = 2.$$

Take two terms for $\frac{1}{31}$ and for $\frac{1}{49}$ and one term for $\frac{1}{161}$. Note that $\sum_{k=n}^{\infty} \frac{y^{2k+1}}{2k+1} < \frac{1}{2n+1} y^{2n+1}(1 - y^2)^{-1}$. Each of the three remainders after multiplication by 14, 10, and 6 resp. contributes at most 10^{-7} to the value of $\log 2 = 0.6973448 \cdots$. **9.** Set $x = b + (a^2 + b^2)s$. This gives $2(a + b)^{-\frac{1}{2}} \int_0^\infty (1 + s^2) = \pi(a + b)^{-\frac{1}{2}}$. **11.** Note that $\arctan a + \arctan b = \arctan \dfrac{a + b}{1 - ab}$. Set $a = x\sqrt{2} + 1$, $b = x\sqrt{2} - 1$ and simplify. **13.** If $Ax + Be^x + Cxe^{-x} = 0$ holds in $(-a, a)$, it holds in particular for $x = 0$. This gives $B = 0$ so that $A + Ce^{-x} = 0$ in $(-a, a)$. If $A \neq 0$, then $e^x = -C/A$ for $-a < x < a$ which is absurd. Hence $A = 0$ and also $C = 0$, i.e., the fcns. are linearly independent. **15.** Form

$$\frac{f'(x)}{g'(x)} = \frac{\sin \frac{1}{x} - \frac{1}{x} \cos \frac{1}{x}}{\frac{1}{2} \operatorname{sgn} x \cdot |x|^{-\frac{1}{2}} + 1} = \frac{|x|^{\frac{1}{2}} \sin \frac{1}{x} - \operatorname{sgn} x\, |x|^{-\frac{1}{2}} \cos \frac{1}{x}}{\frac{1}{2} \operatorname{sgn} x + |x|^{\frac{1}{2}}}.$$

The numerator in the third fraction oscillates between infinite limits of indetermination and the denominator goes to $\pm\frac{1}{2}$ as $x \to 0$. **17.** $\frac{80}{3}$. **19.** 1. **21.** $\sin\{\pi(4n^2 + n)^{\frac{1}{2}}\} = \sin\left\{2n\pi\left(1 + \frac{1}{4n}\right)^{\frac{1}{2}}\right\} = \sin\left\{\left(2n + \frac{1}{2}\right)\pi - \frac{\pi}{64n} + O\left(\frac{1}{n^2}\right)\right\} = \cos\left\{\frac{\pi}{64n} + O\left(\frac{1}{n^2}\right)\right\} \to 1$. **23.** Yes. The sign of the fcn. alternates at the successive multiples of $\frac{1}{5}\pi$, so there are at least 10 real, distinct zeros between 0 and 2π. There are exactly 10 real zeros for, if we set $e^{ix} = t$ in $2 \sin 3x + 10 \cos 5x = 0$, we obtain an algebraic equation in t of degree 10. To the 10 zeros x_j in $(0, 2\pi)$ correspond 10 distinct zeros t_j of the algebraic polyn. so that all roots have been taken into account. **25.** $\dfrac{1}{2}$. **27.** $\dfrac{e - 2}{e - 1}$.

CHAPTER 8

Exercise 8.1. **1.** $TP = \dfrac{y_0}{b^2 x_0}(b^4 x_0^{\,2} + a^4 y_0^{\,2})^{\frac{1}{2}}$, $TX = \left(\dfrac{a y_0}{b x_0}\right)^2$, $XN = \dfrac{b^2 x_0}{a^2}$. Equation of tgt. $b^2 x_0 x - a^2 y_0 y = a^2 b^2$, of normal $b^2 x_0 y + a^2 y_0 x = (a^2 + b^2) x_0 y_0$. **3.** If $P = (x_0, y_0)$, then $N = x_0 + p + q x_0$ and this equals the x-intercept of the normal. **5.** The line $y = mx + b$ is a tgt. to the semi-cubical parabola at (x_0, y_0) if $m = \left(\dfrac{a}{4x_0}\right)^{\frac{1}{3}}$, $b = y_0 - (\frac{1}{4} a x_0^{\,2})^{\frac{1}{3}}$. It is a normal to the parabola at (X_0, Y_0) if $m = -\dfrac{2a}{X_0}$, $b = Y_0 + 2a$. It is both, if $X_0 = -(32 a^2 x_0)^{\frac{1}{3}}$, $Y_0 = 2(\frac{1}{4} a x_0^{\,2})^{\frac{1}{3}}$. Hence, for every value of x_0, there are unique values of X_0, Y_0, i.e., every tgt. of the semi-cubical parabola is a normal to the parabola. **7.** $y - x = 3$. **9.** $a = 1$, b arb. **13.** Pts. of intersection $(a - b, \pm 2\sqrt{ab})$, slopes $\left(\frac{a}{b}\right)^{\frac{1}{2}}$, $-\left(\frac{b}{a}\right)^{\frac{1}{2}}$. **15.** $y = \pm(x + 1)$. The equation has the form $y = -\frac{x_0}{y_0}x + \frac{1}{y_0}$ as tgt. to the circle at (x_0, y_0) and $y = \frac{2}{Y_0}x + \frac{2ax_0}{Y_0}$ as tgt. to the parabola at (X_0, Y_0). Equate slopes and y-intercepts and use $x_0{}^2 + y_0{}^2 = 2$, $Y_0{}^2 = 4X_0$. **17.** The normal at $(\frac{3}{8}, \frac{1}{8})$. **19.** $(aX)^2 - (bY)^2 = (X^2 + Y^2)^2$. An ordinary lemniscate if $a = b$. **21.** $Y^2(X + 2a) + X(X - a)^2 = 0$. Isolated pt. $(a, 0)$. The rest of the curve lies in $-2a < X \leqq 0$. Asymp. $X = -2a$. Vert. tgt. at $(0, 0)$.

Exercise 8.2. **1.** 77. **3.** 30. If $x = 3 \cosh s$, $y = 3 \sinh s$, then $\frac{ds}{dt} = 2$ and $\frac{dy}{dt} = 3 \cosh s \cdot \frac{ds}{dt}$. **5.** $\frac{4}{5}$. **7.** $-0.01a^2$. **9.** $\frac{4\sqrt{3}}{900}m$. **11.** 0.06793.

Exercise 8.3. **1.** $v = 3t^2 - 14t + 8$, $\alpha = 6t - 14$. For $t = 0$, $x = 0$, $v = 8$, $\alpha = -14$. The motion is reversed at $t = \frac{2}{3}$, $x = \frac{68}{27}$. The particle passes the origin at $t = \frac{1}{2}(7 - \sqrt{17})$, reverses at $t = 4$, $x = -16$, passes the origin again at $t = \frac{1}{2}(7 + \sqrt{17})$ and recedes to $+\infty$ when $t \to +\infty$. x is monotone decr. for $t < 0$ and goes to $-\infty$ with t. The acceleration is neg. for $t < \frac{7}{3}$, pos. for $t > \frac{7}{3}$. **3.** If $x^2 = at^2 + 2bt + c$, then $\ddot{x} = (ac - b^2)x^{-3} \neq 0$ by assumption. **5.** If $\frac{dx}{dt} = Cx$, then $x = x_0 e^{Ct}$ which is not 0 for $t = 0$ unless $x_0 = 0$ and $x(t) \equiv 0$. **7.** $(bx)^2 + (ay)^2 = (ab)^2$. Period π. **9.** If $\boldsymbol{x}(t) = x\boldsymbol{i} + y\boldsymbol{j}$, then $x = \alpha \cos kt + \gamma \sin kt$, $y = \delta \sin kt$ so that $(\delta x - \gamma y)^2 + (\alpha y)^2 = (\alpha\delta)^2$ which is an ellipse. The assumption on a is no restriction of the generality. It can be attained by changing the time scale. **11.** $\boldsymbol{x}(t) = 4t^2\boldsymbol{i} + (2t^2 - \frac{3}{2})\boldsymbol{j}$, $\boldsymbol{v}(t) = 8t\boldsymbol{i} + 4t\boldsymbol{j}$, $\boldsymbol{\alpha}(t) = 8\boldsymbol{i} + 4\boldsymbol{j}$. Hodograph $x - 2y = 0$. **13.** $\boldsymbol{x}(t) = R[\cos\theta(t)\boldsymbol{i} + \sin\theta(t)\boldsymbol{j}]$, $\boldsymbol{v}(t) = R[\cos(\theta(t) + \frac{1}{2}\pi)\boldsymbol{i} + \sin(\theta(t) + \frac{1}{2}\pi)\boldsymbol{j}]\,\theta'(t)$, $\boldsymbol{\alpha}(t) = -\boldsymbol{x}(t)[\theta'(t)]^2 + \boldsymbol{v}(t)\frac{\theta''(t)}{\theta'(t)}$. If $\theta'(t) = a$, $\theta(t) = at + b$, $\theta''(t) = 0$.

Exercise 8.4. **1.** (i) incr. in $(-\infty, -1)$, $(1, \infty)$, decr. in $(-1, 1)$; (ii) max. $(-1, 4)$, min. $(1, 0)$; (iii) pt. of infl. $(0, 2)$; (iv) concave downwards or upwards according as $x < 0$ or > 0. **3.** incr. in $(-\infty, -\sqrt{3})$, $(\sqrt{3}, \infty)$, decr. in $(-\sqrt{3}, \sqrt{3})$; (ii) max. $(-\sqrt{3}, 6\sqrt{3})$, min. $(\sqrt{3}, -6\sqrt{3})$; (iii) pts. of infl. $(-\frac{1}{2}\sqrt{6}, \frac{21}{8}\sqrt{6})$, $(0, 0)$, $(\frac{1}{2}\sqrt{6}, -\frac{21}{8}\sqrt{6})$; (iv) concave downwards in $(-\infty, -\frac{1}{2}\sqrt{6})$, $(0, \frac{1}{2}\sqrt{6})$, upwards elsewhere. **5.** (i) decr. for $x < -\frac{1}{2}$, then incr.; (ii) min. $(-\frac{1}{2}, -\frac{1}{64})$; pts. of infl. $(-1, 0)$, $(-\frac{1}{2} - \frac{1}{10}\sqrt{5}, -\frac{1}{125})$, $(-\frac{1}{2} + \frac{1}{10}\sqrt{5}, -\frac{1}{125})$, $(0, 0)$; concave downwards in $(-1, -\frac{1}{2} - \frac{1}{10}\sqrt{5})$ and $(-\frac{1}{2} + \frac{1}{10}\sqrt{5}, 0)$, upwards elsewhere. **7.** (i) incr. in $(-\infty, -1)$ and $(1, \infty)$, decr. in $(-1, 0)$ and $(0, 1)$ but not decr. in $(-1, 1)$ since $x = 0$ is a vert. asymp.; (ii) max. $(-1, -2)$, min. $(1, 2)$; (iii) no pt. of infl.; (iv) concave downwards for $x < 0$, upwards for $x > 0$. Hyperbola. **9.** (i) decr. for $x < -2^{-\frac{1}{3}}$, incr. in $(-2^{-\frac{1}{3}}, 0)$ and $(0, \infty)$, $x = 0$ is vert. asymp.; (ii) min. $(-2^{-\frac{1}{3}}, 2^{\frac{1}{3}} + 2^{-\frac{2}{3}})$; (iii) pt. of infl. $(-1, 0)$; (iv) concave upwards in $(-\infty, 0)$ and $(1, \infty)$, downwards elsewhere. **11.** (i) incr. in $(-\frac{1}{2}\sqrt{2}, \frac{1}{2}\sqrt{2})$, decr. elsewhere; (ii) min. $\left(-\frac{1}{2}\sqrt{2}, -\frac{1}{2}\sqrt{\frac{2}{e}}\right)$, max. $\left(\frac{1}{2}\sqrt{2}, \frac{1}{2}\sqrt{\frac{2}{e}}\right)$; (iii) pts. of infl. $(-\frac{1}{2}\sqrt{6}, -\frac{1}{2}\sqrt{6}e^{-\frac{3}{2}})$, $(0, 0)$, $(\frac{1}{2}\sqrt{6}, \frac{1}{2}\sqrt{6}e^{-\frac{3}{2}})$; (iv) concave downwards in $(-\infty, -\frac{1}{2}\sqrt{6})$ and in $(0, \frac{1}{2}\sqrt{6})$, upwards elsewhere. **13.** (i) decr. in $(0, 1/e)$, then incr.; (ii) min. $(1/e, -1/e)$; (iii) no pts. of infl., vert. tgt. at $(0, 0)$; (iv) concave upwards for $x > 0$. **15.** $y = \sqrt{2}\sin(x + \frac{1}{4}\pi)$. (i) decr. in $((2n + \frac{1}{4})\pi, (2n + \frac{5}{4})\pi)$, n any integer, incr. in each of the remaining intervals; (ii) max. $((2n + \frac{1}{4})\pi, 1)$, min. $((2n + \frac{5}{4})\pi, -1)$; (iii) pts. of infl. $((n + \frac{3}{4})\pi, 0)$; concave downwards in $((2n - \frac{1}{4})\pi, (2n + \frac{3}{4})\pi)$, upwards elsewhere. **17.** (i) incr. in each interval $((n - \frac{1}{2})\pi, (n + \frac{1}{2})\pi)$; (ii) no extrema; (iii) pts. of infl. $(n\pi, 0)$; (iv) concave downwards in $((n - \frac{1}{2})\pi, n\pi)$, upwards elsewhere. **19.** $(1 + 9x^4)^{\frac{3}{2}}(6x)^{-1}$. **21.** $\frac{1}{2}[x^4 + (x^2 - 1)^2]^{\frac{3}{2}}x^{-5}$. **23.** $\cosh^2 x$. **25.** $\dfrac{(1 + 2e^x + 2e^{2x})^{\frac{3}{2}}}{e^x(1 + e^x)}$.

27. $\rho = \dfrac{(1 + \cos^2 x)^{\frac{3}{2}}}{\sin x}$. Min. when $\cos x = 0$ and $\rho = 1$. No finite l.u.b. **31.** $\rho = \infty$. For $\cot x - \dfrac{1}{x} = -\dfrac{1}{3}x + \dfrac{4}{45}x^3 + O(x^5)$ so $f'(0) = -\dfrac{1}{3}$, $f''(0) = 0$. **33.** If $(bx)^2 - (ay)^2 = (ab)^2$, then $y' = \left(\dfrac{b}{a}\right)^2 \dfrac{x}{y}$, $y'' = -\dfrac{b^4}{a^4y^3}$. Hence

$$\rho = -\left(\frac{a}{b}\right)^4 y^3 \left\{1 + \left(\frac{b^2x}{a^2y}\right)^2\right\}^{\frac{3}{2}} \to -\infty,$$

since $\dfrac{x}{y} \to \dfrac{a}{b}$ when $(x, y) \to (+\infty, +\infty)$ along the hyperbola. **35.** Yes.

$$\int_0^\infty (1 + e^{2x})^{-\frac{3}{2}}e^x\,dx = \int_{\pi/4}^{\pi/2} \cos t\,dt = 1 - \tfrac{1}{2}\sqrt{2}$$

by the substitution $e^x = \tan t$. **37.** The curvature is $2(4x + 1)^{-\frac{3}{2}} \to 0$ as $x \to +\infty$. **39.** The radius of curvature at (x_0, y_0) is $2a^{-\frac{1}{2}}(x_0 + a)^{\frac{3}{2}}$. The normal at (x_0, y_0) is $y - y_0 = -\left(\dfrac{x_0}{a}\right)^{\frac{1}{2}}(x - x_0)$ and it intersects the directrix at $\left(-a, y_0 + \left(\dfrac{x_0}{a}\right)^{\frac{1}{2}}(x_0 + a)\right)$ whose distance from (x_0, y_0) is $a^{-\frac{1}{2}}(x_0 + a)^{\frac{3}{2}}$ as asserted. Since the subnormal is constant $= 2a$, the normal through (x_0, y_0) is easily constructed and hence also the center of curvature.

Exercise 8.5. **1.** Rel. min. $(1, 3)$, $(3, 3)$, rel. max. $(2, 4)$. **3.** Rel. min. $(2, \frac{3}{5})$, rel. max. $(0, -1)$. **5.** Rel. max. $(-\frac{1}{3}a, \frac{1}{3}2^{\frac{5}{3}}a)$. **7.** Rel. max. $(e^{1/n}, (ne)^{-1})$. **9.** Rel. max. $((2n + \frac{1}{4})\pi, \frac{1}{2}\sqrt{2}\exp[-(2n + \frac{1}{4})\pi])$, rel. min. $((2n + \frac{3}{4})\pi, -\frac{1}{2}\sqrt{2}\exp[-(2n + \frac{3}{4})\pi])$. **11.** Rel. min. $(a, 2a)$. **13.** Rel. min. $(-2^{\frac{1}{4}}, -2^{\frac{3}{4}})$, rel. max. $(2^{\frac{1}{4}}, 2^{\frac{3}{4}})$. **15.** Rel. max. $(-\frac{1}{2}, 2)$. Node at $(1, -1)$ with slopes $\pm\sqrt{6}$. **17.** $x = \frac{1}{n}(a_1 + a_2 + \cdots + a_n)$. **19.** $f'(x)$ is pos. for $x = [(2n + \frac{3}{4})\pi]^{-1}$, neg. for $[(2n + \frac{1}{4})\pi]^{-1}$, n pos. integer, signs reversed if n is neg. Hence infinitely many rel. max. and min. clustering at $x = 0$. **21.** The extrema correspond to the roots of $\tan\frac{1}{x} = \frac{2}{x}$ which cluster at $x = 0$. $f'(0) = 0$, but $(0, 0)$ is not an extremum since $f(x)$ takes both pos. and neg. values in any neighborhood of $x = 0$. $\lim_{h\to 0} f'(h)$ does not exist. **23.** Rel. max. $(\frac{1}{2}n\pi, 1)$, rel. min. $((\frac{1}{2}n + \frac{1}{4})\pi, \frac{1}{2})$. **25.** Rel. min. $(0, 1)$. **27.** No extrema. **29.** Min. area $2ab$. **31.** Assume sides par. to and symmetric w.r.t. the axes. Max. area $2ab$, sides $\sqrt{2}a$, $\sqrt{2}b$. **33.** The vertices bisect the sides of the triangle when the area has its max. value which is half the area of the triangle. **35.** If base is $2b$, slanting sides a each, and $2a + 2b = p$, then $A = \frac{1}{2}(p)^{\frac{1}{2}}b(p - 4b)^{\frac{1}{2}}$ which is a max. for $b = \frac{1}{6}p$. The triangle is then equilateral. **37.** The max. volume is $\frac{4}{9}$ of that of the cone. The base radius of the cylinder is $\frac{2}{3}$ of that of the cone, the height $\frac{1}{3}$. **39.** If $2b < h$, then $r = \frac{1}{2}\frac{hb}{h - b}$ gives the max. total area $\frac{1}{2}\frac{h^2b}{h - b}\pi$. If $2b \geqq h$, the problem has no actual solution. The total area of an inscribed cylinder has the l.u.b. $2\pi b^2$, which is not reached. **41.** The max. curved area is $2\pi R^2$ for a cylinder of height $\sqrt{2}R$ and base radius $\frac{1}{2}\sqrt{2}R$. **43.** The max. volume is $\frac{2\sqrt{3}}{27}\pi a^3$ for a cone of height $\frac{1}{3}\sqrt{3}a$ and base radius $\frac{1}{3}\sqrt{6}a$. **45.** The derivative is $-\frac{|x|^{\alpha-1}}{|x|^{2\alpha} + 1}\operatorname{sgn} x$. The limit is 0, -1, or $-\infty$ according as $\alpha > 1$, $=1$, <1.

Exercise 8.6. **1.** The extrema of $F = Cf^2$ are given by the extrema and the zeros of f. The latter make F an abs. min. An extremum of f gives an extremum of F of same or opposite type according as f is pos. or neg. at the pt. **3.** The max. distance is $a - b$, reached by the four normals at the pts. $(\pm a^{\frac{3}{2}}(a + b)^{-\frac{1}{2}}, \pm b^{\frac{3}{2}}(a + b)^{-\frac{1}{2}})$. **5.** The fcn. $f(x) = e^{-x} - 1 + x$ has the abs. min. 0 for $x = 0$ since $\operatorname{sgn} f'(x) = \operatorname{sgn} x$. **7.** Horizontal asymp. $y = 1$. Abs. min. $(0, \frac{1}{2})$. Pt. of infl. $(\log 2, \frac{5}{8})$. **9.** If $f(x) = \frac{\sin x}{x}$ then $f'(x) = \frac{\cos x}{x^2}(x - \tan x) < 0$ for $0 < x < \frac{1}{2}\pi$. Hence $f(x) > f(\frac{1}{2}\pi) = \frac{2}{\pi}$ as asserted. **11.** Set $a/b = x$, then

$$(8a^2 + b^2)^2 - 16a^3b = b^4f(x), f(x) = (8x^2 + 1)^2 - 16x^3.$$

Here $f(0) = 1$ and $f'(x) = 16x(16x^2 - 3x + 1) > 0$ for $x > 0$ so f is incr. in $(0, \infty)$. Hence $f(x) > 0$ and the ineq. is proved. **13.** $a = \frac{10}{3}(\sqrt{2} - 1)$. More generally, $\int_\alpha^\beta [f(x) - ag(x)]^2\,dx$ attains its min. value when $a\int_\alpha^\beta [g(x)]^2\,dx = \int_\alpha^\beta f(x)g(x)\,dx$ as is seen by expanding the sq. and differentiating w.r.t. a.

15. $a = -g/\gamma$, $C_1 = -C_2 = (v\gamma + g)/\gamma^2$. $y'(t) = 0$ for $t = \dfrac{1}{\gamma}\log\left(1 + \dfrac{v\gamma}{g}\right)$. Hence $f(v) = \dfrac{v}{\gamma} - \dfrac{g}{\gamma^2}\log\left(1 + \dfrac{v\gamma}{g}\right)$ in agreement with (8.6.1).

Miscellaneous problems for Chapter 8

1. If $f(x) = 2\log x + \dfrac{1}{x} - 1$, the ineq. holds iff $f(x) \geqq 0$. Since $f'(x) = \dfrac{2}{x} - \dfrac{1}{x^2}$, $f(x)$ is decr. in $(0, \frac{1}{2})$, incr. in $(\frac{1}{2}, \infty)$. $f(1) = 0$ gives $f(x) > 0$ in $(1, \infty)$. The equation $f(x) = 0$ has a single root x_0 in $(0, \frac{1}{2})$, x_0 appr. 0.27. The ineq. holds in $(0, x_0]$ and $[1, \infty)$. **3.** The pts. $(\pm 2^{-\frac{3}{2}}, -2^{-\frac{3}{2}})$, $\rho = \frac{3}{2}$. **5.** $b \leqq |v| \leqq a$. **7.** Compare coeffs. in the Maclaurin expansions. Only odd powers of x occur, their coeffs. are pos. and each coeff. on the right is greater than the corresponding coeff. on the left, except the first ones which are equal. Hence the right member is greater than the left for $0 < x < 1$. **9.** P should bisect the line segment cut off from the third line l_3 by the lines l_1 and l_2. Through P draw lines par. to l_1 and l_2. Counting intercepts from the intersection of l_1 and l_2, let a be the intercept on l_1 of the par. to l_2 and b the intercept on l_2 of the par. to l_1. Let l_3 have intercepts x and y on l_1 and l_2 resp. Similar triangles give $y = \dfrac{bx}{x-a}$ and the min. of $xy = \dfrac{bx^2}{x-a}$ is reached for $x = 2a$. Then $y = 2b$ and P bisects the segment of l_3 between l_1 and l_2. **11.** Let $r \leqq R$. Let P be x units from the center of the smaller sphere. The area visible from P is

$$f(x) = 2\pi\left\{R^2\left(1 - \frac{R}{a-x}\right) + r^2\left(1 - \frac{r}{x}\right)\right\},\ r \leqq x \leqq a - R.$$

Here

$$f'(x) = 2\pi\left\{\frac{r^3}{x^2} - \frac{R^3}{(a-x)^2}\right\}$$

which is 0 for $x = x_0$,

$$x_0 = a\frac{r^{\frac{3}{2}}}{r^{\frac{3}{2}} + R^{\frac{3}{2}}}.$$

This gives the max. visible area provided $r \leqq x_0 \leqq a - R$. The first ineq. holds if $ar^{\frac{1}{2}} > r^{\frac{3}{2}} + R^{\frac{3}{2}}$ and if so the second ineq. $x_0 < a - R$ is automatically sat. If $ar^{\frac{1}{2}} \leqq r^{\frac{3}{2}} + R^{\frac{3}{2}}$, then $f(x)$ has its largest value in $[r, a - R]$ at $x = r$, i.e., when P is on the smaller sphere. **13.** Suppose

$$\sum |a_n|\,|e^{\lambda_n z_1}| \equiv \sum u_n \text{ and } \sum |a_n|\,|e^{\lambda_n z_2}| \equiv \sum v_n$$

are cvgt. and set $z = \alpha z_1 + (1 - \alpha)z_2$, $0 < \alpha < 1$. Then

$$\sum |a_n|\,|e^{\lambda_n z}| = \sum u_n^{\alpha} v_n^{1-\alpha} \leqq \alpha \sum u_n + (1 - \alpha)\sum v_n$$

by (8.6.12), so the series cvgs. Hence the domain of abs. cvgce. is convex. **15.** Set $F(x) = f(x) - mx - b$ and suppose $x_1 < x_2 < x_3$ are consecutive zeros of $F(x)$ in (a, b). Then by Rolle's th. there exist pts. x_4 and x_5, $x_1 < x_4 < x_2 < x_5 < x_3$, such that $F'(x_4) = F'(x_5) = 0$. Again by Rolle's th. $F''(x) = f''(x)$ cannot keep a constant sign in (x_4, x_5) and there is at least one pt. x_6, $x_4 < x_6 < x_5$, such that $f''(x_6) = 0$ and $f''(x)$ changes sign when x passes through x_6. Thus $y = f(x)$ has at least one pt. of infl. in (a, b). In the case $y = \dfrac{2}{1 + x^2}$, $y = \frac{1}{5}x + \frac{4}{5}$, we have $x_1 = -3, x_2 = -2, x_3 = 1, x_4 = -2.45 \cdots, x_5 = -0.05 \cdots$, $x_6 = -\frac{1}{3}\sqrt{3} = -0.58 \cdots$. Here $x_7 = \frac{1}{3}\sqrt{3} = 0.58 \cdots$ is also a pt. of infl. in $(-3, 1)$. **17.** Let the line $y - y_0 = m(x - x_0)$ through the pt. $P(x_0, y_0)$ intersect the parabola $y = bx^2$ in the pts. (x_1, y_1), (x_2, y_2), $x_1 < x_2$. The coords. are fcns. of m, but the explicite forms are not needed. Then the area of the segment is

$$\begin{aligned} A(m) &= \int_{x_1}^{x_2} [y_0 + m(x - x_0) - bx^2]\, dx \\ &= y_0(x_2 - x_1) + \tfrac{1}{2}m[(x_2 - x_0)^2 - (x_1 - x_0)^2] - \tfrac{1}{3}b({x_2}^3 - {x_1}^3) \end{aligned}$$

and

$$\begin{aligned} A'(m) = [y_0 + m(x_2 - x_0) - b{x_2}^2]\frac{dx_2}{dm} - [y_0 + m(x_1 - x_0) - b{x_1}^2]\frac{dx_1}{dm} \\ + \tfrac{1}{2}[(x_2 - x_0)^2 - (x_1 - x_0)^2]. \end{aligned}$$

The first two brackets vanish since (x_1, y_1), (x_2, y_2) are the pts. of intersection. Hence $A'(m) = 0$ requires $x_2 - x_0 = \pm(x_1 - x_0)$. Only the minus sign makes sense and gives $x_2 - x_0 = x_0 - x_1$, i.e., P must bisect the segment joining (x_1, y_1) with (x_2, y_2). **19.** If $f(x) = \sin^2 x + \tan^2 x - 2x^2$ then

$$\begin{aligned} f'(x) &= \sin 2x + 2 \sec^2 x \tan x - 4x, \\ f''(x) &= 2 \cos 2x + 2 \sec^4 x + 4 \sec^2 x \tan^2 x - 4. \end{aligned}$$

Thus $f(0) = f'(0) = f''(0) = 0$. In $f''(x)$ express all fcns. in terms of $\cos x$ to obtain

$$f''(x) = (\cos x)^{-4}(1 - \cos^2 x)(6 + 2\cos^2 x - 4\cos^4 x) > 0$$

if $0 < x < \frac{1}{2}\pi$. This shows that $f'(x)$ is incr. and hence pos. in $(0, \frac{1}{2}\pi)$. Hence $f(x)$ is also incr. and pos. there and the inequality is proved. **21.** Since

$$\begin{aligned} f(x) &= x^3(1 - \tfrac{1}{6}x^2 + \cdots)^3 + x^3(1 + \tfrac{1}{3}x^2 + \cdots)^3 - 2x^3 \\ &= x^3[1 - \tfrac{1}{2}x^2 + O(x^4)] + x^3[1 + x^2 + O(x^4)] - 2x^3 \\ &= \tfrac{1}{2}x^5 + O(x^7) \end{aligned}$$

the limit is $\frac{1}{2}$.

CHAPTER 9

Exercise 9.1. **1.** Asymps. $x = -1$, $x = 1$, $y = 1$. The curve is above $y = 1$ for $x > 1$, below for $x < 1$. Max. $(0, -1)$. No pts. of infl. Symmetric with respect to $x = 0$. **3.** Asymps. $x = -1$, $x = 1$, $y = x$. The curve is above the slanting asymp. for $x > 1$, below for $x < -1$. Min. $(\sqrt{3}, \frac{3}{2}\sqrt{3})$, max. $(-\sqrt{3}, -\frac{3}{2}\sqrt{3})$. Pt. of infl. with x-axis as tgt. at $(0, 0)$. **5.** Asymps. $x = 2$, $y = x + 4$. The curve is above the latter for $x > \frac{4}{3}$, below for $x < \frac{4}{3}$. Min. $(6, 13.5)$. Pt. of infl. at $(0, 0)$ with x-axis as tgt. **7.** Asymps. $x = -1$, $x = 1$, asymp. parabola $y = x^2 + 1$.

The curve is above the latter for $x > 1$. Min. at $(\pm\sqrt{2}, 4)$, max. $(0, 0)$. No pt. of infl. Symmetric w.r.t. $x = 0$. **9.** Asymps. $x = 0$, $x = 1$, $y = 0$. The curve is above $y = 0$ for $x > 0$, below for $x < 0$. A max. at the real root of $2x^3 + 9x - 6 = 0$ or appr. 0.60. Corresponding ordinate is about -26.6. No pt. of infl. **11.** Asymps. $x = -1$, $x = 1$, $y = 0$. The curve is above the x-axis. Min. $(0, 4)$. **13.** Asymps. $x = -1$, $x = 1$, $y = -1$, $y = 1$. Excl. regions $-1 < x < 1$, $-1 < y < 1$. No extrema, no pts. of infl. **15.** Asymps. $x = -1$, $x = 1$, asymp. parabola $y^2 = x$. Max. $(3^{\frac{1}{2}}, -2^{-\frac{1}{2}}3^{\frac{3}{4}})$, min. $(3^{\frac{1}{2}}, 2^{-\frac{1}{2}}3^{\frac{3}{4}})$. Cusp at $(0, 0)$. Pts. of infl. for x about 2.5 Excl. regions $x < -1$, $0 < x < 1$. **17.** Asymp. $x = 2$, asymp. parabola $y^2 = x + 4$. Max. $(6, -\frac{3}{2}\sqrt{6})$, min. $(6, \frac{3}{2}\sqrt{6})$. Pts. of infl. for $x = 6 + 4\sqrt{3}$. Cusp at $(0, 0)$. Excl. region $x < 0$. **19.** Asymps. $x = -1$, $x = 1$, $y = x$, $y = -x$. Excl. region $|x| < 1$. Max. at $(\pm\sqrt{2}, -2)$, min. at $(\pm\sqrt{2}, 2)$. No pts. of infl. **21.** Asymps. $x = -1$, $y = 0$. Excl. region $x < -1$. No extrema. Pts. of infl. $(0, \pm 1)$ and $((\frac{4}{5})^{\frac{1}{3}}, \pm\frac{1}{3}\sqrt{5})$. **23.** Slanting asymp. $y + x = 0$. The curve is above this line. Excl. regions $x + y < 0$ and $x + y > 2^{\frac{2}{3}}$. No extrema. Pts. of infl. $(0, 1)$ with horizontal tgt. and $(1, 0)$ with vert. tgt. **25.** $(\frac{4}{3}, \pm\frac{4}{9}\sqrt{3})$. **27.** Asymp. $x = -1$. Excl. regions $x < -1$ and $x > 1$. Node at $(0, 0)$ with slopes ± 1. Max. $(\frac{1}{2}, \frac{1}{6}\sqrt{6})$, min. $(\frac{1}{2}, -\frac{1}{6}\sqrt{6})$. Area of loop $2 - \frac{1}{2}\pi$. Set $\dfrac{1 - x}{1 + x} = t^2$, $t = \tan s$.

Exercise 9.2. **1.** $y^2 = (\frac{1}{4}x^4 + a^2x^2)^{\frac{1}{2}} - \frac{1}{2}x^2$, $2yy' = \dfrac{x^2 + 2a^2}{(x^2 + 4a^2)^{\frac{1}{2}}} - x$. $y' > 0$ since $(x^2 + 2a^2)^2 > x^2(x^2 + 4a^2)$. Hence y is incr. when $(x, y) \to (+\infty, +\infty)$ along the curve in the first quadrant. Since $a^2x^2 > x^2y^2$, $a^2 > y^2$. Since $y^2 = a^2 - \dfrac{y^4}{x^2}$, $\lim y^2 = a^2$ and $\lim y = a$. **3.** No extrema. Pts. of infl. $([1 + \frac{2}{3}\sqrt{3}]^{\frac{1}{2}}, \pm[1 + \frac{2}{3}\sqrt{3}]^{\frac{1}{4}}\frac{2}{3}\sqrt{3})$. **5.** Multiply out and compare with (9.3.13). Factorization gives five excl. regions (i) $x < -1$, $y - x + \frac{1}{2} > 0$, $y + x - \frac{1}{2} < 0$; (ii) $-1 < x < \frac{1}{3}$, $y + x - \frac{1}{2} > 0$; (iii) $-1 < x < \frac{1}{3}$, $y - x + \frac{1}{2} < 0$; (iv) $x > \frac{1}{3}$, $y - x + \frac{1}{2} > 0$, $y + x - \frac{1}{2} < 0$; (v) $y - x + \frac{1}{2} < 0$, $y + x - \frac{1}{2} > 0$. The three asymps. are obtained by equating each of the linear factors on the left to 0. **7.** Excl. regions are (i) $x^2 + y^2 < 1$, $y > 0$; (ii) $x^2 + y^2 > 1$, $y < 0$. Asymp. $x = 0$. The curve is symmetric to the y-axis. It has an infinite branch in the upper half-plane, outside the unit circle, and a closed oval in the lower half of the unit circle provided $0 < \alpha < \frac{2}{9}\sqrt{3}$. The oval reduces to an isolated pt. $(0, -\frac{1}{3}\sqrt{3})$ when $\alpha = \frac{2}{9}\sqrt{3}$ and is missing if $\alpha > \frac{2}{9}\sqrt{3}$. The extremal pts. are on the y-axis at the roots of the equation $y^3 - y - \alpha = 0$. No common pts. **11.** Excl. regions (i) $x + y > 0$, $xy < 1$; (ii) $x + y < 0$, $xy > 1$. The curve is symmetric w.r.t. the line $y = x$. Asymps. $x = 0$, $y = 0$, $x + y = 0$. Three branches, one in the first quadrant above $xy = 1$ which it approaches at both ends. If $0 < \alpha < \frac{4}{9}\sqrt{3}$, the second branch lies below $x + y = 0$ and has this line as an asymp. at both ends while the third lies in the third quadrant above the hyperbola $xy = 1$. If $\alpha = \frac{4}{9}\sqrt{3}$, there is a node at $(-\frac{1}{3}\sqrt{3}, -\frac{1}{3}\sqrt{3})$ where the second and third branches now intersect. For $\alpha > \frac{4}{9}\sqrt{3}$ the second and third branches are symmetric w.r.t. $y = x$ but do not intersect it, both lie below $x + y = 0$ and above $xy = 1$ and have these as asymps. No extrema and no common pts. **13.** Excl. regions (i) $y < x < 0$; (ii) $y < x$, $y^2 < 4x$, (iii) $x > 0$, $y > \max(x, 2\sqrt{x})$. Common pt. (0, 0) where all curves have a vert. tgt. Asymp. $y = x$, asymp. parabola $y^2 = 4x$. Three branches. One goes through (0, 0) and $(-\frac{1}{4}\alpha, 0)$ and is asymp. to the lower halves of the parabola and of the line $y = x$. The second branch is an oval confined to $x < y < 2\sqrt{x}$. The third branch lies in the first quadrant above the parabola and below the line $y = x$ and is asymp. to both. **15.** Excl. regions (i) $\max(-x, +x - \sqrt{x}) \leqq y \leqq x + \sqrt{x}$, (ii) $0 < x$, $y < \min(x - \sqrt{x}, -x)$; (iii) $0 \leqq x + y$, $x \leqq 0$. Common pts. (0, 0), $(\frac{1}{4}, -\frac{1}{4})$. Slope -1 at the former pt. Asymp. $x = 0$, asymp. parabola $(y - x)^2 = x + \alpha$. Two branches. One to the right of the y-axis, above the parabola. The second branch intersects the x-axis at $x = 0$ and $x = \frac{1}{2} \pm (\frac{1}{4} + \alpha)^{\frac{1}{2}}$. It is asymp. to the neg. end of the y-axis and to the lower half of the parabola. **17.** The circles $(x + 1)^2 + y^2 = 3$, $(x - 1)^2 + y^2 = 3$ and the x-axis divide the plane into eight regions, half of which are excl. The region below the x-axis and below the circles is excl. Common pts. $\pm 1 \pm \sqrt{3}$. The curve passes through these pts., is kidney shaped and symmetric w.r.t. the y-axis. There are four extrema, two on the positive y-axis. The position depends on α. The algebra is too complicated and no attempt should be made to find these pts.

Exercise 9.3. **1.** $D_x y = \frac{3}{2}t = \pm\frac{3}{2}\sqrt{x}$, $D_x{}^2 y = \dfrac{3}{4t} = \pm\dfrac{3}{4\sqrt{x}}$.

3. $D_x y = -\dfrac{1 - t^2}{2t} = -\dfrac{x}{y}$, $D_x{}^2 y = -\left(\dfrac{1 + t^2}{2t}\right)^3 = -\dfrac{1}{y^3}$.

5. $D_x y = -\tan t$, $D_x{}^2 y = \dfrac{1}{3a}\sec^4 t \operatorname{cosec} t$. **7.** $D_x y = \dfrac{1 + t^2}{(t + 1)^2(2t - t^2)}$,

$$D_x{}^2 y = -2\,\frac{(t - 1)(2t^5 - 4t^4 + 7t^3 - 11t^2 + t + 1)}{(t + 1)^3(2t - t^2)^3}.$$

9. Substitute! **11.** Excl. regions $0 < x < 4$, fourth quadrant. Asymps. $x = -\frac{1}{2}$, $y = 0$, $y - \frac{1}{2}x + \frac{3}{4} = 0$. No extrema. Vert. tgts. at (0, 0) and $(4, \frac{2}{3})$. Node at $(-1, -1)$. Three branches, two through the node and one to the right of $x = 4$ above the x-axis and below the slanting asymp. One of the branches through the node lies above the slanting asymp. It crosses the vert. asymp. at $(-\frac{1}{2}, -\frac{2}{3})$, passes through (0, 0) and then approaches the vert. asymp. from the right. The other branch through the node lies below the x-axis, to the left of the vert. asymp. It crosses the slanting asymp. at $(-0.9, -1.2)$. The curve is a rational cubic $x(x - 2y - 1)(y + 1) = (x - y)^2$. **13.** The equation in polar coords. $r = 4\dfrac{\cos^2\theta \sin\theta}{\cos 2\theta}$ shows that the curve is unbd. It is symmetric w.r.t. the y-axis. Asymps. $x \pm y + 1 = 0$. The origin is a singular pt. The parametric representation $x = \dfrac{4t}{1 - t^4}$, $y = \dfrac{4t^2}{1 - t^4}$ $(y = tx)$ is useful. At (0, 0) there is a cusp with a vert. tgt. and also an ordinary branch with a horizontal tgt. The infinitary end in the first quadrant lies below the asymp. $x - y + 1 = 0$. The curve crosses this asymp. for $t = -1 + \sqrt{2}$ and then proceeds to (0, 0) where its tgt. is horizontal. The infinitary end in the fourth quadrant lies above the asymp. $x + y + 1 = 0$. This branch intersects the other asymp. for $t = -1 - \sqrt{2}$ and as $t \to -\infty$ the pt. $(x, y) \to (0, 0)$ where the tgt. is vert. The rest follows by symmetry. **15.** The equation in polar coords. $r = a \sin^3 \theta[1 + \frac{1}{2} \sin 2\theta]^{-2}$ shows that $r \leqq 4a$ since $1 + \frac{1}{2} \sin 2\theta \geqq \frac{1}{2}$. The given equation shows that $y \geqq 0$ and that $y = 0$ implies $x = 0$. **19.** $r = 4a \sin^2 (\frac{1}{2}\theta)$. $\tan \psi = \dfrac{1 - \cos \theta}{\sin \theta} = \tan (\frac{1}{2}\theta)$ so $\psi = \frac{1}{2}\theta$. $PN = 4a \sin (\frac{1}{2}\theta) = r \operatorname{cosec} \psi$, $PT = \tan (\frac{1}{2}\theta) \cdot 4a \sin (\frac{1}{2}\theta) = r \sec \psi$. $OT = r \dfrac{r}{r'} = r \tan \psi$. $ON = r' = 2a \sin \theta$.

Exercise 9.4. **1.** $\frac{1}{2}\sqrt{5} + \frac{1}{4} \log (\sqrt{5} + 2)$. **3.** $(a\omega)^{\frac{1}{2}} + \frac{1}{3}a^{-\frac{1}{2}}\omega^{\frac{3}{2}}$. **5.** $\frac{5}{2}a\{\sqrt{3} + \frac{1}{2} \log (2 + \sqrt{3})\}$. **7.** $\log \tan \frac{3}{8}\pi$. **9.** $\frac{1}{2}\omega(\omega^2 + 1)^{\frac{1}{2}} + \frac{1}{2} \log [\omega + (\omega^2 + 1)^{\frac{1}{2}}]$. **11.** Substitute in (9.3.37) for ρ. Further

$$s = 2a \int_{\pi}^{\theta} \sin (\tfrac{1}{2}s) \, ds = -4a \cos (\tfrac{1}{2}\theta) \text{ so } 9\rho^2 + s^2 = 16a^2. \quad \textbf{13.} \int_0^1 |z'(t)| \, dt.$$

Exercise 9.5. **1.** $x(t) = -at(1 + t^2)^{-\frac{1}{2}} \log [t + (1 + t^2)^{\frac{1}{2}}] = -at^2 + O(t^4)$, $y(t) = at - a(1 + t^2)^{-\frac{1}{2}} \log [t + (1 + t^2)^{\frac{1}{2}}] = \frac{2}{3}at^3 + O(t^5)$. Hence a cusp at (0, 0). The curve is in the left half-plane since $x(t) < 0$ for $t \neq 0$. Note

$$\log [-s + (s^2 + 1)^{\frac{1}{2}}] = -\log [s + (s^2 + 1)^{\frac{1}{2}}].$$

3. Use formula (9.4.30). **5.** Use formula (9.4.27). Let φ_1, ρ_1, s_1 be the corresponding quantities for the evolute. Then $\dfrac{ds_1}{ds} = \dfrac{d\rho}{ds}$ by Th. 9.5.2. Further $\varphi_1 = \varphi + \frac{1}{2}\pi$ so $\rho_1(s) = \dfrac{ds_1}{d\varphi_1} = \dfrac{ds_1}{d\varphi} = \dfrac{ds_1}{ds} \cdot \dfrac{ds}{d\varphi} = \rho(s) \dfrac{d\rho}{ds}$ as asserted. For $\rho(s)$ to be differentiable, third order derivatives of $x(s)$ and $y(s)$ must exist.

Exercise 9.6. **1.** $S = \displaystyle\int_0^{2\pi} \{[x'(t)]^2 + [y'(t)]^2\} \, dt = 2R \int_0^{2\pi} \sin (\tfrac{1}{2}t) \, dt = 8R$,

$$A = \int_0^{2\pi} y(t)x'(t) \, dt = 4R^2 \int_0^{2\pi} (1 - \cos t)^2 \, dt = 3\pi R^2.$$

3. Obviously $t = 0$ is a root, if $t = t_1$ is a root so is $t = -t_1$. Since $x''(t) = -2 \sin t < 0$ for $0 < t < \pi$, $x = x(t)$ is concave upwards in $(0, \pi)$ so there can be at most two roots of $t = 2 \sin t$ in $(0, \pi)$. For $t \geqq 2$, $x(t) > 0$, for $t = \frac{1}{2}\pi$, $x(t) < 0$, so there is a unique pos. root $t = t_1$ and $\frac{1}{2}\pi < t_1 < 2$. A table of natural sines gives $t_1 = 1.89 \cdots$. **5.** For the first part see Fig. 9.16. $PT = \rho(t) = a \sinh \left(\dfrac{t}{a}\right)$ by (9.6.5), $TQ = a$, so $\overline{TQ}^2 + \overline{PT}^2 = a^2 \cosh^2 \left(\dfrac{t}{a}\right) = \overline{PQ}^2$.

Hence PQ is perpendicular to the x-axis. $N = \left(t - a \coth\left(\frac{t}{a}\right), 0\right)$ and $NT = a \operatorname{cosech}\left(\frac{t}{a}\right)$ so that $\overline{NT} \cdot \overline{PT} = a^2$. **7.** $\int_0^\infty y(t)x'(t)\,dt = a^2\int_0^\infty \operatorname{sech} s\,(1 - \operatorname{sech}^2 s)\,ds = a^2\int_0^\infty \frac{\sinh^2 s \cosh s}{(\sinh^2 s + 1)^2}\,ds = a^2\int_0^\infty \frac{u^2\,du}{(u^2+1)^2}$. Integrate by parts to obtain $\frac{1}{4}\pi a^2$. **9.** Since $x = 3a\cos t - a\cos 3t$, $y = 3a\sin t - a\sin 3t$, we have $x + iy = a(3e^{it} - e^{3it})$. Set $e^{it} = \frac{1 - is}{1 + is}$. Then

$$x + iy = a\left\{3\,\frac{1 - is}{1 + is} - \left(\frac{1 - is}{1 + is}\right)^3\right\}.$$

Expanding and separating reals and imaginaries we get

$$x = 2a\,\frac{(1 - s^2)(1 + 10s^2 + s^4)}{(1 + s^2)^3}, \qquad y = -a\,\frac{32s^3}{(1 + s^2)^3}.$$

13. $\frac{3}{8}\pi a^2$.

Miscellaneous problems for Chapter 9

1. Excl. region $0 < x \leqq a$. Asymps. $x = a$, $y = \pm(x + \frac{1}{2}a)$. Cusp at $(0, 0)$ with horizontal tgt. Three branches. In the first quadrant the curve lies to the right of the vert. asymp. and above the slanting one. In the second quadrant the curve lies above its asymp. The curve is symmetric w.r.t. the x-axis. **3.** Excl. regions $x < -\frac{3}{2}a$, $\frac{1}{2}a \leqq x$. Vert. asymp. $x = \frac{1}{2}a$. Node at $(0, 0)$ with tgts. $y = \pm\sqrt{3}x$. Vert. tgt. at $(-\frac{3}{2}a, 0)$. Loop in the left half-plane. **5.** Excl. regions $-a \leqq x \leqq a$, $-b \leqq y \leqq b$. Asymp. $x = \pm a$, $y = \pm b$. Equations $\left(\frac{a}{x}\right)^2 + \left(\frac{b}{y}\right)^2 = 1$ and $x = a\,\frac{1 + t^2}{1 - t^2}$, $y = b\,\frac{1 + t^2}{2t}$. Cf. parametric representation of the ellipse. **7.** Equation $x^4 - 2ax^3 + (a^2 - b^2)x^2 = b^2y^2$. For $b < a$, the origin is a node with tgts. $by = \pm(a^2 - b^2)^{\frac{1}{2}}x$. Cusp for $b = a$, isolated pt. for $b > a$. For $b = a$, $r = a(1 + \cos\theta)$ which agrees with the previously given equation for the cardioid if θ is replaced by $\theta + \pi$. **9.** Set $z = x + iy$, $f(s) = x(s) + iy(s)$. Then the imag. part of $f'(s)\overline{f''(s)}$ equals $x'(s)y''(s) - y'(s)x''(s) = K(s)$ by (9.4.29). **11.** $F(y, x) = 0$. **13.** The answer to No. **12** can be written

$$x_a(s) = x(s) + a\sin\varphi(s), \qquad y_a(s) = y(s) - a\cos\varphi(s),$$

where $\varphi(s)$ is the angle which the tgt. of C makes with the x-axis. Hence

$$\begin{aligned}[x_a'(s)]^2 + [y_a'(s)]^2 &= [x'(s) + aK(s)\cos\varphi(s)]^2 + [y'(s) + aK(s)\sin\varphi(s)]^2\\ &= [x'(s)]^2 + [y'(s)]^2 + a^2[K(s)]^2\\ &\quad + 2aK(s)[x'(s)\cos\varphi(s) + y'(s)\sin\varphi(s)]\\ &= [1 + aK(s)]^2.\end{aligned}$$

This is the required expression, since the first member is $[S'(s)]^2$. **15.** Asymps.

$x = \frac{3}{2}$, $y = 0$. Min. $(\frac{1}{2}, -\frac{1}{2}e^{-\frac{1}{4}})$, max. $(1, -e^{-1})$. There are two pts. of infl. **17.** Asymps. $y = x - 1$, $x = 1$. Excl. region $x \leqq 1$. Min. $(2, 3)$. Two branches, one above and the other below the slanting asymp. The latter branch intersects the x-axis at $x = 2^{\frac{2}{3}} + 1$. **19.** Asymps. $x = 0$, $y = x + 5$. Max. $(-2, 2)$. No pts. of infl. The curve is below the slanting asymp. **21.** The curve consists of two symmetric ovals tgt. to each other at $(0, 0)$. The Cartesian equation is $(x^2 + y^2)^3 = x^4$. Extrema $(\pm(\frac{2}{3})^{\frac{3}{2}}, \pm\frac{2}{9}3^{\frac{1}{2}})$. $\rho = \dfrac{1 + 3\sin^2\theta}{3\cos\theta(1 + \sin^2\theta)}$. **23.** $r(\theta) - \rho(\theta) = a\left\{\theta - \dfrac{(\theta^2 + 1)^{\frac{3}{2}}}{\theta^2 + 2}\right\} = \dfrac{a}{2\theta} + O\left(\dfrac{1}{\theta^3}\right) \to 0$ as $\theta \to \infty$. **25.** By No. **16**, Exercise 9.3, $\tan\psi = \dfrac{r}{r'} = \theta$, where ψ is the angle between the curve tgt. and the radius vector. Hence $\tan\alpha = 1/\theta$, if α is the angle between the normal and the radius vector. The coords. of the center of curvature are

$$x = a\theta - \rho(\theta)\cos\alpha = \frac{a\theta}{\theta^2 + 2}, \qquad y = \rho(\theta)\sin\alpha = a\frac{\theta^2 + 1}{\theta^2 + 2},$$

where θ is a multiple of 2π. This gives $\theta = \dfrac{x}{a - y}$ and the centers lie on the ellipse $x^2 + 2y^2 - 3ay + a^2 = 0$. Actually the same ellipse arises, if we take the centers corresponding to intersections of the spiral with an arb. fixed ray, $\theta = \beta$. **27.** By the preceding problem the curvature is a min. at the endpts. of the minor axis where the radius of curvature is a^2/b. The circle of curvature has its center at $\left(0, -\dfrac{a^2 - b^2}{a}\right)$. Solving the equation of the circle for y gives

$$y_c(x) = b - \frac{a^2}{b} + \frac{a^2}{b}\left\{1 - \left(\frac{bx}{a^2}\right)^2\right\}^{\frac{1}{2}},$$

while on the ellipse

$$y_e(x) = b\left\{1 - \left(\frac{x}{a}\right)^2\right\}^{\frac{1}{2}}.$$

Expanding these fcns. in power series gives only even powers of x. The constant terms as well as the terms in x^2 agree. Since $y_c''''(0) = y_e''''(0)$ it follows that the values of the fcns. and their first three derivatives agree for $x = 0$. Hence third order contact. **29.** First order contact since $f(0) = g(0) = 0, f'(0) = g'(0) = 1$, $f''(0) = -2 \neq g''(0) = 0$.

APPENDIX C

Exercise C.1. **1.** 1. **3.** $\frac{1}{2}$. **5.** $\sqrt{2}e^{-\sqrt{2}} - e^{-2}$. **7.** No, for at $x = \dfrac{1}{k}$, $f(x + 0) - f(x - 0) = \dfrac{1}{k}$ and $\Sigma\dfrac{1}{k}$ is dvgt. **9.** $V_a{}^b[f] \leqq M(b - a)$. **11.** $f \in BV[0, \infty]$ iff $f \in BV[0, b]$ for every finite b and $V_a{}^b[f]$ is a bd. fcn. of b. **13.** Yes. **15.** Yes. **17.** $\sum_{n=0}^{\infty} |f_n(x)| \leqq \sum_{n=0}^{\infty} |f_n(a)| + \sum_{n=0}^{\infty} V_a{}^x[f_n] \leqq \sum_{n=0}^{\infty} |f_n(a)| + \sum_{n=0}^{\infty} V_a{}^b[f_n]$. Hence abs. and unif. cvgce. Since $V_a{}^b[\Sigma f_n] \leqq \Sigma V_a{}^b[f_n]$ the sum belongs to $BV[a, b]$.

Exercise C.2. **1.** $V_{1-h}^{1+h} = 1$, $V_{1-h}^{1} = 1$, $V_{1}^{1+h} = 0$ for all $h > 0$ hence also in the limit. **3.** Every sum $\sum_{k=1}^{n} |f(x_k) - f(x_{k-1})|$ is also a sum $\sum_{k=1}^{n} |f[x(t_k)] - f[x(t_{k-1})]|$ and vice versa. Hence $f[x(s)] \in BV[\alpha, \beta]$ iff $f(x) \in BV[a, b]$ and the total variations are the same. **5.** It is enough to assume that the sequence $\{x_n\}$ is monotone. Let $\{x_n\}$ incr. to x_0 and let $\{\xi_n\}$ decr. to x_0. Here $f(x_n) < f(x_{n+1}) < f(\xi_{n+1}) < f(\xi_n)$ so that $\lim f(x_m)$ and $\lim f(\xi_n)$ exist and are equal since $f(x_m) - f(\xi_n)$ is arb. small when m and n are large. The same argument shows that the limit is independent of the choice of sequences. **7.** At the nth stage the sum of the horizontal line segments is $1 - (\frac{2}{3})^n$. Each slanting line segment joins two pts. whose x-coords. differ by 3^{-n} while the y-coords. differ by 2^{-n}. These segments add up to $\{1 + (\frac{4}{9})^n\}^{\frac{1}{2}}$. Thus the length of the polygonal line is $1 - (\frac{2}{3})^n + \{1 + (\frac{4}{9})^n\}^{\frac{1}{2}}$. This is <2 and approaches 2 when $n \to \infty$. Thus $l(C) = 2$.

Exercise C.3. **1.** Yes. For Th. C.3.1 this is obvious since $G(x) = a \cos (b\pi x) + G(bx)$, if $G(x)$ is the sum of the cosine series. For the second th. assume for convenience that b is an odd integer. Take the partition pts. at the multiples of $\frac{1}{2}b^{-n}$. At such pts. $G(b^n x) = 0$. With these changes and obvious modifications of the formulas the argument carries over. **3.** Let k be the least integer such that $1 \leqq k!\, x \leqq k - 1$ and determine c_k as the integral part of $k!\, x$. If c_k, $c_{k+1}, \cdots, c_n$ have been found, determine c_{n+1} as the integral part of

$$(n + 1)! \left\{x - \sum_k^n \frac{c_j}{j!}\right\}.$$

The expansion is non-unique if x is rat. For such a no. has a finite representation. But there is also an infinite expansion since $\sum_m^\infty \dfrac{n-1}{n!} = \dfrac{1}{(m-1)!}$. **5.** If we take $|k| \leqq 4$, then $\dfrac{\sin(\frac{1}{6}k\pi)}{\frac{1}{6}k\pi} \geqq \dfrac{3\sqrt{3}}{4\pi}$. If $c_m = 0, 1, 2, 3, 4, 5$, we choose $k = -1$, $-1, -2, -3, -4, 1$ resp. Then the cosine term is $\geqq \cos \frac{1}{6}\pi = \frac{1}{2}\sqrt{3}$ in abs. value and $A = \dfrac{9}{8\pi}$. **7.** The inequality $|(u + \delta) - (u)| \leqq |\delta|$ for all u and δ is trivial if $|\delta| \geqq \frac{1}{2}$. If u and $u + \delta$ lie in the same interval $(n, n + \frac{1}{2})$ or $(n + \frac{1}{2}, n + 1)$, then equality holds, otherwise ineq. This gives

$$\left|\frac{1}{h}\sum_1^{m-2} a^n\{(b^n x + b^n h) - (b^n x)\}\right| \leqq \sum_1^{m-2} (ab)^n = \frac{(ab)^{m-1}}{ab - 1}.$$

We choose $h = \pm b^{-m}$. Then

$$\frac{1}{h}[K(x + h) - K(x)] = \frac{1}{h}\sum_1^{m-1} a^n\{(b^n x + b^n h) - (b^n x)\}.$$

Further, if $x = \sum_1^\infty c_k b^{-\kappa}$ then

$$(b^{m-1}x + b^{m-1}h) - (b^{m-1}x) = \left(\frac{c_m \pm 1}{b} + \frac{c_{m+1}}{b^2} + \cdots\right) - \left(\frac{c_m}{b} + \frac{c_{m+1}}{b^2} + \cdots\right) = \frac{1}{b}$$

if the sign of h is suitably chosen. The $(m-1)$th term divided by h is then $(ab)^{m-1}$ and dominates the difference quotient provided $ab > 2$. If this is so, then $\limsup \left|\frac{1}{h}[K(x+h) - K(x)]\right| = +\infty$ for all x and $K(x)$ is nowhere differentiable.

Miscellaneous problems for Appendix C

1. The ineq. follows from Th. C.1.2. If $x(t) = t$, $y(t) = f(t)$, the fcn. of No. **7**, Exercise C.2, then $V_0^1[x(t) + iy(t)] = V_0^1[x(t)] + V_0^1[y(t)] = 2$. **3.** By (C.2.7) $V_0^t[z] = \int_0^t |z'(s)|\,ds = \int_0^t (1+s)^{-1}\,ds = \log(1+t) \to \infty$ with t. **5.** The derivative is $-(\delta + i)(1-x)^{\delta-1+i}$, the abs. value of which is $(\delta^2+1)^{\frac{1}{2}}(1-x)^{\delta-1}$, integrable over $[0, 1]$. **7.** The imag. part of $z'(s)\overline{z''(s)}$. **9.** $K(x) = (x) + aK(bx)$. Proceed as in Th. C.3.1. **11.** $\frac{a}{(1-a)b} - \frac{2a}{\pi b} + \frac{1}{\pi}[(-1)^b - 1]\frac{a^2}{b(b-a)}$. **13.** 200. **15.** If x is rat., $x = p/q$, then all terms with $n \geqq q$ are 0. Further,

$$n!\,e = 2 \cdot n! + n(n-1)\cdots 3 + n(n-1)\cdots 4 + \cdots + n + 1 + \frac{1}{n+1} + \sum_2^\infty \frac{1}{(n+1)\cdots(n+k)}.$$

The integral part has the same parity as $n+1$, the infinite series has a sum $\frac{\theta_n}{n^2}$, $1 < \theta_n < 2$. Hence

$$\sin(n!\,\pi e) = (-1)^{n+1}\sin\left(\frac{\pi}{n+1} + \pi\frac{\theta_n}{n^2}\right) = (-1)^{n+1}\frac{\pi}{n+1} + O(n^{-2}),$$

so the series cvgs. Similarly

$$\sin\left(n!\,\pi\frac{2}{e}\right) = (-1)^{n+1}\sin\left(\frac{2\pi}{n+1} - 2\pi\sum_2^\infty \frac{(-1)^k}{(n+1)\cdots(n+k)}\right)$$

$$= (-1)^{n+1}\frac{2\pi}{n+1} + O(n^{-3}),$$

so the corresponding series also cvgs. **17.** $16\alpha^4 - 12\alpha^2 + 1 = 0$, $\alpha = \frac{1}{4}(1+\sqrt{5})$. **19.** The total variation does not exceed $2\sum_0^\infty \exp[-\frac{1}{2}\alpha(2n+1)\pi]$ which cvgs. for $\alpha > 0$.

MISCELLANEOUS PROBLEMS FOR VOLUME I

1. $3(x-1)^{\frac{1}{2}}(x+1)^{-\frac{5}{2}}$. **3.** $\frac{4x}{x^4-1}$. **5.** $\frac{1}{x}\cos(\log x)$. **7.** 2. **9.** $e^x(1+e^{2x})^{-2}$. **11.** $-2\sec x$. **13.** $D[f] = R[f] = (-\infty, \infty)$. (1, 1.) **15.** $a \geqq e$. **17.** $(x-a)^2 + (y-b)^2 + \alpha(bx - ay) = 0$. No circle through

(0, 0). **19.** $y = -\dfrac{b^2}{a^2m}x$, if m is the slope of the lines. **21.** There are no normals perpendicular to the asymps. These excl. values α, β, in general determine three intervals $(-\infty, \alpha)$, (α, β), (β, ∞). Either $(-\infty, \alpha]$ and $[\beta, \infty)$ are excl. or else $[\alpha, \beta]$ depending upon the location of the hyperbola with respect to the asymps. For $xy = a^2$, the interval $[-\infty, 0]$ is excl. **23.** $c^3x^3 - (b^2 - 2bc)cx^2 + (a^2 - 2b)cx - 1 = 0$. **25.** $t^3 - (a + 3d)t^2 + (b + 2ad + 3d^2)t - (c + bd + ad^2 + d^3) = 0$. **27.** $a = 1, x = 1, 1, -1 \pm \sqrt{2}$; $a = -1, x = -1, -1, 1 \pm \sqrt{2}$; $a = \frac{5}{9}\sqrt{3}$, $x = \frac{1}{3}\sqrt{3}$, $\frac{1}{3}\sqrt{3}$, $\frac{1}{3}(-1 \pm \sqrt{10})\sqrt{3}$; $a = -\frac{5}{9}\sqrt{3}$, $x = -\frac{1}{3}\sqrt{3}$, $-\frac{1}{3}\sqrt{3}$, $\frac{1}{3}(-1 \pm \sqrt{10})\sqrt{3}$. **29.** x_n is pos. and decr., hence cvgs. to the pos. root of $x = \frac{1}{2}x + \frac{1}{3}(x^2 + 1)^{\frac{1}{2}}$ which is $\frac{2}{5}\sqrt{5}$. **31.** x_n is incr. and bd. by $1 + \sqrt{2}$ which is $\lim x_n$. **33.** $a^{\frac{1}{3}}$. The fcn. $f(x) = \frac{2}{3}x + \frac{1}{3}ax^{-2}$ takes its min. value $a^{\frac{1}{3}}$ at $x = a^{\frac{1}{3}}$ so that $a^{\frac{1}{3}}$ is a fixed pt. of the mapping $x \to f(x)$. If $1 < a$, the sequence $\{x_n \mid n \geqq 2\}$ is strictly decr. and cvgs. to its g.l.b. $a^{\frac{1}{3}}$. If $0 < a < 1$, the whole sequence is strictly decr. and cvgs. to $a^{\frac{1}{3}}$. If $a = 1$, $x_n = 1$ for all n and $\lim x_n = 1$. We have

$$x_3 = \tfrac{2}{9}(2 + a) + \frac{3a}{(2 + a)^2} < \tfrac{1}{3}(2 + a) = x_2, \qquad 0 < a, \qquad a \neq 1.$$

This follows from $(2 + a)^3 > 27a$, valid for such values of a. Hence $x_3 < x_2$. Now $x_{n+1} - x_n = f(x_n) - f(x_{n-1}) = (x_n - x_{n-1})f'(\xi_n)$ where ξ_n lies between x_n and x_{n-1}. Since $a^{\frac{1}{3}}$ is the min. value of $f(x)$, we conclude that $x_n > a^{\frac{1}{3}}$, for all n if $0 < a < 1$, for $n > 1$ if $a > 1$. It follows that $\xi_n > a^{\frac{1}{3}}$ for $n > 1$. Hence

$$f'(x) = \frac{2}{3}\left(1 - \frac{a}{x^3}\right) > 0$$

for $x = \xi_n$ and $x_{n+1} - x_n$ has the same sign as $x_n - x_{n-1}$ for $n > 2$. Since $x_3 - x_2 < 0$ we have $x_n - x_{n-1} < 0$ for $n > 2$. Hence $\lim x_n$ exists and is $\geqq a^{\frac{1}{3}}$. Equality must hold since the limit is a root of the equation $y = \dfrac{1}{3}\left(2y + \dfrac{a}{y^2}\right)$. This method is used for high-speed computation of cube roots. **35.** Cvgt. **37.** Cvgt. **39.** Cvgt. for $\alpha > 1$, dvgt. for $\alpha \leqq 1$. **41.** $R = \max(1, a)$. **43.** e. **45.** $R = 1$ since $a_n \to 0$ and $\dfrac{a_{n+1}}{a_n} = \dfrac{\log(1 + a_n)}{a_n} \to 1$. **47.** $R = \sqrt[3]{3}$. If $a_n \leqq 1$, then $a_{n+1} \leqq (\frac{2}{3})^{\frac{1}{3}}$ so a_n is bd. by 1 since $a_1 = 1$. Hence $a_{n+1} \leqq (\frac{2}{3})^{\frac{1}{3}}a_n$ or a_n decr. to 0. Then $a_{n+1}/a_n = [\frac{1}{3}(1 + a_n)]^{\frac{1}{3}} \to 3^{-\frac{1}{3}}$ or $R = \sqrt[3]{3}$. **49.** $R = 1$. From $a_{n+2} = \dfrac{n+1}{n+2}a_{n+1} + \dfrac{1}{n+2}a_n$ it follows that if $1 \leqq a_n, a_{n+1} \leqq 2$, then $1 \leqq a_{n+2} \leqq 2$. Hence every a_n lies in the interval $(1, 2)$ and $R = 1$. **51.** $F(z) = \sum_0^\infty (r_1z)^n + \sum_0^\infty (r_2z)^n + \sum_0^\infty (r_3z)^n = (1 - r_1z)^{-1} + (1 - r_2z)^{-1} + (1 - r_3z)^{-1}$. $R = |r_3|^{-1}$ if $|r_1| \leqq |r_2| \leqq |r_3|$. $A = q$, $B = p$, $C = 0$, $D = 1$. $R = 1$ iff either $r_3 = \pm 1$ or $r_3 = \overline{r_2}$ and $|r_2| = |r_3| = 1$. The first case implies $q = \pm(1 + p)$ and $|q| \leqq 1$, the second $p = 1 - q^2$, $|q| \leqq 1$. **53.** $\frac{1}{9}$. **55.** ab. **57.** $\dfrac{m}{n}$.

59. $\frac{1}{6}$. **61.** $e^{\frac{1}{3}}$. **63.** For $|z| < a$ the sum is $a(a - z)^{-1} + b(b - z)^{-1}$. **65.** $(1 - z - z^2)^{-1} = (w + w^{-1})[w(1 + wz)^{-1} + w^{-1}(1 - w^{-1}z)]$ with $w = \frac{1}{2}(\sqrt{5} - 1)$. Expand in geometric series and add the coeffs. to get

$$a_n = (w + w^{-1})[(-1)^n w^{n+1} + w^{-(n+1)}]. \qquad R = w^{-1}.$$

67. $6 + \log \frac{8}{3}$. **69.** 1. **71.** $\frac{1}{5} \log (1 + \frac{2}{3}\sqrt{3})$. **73.** The substitution (7.5.18) leads to $2\int_0^\infty [(a - 1) + (a + 1)t^2]^{-1}\,dt$. Set $(a + 1)^{\frac{1}{2}}t = (a - 1)^{\frac{1}{2}}s$ to obtain the value. **75.** We have $w^{4n+1} + w^{-4n-1} = (w + w^{-1}) \sum_0^{4n} (-1)^k w^{4n-2k}$. Only the term with $k = 2n$ affects the value of the integral. **77.** Set $x = e^{-t}$ obtaining

$$\int_0^\infty \exp(te^{-t} - t)\,dt = \sum_0^\infty \frac{1}{k!} \int_0^\infty t^k e^{-(k+1)t}\,dt = \sum_0^\infty (k + 1)^{-k-1}.$$

Termwise integration is permitted by unif. cvgce. of the series and cvgce. of the result. **79.** $p = -4$, $q = 3$. **81.** The limit is half the sum of the series $\sum_1^\infty (-1)^{n-1} \int_0^\pi (u + n\pi)^{-\frac{1}{2}} \sin u\,du$ which cvgs. since the terms tend to 0 and

$$\int_0^\pi (u + n\pi)^{-\frac{1}{2}} \sin u\,du > \int_0^\pi (u + (n + 1)\pi)^{-\frac{1}{2}} \sin u\,du.$$

83. If $\max |f(x)| = M$ and $\varepsilon > 0$, there is a $\delta > 0$ and an interval of length δ where $|f(x)| \geqq M - \varepsilon$ and

$$(M - \varepsilon)\delta^{1/p} \leqq \left\{\int_a^b |f(x)|^p\,dx\right\}^{1/p} \equiv I_p[f] \leqq M,$$

so that $M - \varepsilon \leqq \liminf I_p[f] \leqq \limsup I_p[f] \leqq M$. Hence $\lim I_p[f] = M$. **85.** (i) The curve exists for $0 \leqq x < 2a$. It is symmetric w.r.t. the x-axis. Vert. tgt. at $(0, 0)$, node at $(a, 0)$ with slopes ± 1, $x = 2a$ is an asymp. (ii) Substitute! (iii) If the pts. lie on the straight line $Ax + By + Ca = 0$, then t_1, t_2, t_3 are roots of $Bt^3 + (2A + C)t^2 - Bt + C = 0$ and $t_1t_2 + t_2t_3 + t_3t_1 = -1$. Conversely, if t_1, t_2, t_3 are nos. with $t_1t_2 + t_2t_3 + t_3t_1 = -1$, then set $B = 1$, $C = -t_1t_2t_3$, $2A = -t_1 - t_2 - t_3 + t_1t_2t_3$ and the straight line is $Ax + By + Ca = 0$. The pts. lie on this line. **87.** If the normal at (x_0, y_0) is taken, then

$$X = \frac{a^2(a^2 + b^2)}{b^4{x_0}^2 + a^4{y_0}^2}\, x_0{y_0}^2, \qquad Y = \frac{b^2(a^2 + b^2)}{b^4{x_0}^2 + a^4{y_0}^2}\, {x_0}^2 y_0.$$

Substitute! **89.** Use polar coord., $r^2 = a^2 \cos 2\theta$, and the formulas of No. 3, Exercise 9.5. This gives $\varphi = \frac{1}{2}\pi + 3\theta$, $\varphi' = 3$, $x' = -a(\cos 2\theta)^{-\frac{1}{2}} \sin 3\theta$, $y' = a(\cos 2\theta)^{-\frac{1}{2}} \cos 3\theta$, and, since $3 \cos 2\theta \cos \theta - \cos 3\theta = 2 \cos^3 \theta$, $3 \cos 2\theta \sin \theta - \sin 3\theta = 2 \sin^3 \theta$, we have $\alpha = \frac{2}{3}a(\cos 2\theta)^{-\frac{1}{2}} \cos^3 \theta$, $\beta = \frac{2}{3}a(\cos 2\theta)^{-\frac{1}{2}} \sin^3 \theta$. Substitute! **91.** With the given choice of coords., $\lambda(s) = \frac{1}{2}y''(0)s^2 + o(s^2)$ with $y''(0) = [\rho(0)]^{-1}$, so that $\lambda(s)s^{-2} \to \frac{1}{2}y''(0)$. Note that $x'(0) = 1$, $x''(0) = 0$, $y'(0) = 0$, $y''(0) \neq 0$ since $\rho(0) \neq \infty$. **93.** The triangle is isosceles. The vertex opposite the base describes an ellipse with center at the midpt. of the base, the endpts. of which are foci of the ellipse. The minor axis is $2(p^2 - ap)^{\frac{1}{2}}$. If the vertex is at (x, y), the area of the triangle is $\frac{1}{2}ay$, a max. when y is a max., i.e., when $y = (p^2 - ap)^{\frac{1}{2}}$ and the vertex is at the endpt. of the minor axis. **95.** If $\frac{1}{2}\sqrt{2} \leqq e < 1$, the minor axis is the longest chord. If $0 < e < \frac{1}{2}\sqrt{2}$, the max. length is $\dfrac{a^2}{(a^2 - b^2)^{\frac{1}{2}}}$ with $\left(\pm \dfrac{a^2}{a^2 - b^2}(a^2 - 2b^2)^{\frac{1}{2}}, -\dfrac{b^3}{a^2 - b^2}\right)$ as lower endpts. of the chord. **97.** (i) If (X, Y) is a pt. on the ellipse, the tgt. of the angle between the tgt. and the radius vector is $\dfrac{a^2b^2}{a^2 - b^2}\dfrac{1}{XY}$. This is a min. when $XY = \dfrac{b}{a}X(a^2 - X^2)^{\frac{1}{2}}$ is a max. This occurs for $X = \frac{1}{2}\sqrt{2}a$, $Y = \frac{1}{2}\sqrt{2}b$. (ii) The intercepts of the normal are $x = \dfrac{a^2 - b^2}{a^2}X$, $y = -\dfrac{a^2 - b^2}{b^2}Y$, so again the problem is to make XY a max. **99.** 2 arc sin $\frac{1}{3}$. **101.** The induction process depends upon the ineq. $\left(\dfrac{3n + 1}{3n + 2}\right)^{\frac{1}{2}} > \dfrac{2n + 1}{2n + 2}$ which is true for $n \geqq 0$ since

$$(2n + 1)^2(3n + 2) = 12n^3 + 16n^2 + 11n + 2 < 12n^3 + 28n^2 + 20n + 4$$
$$= (2n + 2)^2(3n + 1).$$

103. The sum of the series is $\dfrac{1}{x}\log\dfrac{1 + x}{1 - x}$. If x is replaced by $\dfrac{2x}{1 + x^2}$ in the sum fcn. the equation is easily verified. **105.** Set $f(x) = x + x^3 - \tan x$, $g(x) = xf(x)$. Then $f(0) = 0$, $f(\frac{1}{3}\pi) > 0$ and $\cos^2 x\, f'(x) = \cos^2 x\,(1 + 3x^2) - 1 > (1 - \frac{1}{2}x^2)^2(1 + 3x^2) - 1 = x^2(\frac{3}{4}x^4 - \frac{11}{4}x^2 + 2)$. This is 0 for $x = 0$ and 1 and pos. for $0 < x < 1$. Hence $f(x)$ is incr. in $(0, 1]$ and pos. Thus $g(x) > 0$ in $(0, 1]$. But $g'(x) = 2x + 8x^3 - \tan x - x\sec^2 x$ exceeds $10 - \sqrt{3} - \frac{4}{3}\pi > 0$ for $1 \leqq x \leqq \frac{1}{3}\pi$ so $g(x)$ is pos. in $(1, \frac{1}{3}\pi]$. Hence $f(x) > 0$ in $(0, \frac{1}{3}\pi]$. **107.** Set $f(x) = \log x - \text{arc tan } x$. Then $f'(x) = \dfrac{1}{x} - \dfrac{1}{x^2 + 1} = \dfrac{x^2 + 1 - x}{x(x^2 + 1)} > 0$ for $x > 0$ so f is strictly incr. **109.** If $g(x) = x^{-1}f(x)$, then $g'(x) = x^{-2}[xf'(x) - f(x)]$. Now $f(x) = \displaystyle\int_0^x f'(s)\,ds < xf'(x)$ since $f'(s)$ is incr. Hence $g'(x) > 0$ and $g(x)$ is strictly incr. **111.** Since $f(x) - f(0) = \displaystyle\int_0^x f'(s)\,ds < Kx$, $f(x) < f(0) + Kx$. Hence $f(x) - x > 0$ at $x = 0$ and <0 for large x. Thus there is at least one $x > 0$ where $f(x) - x = 0$. If x_0 is the least pos. root and if $x > x_0$, then $f(x) - x = \displaystyle\int_{x_0}^x f'(s)\,ds - (x - x_0) < -(1 - K)(x - x_0) < 0$ so there is no other root. **113.** Use $u^2\displaystyle\int_0^\infty te^{-ut}\,dt = \int_0^\infty se^{-s}\,ds = 1$. For any $\varepsilon > 0$ there is an a such that $|f(t) - At| < \varepsilon t$ for $t \geqq a$. Then

$$\Delta(u) \equiv u^2\int_0^\infty e^{-ut}f(t)\,dt - A = u^2\int_0^\infty e^{-ut}[f(t) - At]\,dt.$$

Here $\lim u^2\displaystyle\int_0^a \cdots = 0$ and $u^2\left|\displaystyle\int_a^\infty \cdots\right| < u^2\displaystyle\int_a^\infty e^{-ut}\varepsilon t\,dt < \varepsilon$. Hence lim sup $\Delta(u) < \varepsilon$ for any ε. This implies $\lim \Delta(u) = 0$. **115.** Let $\varepsilon > 0$ be given. Divide the interval $[0, 1]$ into k equal parts of length $<\varepsilon/(2M)$. Let $x_0 = 0 < x_1 < x_2 < \cdots < x_{k+1} = 1$ be the division pts. There exists an $N(\varepsilon)$ so large that $|f_n(x_j)| < \varepsilon$ for all $n > N(\varepsilon)$ and $j = 0, 1, \cdots, k+1$. Suppose that $x_j < x < x_{j+1}$. Then for $n > N(\varepsilon)$ we have

$$|f_n(x)| < |f_n(x) - f_n(x_j)| + |f_n(x_j)|$$
$$= \left| \int_{x_j}^{x} f_n'(s)\, ds \right| + |f_n(x_j)| \leqq M(x - x_j) + \tfrac{1}{2}\varepsilon < M \cdot \frac{\varepsilon}{2M} + \tfrac{1}{2}\varepsilon = \varepsilon.$$

Hence $|f_n(x)| < \varepsilon$ for all x in $[0, 1]$ and all $n > N(\varepsilon)$. This proves unif. cvgce. to 0 in $[0, 1]$. **117.** If

$$Q_n = n^{-\frac{1}{2}}e^{-n}\int_n^{n+1} x^{\frac{1}{2}}e^x\, dx, \text{ then } \lim Q_n = e - 1.$$

For

$$1 \leqq \left(\frac{x}{n}\right)^{\frac{1}{2}} \leqq \left(1 + \frac{1}{n}\right)^{\frac{1}{2}} \to 1 \text{ as } n \to \infty$$

and

$$e^{-n}\int_n^{n+1} e^x\, dx = e - 1.$$

119. Yes. By Taylor's th. $f\left(\frac{1}{n}\right) - 1 = \frac{1}{2}\frac{1}{n^2}f''\left(\frac{\theta_n}{n}\right)$, $0 < \theta_n < 1$, and since $f''(t)$ is continuous and tends to 2 when $t \to 0$, $\left|f''\left(\frac{\theta_n}{n}\right)\right| < 3$ for all large n so $\Sigma\left[f\left(\frac{1}{n}\right) - 1\right]$ cvgs. as $\Sigma\frac{1}{n^2}$. **121.** For $-1 \leqq x < 1$. Cvgce. by the root test for $-1 < x < 1$. If $x = -1$, note that $\left(-1 + \frac{1}{\sqrt{n}}\right)^n = (-1)^n\left(1 - \frac{1}{\sqrt{n}}\right)^n = (-1)^n\left[\left(1 - \frac{1}{\sqrt{n}}\right)^{\sqrt{n}}\right]^{\sqrt{n}}$. For large n the quantity inside the brackets is appr. e^{-1}, hence less than a fixed number $b < 1$ and $\Sigma\, b^{\sqrt{n}}$ cvgs. The series dvgs. for $x = 1$. **123.** 0. Since $\frac{n}{x+n} \leqq 1$ for all n and all $x \geqq 0$,

$$n\int_0^\infty \frac{e^{-x}}{(x+n)^2}\, dx < \int_0^\infty \frac{e^{-x}}{x+n}\, dx < \frac{1}{n}\int_0^\infty e^{-x}\, dx = \frac{1}{n} \to 0.$$

125. Since $\cot x = \frac{1}{x} - \frac{1}{3}x + O(x^3)$, we have $n \cot\frac{1}{n} = n^2 - \frac{1}{3} + O(n^{-2})$. Hence $a = 1$ and the limit is then $\frac{1}{3}$. **127.** $p = \pm 1$. For $p = 1$, $\frac{e^x - 1 - \sin x}{x \sin x} = \frac{\frac{1}{2}x^2 + O(x^3)}{x^2 + O(x^4)} \to \frac{1}{2}$. For $p = -1$, $\frac{e^x + 1 - \sin x}{x^{-1}\sin x} = \frac{2 + O(x)}{1 + O(x^2)} \to 2$. **129.** $P(x) = \frac{1}{2}x + 2x^2 - \frac{3}{2}x^3$. **131.** $4a$. **133.** $\frac{1}{4}(\sqrt{26} - \sqrt{2}) - \frac{1}{4}\log\frac{5 + \sqrt{26}}{1 + \sqrt{2}}$.

Index

The pages on which the actual theorems are stated are in parenthesis.

Printed in the United States of America